T0180720

Communications in Computer and Information Science   1143

Commenced Publication in 2007
Founding and Former Series Editors:
Phoebe Chen, Alfredo Cuzzocrea, Xiaoyong Du, Orhun Kara, Ting Liu,
Krishna M. Sivalingam, Dominik Ślęzak, Takashi Washio, Xiaokang Yang,
and Junsong Yuan

More information about this series at http://www.springer.com/series/7899

Tom Gedeon · Kok Wai Wong ·
Minho Lee (Eds.)

Neural
Information Processing

26th International Conference, ICONIP 2019
Sydney, NSW, Australia, December 12–15, 2019
Proceedings, Part V

 Springer

Editors
Tom Gedeon 🆔
Australian National University
Canberra, ACT, Australia

Kok Wai Wong
Murdoch University
Murdoch, WA, Australia

Minho Lee 🆔
Kyungpook National University
Daegu, Korea (Republic of)

ISSN 1865-0929   ISSN 1865-0937 (electronic)
Communications in Computer and Information Science
ISBN 978-3-030-36801-2   ISBN 978-3-030-36802-9 (eBook)
https://doi.org/10.1007/978-3-030-36802-9

This Springer imprint is published by the registered company Springer Nature Switzerland AG
The registered company address is: Gewerbestrasse 11, 6330 Cham, Switzerland

Preface

Welcome to the proceedings of the 26th International Conference on Neural Information Processing of the Asia-Pacific Neural Network Society (APNNS 2019), held in Sydney during December 12–15, 2019.

The mission of the Asia-Pacific Neural Network Society is to promote active interactions among researchers, scientists, and industry professionals who are working in Neural Networks and related fields in the Asia-Pacific region. APNNS 2019 had governing board members from 13 countries/regions – Australia, China, Hong Kong, India, Japan, Malaysia, New Zealand, Singapore, South Korea, Qatar, Taiwan, Thailand, and Turkey. The society's flagship annual conference is the International Conference of Neural Information Processing (ICONIP).

The conference had three main themes: "Theory and Algorithms," "Computational and Cognitive Neurosciences," and "Human Centred Computing and Applications." The two CCIS volumes 1142–1143 are organized in topical sections which were also the names of the 12-minute presentation sessions at the conference. The topics were "Adversarial Networks and Learning," "Convolutional Neural Networks," "Deep Neural Networks," "Embeddings and Feature Fusion," "Human Centred Computing," "Human Centred Computing and Medicine," "Human Centred Computing for Emotion," "Hybrid Models," "Artificial Intelligence and Cybersecurity," "Image Processing by Neural Techniques," "Learning from Incomplete Data," "Model Compression and Optimisation," "Neural Network Applications," "Neural Network Models," "Semantic and Graph Based Approaches," "Social Network Computing," "Spiking Neuron and Related Models," "Text Computing using Neural Techniques," "Time-series and Related Models," and "Unsupervised Neural Models."

A Special thanks in particular to the reviewers who devoted their time to our rigorous peer-review process. Their insightful reviews and timely feedback ensured the high quality of the papers accepted for publication. Finally, thank you to all the authors of papers, presenters, and participants at the conference. Your support and engagement made it all worthwhile.

October 2019

Tom Gedeon
Kok Wai Wong
Minho Lee

Organization

Tom Gedeon   The Australian National University, Canberra,
                             Australia
Kok Wai Wong   Murdoch University, Murdoch, Australia
Minho Lee   Kyungpook National University, Daegu, Korea
                             (Republic of)

Program Chairs

Tom Gedeon   The Australian National University, Australia
Kok Wai Wong   Murdoch University, Australia
Minho Lee   Kyungpook National University, Daegu, Korea
                             (Republic of)

Program Committee

Hussein Abbass   UNSW Canberra, Australia
Hosni Adil Imad Eddine   Beijing Institute of Technology, China
Shotaro Akaho   AIST, Japan
Alaa Al-Kaysi   University of Technology, Baghdad-Iraq, Iraq
Bradley Alexander   The University of Adelaide, Australia
Georgios Alexandridis   National Technical University of Athens, Greece
Usman Ali   Shanghai Jiao Tong University, China
Ahmad Ali   Shanghai Jiao Tong University, China
Abdulrahman Altahhan   Leeds Beckett University, UK
Muhamad Erza Aminanto   NICT, Japan
Ali Anaissi   The University of Sydney, Australia
Khairul Anam   University of Jember, Indonesia
Emel Arslan   Istanbul University, Turkey
Sunil Aryal   Deakin University, Australia
Arnulfo Azcarraga   De La Salle University, Philippines
Donglin Bai   Shanghai Jiao Tong University, China
Hongliang Bai   Beijing Faceall Technology Co., Ltd., China
Mehala Balamurali   University of Sydney, Australia
Mohamad Hardyman   Universiti Malaysia Sarawak, Malaysia
  Barawi
Younès Bennani   Université Paris 13 - Sorbonne Paris Cité, France
Christoph Bergmeir   Monash University, Australia
Gui-Bin Bian   Chinese Academy of Sciences, China
Larbi Boubchir   University of Paris 8, France
Amel Bouzeghoub   Télécom SudParis, France
Congbo Cai   Xiamen University, China

Jian Cao	Shanghai Jiaotong University, China
Xiaocong Chen	The University of New South Wales, Australia
Junsha Chen	UCAS, China
Junjie Chen	Inner Mongolia University, China
Qingcai Chen	Harbin Institute of Technology (Shenzhen), China
Gang Chen	Victoria University of Wellington, New Zealand
Junya Chen	Fudan University, USA
Dong Chen	Wuhan University, China
Weiyang Chen	Qilu University of Technology, China
Jianhui Chen	Institute of Neuroscience, Chinese Academy of Science, China
Girija Chetty	University of Canberra, Australia
Sung-Bae Cho	Yonsei University, South Korea
Chaikesh Chouragade	Indian Institute of Science, India
Tan Chuanqi	Tsinghua University, China
Yuk Chung	The University of Sydney, Australia
Younjin Chung	The Australian National University, Australia
Tao Dai	Tsinghua University, China
Yong Dai	Hunan University, China
Popescu Dan	UPB, Romania
V. Susheela Devi	Indian Institute of Science, India
Bettebghor Dimitri	Expleo Group, France
Hai Dong	RMIT University, Australia
Anan Du	University of Technology Sydney, Australia
Piotr Duda	Czestochowa University of Technology, Poland
Pratik Dutta	IIT Patna, India
Asif Ekbal	IIT Patna, India
Mounim El Yacoubi	Télécom SudParis, France
Haytham Elghazel	LIRIS Lab, France
Zhijie Fang	Chinese Academy of Sciences, China
Yuchun Fang	Shanghai University, China
Yong Feng	Chongqing University, China
Raul Fernandez Rojas	UNSW Canberra, Australia
Junjie Fu	Southeast University, China
Bogdan Gabrys	University of Technology Sydney, Australia
Junbin Gao	The University of Sydney, Australia
Guangwei Gao	Nanjing University of Posts and Telecommunications, China
Tom Gedeon	The Australian National University, Australia
Ashish Ghosh	Indian Statistical Institute, India
Heitor Murilo Gomes	The University of Waikato, New Zealand
Iqbal Gondal	Federation University, Australia
Yuri Gordienko	National Technical University of Ukraine, Ukraine
Raju Gottumukkala	University of Louisiana at Lafayette, USA
Jianping Gou	Jiangsu University, China
Xiaodong Gu	Fudan University, China
Joachim Gudmundsson	The University of Sydney, Australia

Xian Guo	Nankai University, China
Jun Guo	East China Normal University, China
Katsuyuki Hagiwara	Mie University, Japan
Sangchul Hahn	Handong Global University, South Korea
Ali Haidar	The University of New South Wales, Australia
Rim Haidar	The University of Sydney, Australia
Fayçal Hamdi	CEDRIC - CNAM Paris, France
Maissa Hamouda	SETIT, Tunisia
Jiqing Han	Harbin Institute of Technology, China
Chansu Han	National Institute of Information and Communications Technology, Japan
Tao Han	Hubei Normal University, China
Jean Benoit Heroux	IBM Research - Tokyo, Japan
Hansika Hewamalage	Monash University, Australia
Md. Zakir Hossain	Murdoch University, Australia
Zexi Hu	The University of Sydney, Australia
Shaohan Hu	Tsinghua University, China
Xiyuan Hu	Chinese Academy of Sciences, China
Gang Hu	Ant Financial Services Group, China
Xinyi Hu	State Key Laboratory of Mathematical Engineering and Advanced Computing, China
Han Hu	Tsinghua University, China
Yue Huang	Xiamen University, China
Shudong Huang	University of Electronic Science and Technology of China, China
Kaizhu Huang	Xi'an Jiaotong Liverpool University, China
Yanhong Huang	East China Normal University, China
Xiaolin Huang	Shanghai Jiao Tong University, China
Chaoran Huang	The University of New South Wales, Australia
Shin-Ying Huang	Institute for Information Industry, Taiwan
Mohamed Ibm Khedher	IRT SystemX, France
Loretta Ichim	UPB, Romania
David Andrei Iclanzan	Sapientia University, Romania
Keiichiro Inagaki	Chubu University, Japan
Radu Ionescu	University of Bucharest, Romania
Masatoshi Ishii	IBM Research - Tokyo, Japan
Masumi Ishikawa	Kyushu Institute of Technology, Japan
Megumi Ito	IBM Research - Tokyo, Japan
Yi Ji	Soochow University, China
Sun Jinguang	Liaoning Technical University, China
Francois Jacquenet	University of Lyon, France
Seyed Mohammad Jafar Jalali	Deakin University, Australia
Zohaib Muhammad Jan	Central Queensland University, Australia
Yasir Jan	Murdoch University, Australia
Norbert Jankowski	Nicolaus Copernicus University, Poland

Sungmoon Jeong	Kyungpook National University, South Korea
Xiaoyan Jiang	Shanghai University of Engineering Science, China
Fei Jiang	Shanghai Jiao Tong University, China
Houda Jmila	Télécom SudParis, France
Tae Joon Jun	Asan Medical Center, South Korea
H. M. Dipu Kabir	Deakin University, Australia
Kyunghun Kang	Kyungpook National University Hospital, South Korea
Asim Karim	Lahore University of Management Sciences, Pakistan
Kathryn Kasmarik	UNSW@ADFA, Australia
Yuichi Katori	Future University Hakodate, Japan
Imdadullah Khan	Lahore University of Management Science, Pakistan
Zubair Khan	Shanghai Jiaotong University, China
Numan Khurshid	Lahore University of Management Sciences, Pakistan
Matloob Khushi	The University of Sydney, Australia
Shuji Kijima	Kyushu University, Japan
Rhee Man Kil	Sungkyunkwan University, South Korea
SangBum Kim	Seoul National University, South Korea
Mutsumi Kimura	Ryukoku University, Japan
Eisuke Kita	Nagoya University, Japan
Simon Kocbek	University of Technology Sydney, Australia
Hisashi Koga	University of Electro-Communications, Japan
Tao Kong	Tsinghua University, China
Nihel Kooli	Solocal, France
Irena Koprinska	The University of Sydney, Australia
Marcin Korytkowski	Czestochwa University of Technology, Poland
Polychronis Koutsakis	Murdoch University, Australia
Aneesh Krishna	Curtin University, Australia
Lov Kumar	BITS Pilani Hyderabad Campus, India
Takio Kurita	Hiroshima University, Japan
Shuichi Kurogi	Kyushu Institute of Technology, Japan
Hamid Laga	Murdoch University, Australia
Keenan Leatham	University of District of Columbia, USA
Xiaoqiang Li	Shanghai University, China
Yantao Li	Chongqing University, China
Ran Li	Shanghai Jiaotong University, China
Jiawei Li	Tsinghua University, China
Tao Li	Peking University, China
Li Li	Southwest University, China
Xiaohong Li	Tianjin University, China
Yang Li	Tsinghua University, China
Nan Li	Tianjin University, China
Mingxia Li	University of Electronic Science and Technology of China, China
Zhixin Li	Guangxi Normal University, China
Zhipeng Li	Tsinghua University, China

Bohan Li	Nanjing University of Aeronautics and Astronautics, China
Mengmeng Li	Zhengzhou University, China
Yaoyi Li	Shanghai Jiao Tong University, China
Yanjun Li	Beijing Institute of Technology, China
Ming Li	Latrobe University, Australia
Mingyong Li	Donghua University, China
Chengcheng Li	Tianjin University, China
Xia Liang	University of Science and Technology, China
Alan Wee-Chung Liew	Griffith University, Australia
Chin-Teng Lin	UTS, Australia
Zheng Lin	Chinese Academy of Sciences, China
Yang Lin	The University of Sydney, Australia
Wei Liu	University of Technology Sydney, Australia
Jiayang Liu	Tsinghua University, China
Yunlong Liu	Xiamen University, China
Yi Liu	Zhejiang University of Technology, China
Ye Liu	Nanjing University of Posts and Telecommunications, China
Zhilei Liu	Tianjin University, China
Zheng Liu	Nanjing University of Posts and Telecommunications, China
Cheng Liu	City University of Hong Kong, Hong Kong, China
Linfeng Liu	Nanjing University of Posts and Telecommunications, China
Baoping Liu	IIE, China
Guiping Liu	Hetao College, China
Huan Liu	Xi'an Jiaotong University, China
Gongshen Liu	Shanghai Jiao Tong University, China
Zhi-Yong Liu	Institute of Automation, Chinese Academy of Science
Fan Liu	Beijing Ant Financial Services Information Service Co., Ltd., China
Zhi-Wei Liu	Huazhong University of Science and Technology, China
Chu Kiong Loo	University of Malaya, Malaysia
Xuequan Lu	Deakin University, Australia
Huimin Lu	Kyushu Institute of Technology, Japan
Biao Lu	Nankai University, China
Qun Lu	Yancheng Institute of Technology, China
Bao-Liang Lu	Shanghai Jiao Tong University, China
Shen Lu	The University of Sydney, Australia
Junyu Lu	University of Electronic Science and Technology of China, China
Zhengding Luo	Peking University, China
Yun Luo	Shanghai Jiao Tong University, China
Xiaoqing Lyu	Peking University, China

Kavitha MS	Hiroshima University, Japan
Wanli Ma	University of Canberra, Australia
Jinwen Ma	Peking University, China
Supriyo Mandal	Indian Institute of Technology Patna, India
Sukanya Manna	Santa Clara University, USA
Basarab Matei	University Paris 13, France
Jimson Mathew	IIT Patna, India
Toshihiko Matsuka	Chiba University, Japan
Timothy McIntosh	La Trobe University, Australia
Philip Mehrgardt	The University of Sydney, Australia
Jingjie Mo	Chinese Academy of Sciences, China
Seyed Sahand Mohammadi Ziabari	Vrije Universiteit Amsterdam, The Netherlands
Rafiq Mohammed	Murdoch University, Australia
Bonaventure Molokwu	University of Windsor, Canada
Maram Monshi	The University of Sydney, Australia
Ajit Narayanan	Auckland University of Technology, New Zealand
Mehdi Neshat	Adelaide University, Australia
Aneta Neumann	The University of Adelaide, Australia
Frank Neumann	The University of Adelaide, Australia
Dang Nguyen	University of Canberra, Australia
Thanh Nguyen	Robert Gordon University, UK
Tien Dung Nguyen	University of Technology Sydney, Australia
Thi Thu Thuy Nguyen	Griffith University, Australia
Boda Ning	RMIT University, Australia
Roger Nkambou	Uqam, Canada
Akiyo Nomura	IBM Research - Tokyo, Japan
Anupiya Nugaliyadde	Murdoch University, Australia
Atsuya Okazaki	IBM Research, Japan
Jonathan Oliver	Trendmicro, Australia
Toshiaki Omori	Kobe University, Japan
Takashi Omori	Tamagawa University, Japan
Shih Yin Ooi	Multimedia University, Malaysia
Seiichi Ozawa	Kobe University, Japan
Huan Pan	Ningxia University, China
Paul Pang	Unitec Institute of Technology, New Zealand
Shuchao Pang	Macquarie University, Australia
Kitsuchart Pasupa	King Mongkut's Institute of Technology Ladkrabang, Thailand
Jagdish Patra	Swinburne University of Technology, Australia
Cuong Pham	Griffith University, Australia
Mukesh Prasad	University of Technology Sydney, Australia
Yu Qiao	Shanghai Jiao Tong University
Feno Heriniaina Rabevohitra	Chongqing University, China
Sutharshan Rajasegarar	Deakin University, Australia

Md Mashud Rana	CSIRO, Australia
Md Mamunur Rashid	CQUniversity, Australia
Pengju Ren	Xi'an Jiaotong University, China
Rim Romdhane	Devoteam, France
Yi Rong	Wuhan University of Technology, China
Leszek Rutkowski	Czestochowa University of Technology, Poland
Fariza Sabrina	CQU, Australia
Naveen Saini	Indian Institute of Technology Patna, India
Toshimichi Saito	Hosei University, Japan
Michel Salomon	University of Bourgogne Franche-Comté, France
Toshikazu Samura	Yamaguchi University, Japan
Naoyuki Sato	Future University Hakodate, Japan
Ravindra Savangouder	Swinburne University of Technology, Australia
Rafal Scherer	Czestochowa University of Technology, Poland
Erich Schikuta	University of Vienna, Austria
Fatima Seeme	Monash University, Australia
Feng Sha	The University of Sydney, Australia
Jie Shao	University of Electronic Science and Technology, China
Qi She	Intel Labs China, China
Michael Sheng	Macquarie University, Australia
Jinhua Sheng	Hangzhou Dianzi University, China
Iksoo Shin	Korea Institute of Science and Technology Information, South Korea
Mohd Fairuz Shiratuddin	Murdoch University, Australia
Hayaru Shouno	University of Electro-Communications, Japan
Sarah Ali Siddiqui	Macquarie University, Australia
Katherine Silversides	The University of Sydney, Australia
Jiri Sima	Czech Academy of Sciences, Czech Republic
Chiranjibi Sitaula	Deakin University, Australia
Marek Śmieja	Jagiellonian University, Poland
Ferdous Sohel	Murdoch University, Australia
Aneesh Srivallabh Chivukula	University of Technology Sydney, Australia
Xiangdong Su	Inner Mongolia University, China
Jérémie Sublime	ISEP, France
Liang Sun	University of Science and Technology Beijing, China
Laszlo Szilagyi	Obuda University, Hungary
Takeshi Takahashi	National Institute of Information and Communications Technology, Japan
Hakaru Tamukoh	Kyushu Institute of Technology, Japan
Leonard Tan	University of Southern Queensland, Australia
Gouhei Tanaka	The University of Tokyo, Japan
Maolin Tang	Queensland University of Technology, Australia
Selvarajah Thuseethan	Deakin University, Australia
Dat Tran	University of Canberra, Australia

Guoxia Xu	Hohai University, China
Jiaming Xu	Institute of Automation, Chinese Academy of Sciences, China
Qing Xu	Tianjin University, China
Li Xuewei	Tianjin University, China
Toshiyuki Yamane	IBM, Japan
Haiqin Yang	Hang Seng University of Hong Kong, Hong Kong, China
Bo Yang	University of Electronic Science and Technology of China, China
Wei Yang	University of Science and Technology of China, China
Xi Yang	Xi'an Jiaotong-Liverpool University, China
Chun Yang	University of Science and Technology Beijing, China
Deyin Yao	Guangdong University of Technology, China
Yinghua Yao	Southern University of Science and Technology, China
Yuan Yao	Tsinghua University, China
Lina Yao	The University of New South Wales, Australia
Wenbin Yao	Beijing Key Laboratory of Intelligent Telecommunications Software and Multimedia, China
Xu-Cheng Yin	University of Science and Technology Beijing, China
Xiaohan Yu	Griffith University, Australia
Yong Yuan	Chinese Academy of Science, China
Ye Yuan	Southwest University, China
Yun-Hao Yuan	Yangzhou University, China
Xiaodong Yue	Shanghai University, China
Seid Miad Zandavi	The University of Sydney, Australia
Daren Zha	Chinese Academy of Sciences, China
Yan Zhang	Tianjin University, China
Xiao Zhang	Huazhong University of Science and Technology, China
Yifan Zhang	CSIRO, Australia
Wei Zhang	The University of Adelaide, Australia
Lin Zhang	Beijing Institute of Technology, China
Yifei Zhang	University of Chinese Academy of Sciences, China
Huisheng Zhang	Dalian Maritime University, China
Gaoyan Zhang	Tianjin University, China
Liming Zhang	University of Macau, Macau, China
Xiang Zhang	The University of New South Wales, Australia
Yuren Zhang	Bytedance.com, China
Jianhua Zhang	Zhejiang University of Technology, China
Dalin Zhang	The University of New South Wales, Australia
Bo Zhao	Beijing Normal University, China
Jing Zhao	East China Normal University, China
Baojiang Zhong	Soochow University, China
Guoqiang Zhong	Ocean University, China

Contents – Part V

Learning from Incomplete Data

Neural Network Models

Semantic and Graph Based Approaches

Social Network Computing

Spiking Neuron and Related Models

Text Computing Using Neural Techniques

Time-Series and Related Models

Unsupervised Neural Models

Image Processing by Neural Techniques

Deep Residual-Dense Attention Network for Image Super-Resolution

Ding Qin and Xiaodong Gu$^{(\boxtimes)}$ (iD)

Department of Electronic Engineering, Fudan University,
Shanghai 200433, China
xdgu@fudan.edu.cn

Abstract. Recently, a great variety of CNN-based methods have been proposed for single image super-resolution. But how to restore more high-frequency details is still an unsolved issue. It is easy to find that the low-frequency information is similar in a pair of low-resolution and high-resolution images. So the model only needs to pay more attention to the high-frequency information to restore more realistic images which have abundant details and meet human visual system better. In this paper, we propose a deep residual-dense attention network (RDAN) for image super-resolution. Specially, we propose a channel attention module to change the weight of each channel and a spatial attention module to rescale the region weight in a channel map, which can make the model focus more on the high-frequency information. Experimental results on five benchmark datasets show that RDAN is superior to those state-of-the-art methods for both accuracy and visual performance.

Keywords: Image super-resolution · Deep convolution neural network · Attention mechanism

1 Introduction

Single image super-resolution (SISR) aims to reconstruct a high-resolution (HR) image I^{SR} from a low-resolution (LR) image I^{LR}. As a low-level computer vision task, image super-resolution has various applications, such as medical image processing and security. While image super-resolution (SR) is an ill-posed problem, one LR image input may correspond to a multitude of SR image outputs, hence plenty of algorithms have been proposed including interpolation-based [2], reconstruction-based [3] and learning-based methods [1, 4, 7, 8].

After Dong et al. [1] first introduced convolutional neural network into image SR to learn the mapping from LR image to HR image, the performance of image SR has been improved significantly. From then on, a variety of network architectures have been proposed to solve this problem. Kim et al. [4] increased the depth of network to 20 in VDSR using skip connection. With the encouragement of the ResNet [5], the network for image SR has deeper structure and more capacity for rich representation of LR images. Ledig et al. [6] used 16 residual blocks and skip connections in their SRResNet model optimized with MSE. Tong et al. [7] introduced dense skip connections into image SR, which enables the previous feature maps can be propagated to each

© Springer Nature Switzerland AG 2019
T. Gedeon et al. (Eds.): ICONIP 2019, CCIS 1143, pp. 3–10, 2019.
https://doi.org/10.1007/978-3-030-36802-9_1

following layers. To make full use of the feature information, Zhang et al. [8] combined residual and dense skip connections and designed the residual dense block for SR to fuse the local features and global features.

The main issue in SISR is how to reconstruct as much high-frequency information as possible. But there is too much low-frequency information and few high-frequency information in a LR image. So the network should pay more attention to the extraction and reconstruction of the high-frequency features. However, all of those methods consider feature maps of each channel are equal and ignore that different channels contain different amounts of the high-frequency information. Furthermore, different regions in one feature map also effect the reconstruction performance in varying degrees.

Based on the above analysis, we propose a deep convolution neural network with channel and spatial attention modules. The channel attention module is applied to learn a specific vector which represents different weights for different channel features. Meanwhile, to obtain the spatial weight map for different regions, we propose the spatial attention module. Through these two modules, our model can pay more attention to the high-frequency details and finally achieve notable reconstruction performance. Overall, our main contributions are as follows.

(1) We propose the deep residual-dense attention network (RDAN) for SISR, which demonstrate attention mechanism can significantly boost the reconstruction performance, especially high-frequency information.
(2) We propose the channel attention module and the spatial attention module in RDAN. These two modules effectively make our network pay more attention to the high-frequency features.
(3) Our RDAN outperforms state-of-the-art methods on five benchmark datasets under the ×4 scale factor and the visual performance is enhanced effectively.

2   Deep Residual-Dense Attention Model

2.1   Network Architecture

Fig. 1. The architecture of Deep Residual-Dense Attention Network (RDAN). Our RDAN is combined with four parts: shallow feature extraction module, residual dense attention blocks (deep feature extraction), upsampling module and reconstruction module (HR space).

In the past methods, LR image is always upsampled into HR space first and then fed into later convolution layers, therefore the most computation is done in the HR feature space, which will consume a lot of computing resources and slow down the training process. Inspired by the architecture of the SRResnet [6], we feed LR image into network directly and place the upsampling module after the feature extraction module as shown in Fig. 1.

To extract and make full use of the deep features, the block layout is employed and all blocks named residual dense attention block (RDAB) have the same structure. As shown in Fig. 2, RDAB consists of dense connections, channel attention module, spatial attention module and residual connection. Through the dense connections, all the feature maps of the preceding layers are concatenated as the input of the next layer.

Channel attention and spatial attention are then used to concentrate on what and where to emphasize or suppress and optimize the weights of feature maps. The feature maps $F \in \mathbb{R}^{C \times H \times W}$ are taken as the input of channel attention module. By using the pooling layer, the spatial information of a feature map is aggregated to a point and these points form a spatial context vector. The vector passes through the nonlinear mapping of the multi-layer perception to obtain the channel weight $W_c \in \mathbb{R}^{C \times 1 \times 1}$. The channel-weighted feature maps are then forwarded into spatial attention module. Different with the channel attention, spatial attention module produces the spatial weight $W_s \in \mathbb{R}^{1 \times H \times W}$. Then the refined feature maps can be obtained through attention mechanism.

$$F_{refined} = W_s \otimes (W_c \otimes F) \tag{1}$$

where $\otimes$ denotes the element-wise multiplication, W_c and W_s represents channel weight and spatial weight respectively.

After the feature extraction of RDABs, fused features are forwarded, including global residual features and local dense features refined by attention module, to the upsampling module. In our RDAN, nearest-neighbor upsampling and convolution are used to upsample features. To fit the requirement of the channel number of the image, two convolution layers are applied in HR space and the last layer outputs 3 feature maps to finish the reconstruction of SR.

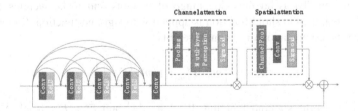

Fig. 2. The architecture of residual dense attention block (RDAB)

2.2   Attention Module

The image in LR space has abundant low-frequency features so how to restore more vivid high-frequency details and improve the reconstruction performance is what we are concerned about. In previous methods, each channel of the feature maps is considered equally and they ignore the inter-channel and inter-spatial relationship of feature maps. In order to make full use of features and restore more high-frequency information, our RDAN is supposed to pay different attention to various channels and locations via attention mechanism.

Channel Attention. The purpose of channel attention is to generate different attention (channel-wise relationship of features) for all channels.

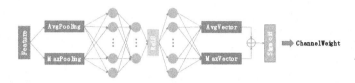

Fig. 3. The architecture of channel attention module

As shown in Fig. 3, the information of each channel is aggregated first through max-polling layer and average-pooling layer to obtain two different channel-wise descriptors: $F_{max}^c, F_{avg}^c \in \mathbb{R}^{C \times 1 \times 1}$, which represent the max-pooled features and average-pooled features.

$$F_{\max}^c = [\max(F_1), \max(F_2), \ldots, \max(F_c)] \tag{2}$$

$$F_{avg}^c = \frac{1}{H \times W} \left[\sum_{i=0}^{H-1}\sum_{j=0}^{W-1} F_1(i,j), \sum_{i=0}^{H-1}\sum_{j=0}^{W-1} F_2(i,j), \ldots, \sum_{i=0}^{H-1}\sum_{j=0}^{W-1} F_c(i,j) \right] \tag{3}$$

where $F_i \in \mathbb{R}^{H \times W}$ denotes the i-th channel of feature maps and $\max(F_i)$ is to get the maximum of the F_i. Then these two kind of features are forwarded to the same multi-layer perception respectively. The multi-layer perception has only one hidden layer and ReLU is used as activation function after the hidden layer. After obtaining $AvgVector \in \mathbb{R}^{C \times 1 \times 1}$ and $MaxVector \in \mathbb{R}^{C \times 1 \times 1}$, these two vectors are then fused through element-wise summation. To avoid the wide range of weights and make weights fall in a reasonable area, the threshold function is applied with sigmoid function. So far we can summarize the computation of channel weights:

$$W_c = \sigma \left\{ mlp\left[F_{\max}^c\right] + mlp\left[F_{avg}^c\right] \right\} \tag{4}$$

where σ denotes the sigmoid function, $W_c \in \mathbb{R}^{C \times 1 \times 1}$ and $F_{max}^c, F_{avg}^c \in \mathbb{R}^{C \times 1 \times 1}$. Then the feature maps can be refined by channel weights adaptively:

$$F_{channel-refined} = W_c \otimes F \tag{5}$$

where $\otimes$ denotes element-wise multiplication and $F \in \mathbb{R}^{C \times H \times W}$.

Spatial Attention. Different from the channel attention, spatial attention focuses on the inter-spatial relationship of a channel feature—'where' contains more information. As shown in Fig. 4, average-pooling and max-pooling are first used along channel dimension to compute the statistics of features and then we get two kinds of spatial feature maps: $F^s_{max}, F^s_{avg} \in \mathbb{R}^{1 \times H \times W}$, where F^s_{max} denotes max-pooled spatial map and F^s_{avg} denotes average-pooled spatial map. It has been proved that spatial attention map obtained via pooling layer along channel axis can improve the network perception and help to focus on the location we are interested in [21].

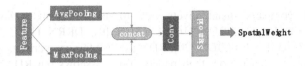

Fig. 4. The architecture of spatial attention module

Then these two maps are concatenated and then forwarded into a basic convolution layer to get the spatial attention map. Same as channel attention, sigmoid activation function is applied to scale the spatial weight to [0, 1]. So the final spatial feature map can be obtained via the following formulas:

$$\left[F^s_{max}, F^s_{avg} \right] = \text{concat}(\text{Avgpool}(F), \text{Maxpool}(F)) \qquad (6)$$

$$W_s = \sigma\left(conv\left(\left[F^s_{max}, F^s_{avg} \right] \right) \right) \qquad (7)$$

where σ represents the sigmoid function and $W_s \in \mathbb{R}^{1 \times H \times W}$.

3   Experiments

3.1   Datasets and Metrics

In our experiment, DIV2k [10] dataset is used as train set, which is proposed for image super-resolution task and contains 800 2k-resolution images. We test the performance on five extensive used benchmark dataset: Set5 [11], Set14 [12], BSD100 [13], Urban100 [14], Manga109 [15]. For image super-resolution task, Peak Signal-to-Noise Ratio (PSNR) and structural similarity index (SSIM) are commonly adopted to measure the reconstruction quality. In our experiment, we evaluate the results with PSNR and SSIM on Y channel of YCbCr space.

3.2   Training Details

Our experiment are performed under the ×4 scale factor. Before training, we need to prepare the LR images corresponding to the HR images. They are first cropped to sub images of size 480 × 480. Then these HR sub images are downsampled by bicubic interpolation to generate corresponding LR sub images in MATLAB.

When training, LR images are randomly cropped to 48 × 48. Adam optimizer is applied to optimize our RDAN with $\beta_1 = 0.9, \beta_2 = 0.999, \epsilon = 10^{-8}$. The initial learning is set as 2×10^{-4} and decays by a factor of 2 after every 2×10^5 iterations. We organize our RDAN in the Pytorch framework with a 2080Ti GPU and train the model for 10^6 iterations in total with L1 loss. Our method works the same for both gray images and color images.

3.3   Comparisons to the State-of-the-Art Methods

Our results are compared quantitatively with 10 state-of-the-art methods, including SRCNN [1], VDSR [4], DRCN [16], LapSRN [9], DRRN [17], MemNet [19], SRDenseNet [7], DSRN [18], IDN [20], RDN [8]. For fair comparisons, all the results of these methods are cited from their papers. Table 1 shows our RDAN achieve the best or the second performance on Set5, Set14, BSD100, Urban100, Manga109 under the ×4 scale factor. Even though our method doesn't perform better than RDN in term of PSNR on Set14, these two methods only differs by 0.01 dB. Meanwhile, our method performs best in term of SSIM. It indicates our attention module improves the reconstruction performance efficiently.

Table 1. Comparisons with PSNR (dB) and SSIM. The best result is in **Bold** and the second is in *Italic*.

Method	Set5		Set14		BSD100		Urban100		Manga109	
	PSNR	SSIM	PSNR	SSIM	PSNR	SSIM	PSNR	SSIM	PSNR	SSIM
Bicubic	28.42	0.8104	26.00	0.7027	25.96	0.6675	23.14	0.6577	24.89	0.7866
SRCNN [1]	30.48	0.8628	27.50	0.7513	26.90	0.7101	24.52	0.7221	27.58	0.8555
VDSR [4]	31.35	0.8838	28.01	0.7674	27.29	0.7251	25.18	0.7524	–	–
DRCN [16]	31.53	0.8854	28.02	0.7670	27.23	0.7233	25.14	0.7510	–	–
LapSRN [9]	31.54	0.885	28.19	0.772	27.32	0.728	25.21	0.756	29.09	0.890
DRRN [17]	31.68	0.8888	28.21	0.7720	27.38	0.7284	25.44	0.7638	–	–
DSRN [18]	31.40	0.883	28.07	0.770	27.25	0.724	25.08	0.747	–	–
MemNet [19]	31.74	0.8893	28.26	0.7723	27.40	0.7281	25.50	0.7630	–	–
IDN [20]	31.82	0.8903	28.25	0.7730	27.41	0.7297	25.41	0.7632	–	–
SRDenseNet [7]	32.02	0.8934	28.50	0.7782	27.53	0.7337	26.05	0.7819	–	–
RDN [8]	*32.47*	*0.8990*	**28.81**	*0.7871*	*27.72*	*0.7419*	*26.61*	*0.8028*	*31.00*	*0.9151*
Ours	**32.59**	**0.9000**	*28.80*	**0.7872**	**27.73**	**0.7420**	**26.65**	**0.8035**	**31.10**	**0.9162**

The visual results of several state-of-the-art methods under the scale factor of ×4 are also illustrated as shown in Fig. 5. For img_74.png and img_005.png from Urban100, it is obvious that the most of compared methods suffer from blurring and over-smoothing artifacts, even distortion in the cropped images. But our RDAN can not only greatly alleviate the artifacts, but also produce sharp edges and abundant details. For 78004.png from BSD100 and Barbara.png from Set14, the right parts of the cropped images have dense horizontal and vertical lines. Unacceptably, the compared methods can't reconstruct these lines properly among the windows and the books. The recovered lines have the wrong directions or even complete distortion. Our RDAN can generate more realistic results.

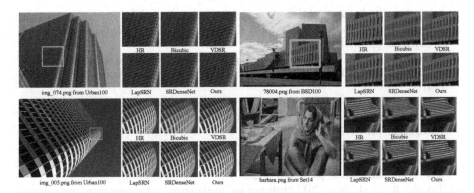

Fig. 5. Visual comparisons with other methods

4 Conclusion

In this paper, we introduce the attention mechanism to single image super-resolution (SISR). Specially, we apply the channel attention module to explore the inter-channel relationship of features and obtain the channel- refined feature maps. Furthermore, the spatial attention module is used to look for the inter-spatial relationship of features. Through the refinement of channel and spatial attention modules, our RDAN is able to focus on more informative features. The evaluations with PSNR and SSIM, as well as the visual performance, indicate our method outperforms those state-of-the-art methods.

Acknowledgments. This work was supported in part by National Natural Science Foundation of China under grants 61771145 and 61371148.

References

1. Dong, C., Loy, C.C., He, K., Tang, X.: Learning a deep convolutional network for image super-resolution. In: Fleet, D., Pajdla, T., Schiele, B., Tuytelaars, T. (eds.) ECCV 2014. LNCS, vol. 8692, pp. 184–199. Springer, Cham (2014). https://doi.org/10.1007/978-3-319-10593-2_13
2. Zhang, L., Wu, X.: An edge-guided image interpolation algorithm via directional filtering and data fusion. IEEE Trans. Image Process. 15(8), 2226–2238 (2006)
3. Zhang, K., Gao, X., Tao, D., Li, X.: Single image super-resolution with non-local means and steering kernel regression. IEEE Trans. Image Process. 21(11), 4544–4556 (2012)
4. Kim, J., Kwon Lee, J., Mu Lee, K.: Accurate image super-resolution using very deep convolutional networks. In: Proceedings of the IEEE Conference on Computer Vision and Pattern Recognition, pp. 1646–1654 (2016)
5. He, K., Zhang, X., Ren, S., Sun, J.: Deep residual learning for image recognition. In: Proceedings of the IEEE Conference on Computer Vision and Pattern Recognition, pp. 770–778 (2016)

6. Ledig, C., et al.: Photo-realistic single image super-resolution using a generative adversarial network. In: Proceedings of the IEEE Conference on Computer Vision and Pattern Recognition, pp. 4681–4690 (2017)

7. Tong, T., Li, G., Liu, X., Gao, Q.: Image super-resolution using dense skip connections. In: Proceedings of the IEEE International Conference on Computer Vision, pp. 4799–4807 (2017)

8. Zhang, Y., Tian, Y., Kong, Y., Zhong, B., Fu, Y.: Residual dense network for image super-resolution. In: Proceedings of the IEEE Conference on Computer Vision and Pattern Recognition, pp. 2472–2481 (2018)

9. Lai, W.S., Huang, J.B., Ahuja, N., Yang, M.H.: Deep laplacian pyramid networks for fast and accurate super-resolution. In: Proceedings of the IEEE Conference on Computer Vision and Pattern Recognition, pp. 624–632 (2017)

10. Timofte, R., et al.: Ntire 2017 challenge on single image super-resolution: methods and results. In: Proceedings of the IEEE Conference on Computer Vision and Pattern Recognition Workshops, pp. 114–125 (2017)

11. Bevilacqua, M., Roumy, A., Guillemot, C., Alberi-Morel, M.L.: Low-complexity single-image super-resolution based on nonnegative neighbor embedding. In: Proceedings British Machine Vision Conference, pp. 135.1–135.10 (2012)

12. Zeyde, R., Elad, M., Protter, M.: On single image scale-up using sparse-representations. In: Boissonnat, J.-D., et al. (eds.) Curves and Surfaces 2010. LNCS, vol. 6920, pp. 711–730. Springer, Heidelberg (2012). https://doi.org/10.1007/978-3-642-27413-8_47

13. Martin, D., Fowlkes, C., Tal, D., Malik, J.: A database of human segmented natural images and its application to evaluating segmentation algorithms and measuring ecological statistics. In: IEEE International Conference on Computer Vision, pp. 416–423 (2001)

14. Huang, J.B., Singh, A., Ahuja, N.: Single image super-resolution from transformed self-exemplars. In: Proceedings of the IEEE Conference on Computer Vision and Pattern Recognition, pp. 5197–5206 (2015)

15. Matsui, Y., et al.: Sketch-based manga retrieval using manga109 dataset. Multimedia Tools Appl. **76**(20), 21811–21838 (2017)

16. Kim, J., Kwon Lee, J., Mu Lee, K.: Deeply-recursive convolutional network for image super-resolution. In: Proceedings of the IEEE Conference on Computer Vision and Pattern Recognition, pp. 1637–1645 (2016)

17. Tai, Y., Yang, J., Liu, X.: Image super-resolution via deep recursive residual network. In: Proceedings of the IEEE Conference on Computer Vision and Pattern Recognition, pp. 3147–3155 (2017)

18. Han, W., Chang, S., Liu, D., Yu, M., Witbrock, M., Huang, T.S.: Image super-resolution via dual-state recurrent networks. In: Proceedings of the IEEE Conference on Computer Vision and Pattern Recognition, pp. 1654–1663 (2018)

19. Tai, Y, Yang, J., Liu, X., Xu, C.: MemNet: a persistent memory network for image restoration. In: Proceedings of the IEEE International Conference on Computer Vision, pp. 4539–4547 (2017)

20. Hui, Z., Wang, X., Gao, X.: Fast and accurate single image super-resolution via information distillation network. In: Proceedings of the IEEE Conference on Computer Vision and Pattern Recognition, pp. 723–731 (2018)

21. Zagoruyko, S., Komodakis, N.: Paying more attention to attention: Improving the performance of convolutional neural networks via attention transfer. arXiv preprint arXiv: 1612.03928 (2016)

Discriminant Feature Learning with Self-attention for Person Re-identification

Yang Li[1], Xiaoyan Jiang[1(✉)], and Jenq-Neng Hwang[2]

[1] School of Electronic and Electrical Engineering,
Shanghai University of Engineering Science,
No. 333 of Longteng Road, Shanghai, China
xiaoyan.jiang@sues.edu.cn
[2] Department of Electrical and Computer Engineering,
University of Washington, Box 352500, Seattle, WA 98195, USA

Abstract. Person re-identification (re-ID) across cameras is a crucial task, especially when cameras' fields of views are non-overlapping. Feature extraction is challenging due to changing illumination conditions, complex background clutters, various camera viewing angles, and occlusions in this case. Moreover, the space mis-alignment of human corresponding regions caused by detectors is a big issue for feature matching across views. In this paper, we propose a strategy of merging attention models with the resnet-50 network for robust feature learning. The efficient self-attention model is used directly on the feature map to solve the space mis-alignment and local feature dependency problems. Furthermore, the loss function which jointly considers the cross-entropy loss and the triplet loss in training enables the network to capture both invariant features within the same individual and distinctive features between different people. Extensive experiments show that our proposed mechanism outperforms the state-of-the-art approaches on the large-scale datasets Market-1501 and DukeMTMC-reID.

Keywords: Person re-identification · Feature extraction · Self-attention · Cross-entropy loss · Triplet loss

1 Introduction

Person re-ID aims to retrieve specific pedestrians in images or video sequences obtained by multiple non-overlapping cameras. Given a query person's image, the task is to find the correct corresponding matches among a set of candidate images captured by different cameras in the gallery. Two crucial issues should be addressed: robust feature extraction and suitable distance metrics. It is a challenging topic because of changing illumination conditions, complex background clutters, various camera viewing angles, and occlusions.

Y. Li and X. Jiang contribute equally to this work.

© Springer Nature Switzerland AG 2019
T. Gedeon et al. (Eds.): ICONIP 2019, CCIS 1143, pp. 11–19, 2019.
https://doi.org/10.1007/978-3-030-36802-9_2

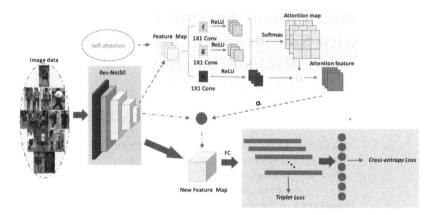

Fig. 1. The proposed network architecture. It consists of three parts: the resnet-50 network, the self-attention model, and the loss function. The self-attention model includes multi-scale transformation spaces and nonlinear combination of feature maps. The loss function considers both intra-person similarity and inter-person distinction.

To overcome the challenges, there have been numerous research efforts on person re-ID, which can be broadly categorized into the following techniques: representation learning, metric learning, local feature-based, generative adversarial networks, and unsupervised learning-based. More specifically, representation learning is a supervised learning relying on feature extraction methods [6–8,28]. Unlike representation learning method, metric learning aims to learn the similarity between two images through the network. Commonly used loss methods for metric learning are contrastive loss [16], triplet hard loss with batch hard mining [4], and margin sample mining loss [20]. The acquisition of person re-ID dataset is another challenge. Therefore, after the emergence of the generative adversarial networks (GAN) [10,19,27,30], the model is used to expand our datasets as much as possible to improve the learning ability of the network. In contrast, the method of supervised learning relies on the demand for a large number of labeled data. Hence, unsupervised learning methods [9,18,22] depending on unlabeled datasets are proposed to cross-view identity-specific information.

We propose a self-attention model extraction framework to obtain discriminant features specifically for person re-ID. Our approach outperforms the state-of-the-art performance on standard datasets evaluated by standard metrics. The main contributions are summarized as following two-folds:

– A multi-scale attention model is jointly with the resnet-50 [3] to solve the spatial mis-alignment and local feature dependency problems of person re-ID. The nonlinear combination of features from multiple scales in the self-attention model merges the global and local features of the image effectively.

– The cross-entropy loss function usually used in multi-classification task is combined with the triplet loss function for person re-ID in the supervised training process of the network. The trained model extracts the similar characteristics of the same individuals better and significantly enlarges feature distinctions between different persons.

2   Proposed Technical Approaches

We propose a deep learning framework to deal with person re-ID as shown in Fig. 1, where images are sequentially fed to the pre-trained resnet-50 [3] network for feature extraction to obtain corresponding feature maps. The feature map is then input to the self-attention network (Sect. 2.1) to generate new discriminative spatial visual features. To better represent intra-person features and distinguish inter-person differences, the loss function fuses cross-entropy and triplet loss (Sect. 2.2) to give reasonable feedback to the network finally. The details of the proposed framework are described as the following.

2.1   Self-attention Model

Refer to the attention mechanism [17], a novel self-attention model is proposed as shown in Fig. 1. We compute the visual space feature of the image $\{f_{n,l}\}_{l=1,\cdots,L}$ by resnet-50, which is simplified to $\mathbf{x} \in \mathbb{R}^{D \times L}$ while keeping the size as the original feature map size (h, w). Then we sent the feature $\mathbf{x}$ into two different feature transformation spaces $\mathbf{f}$ and $\mathbf{g}$ to calculate the attention score, and another different transformation space $\mathbf{h}$ to generate a new feature map. Thus, we have three different feature transformation spaces $\mathbf{k} \in (\mathbf{f}, \mathbf{g}, \mathbf{h})$. We indicate $W_f \in \mathbb{R}^{\bar{D} \times D}, W_g \in \mathbb{R}^{\bar{D} \times D}, W_h \in \mathbb{R}^{\bar{D} \times D}$ to be parameter matrices trained by the network. Dot-product with attention weight is defined by the following formula:

$$f(x) = W_f x, g(x) = W_g x, h(x) = W_h x. \tag{1}$$

Here, we use 1×1 convolution to perform linear combination of feature maps in different scales and different spaces. Since the size of the original feature map is large, the dot-product requires a large memory size. We convert them to a lower-dimensional feature space $\bar{D} = D/8$ by filtering the original feature map using a small number of kernels.

After linear space conversion, the ReLU activation function is applied to perform nonlinear processing to obtain the response $e_{i,j}$, which is corresponding to the original feature map. The characteristic response transformation formula is:

$$e_{i,j} = (max(f(x_i), 0))^T (max(g(x_i), 0)). \tag{2}$$

Based on the obtained response $e_{i,j}$, our attention score $s_{j,i}$ is calculated as follows:

$$s_{j,i} = \frac{exp(e_{i,j})}{\sum_{i=1}^{D} exp(e_{i,j})}, \tag{3}$$

where $s_{j,i}$ indicates the extent to which the model attends to the ith location on the feature map when synthesizing the jth dimension. By definition, each receptive field is a probability mass function since $\sum_{i=1}^{L} \sum_{j=1}^{D} s_{i,j} = 1$.

We denote the output of the attention model layer as $O = (o_1, o_2, \cdots, o_j, \cdots, o_N) \in \mathbb{R}^{D \times L}$, where o_j is defined as

$$o_j = \sum_{i=1}^{L} \sum_{j=1}^{D} s_{i,j} h(x). \tag{4}$$

In addition, we further multiply the output of the attention layer by a scale parameter α, of which the result is added back to the input feature map. Therefore, the final output y_i is given by $y = \alpha o_i + x_i$, where α is initialized as 0. Such parameter setting makes the network to give priority to the local neighborhood and then gradually learn to assign high weights to non-local evidence.

2.2   Loss Function

We combine the triplet loss function with hard mining [4] and the softmax cross-entropy loss function with label smoothing regularization in the training process of the CNN network. The triplet loss function is originally proposed in [4], which is named as Batch Hard triplet loss function. For each batch, we randomly select the number of P different individuals in the dataset. For each person, K tracks are randomly selected. A single track is composed by images captured from different viewing angles or extracted from different sequences. Since the proposed person re-ID approach needs only images not videos, each track here can contains only one image. We set $T = 1$ and each batch contains $P \times K$ images in total.

For each sample a in the batch, the hardest positive and the hardest negative samples within the batch are selected to form the triplet for computing the loss $L_{triplet}$:

$$L_{triplet} = \overbrace{\sum_{i=1}^{p} \sum_{a=1}^{k}}^{\text{all anchors}} [m + \overbrace{\max_{p=1 \cdots K} D(f_a^i, f_p^i)}^{\text{hardest positive}} - \underbrace{\min_{\substack{j=1 \cdots P \\ n=1 \cdots K \\ j \neq i}} D(f_a^i, f_n^i)}_{\text{hardest negative}}]_+. \tag{5}$$

The label-smoothing regularization is used to regularize the model by

$$L'_{softmax} = -\frac{1}{P \times K} \sum_{i=1}^{P} \sum_{a=1}^{K} p_{i,a} log((1 - \varepsilon)q_{i,a} + \frac{\varepsilon}{N}), \tag{6}$$

which is a mixture of the original ground-truth distribution $q_{i,a}$ and the uniform distribution $u(x) = \frac{1}{N}$ with weights 1-ε and ε, respectively. We set $\varepsilon = 0.1$ and N is the number of classes.

The total loss L is the combination of the above mentioned two loss items:

$$L = L_{triplet} + L'_{softmax}. \tag{7}$$

3 Experiment

Our approach is evaluated on two public standard datasets: Market-1501 [26] and Duke MTMC-reID [11], which are collected by multiple cameras and have relatively large-scale sample images. The experimental evaluation demonstrates that our proposed framework obtains better performance than other state-of-the-art methods.

Table 1. Ablation study using the single query.

Database	resnet-50	Spatial-attention	Ours	Evaluation				
				Rank-1	Rank-5	Rank-10	Rank-20	mAP
Market-1501	√	×	×	86.8%	94.0%	96.8%	98.0%	75.6%
	√	√	×	88.6%	93.9%	97.1%	98.0%	76.2%
	√	×	√	**90.2%**	96.7%	98.1%	99.0%	**82.7%**
DukeMTMC-reID	√	×	×	76.8%	88.0%	90.2%	91.6%	74.2%
	√	√	×	78.2%	89.6%	92.4%	92.3%	75.8%
	√	×	√	**81.0%**	92.4%	94.2%	95.9%	**78.0%**

3.1 Evaluation Criteria and Parameter Configuration

Experimental results are evaluated by the accuracy of rank-1, 5, 10, 20 and the mean average precision (mAP) metrics. It is a remarkable fact that all experiments use a single query. The resnet-50 [3] network pre-trained on ImageNet [12] is used to initialize the convolutional layer of the proposed network. The size of all images is set to 256×128. The model is trained for 800 epochs. The starting learning rate is 0.0003 and is reduced by the optimization algorithm Adam with a decay rate of 0.0005 for each 100 epochs. The batch size is 32. The similarity of features is calculated by the Euclidean distance and is normalized to $[0, 1]$.

3.2 Ablation Studies

The experimental results of three different methods are shown in Table 1. The first method is resnet-50-based without using the attention mechanism. The second uses spatial-attention and the third one uses our self-attention. The spatial-attention method directly performs a one-dimensional time-series convolution on each feature map to obtain the corresponding weight value, and each function value is transferred into a vector, which is substituted into the softmax function to obtain the weight value of each feature map. Finally, we fuse multiple feature maps with the last feature map for training based on the weight value. According to the experimental results, the attention mechanism plays a certain role in the two datasets. The Rank-1 accuracy is improved by 3%–5%.

Table 2. Comparison of state-of-the-art methods on the Market-1501.

Database	Market-1501	
Method	Rank-1	mAP
Gated-Sia [16]	65.88	39.55
Spindle [24]	76.90	–
PIE [25]	79.33	55.95
DLPAR [26]	81.00	63.40
Deep Transfer [2]	83.70	65.50
PDC [14]	84.14	63.41
DML [23]	87.70	68.80
JLML [5]	85.10	65.50
PN-GAN [10]	89.43	72.58
Ours	**90.20%**	**82.70%**

Table 3. Comparison of state-of-the-art methods on the DukeMTMC-reID.

Database	DukeMTMC-reID	
Method	Rank-1	mAP
Basel.+LSRO [27]	67.70	47.10
Basel.+OIM [21]	68.10	–
AttIDNet [7]	70.69	51.88
ACRN [13]	72.60	52.00
SVDNet [15]	76.70	56.80
Chen et. al [1]	79.20	60.60
Inception-V3 [29]	80.48	63.27
Ours	**81.00%**	**78.00%**

Table 4. Evaluation on the dataset Market-1501 using different loss function.

Database	Loss		Evaluation				
	Cross-entropy	Triplet	Rank-1	Rank-5	Rank-10	Rank-20	mAP
Market-1501	√		85.3%	94.8%	97.0%	98.3%	76.4%
		√	86.2%	95.1%	97.3%	98.8%	76.8%
	√	√	**90.2%**	96.7%	98.1%	99.0%	**82.7%**

Comparison with the State-of-the-Art. As shown in Tables 2 and 3, we compare our results with the state-of-the-art methods on the dataset Market-1501 and DukeMTMC-reID, respectively. In the comparison, the metric learning based methods include Gated-Sia [16], Basel.+LSRO [27], DML [23], JLML [5], Basel.+OIM [21], and Verif.-Identif.+LSRO [27]. Deep learning based methods include Inception-V3 [29], PDC [14], and Deep Transfer [2].

3.3   Analysis and Visualization

We perform the relevant comparison on the dataset Market-1501, and visualize the results for analysis. In the following section, parameter configuration of the work influences the experimental results.

In the training process, our loss function is composed by the cross entropy and the triplet loss. In order to access the effect of each loss, we did experiments by using only one of the two loss functions for person re-ID as shown in Table 4. The cross entropy loss function mainly focuses on the common features within the class, without paying attention to the differences with other classes, which is beneficial for person re-ID task to learn the characteristics of the same person in different conditions. In contrast, the triplet loss function considers both intra-class features and inter-class properties, helping the network obtaining more invariant features in the training process. Two loss functions are fused in our approach so that they are mutually constrained and interactive with each other.

To analyze the attention model in the network, Fig. 2 shows the effect of the partial kernel matrix on the corresponding original image which is different among various pedestrians. The reason is that the kernel matrix of the attention mechanism is obtained by multiplying the actual position of the corresponding pixel on the feature map. The final highlight is more concentrated on certain parts.

Fig. 2. Visualization of the self-attention model on the Market-1501. Our approach highlights distinctive image regions which are useful for person re-identification. The attention models primarily focus on foreground regions and generally correspond to specific body parts.

4 Conclusion

In this paper, an end-to-end framework for solving person re-ID is proposed. By fusing different spatial feature positions and combining two kinds of loss functions for network training, we obtain more discriminative and representative features. We utilize this attention mechanism to pay more attention to spatial position relationships between local features and global features, and use different nonlinear functions to carry out effective feature combination. Meanwhile, to obtain the similarity and difference characteristics of pedestrians, we combine two high-efficiency loss functions to supervise training of the network to improve the performance. Experiments show that our approach achieves better results than the state-of-the-art methods.

Acknowledgments. The work is supported by the following projects: National Natural Science Foundation of China, Nr.: 61702322, 6177051715, 61831018.

References

1. Chen, Y., Zhu, X., Gong, S.: Person re-identification by deep learning multi-scale representations. In: CVPR (2017)
2. Geng, M., Wang, Y., Xiang, T., Tian, Y.: Deep transfer learning for person reidentification. IEEE TIP (2016)
3. He, K., Zhang, X., Ren, S., Sun, J.: Deep residual learning for image recognition. In: CVPR (2015)
4. Hermans, A., Beyer, L., Leibe, B.: In defense of the triplet loss for person reidentification. arXiv:1703.07737 (2017)
5. Li, W., Zhu, X., Gong, S.: Person re-identification by deep joint learning of multiloss classification. In: IJCAI (2017)
6. Li, W., Zhu, X., Gong, S.: Harmonious attention network for person re-identification. In: CVPR (2018)
7. Lin, Y., Zheng, L., Zheng, Z., Wu, Y., Yang, Y.: Improving person re-identification by attribute and identity learning. In: CVPR (2017)
8. Luo, H., Gu, Y., Liao, X., Lai, S., Jiang, W.: Bag of tricks and a strong baseline for deep person re-identification. In: CVPR (2019)
9. Ma, X., et al.: Personre-identification by unsupervised video matching. Pattern Recogn. **65**, 197–210 (2017)
10. Qian, X., et al.: Pose-normalized image generation for person re-identification. In: CVPR (2018)
11. Ristani, E., Solera, F., Zou, R., Cucchiara, R., Tomasi, C.: Performance measures and a data set for multi-target, multi-camera tracking. In: Hua, G., Jégou, H. (eds.) ECCV 2016. LNCS, vol. 9914, pp. 17–35. Springer, Cham (2016). https://doi.org/10.1007/978-3-319-48881-3_2
12. Russakovsky, O., et al.: Imagenet large scale visual recognition challenge. IJCV **115**(3), 211–252 (2015)
13. Schumann, A., Stiefelhagen, R.: Person re-identification by deep learning attribute-complementary information. In: CVPRW (2017)
14. Su, C., Li, J., Zhang, S., Xing, J., Gao, W., Tian, Q.: Pose-driven deep convolutional model for person re-identification. In: ICCV (2017)
15. Sun, Y., Zheng, L., Deng, W., Wang, S.: Svdnet for pedestrian retrieval. In: ICCV (2017)
16. Varior, R.R., Haloi, M., Wang, G.: Gated siamese convolutional neural network architecture for human re-identification. In: Leibe, B., Matas, J., Sebe, N., Welling, M. (eds.) ECCV 2016. LNCS, vol. 9912, pp. 791–808. Springer, Cham (2016). https://doi.org/10.1007/978-3-319-46484-8_48
17. Vaswani, A., et al.: Attention is all you need. arXiv:1706.03762 (2017)
18. Wang, H., Zhu, X., Xiang, T., Gong, S.: Towards unsupervised open-set person re-identification. In: ICIP (2016)
19. Wei, L., Zhang, S., Gao, W., Tian, Q.: Person transfer gan to bridge domain gap for person re-identification. In: CVPR (2018)
20. Xiao, Q., Luo, H., Zhang, C.: Margin sample mining loss: a deep learning based method for person re-identification. arXiv:1710.00478 (2017)
21. Xiao, T., Li, S., Wang, B., Lin, L., Wang, X.: Joint detection and identification feature learning for person search. In: CVPR (2017)
22. Yu, H., W.Zheng, A, Wu, Guo, X., Gong, S., Lai, J.: Unsupervised person re-identification by soft multilabel learning. In: CVPR (2019)

23. Zhang, Y., Xiang, T., Hospedales, T.M., Lu, H.: Deep mutual learning. arXiv:1706.00384 (2017)
24. Zhao, H., et al.: Spindle net: person re-identification with human body region guided feature decomposition and fusion. In: CVPR (2017)
25. Zheng, L., Huang, Y., Lu, H., Yang, Y.: Pose invariant embedding for deep person reidentification. arXiv:1701.07732 (2017)
26. Zheng, L., Shen, L., Tian, L., Wang, S., Wang, J., Tian, Q.: Scalable person re-identification: a benchmark. In: ICCV (2015)
27. Zheng, Z., Zheng, L., Yang, Y.: Unlabeled samples generated by gan improve the person re-identification baseline in vitro. In: ICCV (2017)
28. Zheng, Z., Zheng, L., Yang, Y.: Pedestrian alignment network for large-scale person re-identification. IEEE Trans. Circ. Syst. Video Technol. (2018)
29. Zhong, Z., Zheng, L., Cao, D., Li, S.: Re-ranking person re-identification with k-reciprocal encoding. In: CVPR (2017)
30. Zhong, Z., Zheng, L., Zheng, Z.: Camera style adaptation for person re-identification. In: CVPR (2018)

SCS: Style and Content Supervision Network for Character Recognition with Unseen Font Style

Wei Tang[1,2,3(✉)], Yiwen Jiang[1,2,3], Neng Gao[3], Ji Xiang[3], Yijun Su[1,2,3],
and Xiang Li[1,2,3]

[1] State Key Laboratory of Information Security, Chinese Academy of Sciences,
Beijing, China
{tangwei,jiangyiwen,suyijun,lixiang9015}@iie.ac.cn
[2] Institute of Information Engineering, Chinese Academy of Sciences, Beijing, China
[3] School of Cyber Security, University of Chinese Academy of Sciences,
Beijing, China
{gaoneng,xiangji}@iie.ac.cn

Abstract. There is a significant style overfitting problem in traditional content supervision models of character recognition: insufficient generalization ability to recognize the characters with unseen font styles. To overcome this problem, in this paper we propose a novel framework named Style and Content Supervision (SCS) network, which integrates style and content supervision to resist style overfitting. Different from traditional models only supervised by content labels, SCS simultaneously leverages the style and content supervision to separate the task-specific features of style and content, and then mixes the style-specific and content-specific features using bilinear model to capture the hidden correlation between them. Experimental results prove that the proposed model is able to achieve the state-of-the-art performance on several widely used real world character sets, and it obtains relatively strong robustness when the size of training set is shrinking.

Keywords: Character recognition · Convolutional neural networks · Style overfitting · Style supervision

1 Introduction

In our daily life, character is the most important information carrier. There are more than tens of thousands of different characters with variable font styles, and most of them can be well recognized by most people. But in the field of artificial intelligence, character recognition is considered as an extremely difficult task due to the very large number of categories, complicated structures, similarity between characters and the variability of font styles. Because of its unique technical challenges and great social needs, there are intensive research in this field and a rapid increase of successful techniques, especially the Convolutional

© Springer Nature Switzerland AG 2019
T. Gedeon et al. (Eds.): ICONIP 2019, CCIS 1143, pp. 20–31, 2019.
https://doi.org/10.1007/978-3-030-36802-9_3

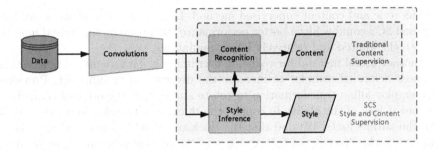

Fig. 1. Different to previous content supervision methods, SCS leverages both style and content supervision simultaneously. SCS takes into account the correlation between style and content in purpose of reducing style overfitting.

Neural Network (CNN), which has played an important role to improve recent Computer Vision studies [4,6,9,14,15].

As CNNs have achieved the great success in general object recognition, face recognition and other image recognition tasks [3,7,11,14,15,22], CNN-based methods also break the bottleneck of character recognition and achieve excellent performance even better than human on several popular large-scale benchmarks such as MNIST and ICDAR [1,2,18,21,24]. But there is a significant style overfitting problem still challenging character recognition: insufficient generalization ability to recognize characters with unseen font styles. For example, a well trained CNN model supervised by content label in a traditional way always perform poorly when test on the characters having such font style that the trained model has never seen [16].

Another limitation of traditional content supervision method is: the generalization ability of the trained model will decline dramatically when the scale of training set is shrinking [16]. Note that people learning new concepts can often generalize successfully from just a single example, yet machine learning algorithms typically require tens or hundreds of examples to perform with similar accuracy, and people can also use learned concepts in richer ways than neural-based algorithms [8]. Unlike human-level character recognition aimed to enhance generalization ability by using limited training set, in traditional applications of character recognition, the training set needs to contain as many font styles as possible to enhance the generalization ability of the trained model. But this naive method of expanding the training set is absolutely costly [6]. In addition, it is contrary to the intention of building a human-level learning model to extract rich concepts from limited data [8].

In this paper, we propose a novel framework named Style and Content Supervision (SCS) network to separate the style-specific and content-specific features of character. Different from traditional approaches only supervised by content labels, SCS simultaneously leverages the style and content supervision and takes into account the hidden correlation of style and content. Figure 1 shows the differences between the traditional content supervised method and our simultaneous style and content supervised method. Leveraging style supervision, the proposed SCS could achieve better performance than content supervised models on both the limited and relatively large training set.

Another useful method for reducing overfitting is data augmentation, which produces deformed samples to increase the diversity of training set. Previous work employ affine transformation to rotate and distort the original characters [23], but it lack of the ability to control the thickness of strokes and hardly deal with the outline fonts. We find that the thickness of strokes is a key point associated with the diversity of font styles, and the outline fonts are more hard to recognize compare with ordinary fonts. According to this situation, we design a simple but efficient data augmentation approach based on morphologic operations such as erosion and dilation to change the thickness of strokes and transfer the input character into outline style.

We conduct extensive experiments on several widely used character sets: Arabic numerals, English letters and simplified Chinese characters. Experimental results demonstrate that the proposed SCS is able to achieve state-of-the-art performance on all of these character sets, and it obtain desirable robustness on both of limited training set and relatively lager training set. And the benefit of the proposed data augmentation method is also be proved in our experiments. Overall, our contributions are as follows:

1. We propose an novel end-to-end trainable framework for character recognition, called Style and Content Supervision (SCS) network, which simultaneously leverages the style and content supervision to recognize character with unseen font style.
2. We verify the effectiveness of data augmentation approach based on erosion and dilation, which alter the thickness of character strokes as well as generate outline fonts to increase the style diversity of the training set for reducing style overffiting.
3. We carry out a series of experiments to evaluate our method, and the experimental results prove that the proposed SCS is able to achieve state-of-the-art performance on multiple character sets even when the scale of training set is changing.

The rest of this paper is organized as follows. Section 2 summarizes the related works. Section 3 introduces the proposed method in details. Section 4 presents the experimental results. Finally in Sect. 5, we conclude our work and discuss the future work.

2 Related Work

In recent years, character recognition has achieved unprecedented success because of the rise of Convolutional Neural Networks. But all these successful models are training on a large scale dataset and test on a relatively small dataset without significant font style variation. Wu et al. [18] propose a character recognition model based on relaxation convolutional neural network, and took the 1st place in ICDAR'13 Handwriting Character Recognition Competition [21]. Meier et al. [2] create a multi-column deep neural network achieving first human-competitive performance on the famous MNIST handwritten digit recognition task. And this neural based model recognize the 3755 classes of handwritten Chinese characters in ICDAR'13 with almost human performance. Zhong et al. [24] design a streamlined version of GoogLeNet for character recognition outperforming previous best result with significant gap. Chen et al. [1] propose a CNN-based character recognition framework employ random distortion [13] and multi-model voting [12]. This classifier perform even better than human on MNIST and Chinese character set. In addition to the above models, the current state-of-the-art character recognition method proposed by Xiao et al. [19] it is also challenged by the style overfitting problem.

A growing number of works are devoted to enhance the generalization ability of character recognition models to make their performance desirable in dealing with unseen font styles. Xu et al. [20] propose a artificial neural network architecture called Cooperative Block Neural Networks to address the variation in the shape of characters by considering only three different fonts. Lv [10] successfully applied the stochastic diagonal Levenberg-Marquardt method to a convolutional neural network to recognize a small set of multi-font characters that consist of the Arabic numerals and English letters without the Chinese characters. Zhong et al. [23] propose a CNN-based multi-font character recognizer using multi-pooling and affine data augmentation achieving acceptable result, but the size of the training set is more than 500% of the test set (240 font styles for training and other 40 for test) and shrinking the size of training set will significantly reduce the effect of the model. Tang et al. [16] propose a specific kernel to extract the marginal distribution of character pixels that takes account the skeleton information to enhance the generalization ability of CNNs when training on a limited data set. But all these previous works are based on a single network supervised by content label and ignore the informative style supervision.

3   Methodology

In this section, we first introduce the overall architecture of the proposed SCS. Then we present our method in detail by successively introducing the modules of data augmentation, shared convolution as well as style and content branches.

3.1   Overall Architecture

An overview of our framework is illustrated in Fig. 2. The backbone of the style inference branch is a full connection network with a softmax classifier supervised by style label, which has a extra mixer to receive the feedback from content branch. And the content recognition branch obtain the same architecture with the style inference branch, which is supervised by content label and also has a mixer to receive the feedback from style branch. We adopt a convolutional network as the main framework of shared convolution module. This shared convolution module is widely used to extract the shared feature of input image. In addition, we use data augmentation to enhance the style diversity of the training data.

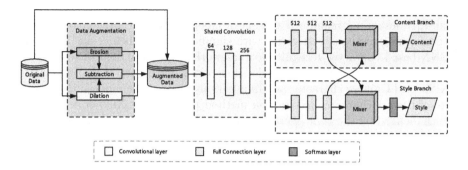

Fig. 2. Overall Architecture of the proposed SCS network. From left to right, first the original training data is augmented to enhance the style diversity; then the augmented data is pass through the shared convolutional layers for extracting texture features; next the shared features is fed to separated supervision branches to learn the hidden representations for different tasks; finally the mixed representations are fed to softmax classifiers to make task-specific predictions. This end-to-end learning approach takes into account the hidden correlations of style and content, and it is supervised by these two factors simultaneously.

3.2 Data Augmentation

It is proved that the diversity of training data is significant for enhancing the generalization ability of deep model [8]. Previous work employ affine transformation to rotate and distort the original characters [23], but it is difficult to control the thickness of character strokes and can not generate outline fonts. However, such style-related factors are typically the important features of character diversity.

According to this situation, we propose an effective data augmentation method in purpose of altering the thickness of character strokes and producing outline fonts, which could considerably increase the style diversity of training data. In our approach, the morphologic operations of dilation and erosion are respectively used to increase and reduce the thickness of character strokes, and then the outline character is produced by the subtraction of the dilated character and the eroded one. This approach do not require any training process and could be a complement to traditional affine transformation. Figure 3 presents some generated samples produced by the proposed data augmentation method.

3.3 Shared Convolution

Convolutional Neural Network (CNN) is widely used to extract the texture feature of image [6,9]. We leverage the strong feature learning power of CNN to

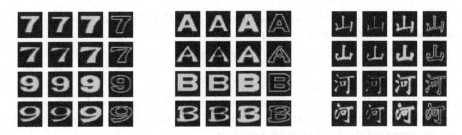

Fig. 3. Samples produced by the proposed data augmentation method. In this figure, the left, center and left parts respectively present several results of Arabic numerals, English letters and Chinese characters. And for each character set, the 1-th, 2-th, 3-th and 4-th columns respectively list the original, the dilated, the eroded and the outline fonts.

capture the shared texture features of input characters, and then feed these shared features into separate learning branches to obtain content-specific feature and style-specific feature.

This shared convolution network consist of a series of Convolution-BatchNorm-LeakyReLU down-sampling blocks that yield the texture features of input images. Here in our model, we totaly stack 3 down-sampling blocks, and in some advanced versions it could be stacked more blocks if necessary. The output channels of convolutional layers are 2, 4, 8 times of 32 respectively. The first convolution layer is with 5×5 kernel and stride 1 and the rest are with 3×3 kernel and stride 1, and all ReLUs are leaky with slope 0.1.

3.4 Style and Content Branches

We design separated learning branches to capture style-specific feature and content-specific feature. In each branch we adopt stacked full connection layers to learn such task-specific features, and then we use the mixers based on bilinear model to communicate each branches for taking into account the correlation between style and content. Finally we employ softmax classifiers to infer the style and recognize the content.

Full connection network is widely used in hidden feature learning on account of their excellent learning ability and desirable scalability [6]. In each branches, we stack 3 full connection layers, and each layers contain 512 cells with dropout rate 0.1. Note that it could be stacked more full connection layers for enhancing the learning ability, but adding more full connection layer could lead to a sharp increase in the number of trainable parameters, which bring high costs.

We combine the style feature and content feature in the mixer, which is a bilinear model. Bilinear model is a two-factor model with the mathematical property of separability: their outputs are linear in either factor when the others held constant, which has been demonstrated that the influences of two factors can be efficiently separated and combined in a flexible representation [17]. The combination function can be formulated as:

$$M_s = F_s W_s F_c$$

$$M_c = F_c W_c F_s$$

where W_s and W_c is the trainable parameters of bilinear models, F_s and F_c denote the feature representations of style and content, M_s and M_c is the mixed feature representations of style and content.

With the mixed feature obtained by the previous bilinear model, we obtain the prediction probability for the style y_s and content y_c of a given character C by softmax classifiers:

$$p(y_s|C) = softmax(V_s M_s)$$

$$p(y_c|C) = softmax(V_c M_c)$$

where V_s and V_c is the trainable parameters, which is responsible for converting the mixed feature representation for each task into predictions through linear transformation.

The goal of traditional character recognition is to infer the contents of characters from their images. It typically minimize the sum of the negative log-likelihoods. In addition to content supervision, here we take into account style supervision. Combined with content loss and style loss, the full loss function is:

$$Loss = - \sum_i (\lambda log p(y_c^i|C^i) + (1 - \lambda) log p(y_s^i|C^i))$$

where the hyper-parameter λ controls the trade-off between content loss and style loss. Considering that content recognition is more important in our experiments, we set λ to be 0.9.

4   Experiments

We conduct extensive experiments to evaluate the effectiveness of our approach. In this section, we present the details of our experiments and analyze the experimental results.

4.1   Data

We build several dataset to evaluate our method, including Arabic numerals (10 content classes), upper-case and lower-case English letters (52 content classes), and simplified Chinese characters (3755 content classes). All of these character sets are extracted from True Type font (TTF) files, which jointly launched by Apple Inc. and Microsoft Corp. as a standard font format file supporting their operating systems. We extract all the character images from 91 TTF files with widely varying font styles. Each character with a certain style and content is presented by a 32×32 PNG image. Figure 4 presents the samples of data. Note that the reason why we abandon the popular benchmarks such as MNIST and ICDAR is that they lack of explicit style labels.

Fig. 4. Samples of data. From left part to right part in this figure, it respectively present several characters of Arabic numerals, English letters and Chinese characters. In each part, there are 91 character images obtain same content but different styles.

4.2 Baseline Models

We compare our models with several baselines, all these models are design for character recognition with unseen styles. The description of these baselines and our model are listed below:

CMP. Convolutional Multiple Pooling network is a CNN-based model adopting multi-pooling and affine transformation. It could achieve desirable performance on a relatively large training set [23].

CSK. Convolutional Skeleton Kernel network is a CNN-based model, which use skeleton kernel to capture the skeletons of characters. It perform well on a limited training set [16].

SCS. Style Content Supervised network is our proposed model. It leverages the style and content supervision simultaneously as well as takes into account the hidden correlations between style and content.

4.3 Implementation Details

We design two experiments for test the baselines and our model training on relatively large dataset and limited dataset respectively. In first case, we randomly choose 70 styles for training and use other 21 styles for test. In second case, the number of training styles: the number of test styles is set to be constantly 1 : 2 and the number of training styles is dynamically changing to verify the robustness of the tested models. In order to get a more convincing conclusion, we conduct each experiment 10 times to calculate the average accuracy and shuffle the datasets after each evaluation.

When training, we use random values drawn from the Gaussian distribution with 0 mean and 0.01 standard deviation to initialize all trainable weights, and all bias are initialized at 0. Adam [5] is used as the optimization algorithm and the mini-batch size is 128. The learning rate is set to be $1e^{-4}$. After each epoch, we shuffle the training data to make different mini-batches. Furthermore, we adopt 2×2 morphology kernel for the proposed data augmentation method, where all the morphology operations are provided by OpenCV.

4.4   Performance Analysis

Large Training Set. Table 1 presents the experimental results of the baselines and our model training on relatively large dataset. Based on these results, we have the following findings:

Table 1. Results of training on relatively large dataset.

Model name	Data augmentation	Character set		
		Arabic numerals	English letters	Chinese characters
CMP	AF[a]	96.63%	92.32%	84.89%
CMP	ED[b]	96.94%	92.85%	85.33%
CMP	AF+ED[c]	97.21%	93.09%	85.58%
CSK	AF	96.41%	92.15%	84.97%
CSK	ED	96.72%	92.65%	85.39%
CSK	AF+ED	97.04%	92.98%	85.62%
SCS	AF	97.12%	93.16%	87.59%
SCS	ED	97.46%	93.54%	87.68%
SCS	AF+ED	**97.69%**	**93.82%**	**88.06%**

[a]Using traditional data augmentation based on affine transformation.
[b]Using proposed data augmentation based on erosion and dilation.
[c]Using two types of data augmentation methods simultaneously.

1. As we emphasized in this paper, the style supervision is important to character recognition. Leveraging this informative style supervision, SCS obtain the ability of style-aware to find that which styles of the input characters are more likely to be. This style inference is a helpful signal for content recognition because the content features of characters often have a rich correlation with their style features. Giving the credit to style supervision, our SCS models outperform the baselines with a significant gap, especially on the hard task of Chinese character recognition.
2. Altering the thickness of character strokes is a helpful approach to reduce style overfitting. The proposed data augmentation method based on erosion and dilation essentially enhance the style diversity of training set. Experimental results proves that the proposed data augmentation method based on erosion and dilation is better than traditional affine transformation on character recognition, and it is also a beneficial complement to affine transformation.

Limited Training Set. Figure 5 shows the results of the baselines and our SCS model training on limited dataset. In this experiment, we abandon data augmentation to evaluate the authentic generalization abilities of all tested models.

When training on limited dataset, our SCS model also outperform all the baselines. Even when the size of training set is changing, SCS maintains the superior position. This evidence demonstrates the robustness of the style supervision in extreme adverse conditions where few styles are available for training.

Fig. 5. Results of training on limited dataset.

5 Conclusion and Future Work

Our purpose is to recognize the character content when the observed characters obtain unseen font styles. It is challenged by style overfitting problem posed by the characters do not have an explicit separation between style and content. To address this issue, we propose a novel model of Style and Content Supervision (SCS) network, which integrate style and content supervision to resist style overfitting. Experimental results demonstrate that SCS achieve the state-of-the-art performance on several widely used character sets.

In the future, we would like to conduct more experiments and find a more effective way to reduce style overfitting. We also hope to verify the effectiveness of our method in a larger number of other tasks, which need more datasets and annotations. More advanced models are expected as some shortcomings are still existed in current models, such as lack of external knowledge and excessive reliance on annotated data.

Acknowledgment. We thank all reviewers for their helpful advice. This work is supported by the National Key Research and Development Program of China, and National Natural Science Foundation of China (No. U163620068).

References

1. Chen, L., Wang, S., Fan, W., Sun, J.: Beyond human recognition: a CNN-based framework for handwritten character recognition. In: IAPR Asian Conference on Pattern Recognition, pp. 695–699 (2015)
2. Dan, C., Meier, U.: Multi-column deep neural networks for offline handwritten Chinese character classification. In: International Joint Conference on Neural Networks, pp. 1–6 (2015)
3. Girshick, R., Donahue, J., Darrell, T., Malik, J.: Rich feature hierarchies for accurate object detection and semantic segmentation. Computer Science, pp. 580–587 (2013)
4. He, K., Zhang, X., Ren, S., Sun, J.: Deep residual learning for image recognition. In: 2016 IEEE Conference on Computer Vision and Pattern Recognition, CVPR 2016, Las Vegas, NV, USA, 27–30 June 2016, pp. 770–778 (2016)

5. Kingma, D.P., Ba, J.: Adam: a method for stochastic optimization. CoRR abs/1412.6980 (2014)
6. Krizhevsky, A., Sutskever, I., Hinton, G.E.: Imagenet classification with deep convolutional neural networks. In: Annual Conference on Neural Information Processing Systems 2012, NIPS 2012, Lake Tahoe, Nevada, United States, 3–6 December 2012, pp. 1106–1114 (2012)
7. Krizhevsky, A., Sutskever, I., Hinton, G.E.: Imagenet classification with deep convolutional neural networks. In: Pereira, F., Burges, C.J.C., Bottou, L., Weinberger, K.Q. (eds.) Advances in Neural Information Processing Systems 25, pp. 1097–1105. Curran Associates, Inc. (2012)
8. Lake, B.M., Salakhutdinov, R., Tenenbaum, J.B.: Human-level concept learning through probabilistic program induction. Science **350**(6266), 1332–1338 (2015)
9. LeCun, Y., Bottou, L., Bengio, Y., Haffner, P.: Gradient-based learning applied to document recognition. Proc. IEEE **86**(11), 2278–2324 (1998)
10. Lv, G.: Recognition of multi-fontstyle characters based on convolutional neural network. In: Fourth International Symposium on Computational Intelligence and Design, pp. 223–225 (2011)
11. Ren, S., He, K., Girshick, R., Sun, J.: Faster R-CNN: towards real-time object detection with region proposal networks. In: International Conference on Neural Information Processing Systems, pp. 91–99 (2015)
12. Schmidhuber, J., Meier, U., Ciresan, D.: Multi-column deep neural networks for image classification. In: Computer Vision and Pattern Recognition, pp. 3642–3649 (2012)
13. Simard, P.Y., Steinkraus, D., Platt, J.C.: Best practices for convolutional neural networks applied to visual document analysis. In: International Conference on Document Analysis and Recognition, p. 958 (2003)
14. Simonyan, K., Zisserman, A.: Very deep convolutional networks for large-scale image recognition. CoRR abs/1409.1556 (2014)
15. Szegedy, C., et al.: Going deeper with convolutions. In: IEEE Conference on Computer Vision and Pattern Recognition, CVPR 2015, Boston, MA, USA, 7–12 June 2015, pp. 1–9 (2015)
16. Tang, W., et al.: CNN-based Chinese character recognition with skeleton feature. In: Cheng, L., Leung, A.C.S., Ozawa, S. (eds.) ICONIP 2018, Part V. LNCS, vol. 11305, pp. 461–472. Springer, Cham (2018). https://doi.org/10.1007/978-3-030-04221-9_41
17. Tenenbaum, J.B., Freeman, W.T.: Separating style and content. In: Annual Conference on Neural Information Processing Systems 1996, NIPS 1996, Denver, CO, USA, 2–5 December 1996, pp. 662–668 (1996)
18. Wu, C., Fan, W., He, Y., Sun, J., Naoi, S.: Handwritten character recognition by alternately trained relaxation convolutional neural network. In: International Conference on Frontiers in Handwriting Recognition, pp. 291–296 (2014)
19. Xiao, X.F., Jin, L., Yang, Y., Yang, W., Sun, J., Chang, T.: Building fast and compact convolutional neural networks for offline handwritten chinese character recognition. Pattern Recogn. **72**, 72–81 (2017)
20. Xu, N., Ding, X.: Printed Chinese character recognition via the cooperative block neural networks. In: IEEE International Symposium on Industrial Electronics, vol. 1, pp. 231–235 (1992)
21. Yin, F., Wang, Q.F., Zhang, X.Y., Liu, C.L.: ICDAR 2013 Chinese handwriting recognition competition (ICDAR), pp. 1464–1469 (2013)

22. Zeiler, M.D., Fergus, R.: Visualizing and understanding convolutional networks. In: Fleet, D., Pajdla, T., Schiele, B., Tuytelaars, T. (eds.) ECCV 2014. LNCS, vol. 8689, pp. 818–833. Springer, Cham (2014). https://doi.org/10.1007/978-3-319-10590-1_53

23. Zhong, Z., Jin, L., Feng, Z.: Multi-font printed Chinese character recognition using multi-pooling convolutional neural network. In: International Conference on Document Analysis and Recognition, pp. 96–100 (2015)

24. Zhong, Z., Jin, L., Xie, Z.: High performance offline handwritten Chinese character recognition using googlenet and directional feature maps. In: International Conference on Document Analysis and Recognition, pp. 846–850 (2015)

T-SAMnet: A Segmentation Driven Network for Image Manipulation Detection

Jie Pan[1,2], Yunshu Chen[1,2], Yue Huang[1,2], Xinghao Ding[1,2], and En Cheng[1,2(✉)]

[1] Key Laboratory of Underwater Acoustic Communication and Marine Information Technology, Xiamen University, Ministry of Education, Xiamen 361005, Fujian, China
{panjie,yschen}@stu.xmu.edu.cn, {yhuang2010,dxh,chengen}@xmu.edu.cn
[2] School of Informatics, Xiamen University, Xiamen 361005, Fujian, China

Abstract. Although most of the current image manipulation detection algorithms gain great breakthrough, it still generally has problems in detecting multiple types of tampering techniques, and in receiving the classification, detection and segmentation results simultaneously. We propose a two streams network driven by segmentation mask, called T-SAMnet, where RGB images and noise images provide semantic features and noise inconsistency features for the network respectively. RGB stream generates tampered region detection bounding box and segmentation mask, and then is fused with the noise stream to generate classification results. The segmentation mask, on the one hand, supervises the characteristics of the network learning tampered region, feeds back as segmentation attention mechanism to constraint detection branch. The experimental results demonstrates that our method achieves state-of-the-art performance on the three standard image manipulation detection datasets.

Keywords: Image manipulation detection · Segmentation attention mechanism · Two-stream network

1 Introduction

The rapid development of digital image editing technology makes it very important to detect the originality of images accurately. So, how to effectively and quickly achieve image forensics and solve the problems of network information security receives more attentions in the academic and industrial circles.

This work was supported in part by the National Natural Science Foundation of China under Grants 61571382, 81671766, 61571005, 81671674, 61671309, 61971369 and U1605252, in part by the Fundamental Research Funds for the Central Universities under Grants 20720160075 and 20720180059, in part by the CCF-Tencent open fund, and the Natural Science Foundation of Fujian Province of China (No. 2017J01126).

© Springer Nature Switzerland AG 2019
T. Gedeon et al. (Eds.): ICONIP 2019, CCIS 1143, pp. 32–40, 2019.
https://doi.org/10.1007/978-3-030-36802-9_4

Image manipulation detection is verifying the authenticity of the images. Systematically, digital image tampering techniques is classified as splicing, copy-move, removal, et al. [1,2]. Removal refers to erasing a specific area in an image. Splicing refers to copying an area from a single image or multiple images and pasting it into another image [3]. Copy-move refers to selecting a specific area and moving this area to other areas in the same image. Since different areas in the same image commonly have similarities in texture, illumination conditions and colors, copy-move is an effective way to tamper images.

At present, image manipulation detection methods are roughly divided into traditional detection algorithm based on image fingerprinting and tampering detection algorithm based on deep learning. The manipulated image partly destroys the internal statistical features of the image, so the tamper sensitive statistical features of the image, such as bicoherence feature [4], color consistency feature [5], can effectively detect whether the image is tampered. Image manipulation detection based on deep learning extracts image features through the deep learning network, implements an end-to-end adaptive learning process through the loss function. Based on the spatial rich model (SRM) [6], Rao [7] utilized deep learning models for image manipulation detection. Zhou [3] proposed a two-stream Faster-RCNN network using RGB image and SRM image.

We adopt a two-stream Faster-RCNN network to analyze visual inconsistency and local noise features in tampering images, prevent network overfitting to image content while extracting tamper information. We add a branch for tamper type classification to network training as an auxiliary information for tamper detection. We extract the features of RGB images and noise images to simultaneously complete the identification, detection and segmentation tasks, and combine the loss of the three tasks in the optimization of network parameters to achieve mutual promotion. There is no supervision added to segmentation information when identifying and detecting. The segmentation information is fed back to the detection branch in the form of attention mechanism, which reduces the interference of the content information of the non-tampered region on the detection of the tampering target.

2   Proposed Method

In this section, we introduce the framework we proposed. The framework consists of two streams. RGB stream extracts the features of the RGB image for detection task and segmentation task. Noise stream extracts the high frequency information of the noise image for classification task. Then we fuse two streams of information to get classification, detection and segmentation results. Finally, we feedback segmentation mask to improve detection accuracy. A summary of our method is shown in Fig. 1.

2.1   Two Streams Image Manipulation Detection Network

As Zhou described in paper [3], tampering information contained in the tampered image represents both by RGB image and noise image. We use a

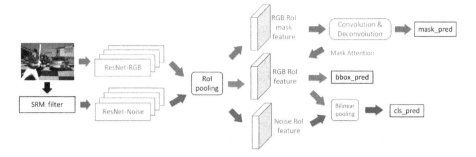

Fig. 1. Our two streams image manipulation detection model based on the segmentation attention mechanism.

two-stream ResNet50 network [8] to learn the characteristics of tampered images. RGB stream acquires the features of high-level semantics, capturing differences in visual aspects of tampered and non-tampered regions, such as unnatural transitions and strong contrast differences around tampered edges. RGB stream is used for bounding box regression and segmentation information extraction.

Individual RGB stream cannot determine the tamper techniques sufficiently. In fact, different tamper techniques have different characteristics, so noise stream provides more detailed information about tampering information. Inspiring by the usage of SRM in image forensics [6], we use SRM filters in the noise stream to extract local noise information to complement the semantic information extracted by RGB stream, which in turns to promote detection tampering targets.

Noise stream and RGB stream share the same Region of Interests (RoI) pooling layer, then we apply compact bilinear pooling [9] on features from two streams for tamper techniques classification. Compact bilinear pooling realizes the suppression of redundant features and the fusion of effective features. Combining two streams features not only preserves spatial information, but also improves the accuracy of the detection.

2.2   Segmentation Module

Currently image manipulation detection algorithms barely have segmentation result for the tampered region. In fact, the segmentation result is an effective promotion method for improving the accuracy of detection and recognition, which adds richer monitoring information to the network. Segmentation of the tampered region depends more on the high level semantic features of the visual level. Therefore, the positioning and segmentation of the tampered region only base on the RoI features generated by the RGB stream. After the RoI pooling of the RGB stream, we add a segmentation module for segmentation.

When network is training, it completes three tasks of tampering with classification, detection, and segmentation. Three tasks provide more supervision information for each other and also restrain each other. The loss function L_{total} of the network is:

$$L_{total} = L_{RPN} + L_{cls}(f_{RGB}, f_N) + L_{bbox}(f_{RGB}) + L_{mask}(f_{RGB\_mask}). \quad (1)$$

where L_{RPN} denotes the loss function of the RPN network. f_{RGB} and f_N denote the ROI features of the RGB stream and noise stream. f_{RGB_mask} denotes the RGB mas feature of the RGB stream. L_{cls} denotes the cross entropy classification loss function used to determine whether the region has been tampered. L_{bbox} denotes the bounding box regression loss function, and L_{mask} denotes the binary cross entropy loss function of the segmentation, both L_{bbox} and Lmask are only determined by the RoI characteristics of the RGB stream.

2.3   Segmentation Mask Attention

The visual attention mechanism is a brain signal processing mechanism unique to human vision. By quickly scanning the whole image, human vision obtains the target area that needs to be focused on, then invests more attention resources in this area to obtain more details of the target while suppressing other useless information [10]. In image manipulation detection, usually the features extracted by network are subtle, tampering information is easily obscured by the content information of the image, resulting in inaccurate detection or even failure to detect tampering areas. Therefore, we apply the attention mechanism in the detection network.

The segmentation feature of the RGB stream characterizes the position distribution of the pixel that may be tampered in the tampered region. We feed it back into the detection branch as a segmentation mask attention mechanism to constrain the detection branch. In our network, we weight the last layer feature of the segmentation branch in the form of dot-multiply to the network of the detection branch, so that the tampering pixel gets a larger response, and the response of the non-tampered region is suppressed, improving the SNR of tampering information, moreover, improving detection accuracy.

3   Experiments

We propose an image manipulation detection model based on the segmentation attention mechanism and two streams Faster-RCNN, which classifies, detects and segments the tampered area concurrently. We compare our two streams attention network with state-of-the-art methods to illustrate the effectiveness of our method.

3.1   Dataset

We evaluate our method on NIST Nimble 2016 [11] (NIST16), CASIA [12,13] and COVER [14]. We create an image manipulation dataset based on the COCO [15], we call it SPCOCO, for network pre-training.

NIST16 provides 564 color images with corresponding tamper target segmentation mask. The tampering techniques includes splicing, copy-move and removal. All the tampered images are blurred and filtered.

CASIA provides two methods of tampering techniques with copy-move and splicing. The dataset divides into version 1.0 and version 2.0. Comparing with CASIA 1.0, images in CASIA 2.0 are filtered, blurred after copy-move and splicing. In this paper we use CASIA 2.0 for training and CASIA 1.0 for testing.

COVER is a lightweight dataset specially for copy-move, which contains only 100 tampered images and 100 original images. The tampered targets are rotated at different angles and scaled at different scales when pasted in the source image. It is worth noting that the tampered area in the image has similar area into the original image, so it is difficult to detect the source and target.

SPCOCO synthesizes by COCO and its annotation information, tampering techniques include copy-move and splicing. SPCOCO consists 12 major categories, each major category of images only appear in the training set or validation set to prevent the network from over-fitting to the image content.

3.2   Comparing with State-of-the-Art Methods

Evaluation Metric. Essentially image manipulation detection is a detection task, so we use the evaluation metric, which is commonly used in the detection task, to evaluate the performance of the model, including Average Precision (AP) and pixel level F1 scores.

Table 1. F1 score comparison on three standard datasets.

	NIST16	COVER	CASIA
ELA	23.6%	22.2%	21.4%
NOI1	28.5%	26.9%	26.3%
CFA1	17.4%	19.0%	20.7%
CFA2	57.1%	-	21.3%
MFCN	57.1%	-	**54.1%**
RGB-N	72.2%	43.7%	40.8%
T-SAMnet	**89.4%**	**66.7%**	49.7%

Baseline Model. To illustrate effectiveness, we compare proposed method with other methods. Baselines are described as below.

ELA [16]. Using error inconsistency calculated by JPEG compression in different images to analyze the tampered area of the image.

NOI1 [17]. Calculating high pass wavelet coefficients to distinguish noise inconsistencies.

CFA1 [18]. Estimating CFA based on the nearby pixel information to analyze the tampering probability of each pixel.

CFA2 [19]. Based on demosaic artifacts to compute a single feature.

MFCN [20]. A FCN networks use edge mask and tamper area mask to jointly detect tamper edges.

RGB-N [3]. Using RGB image and noise image to perform manipulation classification and bounding box regression.

T-SAMnet. Using RGB image and noise image fusion to achieve manipulation classification, bounding box regression and segmentation, the segmentation mask is fed back to the detection branch, our model.

Table 1 shows the F1 score comparison between our method and baselines in three datasets. It can be observed from the table that the image manipulation detection method based on deep learning is better than the traditional method based on manual features in performance. CNN dynamically and efficiently extract the information of tampered image and use this information accurately for the detection of the tampered region. Also, CNN detects tampered areas from the perspective of semantics, instead the information of neighboring pixels, which provides more information for image manipulation detection. At the same time, the model of this paper achieves the highest F1 score on two standard datasets. Although the performance on CASIA is slightly lower than MFCN, our method has better effect and generalization capability.

Table 2. Ablation study on three standard datasets.

AP/F1 score	NIST16	COVER	CASIA
RGB-N	79.2%/72.2%	42.7%/43.7%	28.5%/40.8%
T-Mnet	80.5%/68.2%	55.1%/55.0%	34.4%/44.11%
T-SAMnet	**85.2%/89.36%**	**59.5%/66.67%**	**42.7%/49.71%**

3.3   Ablation Study

We compare our method with the improved method of this paper, and list both AP and F1 score of all the methods, as show in Table 2. T-Mnet uses RGB image and noise image fusion to achieve manipulation classification, bounding box regression and segmentation. It shows that adding a segmentation branch improves the detection result. Multi-task learning helps improve performance of network. T-SAMnet explains that it is effective to feed back the segmentation mask in the form of attention mechanism to the detection branch. Segmentation mask attention mechanism effectively suppresses the effect of non-tampering pixels in bounding box. All the results show the effectiveness of our method.

Table 3. AP and F1 score comparison on multi-class on three datasets.

AP/F1 score	Copy-move	Splicing	Removal
NIST	80.99%/87.17%	89.90%/92.93%	82.33%/89.90%
CASIA	18.99%/28.39%	69.55%/69.62%	-
COVER	59.47%/66.67%	-	-

3.4   Manipulation Technique Detection

For copy-move, splicing and removal, we carry out more specific tests on proposed method. Table 3 shows that splicing achieves the best performance. We believe that splicing generates Unnatural edge and image contrast differences in RGB images, while camera fingerprint features from different images can be extracted on noise residuals, so fusion feature using RGB stream and noise stream can detect more accurately. In addition, on CASIA, result of copy-move performances not good. Because the source and the target are in the same image, therefore, a similar noise distribution is generated on the noise residual to confuse the feature detection of the noise stream. Detection results and segmentation results in three types of tampering techniques in NIST16 are showed in Fig. 2.

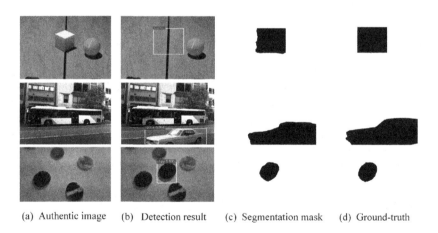

(a) Authentic image   (b) Detection result   (c) Segmentation mask   (d) Ground-truth

Fig. 2. Multi-class image manipulation detection on NIST16 dataset.

4   Conclution

We propose an image manipulation detection network for multiple tampering techniques while completing identification, detection, and segmentation simultaneously. We use RGB stream and noise stream to extract the semantic information and noise inconsistency of tampered images. We add a segmentation branch in detection network, achieving the goal of multi-task learning. Three tasks constrain each other to enhance the extraction of detail features to improve the accuracy of image manipulation detection. Inspiring by the attention mechanism, the segmentation mask feeds back to the detection branch, further constraining the focus of the detection branch. Experiments show that our proposed method achieves state-of-the-art performance on standard datasets and have good generalization performance.

References

1. Farid, H.: Creating and detecting doctored and virtual images: implications to the child pornography prevention act. Department of Computer Science, Dartmouth College, TR2004-518, 13:970 (2004)
2. Wu, Q., Li, G., Tu, D., Sun, S.: A survey of blind digital image forensics technology for authenticity detection. Acta Autom. Sinica **34**(12), 1458–1466 (2008)
3. Zhou, P., Han, X., Morariu, V.I., Davis, L.S.: Learning rich features for image manipulation detection. In: Proceedings of the IEEE Conference on Computer Vision and Pattern Recognition, pp. 1053–1061 (2018)
4. Ng, T.T., Chang, S.F.: A model for image splicing. In: 2004 IEEE International Conference on Image Processing, vol. 2, pp. 1169–1172 (2004)
5. Zhou, Z., Hu, C., Huang, H.: Image blur forgery detection based on color consistency. Comput. Eng. **42**(1), 237–242 (2016)
6. Fridrich, J., Kodovsky, J.: Rich models for steganalysis of digital images. IEEE Trans. Inf. Forensics Secur. **7**(3), 868–882 (2012)
7. Rao, Y., Ni, J.: A deep learning approach to detection of splicing and copy-move forgeries in images. In: 2016 IEEE International Workshop on Information Forensics and Security, pp. 1–6 (2016)
8. He, K., Zhang, X., Ren, S., Sun, J.: Deep residual learning for image recognition. In: Proceedings of the IEEE Conference on Computer Vision and Pattern Recognition, pp. 770–778 (2016)
9. Gao, Y., Beijbom, O., Zhang, N.: Compact bilinear pooling. In: Proceedings of the IEEE Conference on Computer Vision and Pattern Recognition, pp. 317–326 (2016)
10. Fu J., Zheng, H., Mei, T.: Look closer to see better: recurrent attention convolutional neural network for finegrained image recognition. In: Proceedings of the Conference on Computer Vision and Pattern Recognition (2017)
11. Nist nimble 2016 datasets. https://www.nist.gov/itl/iad/mig/nimble-challenge-2017-evaluation/
12. Dong, J., Wang W., Tan, T.: Casia image tampering detection evaluation database (2010). http://forensics.idealtest.org
13. Dong, J., Wang, W., Tan, T.: Casia image tampering detection evaluation database. In: 2013 IEEE China Summit and International Conference on Signal and Information Processing, pp. 422–426 (2013)
14. Wen, B., Zhu, Y., Subramanian, R., Ng, T.-T., Shen, X., Winkler, S.: COVERAGEA novel database for copy-move forgery detection. In: 2016 IEEE International Conference on Image Processing, pp. 161–165 (2016)
15. Lin, T.-Y., et al.: Microsoft COCO: common objects in context. In: Fleet, D., Pajdla, T., Schiele, B., Tuytelaars, T. (eds.) ECCV 2014. LNCS, vol. 8693, pp. 740–755. Springer, Cham (2014). https://doi.org/10.1007/978-3-319-10602-1_48

16. Krawetz, N.: A picture's worth.... Hacker Factor Solutions **6**, 2 (2007)
17. Mahdian, B., Saic, S.: Using noise inconsistencies for blind image forensics. Image Vis. Comput. **27**(10), 1497–1503 (2009)
18. Ferrara, P., Bianchi, T., De Rosa, A.: Image forgery localization via finegrained analysis of CFA artifacts. IEEE Trans. Inf. Forensics Secur. **7**(5), 1566–1577 (2012)
19. Dirik, A.E., Memon, N.: Image tamper detection based on demosaicing artifacts. In: 2009 IEEE International Conference on Image Processing, pp. 1497–1500 (2009)
20. Salloum, R., Ren, Y., Kuo, C.C.J.: Image splicing localization using a multi-task fully convo-lutional network (MFCN). J. Vis. Commun. Image Represent. **51**, 201–209 (2018)

IRSNET: An Inception-Resnet Feature Reconstruction Model for Building Segmentation

Kepeng Xu[1], Li Nie[1], Zhiqiang Zhang[1], Wenxin Yu[1(✉)], Yunye Zhang[1],
Wei Chen[1], Siyuan Li[1], Xinxin Zhou[4,5], Shangwei Deng[1], Pengfei Yu[1],
Yibo Fan[2], Hui Zhang[1], and Valentin Bouillon[3]

[1] Southwest University of Science and Technology, Mianyang, China
star_yuwenxin27@163.com
[2] State Key Laboratory of ASIC and System, Fudan University, Shanghai, China
[3] École internationale des sciences du traitement de l'information, Cergy, France
[4] College of Geographical Science, Nanjing Normal University, Nanjing, China
[5] Key Laboratory of Virtual Geographic Environment, Ministry of Education,
Nanjing Normal University, Nanjing, China

Abstract. Effective analysis of remote sensing images remains a challenging topic in deep learning research field. This paper proposes a semantic segmentation approach based on the inception architecture. Specifically, the combination of residual network and Inception improves the feature extraction capability of encoder network. In addition, this paper discusses the problem of unbalanced information allocation caused by concatenated operation in the up-sampling process of network models such as Unet, and the RCSE module is proposed to complete feature reconstruction to solve this problem. The approach ensures accurate assignment of semantic labels to the buildings in the aerial images. Experiments based on the dataset proposed by CrowdAi certify the effectiveness of our approach with a 3.7% IoU improvement compared to Unet.

Keywords: Aerial image · Building extraction · Convolution
network · Semantic segmentation

1 Introduction

Remote sensing technology has developed rapidly in recent decades, which enabled us to obtain a very large amount of information on the surface of the Earth. Remote sensing image is one of the most important information obtained by remote sensing equipment. At present, the task of labeling building objects in remote sensing images in the industry is still at a very low level of technology, relying on a large amount of manpower for manual labeling. So building a soft system that generates semantic annotations for maps is very important.

© Springer Nature Switzerland AG 2019
T. Gedeon et al. (Eds.): ICONIP 2019, CCIS 1143, pp. 41–48, 2019.
https://doi.org/10.1007/978-3-030-36802-9_5

The goal of semantic segmentation is to assign a corresponding semantic annotation to each pixel in the image. In order to achieve this goal, most of the current network architecture is mainly based on two points to improve: 1. Improve the global information extraction capability of the model, that is, to establish a model that can obtain the global view and detect the building objects. 2. Improve the model's ability to save pixel-level location information.

Remote sensing image building extraction tasks are more complicated than semantic segmentation tasks in conventional scenarios, which can be concluded in the following five points: 1. The shape of a building is varied and very complicated. 2. The styles of buildings in different areas vary greatly. 3. The chromaticity of the acquired images is caused by different atmospheric lighting conditions. 4. The size of the building caused by the difference in sensor height is too large.

This paper proposes an end-to-end semantic segmentation model for building footprint segmentation. The encoder side of the model is based on the inception architecture [9], using the residual connection, as well as the atrous convolution is applied to the encoder. Among them, the receptive field of the convolution kernel is expanded in the last part of the encoder, while in the decoder part of the model, the feature layer is restored to the original image size by stepping up-sampling. Furthermore, in the recovery process, the feature layer in the model encoder is concatenated to the corresponding shape feature layer in the decoder using the shortcut connection [4], which contributes the decoder part of the proposed model to effectively utilize the high-level information and low-level information of the aerial image. Unlike previously, this paper considers a problem that connecting directly to the high and low level of information through the shortcut, though, to some extent, makes the decoder obtain more accurate pixel texture information, the feature layer obtained by the decoder is directly concatenated from the high and low levels and does not undergone a corrective process. And such feature information will interferes with the decoder of the entire semantic recovery to some extent. Considering what has been mentioned above, this paper adds the operation of channel weighting after the up-sampling. Moreover, to concatenate the feature of the layer after each channel gives different weights, RCSE module is proposed in this paper, which can help decoder to complete information reconstruction of the feature layer after upsampling and concatenating. And through such module, the proposed model can reduce the dependence on the information of the redundant feature layer, and finally increase the information recovery ability of decoder, so as to improve the accuracy of the segmentation in the model.

The main contributions of this paper include the following:

- A semantic segmentation model based on inception-resnet for building footprint segmentation is proposed.
- The application of shortcut in the model based on encoder-decoder and its shortcomings are discussed. RCSE module is proposed to reduce the feature redundancy and interference caused by shortcut.
- The proposed model can effectively complete the aerial image building footprint segmentation and achieve 91.2% Iou in the validation set.

2   Related Work

The purpose of image semantic segmentation is to assign pixel-level semantic annotations to images, which is an important part of computer vision. The full convolutional neural network applies convolutional neural networks to semantic segmentation tasks for the first time, and is very effective.

In the past few decades, remote sensing devices have been widely used, and a large amount of remote sensing image data has been collected. The aerial image building footprint extraction has been extensively studied. In the past traditional computer vision technology, researchers extracted features by manually designing feature extractors such as texture and color features, and then through the machine learning classifiers (such as AdaBoost, support vector machine, random forest, etc.) to complete the rough Scene classification, which requires a lot of subsequent processing to optimize the results. With the development of remote sensing equipment production technology, researchers have been able to obtain a large number of high-resolution remote sensing images, however, the traditional methods can't effectively process high-precision images and can't achieve pixel-level segmentation accuracy, but only complete scene-level classification. Therefore, in recent years, scholars have focused on the research of high-resolution image building footprint extraction tasks with convolutional neural networks.

Recently, convolutional neural networks have been rapidly developed in the extraction of building footprints. Zhang et al. proposed a method based on CNN classification, which calculates and classifies scores for each sliding window by using sliding windows of different scales, and then optimizes by using non-maximum suppression result.

Contrast to them, this paper proposes an end-to-end encoder-decoder model. In the encoder, the inception architecture is effectively combined with resnet, and the dilated convolution method is used to enhance the receptive field of the convolution kernel to enhance the coding capability of the model encoder. In the decoder of the model, the squeeze excitation module is combined with the shortcut connection, so that the decoder of the model can assign weights to the feature layer after the shortcut, which enables the model to allocate limited computing resources to the feature layer with the largest amount of information.

3   Methodology

3.1   IRSNET

This section will introduce the model in detail. Figure 1 is the overall architecture of the proposed model, which is divided into two parts: the encoder (Fig. 1 left) decoder (Fig. 1 Right side). The encoder side effectively extracts and abstracts the image pixel information by combining the Inception architecture with the residual connection. In the decoder part, the model gradually restores the feature map to the size of the input picture by upsampling. In order to ensure that the decoder can have pixel-level segmentation accuracy in the case of extracting global information, the proposed method connects the different sizes of the encoder layer directly to the decoder through the shortcut in the upsampling stage.

After each up-sampling process of the model decoder, the feature layer is further modified by the proposed RCSE module. Thus, the interference of the feature layer with less information can be reduced and the feature recovery accuracy of the decoder can be enhanced, which improves the expressive power of the model.

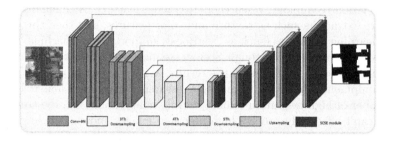

Fig. 1. Model architecture diagram

3.2   Encoder

At the encoder end of the model, for the input of the model, the feature is extracted using three convolution processes in the first two downsampling stages of the encoder. In the first convolution process, we take step size of 2 and complete the down-sampling. To reduce the size of the feature map, the third convolution process uses a hole convolution kernel of rate = 2 for convolution to expand the receptive field of the convolution kernel. During the last three sampling processes of the model encoder, this paper adds the Inception-Resnet structure to the model to improve the feature aggregation ability of the encoder. The Inception-Resnet structure of the three stages is shown in the Fig. 2.

3.3   Decoder

In the decoder part of the model, the feature layer is restored step by step to the original image size by upsampling. In order to enable the decoder to simultaneously acquire the image high-level semantic information and the image pixel position information, the provided method connect directly the corresponding size feature layer in the encoder to the decoder during the upsampling process. And the shape of the feature layer is also directly connected to the decoder. The research process reveals that simply connecting the feature layer and then convoluating allow the model to save both high-level information and low-level pixel information, however, decoder can not effectively distinguish the importance of different feature layers. Therefore, this paper introduces the Squeeze-and-Excitation mechanism into the feature layer concatenation, with different

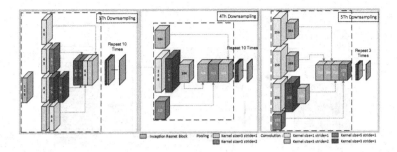

Fig. 2. These are the last three down-sampling model structure diagrams, and the black boxes of each diagram represent the Inception-Resnet structures of each process. These corresponding modules are superimposed several times after each down-sampling, with The Times of (10, 10, 3) respectively. Since in this article, in view of the remote sensing image segmentation task, the image size is small, the down-sampling size is small after the last stage. Although these feature layers extracte the high-level semantic information, lots of convolution padding operations also increase a lot of noise. The last Inception-Resnet has limited effect on improving the performance of the model, which leads that the last model of encoder only had three stages superposition Inception-Resnet structure.

weights assigned to different channels of the concatenated feature layer, which enhances the expressive ability of the decoder and thereby effectively improves the performance of the model.

3.4   RCSE: Refactoring ShortCut Squeeze Excitation Block

Unet segmentation model based on encoder and decoder, in the process of upsampling by Shortcut, can achieve high and low level of information fusion, and the decoder in the process of completing image information recovery can not only complete advanced level of the image object level detection, but be able to complete the accurate using low-level pixel characteristic information to the pixel level distribution of the label.

However, the decoder has the ability of capturing both high and low levels of image feature information at the same time, such simple directly concatenating of feature layers at different levels restricts the expretion ability of decoder. Therefore, we consider a problem that how much higher or lower levels of information contribute to overall model performance improvement after the operation of concatenation. Upon this thinking, we design the RCSE module and add it to the model upsampling process. In the concrete implementation of RCSE each channel of the shortcut feature layer was assigned different weights by SE block [3]. Making the decoder in the process of upsampling different features can extract more directional layer information, improve the efficiency of the decoder. The specific structure of RCSE is shown in Fig. 3.

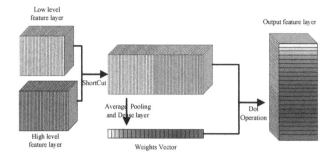

Fig. 3. RCSE architecture. As shown in the diagram, considering the different functions between high and low level feature layers, this paper designs the feature layer after the completion of feature scaling of the RCSE module after concatenation. The proposed module improved model performance with small additional parameters.

4   Experiments

The experiment is based on the Map Challenge data mining competition proposed by Crowdai. The dataset consists of a training set (280,000 300 * 300 aerial images and corresponding manual annotations) and a validation set (60,000 images and annotations). The data set has many advantages, and the semantic labeling is clear. This paper evaluates the experiment based on the dataset and compares it with SegNet, pspnet [11] and Unet, and proves the effectiveness of the proposed method.

4.1   Metrics

Our experiment takes Acc, Iou and F1 Score for evaluation metric.

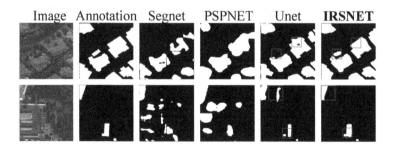

Fig. 4. Experimental result diagram. As shown in the figure, the rightmost side is the output part of the model proposed in this paper. The proposed model improves the extraction accuracy of small buildings and the segmentation effect of special edge structures such as right-angle and straight-line structures (as shown in the red box). (Color figure online)

4.2 Results

The results of the proposed model on the validation set are as follows Fig. 4 and Table 1. From the table, we can see that the performance of our model based on Inception-Resnet is higher to some extent compared with other models, and the RCSE module proposed significantly improves the performance of the model.

Table 1. Comparison table of experimental results

Model	Acc	Iou	F1score
IRSNET	**96.8**	**91.2**	**92.7**
IRSNET (Without RCSE)	**95.7**	**90.2**	**91.7**
Unet (With RCSE)	96	89.0	90.2
Unet	95.1	87.5	89.7
Segnet	87.0	69.13	69.24
pspnet	86.3	68.4	68.9
FCN	83.8	65.8	67.1

4.3 Analysis

The proposed model adds Inception-Resnet to the encoder side of the model, which improves the feature extraction capabilities of the model. In addition, the proposed RCSE block can reconstruct the feature layer formed by short cut after the up-sampling process. This operation enables the model to modify the large difference of eigenvalues between the two ends of short cut, so as to enhance the expressive ability of the model.

Because remote sensing images have the characteristics of great scale changes, PSPNET is chosen as the contrast benchmark to compare the contrast model in the conventional scene. The results show that the model proposed in this paper is much better than the conventional computer vision model.

Compared to Unet, which performs well in the two-category segmentation task, the Inception-resnet based architecture replaces the original simple convolution pooled downsampling process in the encoder. In the decoder, the extrusion excitation module is added after each shortening process, which improves the feature aggregation capability of the model. These allow the model to selectively combine high-level information, rather than simply concatenating high-level and low-level information like traditional shortcut modules, which are the reasons why the performance of the model has been improved.

5 Conclusion

This paper proposes a semantic segmentation model based on Inception-Resnet for aerial image building segmentation in order to solve the problem of unbalance between two ends after the shortcut process. This paper also proposes the RCSE module which can re-correct the feature layer, improving the expressive ability of the model decoder. Furthermore, experiments show that, compared with the previous methods, the proposed method significantly improves the quality of building segmentation in remote sensing images.

Acknowledgement. This research was supported by [2018GZ0517] [2019YFS0146] [2019YFS0155] which supported by Sichuan Provincial Science and Technology Department, [2018KF003] Supported by State Key Laboratory of ASIC & System, Science and Technology Planning Project of Guangdong Province [2017B010110007]. Project S201910619048 supported by Sichuan's Training Program of Innovation and Entrepreneurship for Undergraduate.

References

1. Badrinarayanan, V., Kendall, A., Cipolla, R.: Segnet: a deep convolutional encoder-decoder architecture for image segmentation. IEEE Trans. Pattern Anal. Mach. Intell. **39**, 2481–2495 (2017)
2. Chollet, F.: Xception: Deep Learning with Depthwise Separable Convolutions. arXiv e-prints, October 2016
3. Hu, J., Shen, L., Sun, G.: Squeeze-and-excitation networks. CoRR abs/1709.01507 (2017). http://arxiv.org/abs/1709.01507
4. Huang, G., Liu, Z., van der Maaten, L., Weinberger, K.Q.: Densely connected convolutional networks. In: Proceedings of the IEEE Conference on Computer Vision and Pattern Recognition (2017)
5. Huang, Z., Cheng, G., Wang, H., Li, H., Shi, L., Pan, C.: Building extraction from multi-source remote sensing images via deep deconvolution neural networks. In: 2016 IEEE International Geoscience and Remote Sensing Symposium (IGARSS), pp. 1835–1838, July 2016. https://doi.org/10.1109/IGARSS.2016.7729471
6. Marmanis, D., Schindler, K., Wegner, J.D., Galliani, S., Datcu, M., Stilla, U.: Classification With an Edge: Improving Semantic Image Segmentation with Boundary Detection. arXiv e-prints, December 2016
7. Ronneberger, O., Fischer, P., Brox, T.: U-net: convolutional networks for biomedical image segmentation. In: Navab, N., Hornegger, J., Wells, W.M., Frangi, A.F. (eds.) MICCAI 2015. LNCS, vol. 9351, pp. 234–241. Springer, Cham (2015). https://doi.org/10.1007/978-3-319-24574-4_28
8. Shelhamer, E., Long, J., Darrell, T.: Fully convolutional networks for semantic segmentation. IEEE Trans. Pattern Anal. Mach. Intell. **39**(4), 640–651 (2017). https://doi.org/10.1109/TPAMI.2016.2572683
9. Szegedy, C., Vanhoucke, V., Ioffe, S., Shlens, J., Wojna, Z.: Rethinking the inception architecture for computer vision. CoRR abs/1512.00567 (2015). http://arxiv.org/abs/1512.00567
10. Yang, H., Wu, P., Yao, X., Wu, Y., Wang, B., Xu, Y.: Building extraction in very high resolution imagery by dense-attention networks. Remote Sensing **10**(11) (2018). https://doi.org/10.3390/rs10111768. http://www.mdpi.com/2072-4292/10/11/1768
11. Zhao, H., Shi, J., Qi, X., Wang, X., Jia, J.: Pyramid scene parsing network. In: Proceedings of IEEE Conference on Computer Vision and Pattern Recognition (CVPR) (2017)

Residual CRNN and Its Application to Handwritten Digit String Recognition

Hongjian Zhan, Shujing Lyu$^{(\boxtimes)}$, Xiao Tu, and Yue Lu

Shanghai Key Laboratory of Multidimensional Information Processing,
East China Normal University, Shanghai 200062, China
hjzhan@stu.ecnu.edu.cn, {sjlv,ylu}@cs.ecnu.edu.cn

Abstract. Recently the Convolutional Recurrent Neural Network (CRNN) architecture has shown success in many string recognition tasks and residual connections are applied to most network architectures. In this paper, we embrace these observations and present a new string recognition model named Residual Convolutional Recurrent Neural Network (Residual CRNN, or Res-CRNN) based on CRNN and residual connections. We add residual connections to convolutional layers as well as recurrent layers in CRNN. With residual connections, the proposed method extracts more efficient image features and make better predictions than ordinary CRNN. We apply this new model to handwritten digit string recognition task (HDSR) and obtain significant improvements on HDSR benchmarks ORAND-CAR-A and ORAND-CAR-B.

Keywords: Residual connection · Convolutional recurrent neural network · Handwritten digit string recognition

1 Introduction

Handwritten digit string recognition (HDSR) is one of the main topics in the document analysis community [1]. Due to the connected characters, background noises and high variability of handwriting, recognition of unconstrained handwriting string is still an open and hot area. Many efforts are applied to handle this problem. A straight way is to segment the string image into pieces, and then recognize each piece with a single digit classifier. After that a path-algorithm, such as beam search, is applied to combine the recognition results of previous pieces and output the final global optimal prediction result. These methods are known as over-segmentation strategy. Wu et al. [1] transformed the string image into a sequence of primitive image segments after binarization, then combined these segments to generate candidate character patterns, forming a segmentation candidate lattice. After that a beam search algorithm was used to find an optimal path over the candidate lattice. This method won the first place on the ICFHR2014 HDSR competition [1]. Saabni [2] used sliding window and deep neural network to attain high recognition rates. Gattal et al. [3] applied three segmentation methods to handle handwritten digit strings by combining these

T. Gedeon et al. (Eds.): ICONIP 2019, CCIS 1143, pp. 49–56, 2019.
https://doi.org/10.1007/978-3-030-36802-9_6

segmentation methods depending on the configuration link between digits. But this kind of methods faced many problems in practice that mentioned at the beginning of this paragraph.

An alternative solution is segmentation-free methods. One popular solution is to uss CTC [4] (connectionist temporal classification) to decode the extracted features to generate the final recognition results directly. Messina [5] firstly applied a LSTM-RNN model to off-line Chinese handwritten text recognition. Without well-designed architecture it achieved competitive performance with the state-of-the-art of tradition method [6].

Recent evidences [7,8] reveal that network depth is of crucial importance, and we can gain accuracy from considerably increased depth. But while the networks are going deeper, problems such as vanishing or exploding gradients and degradation problem are been exposed. In order to handle these issues, He et al. [9] introduced residual connections and achieved very good results. Gao Huang et al. [10] further fortified this and proposed the Densenet that connects each layer to every other layer in a feed-forward fashion.

In this paper, we embrace these observations and propose a new network named Residual CRNN (Res-CRNN) based on residual connections and CRNN framework. In Residual CRNN, we apply residual connections between each two convolutional layers, like [11] does, and we also add residual connections in recurrent layers to enhance the information flow. At the top our model, a standard CTC is applied to calculate the loss and yield the recognition result. By taking the advantages of the residual connections, the proposed model work well on HDSR task. Compared with the submitted methods in ICFHR 2014 HDSR competition [1] and other CRNN-based approaches, our network obtains significant improvements.

The remainder of this paper is organized as follows. In Sect. 2 we describe the methods. Then, the details of our experiments are presented in Sect. 3. Section 4 concludes this paper.

2   The Proposed Architecture

The proposed Residual Convolutional Recurrent Neural Network (Res-CRNN) consists of three components, densely residual convolutional layers, residual recurrent layers and Connectionist temporal classification (CTC), which are shown in Fig. 1.

Fig. 1. The architecture of our network.

2.1 Densely Residual Convolutional Layers

Residual connection or ResNet [9] had shown success on many computer vision tasks. [10] embraced this observation and connects each layer to every other layer in a feed-forward fashion with residual connections, which was known as DenseNet. The basic component of DenseNet was a group of such densely connected convolutional layers and named as dense block. In this paper, we use three dense blocks to build the CNN part to extract efficient features.

2.2 Residual Recurrent Layers

Following many existed works, in this paper we use LSTM to build the recurrent layers. Inspire by the super performance of residual in convolutional layers, we also add residual connection in RNNs. The architecture of residual LSTM layers is shown in Fig. 2.

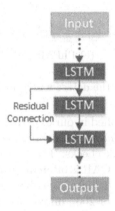

Fig. 2. The structure of residual LSTM layers.

2.3 Connectionist Temporal Classification

Connectionist temporal classification [4] is a kind of output layer, which is usually utilized after the RNN layers to calculate loss and decode the output of RNN layers.

Given the training dataset is $S = (\boldsymbol{X}, \boldsymbol{I})$, where X is the training image and I is the ground truth. The CTC object function $\mathscr{O}(S)$ is defined as the negative log probability of ground truth all the training examples in training set S,

$$\mathscr{O}(S) = - \sum_{(x,i) \in S} \log p(i|y) \tag{1}$$

where y is the sequence produced by the recurrent layers from x. Therefore, the network can be end-to-end trained on pairs of images and sequences.

3 Experiments

In order to evaluate the effectiveness of the proposed model, we conduct two experiments. One is to show the recognition performance on public datasets and the other one is to test the ability of our model on long digit strings recognition.

3.1 Datasets

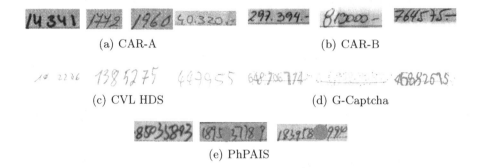

(a) CAR-A   (b) CAR-B

(c) CVL HDS   (d) G-Captcha

(e) PhPAIS

Fig. 3. Samples of the datasets used in our experiments.

Two public datasets are applied in our experiments. The first one is ORAND-CAR, which consists of 11719 images obtained from the Courtesy Amount Recognition (CAR) field of real bank checks. The images in ORAND-CAR has two sources, so in practice ORAND-CAR is often divided into two datasets, ORAND-CAR-A and ORAND-CAR-B, which are abbreviated to CAR-A and CAR-B. The CAR-A database consists of 2009 images for training and 3784 images for testing. The CAR-B database consists of 3000 training images and 2926 testing images. Some samples are shown in Fig. 3(a)–(b).

The other dataset is Computer Vision Lab Handwritten Digit String (CVL HDS) dataset, which is collected from about 300 writers. The CVL HDS dataset has 7960 images, from which 1262 images for training and the other 6698 images for testing. The variability of writers brings high variability with respect to handwritten styles. Some examples are shown in Fig. 3(c) with different written styles.

Images in the two datasets mentioned above mostly have digits less than 7. For string recognition task, the longer string means the harder task. So we use the same dataset G-Captcha used in [12] to show the ability of our model on long digit string recognition. This dataset is generated by using a Python package named 'captcha'[1]. This package can generate arbitrary length captcha images with dirty background and varied writing styles, which are similar to human handwriting. Some examples of G-Captcha are shown in Fig. 3(d).

---

[1] https://pypi.python.org/pypi/captcha/0.1.1.

We also collect a real long digit string dataset named PhPAIS, as shown in Fig. 3(e). The images in this dataset come from the real express form, due to the privacy policy we hide some digits in the samples.

The distribution of the five different datasets with respect to string length are show in Table 1.

Table 1. Distribution of all datasets with respect to string length.

Len	CVL		CAR-A		CAR-B		G-Captcha		PhPAIS	
	Train	Test	Train	Test	Train	Test	Train	Test	Train	Test
2	0	0	22	36	0	0	0	0	0	0
3	0	0	204	387	0	5	0	0	0	0
4	0	0	704	1425	63	69	0	0	0	0
5	125	789	903	1475	1200	1241	0	0	0	0
6	758	4144	145	363	1599	1452	0	0	828	912
7	379	1765	29	87	137	157	0	0	244	350
8	0	0	2	11	1	2	2000	1500	264	379
9	0	0	0	0	0	0	2000	1500	178	313
10	0	0	0	0	0	0	2000	1500	364	586
11	0	0	0	0	0	0	2000	1500	3122	3937
Total	1262	6698	2009	3784	3000	2926	8000	6000	5000	6477

3.2 Pre-processing

The size of images is varying in these datasets, but the fully connected layer in our network requires fixed input dimensions. In this paper, we resize the input images before feeding into the network to address this problem.

3.3 Evaluation Metrics

Two metrics are widely used to measure string recognition performance, a hard metric and a soft metric. Both of them can be defined based on the Levenshtein distance (LD), which is also known as the edit distance. The edit distance between two strings is the minimum number of single-character edits (insertions, deletions or substitutions) required to change one word into the other.

In this paper, we apply the hard metric. Let a_T be a target string (GT) and a_R be the corresponding recognized string. The hard metric, or accuracy rate (AR) is computed by:

$$AR = \frac{N_a}{N} \tag{2}$$

where N is the total number of testing string images and N_a is the number of images that the $LD(a_T, a_R) = 0$. All performance indexes appear in tables are accurate rate.

3.4   Experimental Settings

The proposed network is trained with ADADELTA, and the parameter delta is 10^{-6}. On all datasets we train the network using batch size 128 for 30,000 iterations.

We use three dense blocks in our experiments. The hyper-parameters of three dense blocks are the same. The growth rate is 12 and the number of convolution layer is 16. The number of hidden unit in all recurrent layers is 256. After recurrent layers, the two fully connected layers and the output size is 100 and the second is 11.

We conduct our experiments on a SuperMicro server with Intel Xeon E5-2630 CPU and NVIDIA GeForce 1080TI GPU. We use the Caffe [13] framework with cuDNN V5 accelerated to build our network on Ubuntu 14.04 LTS system.

Table 2. Recognition rates of different models on the datasets described above. (The top five methods are proposed on the HDSRC 2014 [1], the following two are proposed in newest papers. Especially, last but one method uses the ORAND-CAR dataset as a whole.)

Methods	CAR-A	CAR-B	CVLHDS
Singapore [1]	0.5230	0.5930	0.5040
Tebessa I [1]	0.3705	0.2662	0.5930
Pernambuco [1]	0.7830	0.7543	0.5860
Tebessa II [1]	0.3972	0.2772	0.6123
BeiJing [1]	0.8073	0.7013	**0.8529**
FSPP [14]	0.8261	0.8332	0.7923
CRNN [15]	0.8801	0.8979	0.2601
Saabni [2]	0.8580		-
Zhan [12]	0.8975	0.9114	0.2704
Proposed	**0.9197**	**0.9387**	0.2823

3.5   Results and Analysis

The experimental results of our method on public datasets are shown in Table 2, as well as other methods. We can see that the proposed network leading a lot on both ORAND-CAR-A and ORAND-CAR-B than other methods. But it performs very bad on CVL HDS dataset. On the other side, traditional methods performed outstandingly. There are 300 writers that contribute to CVL HDS. For each writer, 26 different digit strings were collected. Only 10 kinds of strings occur in training set. For segmentation methods, this is not a problem because the total categories of numbers are ten. But for methods based on deep learning method which requir, it gets into trouble due to the lack of sample diversity. The CRNN architecture [15] which is also derived from RNN-CTC faces the same problem.

The above experiments demonstrate that the proposed method can work well on handwritten digit string recognition task. Apart from the shorter length strings in public datasets, it can recognize much longer string images. Our model get the highest performance on G-Captcha and PhPAIS datasets, which consists of much longer digit strings than other datasets (Table 3).

Table 3. Accuracy rates of different models on the G-Captcha and PhPAIS datasets.

Methods	G-Captcha	PhPAIS
CRNN [15]	0.9312	0.8792
Zhan [12]	0.9512	0.9092
FSPP [14]	-	0.8472
Proposed	**0.9632**	**0.9366**

4   Conclusion

In this paper, we have presented a novel model based on CRNN architecture and residual connections. By taking the advantages of residual connections, we archive a very good performance on the handwritten digit string recognition benchmarks ORAND-CAR-A and ORAND-CAR-B. The proposed model is a universal architecture, we can apply it to many other sequence labelling tasks such as handwritten Chinese text recognition.

Acknowledgement. The work is supported by Shanghai Natural Science Foundation (No. 19ZR1415900).

References

1. Diem, M., et al.: ICFHR 2014 competition on handwritten digit string recognition in challenging datasets (HDSRC 2014). In: 14th International Conference on Frontiers in Handwriting Recognition (ICFHR), pp. 779–784 (2014)
2. Saabni, R.: Recognizing handwritten single digits and digit strings using deep architecture of neural networks. In: The 3th International Conference on Artificial Intelligence and Pattern Recognition, pp. 1–6 (2016)
3. Gattal, A., Chibani, Y., Hadjadji, B.: Segmentation and recognition system for unknown-length handwritten digit strings. Pattern Anal. Appl. **20**(2), 307–323 (2017)
4. Graves, A., Fernández, S., Gomez, F., Schmidhuber, J.: Connectionist temporal classification: labelling unsegmented sequence data with recurrent neural networks. In: The 23rd International Conference on Machine Learning, pp. 369–376 (2006)

5. Messina, R., Louradour, J.: Segmentation-free handwritten Chinese text recognition with LSTM-RNN. In: 13th IAPR International Conference on Document Analysis and Recognition, pp. 171–175 (2015)
6. Wang, Q.-F., Yin, F., Liu, C.-L.: Handwritten chinese text recognition by integrating multiple contexts. IEEE Trans. Pattern Anal. Mach. Intell. **34**(8), 1469–1481 (2012)
7. Simonyan, K., Zisserman, A.: Very deep convolutional networks for large-scale image recognition, arXiv preprint arXiv:1409.1556 (2014)
8. Szegedy, C., et al.: Going deeper with convolutions. In: Proceedings of the IEEE Conference on Computer Vision and Pattern Recognition, pp. 1–9 (2015)
9. He, K., Zhang, X., Ren, S., Sun, J.: Deep residual learning for image recognition. In: The 29th IEEE Conference on Computer Vision and Pattern Recognition, pp. 770–778 (2016)
10. Huang, G., Liu, Z., Weinberger, K.Q., van der Maaten, L.: Densely connected convolutional networks, arXiv preprint arXiv:1608.06993 (2016)
11. Gao, H., Zhuang, L., Maaten, L.V.D., Weinberger, K.Q.: Densely connected convolutional networks. In: IEEE Conference on Computer Vision and Pattern Recognition (2017)
12. Zhan, H., Wang, Q., Lu, Y.: Handwritten digit string recognition by combination of residual network and RNN-CTC. In: Liu, D., Xie, S., Li, Y., Zhao, D., El-Alfy, E.S. (eds.) ICONIP 2017. LNCS, vol. 10639, pp. 583–591. Springer, Cham (2017). https://doi.org/10.1007/978-3-319-70136-3_62
13. Jia, Y., et al.: Caffe: convolutional architecture for fast feature embedding, arXiv preprint arXiv:1408.5093 (2014)
14. Wang, Q., Lu, Y.: A sequence labeling convolutional network and its application to handwritten string recognition. In: Twenty-Sixth International Joint Conference on Artificial Intelligence, pp. 2950–2956 (2017)
15. Shi, B., Bai, X., Yao, C.: An end-to-end trainable neural network for image-based sequence recognition and its application to scene text recognition. IEEE Trans. Pattern Anal. Mach. Intell. (2016). https://doi.org/10.1109/TPAMI.2016.2646371

G-HAPNet: A Novel Structure for Single Image Super-Resolution

Yan Luo[1,2], Mingyong Zhuang[1,2], Congbo Cai[1,2], Yue Huang[1,2],
and Xinghao Ding[1,2(✉)]

[1] Key Laboratory of Underwater Acoustic Communication and Marine Information
Technology, Ministry of Education, Xiamen University, Xiamen 361005, Fujian, China
{23320171153184,myzhuang}@stu.xmu.edu.cn
{cbcai,yhuang2010,dxh}@xmu.edu.cn
[2] School of Informatics, Xiamen University, Xiamen 361005, Fujian, China

Abstract. Recent investigations on single image super-resolution
(SISR) have progressed with the development of deep convolutional neural networks (DCNNs). However, increasingly complex network designs
cause huge computational budgets. Therefore, a more efficient structure
for SISR task is desirable. In this report, we propose a novel structure,
called G-HAPNet. Specifically, the group-hierarchical atrous pyramid
block (G-HAPB) is built to package as a general block for deeper network
constitution. Firstly, the original features are expanded and grouped in
channel. Then, a atrous pyramid is constructed to extract multi-scale
features from corresponding channels. Besides, we introduce hierarchical
grouping aggregation (HGA) which includes forward aggregation and
backward aggregation by skip connections so that we can achieve hierarchical fusion and information guidance among multi-scale features.
Extensive experiments demonstrate that with the same level depth and
computational budgets, our proposed G-HAPNet has better performance
than state-of-the-art methods on both synthetic datasets and real-world
dataset, which indicates our G-HAPNet is a more efficient and practical
structure for SISR.

Keywords: Single image super-resolution · Deep convolutional neural
networks · Atrous pyramid

1 Introduction

Single image super-resolution (SISR) aims to reconstruct a high-resolution image
(HR) from it's low-resolution (LR) counterpart, which has wide applications in

This work was supported in part by the National Natural Science Foundation of
China under Grants 61571382, 81671766, 61571005, 81671674, 61671309, 61971369 and
U1605252, in part by the Fundamental Research Funds for the Central Universities
under Grants 20720160075 and 20720180059, in part by the CCF-Tencent open fund,
and the Natural Science Foundation of Fujian Province of China (No. 2017J01126).

T. Gedeon et al. (Eds.): ICONIP 2019, CCIS 1143, pp. 57–64, 2019.
https://doi.org/10.1007/978-3-030-36802-9_7

computer vision tasks. Therefore, the SISR has drawn increasing research attention for decades and diverse SISR methods have been proposed. Early methods mainly include bicubic interpolation, Lanczos resampling [3] and more effective methods by statistical image priors is proposed in [9].

In recent years, the deep convolutional neural networks (DCNNs) show power ability in image processing. Thus, a variety of network structures for SISR emerge. Previous networks such as SRCNN [1], FSRCNN [2] adopt relatively shallow network structures. Although significant performance improvement is achieved when compared to no-CNN based methods, they are still inferior when compared to deeper network such as VDSR [7], EDSR [8], etc. The DCNNs are of crucial importance for SISR and the increasing depth bring benefits to representation ability. Subsequently, the network tends to be deeper with tons of parameters (e.g. RCAN [11], DBPN [5]), which is not easily affordable for practical applications. Therefore, the more effective structure for SISR is worth studying.

Research shows that it is of critical importance to enrich multi-scale stimuli for visual cognition tasks [4,5] tries to introduce the dense connection between layers and diverse scales kernels for adapting different SISR tasks. Although multi-scale features are considered in [5], it is still a coarse and complex method. Thus, how to extract more abundant scale features is a core issues for SISR.

To address these issues, we propose a group-hierarchical atrous pyramid net (G-HAPNet) with a cascaded architecture which adopts cascaded scheme with group-hierarchical atrous pyramids blocks (G-HAPB). Support by G-HAPB the more abundant scale features and the finer features aggregation are achieved. Moreover, it is easy to package G-HAPB as a general block for building deeper network.

In general, the contributions of our paper are three-fold.

(1) We build a group-hierarchical atrous pyramid network (G-HAPNet) for highly accurate image SR.
(2) We propose a effective group-hierarchical atrous pyramid block (G-HAPB), a general block for extracting multi-scale features.
(3) The hierarchical grouping aggregation (HGA) is further introduced to helps G-HAPB to achieve effective feature fusion.

2   Proposed Methods

In this section, the progressive deraining networks are presented by G-HAPB and the integration with Modern Methods are also discussed. We can find the overall of G-HAPNet in Fig. 2.

2.1   Group-Hierarchical Atrous Pyramid Block (G-HAPB)

In this paper, we propose a novel structure which is called group-hierarchical atrous pyramid block (G-HAPB) for multi-scale feature extraction. As Fig. 1

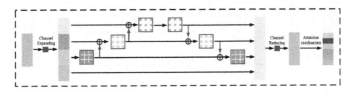

Fig. 1. Group-hierarchical atrous pyramid block (Color figure online)

shows each G-HAPB mainly includes two parts: atrous pyramid and hierarchical grouping aggregation (HGA), in following sections we will introduce their structures, respectively.

Atrous Pyramid: The atrous pyramid is major structure of G-HAPB, which is mainly composed of multiple groups of atrous convolutions in parallel, each group of atrous convolutions have the same rate, but different rates in different groups. The tower combination of convolutions can brings larger receptive field and stronger scale adaptability.

Hierarchical Grouping Aggregation: As red lines show in Fig. 1. HGA is divided into two parts: forward aggregation (FA) and back aggregation (BA). In process of FA, original features are expanded on channels by 1×1 convolution. Then, we evenly split new features into S groups, denoted as x_i, where $i \in \{1, 2, 3, \cdots, s\}$, each feature group x_i has the same spatial size but $1/s$ number of channels compared to original features. Each x_i has it's corresponding atrous convolution, denoted as $K_i(\cdot)$ except x_1, we present Y_i as the output of $K_i(\cdot)$, Meanwhile, in process of BA, we denote the atrous convolution as $F_i(\cdot)$ and L_i is the output of $F_i(\cdot)$. Then, the final output F_{out} of G-HAPB can be defined as:

$$Y_i = \begin{cases} X_i & i = 1, 2, \\ K_i(X_i + Y_{i-1}) & 2 < i \leq s. \end{cases} \tag{1}$$

$$L_i = \begin{cases} X_i & i = 1, \\ F_i(Y_i) & i = s, \\ F_i(Y_i + L_{i+1}) & 1 < i < s. \end{cases} \tag{2}$$

$$F_{out} = concat(L_1, L_2, \cdots, L_{s-1}, L_s) \tag{3}$$

Where the function of concat$(\cdot)$ denotes the concatenation operation. Subsequently, the attention mechanism is adopt to rescale the F_{out} in channel-wise.

2.2 Integration with Modern Methods

Since the G-HAPB is a general module, numerous innovative methods can be integrated into our proposed G-HAPNet easily, such as wide activation and weight normalization [10], attention mechanism [6], etc.

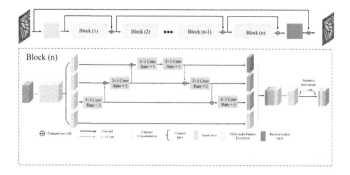

Fig. 2. Network architecture of our proposed G-HAPNet

Wide Activation: Wide activation is proposed in WDSR, which aims to expand features before activation function. In this paper, original feature maps of G-HAPB is expanded to double to implement wider activation.

Weight Normalization: Weight normalization (WN) is a re-parameterization of the weight vectors in a neural network that decouples the length of those weight vectors from their direction, which does not introduce dependencies between the examples in a mini-batch and has the same formulation in training and testing [10]. We can define the Y as the output:

$$Y = W * X + b \tag{4}$$

Where W is a k-dimensional weight vector, b is a scalar bias and x is k-dimensional vector of input. The weight vector is re-parameterized as follow:

$$W = \frac{g}{\|v\|} v \tag{5}$$

Where each weight vector W in terms of a parameter k-dimensional vector v and a scalar g. The $\|v\|$ presents the Euclidean norm of v.

3   Experiments

3.1   Dataset

Datasets: We use Set291 for benchmark to speed up experimental process and save computational budget and four widely used test set including Set5, Set14, BSD100, Urban100 are employed. Meanwhile, a novel dataset of real-world is also provided by NTIRE 2019 challenge (http://www.vision.ee.ethz.ch/ntire19/), which is built by digital single-lens reflex (DSLR) cameras, The whole dataset contains training set (60 images), validation set (20 images). Limited by test set is not available, we set the result of validation as evaluation indicator.

Both the Synthetic datasets and real-world dataset are split into 128 × 128 with stride 64 and the common data augmentation methods including rotation and flip is adopted when generating training set.

3.2   Implementation Detail

For fair comparison, all models are controlled at same depth and same settings. We set the depth of 22 layers to all models. Light-weight models can speed up the experimental process and adapt to the small dataset of real-world (only 60 training images). Besides, since the LR images have the same size as HR images in real-world dataset. We remove the upsampling layer of all models to unite the network structure for synthetic datasets and real-world dataset. In addition, a deeper G-HAPNet+ which has 7 blocks is also trained.

With reference to previous works, the adam optimizer is chosen, the initial learning rate is set to 10^{-3} and 6×10^5 iterations of training are required. In this paper, we set all models with kenerl size of 3×3, 64 channels, except $G-HAPNet$ which has kenerl size of 3×3, 32 channels with rate ranging from 1 to 3. All experiments are conducted by pytorch on a NVIDIA GeForce GTX 1080 with 8 GB GPU memory.

3.3   Evaluation on Synthetic Datasets and Real-World Datasets

In this paper, we compare our methods with several state-of-the-art algorithms including EDSR, WDSR, DBPN, RCAN on both synthetic and real-world datasets. Besides, the benefits of atrous pyramid and HGA is discussed.

Effects Atrous Pyramid and HGA: We train a G-HAPNet_O which use two convolutional layer to replace atrous pyramid as baseline, and a G-HAPNet_AP is trained without HGA. We split HGA into two parts of forward aggregation (FA) and back aggregation (BA) for quantitative analysis, respectively. Table 1 shows, the atrous pyramid is of significant to SISR to bring performance improvement. Besides, HGA is a also essential structure to G-HAPNet.

Result on Synthetic Datasets and Real-World Datasets: Table 2 shows, these state-of-the-art models no longer perform well as the depth reduces, which reveals they are inefficient for SISR. Meanwhile, our G-HAPNet+ achieves the best performance with less than 35% parameters than RCAN, DBPN and EDSR, even G-HAPNet obtain a decent performance with less than 50% parameters. Beside, our models have more obvious superiority on datasets for ×3 and ×4 which illustrates our models have stronger scales adaptability. Figure 3 shows our models can recover more accurate details and produce less blurring.

Table 3 shows, G-HAPNet+ and G-HAPNet obtain the best performance and second one. But RCAN does not have expected performance. It indicates that SISR on real-world images is a more difficult task. In Fig. 4, we can find our proposed models has better visual quality on details reconstruction and texture recovery. Therefore, our proposed model is more suitable for SISR on real-world images, which is of great significance on practical applications.

3.4   Comparison with EDSR on Real-World Dataset

Since EDSR has decent performance on real-world dataset, we choose the EDSR as baseline and adjust the depth of models to observe the performance change.

Table 1. Investigation of atrous pyramid and HGA on Set5 dataset, **Red** indicates best results and *blue* indicates second best

Item	×2	×3	×4
G-HAPNet_O	35.09/0.939011	31.59/0.894037	29.21/0.847830
G-HAPNet_AP	35.14/0.939430	31.69/0.895287	29.45/0.851873
G-HAPNet_AP+BA	*35.15/0.939517*	31.72/0.896067	29.49/0.853288
G-HAPNet_AP+FA	*35.15*/0.939499	*31.73/0.896159*	*29.51/0.853643*
G-HAPNet_AP+HGA	**35.23/0.939961**	**31.76/0.896503**	**29.54/0.854192**

Table 2. Quantitative results on synthetic datasets. **Red** indicates best results and *blue* indicates second best results

Model	Parameters	Scale	Set5		Set14		BSD100		Urban100	
			PSNR	SSIM	PSNR	SSIM	PSNR	SSIM	PSNR	SSIM
Bicubic	-	×2	31.79	0.9088	28.40	0.8427	28.23	0.8298	25.43	0.8275
WDSR [10]	0.44M	×2	34.98	0.9388	30.58	0.8838	30.47	0.8859	28.83	0.8994
DBPN [5]	0.82M	×2	34.81	0.9370	30.43	0.7768	30.37	0.8841	28.65	0.8916
EDSR [8]	0.85M	×2	35.02	0.9388	30.60	0.8841	30.50	0.8863	28.93	0.9003
RCAN [11]	0.86M	×2	35.18	0.9394	*30.71*	0.8848	*30.54*	*0.8868*	*29.05*	0.9015
G-HAPNet	0.39M	×2	*35.23*	*0.9400*	30.65	*0.8854*	30.53	*0.8868*	29.03	*0.9018*
G-HAPNet+	0.53M	×2	**35.28**	**0.9401**	**30.73**	**0.8861**	**30.57**	**0.8873**	**29.11**	**0.9029**
Bicubic		×3	28.63	0.8385	25.87	0.7407	25.90	0.7176	23.03	0.7152
WDSR	0.44M	×3	31.38	0.8920	27.53	0.7949	27.44	0.7810	25.27	0.8037
DBPN	0.82M	×3	31.10	0.8877	27.42	0.7900	27.34	0.7768	25.10	0.7927
EDSR	0.85M	×3	31.52	0.8936	27.64	0.7965	27.49	0.7819	25.42	0.8068
RCAN	0.86M	×3	31.73	0.8955	27.70	0.7971	27.52	0.7828	25.49	0.8075
G-HAPNet	0.39M	×3	*31.76*	*0.8965*	*27.73*	*0.7989*	*27.56*	*0.7835*	*25.59*	*0.8124*
G-HAPNet+	0.53M	×3	**31.89**	**0.8974**	**27.75**	**0.7995**	**27.59**	**0.7845**	**25.64**	**0.8139**
Bicubic		×4	26.69	0.7737	24.39	0.6644	24.66	0.6418	21.71	0.6328
WDSR	0.44M	×4	29.06	0.8447	25.80	0.7229	25.90	0.7019	23.36	0.7205
DBPN	0.82M	×4	28.74	0.8379	25.70	0.7172	25.80	0.6973	23.22	0.7089
EDSR	0.85M	×4	29.19	0.8486	25.89	0.7260	25.94	0.7037	23.49	0.7257
RCAN	0.86M	×4	29.40	0.8507	26.00	0.7270	25.98	0.7053	23.53	0.7263
G-HAPNet	0.39M	×4	*29.54*	*0.8542*	*26.11*	*0.7367*	*26.03*	*0.7068*	*23.66*	*0.7344*
G-HAPNet+	0.53M	×4	**29.63**	**0.8558**	**26.13**	**0.7312**	**26.05**	**0.7082**	**23.68**	**0.7356**

Figure 5 shows, the tendency of EDSR in PSNR suggests that EDSR with deeper depth and tons of parameters may be not suitable on real-world images. On the contrary, G-HAPNet obtains a gradual increase on PSNR as the block goes deeper, In addition, we can find that our models outperforms than EDSR in not matter shallow or deeper networks with few parameters, which proves again that G-HAPNet is an more effective and practical structure for SISR.

Table 3. Quantitative results on real-world dataset. **Red** indicates best results and *blue* indicates second best

Item	LR	DBPN	EDSR	WDSR	RCAN	G-HAPNet	G-HAPNet+
PSNR	27.79	28.72	28.88	28.83	28.86	*29.12*	**28.17**
SSIM	0.8163	0.8437	0.8482	0.8478	0.8463	*0.8568*	**0.8576**

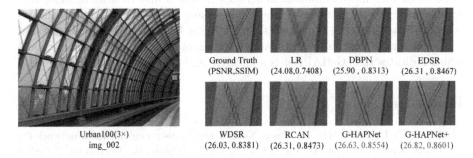

Fig. 3. The visual results comparisons on synthetic dataset for 3× enlargement, **Red** indicates best results and *blue* indicates second best (Color figure online)

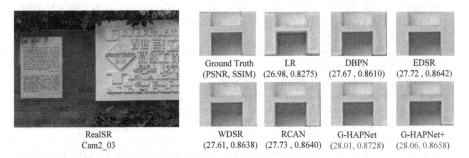

Fig. 4. The visual results comparisons on real-world dataset, **Red** indicates best results and *blue* indicates second best (Color figure online)

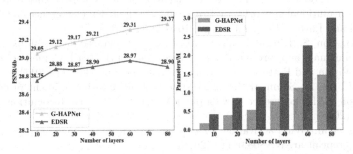

Fig. 5. Comparisons of EDSR and G-HAPNet with different depths on real-world dataset

4   Conclusion

In this paper, we propose G-HAPNet, an effective multi-scale network for SISR task. Specifically, the group-hierarchical atrous pyramid block (G-HAPB) is instituted as a general block to design deeper network. In G-HAPB, the atrous pyramid and hierarchical grouping aggregation (HGA) is proposed to help network to extract more abundant multi-scale features and to achieve more delicate feature fusion. Extensive experiments show that on the same depth and computational budgets, our models can perform better on both synthetic and real-world datasets with fewer parameters than other state-of-the-art methods, which indicates that G-HAPNet is a effective and practical structure for SISR.

References

1. Dong, C., Loy, C.C., He, K., Tang, X.: Learning a deep convolutional network for image super-resolution. In: Fleet, D., Pajdla, T., Schiele, B., Tuytelaars, T. (eds.) ECCV 2014. LNCS, vol. 8692, pp. 184–199. Springer, Cham (2014). https://doi.org/10.1007/978-3-319-10593-2_13
2. Dong, C., Loy, C.C., Tang, X.: Accelerating the super-resolution convolutional neural network. In: Leibe, B., Matas, J., Sebe, N., Welling, M. (eds.) ECCV 2016. LNCS, vol. 9906, pp. 391–407. Springer, Cham (2016). https://doi.org/10.1007/978-3-319-46475-6_25
3. Duchon, C.E.: Lanczos filtering in one and two dimensions. J. Appl. Meteorol. **18**(8), 1016–1022 (1979)
4. Gao, S.H., Cheng, M.M., Zhao, K., Zhang, X.Y., Yang, M.H., Torr, P.: Res2net: a new multi-scale backbone architecture. arXiv preprint arXiv:1904.01169 (2019)
5. Haris, M., Shakhnarovich, G., Ukita, N.: Deep back-projection networks for super-resolution. In: Proceedings of the IEEE Conference on Computer Vision and Pattern Recognition, pp. 1664–1673 (2018)
6. Hu, J., Shen, L., Sun, G.: Squeeze-and-excitation networks. In: Proceedings of the IEEE Conference on Computer Vision and Pattern Recognition, pp. 7132–7141 (2018)
7. Kim, J., Kwon Lee, J., Mu Lee, K.: Accurate image super-resolution using very deep convolutional networks. In: Proceedings of the IEEE Conference on Computer Vision and Pattern Recognition, pp. 1646–1654 (2016)
8. Lim, B., Son, S., Kim, H., Nah, S., Mu Lee, K.: Enhanced deep residual networks for single image super-resolution. In: Proceedings of the IEEE Conference on Computer Vision and Pattern Recognition Workshops, pp. 136–144 (2017)
9. Sun, J., Xu, Z., Shum, H.Y.: Image super-resolution using gradient profile prior. In: 2008 IEEE Conference on Computer Vision and Pattern Recognition, pp. 1–8. IEEE (2008)
10. Yu, J., et al.: Wide activation for efficient and accurate image super-resolution. arXiv preprint arXiv:1808.08718 (2018)
11. Zhang, Y., Li, K., Li, K., Wang, L., Zhong, B., Fu, Y.: Image super-resolution using very deep residual channel attention networks. In: Proceedings of the European Conference on Computer Vision (ECCV), pp. 286–301 (2018)

Inpainting with Sketch Reconstruction and Comprehensive Feature Selection

Siyuan Li[1], Lu Lu[1], Zhijing Li[2], Kepeng Xu[1], Matthieu Claisse[3], Wenxin Yu[1(✉)], Gang He[1], Gang He[4], Yibo Fan[5], and Zhuo Yang[6]

[1] Southwest University of Science and Technology, Mianyang, China
yuwenxin@swust.edu.cn
[2] Accenture Japan Ltd., Tokyo, Japan
[3] Graduate School in Computer Science and Mathematics Engineering France, Pau, France
[4] Xidian University, Xi'an, China
[5] State Key Laboratory of ASIC and System, Fudan University China, Shanghai, China
[6] Guangdong University of Technology, Guangzhou, China

Abstract. With the advent of the convolutional neural network, learning-based image inpainting approaches have received much attention, and most of these methods have been attracted by adversarial learning and various loss functions. This paper focuses on the enhancement of the generator model and guidance of structural information. Hence, a novel convolution block is proposed to comprehensively capture the context information among feature representations. The performance of the proposed method is evaluated on Place2 test dataset, which outperforms the current state-of-the-art inpainting approaches.

Keywords: Image inpainting · Deep learning · Feature selection · Edge guidance

1 Introduction

Image inpainting, also known as image completion, is the process of restoring the missing parts in a damaged image. Because the corrupted region only can be inferred through its neighborhood, it is still a challenging task to recover the details of the corrupted region to completely match the original image. From the tuition of human painting, the edge information in the filled region can guide the inpainting model to produce sharper results and away from the blurred edges, so as to improve inpainting quality and make filled region reasonable. Therefore, if we first paint the missing area with the fine structure or precise edges, the final results guided by the repaired sketch will be greatly improved.

In consideration of these pieces of knowledge, this paper proposes a two-stage, learning-based image inpainting approach with enhanced generator model and a new type of convolution block. Similar to one of the most advanced works [10],

© Springer Nature Switzerland AG 2019
T. Gedeon et al. (Eds.): ICONIP 2019, CCIS 1143, pp. 65–73, 2019.
https://doi.org/10.1007/978-3-030-36802-9_8

the two stages of our proposed method are sketch reconstruction and image completion phases, and both of the two stages also introduce the generative adversarial network (GAN) [3].

The first stage, sketch reconstruction phase, aims to recover the gradient information in the missing area from corrupted sketch maps. In this paper, the Holistically-nested edge detection (HED) [12] is introduced to generate the sketches maps for training and testing. It is worth noting that Kamyar's work [10] has conducted some experiments that also exploited the edge maps generated by HED, however, they assume the edge inpainting process is a relatively easy task and don't put enough attention on the edge generator model. According to their experiments, their edge generator model fails to achieve better accuracy of HED prediction than applying Canny edge detector [1]. The sketch map generated by HED is considered as guiding information in this paper, and we enhance the sketch generator model by increasing the convolution layers to competent the sketch reconstruction task.

In the second phase, the goal of image completion networks is exploiting the repaired sketch maps and corrupted raw RGB image to color the sketch in the filled region. This paper applies a new convolution module named Comprehensive Feature Selection Block (CFS Block) in the second phase to comprehensively capture the saliency of context information among the feature maps. And the weight of the proposed module can be automatically updated during the training phase by the backpropagation.

Although our work is close to the combination of Kamyar's [10] and Yu's [13], our work proposed a novel convolution block to comprehensively select the features among the input features and convolved features in current module meanwhile enhancing the generator model through incorporating lots of popular technology to assure the accuracy of sketch and texture prediction.

This approach we presented is evaluated on the test dataset of Place2 [15]. Compared with those state-of-art inpainting approaches, the produced results quantitatively achieve great improvement. To sum up, our paper makes the following contributions:

- *The introduction of the edge sketch produced by HED, which better represents the rough shape of objects in images.*
- *A reinforced edge generator that can repair or hallucinate the sketch map through rest of sketch.*
- *An integrated inpainting generator with a novel convolution block – CFS Block.*

2   Approach

The proposed scheme divides the learning-based image inpainting process into two phases. At each stage, we create a generator model to produce target image meanwhile establishing a discriminator model that feedback to the generator to help produce high-quality results. In the first phase, the corrupted sketch and grayscale image are concatenated as feature map, the features and the binary mask map (where 1 represent the non-damaged) region are fed into the generator with Partial Convolution, then the generator predicts a complete sketch as output. At the second stage, the restored sketch and damaged RGB image are considered as features together, they are inputted to a new generator built by CFS Blocks, aimimg to get a complete RGB image.

In this section, we describe the detailed architecture of proposed networks in each phase and the detailed design of the Comprehensive Feature Selection Block (CFS Block) and briefly analyze what CFS Block actually doing.

2.1   Networks

The general architecture of the proposed model for each phase is illustrated as Fig. 1. As mentioned in [9], Spectral Normalization can further stabilize the training process through utilizing the maximum singular value of the weight matrix to reduce each weight matrix, which limits the Lipschitz constant of functions to 1. Thus we apply spectral normalization to the generators and discriminators in both phases.

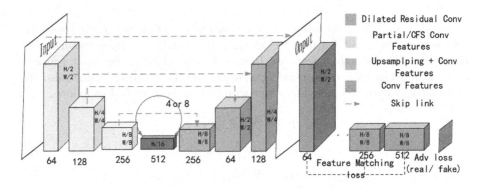

Fig. 1. The design of networks in one of stages.

The design of proposed discriminators in different phases are exactly the same. It's worth to note that the last two convolution layer with the same padding reduce the number of channels to 1 after increasing the number. In sketch reconstruction phase, the task of generator is relatively easy, thus we apply Partial Convolution [8] to each convolution layer in this generator instead of applying CFS Block, another difference is that we set 4 dilated residual block with Partial Convolution in the middle part of generator rather than 8 dilated residual blocks with CFS in image completion generator.

2.2   Comprehensive Feature Selection Block

The inspiration of the front part of CFS Block comes from classical Gated Recurrent Units (GRU) [2] and Gated Convolution [13], whileas the tail of the module directly introduce the Squeeze-and-Excitation [4] block. The proposed integrated CFS Block aims to not only emphasize spatial relationships but also to further concern about channel correlation in feature selection process.

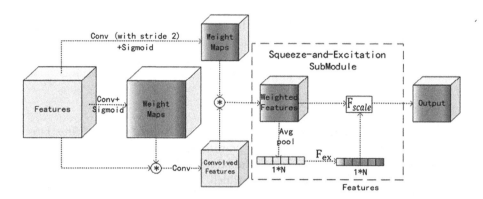

Fig. 2. The internal structure of Comprehensive Feature Selection Block.

As shown in Fig. 2, we adopt two convolution layer followed by sigmoid activation from the input features to calculate the weights of input features and the weights of convoluted features. Before the convolution, the input features are multiplied by their corresponding weights, and the convolved values also are multiplied by their weights. The slider-wise process is described as follows.

$$g = \sigma(W_g^T x + b_g) \tag{1}$$

$$G = \sigma(W_G^T x + b_G) \tag{2}$$

$$f = G \cdot \phi(W_f^T(g \cdot x) + b_f) \tag{3}$$

In which the g represents the gating value of the original feature in one of sliding windows and G is the gating values (weight) of the convolved feature map, the σ denotes sigmoid function and ϕ represents the Leaky ReLU activation with the slope of 0.2. The f corresponds to the weighted feature computed by those foregoing units.

Discussion About CFS Block. All of the parameters in CFS Block (except r) are learnable in the training process, this means that all the gating values (weight maps) in CFS Block can automatically be updated from data, it enables the generator to learn weight maps dynamically thus can select features both in input feature maps and the features in the next level. Because the importance of the damaged image, masks and sketches are obviously different, it is reasonable to set an additional gated convolution to select the input features rather than directly applying cross-level learning pattern [13]. Furthermore, the Squeeze-and-Excitation submodule is set to reinforce the ability of the model to notice some essential channels in the computed features. However, the number of parameters in CFS Block is a bit large, this is the reason why the relatively easy task, the sketch inpainting task, adopts Partial Conv instead of CFS Block.

2.3 Loss Function

For convenience of formulizing, this paper denotes the generator and discriminator in sketch reconstruction as G_s and D_s, denotes the generator and discriminator in image completion as G_i and D_i, and the binary mask that labels the valid pixels as 1 (invalid pixels as 0) is expressed as M, the ground truth images in dataset is represented as $\mathbf{I}_{gt}$, damaged image can be represented as $\acute{\mathbf{I}} = \mathbf{I}_{gt} \odot M$, the complete sketch generated by HED is S_{gt}, the incomplete sketch is $\acute{\mathbf{S}} = \mathbf{S}_{gt} \odot M$, the composite sketch is described as $\mathbf{S}_{comp} = \mathbf{S}_{pred} \odot (1-M) + \mathbf{S}_{gt} \odot M$.

The sketch generator predicts the complete sketch by considering the concatenation of damaged grayscale image and damaged sketch as the features, meanwhile inputting the mask M for Partial Conv. While in the image completion generator, it concatenates the inpainted sketch, damaged image and mask as the input feature.

$$\mathbf{S}_{pred} = G_s\left(\left[\acute{\mathbf{I}}_{gray}, \acute{\mathbf{S}}\right], M\right) \tag{4}$$

$$\mathbf{I}_{pred} = G_I\left(\left[\acute{\mathbf{I}}, \mathbf{S}_{comp}, M\right]\right) \tag{5}$$

On account of the feature-matching loss [11] is introduced as one of loss terms for further optimizing the generator, the total loss for sketch generator is interpreted as Eq. (6).

$$\min_{G_s} \max_{D_s} \mathcal{L}_{G_s} = \min_{G_s}\left(0.1 \max_{D_s}\left(\mathcal{L}_{D_s}\right) + 10\mathcal{L}_{FM}\right) \tag{6}$$

The task of the discriminator is to distinguish whether the input sketch in discriminator belongs to the grayscale image of the corresponding ground truth image, this paper adopts the hinge loss as the target function of discriminator, which train the discriminator more strictly.

$$\begin{aligned}\mathcal{L}_{D_s} = &\ \mathbb{E}_{\mathbf{S}_{gt}}\left[\psi(1 - D_s\left(\mathbf{S}_{gt}, \mathbf{I}_{gray}\right))\right] \\ &+ \mathbb{E}_{\mathbf{S}_{pred}}\left[\psi(1 + D_s\left(\mathbf{S}_{pred}, \mathbf{I}_{gray}\right))\right]\end{aligned} \tag{7}$$

Image completion generator adopts L1 distance and the perceptual loss ($\mathcal{L}_{style}$ and $\mathcal{L}_{perc}$) [6]. It enables the model to learn the high-level representation and remove the checkboard artifacts from the predicted image, the total loss of completion generator is express as

$$\mathcal{L}_{G_i} = \mathcal{L}_{\ell_1} + 0.1\mathcal{L}_{D_i} + 300\mathcal{L}_{perc} + 300\mathcal{L}_{style} + 10\mathcal{L}_{FM} \tag{8}$$

where $\mathcal{L}_{D_i}$ is similar to $\mathcal{L}_{D_s}$ however the input of $\mathcal{L}_{D_i}$ has a slight difference, it just input the $\mathbf{I}_{gt}$ and $\mathbf{I}_{pred}$ without any grayscale images.

3   Experiments

All of the experiments in this paper are conducted in the dataset of Place2 [15], the sketches are inferred from RGB images through HED [12] approach. The irregular mask dataset used in this paper comes from the work of Liu [8]. With these data groups (image sketch mask), We train on single NVIDIA 1080TI with a batch size of 6 until the generators converge using Adam optimizer [7].

Table 1. Comparison of quantitative results, these data are taken from this paper [10].

Mask		CA [14]	GLCIC [5]	PConv [8]	EdgeCnt [10]	Ours
10–20%	PSNR	24.36	23.49	28.02	27.95	**30.85**
	SSIM	0.893	0.862	0.869	0.920	**0.951**
20–30%	PSNR	21.19	20.45	24.90	24.92	**26.87**
	SSIM	0.815	0.771	0.777	0.861	**0.900**
30–40%	PSNR	19.13	18.50	22.45	22.84	**24.17**
	SSIM	0.739	0.686	0.685	0.799	**0.841**
>=40	PSNR	17.75	17.17	20.86	21.16	**21.74**
	SSIM	0.662	0.603	0.589	0.731	**0.7695**

3.1   Quantitative Results

The quantitative results in the test dataset of Place2 are shown in Table 1, this table also shows some results that produced popular inpainting methods in comparison. In the case of all different ratios of the damaged region, the table indicates that our results outperform the others in PSNR (Peak Signal-to-Noise) and SSIM (Structural Similarity) metric, especially in the case of small masks. In addition, the unique sketch prediction task achieved 77% accuracy (better than [10]), it indicates the quantitative improvement is benefited from the enhanced sketch generator.

3.2   Qualitative Results

The EdgeConnect approach [10] has recently shown surprising advancement in image inpainting, therefore we compare our result with this state-of-art work (Figs. 3 and 4). The red box in the figure indicates that the proposed approach with CFS Block can produce a clearer color image with the more obvious edges than EdgeConnect, the 5th column show the generated sketch in the proposed approach, the orange lines in these images represent the inpainted line by generator, it demonstrates the sketch generator with Partial Conv already can handle the sketch reconstruction task.

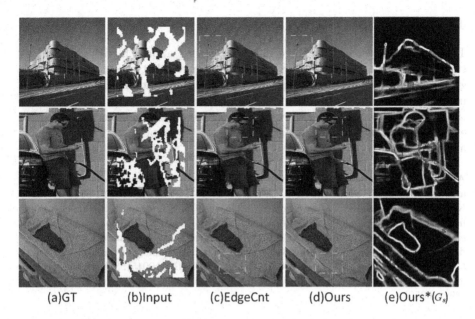

(a)GT   (b)Input   (c)EdgeCnt   (d)Ours   (e)Ours*(G_s)

Fig. 3. The qualitative comparison of results. (a) Ground Truth image. (b) Corrupted image. (c) EdgeConnect [10]. (d) Ours. (e) Restored sketches of Ours. (Color figure online)

(a)Input   (b)EdgeCnt[11]   (c)Ours   (a)Input   (b)EdgeCnt[11]   (c)Ours

Fig. 4. The additional qualitative comparison.

4   Conclusions

This paper proposed a two-stage image inpainting approach with a novel feature selection mechanism – CFS Block, and proves that the enhanced sketch generator and the proposed comprehensive feature selection mechanism can significantly improve the inpainting results. The qualitative comparisons show that the proposed approach produces visually more pleasing results, and the objective evaluations in various sizes of masks demonstrate the superiority of the proposed approach.

Acknowledgements. This research was supported by Sichuan Provincial Science and Technology Department (No. 2018GZ0517,2019YFS0146,2019YFS0155), State Key Laboratory of ASIC & System (No. 2018KF003), Science and Technology Planning Project of Guangdong Province (No. 2017B010110007), National Natural Science Foundation of China (No. 61907009), Natural Science Foundation of Guangdong Province (No. 2018A030313802), Science and Technology Planning Project of Guangdong Province (No. 2017B010110007 and 2017B010110015). Supported by Postgraduate Innovation Fund Project by Southwest University of Science and Technology (19ycx0050).

References

1. Canny, J.: A computational approach to edge detection. In: Readings in computer vision, pp. 184–203. Elsevier (1987)
2. Cho, K., van Merrienboer, B., Gülçehre, Ç., Bougares, F., Schwenk, H., Bengio, Y.: Learning phrase representations using RNN encoder-decoder for statistical machine translation. CoRR (2014). http://arxiv.org/abs/1406.1078
3. Goodfellow, I., et al.: Generative adversarial nets. In: Advances in Neural Information Processing Systems, pp. 2672–2680 (2014)
4. Hu, J., Shen, L., Sun, G.: Squeeze-and-excitation networks. CoRR (2017). http://arxiv.org/abs/1709.01507
5. Iizuka, S., Simo-Serra, E., Ishikawa, H.: Globally and locally consistent image completion. ACM Trans. Graph. (ToG) **36**(4), 107 (2017)
6. Johnson, J., Alahi, A., Fei-Fei, L.: Perceptual losses for real-time style transfer and super-resolution. In: Leibe, B., Matas, J., Sebe, N., Welling, M. (eds.) ECCV 2016. LNCS, vol. 9906, pp. 694–711. Springer, Cham (2016). https://doi.org/10.1007/978-3-319-46475-6_43
7. Kingma, D.P., Ba, J.: Adam: a method for stochastic optimization. arXiv preprint arXiv:1412.6980 (2014)
8. Liu, G., Reda, F.A., Shih, K.J., Wang, T.C., Tao, A., Catanzaro, B.: Image inpainting for irregular holes using partial convolutions. In: Proceedings of the European Conference on Computer Vision (ECCV), pp. 85–100 (2018)
9. Miyato, T., Kataoka, T., Koyama, M., Yoshida, Y.: Spectral normalization for generative adversarial networks. arXiv preprint arXiv:1802.05957 (2018)
10. Nazeri, K., Ng, E., Joseph, T., Qureshi, F., Ebrahimi, M.: Edgeconnect: generative image inpainting with adversarial edge learning. arXiv preprint (2019)
11. Wang, T.C., Liu, M.Y., Zhu, J.Y., Tao, A., Kautz, J., Catanzaro, B.: High-resolution image synthesis and semantic manipulation with conditional GANs. In: Proceedings of the IEEE Conference on Computer Vision and Pattern Recognition, pp. 8798–8807 (2018)

12. Xie, S., Tu, Z.: Holistically-nested edge detection. In: Proceedings of the IEEE International Conference on Computer Vision, pp. 1395–1403 (2015)
13. Yu, J., Lin, Z., Yang, J., Shen, X., Lu, X., Huang, T.S.: Free-form image inpainting with gated convolution. arXiv preprint arXiv:1806.03589 (2018)
14. Yu, J., Lin, Z., Yang, J., Shen, X., Lu, X., Huang, T.S.: Generative image inpainting with contextual attention. In: The IEEE Conference on Computer Vision and Pattern Recognition (CVPR), June 2018
15. Zhou, B., Lapedriza, A., Khosla, A., Oliva, A., Torralba, A.: Places: A 10 million image database for scene recognition. IEEE Trans. Pattern Anal. Mach. Intell. 40(6), 1452–1464 (2017)

Delving into Precise Attention in Image Captioning

Shaohan Hu[1], Shenglei Huang[2], Guolong Wang[1], Zhipeng Li[1], and Zheng Qin[1(✉)]

[1] School of Software, Tsinghua University, Beijing, China
{hush17,wanggl16,lizp14}@mails.tsinghua.edu.cn
qingzh@tsinghua.edu.cn
[2] School of Computer Science, Shanghai Jiao Tong University, Shanghai, China
shengleihuang@sjtu.edu.cn

Abstract. Recent image captioning models usually directly use the output of the last convolutional layer from a pretrained CNN encoder. This intuitive design remains two weaknesses: the top layer feature is not position-sensitive which is harmful for the decoder to generate precise spatial attention for object of interest; irrelevant features will mislead the decoder into focusing irrelevant regions. To tackle these weaknesses, we propose Feature Selection and Fusion Network (FSFN). Specifically, to tackle the first weakness, *Feature Fusion* module is proposed to generate fine-grained and position-sensitive features by fusing multi-scale features. To handle the second weakness, *Feature Selection* module is proposed to select more informative features which will prevent the decoder from focusing on irrelevant regions. Extensive experiments demonstrate that our model has successfully addressed the above two weaknesses and can achieve comparable results with the state-of-the-art under cross entropy loss without any bells and whistles on MSCOCO dataset. Furthermore, our model can improve the performance under different encoders and decoders.

Keywords: Image captioning · Feature selection · Feature fusion

1 Introduction

Image captioning has been well studied in past few years [4,15,21,22] which aims to generate vivid and fluent language descriptions on a given image. The methods to solve the image captioning problem can be divided into two different ways: image CNN feature based method [15,22] and detection based method [1]. The first one only uses the image features from a pretrained classification network and is faster than the latter one which predicts captions using the detection boxes features from Faster RCNN [17]. Our focus is the first one.

© Springer Nature Switzerland AG 2019
T. Gedeon et al. (Eds.): ICONIP 2019, CCIS 1143, pp. 74–82, 2019.
https://doi.org/10.1007/978-3-030-36802-9_9

To predict each word accurately, visual attention mechanism has been utilized [21] by encouraging the model to selectively focus on regions of interest. Though this have greatly boosted the performance, the features input to attention-based decoders are position-insensitive. Existing captioning models usually employ the output feature of the top convolutional layer from a CNN model pretrained on ImageNet classification task as the input of an attention-based decoder. However, this intuitive design remains two weaknesses: (1) The top layer feature of CNN is not position-sensitive [5], which is harmful for the decoder to generate precise spatial attention on object of interest; (2) These models treat all features equal informative, which leads to the irrelevant focus by visual attention mechanism.

Multi-scale context information is widely used in object detection and semantic segmentation tasks which need to regress the precise positions of objects or classify each pixel accurately in the image. Fine-grained and position-sensitive information is missing in deeper layers especially in the top layer due to the large stride factor [10]. Many detection [10,12] or segmentation [23] models choose to fuse multi-scale context from different stages of backbone model to obtain features with high-level semantics and low-level position information, which is already proved to be effective in boosting the performance. In [23], attention mechanism is proved to be efficient in selecting the informative feature and improving the performance.

Motivated by the above analysis, we propose the novel Feature Selection and Fusion Network (FSFN) to address these weaknesses existing in current image captioning models. Our framework, as illustrated in Fig. 1(a), includes a *Feature Selection* and a *Feature Fusion* module between the encoder and decoder, which can effectively transform the original CNN features to fine-grained and position-sensitive features. For the first weakness, we propose *Feature Fusion* module to fuse multi-scale features from different stages: (1) high-level information from top layers (e.g., category); (2) low-level information from lower layers (e.g., object position and boundary). Fusing multi-scale features can generate position-sensitive features, which facilitates the decoder to generate more precise spatial attention for objects of interest. For the second weakness, we propose *Feature Selection* module to select informative features from features on each scale before fusing them. Since some features may be irrelevant and will mislead the decoder into focusing on irrelevant regions, our *Feature Selection* module can effectively suppress the irrelevant features and prevent the decoder from focusing on irrelevant regions. Furthermore, we also propose the *Feature Normalization* module after *Feature Selection* to make all features from different stages to be in the similar numerical scale when feature fusing.

In summary, our main contributions are: (1) We propose Feature Selection and Fusion Network (FSFN) to generate fine-grained and position-sensitive features for the decoder to compute precise spatial attention and better captions. (2) Our model can achieve comparable results with the state-of-the-art under cross entropy loss without any bells and whistles on MSCOCO dataset. (3) Our model can improve the performance under different encoders and decoders.

2   Model

Our goal is to get fine-grained and position-sensitive features for attention-based decoder, by selecting and fusing features from different stages in a pretrained CNN encoder. We will first give an overview of our model, then describe details of each component.

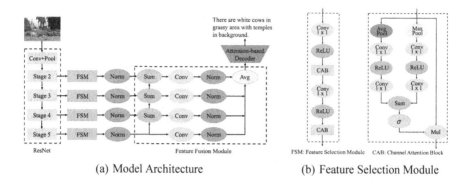

(a) Model Architecture   (b) Feature Selection Module

Fig. 1. Overall model architecture.

2.1   Overview

In our model, we use features from ResNet [6] which is pretrained in ImageNet classification task. We denote the output of the last residual block of each stage in ResNet as $\{C_2, C_3, C_4, C_5\}$, they have strides of $[4, 8, 16, 32]$ with respect to the input image. These features are frozen in our model, which is called "offline features".

The framework we proposed is illustrated in Fig. 1(a). The whole model is composed of three main components: *Feature Selection, Feature Normalization* and *Feature Fusion*. We first use Feature Selection module to select features which could facilitate captioning task. Then we use Feature Normalization to make features from different stages in the same scale before fusing them. Finally, we use a top-down Feature Fusion to make the features from different stages have strong semantics. We aggregate all features through an average process to get a single fine-grained feature containing strong semantic and object position information simultaneously. We also add Feature Normalization before the aggregation.

2.2   Feature Selection

Captioning models usually use the features extracted from CNN encoder which is pretrained in ImageNet classification task as input (i.e. offline features) to attentive decoder. However, the offline features are optimal for the classification task but are not directly designed for the captioning task, there exists some irrelevant or insignificant features [23] which are harmful to compute spatial attention in decoder. Therefore, it is important to select informative and relevant features before fusing features from different stages.

To tackle these weaknesses, we propose a *Feature Selection* module which is composed of two 1×1 convolutional layers, each follows a *channel attention block*, as shown in Fig. 1(b). The two convolutional layers take the offline features $\{C_i\}$ as the generic high-level representation of the input image, to extract more tailored features for captioning task. Since the convolutional operation is unable to do feature extraction and feature selection simultaneously, the channel attention block is designed to model channel interdependency, which makes the model to focus on significant features and suppress irrelevant ones.

We denote the input feature map to *channel attention block* as $F_i \in \mathbb{R}^{C \times H \times W}$, where i means the stage number. Then we aggregate the spatial information of the feature to get channel descriptor of the feature map. We generate two channel descriptors $F_{avg}^i \in \mathbb{R}^{C \times 1 \times 1}$ and $F_{max}^i \in \mathbb{R}^{C \times 1 \times 1}$ by using both global average pooling and max pooling. Then we use two fully connected layers to transform the two channel descriptors to our final channel attention $M_i \in \mathbb{R}^{C \times 1 \times 1}$. Specifically, the final channel attention is computed as:

$$M_i = \sigma(W_1(W_0 F_{avg}^i) + W1(W_0 F_{max}^i)) \tag{1}$$

where σ denotes the sigmoid function since the attention weights should be non-mutually-exclusive. W_0 and W_1 are the weights of the two fully connected layers, each dimension is $C \times C$, where C is the channel dimension of the input feature F^i. Finally, we get our attended feature F_{att}^i as follows:

$$F_{att}^i = F^i \otimes M_i \tag{2}$$

where $\otimes$ denotes element-wise multiplication.

2.3   Feature Normalization

Fusing features from different stages is necessary to get more fine-grained features with high-level semantics and low-level position information simultaneously, which will be discussed in Sect. 2.4. Directly fusing two features would decrease performance marginally due to the numerical scale inconsistency in different stages, because features with large scale would dominate those with small scale. Some works like Feature Pyramid Network (FPN) [12] and Discriminative Feature Network (DFN) [23] apply additional convolutional layers to fuse features from different stages, which plays the same role as normalization before fusion module, but introduces more parameters and computation cost. Furthermore, the scale inconsistency problem is more severe in our captioning model, as the offline feature is not trained for captioning model. Motivated by ParseNet [14], we use L2 Normalization to normalize the feature map along the channel dimension. We denote the feature map in any stage as $X \in \mathbb{R}^{C \times H \times W}$, we reshape it to $[X_1, X_2, ..., X_m]$, where $X_i \in \mathbb{R}^C$ and $m = H \times W$, we do the L2 normalization operation on each $X_i \in \mathbb{R}^C$ as follows:

$$\hat{X}_i = \frac{X_i}{\|X_i\|_2} \tag{3}$$

where $\|X_i\|_2$ is the L2 norm of X_i which is defined as:

$$\|X_i\|_2 = \left(\sum_{k=1}^{C} X_i^k \right)^{\frac{1}{2}} \tag{4}$$

As in Batch Normalization [7], we also introduce a scale parameter γ_k for each channel to increase the feature's representation:

$$Y_i^k = \gamma_k \cdot \hat{X}_i^k \tag{5}$$

2.4   Feature Fusion

To generate precise spatial attention map for decoder to generate high quality description, we propose the Feature Fusion module as shown in Fig. 1(a), which can generate fine-grained and position-sensitive feature map containing high-level semantics and low-level position information simultaneously. We fuse the features in a top-down way. Specifically, we add features from higher stages to feature of current stage to make feature contain semantics at each scale, then we use convolutional layer to the merged feature map. Finally, we use a simple average operation with Feature Normalization to aggregate features from all the stages to get a single fine-grained and position-sensitive feature map with high-level semantics and low-level position information.

3   Experiments

Dataset and Evaluation. We evaluate our model on the largest benchmark dataset, MSCOCO dataset [13], which contains 82,783, 40,504, 40,775 images for training, validation and test, respectively. Each image has 5 human-annotated captions. We follow the data split in [9], which contains 113,287, 5,000, 5,000 for training, validation and test, respectively. We transform all letters into lowercase and replace all words occurring less than 5 times with "UNK". Finally, we get a vocabulary consisting of 9487 words. We use four automatic evaluation metrics - CIDEr [20], BLEU [16], METEOR [2] and ROUGE-L [11].

Encoders and Decoders. To test the compatibility of our model to different encoder-and-decoder combinations, we choose *ResNet34* and *ResNet101* as encoders, and *att2in* [18] and *top-down* [1] as decoders. To test the effectiveness of different multi-scale context, we choose the outputs from *stage2* to *stage5* as our model's inputs, which are denoted as $\{C_2, C_3, C_4, C_5\}$. To simplify computation, all inputs $\{C_2, C_3, C_4, C_5\}$ should be resized to the same fixed spatial size 14×14 through an adaptive average pooling layer. The *top-down* decoder used in our model is slightly different from the original one [1]. Instead of using object RoI Pooling feature from Faster RCNN [17], we use the feature on each grid location of the CNN feature map, but the overall architecture of top-down remains the same.

Baseline and Our Models. We compare our models with several baselines, and we also conduct ablation tests on our model. Baseline models use C_5 as inputs to the decoder directly, and is denoted as *basline{encoder}-{decoder}*. For example, the baseline using *ResNet34* as encoder and *att2in* as decoder is denoted as *baseline34-att2in*. Our models use multi-scale context as inputs to the decoder, and is denoted as *FSFN{encoder}(i)-{decoder}*. For example, our model using *ResNet34* as encoder, *att2in* as decoder and fusing features from C_4 and C_5 is denoted as *FSFN34(4&5)-att2in*. The module notations in ablation test are as follows: "S" denotes the complete *Feature Selection* module, "C" denotes the *Feature Selection* module without *Channel Attention Block*, "N" denotes *Normalization*.

Training Setting. Both LSTM hidden state size and word embedding size are set to 512. The model is trained with Adam optimizer, cross entropy loss and scheduled sampling. The initial learning rate is set to 4×10^{-4} and will decay every 3 epochs by a factor 0.8. The maximum number of training epoch is set to 35 and the model is selected by CIDEr score on validation dataset. Batch size is set to 64. For scheduled sampling, the probability of sampling a token from model increases every 5 epochs by 0.05 until it hits the probability of 0.25. The dropout is applied on the hidden states and is set to 0.5. By default, all models use beam search with size 3.

Table 1. Ablation test results. "dis" = "discriminative supervision", "rl" = "reinforcement learning", "gs" = "greedy search".

Models	C	B-4	M	R-L
baseline34-att2in	97.04	31.83	24.69	53.21
FSFN34(4&5)-att2in	98.74	31.90	24.96	53.71
FSFN34(3~5)-att2in	98.48	31.95	24.84	53.53
FSFN34(2~5)-att2in	98.76	32.18	24.97	53.67

(a)

Models	C	B-4	M	R-L
baseline101-att2in	102.59	33.01	25.55	54.36
baseline101-topdown	104.70	33.83	25.67	54.81
FSFN101(4&5)-att2in+S+N	**108.74**	**34.74**	**26.21**	**55.57**
FSFN101(4&5)-topdown+S+N	105.90	33.89	25.91	54.89

(b)

Models	C	B-4	M	R-L
baseline34-att2in	97.04	31.83	24.69	53.21
baseline34-att2in+C	100.76	32.47	25.24	54.05
baseline34-att2in+S	100.34	32.60	25.24	54.08
FSFN34(2~5)-att2in+S	100.58	32.48	25.19	54.24
FSFN34(3~5)-att2in+S	101.63	32.69	25.27	54.13
FSFN34(4&5)-att2in+S	102.49	33.12	25.36	54.30
FSFN34(4&5)-att2in+S+N	**104.20**	**33.69**	**25.60**	**54.70**
FSFN34(4&5)-att2in+C+N	100.93	32.80	25.17	54.14

(c)

Models	C	B-4	M	R-L
Soft Attention	-	24.3	23.9	-
SCA-CNN	95.2	31.1	25.0	53.1
AdaATT	108.5	33.2	26.6	-
Att2in	101.3	31.3	26.0	54.3
RFNet+gs	105.2	-	-	-
FSFN101(4&5)-att2in+S+N+gs	104.82	32.26	26.01	54.73
FSFN101(4&5)-att2in+S+N	**108.74**	**34.74**	**26.21**	**55.57**
ReviewNet+dis	88.6	29.0	23.7	-
BAM+rl	112.2	35.4	26.5	56.2
RFNet+dis	112.5	35.8	27.4	56.5
RFNet+dis+rl	121.9	36.5	27.7	57.3

(d)

3.1   Quantitative Results

In this section, several ablation tests are conducted to evaluate the effectiveness of our FSFN model.

Ablation on Feature Fusion. To study the effectiveness of *Feature Fusion* module without any feature processing such as selection and normalization, we use *ResNet34* as encoder and *att2in* as decoder, and fuse features of C_2, C_3, C_4, C_5. The results are shown in Table 1(a). Our *Feature Fusion* module can slightly boost the performance across all evaluation metrics, such as the CIDEr score is boosted from 97.04% to 98.76%, BLEU-4 score is boosted from 31.83% to 32.18%. All *FSFN34(C_i)-att2in* models obtain similar results, which demonstrates that fusing $\{C_2, C_3\}$ cannot bring too much improvement. This may due to that features of $\{C_2, C_3\}$ are too shallow which are not suitable to be used for decoder directly.

Ablation on Feature Selection and Normalization. To analyze the importance of *Feature Selection* and *Normalization* before *Feature Fusion* module, we still use *ResNet34* as encoder and *att2in* as decoder. Comparative results are reported in Table 1(b). The results show that fusing features of $\{C_2, C_3\}$ cannot bring improvement even adding Feature Selection module and the reason is discussed before, so we only fuse features $\{C_4, C_5\}$ in later experiments. *FSFN34(4&5)-att2in+S* outperforms *FSFN34(4&5)-att2in* greatly, increasing CIDEr score from 98.74% to 102.49%, which demonstrates that *Feature Selection* is essential to fuse different features and can select more informative feature for captioning. *FSFN34(4&5)-att2in+S+N* achieves the best performance across all metrics, boosting the CIDEr score to 104.2%, which demonstrates that *Feature Selection* and *Feature Normalization* are important before feature fusion procedure and fusing irrelevant features or features with different numerical scales would be harmful to compute precise spatial attention decoder. *FSFN34(4&5)-att2in+C+N* gets worse results, which demonstrates that the increased performance is not due to the two convolutional layers, and it is *Channel Attention Block* that effectively selects the informative feature for captioning. Compared to *baseline34-att2in+C* and *baseline34-att2in+S*, our *Feature Fusion* module can generate fine-grained and position-sensitive features to improve the performance greatly.

Ablation on Encoders and Decoders. Table 1(c) reports the results obtained from different combinations of encoder and decoder. *FSFN101(4&5)-att2in+S+N* outperforms *baseline101-att2in* across all metrics greatly, especially boosting CIDEr score from 102.59% to 108.74%. *FSFN101(4&5)-topdown+S+N* outperforms *baseline101-topdown* across all metrics, and boosts CIDEr score from 104.70% to 105.90%. The results demonstrate that our FSFN model can boost performance on different encoder-decoder combinations.

Comparison with State-of-The-Art. We compare state-of-the-art models with our best model *FSFN101(4&5)-att2in+S+N*. The comparative results are shown in Table 1(d). First, since our model is trained only with cross entropy loss, we compare our model with state-of-the-art models which are also trained with cross entropy loss, including *Soft Attention* [21], *SCA-CNN* [3], *AdaATT* [15], *Att2in* [18] and *RFNet+gs* [8]. Our model *FSFN101(4&5)-att2in+S+N* outperforms these comparative models and achieves comparable results to *RFNet+gs*. Note that *RFNet+gs* uses label-smoothing regularization (LSR) [19] and data augmentation which is not used in our best model. Our *FSFN101(4&5)-att2in+S+N* model outperforms *AdaATT* across all metrics except METEOR, and *AdaATT* used ResNet152 as encoder but we only use ResNet101. More powerful encoder such as ResNet152 can improve the performance greatly as the results reported in Table 1(d). Second, we compare our best model with state-of-the-art models which trained in a more powerful way, including *ReviewNet* [22], *RFNet+dis+rl* [8] and *BAM* [4]. *BAM* is fine-tuned with reinforcement learning. *RFNet+dis+rl* is trained with discriminative supervision [22] and fine-tuned with reinforcement learning, which are not included in our models. Our FSFN model can also be integrated into *RFNet* to get better performance.

4 Conclusion

In this paper, we proposed Feature Selection and Fusion Network to generate fine-grained and position-sensitive features for the decoder to compute precise spatial attention and better captions. Our model can achieve comparable results with the state-of-the-art under cross entropy loss without any bells and whistles on MSCOCO.

References

1. Anderson, P., et al.: Bottom-up and top-down attention for image captioning and visual question answering. In: CVPR (2018)
2. Banerjee, S., Lavie, A.: METEOR: an automatic metric for MT evaluation with improved correlation with human judgments. In: Summarization@ACL (2005)
3. Chen, L., et al.: SCA-CNN: spatial and channel-wise attention in convolutional networks for image captioning. In: CVPR (2017)
4. Chen, S., Zhao, Q.: Boosted attention: leveraging human attention for image captioning. In: ECCV (2018)
5. Dai, J., Li, Y., He, K., Sun, J.: R-FCN: object detection via region-based fully convolutional networks. In: NIPS (2016)
6. He, K., Zhang, X., Ren, S., Sun, J.: Deep residual learning for image recognition. In: CVPR (2016)
7. Ioffe, S., Szegedy, C.: Batch normalization: accelerating deep network training by reducing internal covariate shift. arXiv preprint arXiv:1502.03167 (2015)
8. Jiang, W., Ma, L., Jiang, Y.-G., Liu, W., Zhang, T.: Recurrent fusion network for image captioning. In: Ferrari, V., Hebert, M., Sminchisescu, C., Weiss, Y. (eds.) ECCV 2018. LNCS, vol. 11206, pp. 510–526. Springer, Cham (2018). https://doi.org/10.1007/978-3-030-01216-8_31
9. Karpathy, A., Fei-Fei, L.: Deep visual-semantic alignments for generating image descriptions. IEEE Trans. Pattern Anal. Mach. Intell. **39**(4), 664–676 (2017)

10. Li, Z., Peng, C., Yu, G., Zhang, X., Deng, Y., Sun, J.: DetNet: design backbone for object detection. In: Ferrari, V., Hebert, M., Sminchisescu, C., Weiss, Y. (eds.) ECCV 2018. LNCS, vol. 11213, pp. 339–354. Springer, Cham (2018). https://doi.org/10.1007/978-3-030-01240-3_21

11. Lin, C.Y.: Rouge: a package for automatic evaluation of summaries. In: Text Summarization Branches Out (2004)

12. Lin, T.Y., Dollár, P., Girshick, R., He, K., Hariharan, B., Belongie, S.: Feature pyramid networks for object detection. In: CVPR (2017)

13. Lin, T.-Y., et al.: Microsoft COCO: common objects in context. In: Fleet, D., Pajdla, T., Schiele, B., Tuytelaars, T. (eds.) ECCV 2014. LNCS, vol. 8693, pp. 740–755. Springer, Cham (2014). https://doi.org/10.1007/978-3-319-10602-1_48

14. Liu, W., Rabinovich, A., Berg, A.C.: Parsenet: looking wider to see better. arXiv preprint arXiv:1506.04579 (2015)

15. Lu, J., Xiong, C., Parikh, D., Socher, R.: Knowing when to look: adaptive attention via a visual sentinel for image captioning. In: CVPR (2017)

16. Papineni, K., Roukos, S., Ward, T., Zhu, W.: Bleu: a method for automatic evaluation of machine translation. In: ACL (2002)

17. Ren, S., He, K., Girshick, R.B., Sun, J.: Faster R-CNN: towards real-time object detection with region proposal networks. IEEE Trans. Pattern Anal. Mach. Intell. **39**(6), 1137–1149 (2017)

18. Rennie, S.J., Marcheret, E., Mroueh, Y., Ross, J., Goel, V.: Self-critical sequence training for image captioning. In: CVPR (2017)

19. Szegedy, C., Vanhoucke, V., Ioffe, S., Shlens, J., Wojna, Z.: Rethinking the inception architecture for computer vision. In: CVPR (2016)

20. Vedantam, R., Zitnick, C.L., Parikh, D.: Cider: consensus-based image description evaluation. In: CVPR (2015)

21. Xu, K., et al.: Show, attend and tell: neural image caption generation with visual attention. In: ICML (2015)

22. Yang, Z., Yuan, Y., Wu, Y., Cohen, W.W., Salakhutdinov, R.R.: Review networks for caption generation. In: NIPS (2016)

23. Yu, C., Wang, J., Peng, C., Gao, C., Yu, G., Sang, N.: Learning a discriminative feature network for semantic segmentation. In: CVPR (2018)

Dense Image Captioning Based on Precise Feature Extraction

Zhiqiang Zhang[1], Yunye Zhang[1], Yan Shi[1], Wenxin Yu[1(✉)], Li Nie[1],
Gang He[2], Yibo Fan[3], and Zhuo Yang[4]

[1] Southwest University of Science and Technology, Mianyang, China
zzq.zhangzhiqiang2018@gmail.com, yuwenxin@swust.edu.cn,
star_yuwenxin27@163.com
[2] Xidian University, Xi'an, China
[3] State Key Laboratory of ASIC and System, Fudan University, Shanghai, China
[4] Guangdong University of Technology, Guangzhou, China

Abstract. Image captioning is a challenging problem in computer vision, which has numerous practical applications. Recently, the method of dense image captioning has emerged, which realizes the full understanding of the image by localizing and describing multiple salient regions covering the image. Despite there are state-of-the-art approaches encouraging progress, the ability to position and to describe the target area correspondingly is not enough as we expect. To alleviate this challenge, a precise feature extraction method (PFE) is proposed in this paper to further enhance the effect of dense image captioning. Our model is evaluated on the Visual Genome dataset. It demonstrated that our method is better than other state-of-the-art methods.

Keywords: Dense captioning · Computer vision · Feature extraction · Location and description · Deep learning

1 Introduction

Image captioning has been a great significant research field in computer vision. It has tremendous applications including image retrieval and children's education. In the past few years, the rapid development of deep learning has made a breakthrough in this field. The multimodal Recurrent Neural Network (m-RNN) [1] creatively combined Convolutional Neural Networks (CNN) and Recurrent Neural Networks (RNN), which successfully has achieved efficient image captioning. Vinyals et al. [2] proposed that neural image caption (NIC) further enhances the accuracy of the description. Xu et al. [3] introduced the attention mechanism based on the model of NIC to obtain encouraging results. Wu et al. [4] first used CNN to obtain the advanced semantic concept in the attribute prediction layer, and then put it into LSTM [5] to obtain the final labeling result. Their approaches have achieved remarkable results. Ren et al. [6] achieved better results by using the policy network and value network in reinforcement learning.

© Springer Nature Switzerland AG 2019
T. Gedeon et al. (Eds.): ICONIP 2019, CCIS 1143, pp. 83–90, 2019.
https://doi.org/10.1007/978-3-030-36802-9_10

Despite the acclaimed effects of these approaches in the field of image captioning, there are still some challenges. The biggest one is that the number of generated image descriptions is not enough. In general, only one corresponding description is output for an image.

Recently, Johnson et al. [7] proposed a dense image captioning method to achieve dense descriptions of the image. They conducted regional positioning of the input image based on the faster R-CNN [8], and then put the convolutional features of the localized regions into the LSTM to obtain the corresponding region descriptions. Based on Johnson's work, Yang et al. [9] developed joint inference and context fusion to further improve the accuracy of regional positioning and description. Overall, their results are good but can still be improved.

Through the study of [7] and [9], we found that the method of Johnson et al. and Yang et al. are more concerned with the external network structure (through structural improvement to obtain better results), while there is not much exploration of the internal details. In the task of dense image captioning, the core work is region location and region description. Both are related to regional feature extraction. Accurate feature extraction is conducive to more accurate regional location and better regional description. Based on this, we proposed a precise feature extraction method (PFE) to complete dense image captioning better by extracting precise regional features. The basic structure is still based on faster RCNN. And the method of precise feature extraction is achieved by combining VGG19 [10] with the recently proposed RoIAlign [11].

The contributions to this paper are as follows: (1) Through the internal analysis of dense image captioning, we propose a precise feature extraction method to further improve the quality of the results. Experiments on Visual Genome datasets [12] show that our method is more effective and more outstanding than other existing state-of-the-art methods. (2) A novel combination method, VGG19 + RoIAlign, is used to extract precise features. Experiments show that the combination is efficient.

The rest of this paper is arranged as follows. Section 2 presents our method in detail. Section 3 shows the experimental results and Sect. 4 concludes our work.

2   Our Method

2.1   Basic Idea

Dense captioning mainly consists of two parts i.e. regional positioning and regional description. Both of them are directly related to the convolutional features of the image. On the one hand, the more accurate image feature is extracted, the more accurate the corresponding region location will be obtained. This means that the bounding box can better divide the relevant regions. On the other hand, precise feature extraction will promote image region description, because it will contain more effective region feature information. Based on these, the approach of better feature extraction is focused. The method of VGG combined with RoIAlign is adopted for precise feature extraction.

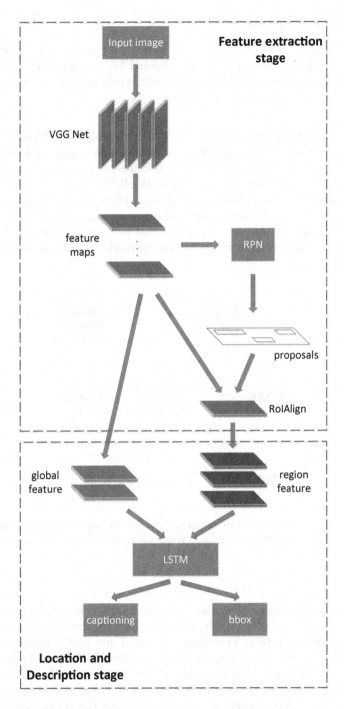

Fig. 1. Our network architecture. The core of the structure is to promote the accuracy of location and description through precise feature extraction.

2.2 Network Structure

Our model structure is shown in Fig. 1. It consists of two parts: one is the feature extraction stage, which mainly extracts regional features from the input image; the other is the location and description stage, which mainly uses the extracted features to locate and describe the region.

Feature Extraction Stage. In the feature extraction stage, VGG Network is used to extract the basic features of the input image to obtain the feature maps. The acquired feature maps are the global features of the image. To acquire the regional features, a Region Proposal Network (RPN) [8] is used to locate the region to obtain the corresponding proposals. Then, regional features extraction is performed from feature maps and regional proposals through RoIAlign [11]. Through such a process, precise regional features are extracted. Experiments show the effectiveness of the feature extraction stage, specifically shown in Sect. 3.1.

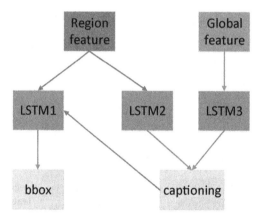

Fig. 2. Specific positioning and description structure. It consists of three LSTMs, one for bbox generation and the others for corresponding text description generation.

Location and Description Stage. In the location and description stage, referring to the idea of Yang et al. [9], the global features are also introduced to obtain a better region description. Therefore, the global and region features acquired during the feature extraction stage are put into the LSTM [5] to obtain the final bounding box and the region description corresponding to the bounding box. The specific structure of this part is shown in Fig. 2. For the acquisition of the captioning, the region features are used to describe the region. In the meantime, to improve the rationality of the regional description, the global features are used to assist the acquisition of regional descriptions. As can be seen in the figure, two LSTMs are used to process the regional features and global features respectively. The advantage of two-LSTMs is that the primary and secondary relationships can be well-formed during training to make the description of the acquisition more precise. For the acquisition of the bounding box, a separate location-LSTM is used together with the previously generated description to generate the final bounding box.

2.3   Details

In the training process, an initial learning rate is set to 0.001, gamma is set to 0.5, and the Stochastic Gradient Descent (SGD) optimization with the momentum is set to 0.98. The learning rate is halved every 100,000 iterations. For the task of captioning, the 10,000 of the most common words as vocabulary are used in this work and the number of hidden layer nodes in LSTM is 512. The batch size is 1 and the total number of iterations is 600,000. In the captioning generation process, the connection method of the results of the LSTM2 and LSTM3 is multiplication. The loss function of training is as follows:

$$Loss = \alpha L_{detection} + \beta L_{bbox} + \gamma L_{caption} \tag{1}$$

where $L_{detection}$, L_{bbox}, $L_{caption}$ represent detection loss, bounding box loss, and caption loss functions, respectively. Bounding box loss function is a smoothed-L1 loss [8], while detection loss and caption loss function are a SoftmaxWithLoss. The values of α, β and γ are 0.1, 0.01 and 1, respectively.

3   Experiments

We validated the effectiveness of our method by quantitative and qualitative results on the Visual Genome dataset [12]. The Visual Genome dataset contains 94,313 images and 4,100,413 region descriptions. In the experiment, we used the same dataset split (77,398 images for training and 5,000 images each for test and validation) and the same evaluation metric of mean Average Precision (mAP) as [7] and [9].

3.1   Qualitative Results

Figure 3 show that our model can locate most of the regions of the image and give corresponding region descriptions. The accuracy of each region description is high. It shows the effectiveness of the PFE method in regional description. At the same time, it can be seen that our method can locate a large number of salient areas, such as various signs, lights, cars, trees and other objects related to the scene.

Figure 4 shows the effectiveness of our method in the precise region location. On the left is the result of not using the PFE method, and on the right is the result of using the PFE method. It can be seen from the figure that our method locates the computer in the figure very accurately. It further defines the specific position of the object. It demonstrates the validity of using the PFE method in regional location on the one hand. And on the other hand, it indirectly shows that the PFE method can extract regional features more accurately.

To further demonstrate the superiority of our method, a comparison with the quantitative results of [7] and [9] is performed, specifically shown in Sect. 3.2.

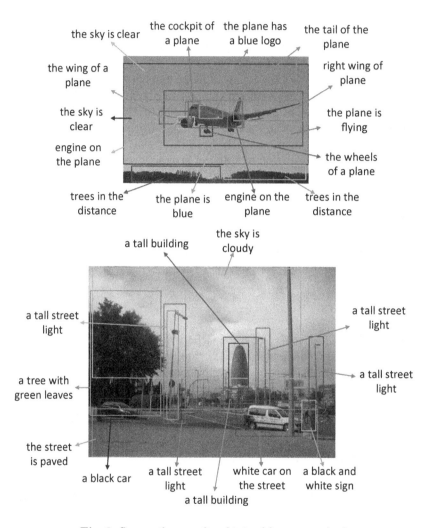

Fig. 3. Some other results obtained by our method.

without PFE   **with PFE**

Fig. 4. The comparison of local result before and after using precise feature extraction method.

3.2   Quantitative Results

The quantitative evaluation material that we use is the mean Average Precision (mAP), which is a comprehensive evaluation of the accuracy of positioning and description. For positioning, the IoU thresholds (0.3, 0.4, 0.5, 0.6, 0.7) are used to calculate its accuracy. For captioning, the Meteor [13] score thresholds (0, 0.05, 0.1, 0.15, 0.2, 0.25) are used to do the calculation. For the input image feature extraction, the VGG16 and VGG19 are used Network respectively (both using RoIAlign). The results are shown in Table 1. Our two methods have achieved better results, and the VGG19 + RoIAlign method has achieved the best results. It is 80.0% and 4.2% higher than other existing state-of-the-art methods [7,9]. The quantitative results show the superiority of our method.

Therefore, the quantitative and qualitative results demonstrate the effectiveness and superiority of our method.

Table 1. The mAP of Johnson et al. [7], Yang et al. [9] and our method, respectively.

Model	mAP
Johnson et al. [7]	5.39
Yang et al. [9]	9.31
PFE (VGG16+RoIAlign)	**9.444**
PFE (VGG19+RoIAlign)	**9.703**

4   Conclusion

In this paper, a precise feature extraction method (PFE) is proposed to promote the effect of dense image captioning. Precise feature extraction can bring better regional features, thus they will promote the better implementation of region positionings and descriptions, and ultimately achieve the enhancement of dense image captioning results. Qualitative and quantitative experimental results demonstrate the effectiveness and superiority of our method. It performs better than the existing state-of-the-art methods.

Acknowledgement. This research was supported by 2018GZ0517, 2019YFS0146, 2019YFS0155, which supported by Sichuan Provincial Science and Technology Department, 2018KF003 Supported by State Key Laboratory of ASIC & System. No. 61907009 Supported by National Natural Science Foundation of China, No. 2018A030313802 Supported by Natural Science Foundation of Guangdong Province, No. 2017B010110007 and 2017B010110015 Supported by Science and Technology Planning Project of Guangdong Province.

References

1. Mao, J., Xu, W., Yang, Y., Wang, J., Yuille, A.L.: Explain Images with Multimodal Recurrent Neural Networks. arXiv:1410.1090 (2014)
2. Vinyals, O., Toshev, A., Bengio, S., Erhan, D.: Show and tell: a neural image caption generator. In: Computer Vision and Pattern Recognition, pp. 3156–3164 (2015)
3. Xu, K., et al.: Show, attend and tell: neural image caption generation with visual attention. In: International Conference on Computer Vision, pp. 2048–2057 (2015)
4. Wu, Q., Shen, C., Liu, L., Dick, A., Hengel, A.: What value do explicit high level concepts have in vision to language problem? In: Computer Vision and Pattern Recognition, pp. 203–212 (2016)
5. Hochreiter, S., Schmidhuber, J.: Long short-term memory. Neural Comput. **9**(8), 1735–1780 (1997)
6. Rennie, S.J., Marcheret, R., Mroueh, Y., Ross, J., Goel, V.: Self-critical sequence training for image captioning. In: Computer Vision and Pattern Recognition, pp. 1179–1195 (2017)
7. Johnson, J., Karpathy, A., Fei-Fei, L.: Densecap: fully convolutional localization networks for dense captioning. In: Computer Vision and Pattern Recognition, pp. 4565–4574 (2016)
8. Ren, S., He, K., Girshick, R.B., Sun, J.: Faster R-CNN: towards real-time object detection with region proposal networks. In: Advances in Neural Information Processing Systems, pp. 91–99 (2015)
9. Yang, L., Tang, K.D., Yang, J., Li, L.: Dense captioning with joint inference and visual context. In: Computer Vision and Pattern Recognition, pp. 1978–1987 (2017)
10. Simonyan, K., Zisserman, A.: Very deep convolutional networks for large-scale image recognition. In: International Conference on Learning Representation (2015)
11. He, K., Gkioxari, G., Dollár, P., Girshick, R.: mask R-CNN. In: International Conference on Computer Vision, pp. 2908–2988 (2017)
12. Krishna, R., et al.: Visual genome: connecting language and vision using crowd-sourced dense image annotations. Int. J. Comput. Vis. **123**(1), 32–73 (2017)
13. Banerjee, S., Lavie, A.: METEOR: an automatic metric for MT evaluation with improved correlation with human judgments. In: ACL Workshop, pp. 65–72 (2005)

Improve Image Captioning
by Self-attention

Zhenru Li, Yaoyi Li, and Hongtao Lu[✉]

Department of Computer Science and Engineering, Shanghai Jiao Tong University,
Shanghai 200240, People's Republic of China
{lethever,dsamuel,htlu}@sjtu.edu.cn

Abstract. The common attention mechanism has been widely adopted
in prevalent image captioning frameworks. In most of the prior work,
attention weights were only determined by visual features as well as the
hidden states of Recurrent Neural Network (RNN), while the interaction
of visual features was not modelled. In this paper, we introduce the
self-attention into the current image captioning framework to leverage
the nonlocal correlation among visual features. Moreover, we propose
three distinctive methods to fuse the self-attention and the conventional
attention mechanism. Extensive experiments on MSCOCO dataset show
that the self-attention can empower the captioning model to achieve
competitive performance with the state-of-the-art methods.

Keywords: Image captioning · Self-attention

1 Introduction

Image captioning, a task to automatically generate natural and complete descriptions of given images, has been an important topic in the field of computer vision.
The capability of describing the content of an image is in high demand in practice, for instance, it could help the visually impaired people better understand the
image. Besides recognizing the objects, finding their attributes and relationships
in the image, image captioning substantially requires to generate syntactically
and semantically correct sentences in human language for the target image.

Owing to the surge of deep learning, there has been remarkable progress
in image captioning in the past few years. Inspired by the success in machine
translation [1], the researchers proposed to follow the encoder-decoder framework
in this task [2]. They employed a deep convolutional neural network (CNN) as an
image "encoder" and a recurrent neural network (RNN) decoder that generated
sentences. Xu *et al.* [4] added an attention mechanism to this framework, which
learned to attend to different subregions of the image when predicting word at
each time step.

Many researches focus on improving the attention mechanism. Earlier models
use the positions in the activation grid of a CNN layer as attention areas, but
they are not flexible and adaptive to the image content. Some methods [6,8,24]

© Springer Nature Switzerland AG 2019
T. Gedeon et al. (Eds.): ICONIP 2019, CCIS 1143, pp. 91–98, 2019.
https://doi.org/10.1007/978-3-030-36802-9_11

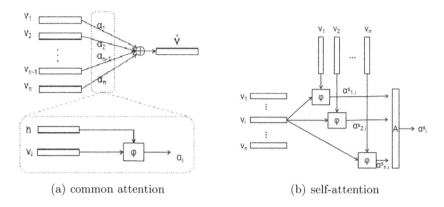

(a) common attention   (b) self-attention

Fig. 1. An overview of the common attention and self-attention mechanism.

use object proposals or salient regions as attention regions, the others [5,10] learn to attend to semantic attributes. As shown in Fig. 1, in previous methods, the common attention mechanism computes weights from the visual feature vector itself and the RNN hidden state, without consideration about the other feature vectors, so we propose to add the self-attention mechanism to model these relations. The self-attention is first proposed in [13], which models the dependencies among all the words in a sequence, we adopt this mechanism from them and study how to combine it with the current attention mechanism.

Our contributions in this paper can be summarized as follows: (1) We propose to add the self-attention mechanism to the current encoder-decoder framework in which we can explicitly model the relation among visual features. To the best of our knowledge, no prior work has considered this before. (2) We conduct a systematic experiment study on the effectiveness of the proposed self-attention module. In addition, we also study the combination of common attention and the self-attention. Experimental results show that the proposed framework based on self-attention mechanism can facilitate image captioning. With self-attention, our method can achieve competitive performance with state-of-the-art approaches on the MSCOCO [22] dataset.

2   Related Work

Other Work on Attention in Captioning. Xu *et al.* [4] proposed a soft and a hard attention mechanism in their work when they introduced attention to this encoder-decoder framework. Soft attention assigned weights to different vectors, and hard attention sampled only one vector according to the weight, which made the process non-differentiable. Adaptive Attention [9] introduced a visual sentinel to determine how much new information to get from the image, because sometimes the RNN language model is enough to predict words like 'a' and 'the'. Besides attention to spatial features, SCA-CNN [7] proposed the channel-wise attention to assign weights to different channels. Some work utilized

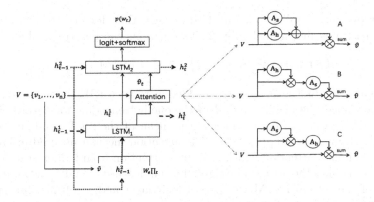

Fig. 2. An overview of the model and proposed three fusion method (on the right). A_s is the self-attention, and A_h is the existing attention mechanism which taking hidden state h as input. These two blocks output a set of attention weights. Final output is obtained by multiplying the weights and summing them up.

better image features. Semantic attention [10] used a set of attribute detectors to get a list of visual attributes or concepts and applied attention to these visual features. [12] exploited image-level features and object-level features to address the object missing and misprediction problem. Areas of Attention [11] used spatial transformer network to extract image features and proposed a new attention mechanism to model the three pairwise interaction among the RNN state, image features and word embedding vectors, but they did not model the interplay among image features themselves. Anderson *et al.* [24] applied attention to salient image regions by leveraging Faster R-CNN.

Self-attention. Self-attention was first introduced in [13]. In our method, we only adopt part of the scaled dot-product to implement self-attention. Their method computed self-attention on hidden states or the input word vectors in the sequence, while we applied self-attention to spatial features. Another work related to ours is the non-local neural network [14], which proposed the non-local operations to model long-range dependencies so that response at one position is a weighted sum of features at all positions. Different from their method, we apply self-attention in image captioning and study how to combine it with existing attention method.

3   Method

3.1   Baseline

For our baseline model,we follow the Up-Down (UD) [24] to encode the image into a possibly variably-sized set of vectors, $V = \{v_1, v_2, ..., v_n\}$, where each vector represents a salient region of the image, then for decoder we utilize a two-layer lstm as shown in the Fig. 2, the functions can be expressed as

$$h_t^1 = \mathrm{LSTM}_1(\bar{v}, W_e \Pi_t, h_{t-1}^2, h_{t-1}^1); h_t^2 = \mathrm{LSTM}_2(\hat{v}_t, h_t^1, h_{t-1}^2) \qquad (1)$$

The input of the first layer is the embedded word vector $W_e \Pi_t$, where W_e is an learnable embedding matrix, and Π_t is the one-hot encoding for a word at time t. The global visual vector $\bar{v}$ which is the mean of the set V, and the hidden state h_{t-1}^2 from second upper layer. The attention mechanism used in baseline model is common attention which computes attention from hidden state h. The second layer takes the attention weighted visual feature $\hat{v}_t$ and h_t^1 as input. The output of the second LSTM will pass a logit layer and a softmax function to predict the probability distribution.

3.2   Self-attention

In general formulation, to compute attention we have a set of query vectors $Q = \{q_1, ..., q_m\}$, a set of key vectors $K = \{k_1, ..., k_n\}$, and a set of value vectors to assign weight to $\dot{V} = \{\dot{v}_1, ..., \dot{v}_n\}$. In our self-attention, Q, K are calculated from the same set of visual vectors V so that $m = n$, where n is the number of visual features for the current input image. And $\dot{V}$ is V without any transformation so we use same notation V later.

To combine with existing attention mechanism, we only use the relevance matrix $R = QK^T$, where $Q \in \mathbb{R}^{n \times d_1}$, $K \in \mathbb{R}^{n \times d_1}$, and each element $R_{i,j} = f_q(v_i) \cdot f_k(v_j)$, where v_i and v_j are visual features, f_q and f_k are transformation functions, here we adopt two linear transformations to project the visual features to a common embedding space and compute the dot product as the relevance. After applying the softmax function to the relevance matrix R, each row of it is a set of normalized weights, we can get a new set of vector $V^s = \{v_1^s, ..., v_n^s\}$, where

$$\dot{R}_{i,*} = softmax(R_{i,*}); v_i^s = \sum_{j=1}^{n} \dot{R}_{i,j} v_j \qquad (2)$$

and $v_i^s \in \mathbb{R}^{d_2}$ is a weighted sum of the original whole visual features. To further get a single representation, we take the average or the summation of V^s.

3.3   Fusion Method

We consider how to combine common attention and self-attention, which is illustrated in Fig. 2. We propose 3 kinds of combinations, which are A, B, C in the right part of Fig. 2. We denote the traditional method by A_h, and self-attention by A_s. Both blocks output a weight vector. Fused-att-A(A) applies the two mechanism in two branches and add the results together. Fused-att-B(B) employs A_h first, and multiplies the assigned weight to each visual feature to get a new set of attention weighted visual features $\tilde{V}$. In experiment we find passing it through a linear layer will improve the performance, which is not shown in the figure. Then fused-att-B applies A_s" to $\tilde{V}$ to obtain a new weight vector. In the end the whole module outputs a vector $\hat{v}$, which is the weighted sum of V. Fused-att-C(C) just changes the order of the two mechanisms in B.

4   Experiment

4.1   Dataset and Evaluation Metrics

To evaluate our proposed captioning model, we conduct experiments on the MSCOCO 2014 caption dataset [22]. To make the results comparable to others, we follow the Karpathy's splits [3] and use 113287 training images, 5000 images for validation, and 5000 for testing. The metrics we use are the standard automatic evaluation metrics, including BLEU, METEOR, ROUGE, CIDEr and SPICE.

4.2   Implementation Details

We adopt the model in [24], for encoder they leverage Faster R-CNN with ResNet-101 pretrained for classification on ImageNet, and train this bottom-up attention model on Visual Genome dataset. We just adopt the extracted image features from their released code. Each feature represents a region from detection result, the dimension of it is 2048. For decoder, we utilize two LSTMs and both the word embedding size and the hidden sate size are 1024. In attention module, the image features and sentences features are projected to the joint embedding space of dimension 512. The similarity is measure by dot product on the normalized feature vectors.

4.3   Results

As shown in Table 1, our baseline is very high because we have same image features as UD [24], our encoder includes more information than that trained by classification on ImageNet, and the extracted features are more flexible and adaptive to image subregions. Our baseline is our reproduced UD model. Our results (fusion method A) are slightly better than UD when trained for cross entropy loss, and are much higher than UD when optimized for CIDEr score. Our proposed model obtains a better result on all metrics compared to baseline, which prove the self-attention can effectively improve the baseline model. Our best single model performs on par with state-of-the-art methods.

4.4   Fusion Experiments

From Table 2 we find that, all of the proposed methods improve the baseline more or less. And the best way to put self-attention into traditional paradigm is to apply them in parallel]. Compared to sequential methods B and C, there is less interaction between self-attention and common attention in method A, maybe the sequential method makes it harder to learn the two mechanism well at the same time.

Table 1. Performance Comparisons on the test set of Karpathy's split [3], where B@N, M, R, C, S are short for BLEU@N, METEOR, ROUGE-L, CIDEr, SPICE. $^\Sigma$ indicates an ensemble, † indicates using image features from ResNet-101, and "-" indicates not reported. All values are reported as percentage (%), with the highest value of each entry highlighted in boldface

Method	Cross entropy loss						CIDEr optimization					
	B@1	B@4	M	R	C	S	B@1	B@4	M	R	C	S
SCST†[18]	-	30.0	25.9	53.4	99.4	-	-	34.2	26.7	55.7	114.0	-
UD [24]	77.2	36.2	27.0	56.4	113.5	20.3	79.8	36.3	27.7	56.9	120.1	21.4
UD-GAN$^\Sigma$[19]	-	-	-	-	-	-	**81.8**	**39.6**	**28.9**	59.1	125.6	**22.3**
DISC-cap [21]	-	-	-	-	-	-	-	36.1	27.4	57.3	114.3	21.1
RFNET†[17]	**77.4**	37.0	27.9	**57.3**	116.3	20.8	80.4	37.9	28.3	58.3	125.7	21.7
GCN†[16]	**77.4**	**37.1**	**28.1**	57.2	**117.1**	21.1	80.9	38.3	28.6	58.5	**128.7**	22.1
STACK†[15]	76.2	35.2	26.5	-	109.1	-	78.6	36.1	27.4	**59.6**	120.4	20.9
SCST-n†[20]	-	-	-	-	-	-	77.9	35.0	26.9	56.3	115.2	20.4
Baseline	77.0	36.3	27.8	56.9	114.4	21.0	79.9	37.5	28.3	58.0	127.0	21.8
Ours	77.2	36.8	28.0	57.1	116.3	**21.2**	80.2	38.0	28.6	58.4	128.6	22.1

Table 2. Ablation study and fusion results

Method	att_h	att_s	B@1	B@2	B@3	B@4	M	R	C	S
ablation	×	×	73.67	56.82	42.13	31.00	25.50	53.99	99.03	18.56
	✓	×	77.02	61.14	47.28	36.34	27.82	56.90	114.4	20.95
	×	✓	76.46	60.30	46.20	35.29	27.37	56.29	111.7	20.37
	✓	✓	**77.16**	**61.46**	**47.70**	36.81	**27.98**	**57.06**	**116.3**	**21.15**
fused-att-A	✓	✓	77.16	**61.46**	**47.70**	36.81	**27.98**	**57.06**	**116.3**	**21.15**
fused-att-B	✓	✓	77.28	61.36	47.67	**36.92**	27.93	56.97	115.3	21.03
fused-att-C	✓	✓	**77.31**	**61.46**	47.64	36.71	27.86	57.05	115.5	20.94

4.5   Ablation Study

We run ablation study to examine whether the self-attention works. And as Table 2 suggests, self-attention (att_s) do improves the model as an attention mechanism, but the progress is less than common attention (att_h), and our proposed model can take advantage of both to obtain best results. Self-attention gets the same input each time during generating the whole word sequence, it's not dynamic as the common attention which takes hidden state h_t as input, which is the limit of this method, but it's still better than the averaged feature $\bar{v}$.

4.6   Qualitative Analysis

Table 3 shows qualitative captioning results. Our proposed model can better capture the details in the target image, as self-attention leads the model to find relationship between the visual features. For instance, the "baby elephant",

Table 3. Qualitative results by baseline (B), ours (O), and the ground truth caption (G)

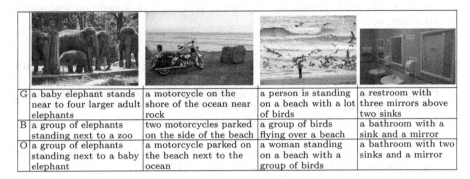

G	a baby elephant stands near to four larger adult elephants	a motorcycle on the shore of the ocean near rock	a person is standing on a beach with a lot of birds	a restroom with three mirrors above two sinks
B	a group of elephants standing next to a zoo	two motorcycles parked on the side of the beach	a group of birds flying over a beach	a bathroom with a sink and a mirror
O	a group of elephants standing next to a baby elephant	a motorcycle parked on the beach next to the ocean	a woman standing on a beach with a group of birds	a bathroom with two sinks and a mirror

"ocean", "woman" in the positive examples, they are missed by the baseline model. However, our model might fail to figure out the right number of the object, such as "three mirrors" in the fourth image. But compared to baseline, our model gives the right number of sinks in the four image. This could be understood because there is not enough supervised information about the concept of number in training data.

5   Conclusion

In this paper, we propose an image captioning model by introducing the self-attention into current framework and study the combination with existing attention mechanism. The self-attention models the relation among the visual features themselves, which is not considered in the common attention mechanism. Results show that self-attention works effectively as an attention mechanism though it is static, combining it with the common attention brings enhancement to the baseline model and we achieve competitive performance with state-of-the-art methods. Better ways to combine and exploit these two mechanism will be addressed in our future work.

Acknowledgement. This paper is supported by NSFC (No. 61772330, 61533012, 61876109), the pre-research project (no.61403120201), Shanghai authentication key Lab. (2017XCWZK01), and Technology Committee the interdisciplinary Program of Shanghai Jiao Tong University (YG2019QNA09).

References

1. Bahdanau, D., Cho, K., Bengio, Y.: Neural machine translation by jointly learning to align and translate. In: ICLR (2015)
2. Vinyals, O., Toshev, A., Bengio, S., Erhan, D.: Show and tell: a neural image caption generator. In: CVPR (2015)

3. Karpathy, A., Fei-Fei, L.: Deep visual-semantic alignments for generating image descriptions. In: CVPR (2015)
4. Xu, K., et al.: Show, attend and tell: neural image caption generation with visual attention. In: ICML (2015)
5. Wu, Q., Shen, C., Liu, L., Dick, A., van den Hengel, A.: What value do explicit high level concepts have in vision to language problems? In: CVPR (2016)
6. Yao, T., Pan, Y., Li, Y., Qiu, Z., Mei, T.: Boosting image captioning with attributes. In: ICCV (2017)
7. Chen, L., et al.: SCA-CNN: spatial and channel-wise attention in convolutional networks for image captioning. In: CVPR (2017)
8. Wang, J., Madhyastha, P.S., Specia, L.: Object counts! Bringing explicit detections back into image captioning. In: NAACL-HLT (2018)
9. Lu, J., Xiong, C., Parikh, D., Socher, R.: Knowing when to look: adaptive attention via a visual sentinel for image captioning. In: CVPR (2017)
10. You, Q., Jin, H., Wang, Z., Fang, C., Luo, J.: Image captioning with semantic attention. In: CVPR (2016)
11. Pedersoli, M., Lucas, T., Schmid, C., Verbeek, J.: Areas of attention for image captioning. In: ICCV (2017)
12. Li, L., Tang, S., Deng, L., Zhang, Y., Tian, Q.: Image caption with global-local attention. In: AAAI (2017)
13. Vaswani, A., et al.: Attention is all you need. In: NIPS (2017)
14. Wang, X., Girshick, R.B., Gupta, A., He, K.: Non-local Neural Networks. In: CVPR (2018)
15. Gu, J., Cai, J., Wang, G., Chen, T.: Stack-captioning: coarse-to-fine learning for image captioning. In AAAI (2018)
16. Yao, T., Pan, Y., Li, Y., Mei, T.: Exploring visual relationship for image captioning. In: Ferrari, V., Hebert, M., Sminchisescu, C., Weiss, Y. (eds.) Computer Vision – ECCV 2018. LNCS, vol. 11218, pp. 711–727. Springer, Cham (2018). https://doi.org/10.1007/978-3-030-01264-9_42
17. Jiang, W., Ma, L., Jiang, Y.-G., Liu, W., Zhang, T.: Recurrent fusion network for image captioning. In: Ferrari, V., Hebert, M., Sminchisescu, C., Weiss, Y. (eds.) ECCV 2018. LNCS, vol. 11206, pp. 510–526. Springer, Cham (2018). https://doi.org/10.1007/978-3-030-01216-8_31
18. Rennie, S.J., Marcheret, E., Mroueh, Y., Ross, J., Goel, V.: Self-critical sequence training for image captioning. In: CVPR (2017)
19. Chen, C., et al.: Improving image captioning with conditional generative adversarial nets. In: AAAI (2019)
20. Gao, J., Wang, S., Wang, S., Ma, S., Gao, W.: Self-critical n-step training for image captioning. In: CVPR (2019)
21. Luo, R., Price, B.L., Cohen, S., Shakhnarovich, G.: Discriminability objective for training descriptive captions. In: CVPR (2018)
22. Lin, T.-Y., et al.: Microsoft COCO: common objects in context. In: Fleet, D., Pajdla, T., Schiele, B., Tuytelaars, T. (eds.) ECCV 2014. LNCS, vol. 8693, pp. 740–755. Springer, Cham (2014). https://doi.org/10.1007/978-3-319-10602-1_48
23. Ren, Z., Wang, X., Zhang, N., Lv, X., Li, L.J.: Deep reinforcement learning-based image captioning with embedding reward. In: CVPR (2017)
24. Anderson, P., et al.: Bottom-up and top-down attention for image captioning and VQA. In: CVPR (2018)

Dual-Path Recurrent Network for Image Super-Resolution

Xinyao Li, Dongyang Zhang, and Jie Shao[✉]

Center for Future Media, School of Computer Science and Engineering, University of
Electronic Science and Technology of China, Chengdu 611731, China
{xinyaoli,dyzhang}@std.uestc.edu.cn, shaojie@uestc.edu.cn

Abstract. Recently, significant progress has been made for the task
of image super-resolution (SR). However, existing methods show that
stacking convolution layers blindly leads to overwhelming parameters and
high computational complexities. Besides, the conventional feed-forward
architectures can hardly fully exploit the mutual dependencies between
low- and high-resolution images. Motivated by these observations, we
first propose a novel architecture by taking advantage of recursive learning. Based on dual-path block (DPB), we enhance the network performance simply by increasing the number of DPB recursions and avoiding
additional parameters. Moreover, in contrast to most convolutional neural network (CNN) methods with one state, our network is endowed with
two states (low-resolution state and high-resolution state) transformed
mutually. We exploit the relationship between the two states by introducing a back-projection operation which calculates the differences between
the two states for a better result. Extensive experiments show that our
method outperforms the state-of-the-art methods.

Keywords: Super-resolution · Recursive learning · Dual-path

1 Introduction

Image super-resolution (SR) is of great importance in image processing and
computer vision, which aims to reconstruct high-resolution (HR) images from
low-resolution (LR) counterparts. Since the nature of SR is to improve perceptual
quality by restoring the missing high-frequency part in LR images, it has a
great demand in the applications where the acquisition of high-quality images is
arduous and expensive. SR also does a favour for other real-world tasks, such as
video surveillance, object detection and action recognition. However, image SR
is an inherently underdetermined inverse problem. For a given LR image input,
there could be various reasonable HR image outputs. To solve this problem,
numerous learning-based SR methods have been proposed.

In recent years, based on deep learning, super-resolution convolutional neural network (SRCNN) [3] and then more advanced SR methods [11,12,16,19–22]
which achieved great performance have been proposed. In order to recreate HR

© Springer Nature Switzerland AG 2019
T. Gedeon et al. (Eds.): ICONIP 2019, CCIS 1143, pp. 99–108, 2019.
https://doi.org/10.1007/978-3-030-36802-9_12

of high quality, increasing the depth of network is the first approach to be considered. However, network depth increase may cause gradient vanishing or gradient exploding. The residual network (ResNet) [9] is an effective strategy for deepening the depth and reducing the overfitting. The second approach is increasing the width of network, which contributes to learn more features. Enhanced deep super-resolution network (EDSR) [16] removes batch normalization and learns multi-scale mapping functions via parameter sharing, and it indicates that network width is important for SR. Li et al. [15] propose multi-scale residual network (MSRN), which can not only adaptively detect the image features, but also achieve feature fusion at different scales.

In fact, a majority of satisfactory results are achieved by very deep models which follow a simple principle, the deeper the better [11]. By simply stacking convolution layers, these deep models are endowed with the ability to learn complex non-linear mapping functions for recreating better upscaling images, but this unavoidably leads to overwhelming parameters and high computational complexities. Recurrent neural network (RNN) is an effective way to solve these problems, which utilizes recursive learning to share parameters across various blocks.

Most existing SR methods are based on single-state, e.g., deeply-recursive convolutional network (DRCN) [12] and deep recursive residual network (DRRN) [19] take upscaled LR image by bicubic interpolation as input and output HR counterparts directly where the spatial resolution is not changed during recursion. The essence of SR is to explore the relationship between LR-HR image pairs. Thus, multi-state is better than single-state in RNN architecture from the perspective of feature engineering. Dual-state recurrent network (DSRN) [7] adopts dual-state design where the two kinds of state signals are exchanged bidirectionally, LR to HR and vice versa. However, we argue that the state transformation in DSRN is not fully exploited, and the mutual dependencies of the two states can be further utilized.

Normally, the feed-forward architectures and the non-linear mapping from low-resolution input to high-resolution output are used for SR. However, this method cannot completely address the interdependence between low-resolution and high-resolution images. In deep back-projection network (DBPN) [8], an iterative up- and down-sampling approach is proposed where two kinds of feature maps with different spatial resolutions (two states) are exploited during feed-forward pass. In this work, we explore a new dual-path recurrent network (DPRN) to tackle the above issue for SR, with recursive learning and back-projection [8]. Similar to DSRN [7], we also implement our network with two states which operate at different spatial resolutions. The most important part of our network is dual-path block (DPB), which contains two main modules. One is back-projection block that includes a downsampling unit (HR to LR) and an upsampling unit (LR to HR). The other is transmission block (forward propagation between identical states). Inspired by DBPN [8], we impose the back-projection on downsampling and upsampling units to further calculate the differences between state transformation, which is simpler than DBPN and also provides an error feedback mechanism during recursions to better guide the learning of feature representation. To some extent, the proposed DPRN can be regarded as the RNN version of DBPN with finite unfolding. To sum up, the main contributions of our work are three-fold:

- *Recursive learning*: by introducing the recursive learning in the network, we propose a compact recurrent neural network with two states, which renders our model lightweight and efficient.
- *Back projection*: for the first time, we integrate the back-projection into RNN architecture to fully exploit the dependency between LR and HR states.
- We provide a comprehensive comparison with recent methods on common benchmarks, in which our method outperforms previous methods based on recursive architecture.

The rest of this paper is organized as follows. Section 2 reviews the related work, and Sect. 3 introduces the proposed method. Section 4 presents the experimental study. Finally, Sect. 5 concludes.

2   Related Work

2.1   Image Super-Resolution

Image super-resolution (SR), referring to restore an HR image from a single LR observation, has achieved increasing research attention as the boom of deep learning. Lots of literatures contribute to address this ill-posed problem. In this section, we give a brief description on image SR.

Until now, most existing methods deal with the SR task in a supervised manner, i.e., with both LR images and ground-truth HR counterparts. SRCNN [3] is a seminal work in the line of SR research, which proposes a simple but efficient CNN based architecture with three layers for the first time. Subsequent works such as [8] gradually increase the network depth and width to improve the capacity of model, which is the key factor for the progress in SR.

Fast super-resolution convolutional neural networks (FSRCNN) [4] does not have to process the upscaled LR images, and it can utilize a more efficient mapping layer. In very deep convolutional networks (VDSR) [11], the residual learning is conducted on the LR and HR images to learn the damage in image restoration.

Based on the previous Laplacian pyramid super-resolution network (Lap-SRN) [13], multi-scale Laplacian pyramid network (MS-LapSRN) [14] proposes the parameter sharing among different pyramid layers with recursive convolution network structure. In addition, cascading residual network (CARN) [1] transforms the residual block of ResNet [9] into a cascading residual network by adding local and global cascading modules, which can reflect inputs of different levels so as to receive more information.

2.2 Dual-State

Most state-of-the-art deep architectures for SR can be translated into dealing with feature in a single state. For example, in SRCNN [3] features are upsampled before mapping, which causes high computational complexity. Besides, various methods such as multi-scale deep super-resolution (MDSR) [16], VDSR [11] and FSRCNN [4] are proposed to set the original LR as the input, and upsample features at the end of the network. Different from these single state methods above, dual-state recurrent network (DSRN) is proposed in [7], where the HR state and the LR state are transformed bidirectionally during recursions. Consequently, DSRN greatly reduces model size up to 8 times at the expense of a small performance loss.

2.3 Recursive Learning

RNN is a special neural network structure, whose inspiration comes from the idea that human cognition is based on experience and memory of the past. It differs from deep neural network (DNN) and convolutional neural network (CNN) in that RNN not only considers current input, but also endows the network with a memory function for the previous content. The methods based on recursive learning are most related to ours. Kim et al. [12] proposed the 16-recursion DRCN employing a single convolution layer as the recursive unit. Moreover, they proposed two techniques (recursive-supervision and skip-connection) to ease the difficulty of training. Afterwards, some other variants performing recursions on residual block are proposed, such as DRRN [19] and CARN [1]. Different from these works, in order to fully explore the deep relationships between LR-HR image pairs, Han et al. [7] proposed dual-state recurrent network (DSRN). However, DSRN ignores the transformations between the two states, and with the proposed dual-path block our DPRN model can further utilize such mutual dependencies. We impose the back-projection on down- and up-sampling units to calculate the differences between state transformation, which provides an error feedback mechanism during recursions to better guide the learning of feature representation.

3 Proposed Method

3.1 Dual-Path Recurrent Network

As shown in Fig. 1, our dual-path recurrent network consists of three parts: feature extraction (FE), dual-path block (DPB), and upscale layer (UL).

We represent L as the input LR image, and S as the output reconstructed SR image. At the beginning, we firstly use feature extraction to extract the LR features F_0.

$$F_0 = f_{FE}(L), \tag{1}$$

where f_{FE} denotes the operations of FE.

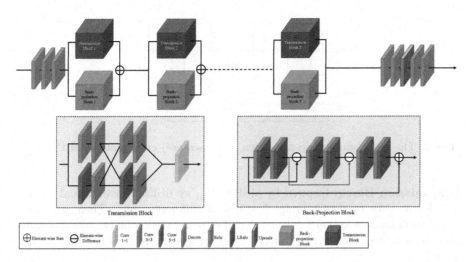

Fig. 1. The architecture of our DPRN model, and the details of transmission block and back-Projection block.

F_0 is used as the input of DPB. More details about DPB will be given in Sect. 3.2. The output of DPB can be formulated as

$$F_L = f_{DPB}(F_0), \tag{2}$$

where f_{DPB} denotes the operations of DPB.

After recursive learning, we utilize ESPCN [18] followed by one convolution layer as UL. The output SR image of DPRN can be represented as

$$S = f_{DPRN}(L) = f_{UL}(F_L), \tag{3}$$

where f_{DPRN} denotes the operations of DPRN, and f_{UL} denotes the function of upscale layer.

3.2 Dual-Path Block (DPB)

Now we give more details of DPB. As shown in Fig. 1, a DPB contains T dual-path groups (each contains one transmission block and one back-projection block) and a global feature concatenation. After each dual-path group (except the T-th dual-path group), we use an element-wise sum to get the local fusion features. We suppose that the input of the i-th dual-path group is F_{i-1} and the output of its sum is F_i which can be represented as

$$F_i = f_t(F_{i-1}) + f_{bp}(F_{i-1}), \tag{4}$$

where f_t denotes the operations of transmission block of the i-th dual-path group, and f_{bp} denotes the function of back-projection block of the i-th dual-path group.

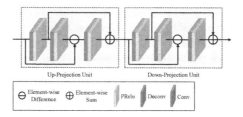

Fig. 2. The architecture of up-projection and down-projection units in DBPN [8].

Transmission Block: Inspired by MSRN [15], we utilize convolutional kernels of different sizes to detect image features at different scales. Then, we use a convolution layer of kernel 1×1 at the end of transmission block, which obtains multi-scale information of two convolutional kernels and reduces network parameters.

$$M_{i1} = [\sigma(f_{3\times3}(F_{i-1})), \sigma(f_{5\times5}(F_{i-1}))], \tag{5}$$

$$M_{i2} = [\sigma(f_{3\times3}(M_{i1})), \sigma(f_{5\times5}(M_{i1}))], \tag{6}$$

$$F_i^t = f_{1\times1}(M_{i2}), \tag{7}$$

where σ denotes Relu activation function, and $f_{a\times a}$ is the weight of the convolution layer of kernel $a \times a$.

Back-Projection Block: As shown in Figs. 1 and 2, our back-projection block is much simpler than DBPN [8] and it also considers the difference between LR and HR, which uses mutually connected down- and up-sampling units to represent the relation of LR and HR images. After connecting three down- and up-sampling units, we conduct a short skip connection that performs residual learning on F_{i-1} and X_i (the difference between the two states) to solve the problem of degradation caused by increasing network depth. The operation can be defined as

$$E_{i1} = \varphi(f_{up}(\varphi(f_{down}(F_{i-1})))) - \varphi(f_{down}(F_{i-1})), \tag{8}$$

$$E_{i2} = \varphi(f_{up}(\varphi(f_{down}(E_{i1})))) - \varphi(f_{down}(E_{i1})), \tag{9}$$

$$X_i = \varphi(f_{up}(\varphi(f_{down}(E_{i2})))), \tag{10}$$

$$F_i^{bp} = F_{i-1} + X_i, \tag{11}$$

where φ denotes Leaky Relu (LRelu) activation function, and f_{down} (or f_{up}) is the weight of the convolution (or deconvolution) layer.

After the T-th dual-path group, we utilize the concatenation which can reuse local features to widen the channel of features. Then, the global fusion feature can be represented as

$$F_L = [f_t(F_{T-1}), f_{bp}(F_{T-1})]. \tag{12}$$

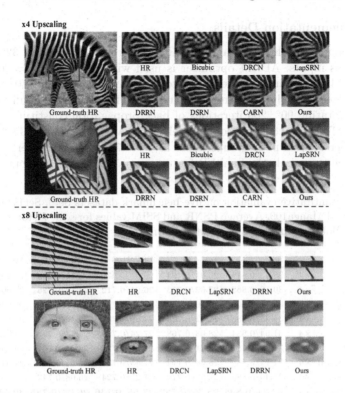

Fig. 3. Visual comparison with other methods. The two groups show the output images of our DPRN with the scale factors of ×4 and ×8.

Given a training set $\{L^{(i)}, H^{(i)}\}_{i=1}^{N}$, where N is the number of training images and $H^{(i)}$ is the ground-truth HR image of the LR image $L^{(i)}$, the loss function of DPRN is

$$Loss(\theta) = \frac{1}{N} \sum_{i=1}^{N} \|H^{(i)} - F(L^{(i)})\|^2, \tag{13}$$

where θ denotes the parameter set, and $F(\cdot)$ denotes the network reconstruction function.

4 Experiments

4.1 Datasets

For training, we use the DIV2K dataset for image restoration applications. DIV2K consists of 800 images for training, 100 images for validation, and 100 images for testing. We train our model with 800 training images. For testing, we use 5 standard benchmark datasets: Set5 [2], Set14 [6], Urban100 [10], Manga109 [5] and BSDS100 [17]. Our method can deal with large scale factors, so the experiments are performed with ×4 upscaling and ×8 upscaling between LR and HR images.

4.2   Implementation Details

We set training batch as 16, and 16 LR patches with the size of 32×32 are extracted as inputs. ADAM optimizer is used with $\beta_1 = 0.9, \beta_2 = 0.009, \epsilon = 10^{-8}$. The initial learning rate is set to 10^{-4} and decayed by 10 every 200 epoches. We randomly augment the patches by flipping horizontally or vertically and rotating $90°$. 200 iterations of back-propagation constitute an epoch. We use PyTorch to construct our model with NVIDIA GeForce GTX 1080Ti. The number of dual-path groups is set as $T = 7$. In upscale layer, we utilize ESPCN [18] to upscale the final LR feature maps into the HR output.

Table 1. Quantitative evaluation on five benchmark datasets Set5, Set14, BSDS100, Urban100 and Manga109. Average PSNR and SSIM values for scale factors $\times 4$ and $\times 8$ are shown. Bold text indicates the best performance.

Method	Scale	Set5	Set14	BSDS100	Urban100	MANGA109
Bicubic	$\times 4$	28.43/0.811	26.01/0.704	25.97/0.670	23.15/0.660	24.89/0.786
SRCNN [3]	$\times 4$	30.50/0.863	27.52/0.753	26.91/0.712	24.53/0.725	27.66/0.858
FSRCNN [4]	$\times 4$	30.72/0.866	27.70/0.755	26.98/0.715	24.62/0.728	27.89/0.859
VDSR [11]	$\times 4$	31.35/0.883	28.02/0.768	27.29/0.726	25.18/0.754	28.82/0.886
DRCN [12]	$\times 4$	31.54/0.884	28.03/0.768	27.24/0.725	25.14/0.752	28.97/0.886
DRRN [19]	$\times 4$	31.68/0.888	28.21/0.772	27.38/0.728	25.44/0.764	29.46/0.896
MS-LapSRN [14]	$\times 4$	31.74/0.889	28.26/0.774	27.43/0.731	25.51/0.768	29.54/0.897
DSRN [7]	$\times 4$	31.40/0.883	28.07/0.770	27.25/0.724	25.08/0.747	–
DPRN (ours)	$\times 4$	**32.21/0.898**	**28.75/0.782**	**27.58/0.736**	**26.08/0.785**	**30.52/0.908**
Bicubic	$\times 8$	24.40/0.658	23.10/0.566	23.67/0.548	20.74/0.516	21.47/0.649
SRCNN [3]	$\times 8$	25.33/0.690	23.76/0.591	24.13/0.566	21.29/0.544	22.37/0.682
FSRCNN [4]	$\times 8$	25.41/0.682	23.93/0.592	24.21/0.567	21.32/0.537	22.39/0.672
VDSR [11]	$\times 8$	25.72/0.711	24.21/0.609	24.37/0.576	21.54/0.560	22.83/0.707
DRRN [19]	$\times 8$	26.18/0.738	24.42/0.622	24.59/0.587	21.88/0.583	23.60/0.742
DRCN [12]	$\times 8$	25.93/0.723	24.25/0.614	24.49/0.582	21.71/0.571	23.02/0.724
MS-LapSRN [14]	$\times 8$	26.34/0.753	24.57/0.629	24.65/**0.592**	**22.06/0.598**	23.90/0.759
DPRN (ours)	$\times 8$	**26.55/0.758**	**24.61/0.633**	**24.68/0.592**	22.04/**0.598**	**23.94/0.762**

4.3   Results

In this section, we evaluate the performance of our dual-path recurrent network (DPRN), and compare it with that of related state-of-the-art methods. Our method is compared both numerically and qualitatively with Bicubic, SRCNN [3], FSRCNN [4], VDSR [11], DRCN [12], DRRN [19], and MS-LapSRN [14]. Quantitative results on Set5, Set14, Urban100, Manga109 and BSDS100 in terms of average peak signal-to-noise (PSNR) and structural similarity (SSIM) are shown in Table 1. Our method outperforms most of other methods considered both numerically and more importantly visually. As expected, the Bicubic method for the scale factor $\times 4$ does not perform well and neither does the scale factor $\times 8$. Overall, the results obtained by these methods are relatively noisy and blurry compared with DPRN.

In addition, we also illustrate a visual quality comparison to show the reconstructed details on images from Set5, Set14 and Urban100 datasets. By taking advantage of recursive learning and fully exploiting the mutual dependencies between low- and high-resolution images, Fig. 3 shows that our method can produce the most appealing visual results. Both results show that, in contrary to all other methods considered, our model can ontain more clear straight lines and sharper edges, while the results of other methods are blurrier.

5   Conclusion

In this paper, we introduce a dual-path recurrent network (DPRN) for highly accurate image SR of ×4 and ×8 scale factors. Based on dual-path block, our DPRN focuses on both HR and LR states, and shows mutual dependencies between them. In addition, recursive learning is used for parameter sharing in order to reduce model size. We show that our method can be used to effectively increase the quality of image SR. Extensive experiments on benchmark datasets demonstrates that our DPRN performs better than most of the state-of-the-art methods.

Acknowledgments. This work is supported by National Natural Science Foundation of China (grants No. 61672133 and No. 61832001).

References

1. Ahn, N., Kang, B., Sohn, K.-A.: Fast, accurate, and lightweight super-resolution with cascading residual network. In: Ferrari, V., Hebert, M., Sminchisescu, C., Weiss, Y. (eds.) ECCV 2018. LNCS, vol. 11214, pp. 256–272. Springer, Cham (2018). https://doi.org/10.1007/978-3-030-01249-6_16
2. Bevilacqua, M., Roumy, A., Guillemot, C., Alberi-Morel, M.: Low-complexity single-image super-resolution based on nonnegative neighbor embedding. In: British Machine Vision Conference, BMVC 2012, pp. 1–10 (2012)
3. Dong, C., Loy, C.C., He, K., Tang, X.: Learning a deep convolutional network for image super-resolution. In: Fleet, D., Pajdla, T., Schiele, B., Tuytelaars, T. (eds.) ECCV 2014. LNCS, vol. 8692, pp. 184–199. Springer, Cham (2014). https://doi.org/10.1007/978-3-319-10593-2_13
4. Dong, C., Loy, C.C., Tang, X.: Accelerating the super-resolution convolutional neural network. In: Leibe, B., Matas, J., Sebe, N., Welling, M. (eds.) ECCV 2016. LNCS, vol. 9906, pp. 391–407. Springer, Cham (2016). https://doi.org/10.1007/978-3-319-46475-6_25
5. Fujimoto, A., Ogawa, T., Yamamoto, K., Matsui, Y., Yamasaki, T., Aizawa, K.: Manga109 dataset and creation of metadata. In: Proceedings of the 1st International Workshop on coMics ANalysis, Processing and Understanding, MANPU@ICPR 2016, pp. 2:1–2:5 (2016)
6. Glorot, X., Bengio, Y.: Understanding the difficulty of training deep feedforward neural networks. In: Proceedings of the Thirteenth International Conference on Artificial Intelligence and Statistics, AISTATS 2010, pp. 249–256 (2010)

7. Han, W., Chang, S., Liu, D., Yu, M., Witbrock, M., Huang, T.S.: Image super-resolution via dual-state recurrent networks. In: 2018 IEEE Conference on Computer Vision and Pattern Recognition, CVPR 2018, pp. 1654–1663 (2018)

8. Haris, M., Shakhnarovich, G., Ukita, N.: Deep back-projection networks for super-resolution. In: 2018 IEEE Conference on Computer Vision and Pattern Recognition, CVPR 2018, pp. 1664–1673 (2018)

9. He, K., Zhang, X., Ren, S., Sun, J.: Deep residual learning for image recognition. In: 2016 IEEE Conference on Computer Vision and Pattern Recognition, CVPR 2016, pp. 770–778 (2016)

10. Huang, J., Singh, A., Ahuja, N.: Single image super-resolution from transformed self-exemplars. In: IEEE Conference on Computer Vision and Pattern Recognition, CVPR 2015, pp. 5197–5206 (2015)

11. Kim, J., Lee, J.K., Lee, K.M.: Accurate image super-resolution using very deep convolutional networks. In: 2016 IEEE Conference on Computer Vision and Pattern Recognition, CVPR 2016, pp. 1646–1654 (2016)

12. Kim, J., Lee, J.K., Lee, K.M.: Deeply-recursive convolutional network for image super-resolution. In: 2016 IEEE Conference on Computer Vision and Pattern Recognition, CVPR 2016, pp. 1637–1645 (2016)

13. Lai, W., Huang, J., Ahuja, N., Yang, M.: Deep Laplacian pyramid networks for fast and accurate super-resolution. In: 2017 IEEE Conference on Computer Vision and Pattern Recognition, CVPR 2017, pp. 5835–5843 (2017)

14. Lai, W., Huang, J., Ahuja, N., Yang, M.: Fast and accurate image super-resolution with deep Laplacian pyramid networks. CoRR abs/1710.01992 (2017)

15. Li, J., Fang, F., Mei, K., Zhang, G.: Multi-scale residual network for image super-resolution. In: Ferrari, V., Hebert, M., Sminchisescu, C., Weiss, Y. (eds.) ECCV 2018. LNCS, vol. 11212, pp. 527–542. Springer, Cham (2018). https://doi.org/10.1007/978-3-030-01237-3_32

16. Lim, B., Son, S., Kim, H., Nah, S., Lee, K.M.: Enhanced deep residual networks for single image super-resolution. In: 2017 IEEE Conference on Computer Vision and Pattern Recognition Workshops, CVPR Workshops 2017, pp. 1132–1140 (2017)

17. Martin, D.R., Fowlkes, C.C., Tal, D., Malik, J.: A database of human segmented natural images and its application to evaluating segmentation algorithms and measuring ecological statistics. In: Proceedings of the Eighth International Conference On Computer Vision (ICCV-01) - Volume 2, pp. 416–425 (2001)

18. Shi, W., et al.: Real-time single image and video super-resolution using an efficient sub-pixel convolutional neural network. In: 2016 IEEE Conference on Computer Vision and Pattern Recognition, CVPR 2016, pp. 1874–1883 (2016)

19. Tai, Y., Yang, J., Liu, X.: Image super-resolution via deep recursive residual network. In: 2017 IEEE Conference on Computer Vision and Pattern Recognition, CVPR 2017, pp. 2790–2798 (2017)

20. Tong, T., Li, G., Liu, X., Gao, Q.: Image super-resolution using dense skip connections. In: IEEE International Conference on Computer Vision, ICCV 2017, pp. 4809–4817 (2017)

21. Zhang, D., Shao, J., Hu, G., Gao, L.: Sharp and real image super-resolution using generative adversarial network. In: Liu, D., Xie, S., Li, Y., Zhao, D., El-Alfy, E.S. (eds.) Neural Information Processing - 24th International Conference. ICONIP 2017, Proceedings, Part III, vol. 10636, pp. 217–226. Springer, Cham (2017). https://doi.org/10.1007/978-3-319-70090-8_23

22. Zhang, Y., Tian, Y., Kong, Y., Zhong, B., Fu, Y.: Residual dense network for image super-resolution. In: 2018 IEEE Conference on Computer Vision and Pattern Recognition, CVPR 2018, pp. 2472–2481 (2018)

Attention-Based Image Captioning
Using DenseNet Features

Md. Zakir Hossain[1(✉)], Ferdous Sohel[1], Mohd Fairuz Shiratuddin[1],
Hamid Laga[1], and Mohammed Bennamoun[2]

[1] Murdoch University, Perth, WA 6155, Australia
{MdZakir.Hossain,F.Sohel,F.Shiratuddin,H.Laga}@murdoch.edu.au
[2] The University of Western Australia, Perth, WA 6009, Australia
mohammed.bennamoun@uwa.edu.au

Abstract. We present an attention-based image captioning method
using DenseNet features. Conventional image captioning methods
depend on visual information of the whole scene to generate image captions. Such a mechanism often fails to get the information of salient
objects and cannot generate semantically correct captions. We consider
an attention mechanism that can focus on relevant parts of the image to
generate fine-grained description of that image. We use image features
from DenseNet. We conduct our experiments on the MSCOCO dataset.
Our proposed method achieved 53.6, 39.8, and 29.5 on BLEU-2, 3, and 4
metrics, respectively, which are superior to the state-of-the-art methods.

Keywords: Image captioning · DenseNet · Attention

1 Introduction

Image captioning is the task of describing an image with natural language. Automatic image captioning has many applications such as helping visually impaired
people to understand their surroundings and automatic image indexing.

Image captioning has been extensively studied in the literature. It has been
addressed using both traditional techniques and deep learning techniques [17].
Deep learning-based techniques such as Convolutional Neural Network (CNN),
Recurrent Neural Network (RNN), and Long Short-Term Memory (LSTM) have
been widely used as they are capable of handling the complexities and challenges
of image captioning.

Different CNNs such as VGGNet [14], ResNet [3], and DenseNet [6] have
their own strengths and weaknesses. It is generally accepted that the deeper
is the network, the more relevant are the learned features [3]. However, if the
depth of the network exceeds a certain number, one may obtain the opposite
effect, i.e., a decline in performance. There are two main reasons behind this
fact: (i) The vanishing-gradient problem and (ii) the degradation problem. This
problem has been addressed in the literature by using residual learning mechanisms such as ResNet. However, the element-wise addition is used in identity

T. Gedeon et al. (Eds.): ICONIP 2019, CCIS 1143, pp. 109–117, 2019.
https://doi.org/10.1007/978-3-030-36802-9_13

mapping in ResNet is computationally expensive during training. On the other hand, in DenseNet, each layer has connections with every other layer in the network in a feed-forward manner. The network reuses the feature-maps and uses concatenation for various operations instead of addition. Therefore, it can reduce the number of parameters and it can be memory efficient. Moreover, since each layer of DenseNet receives feature maps from all previous layers, it gets diversified features and tends to have rich patterns. For this reason, we use DenseNet for extracting features from images.

Most existing image captioning methods including deep learning-based techniques focus only on factual description of an image [17]. During feature learning, these methods compress the entire scene into a fixed vector representations. As a result, they often lose information of the prominent objects in the scene. Attention mechanisms [5] can focus on the parts of the image that are relevant, for a period of time, similar to the human visual system. Simultaneously, they can discard irrelevant information.

In this paper, we propose an image captioning method where attention is a key mechanism to describe the important objects in a scene. Overall, the contributions of the paper are:

- We use DenseNet for extracting image features as it can extract diversified and rich feature patterns.
- We use an attention mechanism in our image captioning method that can focus on the salient parts of the image and describe fine-grained captions.

2   Related Work

This section is divided into two major parts; (i) image captioning and (ii) attention in image captioning.

2.1   Image Captioning

In the last few years, with the advancements in deep neural network models for Computer Vision (CV) and Natural Language Processing (NLP), automatic image captioning has become a promising research area. Hossain et al. [4] present a comprehensive survey of the topic. They grouped the methods into a number of categories. They include template-based image captioning, retrieval-based image captioning, and novel caption generation. Template-based approaches have fixed templates with a number of blank slots to generate captions. In retrieval-based approaches, captions are retrieved from a set of existing captions. These methods produce general and syntactically correct captions. However, they cannot produce syntactically correct image-specific captions. Novel caption generation methods first analyze the visual content of the image first and then generate image captions from the visual content using a language model. These methods can generate image captions that are semantically more accurate than the aforementioned approaches. However, these methods have problems in identifying prominent objects from image.

2.2 Attention in Image Captioning

Most existing image captioning methods consider the scene as a whole at the time of generating captions. These methods cannot analyze the image over time while they generate the descriptions. Attention-based image captioning methods can dynamically focus on the relevant parts of the image while the output sequences are being produced. Such methods are now getting increasingly popular in deep learning as well as in image captioning. The first attention-based image captioning method was proposed by Xu et al. [17]. This method can automatically describe the salient contents of an image. It introduces two attention-based generators: stochastic hard attention and deterministic soft attention to describe the main parts of the image. A different type of attention-based image captioning method was introduced by Park et al. [13], which addresses the personal issues of an image. In fact, nowadays, people share a lot of photos on social media. It mainly focuses on two tasks: hashtag prediction and post generation. However, these methods have problems in recognizing correct objects from an image.

3 The Proposed Architecture

The proposed model consists of three main parts: a CNN encoder (i.e., DenseNet) to extract image features, an attention module (i.e., Soft and Hard attention) to dynamically focus on the relevant parts of the image and a decoder LSTM to generate image captions with the information of salient objects. The overall architecture of our image captioning method is shown in Fig. 1. In Fig. 1, $\hat{z}_t$, h_t and s_t refer to the context vector, the LSTM hidden state vector, and the previously generated word, respectively. The LSTM is trained to compute the output word (s_t) probability condition on the context vector $(\hat{z}_t)$ and the previously generated word s_t at time t. It is given as:

$$P(s_0, s_1, \ldots, s_m) = \prod_{i=0}^{m} P(s_i | \hat{z}, s_0, s_1 \ldots, s_m), \tag{1}$$

3.1 Image Encoder

Traditional convolutional networks with L layers have L connections. However, DenseNet has $L(L+1)/2$ direct connections. As a result, the feature-maps of all preceding layers are used as inputs of the current layer, and its own feature-maps are used as inputs into all subsequent layers. Suppose an image I_0 is passed through a CNN. Consider a CNN network that has L layers and each layer l applies a non-linear transformation $H_l(.)$. The transformation function for traditional CNNs can be written as:

$$I_l = H_l(I_{l-1}), \tag{2}$$

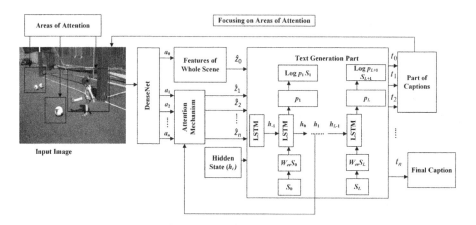

Fig. 1. The overall architecture of the proposed image captioning method. The method uses DenseNet to obtain the image features and an attention mechanism to selectively focus on relevant parts of the image.

However, the transformation mechanism is different in ResNets. Such a network adds a skip-connection using an identity function for transformation:

$$I_l = H_l(I_{l-1}) + I_{l-1}, \tag{3}$$

In ResNets, the identify function is used to flow gradient from later layers to the earlier layers. Since summation is used here to merge the output of H_l and the identity function, it may delay the information flow to the network. Consequently, the l-th layer of DenseNet receives the feature-maps of all the previous layers, $I_0, I_1, \ldots, I_l$ as input. The transformation function for DenseNet is:

$$I_l = H_l([I_0, I_1, \ldots, I_{l-1}]), \tag{4}$$

where $[I_0, I_1, \ldots, I_{l-1}]$ refers to the concatenation of the feature-maps generated in layers $0, 1, \ldots, l - 1$ and $H_l(.)$ is a composite function.

3.2   Attention Models

The attention-based network can recompute its attention for the relevant parts of the image according to the perceived importance from LSTM. This recomputed image feature is a dynamic representation of the relevant parts of the image and is called a context vector $(\hat{z}_t)$. Such a vector is computed from the annotation vector a_i defined in Eq. (5) and the attention weight (α_{ti}). The attention weight is obtained from the alignment score (e_{ti}). The score defines how well each annotation vector matches with the previous hidden state output (h_{t-1}) of the LSTM decoder. Such an alignment score is computed by applying an attention function (f_{att}):

$$e_{ti} = f_{att}(a_i, h_{t-1}), \tag{5}$$

Next, the attention weight is obtained by normalizing e_{ti} using a Softmax function:

$$\alpha_{ti} = \frac{\exp(e_{ti})}{\sum_{k=1}^{L} \exp(e_{tk})}, \tag{6}$$

Then we compute the context vector $(\hat{z}_t)$ using Eqs. (5) and (6) as follows:

$$\hat{z}_t = \phi(\{a_i\}, \{\alpha_i\}), \tag{7}$$

Deterministic Soft Attention: We compute (α_i) for each image region (x_i) and then we calculate the weighted average for (x_i) to use it as the input of LSTM. Hence the context vector $\hat{z}_t$ for soft attention can be calculated as:

$$E_{p(x_t|a)}[\hat{z}_t] = \sum_{i=1}^{L} \alpha_i, a_i, \tag{8}$$

Stochastic Hard Attention: In hard attention, instead of a weighted average, we use α_i in a stochastic manner to pick up one x_i. We compute $\hat{z}_t$ for hard attention as follows:

$$\hat{z}_t = \sum_{i=1}^{t} s_{t,i}, a_i, \tag{9}$$

3.3 Language Decoder

LSTM is a type of RNN that works well on temporal and sequential data. RNNs are similar to the feed forward artificial neural network except that they can feed outputs back to the input. In our model, LSTM takes context vector $(\hat{z}_t)$ and the hidden state vector (h_t) as input at each time step and generates a word as output.

4   Experiments

In this section, we demonstrate our proposed method using the MSCOCO [10] dataset and commonly used evaluation metrics such as BLEU [12], METEOR [1], ROUGE [9], and CIDEr [15] for image captioning. We implement both stochastic hard attention and deterministic soft attention.

4.1   Dataset and Experimental Setup

MSCOCO: Microsoft COCO Dataset is a large and popular dataset for object recognition, segmentation, and image captioning. The dataset consists of 82,783 training and 40,504 validation images. Each image has at least 5 human annotated ground-truth captions.

Implementation Details: In our framework, we use DenseNet121 [6] with fully connected layers for obtaining image features. The DenseNet121 is pre-trained on

ImageNet dataset. We apply the fc7 feature map to compute attention features. The dimension of our feature map is 1×1024. The size of the hidden layer in the prediction module is 1024. We apply dropout, learning rate to 0.001 and use a linear layer to obtain a 512-dimensional word embedding. We also apply Adam optimizer with mini-batch size 16 to train the model. Then we upsample the word embedding vector via ReLU activation on the fully connected layer, and pass it through a softmax to obtain the output word probabilities $P_{i,w}(y_i|y_{<i}, I)$. Our method was trained for 20 epochs and we evaluate the metrics on the validation dataset, after every epoch, to pick the best model. The model was implemented in Tensorflow 1.2.

Evaluation Metrics: A number of evaluation metrics such as BLEU [12], METEOR [1], ROUGE [9], and CIDEr [15] have widely been used to measure the quality of the generated image captions compared to the ground truth. Each metric applies its own technique for computation and has distinct advantages. In this experiment, we consider BLEU-1, BLEU-2, BLEU-3, BLEU-4, METEOR, ROUGE, and CIDEr to evaluate our method. For all metrics, higher values indicate better performance.

Table 1. Performance of our method on MSCOCO dataset. M, R, C stand for METEOR, ROUGE, and CIDEr, respectively (Bold indicates the best result and a dash(-) indicates results are unavailable).

Model	BLEU-1	BLEU-2	BLEU-3	BLEU-4	M	R	C
DeepVS [8]	62.5	45.0	32.1	23.0	19.5	–	66.0
m-RNN [11]	67.0	49.0	35.0	25.0	–	–	–
NIC [16]	66.6	46.1	32.9	24.6	–	–	–
g-LSTM [7]	67	49.1	35.8	26.4	23.9	–	–
LRCN [2]	69.7	51.9	38.0	27.8	22.9	50.8	83.7
Hard-ATT [17]	**71.8**	50.4	35.7	25.0	23.0	–	–
Soft-ATT [17]	70.7	49.2	34.4	24.3	**23.9**	–	–
ConvCap [13]	69.3	51.8	37.4	26.8	23.8	51.1	**85.5**
Ours-Dense (Soft-ATT)	68.0	47.4	32.5	22.9	22.6	**53.0**	74.3
Ours-Dense (Hard-ATT)	70.3	**53.6**	**39.8**	**29.5**	–	–	–

4.2 Analysis of Results on MSCOCO Dataset

Quantitative Evaluation Results: Table 1 shows the results on the MSCOCO dataset. Note that all methods only use the image information without semantics or attributes boosting. The methods use different CNN encoder for image representation. NIC [16] and g-LSTM [7] exploit GoogleNet to extract image features. LRCN [2] utilizes AlexNet to obtain the features. DeepVS [8], m-RNN [11], and

Soft/Hard attention [17] use VGGNet to get image-level representation. However, we use DenseNet in our method for the task. In terms of the BLEU-1 score, which only considers bigrams, the methods NIC [16], m-RNN [11], g-LSTM [7] and DeepVS [8] achieved 66.6, 67, 67, and 62.5, respectively. In contrast, our method (Soft) achieved 68.0 and (Hard) achieved 70.3 on BLEU-1, which are higher than these methods. The scores for Hard-ATT [17] and Soft-ATT [17] are 71.8 and 70.7, respectively. In terms of score, we are slightly inferior to this popular method. However, our method has superior performance to all the methods on BLEU-2, 3, and 4 metrics. The METEOR metric considers the precision, recall, and the alignment of the matched tokens. The results show that our (Soft) method has better precision and recall accuracy than Karpathy's DeepVS method. ROUGE evaluates the adequacy and fluency of the generated captions, whereas CIDEr focuses grammaticality and saliency. Our method achieved the best result in terms of adequacy, fluency, and saliency.

Table 2. Comparison between our methods and a baseline line method on generated captions. The sample images and their ground-truth captions are collected from MS COCO dataset (Images are best viewed in colour).

Image	Captions
	Ground-Truth Captions: A dog and his owner on the back of a boat. **Generated Captions:** *(VGGNet-ATT):* *A woman holding a piece of pizza.* *(Ours DenseNet-ATT (Soft)):* *A man holding a dog with a surfboard.* *(Ours DenseNet-ATT (Hard)):* *A man sitting on a boat with a dog*
	Ground-Truth Captions: Four airplanes are shown flying through a cloudy sky. **Generated Captions:** *(VGGNet-ATT):* *A group of people flying kites in a field.* *(Ours DenseNet-ATT (Soft)):* *A flock of airplanes flying in the sky.* *(Ours DenseNet-ATT (Hard)):* *A plane flying through a cloudy sky.*

Qualitative Evaluation Results: We choose some sample images and their ground-truth captions from popular MSCOCO dataset. We generate captions using our method and a baseline attention-based method and show the comparison on the generated captions in Table 2. It is seen that the encoder CNN has a big influence on the overall performance of image captioning. Table 2 shows some examples of generated captions by different models. It is easy to see that most of the VGGNet-ATT generated image captions are somewhat relevant to the images. However, in some cases, it cannot predict visual attributes properly. For example, in the first image, "dog" and "man" are considered an important object even in the ground-truth captions. However, VGGNet-ATT cannot recognize them properly. VGGNet-ATT also has problems in distinguishing "airplanes" and "kites". However, in each case, DenseNet-ATT can recognize the objects properly and include them successfully in the generated captions.

Visualization of Attention Probabilities: We visualize attention probabilities generated by DenseNet-ATT in Fig. 2. We see that the word "dog" and "laying" mostly focus on the head and the body parts of the image. However, when the network generates the word "in", it shifts its attention to other regions of the image.

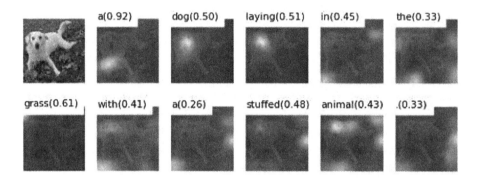

Fig. 2. Attention visualization generated by DenseNet-ATT.

5   Conclusion

We have proposed an attention-based image captioning method that uses DenseNet features and evaluated its performance on the MSCOCO dataset. DenseNet can extract rich image feature maps and attention mechanism can selectively focus on relevant image features. We have reported our results on commonly used evaluation metrics such as BLEU, METEOR, ROUGE, and CIDEr, and show that our proposed method achieved better results compared to all the other methods on BLEU-2, 3, and 4 metrics and the third best result on BLEU-1. We have also shown the generated captions by our methods and VGGNet-ATT methods. In some cases, our method generates semantically richer captions than VGGNet-ATT. Finally, we have presented the attention visualization details and described how attention shifts from one image region to another based on generated words at each time step.

References

1. Banerjee, S., Lavie, A.: METEOR: an automatic metric for MT evaluation with improved correlation with human judgments. In: The ACL Workshop on Intrinsic and Extrinsic Evaluation Measures for Machine Translation and/or Summarization, vol. 29, pp. 65–72 (2005)
2. Donahue, J., et al.: Long-term recurrent convolutional networks for visual recognition and description. In: Computer Vision and Pattern Recognition, pp. 2625–2634 (2015)
3. He, K., Zhang, X., Ren, S., Sun, J.: Deep residual learning for image recognition. In: Computer Vision and Pattern Recognition, pp. 770–778 (2016)
4. Hossain, M.Z., Sohel, F., Shiratuddin, M.F., Laga, H.: A comprehensive survey of deep learning for image captioning. ACM Comput. Surv. (CSUR) **51**, 118 (2019)
5. Hossain, M.Z., Sohel, F., Shiratuddin, M.F., Laga, H., Bennamoun, M.: Bi-san-cap: bi-directional self-attention for image captioning. In: Accepted in Digital Image Computing: Techniques and Applications (DICTA) (2019)
6. Huang, G., Liu, Z., Van Der Maaten, L., Weinberger, K.Q.: Densely connected convolutional networks. In: Computer Vision and Pattern Recognition (CVPR), pp. 2261–2269. IEEE (2017)
7. Jia, X., Gavves, E., Fernando, B., Tuytelaars, T.: Guiding the long-short term memory model for image caption generation. In: International Conference on Computer Vision, pp. 2407–2415 (2015)
8. Karpathy, A., Fei-Fei, L.: Deep visual-semantic alignments for generating image descriptions. In: Computer Vision and Pattern Recognition, pp. 3128–3137 (2015)
9. Lin, C.Y.: ROUGE: a package for automatic evaluation of summaries. In: Text Summarization Branches Out: Proceedings of the ACL-04 Workshop, Barcelona, Spain, vol. 8 (2004)
10. Lin, T.-Y., et al.: Microsoft COCO: common objects in context. In: Fleet, D., Pajdla, T., Schiele, B., Tuytelaars, T. (eds.) ECCV 2014. LNCS, vol. 8693, pp. 740–755. Springer, Cham (2014). https://doi.org/10.1007/978-3-319-10602-1_48
11. Mao, J., Xu, W., Yang, Y., Wang, J., Huang, Z., Yuille, A.: Deep captioning with multimodal recurrent neural networks (m-RNN). In: International Conference on Learning Representations (2015)
12. Papineni, K., Roukos, S., Ward, T., Zhu, W.J.: BLEU: a method for automatic evaluation of machine translation. In: Association for Computational Linguistics, pp. 311–318 (2002)
13. Park, C.C., Kim, B., Kim, G.: Attend to you: personalized image captioning with context sequence memory networks. In: Computer Vision and Pattern Recognition, pp. 6432–6440 (2017)
14. Simonyan, K., Zisserman, A.: Very deep convolutional networks for large-scale image recognition. In: International Conference on Learning Representations (2015)
15. Vedantam, R., Lawrence Zitnick, C., Parikh, D.: CIDEr: consensus-based image description evaluation. In: Computer Vision and Pattern Recognition, pp. 4566–4575 (2015)
16. Vinyals, O., Toshev, A., Bengio, S., Erhan, D.: Show and tell: a neural image caption generator. In: Computer Vision and Pattern Recognition, pp. 3156–3164 (2015)
17. Xu, K., et al.: Show, attend and tell: neural image caption generation with visual attention. In: International Conference on Machine Learning, pp. 2048–2057 (2015)

High-Performance Light Field Reconstruction with Channel-wise and SAI-wise Attention

Zexi Hu[1(⊠)], Yuk Ying Chung[1], Seid Miad Zandavi[1], Wanli Ouyang[2], Xiangjian He[3], and Yuefang Gao[4]

[1] School of Computer Science, University of Sydney, Sydney, Australia
huzexi@outlook.com, {vera.chung,miad.zandavi}@sydney.edu.au
[2] School of Electrical and Information Engineering, University of Sydney, Sydney, Australia
wanli.ouyang@sydney.edu.au
[3] School of Computing and Communications, University of Technology Sydney, Sydney, Australia
Xiangjian.He@uts.edu.au
[4] College of Mathematics and Informatics, South China Agricultural University, Guangzhou, China
gaoyuefang@scau.edu.cn

Abstract. Light field (LF) images provide rich information and are suitable for high-level computer vision applications. To acquire capabilities of modeling the correlated information of LF, most of the previous methods have to stack several convolutional layers to improve the feature representation and result in heavy computation and large model sizes. In this paper, we propose channel-wise and SAI-wise attention modules to enhance the feature representation at a low cost. The channel-wise attention module helps to focus on important channels while the SAI-wise attention module guides the network to pay more attention to informative SAIs. The experimental results demonstrate that the baseline network can achieve better performance with the aid of the attention modules.

Keywords: Light field · Image processing · Deep learning

1 Introduction

Light field (LF) images can provide both angular and spatial information by capturing the appearance of objects from several angles in one shot. Compared with regular images, such a feature realizes many high-level computer vision applications such as depth extraction [5,6,8,13,14], refocusing [4] and material classification [10,15]. With the emergence of commercial and industrial LF cameras, *e.g.* Lytro and RayTrix, LF has drawn more attention. However, LF cameras suffer from the inherent trade-off between angular and spatial resolution caused by the limitation of sensor space.

© Springer Nature Switzerland AG 2019
T. Gedeon et al. (Eds.): ICONIP 2019, CCIS 1143, pp. 118–126, 2019.
https://doi.org/10.1007/978-3-030-36802-9_14

LF reconstruction is adopted for alleviating this dilemma which focuses on boosting the angular resolution, *i.e.* the number of SAIs. For example, a densely sampled 8×8 LF is possible to be reconstructed from a sparsely sampled 2×2 LF. With the introduction of deep learning, the performance is significantly improved by the powerful feature representation learned from training samples. In [9], Kalantari *et al.* have proposed the first deep learning-based method to tackle this task by extracting disparity maps using a disparity network, and then warping the input SAIs into the novel SAIs by a color network. With the deep features, it has achieved state-of-the-art performance, nevertheless, it has suffered from intensive computation as it has to reconstruct SAIs separately. Yeung *et al.* [16] proposed a fully convolutional network (FCN) where the 4D LF image can be processed jointly. To mitigate the intensive computational problem of directly operating convolution on the 4D LF data, they proposed a Spatial-Angular Alternating convolutional layer to approximate the 4D convolution by a spatial convolution and an angular convolution. However, to acquire higher-level features and a larger receptive field, these methods have to stack more convolutional layers, which leads to heavy computation and large model size. On the other hand, they treat the features in different locations equally. It is possible that some of the learned features are more important and should not be paid the same attention as other features.

In this paper, we propose two different kinds of attention modules, namely channel-wise and SAI-wise attention for LF. We have assumed that in the processing of LF reconstruction, some of channels and SAIs may carry more important information and to enhance the feature representation these components should be reinforced while the other trivial components should be suppressed. To verify this hypothesis, we apply the two attention modules on a baseline network and propose a novel Channel-wise and SAI-wise Attention (CSA) network which can reconstruct the LF with high quality at a low cost of computation.

2   Proposed Method

The proposed SAI-wise and channel-wise attention modules will be elaborated in Sect. 2.1 and Sect. 2.2 correspondingly. We adopt the skeleton in [16] as our baseline network in Fig. 1 where the reconstruction network is firstly employed to reconstruct the coarse novel SAIs, and then these intermediate SAIs will be refined in the following refinement network. The stacked layers in the reconstruction network are replaced with U-Net to extract the features from input SAIs. U-Net has been proved to be capable to extract multi-level features carrying hierarchical information [5,6,12]. The proposed modules will be applied to the baseline network to demonstrate their impact.

2.1   SAI-wise Attention

In a LF image, even though all SAIs share a major proportion of information in common, details still vary in different SAIs, especially when occlusion happens.

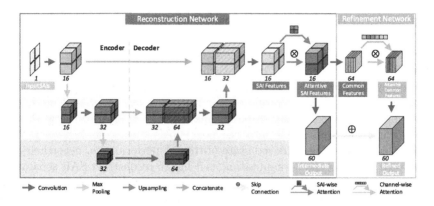

Fig. 1. The illustration of the proposed network architecture. Layers with SAI-wise and channel-wise attention are colored in purple and green respectively. (Color figure online)

Occlusion often plays a challenging role to cause artifacts. Therefore, paying more attention to the SAIs with more details can be beneficial. To this end, we propose SAI-wise attention as shown in Fig. 2(a). For simplicity, some layers are omitted.

Let x be the 4D input tensor of size (U, V, W, H, C) where U and V are the angular dimensions, W and H are the spatial dimensions and C is the size of the channel space. At first, the dimensions (W, H, C) are shrunk into one dimension by global average pooling to get SAI-wise statistic z^{SAI}, in which s-th element is calculated as

$$z_s^{SAI} = f_{GAP}(x_s) = \frac{1}{W \times H \times C} \sum_i^W \sum_j^H \sum_k^C x_s(i, j, k) \qquad (1)$$

where x_s denotes the corresponding SAI. Afterwards, $z^{SAI} \in \mathbb{R}^{(U \times V)}$ is vectorized and fed to three fully connected (FC) layers with n_1, n_2 and $U \times V$ neuron units subsequently, the first two FC layers are followed by ReLU [11] functions and the third one is followed by Sigmoid gating function. Formally, SAI-wise attention is obtained as

$$s^{SAI} = Sigmoid(\mathbb{W}_3 \cdot ReLU(\mathbb{W}_2 \cdot ReLU(\mathbb{W}_1 \cdot z^{SAI}))) \qquad (2)$$

where $\mathbb{W}_1$, $\mathbb{W}_2$ and $\mathbb{W}_3$ are trainable parameters of the three FC layers. Finally, the attentive features of $\hat{x}_s$ are obtained as

$$\hat{x}_s = s_s^{SAI} \cdot x_s \qquad (3)$$

where $\hat{x}_s$ and s_s^{SAI} indicate the s-th elements of $\hat{x}$ and s^{SAI}.

2.2   Channel-wise Attention

Most of the previous deep learning-based LF methods treat the channels equally, which hinders the feature representation. In the refinement network, the features

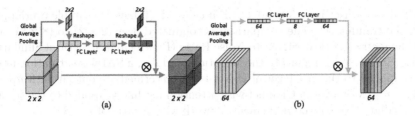

Fig. 2. The illustration of (a) SAI-wise attention and (b) Channel-wise attention.

are convoluted from 4D to 3D which can help to further extract common features and refine the reconstructed SAIs. Inspired by [7], we presume that the channels of the learned common features are not always informative for refinement in some cases and introduce the squeeze-and-excitation channel-wise attention module for our LF reconstruction which is shown in Fig. 2(b).

Given x indicating the 3D input tensor, each channel is denoted as $x_c \in (W, H, F)$ where W and H are spatial width and height and F is the vectorized SAI dimension. A global average pooling layer f_{GAP} is operated on x to get channel-wise statistic $z^{channel} \in \mathbb{R}^C$. Formally, c-th element of $z^{channel}$ is calculated as

$$z_c^{channel} = f_{GAP}(x_c) = \frac{1}{W \times H \times F} \sum_i^W \sum_j^H \sum_k^F x_c(i, j, k). \tag{4}$$

Then, the shrunk features are fed into two FC layers in succession to calculate the attention weights as

$$s^{channel} = Sigmoid(\mathbb{W}_E \cdot ReLU(\mathbb{W}_S \cdot z^{channel})) \tag{5}$$

where the first FC layer $\mathbb{W}_E$ squeezes the features by ratio r and the second FC layer $\mathbb{W}_S$ serves as excitation from the C/r neuron units back to the C neuron units. ReLU and Sigmoid gating functions are applied to the two FC layers respectively. $s^{channel} \in \mathbb{R}^C$ is the set of weights which will be multiplied with x to obtain the attentive features

$$\hat{x}_c = s_c^{channel} \cdot x_c \tag{6}$$

where $\hat{x}_c$ and $s_c^{channel}$ are the c-th channel of the attentive features $\hat{X}$ and channel-wise attention weights $s^{channel}$ respectively.

3   Experiments

3.1   Implementation and Evaluation Details

We implement our proposed CSA method using the deep learning library Keras [3] with Tensorflow [2] backend. To evaluate the proposed method, the experiments have been carried out on extensive LF datasets *30 Scenes* [9] and *Occlusions* [1] which are captured with the Lytro Illum camera. The *30 Scenes* and

Occlusions dataset have 30 and 43 LF images which have no sample overlapped with the training set. The comparison is conducted over (2×2) to (8×8) task and the metrics of Peak Signal-to-Noise Ratio (PSNR) and Structural Similarity Index (SSIM). With regard to the attention modules, a SAI-wise attention module is applied to the last layer of U-Net before intermediate output with $n_1 = 4$ and $n_2 = 4$. Moreover, a Channel-wise attention module is applied to the second intermediate layer of the refinement network with a ratio of $r = 8$.

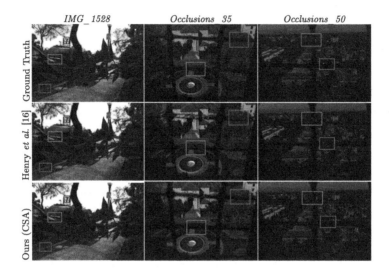

Fig. 3. Visualization of reconstruction.

Table 1. Comparison of overall performance. PSNR and SSIM scores are demonstrated as the metrics of reconstruction quality. Bold scores indicate the best results.

Method	*30 Scenes*	*Occlusions*
Kalantari et al.	37.97/0.9725	31.70/0.8915
Yeung et al.	38.85/0.9759	32.52/0.9029
Ours (CSA)	**39.03/0.9762**	**33.15/0.9067**

3.2   Comparison with State-of-the-Art

We compare our method with 2 state-of-the-art methods, Kalantari *et al.* [9] and Yeung *et al.* [16], and the results are shown in Table 1. In the *30 Scenes* dataset, our method achieves the best result outperforming Yeung *et al.* by 0.18 db PSNR. The edge extends to more than 0.50 db in *Occlusions* dataset which features the challenging scenarios of occlusion. To demonstrate the algorithm efficiency, we

Table 2. Comparison of number of trainable parameters and running speed.

Method	# Trainable Parameters	Speed (Seconds per sample)
Kalantari et al.	1644204	593.47
Yeung et al.	1498752	30.23
Ours (CSA)	**719198**	**13.16**

compare the number of trainable parameters and the running speed in Table 2. The running speed is measured by executing the methods in CPU-only mode. It is observed that our method has achieved better performance with a substantially smaller model and higher speed. The method of Yeung *et al.* suffers from the giant network with the 16 stacked alternating filters while our method comes with less than half of the size and runs at 2 times faster speed.

Visualization of some examples is demonstrated in Fig. 3. In *IMG_1528*, it is observed that the improvement comes from the area with reflective surfaces, *e.g.* the surface of the vehicle's back in the red box, where the result of Henry *et al.* is light leaking while CSA is reconstructed perfectly. More significant differences are observed in the occluded areas, *e.g.* in *Occlusion_35* and *Occlusion_50*, the window frames in the red and blue boxes are blurred while CSA reconstructs these parts more completely. In *IMG_1528*, the edge occluded by the leaf is also more clear in the result of CSA than the other method.

Fig. 4. Visualization of SAI-wise attention on *IMG_1555*. The input SAIs placed by (2×2) in the first two columns annotated with attentive weights and arrows indicating their locations. Selected reconstructed SAIs of Henry *et al.* [16] and ours (CSA) are shown in the third column. Selected regions are annotated with the red boxes in the examples and zoomed in the last row. (Color figure online)

Table 3. Ablation study of the attention modules.

SAI-wise attention	Channel-wise attention	*30 Scenes*	*Occlusions*
✗	✗	38.94/0.9749	32.95/0.9054
✓	✗	38.99/0.9750	33.09/0.9060
✗	✓	39.02/0.9754	33.13/0.9064
✓	✓	**39.03/0.9762**	**33.15/0.9067**

3.3   Study of Attention Modules

In order to investigate the contribution of our proposed attention modules, in this section, an ablation study is conducted by evaluating the model without the attention modules. As shown in Table 3, the baseline model without SAI-wise and channel-wise attention modules produces just slightly better performance than Yeung *et al.* [16]. With SAI-wise attention only, 0.05 db and 0.14 db improvement is obtained in *30 Scenes* and *Occlusions* correspondingly. A similar improvement is observed with channel-wise attention solely as 0.08 db and 0.18 db improvement is obtained. Such results demonstrate that the two proposed attention modules are beneficial to the feature representation when they are working separately. If combining these two modules, the performance gains a little bit better by 0.09 db and 0.20 db, meaning these two attention modules have a similar effect on feature representation, hence the improvement has saturated.

3.4   Study of SAI-wise Attention

To further understand how SAI-wise attention module contributes to the performance, we visualize the input SAIs and the corresponding attention weights of selected samples in Fig. 4. It is observed that some SAIs are weighted higher than others. In *IMG_1555*, some details are not occluded in the top left SAI such as the window frame of the house behind the flower annotated by the red box. In the bottom right SAI which is with the lowest weight, the junction of the window is fully occluded. It is possible that treating SAIs equally may cause artifacts because of mixing visible and occluded details. With the feature reinforcement by SAI-wise attention, the window frame is reconstructed by CSA compared to Henry *et al.* which gets distorted content. In terms of the results, SAI-wise attention has successfully learned to weight the SAIs and reinforced the ones with informative details.

4   Conclusion

In this paper, we have presented two attention modules for LF reconstruction which enhance the channel-wise and SAI-wise feature representation respectively. Our experimental results have demonstrated that these two attention modules have succeeded to guides the baseline network to focus on these informative channels and SAIs, and the proposed CSA network can achieve the state-of-the-art performance at a low cost.

References

1. Stanford Lytro Light Field Archive. http://lightfields.stanford.edu/LF2016.html
2. Abadi, M., et al.: TensorFlow: a system for large-scale machine learning. In: 12th *USENIX* Symposium on Operating Systems Design and Implementation OSDI 2016, pp. 265–283 (2016)
3. Chollet, F., et al.: Keras (2015). https://keras.io
4. Fiss, J., Curless, B., Szeliski, R.: Refocusing plenoptic images using depth-adaptive splatting. In: 2014 IEEE International Conference on Computational Photography (ICCP), pp. 1–9. IEEE (2014)
5. Heber, S., Yu, W., Pock, T.: U-shaped networks for shape from light field. In: Proceedings of the British Machine Vision Conference 2016, vol. 1, pp. 37.1–37.12 (2016)
6. Heber, S., Yu, W., Pock, T.: Neural EPI-volume networks for shape from light field. In: Proceedings of the IEEE International Conference on Computer Vision, vol. 2017-October, pp. 2271–2279, October 2017
7. Hu, J., Shen, L., Sun, G.: Squeeze-and-excitation networks. In: 2018 IEEE/CVF Conference on Computer Vision and Pattern Recognition, pp. 7132–7141. IEEE, June 2018
8. Jeon, H.G., et al.: Depth from a light field image with learning-based matching costs. IEEE Trans. Pattern Anal. Mach. Intell. **41**(2), 297–310 (2019)
9. Kalantari, N.K., Wang, T.C., Ramamoorthi, R.: Learning-based view synthesis for light field cameras. ACM Trans. Graph. **35**(6), 193 (2016). (Proceedings of SIGGRAPH Asia 2016)
10. Lu, Z., Yeung, H.W.F., Qu, Q., Chung, Y.Y., Chen, X., Chen, Z.: Improved image classification with 4D light-field and interleaved convolutional neural network. Multimedia Tools Appl. **78**(20), 29211–29227 (2019)
11. Nair, V., Hinton, G.E.: Rectified linear units improve restricted Boltzmann machines. In: Proceedings of the 27th International Conference on Machine Learning (ICML-10), pp. 807–814 (2010)
12. Ronneberger, O., Fischer, P., Brox, T.: U-net: convolutional networks for biomedical image segmentation. In: Navab, N., Hornegger, J., Wells, W.M., Frangi, A.F. (eds.) MICCAI 2015. LNCS, vol. 9351, pp. 234–241. Springer, Cham (2015). https://doi.org/10.1007/978-3-319-24574-4_28
13. Shin, C., Jeon, H.G., Yoon, Y., So Kweon, I., Joo Kim, S.: EPINET: a fully-convolutional neural network using epipolar geometry for depth from light field images. In: The IEEE Conference on Computer Vision and Pattern Recognition (CVPR), June 2018
14. Wang, T.C., Efros, A.A., Ramamoorthi, R.: Occlusion-aware depth estimation using light-field cameras. In: Proceedings of the IEEE International Conference on Computer Vision, pp. 3487–3495 (2015)

15. Wang, T.-C., Zhu, J.-Y., Hiroaki, E., Chandraker, M., Efros, A.A., Ramamoorthi, R.: A 4D light-field dataset and CNN architectures for material recognition. In: Leibe, B., Matas, J., Sebe, N., Welling, M. (eds.) ECCV 2016. LNCS, vol. 9907, pp. 121–138. Springer, Cham (2016). https://doi.org/10.1007/978-3-319-46487-9_8
16. Yeung, H.W.F., Hou, J., Chen, J., Chung, Y.Y., Chen, X.: Fast light field reconstruction with deep coarse-to-fine modeling of spatial-angular clues. In: Ferrari, V., Hebert, M., Sminchisescu, C., Weiss, Y. (eds.) ECCV 2018. LNCS, vol. 11210, pp. 138–154. Springer, Cham (2018). https://doi.org/10.1007/978-3-030-01231-1_9

Image Generation Framework for Unbalanced License Plate Data Set

Ming Sun, Fang Zhou(✉), Chun Yang, and Xucheng Yin

Department of Computer Science, University of Science and Technology Beijing,
Beijing, China
mingsun@xs.ustb.edu.cn, zhoufang@ies.ustb.edu.cn,
{chunyang,xuchengyin}@ustb.edu.cn

Abstract. Deep learning based methods have achieved promising performance in the image fields, but they are significantly sensitive to the distribution of data. To obtain balanced data, the acquisition of annotation data is very time-consuming and laborious. Generative Adversarial Networks have made a lot of progress in data generation but may degrade greatly in the unbalanced data set. In this paper, we propose an image generation framework to generate photo-realistic, various, and balanced images of text. Specifically, we add the module for training unpaired images (U-module) and target selector to our framework, and the target selector uses text string to select images in the extended real images, which contain real images and the generated images by the U-module, in addition, global generator and local enhancer network are applied to improve the quality of the generated images. We demonstrate our method on the Chinese license plates, and the unbalance of the license plate data set is shown in the provincial category and the special license plates. The Inception Score and the FID Score are used as metrics to validate our method. The experimental results show that the Inception Score of the generated images is close to that of the real images, and our method achieves the lower FID Score than the state-of-the-art. In the SYSU-ITS dataset, the accuracy of license plate recognition has been largely improved, especially for the provinces with no or few real samples.

Keywords: Unbalanced data · Image generation · Generative adversarial networks · License plate recognition

1 Introduction

License plate recognition plays an important role in intelligent traffic management systems. Along with the great development of deep neural networks, deep learning based license plate recognition algorithms achieve promising performance. However, it degrades in unbalanced license plate data set (the number of categories is extremely uneven).

The license plate may involve personal privacy as well as multiple types, and the cost of collection and annotation is high. Moreover, license plate recognition is a task with regional characteristics [12]. In different countries or regions,

© Springer Nature Switzerland AG 2019
T. Gedeon et al. (Eds.): ICONIP 2019, CCIS 1143, pp. 127–134, 2019.
https://doi.org/10.1007/978-3-030-36802-9_15

the definition rules of license plate numbers are variant. For example, the first characters of license plates are different between each region in China, so it is necessary to collect license plates for each region. In natural scenes, there are various license plates such as night, tilt angle, low resolution [6,7], motion blur and so on, which further increases the difficulty of collecting license plate data. In previous work, computer graphics scripts are used to synthesize license plate images according to the font, color, and composition rules [13]. However, it is not easy to generate photo-realistic images solely upon rule-based image generation. As Fig. 1(a, b) shows, there is a big gap between the images generated based on rules and the real ones, which greatly limits the applications of these synthetic data.

Recently, Generative Adversarial Networks (GANs) [2] has become popular research in many fields with its superior generating ability. But the existing GANs algorithms may have poor performance in the case of unbalanced real samples. As shown in Fig. 1(d), in license plate images generated by pix2pix [5], and the experimental results show that the Chinese character part is more ambiguous than other characters. CycleGAN [15] can be trained with unpaired datasets, but the generated images can only fall into a kind of domain as shown in Fig. 1(c).

(a) Synthetic Images   (b) Real Images   (c) CycleGAN   (d) Pix2pix

Fig. 1. (a) The synthetic images synthesized by the scripts. (b) Real license plates. (c) The examples of license plates generated by the CycleGAN. (d) The examples of license plates generated by the pix2pix.

To generate more realistic images with various background and text noise, we propose a unified framework combining two modules, which utilizing unpaired and paired images simultaneously. We demonstrate our method in the context of automotive license plates. The generation of license plates has important consequences for the training of license plate recognition systems.

The main contributions of this paper can be summarized as follows:

- We propose an image generation framework combining two modules to solve the problem of image generation in the unbalanced data set, which utilizing unpaired and paired images simultaneously.
- We design the target selector to combine two modules, it uses text string to select images in the extended real images, which contain real images and the generated images by the module for training unpaired images.
- We implement a novel pipeline with the proposed framework for a typical unbalanced data set, the license plate. The experimental results show that

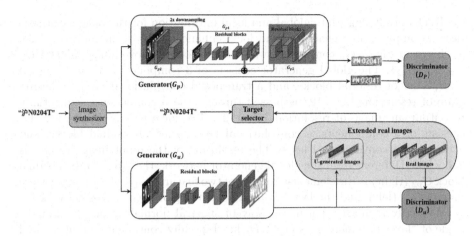

Fig. 2. The whole proposed framework. Synthetic data is generated by the image synthesizer. G_p and D_p are generator and discriminator of P-module respectively. G_u and D_u are generator and discriminator of U-module respectively. The target selector accepts text string and selects the appropriate image from the extended real images. The real images are the unbalanced license plate data set, and some provinces are missing.

the Inception Score of the generated image is close to that of the real images, and our method achieves the lower FID Score than the state-of-the-art.

The rest of this paper is organized as follows. Section 2 introduces some approaches relevant to our work. In Sect. 2.1, we describe the details of networks used in our approach. Experimental results are provided in Sect. 3. Finally, we conclude the work in Sect. 4.

2   Our Approach

2.1   Proposed Framework

The proposed framework consists of an image synthesizer, a target selector, a module for training paired images (P-module) and a module for training unpaired images (U-module), as shown in Fig. 2. We discuss each of these components and their interactions in the following.

We use a standard license plate synthesizer to synthesize a license plate image $x = Synthesizer(s, z)$ with a license plate string s and rendering parameters z. z describes how nuisance factors are added in the synthetic image, and is drawn randomly from a distribution covering the combinations of various factors including foot, outline, color, shading, background, perspective warping, and imaging noise. In our case, z corresponds to a standard font and zero noise perturbation, yielding a clean image x, as shown in Fig. 4. The image synthesizer provides the prototype of the license plate, and it is not trainable.

To effectively aggregate global and local information for the image synthesis task, we apply global generator and local enhancer network in the P-module, as shown in Fig. 2. G_p consists of two generators to improve image fidelity, G_p = $\{G_{p1}, G_{p2}\}$. The global generator consists of 3 components: a convolutional layer, a set of residual blocks, and a transposed convolutional layer. A license plate of resolution 128×128 is passed through the 3 components sequentially to output an image of resolution 128×128. The local enhancer network also consists of 3 components: a convolutional layer, a set of residual blocks, and a transposed convolutional layer. The resolution of the input image to G_{p2} is 256×256. Different from the global generator network, the input to the residual block G_{p2} (right) is the element-wise sum of two feature maps: the output feature map of G_{p2} (left), and the last feature map of the global generator network.

We apply regularization in the way of spectral normalization [9] from the angle of "layer parameters", so that D_p has Lipschitz continuous condition, and D_p modified from pix2pix. The least squares loss [8] and the feature matching loss are used as adversarial loss and the penalty respectively. The experiments show that P-module has excellent network generation capabilities but it works poorly on the unbalanced data set such as the Chinese license plate. We add U-module in our framework, which solves the training problem of unpaired data through reconstruction loss:

$$\mathcal{L}_{res}(G_x, G_{y_{real}}) = \lambda_1 \mathbb{E}_{y_{real} \sim \mathbb{P}_{y_{real}}}[\|G_x(G_{y_{real}}(y_{real})) - y_{real}\|_1]$$
$$+ \lambda_2 \mathbb{E}_{x \sim \mathbb{P}_x}[\|G_{y_{real}}(G_x(x)) - x\|_1] \tag{1}$$

where $\mathbb{P}_x$ and $\mathbb{P}_{y_{real}}$ are the data distribution of synthetic data and real data respectively, generator G_x: $x \to y_{real}$, generator $G_{y_{real}}$: $y_{real} \to x$, λ_1 and λ_2 control the relative importance of two objectives.

Traditional CycleGAN uses least squares loss, we applied WGAN-GP [3] to CycleGAN to improve the stability of training and quality of generated images. WGAN-GP applies gradient penalty to solve the problem of gradient disappearance and gradient explosion. The loss function of G_u in U-module is:

$$\min_{G_u} \mathcal{L}(G_u) = -\mathbb{E}_{y_{real} \sim \mathbb{P}_{y_{real}}}[D_{y_{real}}(G_{y_{real}}(y_{real}))] - \mathbb{E}_{x \sim \mathbb{P}_x}[D_x(G_x(x))]$$
$$+ \mathcal{L}_{res}(G_x, G_{y_{real}}) \tag{2}$$

where $D_{y_{real}}$ discriminates whether $x = G_{y_{real}}(y_{real})$ is true or not, and D_x discriminates whether $y_{real} = G_x(x)$ is true or not.

We design the target selector to connect the P-module and the U-module. The P-module has excellent network generation capabilities but it works poorly on the unbalanced data set. The U-module can solve the training problem of unpaired data because of the weak constraint structure, however, it generates low-resolution images. The target selector can well combine the P-module and the U-module to generate high-quality images on the unbalanced data set. Its input includes license plate string s and extended real images, as illustrated in Fig. 2. It includes a total of two parts of the function as:

$$Y_{data}(s) = y_{real}(s) \cup G_u(Synthesizer(s, z)) \tag{3}$$

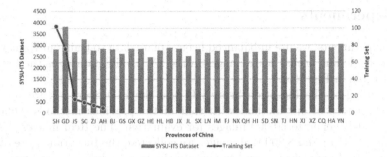

Fig. 3. The abscissa is the abbreviation of provinces in China, the ordinate is the number of license plates. The distribution of SYSU-ITS dataset and training set are described.

where $y_{real}(s)$ is the set of the real images, whose text string equal to s, and G_u is the generator of the U-module.

Traditional methods without the U-module directly take data from $y_{real}(s)$ for training. However, if $y_{real}(s)$ is an empty set, the image corresponding to this text string s cannot be trained. In order to solve the training problem of the P-module and play its own advantages, we add the U-module and the target selector to our framework. For each batch, $y_{data}(s)$ randomly selects data from $Y_{data}(s)$, in the case where $y_{real}(s)$ is an empty set, $y_{data}(s)$ can select data from $G_u(Synthesizer(s,z))$. Adding U-module and target selector to assist P-module in generating images on the unbalanced data set. The full objective combines both GANs loss and feature matching loss as:

$$\mathcal{L}(G_p, D_p, x, y_{data}) = \mathcal{L}_{GAN}\Big(G_p(x), D_p\big(x, G_p(x), y_{data}\big)\Big)$$
$$+ \lambda \mathcal{L}_{FM}\Big(D_p\big(x, G_p(x), y_{data}\big)\Big) \tag{4}$$

where λ controls the importance of the two terms, x is synthetic data by the image synthesizer, $x = Synthesizer(s,z)$, G_p and D_p are generator and discriminator of the P-module respectively.

2.2 Dataset

We evaluated the performance of our approach on the SYSU-ITS [14] dataset, which contains balanced license plate data for 31 provinces, as illustrated in Fig. 3 (SYSU-ITS). Since this database does not provide a rectangle box for the license plate, we do not consider whether the license plate image can be recalled correctly by the detection network, which is similar to Cao et al. [1].

The second data set is the license plate dataset for six provinces, as is shown in Fig. 3 (training set). The data set is collected in natural scenes, totaling more than 8,200. However, the data set is unbalanced, such as the Shanghai and the Guangdong account for a large proportion, many provinces have no data.

3 Experiments

In our experiments, we use unbalanced license plate data to train our framework, and the single-line license plate images in the SYSU-ITS dataset as the test set, as illustrated in Fig. 3(b).

According to the Inception Score [10] and the FID Score [4], we compare the performance of our method with that of the previous methods. Different GANs models generate 60,000 images for evaluation. The Real images dataset contains SYSU-IS and 8,200 real images we collected, the Inception score of it is 2.27. In order to explore the performance of the two modules, we evaluate them separately. Then, we add the U-module and the target selector to our framework, and the performance of ours (P-module+U-module) is evaluated. According to Table 1, the experimental results show that the Inception Score of the generated image is close to that of the real images, and our method achieves the lower FID Score than the previous methods.

In recognition task, we train the license plate recognition network to use the synthetic image, real images or images generated by different GANs as shown in Table 1, and the test set we use the SYSU-ITS dataset. The model has poor performance trained by synthetic data. We use augmented license plate data to compare CycleGAN, pix2pix, and our method's performance in recognition experiments. We use real license plate data from few provinces, utilize our method to generate data and then train recognition network. Moreover, The recognition accuracy of our method can reach 98.6% without Chinese character and 94.3% with Chinese character. Our method has greatly improved the recognition accuracy of license plates with Chinese character. The U-module guides the generation process of the P-module well.

To further evaluate the performance of our method in the unbalanced license plate data set, we conducted separate tests on other provinces that did not appear in the training of real data, as shown in Table 2. It performs poorly in these provinces, and the recognition network trained by synthetic data. The license plate data generated by CycleGAN has slightly improved in these provinces, but pix2pix does not perform well in these provinces, and there are

Table 1. Comparison experiments with different GAN models, "Accuracy-NC" is the recognition accuracy (%) of the letters and numbers of the last six characters.

Methods	Inception Score	FID	Accuracy-NC (%)	Accuracy (%)
Real images	2.27	–	85.0	83.3
CycleGAN [15]	1.43	67.19	95.4	88.0
pix2pix [5]	1.73	39.65	93.6	86.8
pix2pixHD [11]	1.77	36.76	94.4	87.0
Ours (U-module)	1.51	59.04	96.7	90.5
Ours (P-module)	1.86	35.09	96.3	89.0
Ours (P-module+U-module)	**2.11**	**33.59**	**98.6**	**94.3**

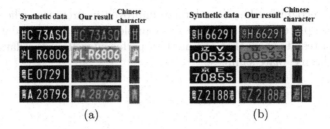

| Synthetic data | Our result | Chinese character | | Synthetic data | Our result | Chinese character |

(a)   (b)

Fig. 4. (a) Single-line license plates generated by our method, and (b) double-line license plates and special license plates generated by our method.

Table 2. Comparison of license plate recognition experiments using different GANs models in some provinces.

Province	Accuracy (%)			
	Script	CycleGAN [15]	pix2pix [5]	Our method
Shandong	69.1	73.0	69.2	**87.5**
Gansu	78.5	96.7	80.5	**98.6**
Anhui	78.2	83.3	76.1	**90.1**
Yunnan	73.5	90.4	73.5	**97.0**
Guangxi	60.5	64.6	60.8	**96.4**
Qinghai	67.3	83.4	67.5	**85.1**
Chongqing	67.1	70.3	67.9	**99.8**
Heilongjiang	85.1	91.3	84.5	**97.9**

still some declines in some provinces. Our method has been greatly improved in these provinces. The experimental comparison in Table 2 shows that our method performs well in unbalanced license plate data.

As shown in Fig. 4, our method is a general framework, The blurring in the Chinese character part has been well solved, and the generated image has high quality, rich background, and multiple styles. Although we present license plates as an application of our method, the same method also works for generating any kind of stylized text in the unbalanced data set.

4   Conclusion

In this paper, we propose an image generation framework that designs the target selector to combine the U-module and the P-module, generating photo-realistic, various, and balanced datasets. Adding U-module and target selector to assist P-module in generating images on the unbalanced data set. The experimental results show that the Inception Score of the generated image is 2.11, which is close to that of the real image, and the FID Score of our method is 33.59, which is lower than the start-of-the-art. In the SYSU-ITS dataset, the accuracy of license plate recognition has been largely improved in provinces where there is no real license plate data or very few real samples for the training dataset. The experimental results show that our method performs well on the datasets with unbalanced license plate samples.

References

1. Cao, Y., Fu, H., Ma, H.: An end-to-end neural network for multi-line license plate recognition. In: ICPR, pp. 3698–3703. IEEE Computer Society (2018)
2. Goodfellow, I.J., et al.: Generative adversarial networks. CoRR abs/1406.2661 (2014)
3. Gulrajani, I., Ahmed, F., Arjovsky, M., Dumoulin, V., Courville, A.C.: Improved training of Wasserstein GANs. In: NIPS, pp. 5769–5779 (2017)
4. Heusel, M., Ramsauer, H., Unterthiner, T., Nessler, B., Hochreiter, S.: GANs trained by a two time-scale update rule converge to a local Nash equilibrium. In: NIPS, pp. 6629–6640 (2017)
5. Isola, P., Zhu, J., Zhou, T., Efros, A.A.: Image-to-image translation with conditional adversarial networks. In: CVPR, pp. 5967–5976. IEEE Computer Society (2017)
6. Liu, X., Liu, W., Mei, T., Ma, H.: A deep learning-based approach to progressive vehicle re-identification for urban surveillance. In: Leibe, B., Matas, J., Sebe, N., Welling, M. (eds.) ECCV 2016. LNCS, vol. 9906, pp. 869–884. Springer, Cham (2016). https://doi.org/10.1007/978-3-319-46475-6_53
7. Liu, X., Liu, W., Mei, T., Ma, H.: PROVID: progressive and multimodal vehicle reidentification for large-scale urban surveillance. IEEE Trans. Multimedia **20**(3), 645–658 (2018)
8. Mao, X., Li, Q., Xie, H., Lau, R.Y.K., Wang, Z., Smolley, S.P.: Least squares generative adversarial networks. In: ICCV, pp. 2813–2821. IEEE Computer Society (2017)
9. Miyato, T., Kataoka, T., Koyama, M., Yoshida, Y.: Spectral normalization for generative adversarial networks. In: ICLR (2018). OpenReview.net
10. Salimans, T., Goodfellow, I.J., Zaremba, W., Cheung, V., Radford, A., Chen, X.: Improved techniques for training GANs. In: NIPS, pp. 2226–2234 (2016)
11. Wang, T., Liu, M., Zhu, J., Tao, A., Kautz, J., Catanzaro, B.: High-resolution image synthesis and semantic manipulation with conditional GANs. In: CVPR, pp. 8798–8807. IEEE Computer Society (2018)
12. Wang, X., You, M., Shen, C.: Adversarial generation of training examples for vehicle license plate recognition. CoRR abs/1707.03124 (2017)
13. Wu, C., Xu, S., Song, G., Zhang, S.: How many labeled license plates are needed? In: Lai, J.-H., et al. (eds.) PRCV 2018. LNCS, vol. 11259, pp. 334–346. Springer, Cham (2018). https://doi.org/10.1007/978-3-030-03341-5_28
14. Zhao, Y., Yu, Z., Li, X., Cai, M.: Chinese license plate image database building methodology for license plate recognition. J. Electron. Imaging **28**(1), 013001 (2019)
15. Zhu, J., Park, T., Isola, P., Efros, A.A.: Unpaired image-to-image translation using cycle-consistent adversarial networks. In: ICCV, pp. 2242–2251. IEEE Computer Society (2017)

Cross-Domain Scene Text Detection via Pixel and Image-Level Adaptation

Danlu Chen, Lihua Lu, Yao Lu$^{(\boxtimes)}$, Ruizhe Yu, Shunzhou Wang, Lin Zhang, and Tingxi Liu

Beijing Laboratory of Intelligent Information Technology, School of Computer Science, Beijing Institute of Technology, Beijing, China
{chendanlu,lulihua,vis_yl,2120171090,shunzhouwang, zhanglin,liutx}@bit.edu.cn

Abstract. Building a robust text detector suitable for different kinds of scene images is a challenging task because the performance of the text detector will be degraded due to the domain shift between different scenes. In this paper, we propose a cross-domain scene text detection method, consisting of pixel-level domain adaptation component (PDA) and image-level domain adaptation component (IDA), to enhance the robustness of scene text detection cross domains. We implement the PDA and IDA by a Gradient Reverse Layer (GRL) and a domain classifier that can distinguish the input scene comes from the source domain or target domain. Besides, to encourage the model to extract domain-invariant features, we introduce the adversarial training to the GRL block. Thus the detector trained with source data can produce more generalized results on the target domain. Experimental results on two pairs of datasets, including MTWI2018 and MSRA-TD500, ICDAR2015 and ICDAR2013, indicate that our proposed method is effective in cross-domain scene text detection and outperforms the methods without domain adaptation.

Keywords: Scene text detection · Domain adaptation · Unsupervised learning

1 Introduction

As an essential step prior to text recognition, scene text detection has been widely used in many applications, such as scene parsing, autonomous driving, instant translation, product identification and so on. Many scene text detection approaches [2,9,11,12,17] have yielded significant performance. The previous methods always assume that training data and test data are in the same distribution. Nevertheless, this assumption is not necessarily true in reality. The disparity between datasets may cause a considerable domain shift, which has been proved to result in performance degradation [4].

Although increasing the diversity of training data may alleviate the effect of domain shift, but it is difficult to collect enough data. Recently, an unsupervised domain adaptation method is proposed in [1] to address the domain

© Springer Nature Switzerland AG 2019
T. Gedeon et al. (Eds.): ICONIP 2019, CCIS 1143, pp. 135–143, 2019.
https://doi.org/10.1007/978-3-030-36802-9_16

Fig. 1. Representative images from different scene text detection datasets: (a) a synthetic images from Internet gathered by MTWI2018; (b) a photo captured in outdoor scene from MSRA-TD500; (c) a focused scene text image provided by ICDAR2013; (d) a incidental scene text image from ICDAR2015.

shift problem in object detection task. Their proposed method based on two-stage network tackles the instance shift and the image shift by introducing two domain components and a consistency regularizer. In this paper, we present an unsupervised domain adaptation method to alleviate the performance degradation caused by domain shift, targeting on scene text detection task. Different from the work [1], we apply a single-stage detection network, which predicts the text instance densely without adopting region proposal network. Therefore, we handle the domain shift problem not on the instance level but the pixel level, focusing on the feature representation of each pixel (Fig. 1).

Considering the domain shift often occurs on pixel level (such as the shift of text and background) and image-level (such as text style and image scale), we design the domain adaptation with two blocks: pixel-level domain adaptation block (PDA) and image-level domain adaptation block (IDA), shown in Fig. 2b. Through the adversarial training, the representation space is updated according to the inverse gradient generated by the domain classifier [3,13], therefore, the discrepancy between domains would be minimize.

The contribution of our work can be summarized as follows: (a) We firstly propose an unsupervised domain adaptation scene text detection method, to enhance the robustness of scene text detection cross domains. (b) This paper designs two domain adaptation components: PDA and IDA, to reduce the domain shift by introducing adversarial training. (c) The proposed method yields better performance on the cross-domain scene text detection.

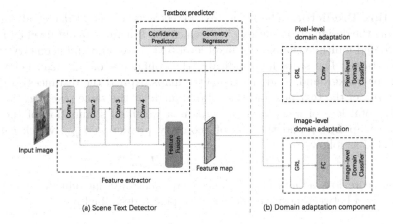

Fig. 2. The architecture of our method can be decomposed into two parts: the scene text detector and the domain adaptation component. The scene text detector predicts the bounding box coordinates of text in the image, and the domain adaptation component, which includes IDA and PDA, distinguishes which domain the feature map belongs to. In the adversarial training stage, the domain adaptation component encourages FE to extract domain-invariant feature for scene text detector.

2   Approach

Figure 2 illuminates the entire procedure of our method. The input image will firstly be sent to FE to obtain the corresponding feature map, which is then fed to both TBP and domain adaptation components. In the training stage, PDA will do pixel-wise classification, while IDA will just output an image-level classification result, with respect to the input feature map extracted before.

Let us denote the detection loss as L_{det}, pixel-level domain classification loss as L_{pxl}, and image-level domain classifier loss as L_{img}. The final training loss is a weighted summation of each component, given by:

$$L = L_{det} + \lambda_{pxl}L_{pxl} + \lambda_{img}L_{img} \qquad (1)$$

where λ_{pxl} and λ_{img} respectively are the balance weights of image-level and pixel-level domain classifier loss.

2.1   Scene Text Detector

Feature Extractor. The convolution neural network extracts the feature from the input image. We choose the ResNet50 [5] as the backbone. Since the size of text regions varies tremendously, the network must use a series of feature maps having various sizes of the receptive field and different levels of visual information. Inspired by EAST [17], we gradually merge the feature from four different layers into a feature map. Therefore, it utilizes different level features and avoids a large computation overhead at the same time.

Text Box Predictor. The scene text detector is a general detect algorithm following the design of DenseBox [7], in which the feature map obtained by FE is fed into and multiple channels of pixel-level text score map and geometry maps are generated. The score map indicates the confidence of the pixel belonging to text or background. And the geometry maps express the distances to the 4 boundaries of the text box and the rotated angle at the same pixel location.

The total loss of scene text detector is a weighted summation of the loss of score map L_{score} and the loss of geometry prediction L_{geo}, which can be formulated as follows:

$$L_{det} = L_{score} + \lambda_{geo} L_{geo} \tag{2}$$

The score loss adopts class-balanced cross-entropy, introducing a balance parameter β between positive and negative pixels, described by:

$$L_{score} = -\beta S \log \hat{S} - (1 - \beta)(1 - S) \log(1 - \hat{S}) \tag{3}$$

where S and $\hat{S}$ denote the ground truth and prediction of score map.

The loss of geometry L_{geo} contains two parts:

$$L_{geo} = -\log IoU(B, \hat{B}) + \lambda_{angle}[1 - \cos(\theta - \hat{\theta})] \tag{4}$$

The box loss computes the IoU of prediction of rectangular text boxes $\hat{B}$ and the corresponding ground truth B. And the angle loss is computed the cosine value of angular offset between ground truth θ and prediction $\hat{\theta}$.

2.2   Domain Adaptation Components

As shown in Fig. 2b, we build two domain adaptation components on pixel-level and image-level in the remedy of the feature gap between domains. Each adaptation component is implemented by a domain classifier with a GRL ahead. We seek to train the domain classifiers to recognize the difference between domains and guide the distance minimization of the source and target feature representations. With the adversarial learning manner, the FE would be optimized and guided toward extracting more generalized convolutional features among both source and target domain.

Pixel-Level Domain Adaptation. We directly provide PDA with the feature map as same as the one used to produce the final dense prediction by the detector. This supplies a meaningful view of the pixel-level feature representation distribution distance of two domains which needs to be minimized. The PDA executes a pixel-wise binary classification and outputs a 1-channel predictive map whose pixel values are in the range of [0, 1].

In particular, we denote the domain label of $k-th$ training image I_k by $D_k \in$ [0,1], where 0 denotes to the source domain, and 1 denotes to the target domain.

And $\hat{D}_k^{pxl}(i,j)$ denotes to the prediction of pixel-level domain classifier at the location of (i,j). The pixel-level adaptation loss L_{pxl} can be given by:

$$L_{pxl} = -\sum_{k,i,j}[D_k \log \hat{D}_k^{pxl}(i,j) + (1 - D_k) \log(1 - \hat{D}_k^{pxl}(i,j))] \qquad (5)$$

In order to enhance the robustness of feature representation, we need to enforce FE to output the feature that minimize the domain distance. This optimization term can be convert to maximize the domain classification loss, which is implemented by introducing GRL. During the forward process of adversarial training, GRL would not work. However, in the backward stage, the sign of the domain classifier gradient will be reversed when propagating through GRL back to FE. The parameters of PDA would be updated by minimizing the pixel-level domain classification loss L_{pxl}. Meanwhile, the parameters of FE are optimized by and maximizing L_{pxl}, which attempts to align the two domains.

Image-Level Domain Adaptation. To further minimize the domain bias as well as to solve the global semantic missing problem, the IDA component is added. By aligning the feature representation on image-level, the model can reduce the domain difference caused by the global image difference, such as illumination, background, text style, etc. The input of IDA refers to the whole feature map extracted by FE. And IDA outputs the binary classification result representing which domain the current image-level feature map comes from.

As the same with pixel-level, D_k denote to the domain label of image I_k, and $\hat{D}_k^{img}$ represents the corresponding image-level domain classifier output. The loss function of the pixel-level classifier can be written as:

$$L_{img} = -\sum_{k}[D_k \log \hat{D}_k^{img} + (1 - D_k) \log(1 - \hat{D}_k^{img})] \qquad (6)$$

There is also a GRL for IDA to apply adversarial training. Thus, the FE is updated by simultaneously minimizing L_{det} and maximizing two domain classifier losses L_{pxl} and L_{img}.

3 Experiments

We evaluate our proposed method on two domain shift scenarios: (1) MTWI2018 to MSRA-TD500, the source data comes from the Internet, while the target data is captured in nature scene; (2) ICDAR2015 to ICDAR2013, the source data and target data are captured with different focus.

3.1 Implementation Details

Follow the common setting of unsupervised domain adaptation, in the training stage, only the detection labels of source dataset are given, while the ground

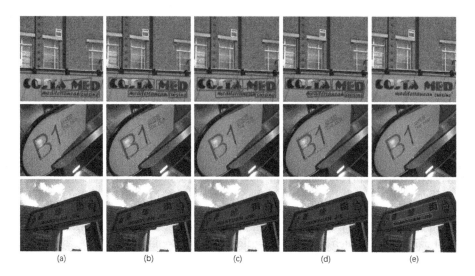

Fig. 3. A comparison of the detection results of the target datasets: ICDAR2013 (the 1st line) and MSRA-TD500 (the 2nd and 3rd line). (a) The baseline result; (b) Only use the PDA component; (c) Only adopt the IDA component; (d) Both PDA and IDA are employed; (e) The ground truth.

Table 1. Detection results evaluated on MSRA-TD500 with MTWI2018 as source dataset, and evaluated on ICDAR2013 with ICDAR2015 as source dataset.

Method	Setting		MSRA-TD500			ICDAR2013		
	PDA	IDA	Precision	Recall	F-score	Precision	Recall	F-score
Baseline			55.8%	55.3%	57.0%	77.5%	61.9%	68.8%
Ours	√		64.1%	**67.1%**	65.6%	**89.2%**	62.3%	73.3%
Ours		√	59.2%	66.8%	62.8%	75.8%	**66.1%**	70.6%
Ours	√	√	**65.5%**	66.3%	**65.9%**	87.0%	65.6%	**74.8%**

truth of the target domain is not accessible. In training, our model is optimized by using a mini-batch stochastic gradient descent algorithm (SGD) with back-propagation with a momentum of 0.9 and a weight decay of 0.0005. We uniform the training samples in the shape, forming a mini-batch of size 16. We use ResNet50 pre-trained on ImageNet to initialize the parameters of the convolutional layers. First, we use the learning rate with 10^{-4} to train the model for 10 epochs, and then the learning rate decays by a weight 0.1 every 20 epochs, stops at 10^{-6}. And the training is stopped when the loss no longer decreases.

3.2   Experimental Results

MTWI2018 to MSRA-TD500. We summarize the results of different experimental settings in Table 1. The baseline trained on the MTWI2018 dataset, has the F-score of 57.0% on MSRA-TD500. Specifically, the PDA module and IDA module can respectively bring 8.6% and 5.8% gains onto the baseline. When integrating the both two components, the F-score can get an 8.9% increase, slightly better than only applying one of them. We give some visual results shown in Fig. 3. With the employment of IDA and PDA, the false negative detections of baseline have been reduced while detections of dense-text area are more accurate. We believe that the performance improvement can be attributed to the adversarial training manner as well as the supervision of IDA and PDA. Compared with other state-of-the-art methods, shown in Table 2, our method is also comparable with those methods trained with the labeled target data.

Table 2. State-of-the-art detection performance on MSRA-TD500.

Method	Performance		
	Precision	Recall	F-score
Yao et al. [15]	63%	63%	60%
Yin et al. [16]	71%	61%	65%
Kang et al. [8]	71%	62%	66%
Zhou et al. [17]	**85.7%**	64.1%	73.3%
He et al. [6]	77%	**70%**	**74%**
Liao et al. [10]	82%	68%	**74%**
Wang et al. [14]	80.3%	65.6%	72.2%
Our Method	65.5%	66.3%	65.9%

ICDAR2015 to ICDAR2013. We present our experimental results conducted on the dataset pairs of ICDAR2015 to ICDAR2013 in Table 1. And the visual result is also shown in Fig. 3. The baseline is trained on the source ICDAR2015 dataset and evaluated on the target ICDAR2013 dataset, which achieves F-score of 68.8%. Our method combining all components gains 6.0% performance improvement compared with the baseline. PDA and IDA improve the F-score by 4.5% and 1.8% respectively. The above comparisons demonstrate the effectiveness of our proposed adversarial training manner and indicate that our method enhances the model adaptability and handles the domain shift efficiently.

4   Conclusion

In this paper, we propose an unsupervised domain adaptation method for scene text detection. The model, obtained by introducing PDA and IDA, is more robust on the task of cross-domain scene text detection. To be specific, the model is gained only by labeled source training data, without using manually annotated data on the target domain. Based on adversarial learning method, the model is trained to extract the invariant feature on both source and target domain, effectively reducing the domain gap. To demonstrate the capability, we validate our approach on several scenarios of domain shift and it achieves better performance on the target domain than the baseline without adopting domain adaptation. In future work, we will further introduce the domain adaptation components into other researches in scene text detection task.

Acknowledgement. This work is in part supported by the National Nature Science Foundation of China (No. 61273273), by the National Key Research and Development Plan, China (No. 2017YFC0112001).

References

1. Chen, Y., Li, W., Sakaridis, C., Dai, D., Van Gool, L.: Domain adaptive faster R-CNN for object detection in the wild. In: Proceedings of the IEEE Conference on Computer Vision and Pattern Recognition, pp. 3339–3348 (2018)
2. Deng, D., Liu, H., Li, X., Cai, D.: Pixellink: detecting scene text via instance segmentation. In: AAAI Conference Artificial Intelligence (2018)
3. Ganin, Y., et al.: Domain-adversarial training of neural networks. J. Mach. Learn. Res. **17**(1), 2096, 2030 (2016)
4. Gopalan, R., Li, R., Chellappa, R.: Domain adaptation for object recognition: an unsupervised approach. In: 2011 International Conference on Computer Vision, pp. 999–1006. IEEE (2011)
5. He, K., Zhang, X., Ren, S., Sun, J.: Deep residual learning for image recognition. In: Proceedings of the IEEE Conference on Computer Vision and Pattern Recognition, pp. 770–778 (2016)
6. He, W., Zhang, X.Y., Yin, F., Liu, C.L.: Deep direct regression for multi-oriented scene text detection. In: Proceedings of the IEEE International Conference on Computer Vision, pp. 745–753 (2017)
7. Huang, L., Yang, Y., Deng, Y., Yu, Y.: DenseBox: unifying landmark localization with end to end object detection. arXiv preprint arXiv:1509.04874 (2015)
8. Kang, L., Li, Y., Doermann, D.: Orientation robust text line detection in natural images. In: Proceedings of the IEEE Conference on Computer Vision and Pattern Recognition, pp. 4034–4041 (2014)
9. Liao, M., Shi, B., Bai, X., Wang, X., Liu, W.: TextBoxes: a fast text detector with a single deep neural network. In: AAAI Conference Artificial Intelligence (2017)
10. Liao, M., Zhu, Z., Shi, B., Xia, G.S., Bai, X.: Rotation-sensitive regression for oriented scene text detection. In: Proceedings of the IEEE Conference on Computer Vision and Pattern Recognition, pp. 5909–5918 (2018)
11. Shi, B., Bai, X., Belongie, S.: Detecting oriented text in natural images by linking segments. In: IEEE Conference Computer Vision and Pattern Recognition, pp. 2550–2558 (2017)
12. Tian, Z., Huang, W., He, T., He, P., Qiao, Y.: Detecting text in natural image with connectionist text proposal network. In: Leibe, B., Matas, J., Sebe, N., Welling, M. (eds.) ECCV 2016. LNCS, vol. 9912, pp. 56–72. Springer, Cham (2016). https://doi.org/10.1007/978-3-319-46484-8_4
13. Tzeng, E., Hoffman, J., Darrell, T., Saenko, K.: Simultaneous deep transfer across domains and tasks. In: Proceedings of the IEEE International Conference on Computer Vision, pp. 4068–4076 (2015)

14. Wang, F., Zhao, L., Li, X., Wang, X., Tao, D.: Geometry-aware scene text detection with instance transformation network. In: Proceedings of the IEEE Conference on Computer Vision and Pattern Recognition, pp. 1381–1389 (2018)
15. Yao, C., Bai, X., Liu, W., Ma, Y., Tu, Z.: Detecting texts of arbitrary orientations in natural images. In: 2012 IEEE Conference on Computer Vision and Pattern Recognition, pp. 1083–1090. IEEE (2012)
16. Yin, X.C., Yin, X., Huang, K., Hao, H.W.: Robust text detection in natural scene images. IEEE Trans. Pattern Anal. Mach. Intell. 36(5), 970–983 (2013)
17. Zhou, X., et al.: EAST: an efficient and accurate scene text detector. In: IEEE Conference Computer Vision and Pattern Recognition, pp. 5551–5560 (2017)

Modified Adaptive Implicit Shape Model
for Object Detection

Ziyan Xu[1], Shujing Lyu[1(✉)], Weiping Jin[2], and Yue Lu[1]

[1] Shanghai Key Laboratory of Multidimensional Information Processing,
East China Normal University, Shanghai 200062, China
{sjlv,ylu}@cs.ecnu.edu.cn
[2] Shanghai Simba Automation Technology Ltd., Shanghai 201803, China

Abstract. Automated threat object detection in X-ray images is needed urgently in baggage inspection at airports, railway stations and other public places. However, the works on object detection are still very limited to meet the needs of practical application. In this paper, we propose a modified adaptive implicit shape model (MAISM) to detect threat objects in X-ray images, in which the triangle patches are used to compute occurrence of the centroid of object instead of keypoints. This model is adaptive for object detection in images of variable scales through triangle patch matching. Experiments on three different threat objects images (razor blades, shuriken, handguns) of various scales demonstrate the effectiveness of the proposed method.

Keywords: Object detection · X-ray images · MAISM · Triangle patch matching

1 Introduction

With the rapid development of transportation, the passenger throughput of airports, subways, railway stations and other hub stations has increased a lot, which results in increasing focus on transportation security. Therefore, it is of vital importance for baggage inspection to prevent passengers from carrying threat objects such as guns, knives, inflammable explosives, etc.

X-ray scanning is widely used for baggage inspection in recent years. With the help of X-ray machines, human inspectors detect threat objects through observing X-ray images. However, it is challenged to stare at the screening all the time. Furthermore, it can be stressful, for human inspectors have to examine baggage in few seconds at the peak time. Therefore, automated object detection in X-ray images is important and urgently needed.

Over the past decades, some object detection algorithms on X-ray baggage images [1, 2] have been proposed, which include traditional algorithms and deep learning algorithms. Most of traditional algorithms are based on Bag of Visual Words (BoVW) model [3–9] although there are also researches using other techniques such as sparse representation [10] and adaptive implicit shape model (AISM) [11]. Bastan et al. [8] utilized the extended SPIN color descriptor and linear structural SVM classifier for handguns detection. Mikolaj et al. [9] employed FAST-SURF descriptor and SVM

© Springer Nature Switzerland AG 2019
T. Gedeon et al. (Eds.): ICONIP 2019, CCIS 1143, pp. 144–151, 2019.
https://doi.org/10.1007/978-3-030-36802-9_17

classifier for firearms detection. Merry et al. [10] proposed a method that extracted image patches to construct representative dictionaries in order to classify the testing image with sparse representation. Riffo et al. [11] proposed Adaptive Implicit Shape Model (AISM) based on Implicit Shape Model that used the concept of visual vocabulary and occurrence of all feature keypoints in every training image in order to find all candidate centroids of threat object in testing image with a probabilistic framework. In addition, deep network is applied for object detection in X-ray images. Akcay et al. [12] used YoloV2 for two-class firearms detection that improved the method [13] based on CNN with transfer learning. But the performance of it is not very well due to the quantitative limitation of X-ray images.

AISM performs well in detecting objects with symmetrical shapes and is worthy of aiding in threat object detection in X-ray images. However, it is not adaptive to scale changes of object in different images. If the size of object in a testing image is different from that of object in training image, AISM does not perform well in detecting it. In practical application, the size of object is various owing to the different distances between the object and X-ray emitter. Thus, we propose modified adaptive implicit model (MAISM) for threat object detection, which modifies AISM through triangle patches. In a similar vein to AISM, MAISM uses exemplar images with different poses to represent a certain threat object and figures out the centroid of threat object in testing image through occurrence that is a set of relative coordinates. Differently, instead of keypoints, our proposed approach introduces triangle patches to compute occurrence that leads to much more matching constraints and provides scale invariance for object detection. Our experiments on the images with different scales yielding 90.60% TPR and 4.97% FPR have demonstrated the effectiveness of our proposed MAISM for detecting threat objects in X-ray images.

The remainder of this paper is organized as following. Section 2 introduces the details of our proposed MAISM. The experimental results are given in Sect. 3. And Sect. 4 offers conclusions related to this paper.

2 MAISM

2.1 Exemplar Representation

2.1.1 Keypoints Extraction

We use the well-known SIFT to extract keypoints and their local visual descriptors from every exemplar image. As shown in Fig. 1(a), the keypoints are marked with solid points on the threat object.

2.1.2 Triangle Patches Representation

After extracting the keypoints, we make up triangle patches with keypoints as vertices and their connections as edges in every exemplar image. If an exemplar image has N keypoints, then there are $C_N^3 = N(N-1)(N-2)/6$ triangle patches in it, as shown in Fig. 1(b).

A triangle patch can be represented with a triple constituting vertex attribute V_j^i with edge attribute E_j^i and occurrence Z_j^i, denoted as (V_j^i, E_j^i, Z_j^i), where $i = 1, \ldots M, M$

is the total number of exemplar images, $j = 1, ..., T^i$, T^i is the total number of triangle patches in exemplar image i. Vertex attribute includes descriptors of three keypoints in each triangle patch. Edge attribute includes edge length information of three edges. Occurrence is the centroid coordinate of the triangle patch with respect to the centroid of the object in the exemplar image.

2.2   Object Detection

2.2.1   Triangle Patches Generation

We also apply SIFT to X-ray testing image to extract all keypoints and descriptors f. Generally, the X-ray testing image provides with a large number of descriptors, while most of them are not useful for object detection. Therefore, we remove unnecessary descriptors from the testing image by performing descriptor matching between all descriptors f^i of exemplar image i and the descriptors f of the testing image using the expression: $d_E(f, f^i) < \theta$, where θ is the minimum distance allowed between f and f^i. The descriptors in testing image fulfil this expression are denoted as $\hat{f}$. As shown in Fig. 2(a), the solid points represent the keypoints remained in the testing image. All triangle patches are generated by using the remained keypoints as vertices and connections as edges. Then, the vertex attribute and edge attribute of triangle patch k in the testing image are denoted as $\hat{V}_k = \left(\hat{f}_{k1}, \hat{f}_{k2}, \hat{f}_{k3}\right)$ and $\hat{E}_k = \left(\hat{e}_{k1}, \hat{e}_{k2}, \hat{e}_{k3}\right)$, respectively. However, occurrence $\hat{Z}_k = \left(\hat{X}_k, \hat{Y}_k\right)$ is unknown due to lack of information of the centroid of the object in the testing image.

 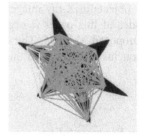

(a) SIFT keypoints   (b) Triangle Patches

Fig. 1. Keypoints extraction in exemplar X-ray image

2.2.2   Triangle Patch Matching

MAISM tries to find out the candidate centroids of the object by triangle patch matching. The triangle patch matching rules can be summarized as vertex similarity and edge similarity.

Firstly, vertex similarity is defined as each vertex of triangle patch k $\hat{P}_k$ in the testing image matching with that of triangle patch j P_j^i in exemplar image i by comparing the distance of the closest neighbor to that of the second-closest neighbor, where $k = 1, ..., \hat{T}$, $\hat{T}$ is the number of triangle patches in testing image, $i = 1, ..., M$, M is the number of exemplar images and $j = 1, ..., T^i$, T^i is the number of triangle patches in exemplar image i. If vertex $\hat{f}_{kn}$ of $\hat{P}_k$ and vertex f_{jn}^i of P_j^i ($n = 1, 2, 3$) fulfill the expressions, we consider that they are matched:

$$d_E\left(\hat{f}_{kn}, f_{jn}^i\right) = min\, d_E\left(\hat{f}_{kn}, f^i\right) \tag{1}$$

$$\frac{min\, d_E\left(\hat{f}_{kn}, f^i\right)}{sencond - min\, d_E\left(\hat{f}_{kn}, f^i\right)} < \theta_{match} \tag{2}$$

Where f^i is the set of all descriptors in exemplar image i.

Secondly, edges similarity means that the edge length ratios ELR_{ij}^k of edges in $\hat{P}_k$ to P_j^i are numerically close:

$$ELR_{ij}^k = \frac{\hat{e}_{k1}}{e_{j1}^i} = \frac{\hat{e}_{k2}}{e_{j2}^i} = \frac{\hat{e}_{k3}}{e_{j3}^i} \tag{3}$$

If $\hat{P}_k$ and P_j^i are satisfied with the matching rules, then we consider ELR_{ij}^k as the scaling ratio of $\hat{P}_k$ to P_j^i.

What's more, we can calculate how many degrees $\hat{P}_k$ rotated counterclockwise comparing to P_j^i by affine transformation of triangle. The rotation angle is denoted as θ_{rotate}^k. As shown in Fig. 2(b), similar vertices are connected with solid lines in the matched triangle patches.

2.2.3 Voting for Candidate Centroids

Firstly, with the information of scaling ratio ELR_{ij}^k, rotation angle θ_{rotate}^k of triangle patch k $\hat{P}_k$ in testing image and occurrence $Z_{ij} = \left(X_{ij}, Y_{ij}\right)$ of triangle patch j P_j^i in exemplar image i, we can deduce occurrence $\hat{Z}_k = \left(\hat{X}_k, \hat{Y}_k\right)$ of triangle patch $\hat{P}_k$ from all these information. The occurrence $\hat{Z}_k$ can be calculated as:

$$\begin{bmatrix} \hat{X}_k \\ \hat{Y}_k \end{bmatrix} = ELR_{ij}^k * \begin{bmatrix} \cos\left(\theta_{rotate}^k\right) & -\sin\left(\theta_{rotate}^k\right) \\ \sin\left(\theta_{rotate}^k\right) & \cos\left(\theta_{rotate}^k\right) \end{bmatrix} \begin{bmatrix} X_{ij} \\ Y_{ij} \end{bmatrix} \tag{4}$$

Next, we can get the candidate centroid $\left(\hat{C}_x, \hat{C}_y\right)$ of target object in the testing image as follows:

$$\left(\hat{C}_x, \hat{C}_y\right) = \left(\hat{T}_x^k - \hat{X}_k, \hat{T}_y^k - \hat{Y}_k\right) \tag{5}$$

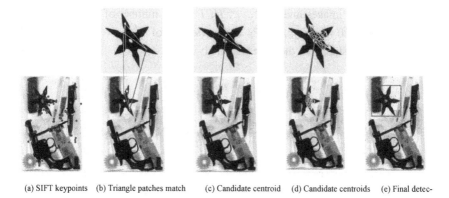

(a) SIFT keypoints   (b) Triangle patches match   (c) Candidate centroid   (d) Candidate centroids   (e) Final detec-

Fig. 2. Object detection with MAISM

Where $\left(\hat{T}_x^k, \hat{T}_y^k\right)$ is the centroid of triangle patch $\hat{P}_k$. As shown in Fig. 2(c), the candidate centroid of the object is marked with solid red point.

Finally, MAISM votes for every candidate centroid with the following score:

$$\text{score}\left(\hat{C}_x, \hat{C}_y\right) = \sum_{i=1}^{M} \sum_{j=1}^{\hat{T}_i} \frac{1}{T_i} \tag{6}$$

Where $\hat{T}_i$ is the total number of triangle patches that matching with triangle patches in exemplar image i, T_i is the total number of triangle patches in exemplar image i, M is the total number of exemplar images. As shown in Fig. 2(d), all matched triangle patches are marked with triangles and the candidate centroids are connected with centroid of the object in exemplar image.

2.2.4   Final Detection

In order to be robust to interclass variation, we have to ignore small shape deformation by merging the candidate centroids that locate closely and adding up their scores. In addition, MAISM keeps the candidate centroids whose scores are greater than threshold θ_s for final object detection.

The final detection is obtained as follows:

If there is no candidate centroid whose score is greater than θ_s, then no potential object is detected in the testing image. If there exist candidate centroids meeting the condition, then we take them as the centers of the detected bounding boxes and draw detection windows. If the detection windows are connected or overlapped, we exploit the non-maximum suppression algorithm in order to remove redundant detection windows and retain the best one. As shown in Fig. 2(e), the small points are all candidate centroids of the object and the biggest solid point is the final centroid of the object and the box is the final detection window of the testing image.

3 Experimental Results

3.1 Dataset

We conduct three experiments for detecting three different kinds of threat objects (razor blade, shuriken and handgun) in X-ray baggage images. Therefore, the datasets in our experiments consist of three kinds of threat objects, which are provided by D. Mery in GDXray [12]. Every dataset has both exemplar images and testing images.

Since object has different poses in baggage in reality, the exemplar images should cover as many poses as possible. The number of exemplar images in each experiment is: 100 for razor blades, 100 for shuriken, 200 for handguns.

In GDXray, the original testing set for each experiment consists of 150 X-ray baggage images and the original size of testing images is 2208 * 2688 pixels. In order to verify the sale invariance of MAISM for object detection, we expand the testing set by reducing the size of all testing images to 2016 * 1656 pixels and 1344 * 1104 pixels. So, we have three testing sets consisting of X-ray baggage images of different sizes. Every image contains only 1 threat object and 18 non-threat objects like screws, clips, pens, etc. Therefore, there are $N_p = 150$ threat object (positive) and $N_n = 18 * 150 = 2700$ non-threat objects (negative) in every testing set.

Besides, we also use the ground truth bounding box that manually annotated by D. Mery to outline the position of threat object in X-ray testing image, denoted as BB_{gt}.

3.2 Evaluation Methodology

Performance of experiments is evaluated by the standard of PASCAL VOC object detection challenge. Detection judged to be correct or false is using the measurement of the overlapping area, denoted as a_0:

$$a_0 = \frac{area\left(BB_d \cap BB_{gt}\right)}{area\left(BB_d \cup BB_{gt}\right)} \tag{7}$$

Where BB_d is the detecting bounding box also referred as detection window in Sect. 2.2.4. In our experiment, we consider the detection to be correct if $a_0 \geq 0.5$.

The number of correct detecting bounding boxes is denoted as TP and the number of false detecting bounding box is denoted as FP. Ideally, $TP = 150$, $FP = 0$ for detecting threat object in every testing set. The true positive rate $TPR = TP/N_p$ and the false positive rate $FPR = FP/N_n$ are used for evaluating the performance. The evaluation standard is the same as AISM adopts.

3.3 Results

The TPR and FPR for detecting razor blades, shuriken and handguns in X-ray images with different sizes are shown in Table 1.

On the one hand, MAISM achieves 94.89% TPR and 3.27% FPR for razor blades detection, 93.57% TPR and 6.63% TPR for shuriken detection, 83.33% TPR and 5% FPR for handguns detection in all testing sets, all these TPRs are higher than AISM and all these FPRs are lower than AISM, which demonstrates that MAISM performs better than AISM.

On the other hand, with the size of image is getting smaller, TPR of MAISM is reduced and FPR of MAISM is increased in a small range, while AISM changes a lot. The average TPR of AISM reduces from 95.11% to 18.89%, which is a very bad result, while that of MAISM keeps is still above 90%. So, we can find that AISM performs badly in detecting objects within various images and MAISM is obviously much better than AISM.

From the above experiments, we draw conclusions that MAISM performs well in detecting objects with variable scales and it is a promising method for object detection.

Table 1. TPR (%) and FPR (%) in detecting image with different sizes

Size	Method	Razor blade		Shuriken		Handgun		Average	
		TPR	FPR	TPR	FPR	TPR	FPR	TPR	FPR
1344 * 1104	AISM	48.67	0.19	8.00	0.11	0.00	21.29	18.89	7.20
	MAISM	86.67	6.15	94.67	10.3	88.67	6.07	90.00	7.51
2016 * 1656	AISM	90.00	1.70	61.33	1.15	61.33	44.67	70.87	15.84
	MAISM	98.67	2.41	92.00	5.85	80.00	5.15	90.22	4.47
2688 * 2208	AISM	98.66	2.00	97.33	6.00	89.33	17.00	95.11	8.33
	MAISM	99.33	1.26	94.00	3.74	81.33	3.80	91.55	2.93
Average	AISM	79.11	3.89	55.55	7.26	50.22	27.65	61.63	12.93
	MAISM	94.89	3.27	93.57	6.63	83.33	5.00	90.60	4.97

4   Conclusion

In this work, we propose MAISM for detecting threat objects in X-ray images. We assign three attributes to each triangle patch generated from exemplar images and use triangle patch matching to find all candidate centroids of object in testing image. Using MAISM for detecting different threat objects (razor blades, shuriken, handguns) yields 90.6% TPR and 4.97% FPR, which is much better than AISM that only achieves 61.63% TPR and 12.93% FPR. The result demonstrates that MAISM is valid for scale invariance in object detection.

References

1. Abidi, B., Zheng, Y., Gribok, A., et al.: Improving weapon detection in single energy X-ray images through pseudocoloring. IEEE Trans. Syst. Man Cybern. Part C (Appl. Rev.) **36**(6), 784–796 (2006)

2. Chen, Z., Zheng, Y., Abidi, B.: A combinational approach to the fusion, de-noising and enhancement of dual-energy X-ray luggage images. In: IEEE Computer Society on Computer Vision and Pattern Recognition Workshops. IEEE (2006)

3. Rosten, E., Porter, R., et al.: Faster and better: a machine learning approach to corner detection. IEEE Trans. Pattern Anal. Mach. Intell. **32**(1), 105–119 (2008)

4. Franzel, T., Schmidt, U., Roth, S.: Object detection in multi-view X-ray images. In: Pinz, A., Pock, T., Bischof, H., Leberl, F. (eds.) DAGM/OAGM 2012. LNCS, vol. 7476, pp. 144–154. Springer, Heidelberg (2012). https://doi.org/10.1007/978-3-642-32717-9_15

5. Baştan, M., Yousefi, M.R., Breuel, T.M.: Visual words on baggage X-ray images. In: Real, P., Diaz-Pernil, D., Molina-Abril, H., Berciano, A., Kropatsch, W. (eds.) CAIP 2011. LNCS, vol. 6854, pp. 360–368. Springer, Heidelberg (2011). https://doi.org/10.1007/978-3-642-23672-3_44

6. Turcsany, D., Mouton, A., Breckon, T.P.: Improving feature-based object recognition for X-ray baggage security screening using primed visual words. In: Proceedings International Conference on Industrial Technology, pp. 1140–1145. IEEE (2013)

7. Bastan, M., Byeon, W., Breuel, T.M.: Object recognition in multi-view dual energy X-ray images. In: BMVC, vol. 1, no. 2, p. 11 (2013)

8. Baştan, M.: Multi-view object detection in dual-energy X-ray images. Mach. Vis. Appl. **26**(7-8), 1045–1060 (2015)

9. Kundegorski, M.E., Akcay, S., et al.: On using feature descriptors as visual words for object detection within X-ray baggage security screening. In: International Conference on Imaging for Crime Detection and Prevention (2016)

10. Mery, D., Svec, E., Arias, M.: Object recognition in baggage inspection using adaptive sparse representations of X-ray images. In: Bräunl, T., McCane, B., Rivera, M., Yu, X. (eds.) PSIVT 2015. LNCS, vol. 9431, pp. 709–720. Springer, Cham (2016). https://doi.org/10.1007/978-3-319-29451-3_56

11. Riffo, V., Mery, D.: Automated detection of threat objects using adapted implicit shape model. IEEE Trans. Syst. Man Cybern. Syst. **46**(4), 472–482 (2016)

12. Akcay, S., Kundegorski, M.E., et al.: Using deep convolutional neural network architectures for object classification and detection within X-ray baggage security imagery. IEEE Trans. Inf. Forensics Secur. **13**(9), 2203–2215 (2018)

13. Akcay, S., Kundegorski, M.E., et al.: Transfer learning using convolutional neural networks for object classification within X-ray baggage security imagery. In: IEEE International Conference on Image Processing, pp. 1056–1061 (2016)

Learning from Incomplete Data

An Expert Validation Framework for Improving the Quality of Crowdsourced Clustering

Liu Jiang, Zheng Qin$^{(\boxtimes)}$, Zhipeng Li, Pengbo Shen, and Shaohan Hu

School of Software, Tsinghua University, Beijing, China
{jiangl16,lizp14,spb17,hush17}@mails.tsinghua.edu.cn,
qingzh@tsinghua.edu.cn

Abstract. Crowdclustering is a cost-effective mechanism that learns a cluster structure from data and crowdsourced human pairwise labels. Though some initial efforts have shown some effectiveness of crowdclustering, performing a reliable crowdclustering is inherently challenging due to the noisy and uncertain nature of crowdsourced labels; the consistency of crowdclustering quality is also not guaranteed across datasets. To improve the quality of crowdsourced clustering, we argue for the need of expert validations for post-processing clustering results. To this end, we establish a novel expert validation framework comprised of a dynamic multi-criteria-based pair selection component to actively select most informative data pairs, and a pairwise label propagation component for enhancing the expert influence and incorporating them to guide the crowdclustering. Both components serve to minimize expert validation efforts. Experimental results on six real-world and synthetic datasets show the effectiveness of our overall approach and its key components, respectively.

Keywords: Crowdsourcing · Clustering · Expert validation

1 Introduction

Cluster analysis is aimed at partitioning data into multiple groups, where intra-group instances are similar and inter-group instances are not. Either as an important task of unsupervised learning or as a common technique for knowledge discovery, clustering has been used in a wide variety of fields [2,6]. The key concept in a clustering model is the notion of similarity between instances. Most of the existing clustering methods base themselves on some general similarity measures defined on features of data instances. This, however, leads to unstable clustering results given various measures, and oftentimes makes it difficult to fit a particular task of the user. A common method to address this is to elicit human labels[1], to supervise the clustering model [23]. To collect human supervision, crowdsourcing

---

[1] Labels are usually in the form of must- or cannot-link constraints.

© Springer Nature Switzerland AG 2019
T. Gedeon et al. (Eds.): ICONIP 2019, CCIS 1143, pp. 155–167, 2019.
https://doi.org/10.1007/978-3-030-36802-9_18

paradigm has been recently shown to be efficient and cost-effective on acquiring manual annotations on pairwise labels between instances. Various *crowdclustering* models for learning data clusters by incorporating these labels from crowd workers are proposed [7,16,25]. Despite some effectiveness of crowdclustering, it is still challenging to learn a reliable clustering from noisy and uncertain crowdsourced pairwise labels, due to large variations in knowledge background and levels of expertise of crowd workers. For example, spammers often give random annotations; sloppy workers may return wrong answers to tasks, while an experienced worker tend to correctly label most of the data. Also, since crowdclustering models heavily rely on model assumptions, the quality of the clustering result is practically inconsistent across datasets.

Against this background, we argue that the performance of crowdclustering is limited due to the lack of expert guidance; a post-processing expert validation stage where experts input a few ground truth labels to verify or correct crowdsourced labels, is required to improve the quality of clustering results. This is motivated by constrained clustering work where an oracle provides pairwise constraints to guide the model towards a better solution. More broadly, this is grounded by early efforts on aggregating classification labels from crowds and experts [9–11]. To the best of our knowledge, the expert validation paradigm is less explored in a crowdclustering setting.

This paper presents a first step towards improving crowdclustering through expert validation. The key components of our proposed expert validation framework are **pair selection** and **pairwise label propagation**. Due to the high cost incurred by querying expert, the pair selection aims to select most informative instance pairs to be validated. To this end, we develop a dynamic multi-criteria-based two-phase selection strategy. In the first phase, we propose to guide the important skeleton exploration by utilizing expert validation. In the second phase, based on the principle of learning with uncertainty [24], our method focuses on making the skeleton outlined in the first phase more solid by comprehensively addressing uncertainty associated with *crowd workers, data instances* and *crowdclustering model*. The pairwise label propagation strives to maximize the influence of experts by propagating their labels to instance pairs not validated. The key challenge is to maximally but precisely augment the set of expert labels and seamlessly integrate them into the crowdclustering model. We accomplish this by a constraint expansion method and a Bayesian formulation of the crowdclustering model for enforcing expert constraints. Extensive experiments on six real-world and synthetic datasets demonstrate the effectiveness of our approach as well as its superiority over competing alternatives.

2   Probabilistic Discriminative Clustering Model

To perform expert-guided crowdclustering, we use the probabilistic discriminative clustering model (PDCM) [1] as our base representation. Given a set of n data instances $\mathbf{X} = [\mathbf{x_1}, ..., \mathbf{x_n}]^T \in \mathbb{R}^{n \times d}$, where d is the feature dimension, and pairwise labels $\mathbf{S}$ collected from M crowd workers, PDCM clusters the

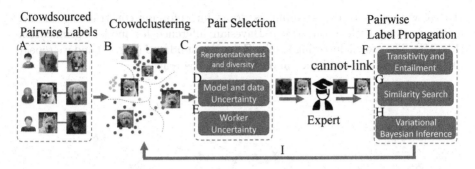

Fig. 1. Our expert validation framework for improving the quality of crowdclustering.

data into K groups by jointly learning a discriminative logistic regressor and a worker uncertainty model. The logistic regressor are parameterized by matrix $\mathbf{W} = [\mathbf{w_1}, ..., \mathbf{w_K}] \in \mathbb{R}^{d \times K}$, and vector $\mathbf{b} = [b_1, ..., b_K] \in \mathbb{R}^{K \times 1}$. The worker uncertainty model [5] associates each worker m with sensitivity and specificity parameters $\alpha_m, \beta_m \in [0, 1]$, and assumes the pairwise labels $\mathbf{S}$ are generated as follows: $p(S_{ij}^{(m)} = 1 | z_i = z_j) = \alpha_m$, $p(S_{ij}^{(m)} = 0 | z_i \neq z_j) = \beta_m$, where z_i is the cluster assignment of instance $\mathbf{x}_i$, $S_{ij}^{(m)}$ is the pairwise label annotated by worker m (must- or cannot-link). The worker uncertainty model and discriminative logistic regressor share the latent variables $\mathbf{Z}$ (cluster assignments). Hence, EM algorithm is exploited to train them jointly by maximizing the likelihood of the crowdsourced labels with respect to model parameters $\theta = \{\mathbf{W}, \mathbf{b}, \alpha, \beta\}$.

3   Framework Overview

This section presents an overview of the proposed expert validation framework. The overall pipeline is shown in Fig. 1. Given the data matrix $\mathbf{X}$ and crowdsourced pairwise labels $\mathbf{S}$ (Fig. 1A), we generate initial cluster assignments $\mathbf{Z}$ using PDCM crowdclustering model (Fig. 1B). We perform a post-processing expert validation on the clustering result $\mathbf{Z}$. First, we propose a dynamic multi-criteria-based strategy for actively selecting instance pairs (Fig. 1C). The selection method is aimed at maximizing representativeness and diversity of the selected instances, and comprehensively reducing the uncertainties associated with *model, data* and *worker* (Fig. 1D, E). Second, upon the acquired expert validation on selected instance pairs, we propose a pairwise label propagation approach (Fig. 1F) to maximize the expert influence. It contains a constraint expansion preprocessing based on transitivity & entailment properties of pairwise constraints[2], and a similarity search for further constraint augmentation (Fig. 1G). To incorporate the constraints into PDCM model, we develop a Bayesian variant

---

[2] The pairwise labels of experts are intrinsically the constraints to instance pairs.

of PDCM that takes expert input as priors for enforcing constraints in model learning, along with a variational Bayesian inference for model optimization (Fig. 1H). The crowdclustering results **Z** are then updated (Fig. 1I) and we repeat the above iterations until the validation budget is reached.

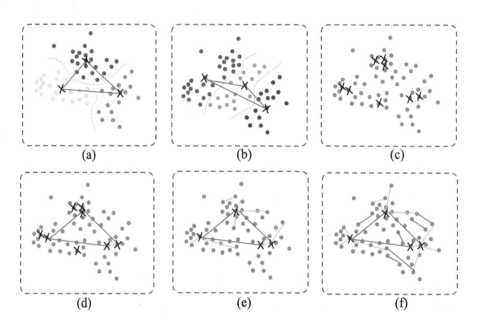

Fig. 2. (a) Ground truth cluster distribution of a synthetic dataset, the cross marks represent the cluster centroids; (b) Initial inferred cluster distribution by PDCM, which greatly deviates from the ground truth; (c) Selecting representatives based on density-biased sampling; (d) Skeleton exploration results, red color encodes cannot-link, green encodes must-link; (e, f) Uncertainty-driven phase progressively strengthen the skeleton and reducing the uncertainty regarding the cluster assignments. (Color figure online)

4   Pair Selection Method

The pair selection method is focused on selecting the most informative *instance pairs* for minimizing expert validation efforts. Overall, it is a dynamic multi-criteria-based strategy comprised of skeleton exploration and uncertainty-driven phases.

4.1   Skeleton Exploration Phase

The cluster skeleton greatly affects the overall clustering quality since the subsequent iterations of clustering optimization are largely based on the notion of the cluster skeleton learned in the early stage. To build a viable skeleton, we propose to construct a skeleton comprised of data points with high representativeness and diversity.

Representativeness. To select representative instances, a straightforward way is to select cluster centroids by checking the average similarity of an instance to others within the same cluster based on the cluster assignments estimated by the PDCM model. However, this approach may exacerbate the potentially incorrect cluster distributions learned by the model. To illustrate this, Fig. 2(a) shows the ground truth cluster distribution of a synthetic three-blob dataset. Figure 2(b) displays the initial cluster assignments learned by PDCM model, which greatly deviates from the true cluster distribution. Simply selecting the centroids of clusters based on cluster assignments in Fig. 2(b) leads to an incorrect skeleton and degrades the clustering performance. To tackle this, we adopt the idea of density-biased sampling [12], i.e., sampling each instance with the probability biased by its local density, $prob(i) \propto \frac{|\{j|\mathbf{x}_j \in neighborhood\,(\mathbf{x}_i,R)\}|^\gamma}{R^d}$, where R specifies the radius of spherical neighborhood centered on $\mathbf{x}_i$ and γ controls the degree of the bias caused by local density. Intuitively, the above equation suggests a higher sampling rate for dense regions and lower for sparse regions, which fits our objective to select representatives. We show an example of sampling result in Fig. 2(c).

Diversity. Given the representative instances, our diversity-driven method aims to build a skeleton that has the largest coverage of the data distribution, or equivalently, maximizing the diversity of data instances contained therein. To formalize this, given the number of cluster K, we strive to search K-disjoint cliques $\mathcal{N} = \{\mathcal{N}(1), ..., \mathcal{N}(K)\}$; instances within a clique are "must-linked", while instances between cliques are "cannot-linked". Since exhaustively querying all data pairs is not practical, we employ the farthest-first-traversal approach used in the classic K-center optimization problem [8], greedily selecting the data instance to grow the K-disjoint cliques. Let $\mathcal{D}$ denotes the set of representative instances, $\mathcal{N}_t$ denotes the disjoint cliques built in t-th iteration. We select an instance $\mathbf{x}_i$ that is farthest to the existing disjoint cliques, $\mathbf{x}_i = \arg\max_{\mathbf{x}_j \in \mathcal{D}} \min_{\mathcal{N}(k) \in \mathcal{N}_t} dist(\mathbf{x}_j, \mathcal{N}(k))$, where $dist(\cdot, \cdot)$ measure the Euclidean distance between an instance and a set. Then we grow the disjoint cliques by querying expert about the relations between instance $\mathbf{x}_i$ and each clique in the disjoint cliques. The growing of disjoint cliques terminates once its cardinality reaches K. An example of the built K-disjoint cliques ($K = 3$) is demonstrated in Fig. 2(d).

4.2 Uncertainty-Driven Phase

Based on the built skeleton, we aim to resolve the uncertainty in crowdclustering, consolidating the skeleton to further reveal fine-grained latent cluster structure. We propose to assess the *data*, *model* and *worker* uncertainty and dynamically weight them to drive our selection method.

Data Uncertainty. In our crowdclustering setting, we measure the uncertainty associated with data by calculating the information entropy of the probabilistic distribution of pairwise relations. Formally, data uncertainty over an instance pair $(\mathbf{x}_i, \mathbf{x}_j)$, is computed as $U_d(i,j) = -p(z_i = z_j)\log p(z_i = z_j) - (1 - p(z_i = z_j))\log(1 - p(z_i = z_j))$, where z_i is the estimated cluster that instance $\mathbf{x}_i$ is affiliated to, $p(z_i = z_j) = \sum_k p(z_i = k|\mathbf{x}_i) \cdot p(z_j = k|\mathbf{x}_j)$ measures how likely $\mathbf{x}_i$ and $\mathbf{x}_j$ belong to a same cluster.

Model Uncertainty. We associate each individual data instance with a point-wise uncertainty to inherently disclose the model confidence on the decision of cluster assignments. To elaborate on this, define the overall model uncertainty as $U_m = \sum_i U_m(i) = \sum_i \sum_k -p(z_i = k|\mathbf{x}_i)\log p(z_i = k|\mathbf{x}_i)$, where $p(z_i = k|\mathbf{x}_i)$ is the posterior probability of $\mathbf{x}_i$ being assigned cluster k, $U_m(i)$ gauges the confidence of PDCM model regarding pointwise decision on $\mathbf{x}_i$. We use the sum of the pointwise $U_m(i)$ as an intrinsic *surrogate* for model uncertainty. Intuitively, a lower U_m indicates a large separation between clusters, i.e., decision boundaries for discriminating clusters have large margins [13]. To obtain $p(z_i = k|\mathbf{x}_i)$, we use Monte Carlo sampling to approximate the computationally intractable integral $p(z_i = k|\mathbf{x}_i) = \int_W \int_b p(z_i = k|\mathbf{W}, \mathbf{b}; \mathbf{x}_i)q(\mathbf{W})q(\mathbf{b})d\mathbf{W}d\mathbf{b}$, where $q(\mathbf{W}), q(\mathbf{b})$ are posterior distributions of logistic regression parameters.

Worker Uncertainty. To measure the uncertainty introduced by crowd workers, we adopt the idea of a spammer-detection method by Hung et al. [10]. Specifically, in our scenario, each worker m is characterized by its accuracy parameters α_m, β_m. We build a confusion matrix for each worker, $CF_m = \begin{pmatrix} \alpha_m, & 1 - \alpha_m \\ 1 - \beta_m, & \beta_m \end{pmatrix}$. Based on this, we use Hung et al.'s method to estimate the spammer score $SP(m)$ of worker m. The worker uncertainty associated in each data pair is then evaluated as $U_w(i,j) = \sum_{m \in \mathcal{W}(i,j)} SP(m)$, where $\mathcal{W}(i,j)$ is the set of workers who label the instance pair $(\mathbf{x}_i, \mathbf{x}_j)$.

Dynamic Weighting Scheme. We exploit a dynamic weighting scheme to combine the uncertainties discussed above. The basic idea of our scheme is that high ratio of spammers and high querying error rate[3] calls for a worker-uncertainty-driven method, otherwise data- and model-uncertainty-driven method are favored. In the initial model building phase, since the PDCM may mis-detect spammers, we allocate small weight to spammer ratio; the weight is increased as model optimization proceeds. We show an example of pair selection by our uncertainty-driven method in Fig. 2(e-f), in which the skeleton is progressively consolidated and ambiguous data pairs are selected for reducing the overall uncertainty in crowdclustering.

5   Pairwise Label Propagation

After the pairwise labels by experts are solicited, we aim to maximize the influence of experts. To this end, we develop a pairwise label propagation approach

---

[3] The querying error is the ratio of pairwise labels identified by experts as incorrect.

based on two important components, i.e., *constraint expansion* and *variational Bayesian inference*.

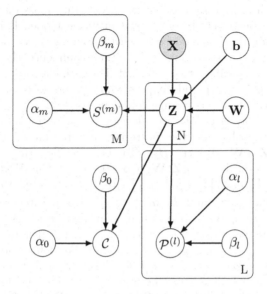

Fig. 3. Overall graphical model of BayesPDCM

5.1   Constraint Expansion

Since the expert label is a must-link or cannot-link constraint to a data pair $(\mathbf{x}_i, \mathbf{x}_j)$, we take advantage of their transitivity and entailment properties [23] to derive more constraints. For example, if $\mathbf{x}_i$ has a must-link to $\mathbf{x}_j$, and $\mathbf{x}_j$ has a must-link to $\mathbf{x}_k$, we can infer that $\mathbf{x}_i$ also has an implicit must-link to $\mathbf{x}_k$. Let $\mathcal{C}$ denote the constraint set expanded by transitivity and entailment, we use similarity search to further find a pseudo constraint set $\mathcal{P}$. For each constraint tuple $((\mathbf{x}_i, \mathbf{x}_j), relation_{ij})$ in $\mathcal{C}$, we retrieve a set of similar data pairs to $(\mathbf{x}_i, \mathbf{x}_j)$, from the instances not validated by experts. The similarity between data pairs is calculated by $\max(sim(i,k) \cdot sim(j,l), sim(i,l) \cdot sim(j,k))$, where $sim(i,k) = \exp(-\frac{\|\mathbf{x}_i - \mathbf{x}_k\|_2^2}{2\sigma^2})$ We regard it as the confidence of $(\mathbf{x}_k, \mathbf{x}_l)$ having the same constraint as $(\mathbf{x}_i, \mathbf{x}_j)$. To achieve an accurate pair retrieval, we filter out data pairs $(\mathbf{x}_k, \mathbf{x}_l)$ that has a low confidence.

5.2   Variational Bayesian Inference

The "golden" constraint set $\mathcal{C}$ and "pseudo" constraint set $\mathcal{P}$, are both exploited to guide the crowdclustering. To seamlessly incorporate them into the PDCM model, we propose a Bayesian variant of PDCM that takes expert input as priors for enforcing the constraints into the crowdclustering model (referred to as BayesPDCM). The overall graphical model is shown in Fig. 3. The novel part of BayesPDCM is that it adds dependencies between cluster assignments $\mathbf{Z}$ and the constraints derived from experts, $\mathcal{C}$ and $\mathcal{P}$. To enforce that the cluster assignments inferred by BayesPDCM respect the constraints in $\mathcal{C}$, we set the prior of the accuracy parameters α_0, β_0 as a "sharp" beta distribution specified by $Beta(50, 1)$, which imposes almost all the probability mass on $\alpha_0, \beta_0 = 1$. We regard $\mathcal{P}$ as soft constraints to our crowdclustering model. Particularly, recall that each constraint in $\mathcal{P}$ is associated with a confidence, we use the K-means algorithm to cluster the constraints in $\mathcal{P}$ by their confidence; the number of cluster L is determined by silhouette coefficient. For each constraint cluster $\mathcal{P}^{(l)}$, we assume it is labeled by a pseudo expert whose priors of accuracy parameters α_l, β_l are proportional to the average confidence of $\mathcal{P}^{(l)}$.

To optimize BayesPDCM, we develop a variational Bayesian inference approach to solve for the posterior distribution of the latent variables $h = \{\mathbf{Z}, \mathbf{W}, \mathbf{b}, \boldsymbol{\alpha}, \boldsymbol{\beta}\}$. To be specific, the variational distribution is factorized as $q(\mathbf{Z}, \mathbf{W}, \mathbf{b}, \boldsymbol{\alpha}, \boldsymbol{\beta}) = q(\mathbf{Z})q(\mathbf{W})q(\mathbf{b})q(\boldsymbol{\alpha})q(\boldsymbol{\beta}) = \prod_m [q(\alpha_m)q(\beta_m)] \prod_i q(z_i) \prod_{k,d} q(W_{kd}) \prod_k q(b_k)$, where $q(\alpha_m), q(\beta_m)$ are beta distributions $Beta(\tau_\alpha^1, \tau_\alpha^2)$ and $Beta(\tau_\beta^1, \tau_\beta^2), q(z_i)$ is a categorical distribution specified by $Cat(\eta_i)$, $q(W_{kd})$ follows Gaussian distribution $\mathcal{N}(\mu_{W_{ij}}, \sigma_{W_{ij}})$ and $q(b_k)$ follows Gaussian distribution $\mathcal{N}(\mu_{b_i}, \sigma_{b_i})$. Let θ denotes the variational parameters $\{\tau_\alpha^1, \tau_\alpha^2, \tau_\beta^1, \tau_\beta^2, \eta_i, \mu_{W_{ij}}, \sigma_{W_{ij}}, \mu_{b_i}, \sigma_{b_i}\}$. We define the variational distribution parameterized by θ as $q(h; \theta)$. The optimal θ can be obtained by minimizing the KL-divergence between $q(h; \theta)$ and the joint distribution $p(h, \mathbf{X}, \mathbf{S}, \mathcal{P}, \mathcal{C}) = p(\mathbf{S}, \mathcal{P}, \mathcal{C}|\mathbf{Z}, \boldsymbol{\alpha}, \boldsymbol{\beta})p(\mathbf{Z}|\mathbf{X}, \mathbf{W}, \mathbf{b})p_0(\mathbf{W}, \mathbf{b}, \boldsymbol{\alpha}, \boldsymbol{\beta})$. To optimize the KL-divergence, we use a gradient-based approach [17] to find its minimum.

6   Evaluation

We empirically evaluate the overall performance of our framework and respectively analyze its key components. The experiments show encouraging results of our framework on improving crowdclustering through expert validation.

6.1   Experimental Setup

Datasets. The experiments are conducted on six datasets widely used for cluster analysis. Dermatology, seeds and newthyroid datasets are acquired from UCI machine learning repository [15]; they have moderate number of features (5 to 35), data instances (200 to 400) and clusters (3 to 6). Face dataset [20] contains 640 face images of people taking 4 different poses. For an efficient cluster analysis, we use PCA to reduce its data dimensionality to 50. WebKB [4] is comprised of 1041 webpages hosted by universities, which can be categorized into 4 classes: course, faculty, project, student. Each webpage is represented by its top 200 frequently occurring words. MNIST [14] is a classic dataset for handwritten digits recognition. We sample 2000 images to be used in our experiments. The feature of each sample is the 50 PCA components of its pixel representation.

Generating Crowd Worker Labels. For the MNIST dataset, we use the crowd worker labels collected from real-world crowdsourcing platforms [16].

For the rest of our datasets, we follow the protocol of [1] to generate synthetic crowdsourced pairwise labels. Specifically, we simulate crowd workers with varying expertise (accuracy parameters α, β) from 0.5 to 0.95; each worker annotate 200 instance pairs randomly chosen from the whole dataset.

Evaluation Metric. The commonly-used normalized mutual information (NMI) metric [19] (ranging from 0 to 1) is adopted to evaluate the clustering performance. NMI measures the matchness between inferred cluster assignments and ground truth (higher the better).

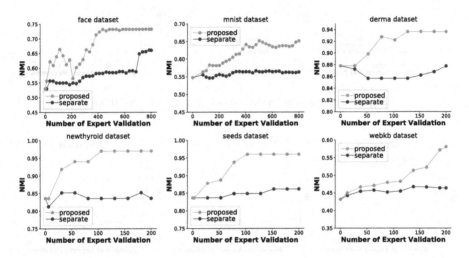

Fig. 4. Overall performance comparison of the proposed method against a baseline.

6.2 Experimental Results

Overall Performance. Our proposed framework is inherently an expert-guided clustering paradigm that learns from **combined** noisy crowd labels and golden expert answers. To verify this way of integrating the expert as an oracle, we compare our overall performance against a common "**separate**" strategy that regards experts as crowd workers. The results are shown in Fig. 4, in which our method consistently outperforms the separate strategy. The separate strategy can improve the clustering performance to some extent (e.g., on face dataset) because it feeds more data to crowdclustering model (PDCM) by adding expert answers to crowd labels. However, its improvement is very limited due to its lack of effective mechanism for seeking and incorporating expert validation. In contrast, our method achieve a significant NMI improvement since we actively select data to be validated and enforce expert-provided constraints to guide the clustering model.

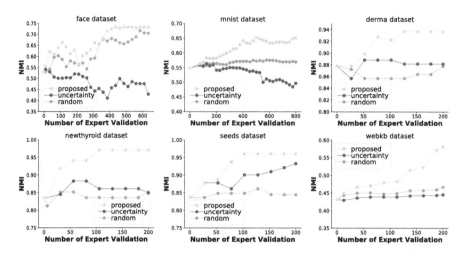

Fig. 5. Comparative evaluation of pair selection method.

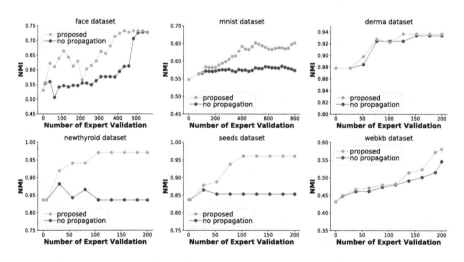

Fig. 6. The benefits of pairwise label propagation.

Pair Selection. We conduct a comparative evaluation of our pair selection method. The baseline selection methods are chosen as: **uncertainty** method that drives the pair selection by minimizing the decision uncertainties on instance pairs; **random** method that introduces no bias to favor any particular instance pair. As shown in Fig. 5, the proposed method makes more gain than baselines across all numbers of expert validation. Given an amount of expert validation budget (e.g., 200 queries), our pair selection can maximally achieve a 0.15 NMI improvement over baselines. This is due to that our dynamic two-phase strategy can select most informative instance pairs with respect to the learning progress of the PDCM model. Interestingly, we find that purely based on uncertainty method may sometimes harm the PDCM model (e.g., on face and mnist dataset). We speculate that uncertainty method tends to find too many data pairs lying far away from cluster centers which fits the noise (e.g., outliers) in the data, making the PDCM model hard to find a consistent and solid cluster structure.

Pairwise Label Propagation. To demonstrate the benefits of our pairwise label propagation, we remove the constraint expansion component and directly input the acquired expert labels to the BayesPDCM. We report the results in Fig. 6, which highlights the superiority of label propagation on all datasets. For most cases, the label propagation can effectively maximize the influence of expert and hence save the expert validation efforts. For example, on face dataset, to achieve a NMI goal of 0.65, our method only consumes about 100 expert queries, while it takes 500 queries without label propagation. The benefit of label propagation is mainly credited to our constraint expansion that augments the expert-provided constraints by deriving its deduced constraints and searching "pseudo" constraints (with high confidence similar to the golden ones).

7   Related Work

This paper focuses on crowdclustering, which learns a data cluster structure from crowdsourced labels. The labels often take the form of pairwise comparisons derived from keyword annotations [25], partial clusterings [7,16] or edge queries [1,21]. The seminal work on crowdclustering [7] proposes a generative model that takes each worker as a pairwise classifier. Their main drawback is requiring each instance to be included in at least one labeled pair, which incurs high labeling cost. To address this, Yi et al. [25] propose a metric learning method to generalize crowdclustering to the setting that requires only a small fraction of instances being labeled. Recently, researchers present different methods to model worker behaviors. Among them, most popular ones are DS model [1,16] and stochastic block model [21], mainly describing worker confusion patterns. Various mappings between data features and latent cluster assignments, such as logistic regression and GMM [1,16], are also studied in the context of crowd-clustering. Besides, there are some initial efforts on discovering clusters from crowdsourced labels beyond pairwise comparisons, e.g., partition labels [3] and triplet comparisons [22]. We adopt PDCM [1] as our base crowdclustering model because it is compact and flexible; our expert validation framework can be easily extended to other base models. Our work is also related to active learning [18]. The proposed pair selection is based on the important concepts mentioned in active learning community, i.e., uncertainty, representativeness and diversity. We extend their basic ideas to our crowdclustering scenario and organically combine them to design our dynamic two-phase pair selection. Particularly, the uncertainty introduced by data, model and worker in crowdclustering is comprehensively evaluated.

8   Conclusion and Future Work

This paper contributes a novel expert validation framework for quality control of crowdsourced clustering. The framework mainly features a pair selection and a pairwise label propagation. To select valuable instance pairs, we adapt the widely-used uncertainty sampling approach to our crowdclustering task, and systematically combine it with the representativeness and diversity measure to maximize the gain of the pair selection. The expert labels are propagated via the constraint expansion method and integrated into our crowdclustering model through prior enforcing and variational Bayesian inference techniques. Experiments show the effectiveness of our approach on actively seeking and seamlessly incorporating expert validations. Interesting avenues for future work include exploring sequential Bayesian updating techniques to online update the BayesPDCM model for a more efficient computation in the large-scale crowd-clustering setting.

References

1. Chang, Y., Chen, J., Cho, M., Castaldi, P., Silverman, E., Dy, J.: Clustering from multiple uncertain experts. In: AISTATS, pp. 28–36 (2017)
2. Charikar, M., Chekuri, C., Feder, T., Motwani, R.: Incremental clustering and dynamic information retrieval. SIAM J. Comput. 33(6), 1417–1440 (2004)
3. Chen, J., Chang, Y., Castaldi, P., Cho, M., Hobbs, B., Dy, J.: Crowdclustering with partition labels. In: AISTATS, pp. 1127–1136 (2018)
4. CMU: Webkb dataset (1998). http://www.cs.cmu.edu/~webkb/
5. Dawid, A.P., Skene, A.M.: Maximum likelihood estimation of observer error-rates using the EM algorithm. J. Roy. Stat. Soc. 28(1), 20–28 (1979)
6. Frigui, H., Krishnapuram, R.: A robust competitive clustering algorithm with applications in computer vision. IEEE TPAMI 5, 450–465 (1999)
7. Gomes, R.G., Welinder, P., Krause, A., Perona, P.: Crowdclustering. In: NeurIPS, pp. 558–566 (2011)
8. Hochbaum, D.S., Shmoys, D.B.: A best possible heuristic for the k-center problem. Math. Oper. Res. 10(2), 180–184 (1985)
9. Hu, Q., He, Q., Huang, H., Chiew, K., Liu, Z.: Learning from crowds under experts' supervision. In: Tseng, V.S., Ho, T.B., Zhou, Z.-H., Chen, A.L.P., Kao, H.-Y. (eds.) PAKDD 2014. LNCS (LNAI), vol. 8443, pp. 200–211. Springer, Cham (2014). https://doi.org/10.1007/978-3-319-06608-0_17
10. Hung, N.Q.V., Thang, D.C., Weidlich, M., Aberer, K.: Minimizing efforts in validating crowd answers. In: SIGMOD, pp. 999–1014 (2015)
11. Kajino, H., Tsuboi, Y., Sato, I., Kashima, H.: Learning from crowds and experts. In: AAAI Workshop on Human Computation (2012)
12. Kollios, G., Gunopulos, D., Koudas, N., Berchtold, S.: Efficient biased sampling for approximate clustering and outlier detection in large data sets. IEEE TKDE 15(5), 1170–1187 (2003)
13. Krause, A., Perona, P., Gomes, R.G.: Discriminative clustering by regularized information maximization. In: NeurIPS, pp. 775–783 (2010)
14. LeCun, Y.: MNIST (1998). http://yann.lecun.com/exdb/mnist/
15. Lichman, M., et al.: UCI machine learning repository (2013)

16. Luo, Y., Tian, T., Shi, J., Zhu, J., Zhang, B.: Semi-crowdsourced clustering with deep generative models. In: NeurIPS, pp. 3212–3222 (2018)
17. Maclaurin, D., Duvenaud, D., Adams, R.P.: Autograd: reverse-mode differentiation of native python. In: ICML workshop on Automatic Machine Learning (2015)
18. Settles, B.: Active learning literature survey. Technical report, University of Wisconsin-Madison (2009)
19. Strehl, A., Ghosh, J.: Cluster ensembles–a knowledge reuse framework for combining multiple partitions. JMLR **3**, 583–617 (2002)
20. Tom, M.: Face dataset (1996). http://www.cs.cmu.edu/~tom/faces.html
21. Vinayak, R., Hassibi, B.: caltech.edu: Clustering by comparison: stochastic block model for inference in crowdsourcing. In: ICML Workshop on Machine Learning and Crowdsourcing (2016)
22. Vinayak, R.K., Hassibi, B.: Crowdsourced clustering: querying edges vs triangles. In: NeurIPS, pp. 1316–1324 (2016)
23. Wagstaff, K., Cardie, C., Rogers, S., Schrödl, S., et al.: Constrained k-means clustering with background knowledge. In: ICML, vol. 1, pp. 577–584 (2001)
24. Wang, X., Zhai, J.: Learning with Uncertainty. CRC Press, Boca Raton (2016)
25. Yi, J., Jin, R., Jain, S., Yang, T., Jain, A.K.: Semi-crowdsourced clustering: deneralizing crowd labeling by robust distance metric learning. In: NeurIPS, pp. 1772–1780 (2012)

A Label Embedding Method
for Multi-label Classification via
Exploiting Local Label Correlations

Xidong Wang, Jun Li, and Jianhua Xu[✉]

School of Computer Science and Technology, Nanjing Normal University,
Nanjing 210023, Jiangsu, China
1529980236@qq.com, {lijuncst,xujianhua}@njnu.edu.cn

Abstract. Multi-label learning has attracted more attention recently due to many real-world applications (e.g., text categorization and scene annotation). As the dimensionality of label space increases, it becomes more difficult to deal with this kind of applications. Therefore, dimensionality reduction techniques originally for feature space is also applied to label space, one of which is label embedding strategy which converts the high-dimensional label space into a low-dimensional reduced one. So far, existing label embedding methods mainly investigate the global recoverability between original labels and reduced labels, dependency between original features and reduced labels, or both. It is widely recognized that local label correlations could improve multi-label classification performance effectively. In this paper, we construct a trace ratio minimization problem as a novel label embedding criterion, which not only includes the global label recoverability and dependency, but also exploits the local label correlations as a local recoverability factor. Experiments on four benchmark data sets with more than 100 labels demonstrate that our proposed method is superior to four state-of-the-art techniques, according to two performance metrics for high-dimensional label space.

Keywords: Multi-label learning · Dimensionality reduction · Global recoverability · Local recoverability · Label correlation

1 Introduction

Traditional single-label learning deals with binary and multi-class supervised classification problems where each instance is only associated with one of several class labels [2]. When some instances are annotated by multiple labels at the same time, the multi-class problem is extended into a multi-label one, which induces a novel learning paradigm: multi-label learning [8]. Nowadays, there are many multi-label learning applications, for example, text categorization, scene semantic annotation, music emotion classification and so on [8]. Therefore, more

This work was supported by the Natural Science Foundation of China (NSFC) under grants 61273246 and 61703096.

attention has been paid to multi-label learning recently in machine learning, pattern recognition, data mining, neural networks, deep learning and data science.

In single-label learning, besides large instance size, we usually have to cope with the high-dimensional feature problem. However, the high-dimensional label problem also occurs frequently in many multi-label applications. In this case, for multi-label learning tasks, the widely-used dimensionality reduction strategy for feature space originally in single-label learning is also applied to label space, to reduce computational burdens for time and memory [8].

With label space dimensionality reduction, multi-label classification techniques consist of two more complicated training and predicting stages [3,8]. The training procedure is, (a) to design an encoding operator and/or a decoding one, (b) to reduce the high-dimensional binary label space into a low-dimensional binary or real sub-space via the encoding operator, and (c) to train a multi-output regressor or a multi-label classifier in the reduced label space. In the predicting procedure, for a test instance, its low-dimensional predicted label vector is estimated using a trained regressor or classifier, and then is recovered back to a high-dimensional binary label vector through the designed decoding operator. In this case, how to design the encoding and decoding operators becomes a crucial issue for this kind of multi-label classification algorithms.

Existing label space dimensionality reduction techniques are divided into two categories: label selection and label embedding [19], which are similar to feature selection and feature extraction in feature space. However, the fatal difference is that an additional reconstruction procedure is needed for label space.

In label selection methods, the encoding operator is to select an informative label subset directly from the original label set, which is realized by solving an L1 and L1,2 regularized least square regression in [1], and using an associated rule mining algorithm in [4]. Their decoding operators are based on a two-stage maximum posterior probability model in [1], and an iterative label search way in [4]. Such two methods would result in a reduced multi-label classification problem, which means that some label correlation information still remains. Additionally, it is observed that their decoding procedures are relatively complicated.

Label embedding methods design their linear and nonlinear encoding and decoding operators simultaneously [16,17]. In this study, we focus on our research work on linear label embedding methods, which usually could satisfy some particular requirements, e.g., orthonormal projection directions [17] and orthonormal reduced labels [15]. The existing methods mainly investigate the recoverability from reduced labels to original ones, the dependency between reduced labels and original features, or both, which are categorized into three sub-groups:

(a) Recoverability-based methods are only to maximize recoverability between original labels and reduced labels, which essentially exploit global label correlations in their encoding and decoding operators. Their representatives are based on compressive sensing (ML-CS) [9], principal component analysis (PLST) [17], binary matrix decomposition (MLC-BMaD) [11], compressed labeling (ML-CL) [22], binary linear compression (BILC) [23]. Specially, PLST could create a set of orthonormal projection directions.

(b) Dependency-based methods are to maximize correlations between reduced labels and original features. In [5], as a byproduct, the conventional canonical correlation analysis (CCA) is generalized to construct its orthonormal version (OCCA). The Hilbert-Schmidt independence criterion (HSIC) [7] is to construct a label compression coding approach (LCCMD) in [3]. All projection directions from such two techniques satisfy the orthonormal constraints.

(c) Hybrid methods are to maximize both recoverability and dependency simultaneously. When the aforementioned OCCA is combined with PLST, a conditional PLST is proposed (i.e., CPLST) in [5]. Essentially, dependence maximization based on label space dimension reduction (DMLR) is to integrate PLST with LCCMD linearly [20]. These two techniques could provide orthonormal projection directions. In [15], a feature-aware implicit label space coding method (FaIE) is proposed via combining PLST and OCCA linearly with orthonormal projection labels, which then is extended into several variants in [14].

In [5,11,14,15,17,20,23], the recoverability is described using the entire original label data (i.e., global label correlations). However, it is widely recognized that the local label correlations are also very helpful to boost multi-label classification performance [10,21]. In this paper, we integrate global and local recoverability, and dependency to build a new label encoding method, where the local consistency in [18] originally for feature space is used to characterize local label recoverability for label space. Experiments on four data sets show that our proposed method performs the best, compared with four existing techniques in [5,11,15,17].

2   Preliminaries

Let a multi-label training set of size N be

$$\{(\mathbf{x}_1, \mathbf{y}_1), ..., (\mathbf{x}_i, \mathbf{y}_i), ..., (\mathbf{x}_N, \mathbf{y}_N)\} \tag{1}$$

where for the i-th instance, the $\mathbf{x}_i \in \mathcal{R}^D$ and $\mathbf{y}_i \in \{0,1\}^L$ denote its D-dimensional feature column vector and L-dimensional binary label column vector. For convenience of formula representation, we also define the feature and label matrices

$$\mathbf{X} = [\mathbf{x}_1, ..., \mathbf{x}_N]; \ \mathbf{Y} = [\mathbf{y}_1, ..., \mathbf{y}_N]. \tag{2}$$

Using a centering matrix of size $N \times N$: $\mathbf{H}^{(N)} = \mathbf{I} - \mathbf{1}\mathbf{1}^T/N$, where $\mathbf{I}$ indicates the identity matrix of size $N \times N$ and $\mathbf{1}$ is the column vector with N one elements, we build two centered feature and label matrices

$$\bar{\mathbf{X}} = \mathbf{X}\mathbf{H}^{(N)}; \ \bar{\mathbf{Y}} = \mathbf{Y}\mathbf{H}^{(N)}. \tag{3}$$

The traditional multi-label classification is to learn a classifier $\mathbf{y} = f(\mathbf{x})$: $\mathcal{R}^D \to \{1, 0\}^L$, according to the above training set (1), which then is used to predict the binary label vectors for unseen instances [8].

When a proper label embedding method is applied to multi-label classification, we need to find out a label encoding operator ϕ and its decoding operator ϕ^{-1}. Via ϕ, each original label binary vector $\mathbf{y} \in \{1, 0\}^L$ is transformed into a l-dimensional real label vector $\mathbf{z} \in \mathcal{R}^l$ or binary one $\mathbf{z} \in \{1, 0\}^l$, where $l < L$. Then a proper classifier ($\mathbf{z} \in \{1, 0\}^l$) or regressor ($\mathbf{z} \in \mathcal{R}^l$) $\mathbf{z} = g(\mathbf{x})$ is trained, which is used to estimate a predicted label $\hat{\mathbf{z}}$ for a testing instance $\mathbf{x}$. According to ϕ^{-1}, we recover $\hat{\mathbf{z}}$ back to an L-dimensional binary predicted label vector $\hat{\mathbf{y}}$. When $\bar{\mathbf{Y}}$ is used, a de-centering step is added to the decoding operator. Additionally, for most of linear embedding methods, their encoding and decoding operators are described as two matrices, i.e., $\phi = \mathbf{P} \in \mathcal{R}^{L \times l}$ and $\phi^{-1} = \mathbf{Q} \in \mathcal{R}^{l \times L}$. In this case, we formally have the following reduced and reconstructed label matrices

$$\mathbf{Z} = [\mathbf{z}_1, ..., \mathbf{z}_N] = \mathbf{P}^T \bar{\mathbf{Y}}; \; \hat{\mathbf{Y}} = [\hat{\mathbf{y}}_1, ..., \hat{\mathbf{y}}_N] = \mathbf{Q}^T \mathbf{Z}. \tag{4}$$

3   A Novel Label Embedding Approach via Exploiting Local Label Correlations

In this section, we review maximum global recoverability and dependency criteria which will be involved in our new label embedding objective first. Then a local label correlation measure is introduced. Finally, we define a trace ratio minimization problem and design its efficient solution method to construct our novel label embedding technique.

3.1   Maximum Global Recoverability Criterion

The maximum global recoverability is to minimize the loss between the original label matrix $\bar{\mathbf{Y}}$ and reconstructed one $\mathbf{Q}^T \mathbf{Z} = \mathbf{Q}^T \mathbf{P}^T \bar{\mathbf{Y}}$. Nowadays, the widely-used form is formulated as follows

$$\min \left\| \bar{\mathbf{Y}} - \mathbf{Q}^T \mathbf{P}^T \bar{\mathbf{Y}} \right\|_F^2 = \left\| \bar{\mathbf{Y}} - \mathbf{P} \mathbf{P}^T \bar{\mathbf{Y}} \right\|_F^2, \text{ s.t. } \mathbf{P}^T \mathbf{P} = \mathbf{I} \tag{5}$$

where $\|\cdot\|_F$ indicates the Frobenius norm of matrix, and $\mathbf{Q} = \mathbf{P}^T$ due to $\mathbf{P}^T \mathbf{P} = \mathbf{I}$, which can be converted into

$$\max tr(\mathbf{P}^T \bar{\mathbf{Y}} \bar{\mathbf{Y}}^T \mathbf{P}), \text{ s.t. } \mathbf{P}^T \mathbf{P} = \mathbf{I} \tag{6}$$

where "tr" is the trace operation of matrix. This objective in (6) is also regarded to maximize variances of matrix, as in principal component analysis [2].

In PLST [17], the above (6) is solved to construct a label encoding operator $\mathbf{P}$. In CPLST [5], DMLR [20] and FaIE [15] and its variants [14], the objective function in (5) is to characterize the global recoverability only.

3.2 Maximum Dependency Criterion

The maximum dependency is to maximize the correlations between the original feature matrix $\bar{\mathbf{X}}$ and the reduced label matrix $\mathbf{Z}$. Hilbert Schmidt Independence criterion (HSIC) [7] is a non-parametric dependence index based on kernels and reproduced kernel Hilbert space to measure dependence between two sets of random variables. When the linear kernel is applied in both feature space and reduced label space, the empirical estimator of HSIC is

$$\text{HSIC} = tr(\bar{\mathbf{X}}^T \bar{\mathbf{X}} \mathbf{Z}^T \mathbf{Z}) = tr(\mathbf{P}^T \bar{\mathbf{Y}} \bar{\mathbf{X}}^T \bar{\mathbf{X}} \bar{\mathbf{Y}}^T \mathbf{P}). \tag{7}$$

With the orthonormal constraints $\mathbf{P}^T \mathbf{P} = \mathbf{I}$, we constructed a label embedding method (LCCMD) [3] via the following optimization problem

$$\max tr(\mathbf{P}^T \bar{\mathbf{Y}} \bar{\mathbf{X}}^T \bar{\mathbf{X}} \bar{\mathbf{Y}}^T \mathbf{P}), \text{ s.t. } \mathbf{P}^T \mathbf{P} = \mathbf{I}. \tag{8}$$

In DMLR [20], the above HSIC (7) is used to describe the dependency between original feature matrix $\bar{\mathbf{X}}$ and reduced label matrix $\mathbf{Z}$.

3.3 Minimum Local Recoverability Criterion

In [18], the local information in feature space is investigated to improve the unsupervised dimensionality reduction performance. In this paper, we apply its idea to characterize the local label correlations. For the i-th instance label vector $\mathbf{y}_i$, we search for its k-nearest neighbours in label space, denoted by the index set $\mathcal{N}_i$, and then build a corresponding label sub-matrix of size $L \times (k+1)$

$$\mathbf{Y}_i = [\bar{\mathbf{y}}_i, \bar{\mathbf{y}}_j | j \in \mathcal{N}_i]. \tag{9}$$

Similarly, we could achieve a centered local label sub-matrix

$$\bar{\mathbf{Y}}_i = \mathbf{Y}_i \mathbf{H}^{(k+1)} \tag{10}$$

and its local covariance sub-matrix

$$\mathbf{C}_i = \bar{\mathbf{Y}}_i \bar{\mathbf{Y}}_i^T. \tag{11}$$

Then an overall local covariance matrix is defined as

$$\mathbf{C} = \sum_{i=1}^{N} \mathbf{C}_i = \sum_{i=1}^{N} \bar{\mathbf{Y}}_i \bar{\mathbf{Y}}_i^T. \tag{12}$$

Finally, we construct the following optimization task which minimizes the variances of matrix $\mathbf{C}$

$$\min tr\left(\mathbf{P}^T \left(\sum_{i=1}^{N} \bar{\mathbf{Y}}_i \bar{\mathbf{Y}}_i^T\right) \mathbf{P}\right), \text{ s.t. } \mathbf{P}^T \mathbf{P} = \mathbf{I} \tag{13}$$

to minimize the local label correlations, in this paper.

Algorithm 1. Pseudo-code of our ML-mLV

TRAINING STAGE:

Input:

 $\mathbf{X}$ and $\mathbf{Y}$: feature and label matrices.

 l: reduced dimensionality.

 ϵ: stopping condition.

Process:

 To centralize feature and label matrices to obtain $\bar{\mathbf{X}}$ and $\bar{\mathbf{Y}}$.

 To construct two proxy matrices ($\mathbf{S}_L$ and $\mathbf{S}_G$) using (15).

 To initialize the $\mathbf{P}_0$ randomly and $t = 1$.

  To execute the following sub-steps until $|\rho_t - \rho_{t-1}| \leq \epsilon$ is satisfied.

   To estimate ρ_t according to (17).

   To achieve $\mathbf{P}_t$ via (18).

   $t \leftarrow t + 1$.

 To determine encoding matrix $\mathbf{P}$ with the l columns of $\mathbf{P}_t$.

 To calculate reduced label matrix $\mathbf{Z} = \mathbf{P}^T \bar{\mathbf{Y}}$.

 To learn a regressor $\mathbf{z} = g(\mathbf{x})$.

Output:

 $\mathbf{Q} = \mathbf{P}^T$: a decoding matrix of size $l \times L$.

 $g(\mathbf{x})$: a trained regressor.

TESTING STAGE:

Input:

 $\mathbf{x}$: a testing instance vector.

Process:

 To calculate the l-dimensional real label vector $\mathbf{z} = g(\mathbf{x})$.

 To reconstruct the L-dimensional real label vector $\hat{\mathbf{y}} = \mathbf{Q}^T \mathbf{z}$.

 To de-centering operation to achieve a real label vector $\tilde{\mathbf{y}}$.

 To detect a binary label vector $\mathbf{y}$ using round operation on $\tilde{\mathbf{y}}$.

Output:

 $\mathbf{y}$: a predicted high-dimensional binary label vector.

3.4 Novel Label Embedding Criterion

In the above sub-sections, we introduce two existing global label embedding criteria (global recoverability and dependency-based one), and a new local label recoverability criterion. Here, we define a trace ratio optimization as follows

$$\min \frac{tr\left(\mathbf{P}^T\left(\sum_{i=1}^N \bar{\mathbf{Y}}_i \bar{\mathbf{Y}}_i^T\right)\mathbf{P}\right)}{tr\left(\mathbf{P}^T \bar{\mathbf{Y}}\bar{\mathbf{Y}}^T\mathbf{P}\right)+\mu tr\left(\mathbf{P}^T \bar{\mathbf{Y}}\bar{\mathbf{X}}^T\bar{\mathbf{X}}\bar{\mathbf{Y}}^T\mathbf{P}\right)}, \text{ s.t. } \mathbf{P}^T\mathbf{P} = \mathbf{I} \tag{14}$$

where μ is a tunable parameter to control the balance of two terms in the denominator. Our objective is to maximize the global label recoverability and dependency, and minimize the local label recoverability.

After defining two proxy matrices

$$\mathbf{S}_L = \sum_{i=1}^N \bar{\mathbf{Y}}_i \bar{\mathbf{Y}}_i^T; \ \mathbf{S}_G = \bar{\mathbf{Y}}\bar{\mathbf{Y}}^T + \mu \bar{\mathbf{Y}}\bar{\mathbf{X}}^T\bar{\mathbf{X}}\bar{\mathbf{Y}}^T \tag{15}$$

Table 1. Four benchmark multi-label data sets used in our experiments.

Dataset	Train	Test	Features	Labels	Cardinality
Bibtex	4480	2515	1836	159	2.40
Cal500	300	202	68	174	2.64
EUR-Lex1	15539	3809	5000	412	1.92
RCV1v2(s1)	3000	3000	47236	101	2.88

Table 2. The number of wins for each method and metric across four data sets

Metric	PLST	CPLST	MLC-BMaD	FaIE	ML-mLV
Precision@1	0	1	3	1	35
Precision@3	1	3	0	1	35
Precision@5	1	1	0	1	37
DCG@1	0	1	3	1	35
DCG@3	1	1	10	11	17
DCG@5	1	2	0	2	35
Total wins	4	9	16	17	194

we have a concise form for (14)

$$\min \frac{tr(\mathbf{P}^T\mathbf{S}_L\mathbf{P})}{tr(\mathbf{P}^T\mathbf{S}_G\mathbf{P})}, \text{ s.t. } \mathbf{P}^T\mathbf{P} = \mathbf{I}. \tag{16}$$

Note that this problem could not converted into an eigenvalue decomposition task.

3.5   Efficient Solution Method

In this sub-section, we apply an efficient solution used in [13,18] to solve (16). After the initial orthonormal $\mathbf{P}_0$ is created randomly, we first compute the trace ratio ρ_t $(t = 1, 2, 3, ...)$

$$\rho_t = \frac{tr\left(\mathbf{P}_{t-1}^T\mathbf{S}_L\mathbf{P}_{t-1}\right)}{tr\left(\mathbf{P}_{t-1}^T\mathbf{S}_G\mathbf{P}_{t-1}\right)} \tag{17}$$

and then update the $\mathbf{P}_t$ via solving the following minimization problem

$$\min tr\left(\mathbf{P}_t^T(\mathbf{S}_L - \rho_t\mathbf{S}_G)\mathbf{P}_t\right), \text{ s.t. } \mathbf{P}_t^T\mathbf{P}_t = \mathbf{I} \tag{18}$$

according to eigenvalue decomposition. When $|\rho_t - \rho_{t-1}| \leq \epsilon$ (a pre-defined threshold), we stop the iterative procedure and construct $\mathbf{P}$ with the l column vectors corresponding to the l smallest eigenvalues from $\mathbf{P}_t$. It has been illustrated that this iterative process is convergent [18].

Finally, we summarize the training and testing procedures of our proposed approach in **Algorithms** 1, which is referred to as a novel label embedding algorithm via maximizing global recoverability and dependency, and minimizing local variances, or simply ML-mLV.

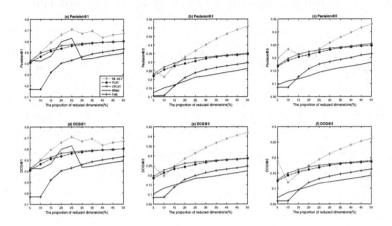

Fig. 1. Two metrics (at $n = 1$, 3 and 5) from five methods on Bibtex

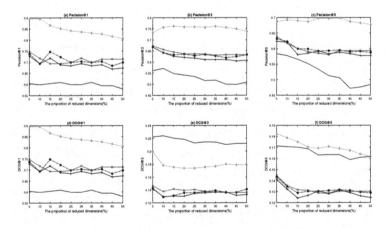

Fig. 2. Two metrics (at $n = 1$, 3 and 5) from five methods on CAL500

4   Experiments

In this section, we compare our proposed ML-mLV with PLST [17], CPLST [5], MLC-BMaD [11] and FaIE [15] experimentally, on four data sets and two metrics.

4.1   Benchmark Data Sets and Evaluation Metrics

In this study, we select four multi-label benchmark data sets from[1], as shown in
Table 1. These sets have been split into training and testing sub-sets. In Table 1,
some useful statistics are provided, e.g., the size of training and testing sets, the
dimensionality of features, the dimensionality of labels, and cardinality.

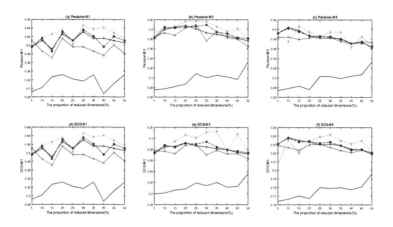

Fig. 3. Two metrics (at $n = 1$, 3 and 5) from five methods on EURLex

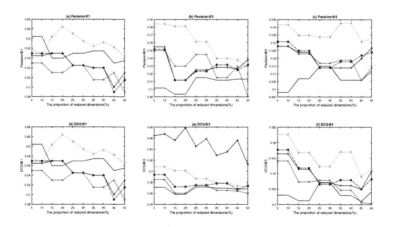

Fig. 4. Two metrics (at $n = 1$, 3 and 5) from five methods on RCV1v2(s1)

---

[1]   http://mulan.sourceforge.net/datasets-mlc.html.

Two evaluation metrics for large-scale label sets are adopted as our evaluation indexes [12]: precision@n and (DisCounted Gain) DCG@n ($n = 1, 2, 3, ...$). For a testing instance $\mathbf{x}$, its ground label vector is $\mathbf{y} = [y_1, ..., y_i, ...y_L]^T$ and predicted function values $\hat{\mathbf{y}} = [\hat{y}_1, ..., \hat{y}_i, ..., \hat{y}_L]^T$, these two metrics are defined as follows.

$$\text{Precision@}n = \frac{1}{n} \sum_{i \in rank_n(\hat{\mathbf{y}})} y_i; \ \text{DCG@}n = \frac{1}{n} \sum_{i \in rank_n(\hat{\mathbf{y}})} \frac{y_i}{\log_2(i+1)} \quad (19)$$

where $rank_n(\hat{\mathbf{y}})$ returns the top n label indexes of $\hat{\mathbf{y}}$. Finally, their average values are calculated across all testing instances.

4.2 Compared Methods and Experimental Settings

In our experiments, we validate our method ML-mLV and four existing methods: PLST [17], CPLST [5], MLC-BMaD [11] and FaIE [15]. Here, the linear regressor is chosen as our baseline for all compared methods.

In order to investigate how the reduced dimensionality l affects the classification performance, we consider different reduced proportions from 5% to 50% of original label dimensionality L with a step 5%. Additionally, $k = 10$ for (9), $\mu = 1$ for (15), and $n = 1$, 3 and 5 for (19).

4.3 Performance Evaluation and Analysis

In this sub-section, we compare our ML-mLV with existing four methods (PLST, CPLST, MLC-BMaD and FaIE), as shown in Figs. 1, 2, 3 and 4, where each metric is regarded as a function of the propositions of reduced dimensionality from 5% to 50% with an increment 5% for each data set.

From these four figures, it is observed that at most of proportions of reduced dimensionality, our method ML-mLV performs the best, compared with the other approaches. In order to evaluate these compared methods comprehensively, we introduce the "Win" index in [6], which denotes how many times each method achieves the best metric values for all data sets and all proportions, as shown in Table 2. It is found out that our ML-mLV achieve 194 wins among total 240 ones, which is greater than those of the other approaches.

Finally, given 30% reduced dimensionality in label space, we list two metrics with $n = 1$, 3 and 5 for five methods and four data sets, as shown in Table 3. Our ML-mLV achieves 22 best metric values and 2 second top ones. MLC-BMaD and FaIT obtain one best DCG@3 on CAL500 and RCV1v2(s1), respectively.

The above experimental results demonstrate that our proposed ML-mLV performs the best, compared with four existing state-of-the-art techniques, according to two performance metrics for large-scale label sets.

Table 3. Two metrics from five methods and four data sets with 30% reduced labels

Metric	PLST	CPLST	MLC-BMaD	FaIE	ML-mLV
Bibtex					
Precision@1	0.5734	0.5817	0.4354	0.4596	**0.6713**
Precision@3	0.3207	0.3287	0.2205	0.2562	**0.4032**
Precision@5	0.2308	0.2367	0.1531	0.1847	**0.2581**
DCG@1	0.5734	0.5817	0.4354	0.4354	**0.6713**
DCG@3	0.2664	0.2721	0.1891	0.2125	**0.3438**
DCG@5	0.1757	0.1795	0.1220	0.1402	**0.2130**
CAL500					
Precision@1	0.7030	0.7178	0.5990	0.6931	**0.8366**
Precision@3	0.6337	0.6353	0.5165	0.6106	**0.7574**
Precision@5	0.5931	0.5802	0.5129	0.5832	**0.6970**
DCG@1	0.7030	0.7178	0.5990	0.6931	**0.8366**
DCG@3	0.1374	0.1377	**0.2146**	0.1356	0.1757
DCG@5	0.1309	0.1282	0.1636	0.1284	**0.1708**
EURLex					
Precision@1	0.4356	0.3960	0.3168	0.4307	**0.4459**
Precision@3	0.4092	0.3762	0.3003	0.3927	**0.4159**
Precision@5	0.3851	0.3832	0.303	0.3822	**0.3946**
DCG@1	0.4356	0.3960	0.3168	0.4307	**0.4459**
DCG@3	0.2939	0.2701	0.2142	0.2846	**0.3112**
DCG@5	0.2330	0.2267	0.1790	0.2306	**0.2415**
RCV1v2(s1)					
Precision@1	0.5446	0.5446	0.5693	0.5446	**0.5945**
Precision@3	0.5231	0.5446	0.5149	0.5248	**0.5611**
Precision@5	0.5139	0.5168	0.5149	0.5139	**0.5337**
DCG@1	0.5446	0.5446	0.5693	0.5446	**0.5945**
DCG@3	0.3746	0.3740	0.3713	**0.4589**	0.3861
DCG@5	0.3065	0.3078	0.3071	0.3070	**0.3122**

5   Conclusions

In this paper, we proposed a novel label embedding method for multi-label classification, which not only investigates the global label recoverability and dependency, but also exploits the local label recoverability, resulting into a more competitive approach for multi-label applications. Our detailed experimental results demonstrate that our proposed technique is superior to four state-of-the-art methods, according to precision and discounted gain (at $n = 1$, 3 and 5). In the future, we will validate our proposed method with more benchmark data sets.

References

1. Balasubramanian, K., Lebanon, G.: The landmark selection method for multiple output prediction. In: 29th International Conference on Machine Learning, pp. 983–990. OmniPress, Madison (2012)
2. Bishop, C.M.: Pattern Recognition and Machine Learning. Springer, Singapore (2006)
3. Cao, L., Xu, J.: A label compression coding approach through maximizing dependence between features and labels for multi-label classification. In: 27th International Joint Conference on Neural Networks, pp. 1–8. IEEE Press, New York (2015)
4. Charte, F., Rivera, A.J., del Jesus, M.J., Herrera, F.: LI-MLC: a label inference methodology for addressing high dimensionality in the label space for multilabel classification. IEEE Trans. Neural Netw. Learn. Syst. 25(10), 1842–1854 (2014)
5. Chen, Y., Lin, H.: Feature-aware label space dimension reduction for multi-label classification. In: 26th Annual Conference on Neural Information Processing Systems, pp. 1538–1546. MIT Press, Cambridge (2012)
6. Demsar, J.: Statistical comparisons of classifiers over multiple data sets. J. Mach. Learn. Res. 7, 1–30 (2006)
7. Gretton, A., Bousquet, O., Smola, A., Schölkopf, B.: Measuring statistical dependence with Hilbert-Schmidt norms. In: Jain, S., Simon, H.U., Tomita, E. (eds.) ALT 2005. LNCS (LNAI), vol. 3734, pp. 63–77. Springer, Heidelberg (2005). https://doi.org/10.1007/11564089_7
8. Herrera, F., Charte, F., Rivera, A.J., del Jesus, M.J.: Multilabel Classification: Problem Analysis, Metrics and Techniques. Springer, Switzerland (2016). https://doi.org/10.1007/978-3-319-41111-8
9. Hsu, D., Kakade, S., Langford, J., Zhang, T.: Multi-label prediction via compressed sensing. In: 23rd Annual Conference on Neural Information Processing Systems, pp. 772–780. MIT Press, Cambridge (2009)
10. Huang, S.J., Zhou, Z.H.: Multi-label learning by exploting label correlations locally. In: 26th AAAI Conference on Artificial Intelligence, pp. 949–955. AAAI Press, New York (2012)
11. Wicker, J., Pfahringer, B., Kramer, S.: Multi-label classification using Boolean matrix decomposition. In: 27th Annual ACM Symposium on Applied Computing, pp. 179–186. ACM Press, New York (2012)
12. Jain, H., Prabhu, Y., Varma, M.: Extreme multi-label loss functions for recommendation, tagging, ranking and other missing label applications. In: 22nd ACM SIGKDD Conference on Knowledge Discovery and Data Mining, pp. 435–944. ACM Press, New York (2016)
13. Jia, Y., Nie, F., Zhang, C.: Trace ratio problem revisited. IEEE Trans. Neural Netw. 20(4), 729–735 (2009)
14. Lin, Z., Ding, G., Han, J., Shao, L.: End-to-end feature-aware label space encoding for multilabel classification with many classes. IEEE Trans. Neural Netw. Learn. Syst. 29(6), 2472–2487 (2018)
15. Lin, Z., Ding, G., Hu, M., Wang, J.: Multi-label classification via feature-aware implicit label space encoding. In: 31st International Conference on Machine Learning, pp. 325–333. JMLR Website (2014)
16. Luo, J., Cao, L., Xu, J.: A non-linear label compression coding method based on five-layer auto-encoder for multi-label classification. In: Hirose, A., Ozawa, S., Doya, K., Ikeda, K., Lee, M., Liu, D. (eds.) ICONIP 2016. LNCS, vol. 9949, pp. 415–424. Springer, Cham (2016). https://doi.org/10.1007/978-3-319-46675-0_45

17. Tai, F., Lin, H.: Multilabel classification with principal label space transformation. Neural Comput. **24**(9), 2508–2542 (2012)
18. Wang, H., Nie, F., Huang, H.: Globally and locally consistent unsupervised projection. In: 28th AAAI Conference on Artificial Intelligence, pp. 1328–1333. AAAI Press, New York (2014)
19. Xu, J., Mao, Z.H.: Multilabel feature extraction algorithm via maximizing approximated and symmetrized normalized cross-covariance operator. IEEE Trans. Cybern. 1–14 (2019, in press)
20. Zhang, J.J., Feng, M., Wang, H., Li, X.: Depedence maximization based label space dimension reduction for multi-label classification. Eng. Appl. Artif. Intell. **45**, 453–463 (2015)
21. Zhang, M.L., Zhang, K.: Multi-label learning by exploiting label dependency. In: 16th ACM SIGKDD Conference on Knowledge Discovery and Data Mining, pp. 999–1007. ACM Press, New York (2010)
22. Zhou, T., Tao, D., Wu, X.: Compressed labeling on distilled labelsets for multi-label learning. Mach. Learn. **88**(1–2), 69–126 (2012)
23. Zhou, W.J., Yu, Y., Zhang, M.L.: Binary linear compression for multi-label classification. In: 26th International Joint Conference on Artificial Intelligence, pp. 3546–3552. IJCAI Website (2017)

The Inadequacy of Entropy-Based Ransomware Detection

Timothy McIntosh[1(✉)], Julian Jang-Jaccard[1], Paul Watters[2], and Teo Susnjak[1]

[1] Massey University, Auckland 0632, New Zealand
t.mcintosh@massey.ac.nz
[2] La Trobe University, Bundoora, VIC 3086, Australia

Abstract. Many state-of-the-art anti-ransomware implementations monitoring file system activities choose to monitor file entropy-based changes to determine whether the changes may have been committed by ransomware, or to distinguish between compression and encryption operations. However, such detections can be victims of spoofing attacks, when attackers manipulate the entropy values in the expected range during the attacks. This paper explored the limitations of entropy-based ransomware detection on several different file types. We demonstrated how to use Base64-Encoding and Distributed Non-Selective Partial Encryption to manipulate entropy values and to bypass current entropy-based detection mechanisms. By exploiting this vulnerability, attackers can avoid entropy-based detection or degrade detection performance. We recommended that the practice of relying on file entropy change thresholds to detect ransomware encryption should be deprecated.

Keywords: Ransomware · Entropy · Encryption · File integrity

1 Introduction

Crypto-ransomware has recently drawn much attention as the most damaging ransomware often with disastrous consequences [2]. It has been found to exploit multiple infection vectors of varying complexity to aggressively spread into and infect victims' systems [2,15]. Ransomware differs to traditional malware in that it attacks user files often without modifying operating system kernels, its attacks are often sudden, and the file damages due to encryption are often irreversible unless victims pay the ransom to cybercriminals [15].

Many studies have been conducted to propose new ransomware detection or prevention solutions. Because ransomware must attack the file systems [15], the implementations monitoring file system activities [2,9,13,16,17] appear more promising, offering high detection rates and/or file recovery mechanisms. Other studies have attempted to analyze ransomware according to the network traffic activities [1,18], features in the ransomware executables or payloads [7,18], modification of Windows Registry values [18], usage of Windows Cryptographic APIs

© Springer Nature Switzerland AG 2019
T. Gedeon et al. (Eds.): ICONIP 2019, CCIS 1143, pp. 181–189, 2019.
https://doi.org/10.1007/978-3-030-36802-9_20

[11,18], and access of honeypot resources [5] *etc.* Although those studies discovered and summarized common features of the ransomware samples they studied, such features may not apply to all ransomware samples. Some ransomware variants obscure their communication to Command and Control Centers (C&C) [9], use embedded or derived encryption keys offline [4], apply sophisticated obfuscation and packing [9], or use own cryptographic libraries [11].

Problem Statement. While the recently proposed implementations to monitor file system activities by ransomware have made significant progress in ransomware detection, many of them decided to monitor file entropy-based changes as possible indicators of ransomware cryptographic activities. Although no ransomware has been found to manipulate the entropy values of encrypted content to date, it is possible to perform this type of activity to beat entropy-based detection. In this study, we evaluate the entropy-based changes of different file types upon Base64-Encoding and partial or full encryption. By proving that there are at least two methods to manipulate file entropy values, we question the effectiveness of entropy-based ransomware detection.

Summary of Original Contributions

- We found that files of naturally low or high entropy could respond differently to further encryption or compression.
- We demonstrated how to apply Base64-Encoding to reduce entropy values of encrypted content, and how to perform distributed non-selective partial encryption to adequately corrupt file data and structures while introducing small entropy changes.
- We recommend that the practice of relying on file entropy change thresholds to detect ransomware encryption should be deprecated.

2   Related Work

Although previous studies used entropy of malware executables to classify the malware, our focus is on data transformation of user files encrypted by ransomware. Files with high entropy values could indicate the stored information is encrypted and/or compressed, both of which could increase the randomness of file data with typically few repeated patterns [2,19]. [19] found that on Windows platforms, older document formats (TXT, RTF, DOC, PPT, XLS *etc.*) exhibited lower mean entropy values between 3.822 and 5.989 bits/byte; media (MP4, M4A *etc.*) and modern document formats (DOCX, PPTX, XLSX, PDF *etc.*) exhibited higher mean entropy values between 7.270 and 7.981 bits/byte, possibly due to improved compression algorithms [19].

Entropy-based file changes have been chosen as possible indicators of files being encrypted with strong cryptographic algorithms [2,9,13,16,17]. In [2], the entropy changes of I/O Request Packets (IRPs) was used to check the legality of the IRPs. In [17], if the difference in write entropy and read entropy of atomic read/write operations exceeded a pre-defined threshold of 0.1 (out of 0–8 range), the *Write* behavior was considered suspicious. In [9], entropy ratios of data blocks were calculated in a similar way to that of [2], and a malice score was calculated to estimate the probability of ransomware attack, although the entropy change threshold applied was not provided. In [16], it was assumed that encrypted files would have entropy values close to 8. [13] demonstrated that the entropy value was not always a good distinguisher for encryption on JPG files. [6] used a threshold of 0.04 for standard deviation of entropy values of file data blocks in 2048 bytes; compressed files had standard deviation of entropy values larger than 0.04 whereas encrypted files had values below 0.04.

Those ransomware studies [2,6,9,13,16,17] all measured file entropy-based changes as possible indicators of ransomware encryption activities, yet all but [13] assumed that ransomware encrypted the entire content of user files and that files with lower entropy responded to encryption similarly as files with higher entropy did. In the next section, we will demonstrate that it is possible to attack entropy-based ransomware detection by slightly changing the way encryption is performed.

3   Attacking Entropy-Based Ransomware Detection

In this section, proof-of-concept attacks are conducted on a few different file types in controlled environments, to analyze the validity of monitoring file entropy-based changes as an indicator of ransomware activities and to demonstrate the possibility of exploiting this popular detection mechanism.

3.1   Selection of File Samples

To adequately test on file samples and to produce convincing and reproducible results, we downloaded file samples from the file reference corpora collection, which contained one million random representative files of different types and sizes, cleared for privacy and copyright issues [3]. The PDF, JPG and Microsoft Office file types were selected for testing as they were some of the most commonly attacked file types [10,15]. The CSV and TXT files were selected as pure text files. Due to the large quantity of files in some groups, we randomly selected files from each group of those file types to achieve 99% confidence level with 1% margin of error. Because there were limited number of DOCX, XLSX and PPTX files provided for analysis, all such files were selected. The entropy values of selected file samples and the standard deviations are displayed in Fig. 1. It was noticed that:

- Pure-text based files (CSV and TXT) have naturally lower entropy values than those of binary files, and the differences between entropy values of different files are large.
- Modern XML-based file formats (DOCX, XLSX and PPTX) have higher average entropy values, possibly due to enhanced compression algorithms, than the older binary file formats (DOC, XLS and PPT).
- Naturally compressed files (JPG) have higher average entropy values.

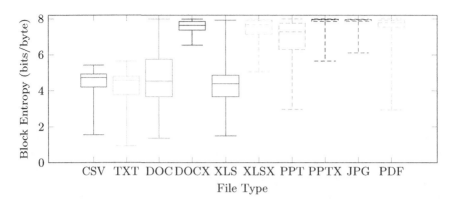

Fig. 1. The file entropy values of different file samples by file type

3.2 Applying Base64 Encoding to Reduce Entropy Values in Encrypted Files

Base64-Encoding. Base64-Encoding is a binary-to-text encoding scheme to represent binary data into an radix-64 representation of ASCII string, and is often used to facilitate easier storage and transfer of binary data over media designed to store textual data [8]. To our best knowledge, no ransomware has attempted to combine Base64-Encoding with encryption to attack user files.

Performing Base64-Encoding of Encrypted Contents. We implemented a simple prototype to fully encrypt files and to encode the encrypted files with Base64-Encoding. The average entropy values of each file type before and after encryption and Base64-Encoding were illustrated in Fig. 2. After being fully encrypted, files of all file types achieved entropy values close to 8.0. However, once the encrypted files were further Base64-Encoded, the entropy values decreased to close to 6.0. If ransomware encrypts user files and converts them with Base64-Encoding, before writing them back to the disk, it could potentially defeat detection based on entropy value increases.

3.3 Performing Partial Encryption to Attack File Entropy Change Thresholds

Partial Encryption Attacks. *Partial Encryption* is when a secure encryption algorithm is applied to encrypt only part of the file targets instead of the whole files [13]. While most known ransomware samples encrypted the full content of user files, we implemented a Distributed Non-Selective Partial Encryption with no intention of preserving file formats or data. It distributed encrypted pockets of data throughout the file content to maximize the corruption of file data, while introducing smaller local entropy change (Partial), and to prevent content recovery (Distributed). Partial encryption could encrypt parts of the files into

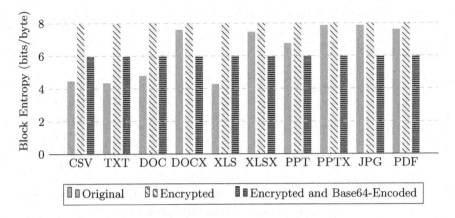

Fig. 2. Average entropy values upon encryption and Base64-Encoding

random data without the need to understand the file structure, schemata or encoding schemes (Non-Selective). Although the rest of a partially encrypted file could still be opened in binary editors, the encrypted file contents could be difficult to recover and restore.

Effects of Partial Encryption on Files of Higher Entropy. Partial encryption on binary files of higher entropy could successfully corrupt file structures while introducing minimal entropy changes (Fig. 3). Many PPTX and JPG files could be fully encrypted while still producing entropy increases less than 0.1, the threshold used by [17]. The larger DOCX files had higher initial entropy values but exhibited smaller increases. PDF file entropy values varied depending on the contents embedded in the documents and did not always correlate with the PDF file sizes; files with more embedded non-text contents had higher initial entropy values and exhibited smaller increases. Again, a 10% partial encryption result in many files with entropy increases below 0.1. JPG and DOCX files exhibited less entropy changes upon encryption than PDF files.

Effects of Partial Encryption on Files of Lower Entropy. Files of Lower Entropy (CSV, TXT, DOC, XLS and PPT files) initially had lower entropy values when partially encrypted (Fig. 4). As the percentage of partial encryption increased, the entropy values rose faster than those of files of higher entropy. It was noticed that almost all files reached entropy values of approximately 8 bits/byte when 100% encrypted, which is consistent with previous findings [19].

In summary, a low percentage partial encryption could be an effective strategy to be employed by crypto-ransomware to attack already compressed or optimized file types while minimizing the risk of triggering entropy change monitor alarms. Using file entropy change as an indicator for crypto-ransomware attacks could be ineffective on file types that have already been compressed or optimized.

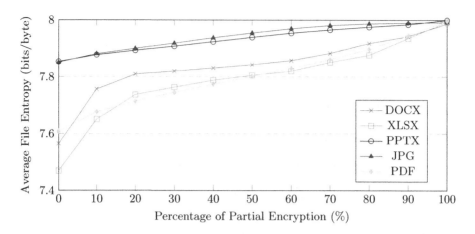

Fig. 3. Partial encryption on files of higher entropy

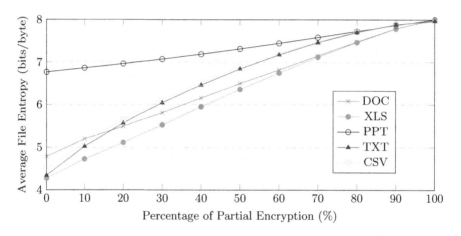

Fig. 4. Partial encryption on files of lower entropy

4   Discussion

In our previous experiment, we demonstrated that checking changes of entropy values of encrypted files was not always a reliable method to detect malicious encryption by ransomware attacks, and was not always a reliable indicator of the actual file types, file contents or file integrity.

4.1   Ransomware-Initiated File Encryption Corrupts File Integrity

File integrity is defined as data stored in files with internal consistency without corruption, and is already used to verify whether selected files have been tampered with [12]. Ransomware-initiated file encryption differs from benign file modifications in that ransomware usually compromises file integrity [15], because:

- Each ransomware sample usually uses one type of encryption algorithm for all files, possibly to reduce design complexity for a quicker profit gain.
- File input data are treated as raw binary inputs regardless of whether the files are text-based or not.
- Other than pure encryption, there is usually no additional processing to preserve file schemata or encoding schemes if they exist.

4.2   File Validation with File Type Identification as a Possible Mitigation Strategy

File Type Identification aims to correctly identify the file types based on clues in the raw file data [12]. *File Validation* is the process of validating the logical integrity of a file, and whether its data clusters violate the structure or rules required for the specified file type [14]. An issue with existing entropy-based ransomware detection is that it does not perform *File Type Identification* or *File Validation*. Anything could be written by any program into any files, when the file content does not have to be consistent with the file type stated in its file extension name [14]. Relying on entropy-based detection could leave the detection mechanisms vulnerably to exploitation. *File Validation* combined with *File Type Identification* could be a possible simple mitigation strategy, to detect possible ransomware encryption that corrupts file structures. Further inspection is required to determine whether the encryption is benign or malicious.

5   Conclusion

This paper presents an important vulnerability of entropy-based ransomware detection. We believe applying a universal model to evaluate entropy-based changes on all file types should be deprecated, as it could be easily attacked by ransomware in multiple means to bypass entropy-based detection. As crypto-ransomware usually encrypts target files and corrupts the file schemata or encoding schemes of formatted files, checking for possible file structural corruption by applying *File Validation* and *File Type Identifcation* on formatted file types could be an alternative approach. This is a step forward towards building more advanced dynamic analysis that could detect crypto-ransomware that seek to perform less noticeable encryption to user files.

Acknowledgment. This work was made possible by the support of a grant (UOCX1720) from the Ministry of Business, Innovation and Employment of New Zealand, September 2017 Catalyst: Strategic Investment Round.

References

1. Ahmadian, M.M., Shahriari, H.R., Ghaffarian, S.M.: Connection-monitor & connection-breaker: a novel approach for prevention and detection of high survivable ransomwares. In: 2015 12th International Iranian Society of Cryptology Conference on Information Security and Cryptology (ISCISC), pp. 79–84. IEEE (2015)
2. Continella, A., et al.: Shieldfs: a self-healing, ransomware-aware file system. In: Proceedings of the 32nd Annual Conference on Computer Security Applications, pp. 336–347. ACM (2016)
3. Garfinkel, S., Farrell, P., Roussev, V., Dinolt, G.: Bringing science to digital forensics with standardized forensic corpora. Digit. Invest. **6**, S2–S11 (2009)
4. Genç, Z.A., Lenzini, G., Ryan, P.Y.: Security analysis of key acquiring strategies used by cryptographic ransomware. In: Proceedings of the Central European Cybersecurity Conference 2018, p. 7. ACM (2018)
5. Gómez-Hernández, J., Álvarez-González, L., García-Teodoro, P.: R-locker: Thwarting ransomware action through a honeyfile-based approach. Comput. Secur. **73**, 389–398 (2018)
6. Held, M., Waldvogel, M.: Fighting ransomware with guided undo. In: Proceedings of the 11th Norwegian Information Security Conference (2018)
7. Homayoun, S., Dehghantanha, A., Ahmadzadeh, M., Hashemi, S., Khayami, R.: Know abnormal, find evil: Frequent pattern mining for ransomware threat hunting and intelligence. IEEE Trans. Emerg. Top. Comput. **PP**, 1 (2017)
8. Josefsson, S.: The base16, base32, and base64 data encodings (2006)
9. Kharraz, A., Kirda, E.: Redemption: real-time protection against ransomware at end-hosts. In: Dacier, M., Bailey, M., Polychronakis, M., Antonakakis, M. (eds.) RAID 2017. LNCS, vol. 10453, pp. 98–119. Springer, Cham (2017). https://doi.org/10.1007/978-3-319-66332-6_5
10. Kirda, E.: Unveil: a large-scale, automated approach to detecting ransomware (keynote). In: 2017 IEEE 24th International Conference on Software Analysis, Evolution and Reengineering (SANER), p. 1. IEEE (2017)
11. Kolodenker, E., Koch, W., Stringhini, G., Egele, M.: Paybreak: defense against cryptographic ransomware. In: Proceedings of the 2017 ACM on Asia Conference on Computer and Communications Security, pp. 599–611. ACM (2017)
12. Li, W.J., Wang, K., Stolfo, S.J., Herzog, B.: Fileprints: identifying file types by n-gram analysis. In: Proceedings from the Sixth Annual IEEE SMC Information Assurance Workshop, IAW 2005, pp. 64–71. IEEE (2005)
13. Mbol, F., Robert, J.-M., Sadighian, A.: An efficient approach to detect torrentlocker ransomware in computer systems. In: Foresti, S., Persiano, G. (eds.) CANS 2016. LNCS, vol. 10052, pp. 532–541. Springer, Cham (2016). https://doi.org/10.1007/978-3-319-48965-0_32
14. McDaniel, M., Heydari, M.H.: Content based file type detection algorithms. In: Proceedings of the 36th Annual Hawaii International Conference on System Sciences, p. 10. IEEE (2003)
15. McIntosh, T.R., Jang-Jaccard, J., Watters, P.A.: Large scale behavioral analysis of ransomware attacks. In: Cheng, L., Leung, A.C.S., Ozawa, S. (eds.) ICONIP 2018. LNCS, vol. 11306, pp. 217–229. Springer, Cham (2018). https://doi.org/10.1007/978-3-030-04224-0_19

16. Mehnaz, S., Mudgerikar, A., Bertino, E.: RWGuard: a real-time detection system against cryptographic ransomware. In: Bailey, M., Holz, T., Stamatogiannakis, M., Ioannidis, S. (eds.) RAID 2018. LNCS, vol. 11050, pp. 114–136. Springer, Cham (2018). https://doi.org/10.1007/978-3-030-00470-5_6
17. Scaife, N., Carter, H., Traynor, P., Butler, K.R.: Cryptolock (and drop it): stopping ransomware attacks on user data. In: 2016 IEEE 36th International Conference on Distributed Computing Systems (ICDCS), pp. 303–312. IEEE (2016)
18. Verma, M., Kumarguru, P., Deb, S.B., Gupta, A.: Analysing indicator of compromises for ransomware: Leveraging IOCs with machine learning techniques. In: 2018 IEEE International Conference on Intelligence and Security Informatics (ISI), pp. 154–159. IEEE (2018)
19. Weston, P., Wolthusen, S.D.: Forensic entropy analysis of microsoft windows storage volumes. SAIEE Afr. Res. J. **105**(2), 63–70 (2014)

Attention Network for Product Characteristics Prediction Based on Reviews

Wan Li$^{(\boxtimes)}$, Lei Zhang, and Si Chen

Beijing University of Posts and Telecommunications, Beijing, China
{liwan,zlei}@bupt.edu.cn,chens_bupt@163.com

Abstract. Online reviews are now widely used in e-commerce and other websites, which can provide rich information about products to consumers or potential consumers. In reality, consumers often get more information about products from reviews and find the more suitable products for themselves. However, reading a lot of reviews is obviously a waste of time for customers. In recent years, researchers have begun to study how to extract critical information from reviews through machine learning as reference information for products, instead of viewing lots of reviews by customers themselves. However, it is limited to emotional judgment. There are more product characteristics which can be extracted to form more comprehensive information for consumers. Meanwhile, existing methods all ignore the descriptions of the product itself given by sellers. By above motivations, we propose a new network based on attention mechanism (RAN for short) to predict more product characteristics (e.g., short or horror). Experimental results show that our method achieves state-of-the-art result on two real-world datasets.

Keywords: Review mining · Product attention · User attention

1 Introduction

With the development of Internet technology, various e-commerce and group-buying websites have emerged. In reality, the descriptions of the product are generally short and incomplete. Consumers usually judge the quality of products and get more information through reviews of other consumers to find the more suitable products for themselves. However, at the same time, there may be a lot of reviews on each product. Reading a lot of reviews is obviously a waste of time for customers. Fortunately, in recent years, researchers have begun to study how to extract critical information from reviews through machine learning as reference information for products, instead of viewing lots of reviews by customers themselves. Previous works in this area always focus on sentiment classification [2,3] based on user-generated reviews. For example, Wu et al. proposed a novel framework to encode reviews for sentiment classification [12] and Chen et al. proposed an effective neural classification model by taking global

© Springer Nature Switzerland AG 2019
T. Gedeon et al. (Eds.): ICONIP 2019, CCIS 1143, pp. 190–197, 2019.
https://doi.org/10.1007/978-3-030-36802-9_21

user and product information into consideration [2]. However, existing models always give the sentiment classification result of certain review of product [3,12]. Users still need to view the analysis results of each review to deduce a conclusion and the conclusion is limited to the emotional analysis. What's more, existing models all ignore the descriptions of the product itself given by sellers. They tend to use the random initialization vector as the product representations such as [2]. Although the descriptions of the product itself may be incomplete, we firmly believe that it has significant influence on mining more information of the product from reviews. For example, people can obtain basic information through incomplete product descriptions although they know nothing about products.

By above motivation, in this paper, we propose a new deep learning model (RAN for short) to mine the rich hidden information of products. Our model classification results are not limited to the emotional judgment of reviews, but extract more product characteristics (e.g., short or horror) from lots of reviews to form a more objective result. We formalize this problem into a multi-label classification task, because a product may have more than one characteristics at the same time. Specifically, we employ a user attention layer to obtain user-generated reviews which show strong user preferences by multiple reviews of one product. Then we employ a product attention layer to introduce the descriptions of product itself. Finally, we predict the product characteristics by the features mined objectively and fully after above steps.

Overall, the major contributions of our work are as follows:

- To the best of our knowledge, this is the first work to focus on extracting more product characteristics which are not limit to emotion analysis from multiple reviews.
- We propose a new deep learning model combining two attention mechanism and introduce the descriptions of product itself to predict the product characteristics.
- We conduct experiments on two real-world datasets to verify the effectiveness of our model. The experimental results demonstrate that our model significantly and consistently outperforms other state-of-the-art models.

The rest of this paper is organized as follows. Section 2 will introduce our model structure and the specific details of method. The thorough experimental evaluation of our methods will be discussed in Sect. 3. We will give a brief summary of this work in Sect. 4.

2   Method

In this section, we first introduce the formalization of product characteristics classification based on multi-reviews. Afterwards, we discuss how to obtain n-gram word representation via the convolution and how to use user attention and the product attention to enhance review representations. Finally, we introduce our prediction and optimization functions. The overall architecture of RAN has been shown in Fig. 1.

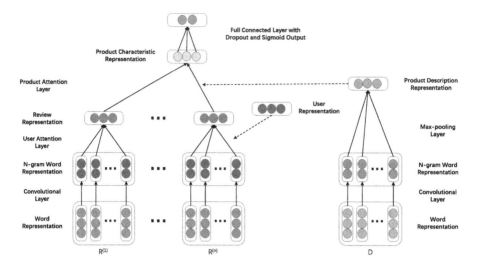

Fig. 1. The architecture of Review Attention Network.

2.1 Formalization

We define a product as $P = \{R^{(1)}, R^{(2)}, \ldots, R^{(m)}\}$, and the truth characteristics distributions of P as $y = \{y_1, y_2, \ldots, y_C\}$, where $R^{(i)}$ means the i-th review owned by the product and m means the number of reviews. We define C as the number of product characteristics. Each review is defined as $R^{(i)} = \{x_1^i, x_2^i, \cdots, x_{n^i}^i\}$, where x_j^i means the j-th word embedding in $R^{(i)}$ and n^i is the length of i-th review. The i-th review is provided by a specific consumer and we defined this consumer as u^i. The descriptions of product which is defined as D consist of l words. We treat D as $\{x_1, x_2, \cdots, x_l\}$.

Next, for a product, our object is to learn the predicted product characteristics distributions $o = \{o_1, o_2, \ldots, o_C\}$ according to its description and reviews, where C is the number of product characteristics we define. Each $o_i \in \{0, 1\}$ indicates whether the i-th characteristic is hit or not.

2.2 N-Gram Word Representation

In our model, in order to better reflect the semantic information, we use the convolution layer of TextCNN [5] to learn the n-gram word representation as our input. $x_j^i \in R^d$ is the embedding vector of the j-th word in i-th review of a product. N-gram word representation is calculated as follow:

$$\hat{x}_{jk}^i = f(w_k \cdot x_{j-h_k:j+h_k}^i + b_k) \tag{1}$$

where $w_k \in R^{(2h_k+1)t}$ is the k-th filter weight and $b_k \in R$ is the k-th bias. $\hat{x}_{jk}^i$ is the feature value of the k-th filter and $\hat{x}_j^i \in R^{d_f}$ means combination of all filters, where d_f is the number of filters.

In order to make the length of the vector encoded by each filter equal to the length of the review, for the k-th filter, we add h_k padding vectors at both the beginning and the end of $R^{(i)}$.

2.3   User Attention

It is obvious that not all words contribute equally to the review meaning for different users. Hence, in word level, instead of feeding $\hat{x}_j^i$ to an average pooling layer, we adopt a user attention mechanism to extract specific words that are important to the meaning of reviews. Finally, we aggregate the representations of those informative words to form the review representation.

Formally, the review representation is a weighted sum of n-gram feature vectors as:

$$r^i = \sum_{j=1}^{n^i} \alpha_j^i \cdot \hat{x}_j^i, \tag{2}$$

where α_j^i is the attention weight of $\hat{x}_j^i$ which can measure the importance of the j-th word for current user. We use a continuous and real-valued vector $u^i \in R^{d_u}$ to represent a user, where d_u is the dimension of user embeddings. The attention weight α_j^i can be calculated as follow:

$$\alpha_j^i = \frac{exp(e(\hat{x}_j^i, u^i))}{\sum_{k=1}^{n^i} exp(e(\hat{x}_k^i, u^i))}, \tag{3}$$

where $e(\cdot)$ is a score function which scores the importance of words for review representation about current user. The score function $e(\cdot)$ is defined as:

$$e(\hat{x}_j^i, u^i) = v_u^T tanh(\mathbf{W}_w \hat{x}_j^i + \mathbf{W}_u u^i + b_u), \tag{4}$$

where v_u^T is the transpose vector of weight vector v_u. $\mathbf{W}_w$ and $\mathbf{W}_u$ are weight matrices, and b_u is the bias vector.

2.4   Product Attention

Similarly, for different products, each review contributes different implicit information to the product characteristics. Based on common sense, the product attention mechanism introduce product information into review representations to generate the product characteristics representation $\hat{p}$, which is calculated as follows:

$$e(r^i, p) = v_p^T tanh(\mathbf{W}_r r^i + \mathbf{W}_p p + b_p), \tag{5}$$

$$\beta^i = \frac{exp(e(r^i, p))}{\sum_{k=1}^{m} exp(e(r^k, p))}, \tag{6}$$

$$\hat{p} = \sum_{i=1}^{m} \beta^i \cdot r^i, \tag{7}$$

where β^i is the weight of i-th review representation r^i and p is the product description representation generated by TexCNN [5] from the textual descriptions of the product itself. For each feature value p_j in product description representation p, it is obtained by convolution operation and max-pooling operation after inputting the word embeddings in the product description. $\mathbf{W}_r$ and $\mathbf{W}_p$ are weight matrices, and b_p is bias vector.

2.5   Prediction and Optimization

After introducing the description of the product and the review information of the product into $\hat{p}$, we use a linear layer to predict product characteristics distribution o.

$$o = \mathbf{W}_c\hat{p} + b_c. \tag{8}$$

Since binary cross-entropy (BCE) loss over sigmoid activation has better optimization effects when applied to multi-label classification datasets [8], so we use BCE loss as the loss function of the classifier:

$$L = -\sum_{d\in T}\sum_{c=1}^{C}[y_c(d)log(\sigma(o_c(d))) + (1 - y_c(d))log(1 - \sigma(o_c(d)))], \tag{9}$$

where $y_c(d)$ indicates whether d exhibits characteristic c, and T represents the training set. $\sigma(\cdot)$ is a sigmoid function.

3   Experiments

In this section, we first introduce the datasets and baselines used in our experiments. Then we give the experimental hyper-parameter settings. Finally, we analyze the experimental results to verify the performance of our model.

3.1   Dataset

We evaluate our model with the following datasets. The characteristic of a product is the category definition of the product on the website. IMDb: We select lots of customer reviews of movies from IMDb as a dataset. One movie may have one or more characteristics. Douban: This dataset is collected from Douban, which provides products about books, movies, music, games, etc. The descriptions and reviews of products are generated by the users.

The detail information about datasets are listed in Table 1. T is the total number of reviews, C is the total number of characteristics, P is the total number of products, U is the total number of user, $\hat{C}$ is the average number of characteristics per product, $\hat{P}$ is the average number of reviews per product, and $\hat{U}$ is the average number of reviews per user.

Table 1. Datasets information.

Datasets	T	C	P	U	$\hat{C}$	$\hat{P}$	$\hat{U}$
IMDb	887,902	27	16,376	123,460	2.34	54.22	7.19
Douban	5381,586	39	24497	627,181	3.21	219.68	8.58

3.2 Baseline

We adopt the following representative methods of text classification as baselines to compare with our model.

MLKNN [13] is one of the most representative traditional machine learning models in multi-label classification tasks. SVM [1] is an another representative traditional model for classification. We use the TF-IDF vector as the feature vector of the texts. FastText [4] is a simple neural network by averaging all word embeddings of a text as feature vectors. TextCNN [5] applies the CNN to the text classification task, using multiple kernels of different sizes to extract key information in the text. TextRNN [10] uses the hidden state of the last time step to encode the text. TextRCNN [7] apply a recurrent structure and a max-pooling layer to automatically judge which words play key roles in text classification.

3.3 Evaluation Metrics

We employ macro-precision (MacroP), macro-recall (MacroR) and macro-F1 (MacroF1) which are widely used in the classification task as evaluation metrics to evaluate the performance of the model on product characteristics prediction. The macro- precision/recall/F1 are calculated by averaging the precision/recall/F1 of each category same as [11].

3.4 Experiments Settings

For traditional machine learning models (SVM, MLKNN), we use TF-IDF as the feature extraction method for text representations. Meanwhile, we set k to 7 when training for MLKNN. For all other models, we adopt pre-train 300-dimensional Glove vectors [9] on IMDb dataset, and random initialization 300-dimensional vectors on Douban dataset. Furthermore, for TextCNN, we set the filter widths to (3, 4, 5) with each filter size to 128 for consistency, and for TextRNN and TextRCNN, we set the hidden vector size to 100.

In our model, we set the filter widths to (3, 5, 7) with each filter size to 100 for consistency when we generate the n-gram word representations and product description representations. Finally, the learning rate of Adam [6] optimizer is 10^{-3}, and the dropout probability is 0.5. The batch size is set to 128 and L2 regularization is set to 0.005.

Specially, the review number m is set to 32. For FastText, TextCNN, TextRNN and TextRCNN, we average the feature vectors of each review to represent the final feature vector.

Table 2. Product characteristics classification results. Marked black denotes the best results on each dataset.

Datasets	IMDb			Douban		
Metrics	MicroP	MicroR	MicroF1	MicroP	MicroR	MicroF1
MLKNN	0.8178	0.6519	0.7255	0.7616	0.6032	0.6732
SVM	0.8359	0.7319	0.7805	0.7778	0.6798	0.7255
FastText	0.9772	0.9084	0.9415	0.9223	0.8487	0.8840
TextRNN	0.9815	0.9146	0.9469	0.9189	0.8472	0.8816
TextCNN	0.9829	0.9156	0.9481	0.9245	0.8512	0.8863
TextRCNN	0.9832	0.9215	0.9514	0.9286	0.8506	0.8879
RAN	**0.9896**	**0.9575**	**0.9733**	**0.9334**	**0.8721**	**0.9017**
Improvement	0.0064	0.0360	0.0219	0.0048	0.0209	0.0138

3.5   Result Analysis

As shown in the Table 2, we can see that our model significantly and consistently outperforms all the baselines and has a huge advantage over two datasets which achieves promising improvements ($0.48\% \sim 0.64\%$, $2.09\% \sim 3.60\%$, and $1.38\% \sim 2.19\%$).

We further compare our model to other baseline methods for the in-depth analysis. Compared against traditional models, manually defined feature extraction methods cause the extracted features limited. Hence, it is not surprising to see that traditional learning methods obtain the worst or the second worst performance in terms of all the evaluating indicator. Deep learning models are superior to the traditional learning methods. However, they all ignore the product descriptions and user information, so they are not better than our model. Drawing conclusion from the above analysis, our method has its own unique advantages.

4   Conclusion

In this paper, we propose a new framework that combines user information and product information to mine product characteristics. Unlike previous reviews mining tasks, we use two attention mechanisms and fuse multiple reviews of a product to enhance the ability of the model to represent the product characteristics, and also we employ the description of product itself. We evaluate our model on the two datasets. The experimental results show that our model achieves significant improvements after introducing the user attention and the product attention, compared to other state-of-the-art models.

References

1. Cao, P., Liu, X., Zhao, D., Zaiane, O.: Cost sensitive ranking support vector machine for multi-label data learning. In: Abraham, A., Haqiq, A., Alimi, A.M., Mezzour, G., Rokbani, N., Muda, A.K. (eds.) HIS 2016. AISC, vol. 552, pp. 244–255. Springer, Cham (2017). https://doi.org/10.1007/978-3-319-52941-7_25

2. Chen, H., Sun, M., Tu, C., Lin, Y., Liu, Z.: Neural sentiment classification with user and product attention. In: Proceedings of the 2016 Conference on Empirical Methods in Natural Language Processing, EMNLP 2016, Austin, Texas, USA, pp. 1650–1659, 1–4 November 2016

3. Dong, R., O'Mahony, M.P., Schaal, M., McCarthy, K., Smyth, B.: Combining similarity and sentiment in opinion mining for product recommendation. J. Intell. Inf. Syst. **46**(2), 285–312 (2016). https://doi.org/10.1007/s10844-015-0379-y

4. Grave, E., Mikolov, T., Joulin, A., Bojanowski, P.: Bag of tricks for efficient text classification. In: Proceedings of the 15th Conference of the European Chapter of the Association for Computational Linguistics, EACL 2017, Valencia, Spain, 3–7 April 2017, Volume 2: Short Papers, pp. 427–431 (2017)

5. Kim, Y.: Convolutional neural networks for sentence classification. In: Proceedings of the 2014 Conference on Empirical Methods in Natural Language Processing, EMNLP 2014, 25–29 October 2014, Doha, Qatar, A meeting of SIGDAT, a Special Interest Group of the ACL, pp. 1746–1751 (2014)

6. Kingma, D.P., Ba, J.: Adam: a method for stochastic optimization. In: 3rd International Conference on Learning Representations, ICLR 2015, San Diego, CA, USA, 7–9 May 2015. Conference Track Proceedings (2015)

7. Lai, S., Xu, L., Liu, K., Zhao, J.: Recurrent convolutional neural networks for text classification. In: Proceedings of the Twenty-Ninth AAAI Conference on Artificial Intelligence, 25–30 January 2015, Austin, Texas, USA, pp, 2267–2273 (2015)

8. Nam, J., Kim, J., Loza Mencía, E., Gurevych, I., Fürnkranz, J.: Large-scale multi-label text classification—revisiting neural networks. In: Calders, T., Esposito, F., Hüllermeier, E., Meo, R. (eds.) ECML PKDD 2014. LNCS (LNAI), vol. 8725, pp. 437–452. Springer, Heidelberg (2014). https://doi.org/10.1007/978-3-662-44851-9_28

9. Pennington, J., Socher, R., Manning, C.D.: Glove: global vectors for word representation. In: Proceedings of the 2014 Conference on Empirical Methods in Natural Language Processing, EMNLP 2014, 25–29 October 2014, Doha, Qatar, A meeting of SIGDAT, a Special Interest Group of the ACL, pp. 1532–1543 (2014)

10. Tang, D., Qin, B., Liu, T.: Document modeling with gated recurrent neural network for sentiment classification. In: Proceedings of the 2015 Conference on Empirical Methods in Natural Language Processing, EMNLP 2015, Lisbon, Portugal, 17–21 September 2015, pp. 1422–1432 (2015)

11. Wang, P., Yang, Z., Niu, S., Zhang, Y., Zhang, L., Niu, S.: Modeling dynamic pairwise attention for crime classification over legal articles. In: The 41st International ACM SIGIR Conference on Research & Development in Information Retrieval, SIGIR 2018, Ann Arbor, MI, USA, 08–12 July 2018, pp. 485–494 (2018). https://doi.org/10.1145/3209978.3210057

12. Wu, Z., Dai, X., Yin, C., Huang, S., Chen, J.: Improving review representations with user attention and product attention for sentiment classification. In: Proceedings of the Thirty-Second AAAI Conference on Artificial Intelligence (AAAI-18), the 30th innovative Applications of Artificial Intelligence (IAAI-18), and the 8th AAAI Symposium on Educational Advances in Artificial Intelligence (EAAI-18), New Orleans, Louisiana, USA, 2–7 February 2018, pp. 5989–5996 (2018)

13. Zhang, M., Zhou, Z.: ML-KNN: a lazy learning approach to multi-label learning. Pattern Recogn. **40**(7), 2038–2048 (2007). https://doi.org/10.1016/j.patcog.2006.12.019

Semi-supervised Regularized Coplanar Discriminant Analysis

Rakesh Kumar Sanodiya[1(✉)], Michelle Davies Thalakottur[3], Jimson Mathew[1], and Matloob Khushi[2]

[1] Indian Institute of Technology Patna, Patna, India
rakesh.pcs16@iitp.ac.in
[2] School of Computer Science, The University of Sydney, Sydney, Australia
[3] MKSSS Cummins College of Engineering, Pune, India

Abstract. Dimensionality Reduction is a widely used method of removing redundant features and data compression. Dimensionality reduction usually occurs in a supervised setting in which all the samples are labelled. However, aspects of spectral clustering and semi-supervised learning can be used in Dimensionality reduction to ensure minimum loss of important data while projecting the high-dimensional data into lower dimensions. We have proposed a novel framework called Semi-supervised Regularized Co-planar Discriminant Analysis (SRCDA) that creates a graph of labelled and unlabelled data and uses label propagation to predict the classes of the unlabelled data. Additionally, we introduce a regularized term which is used to prevent overfitting. The proposed algorithm is evaluated against several other state-of-the-art algorithms with benchmark datasets including PIE Face, ORL and Yale Dataset. The proposed algorithm shows higher accuracies compared to the other algorithms and can be used in real-life datasets where the unlabelled data is vastly greater than the labelled samples. We have also conducted a statistical significance test to verify the results obtained.

Keywords: Semi-supervised learning · Dimensionality Reduction · Regularization

1 Introduction

Dimensionality Reduction (DR) is the cornerstone when it comes to analyzing high dimensional data. In the real-world where high dimensional data is being generated at ever increasing rates, the usefulness on Dimensionality Reduction techniques cannot be underestimated. They allow data in high dimensions to be projected to a lower dimensional subspace while attempting to preserve some feature or structure of the data. Dimensionality Reduction can be used for data visualization and data compression and can improve the performance of a model by taking care of multi-collinearity by removing redundant features and make it more efficient by reducing the computing time. Often, many features are redundant or are correlated instead of being independent. Hence, ultimately there

© Springer Nature Switzerland AG 2019
T. Gedeon et al. (Eds.): ICONIP 2019, CCIS 1143, pp. 198–205, 2019.
https://doi.org/10.1007/978-3-030-36802-9_22

is only a small subspace of the original representation that has features relevant to the task at hand. In such a case, Dimensionality Reduction enables low dimensional representation while ensuring minimum information loss.

The use of entirely labelled information can be called supervised learning while if only unlabelled information is used to train the model, it is called unsupervised learning [1]. The use of large amounts of unlabelled information in conjunction with few labelled points is referred to as semi-supervised learning [2]. Supervised learning holds the disadvantage that supervised models require large amounts of labelled information that is time-consuming and expensive to annotate. Additionally, unlabelled information is more readily available in the real-world. Dimensionality Reduction has traditionally been a unsupervised learning technique, but can hold the disadvantage that label information is not used at all, when the labels can also provide useful information that can be propagated to the unlabelled data via the geometry of the data distribution.

In this paper, we have proposed Semi-supervised Regularized Coplanar Discriminant Analysis (SRCDA) which adds a regularization term to the original criteria of RCDA which is based on the prior knowledge of the distribution geometry provided by both the labeled and unlabeled data. This is done by constructing a graph using similarity between embeddings and finding its Laplacian. Our proposed algorithm is shown to perform better than most state-of-the-art Dimensionality Reduction algorithms when evaluated on various real-world datasets.

The major contributions of the current work are that we have utilized both labelled and unlabelled information instead of relying on one over another. A graph is constructed which allows for label information propagation which helps in classifying the unlabelled information more accurately. Additionally, we have used learnt knowledge of the manifold in which the data exists to add a regularization term that prevents overfitting of the model to the few labelled samples as well as giving better accuracies.

The rest of the paper is organized as follows: Sect. 2 explores the relevant related work after going through various literature surveys. In Sect. 3 we provide a detailed overview of RCDA and the specifications of our proposed method. Experimental Results are discussed in Sect. 4 and we conclude and provide suggestions for future work in Sect. 5.

2   Related Work

We shall first discuss some relevant Dimensionality Reduction algorithms postulated from literature surveys [3]. Principal Component Analysis (PCA) was developed by Pearson [4] in 1901. PCA transforms the input data by projecting it to a new coordinate system such that the variance along the principal components is maximized. PCA has been generalized for non-linear data in Kernel PCA [5] by using a kernel function to project the lower-dimensional data to higher dimensions where it is easier to work with it. However, PCA is not without its faults. It fails to capture the intrinsic geometry of the manifold in which the data exists while Kernel PCA can suffer from the curse of dimensionality with the model requiring more data samples.

Linear Discriminant Analysis (LDA) [6] tries to find a linear transformation such that the class structure and differences of the data distribution remain conserved. To this end, it calculates a within-class scatter matrix, which it tries to minimize, and a between-class scatter matrix, which it tries to maximize, in order to find the projection matrix for the data. LDA has a few constraints which include that each data point must be labeled with a single class and the class boundaries are taken to be linear. Hence, LDA does not allow for outliers that do not belong to any class or samples that belong to more than one class.

Locally Linear Embedding (LLE) [7] approaches dimensionality reduction by considering neighbour information instead of the global manifold. It computes a lower dimensional embedding through weights learnt from its neighbours that allow the data point to be reconstructed. While LLE might be favoured by researchers due to its non-iterative approach to finding low-dimensional embeddings, it is sensitive to noise and is unable to deal with outliers.

Like LLE, Locality Preserving Projections (LPP) [8] also considers local neighbour information to calculate the Laplacian of a graph whose edges are assigned weights in order to preserve the local geometry. The projection will then use these weights to preserve the local manifold structure. While LPP is very good at capturing the geometry of the neighbourhood, it is not able to capture variable relationships which can then be lost after projection. Related to LPP, Globality-locality Preserving Projections (GLPP) [9] tries to preserve the sample relationships as well as the local manifold structure by computing a Laplacian on the subject invariant part of the samples and another Laplacian on the intra-subject part of the sample. It then jointly learns a common Laplacian which is more robust to noise and is able to preserve the geometry better.

Regularized Coplanar Discriminant Analysis (RCDA) [10] uses coplanarity of samples to preserve class information while projecting the data to lower dimensions. It is discussed in more detail in Sect. 3.1.

Semi-supervised Discriminant Analysis (SDA) [11] is an extension of LDA which uses a graph Laplacian to learn the structure of the data distribution in order to introduce some distribution appropriate regularization term that helps the model to predict labels more accurately.

3   The Proposed Method

3.1   An Overview of RCDA

RCDA is a Dimensionality Reduction algorithm that calculates a projection matrix such that when the high dimension data is projected to lower dimensions, the between-class variance is preserved in the new manifold while the within-class variance is reduced [10]. Thus, the class information is preserved while projecting to a lower dimension.

Linear projection directions are found such that the coplanarity of samples from the same class are maximized while samples from different classes are made noncoplanar. For this, a within-class coplanar compactness and between-class coplanar separability is calculated after projection of the high dimensional feature matrix to lower dimensions. The within-class coplanar compactness is used to measure the error in the within-class linear representation, in order to minimize it, while the between-class coplanar separability is used to measure the error in the between-class linear representation in order to maximize it. Firstly, mean normalization is performed on the projected lower dimension data after which the within-class coplanar compactness is defined as the sum of errors of the within-class linear representation. It is represented by the following term:

$$\sum_{i=1}^{n}(\|W^T x_i - W^T X_i \beta_i^W\|_2^2 + \lambda\|\beta_i^W - \tilde{\beta}_i^W\|_2^2)$$

$$= \mathrm{Tr}(W^T S_W W) + \lambda\|B_w - \tilde{B}_W\|_F^2 \tag{1}$$

Similarly, after mean-normalization, the within-class coplanar compactness is defined by the sum of the errors of within-class linear representation.

$$\sum_{i=1}^{n}\sum_{c=1}^{C}(\|W^T x_i - W^T X_c \beta_{i,c}^b\|_2^2 + \lambda\|\beta_{i,c}^b - \tilde{\beta}_i^b\|_2^2)$$

$$= \mathrm{Tr}(W^T S_b W) + \lambda\sum_{c=1}^{C}\|B_c - \tilde{B}_W\|_F^2 \tag{2}$$

Using the within-class coplanar compactness and the between-class coplanar separability terms, RCDA defines its optimization function as follows,

$$\min_{W}\frac{\min_{\beta^W}\mathrm{Tr}(W^T S_W W) + \lambda\|B_W - \tilde{B}_W\|_F^2}{\min_{\beta^b}\mathrm{Tr}(W^T S_b W) + \lambda\sum_{c=1}^{C}\|B_c - \tilde{B}_c\|_F^2} \tag{3}$$

A linear projection matrix W, the within-class linear representation coefficient β^W and the between-class linear representation coefficient β^b is learnt simultaneously.

3.2 Semi-supervised Setting

We have considered the case of RCDA in a semi-supervised setting and have modified the objective function so that the algorithm performs better for datasets with large amounts of unlabelled information and a few labelled data points. We can rewrite Eq. 3 and introduce a Regularization term $J(a)$ to form our objective function. Here, a is the eigen vector corresponding to non-zero eigenvalue of the optimization problem in Eq. 4. Due to the lack of this regularizer term, RCDA suffers from over-fitting when there is insufficient training examples. Based on the a priori information of the manifold structure, we can determine the value of the term regularization, thereby merging the manifold structure to improve accuracy.

$$\min_{W} \frac{\min_{\beta w} Tr(W^T S_W W) + \lambda \|B_W - \tilde{B}_W\|_F^2 + J(a)}{\min_{\beta b} Tr(W^T S_b W) + \lambda \sum_{c=1}^C \|B_c - \tilde{B}_c\|_F^2} \tag{4}$$

We use graph based semi-supervised learning, in which labeled data instances information can be propagated to unlabeled instances by using the similarity matrix. In another word, we can say that similar data instances in low dimensional manifold will have same class label because of their similar embedding. Therefore, the relationship between the instances in the given data-set can be represented as a weight graph or matrix G. For example, the relationship between data instances x_i and x_j is represented by an edge if they are the similar else there is no edge between them. Thus, the weight matrix G can be computed as follows,

$$G_{ij} = \begin{cases} 1, & \text{if } x_i \in N_k(x_j) \mid x_j \in N_k(x_i) \\ 0, & \text{otherwise} \end{cases}$$

where, $N_k(x_j)$ is the set of k nearest neighbours of x_j. From G, Laplacian matrix Matrix L can be calculated as $L = D - G$ where, D is the diagonal matrix whose entries are the column sums of G. Thus we can define a regularizer as follows,

$$\begin{aligned} J(a) &= \sum_{ij} \left(a^T x_i - a^T x_j\right)^2 S_{ij} = 2\sum_i a^T x_i D_{ii} x_i^T a - 2\sum_{ij} a^T x_i G_{ij} x_j^T a \\ &= 2a^T X(D-G)X^T a = 2a^T X L X^T a \end{aligned} \tag{5}$$

Thus, the Objective function for our proposed model can be written as,

$$\min_{W} \frac{\min_{\beta w} Tr(W^T S_W W) + \lambda \|B_W - \tilde{B}_W\|_F^2 + 2a^T X L X^T a}{\min_{\beta b} Tr(W^T S_b W) + \lambda \sum_{c=1}^C \|B_c - \tilde{B}_c\|_F^2} \tag{6}$$

This optimization problem can be solved by the maximum eigenvalue solution to the generalized eigenvalue problem.

4   Experimental Results

4.1   Experimental Setup and Benchmark Datasets

We have evaluated our proposed algorithm, SRCDA, against 4 other algorithms including, RCDA [10], LDA [6], PCA [4] and a NN (Nearest Neighbour) on the Cambridge ORL Database of Faces [12], Yale Face Database [13], PIE Face Dataset [14] and the COIL Dataset [15]. For our experiments, we have chosen a varying number (1–5) of samples per class as our labelled dataset and taken the rest of dataset, after discarding their labels as the unlabelled dataset. We have then performed PCA on the dataset to reduce its dimensionality. Datasets like the PIE Face Dataset have 1024 features per sample and performing the experiment on such a large feature matrix would be computationally inefficient. Hence we project the data onto a lower dimension (100) and then evaluate the algorithm on the projected matrix. We use a kNN classifier evaluate the performance of the algorithm. We run the experiment 30 times and compute the average result which has been shown in given in Table 1.

The datasets used in evaluating the algorithm have been discussed below:

- Cambridge ORL [12]: This database, compiled by AT&T Laboratories Cambridge, consists of 40 subjects against a dark homogeneous background in an upright, frontal position. Each subject has 10 images each with varying lighting, facial expressions and facial details.
- Yale [13]: This database has 10 subjects with images captured with 9 poses, under 64 illumination conditions and 1 image of the subject with an ambient illumination for each pose. There are a total of 5850 images in the database. The size of each image is 640 × 480.
- PIE Face [14]: The CMU Pose, Illumination, and Expression (PIE) database has 68 subjects captured across 13 poses, under 43 illumination conditions and with 4 different expressions. It has over 40,000 images. The images are all high quality images captured using 13 Sony DXC 9000 camera's.
- COIL [15]: Columbia Object Image Library (COIL-20) is a database of grayscale images of 20 subjects with 72 images each. Images of the subject were taken at pose intervals of 5° over 360°, against a black background. The database has two sets of images, the first of which contains 720 unprocessed images of 10 subjects while the second set contains 1,440 size normalized images of 20 objects.

4.2   Results and Discussions

In Table 1 we have reported the values obtained in our experiments and have plotted the values of a few of the datasets in Fig. 1. In the table the highest accuracies have been shown in bold text. We see that SRCDA shows a marked improvement over RCDA and this trend is seen in every dataset. This is especially true when the number of labelled samples per class is small. As the number of labelled samples per class increases it is seen that the performance of RCDA becomes comparable to that of SRCDA in some datasets. If the dataset has a large number of samples per class, then the performance may not be as high as that as SRCDA until the number of labelled samples per class is equal to the total number of labelled samples per class, i.e., the problem is converted into a supervised learning problem. Hence, SRCDA would be able to generalize well on any type of dataset and perform well.

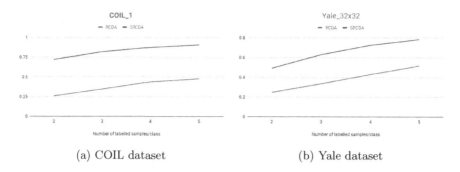

(a) COIL dataset   (b) Yale dataset

Fig. 1. Comparison of performance of SRCDA vs RCDA on COIL and Yale dataset

Table 1. Comparison of SRCDA with various state-of-the-art algorithms

ORL Dataset, Task: ORL_32x32						Yale Dataset, Task: Yale_32x32				
Algorithms	1	2	3	4	5	Algorithms	1	2	3	4	5
RCDA	0.1417	0.3067	0.8325	0.9327	0.9579	RCDA	0.1352	0.2461	0.3343	0.4295	0.5184
LDA	N/A	0.4439	0.7385	0.9384	**0.9707**	LDA	N/A	0.2489	0.3389	0.4303	0.5269
PCA	0.6036	0.7640	0.8523	0.9018	0.9407	PCA	**0.4189**	0.5541	0.6440	0.7147	0.7699
NN	0.5949	0.7548	0.8492	0.9058	0.9400	NN	0.4145	**0.5569**	**0.6503**	0.7145	0.7683
SRCDA	**0.6123**	**0.8218**	**0.9093**	**0.9458**	0.9632	SRCDA	0.3129	0.4925	0.6311	**0.7274**	**0.7855**
PIE Face Dataset, Task: PIE05						PIE Face Dataset, Task: PIE07				
Algorithms	1	2	3	4	5	Algorithms	1	2	3	4	5
RCDA	0.0789	0.6992	0.7368	0.7066	0.6969	RCDA	0.0997	0.7281	0.8196	0.8286	0.8235
LDA	N/A	0.5404	0.7795	0.8350	**0.8727**	LDA	N/A	0.6176	0.8211	0.8782	**0.9067**
PCA	0.1559	0.2550	0.3359	0.4088	0.4881	PCA	0.1835	0.3037	0.4046	0.4854	0.5568
NN	0.1654	0.2705	0.3593	0.4329	0.4947	NN	0.1858	0.3062	0.4047	0.4927	0.5656
SRCDA	**0.5327**	**0.7289**	**0.8021**	**0.8393**	0.8685	SRCDA	**0.5452**	**0.7396**	**0.8373**	**0.8812**	0.9044
PIE Face Dataset, Task: PIE09						PIE Face Dataset, Task: PIE27				
Algorithms	1	2	3	4	5	Algorithms	1	2	3	4	5
RCDA	0.0984	0.7285	0.8053	0.8189	0.8292	RCDA	0.0762	0.7065	0.7644	0.7571	0.7525
LDA	N/A	0.6043	0.8181	0.8701	**0.9054**	LDA	N/A	0.5452	0.7844	0.8389	**0.8751**
PCA	0.1925	0.3176	0.4204	0.5096	0.5826	PCA	0.1559	0.2550	0.3359	0.4088	0.4678
NN	0.1938	0.3234	0.4276	0.5154	0.5963	NN	0.1579	0.2576	0.34289	0.4161	0.4778
SRCDA	**0.5482**	**0.7466**	**0.8294**	**0.8767**	0.9044	SRCDA	**0.5435**	**0.737**	**0.7954**	**0.8341**	0.863
PIE Face Dataset, Task: PIE29						COIL Dataset, Task: COIL_1				
Algorithms	1	2	3	4	5	Algorithms	1	2	3	4	5
RCDA	0.1117	0.7205	0.8136	0.8246	0.8263	RCDA	0.0797	0.2591	0.3443	0.4365	0.4772
LDA	N/A	0.6056	**0.8148**	**0.8683**	**0.9033**	LDA	N/A	0.3823	0.4560	0.5355	0.6492
PCA	0.2058	0.3337	0.4357	0.5257	0.5918	PCA	**0.6903**	**0.7858**	**0.8414**	**0.8802**	0.9019
NN	0.2117	0.3358	0.4434	0.5312	0.6019	NN	0.2986	0.3631	0.3946	0.4215	0.8999
SRCDA	**0.5260**	**0.7229**	0.8121	0.8659	0.8946	**SRCDA**	0.6155	0.7211	0.8205	0.8753	**0.9085**

5   Conclusion

In this paper we have proposed a novel approach to Semi-supervised Dimensionality Reduction and have evaluated the proposed algorithm against several other state-of-the-art algorithms with various datasets. We have used label propagation with the help of a graph to better predict the labels of unlabelled data.

Additionally we have added a regularization term that helps incorporate the manifold structure in the objective function. In future we aim to implement the method on medical data [16, 17].

References

1. Bishop, C.M.: Pattern Recognition and Machine Learning. Springer, New York (2006)
2. Zhu, X.J.: Semi-supervised learning literature survey. Technical report, University of Wisconsin-Madison Department of Computer Sciences (2005)
3. Cunningham, J.P., Ghahramani, Z.: Linear dimensionality reduction: survey, insights, and generalizations. J. Mach. Learn. Res. **16**(1), 2859–2900 (2015)
4. Pearson, K.: LIII. On lines and planes of closest fit to systems of points in space. London Edinburgh Dublin Philos. Mag. J. Sci. **2**(11), 559–572 (1901)
5. Schölkopf, B., Smola, A., Müller, K.-R.: Kernel principal component analysis. In: Gerstner, W., Germond, A., Hasler, M., Nicoud, J.-D. (eds.) ICANN 1997. LNCS, vol. 1327, pp. 583–588. Springer, Heidelberg (1997). https://doi.org/10.1007/BFb0020217
6. Mika, S., Ratsch, G., Weston, J., Scholkopf, B., Mullers, K.R.: Fisher discriminant analysis with kernels. In: Neural Networks for Signal Processing IX: Proceedings of the 1999 IEEE Signal Processing Society Workshop (cat. no. 98th8468), pp. 41–48. IEEE (1999)
7. Roweis, S.T., Saul, L.K.: Nonlinear dimensionality reduction by locally linear embedding. Science **290**(5500), 2323–2326 (2000). https://doi.org/10.1126/science.290.5500.2323
8. He, X., Niyogi, P.: Locality preserving projections. In: Advances in Neural Information Processing Systems, pp. 153–160 (2004)
9. Huang, S., Elgammal, A., Huangfu, L., Yang, D., Zhang, X.: Globality-locality preserving projections for biometric data dimensionality reduction. In: The IEEE Conference on Computer Vision and Pattern Recognition (CVPR) Workshops, June 2014
10. Huang, K.K., Dai, D.Q., Ren, C.X.: Regularized coplanar discriminant analysis for dimensionality reduction. Pattern Recogn. **62**, 87–98 (2017). https://doi.org/10.1016/j.patcog.2016.08.024
11. Cai, D., He, X., Han, J.: Semi-supervised discriminant analysis (2007)
12. Samaria, F.S., Harter, A.C.: Parameterisation of a stochastic model for human face identification. In: Proceedings of 1994 IEEE Workshop on Applications of Computer Vision, pp. 138–142. IEEE (1994)
13. Yale face database b. http://cvc.yale.edu/projects/yalefaces/yalefaces.html
14. Sim, T., Baker, S., Bsat, M.: The CMU pose, illumination, and expression (PIE) database. In: Proceedings of Fifth IEEE International Conference on Automatic Face Gesture Recognition, pp. 53–58. IEEE (2002)
15. Nene, S.A., Nayar, S.K., Murase, H., et al.: Columbia object image library (COIL-20) (1996)
16. Barlow, H., Mao, S., Khushi, M.: Predicting high-risk prostate cancer using machine learning methods. Data **4**(3), 129 (2019)
17. Khushi, M., Dean, I.M., Teber, E.T., et al.: Automated classification and characterization of the mitotic spindle following knockdown of a mitosis-related protein. BMC Bioinform. **18**(16), 566 (2017)

Multi-object Tracking with Conditional Random Field

Xianming Zeng[1,2], Song Wu[1,2], and Guoqiang Xiao[1,2(✉)]

[1] College of Computer and Information Science, Southwest University,
Chongqing, China
`fuuko@email.swu.edu.cn`, {`songwu,gqxiao`}`@swu.edu.cn`
[2] National and Local Joint Engineering Laboratory of Intelligent Transmission
and Control Technology (Chongqing), Chongqing, China

Abstract. The goal of online multi-object tracking (MOT) is to estimate the tracks of multiple objects instantly with each incoming frame
using up-to-the-moment information. Conventional approaches to MOT
do not account for motion of the camera or background. Also, traditional
optical flow features or color features do not produce satisfactory discrimination. This paper attempts to deal with such problems by turning the
online MOT problem into the problem of minimizing the global energy
by constructing a conditional random field model. At the same time,
structural information and discriminative deep appearance features are
integrated together to make data association more accurate. In this way,
individual objects are not only more precisely associated across frames,
but are also dynamically constrained with each other in a global manner.
We conduct experiments using the MOT challenge benchmark to verify
the effectiveness of our method and achieve competitive results.

Keywords: Conditional random field · Deep appearance learning ·
Structural information · Multi-object tracking

1 Introduction

Multiple Object Tracking is emerging technology employed in many real world
applications. Given an input video, the task of MOT is largely partitioned to
locate multiple objects, maintaining their identities, and yielding their individual trajectories. Regarding the association strategy, frame-by-frame association
method that evaluates the affinity between a pair of tracklets, seems to be the
most important one in MOT. From the theory perspective, such a method is
able to obtain a good performance with lower complexity in simple scenarios
and situations when only a single motion pattern is involved. However, it tends
to produce more identity switch errors in complex scenarios, especially for intelligent vehicles, in which the online MOT algorithms suffer from various motion
patterns and severe occlusions.

© Springer Nature Switzerland AG 2019
T. Gedeon et al. (Eds.): ICONIP 2019, CCIS 1143, pp. 206–214, 2019.
https://doi.org/10.1007/978-3-030-36802-9_23

Deep learning based models have emerged as an extremely powerful framework to deal with different kinds of vision problems including image classification, object detection [1], and more relevantly single object tracking [2]. For the MOT problem, the strong observation model provided by the deep learning model for target detection can boost the tracking performance significantly [3]. The formulation and modeling of the target association problem using deep neural networks need more research efforts, although the first attempt to employ sequential neural networks for online MOT was made very recently.

In this paper, to overcome interference caused by camera or background motion and cope with a target's motion information that is too complex to calculate, we introduce the advantage of position invariability of structural information to design the algorithm. At the same time, with the CRF model, we are no longer limited to the difference between two sequential frames. By doing so, our method is capable of handling much more complicated motion patterns. Accordingly, we use a series of reasonable evaluation criteria for the object tracker, which provides reliable evidence for the experiment results.

The main contributions of this paper are threefold:

(1) We present a framework in which structural information is used in preliminary data association, and thus improve the association of object trajectories.
(2) In order to make the classification more effective, by amplifying the samples, we introduce a deep learning method in which the original images are projected into the feature space by metric learning, so as to obtain more discriminating features.
(3) We propose a new model for evaluating the reliability of trajectories based on the concept of a CRF.

The rest of this paper is organized as follows. Related work is reviewed in Sect. 2. In Sect. 3, we detail the proposed method. Section 4 shows results of experiments on challenging MOT datasets, while Sect. 5 concludes the paper.

2   Related Work

Most of the current multi-target tracking methods are based on the tracking-by-detection principle. A majority of the batch methods [4] formulate MOT as a global optimization problem in a graph-based representation, while online methods solve the data association problem either probabilistically or deterministically. Yang et al. [5] present a hybrid data association framework with a min-cost, multi-commodity network flow for robust online MOT. Li et al. [6] proposed a solution "designed in the labeled random finite set framework, using the product styled representation of labeled multi-object densities."

In the area of MOT, the deep-structured CRF model proposed by Shafiee et al. [7] "consists of a series of state layers, where each state layer spatially characterizes the object silhouette at a particular point in time." In their work,

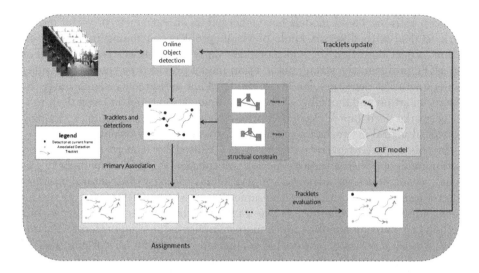

Fig. 1. Proposed framework for online MOT. For each frame of input, online multi-target detection is carried out first, and then the structural information is used to constrain the current frame and historical frame to produce a variety of possible data association schemes (primary association). Finally, the optimal scheme is discovered by constructing a conditional random field (CRF) model (tracklets evaluation), so as to realize tracking.

"the interactions between adjacent state layers are established by interlayer connectivity dynamically determined based on interframe optical flow." Azimifar et al. [8] developed a CRF-based predictor, which proposed a set of "new temporal relations for object tracking, with feature functions such as optical flow (calculated among consequent frames) and line filed features." The work presents a promising result, but the model is complicated by joint inference on both the detection and tracklet association. In contrast, our CRF method provides a discriminative affinity measurement that is generally applicable in different data sets.

3   Multi-object Tracking Framework

The trajectory of an object is represented by a sequence of states denoting the position and size of an object over a period of time. Figure 1 shows the detailed process. In order to better express the state of the target, we added the conception of velocity $(\dot{x}_t^i, \dot{y}_t^i)$. We denote the state of object i at frame t as $s_t^i = [x_t^i, y_t^i, \dot{x}_t^i, \dot{y}_t^i, w_t^i, h_t^i]$, respectively representing the coordinate (x_t^i, y_t^i), velocity and size of bounding box (w_t^i, h_t^i). The states of all target in frame t as $S_t(s_t^i \in S_t)$, $i \in N_t$, N_t is the number of targets. The structural information between any two targets is represented by relative position and relative speed:

$$e_t^{i,j} = [x_t^{i,j}, y_t^{i,j}, \dot{x}_t^{i,j}, \dot{y}_t^{i,j}]^T = [x_t^i - x_t^j, y_t^i - y_t^j, \dot{x}_t^i - \dot{x}_t^j, \dot{x}_t^i - \dot{x}_t^j] \tag{1}$$

$\varepsilon_t^i = \{e_t^{i,j} | \forall j \in N_t\}$ represents the correlation of all targets in frame t, $\varepsilon_t = \{\varepsilon_t^{i,j} | \forall i \in N_t\}$ represents the interrelation of all targets in a video. We also introduce an association matrix to represent the relationship between the target in the previous frame and the detection in the current frame. $A = \{a^{i,k} | i \in N_s, k \in N_d\}$ represents the incidence matrix. N_s and N_d, respectively, represent the number of targets and detections. If detection k is assigned to object i, the assignment is denoted $\{a^{i,k} = 1\}$. Otherwise, it is denoted $\{a^{i,k} = 0\}$.

3.1   Tracklet Generation

In order to adapt to the situation where the camera moves a lot, based on the theory that the mutual structural relations between objects in the continuous frame images do not change much, we add the structural information into the data association of the target, hoping that assignment can keep the structural information. Based on the theory that the motion of each target is equivalent to the superposition of multiple gaussian distributions. When the anchor points are matched, the whole trajectory should be satisfied:

$$\{T^i | \frac{\hbar}{2} \leq \sigma_{\hat{e}_t^i} \sigma_{\hat{s}_t^i} \leq s\frac{\hbar}{2}\} \tag{2}$$

Where $\hbar$ is the Planck's constant, $\sigma_{\hat{e}_t^i}$ and $\sigma_{\hat{s}_t^i}$ are the variance of relative position and size, respectively, and s is the scaled factor [9]. After the anchor points are aligned, structural information is used to match the remaining targets and observations. then we use the following function to determine the approximate position of other targets in the current frame:

$$d^{(i,k),j} = [x_d^k, y_d^k, 0, 0]^\top + [x^{j,i}, y^{j,i}, w^j, h^j] \tag{3}$$

We use the structural $e^{j,i}$ constraints between the position of k corresponding to the anchor point i and the target in the previous frame. Similarity measurement is represented by the following:

$$F(d^{(i,k),j}, d^q) = -\ln(\frac{area(B(d^{(i,k),j}) \cap B(d^q))}{area(B(d^{(i,k),j}) \cup B(d^q))}) \tag{4}$$

The similarity uses the overlap ratio, and $B(.)$ is the bounding box. In this way, for each pair of anchor points identified, an assignment can be obtained.

3.2   Tracklet Evaluation Based on Conditional Random Fields

In the CRF model, C^{*t-1} represents the tracklet that has been tracked before frame t. This part is considered accurate and will not be changed. $H^t(x)$ represents the possible data association in the structure constraint of the previous step. After matching tracklet C_m^{*t-1} with a certain detection d_i, a unique label is assigned to it. The energy function E is then constructed based on the CRF:

$$E(C^{*t-1}, H^t(x)) = \sum_m \Psi(C_m^{*t-1}, H_m^t(x)) + \sum_{m,l} \Phi(C_m^t, C_l^t) \tag{5}$$

$$\sum_m \Psi(C_m^{*t-1}, H_m^t(x)) = \sum_{i \in H_{m,x_m}^t} \psi_u(C_m^{*t-1}, d_i) + \psi_h(C_m^{*t-1}, H_{m,x_m}^t) \quad (6)$$

The first term $\Psi(.)$ represents the cost in the trajectory, and the second term $\Phi(.)$ represents the correlation between the trajectories after data association. The first term of the energy function consists of two parts, the overall motion smoothness and apparent consistency after assigning the current frame detection to the trajectory, noting that the objective function is minimized and that some of the expressions contain negative signs.

Motion Smooth Potential Energy Function. The motion smooth potential energy function is expressed as follows:

$$\psi_u(C_m^{*t-1}, d_i) = \begin{cases} \infty, if o^2(p(C_m^{*t-1}, t_i), d_i) < 0.5 \\ -\eta o^2(p(C_m^{*t-1}, t_i), d_i), otherwise \end{cases} \quad (7)$$

This reformulation mainly considers the similarity obtained by motion estimation, where $p(C_m^{*t-1}, t_i)$ denotes the position and size of the fitting curve at the moment t_i using least squares to estimate the trajectory. The fitting curve's order is different in different data sets; the MOT data set uses the first order, a straight line. The KITTI dataset uses the second order. $o^2(d_i, d_j)$ represents the Intersection over Union (IoU) of two detections. η is the decay factor.

Apparent Uniform Potential Energy Function. The apparent uniform potential energy function,

$$\psi_h(C_m^{*t-1}, H_{m,x_m}^t) = \gamma \cdot \sum_{i \in H_{m,x_m}^t} \xi(C_m^{*t-1}, d_i) + \varepsilon \cdot \sum_{(i,j) \in C_m^{*t-1} \cup H_{m,x_m}^t} \theta - K(d_i, d_j)$$

$$(8)$$

measures the apparent consistency of the trajectory. Here, γ, ε are scalar parameters. $\xi(d_i, d_j)$ denotes the square of the distance between two detections. The first item of the formula is the difference between the predicted detection and the detected detection. θ represents the difference in the ratio of width to height. K describes the apparent similarity of two detections, which is calculated using a depth feature proposed in Subsect. 3.3.

Mutually Exclusive Potential Function. In the mutually exclusive potential function,

$$\Phi(C_m^t, C_l^t) = \sum_{m,l} \alpha \cdot o^2(H_m^t, H_l^t) + \beta \cdot \rho(H_m^t, H_l^t) \quad (9)$$

the first term indicates the overlap ratio between the two hypotheses $o^2(H_m^t, H_l^t) = 2 * IoU(H_m^t, H_l^t)$. $\rho(H_m^t, H_l^t)$ is an indicator function that determines whether H_m^t and H_l^t are the same target. $\alpha = 0.5, \beta = 100$, respectively, represent the penalty coefficient. It is clear that there is zero tolerance for the same detection in the assignment hypothesis. The final loss function is expressed as

$$\hat{H} = \arg \min_{H} E(C^{*t-1}, H^t(x)) \tag{10}$$

Therefore, the target tracking is realized by finding the hypothesized scheme that minimizes the energy function.

3.3 Deep Metric Learning Feature Descriptors

The deep network consists of eight layers: three convolutional, three pooling, and two fully connected layers. The term Lower level (conv, max pooling) is used to calculate lower level features, and High-level (fc) is used to compute high-level features, so the complexity of computation is reduced dramatically. The trained VGG-M [10] network was used to adjust the parameters. Given image pairs of objects, x_p and x_q, we can extract their high-level features x_p^l and x_q^l by passing them to the network. We then define the similarity between x_p^l and x_q^l as the distance between their output features using the square of the L2 norm as

$$D_g^2(x_p, x_q) = \left\| x_p^L - x_q^L \right\|_2^2 \tag{11}$$

Since our goal is to build a deep network that can discriminate between the appearances of object image pairs, we impose $D_g^2(x_p, x_q) \leq \tau$ with a threshold τ for positive pairs, whereas $D_g^2(x_p, x_q) > \tau$ for negative pairs, when training the network. Based on this pairwise constraint, we define a loss function and learn parameters by minimizing the function defined as

$$\arg \min_{\{w^l, b^l\}} F = \frac{1}{2} \sum_{p,q} \Omega(y_{p,q}(D_g^2(x_p, x_q) - \tau)) \tag{12}$$

$$+ \frac{\lambda}{2} \sum_{l=1}^{L} (\left\| \mathbf{w}^l \right\|_F^2 + \left\| \mathbf{b} \right\|_2^2), l = 1, ..., L$$

where $\mathbf{w}^l = \{w_{ij}^l\}$ and $\mathbf{b}^l = \{b_j^l\}$ for $l = 1, 2, 3$. $\Omega(e) = \frac{1}{\beta} \log(1 + \exp(\beta e))$ is the generalized logistic loss function, which is the smooth approximation of the hinge loss function. If x_p and x_q are a positive pair, $y_{pq} = 1$. Otherwise, $y_{pq} = -1$. The first term therefore penalizes a pair that does not follow the pairwise constraint. On the other hand, the second term penalizes the large weights and biases with the parameter λ for preventing overfitting to the training data.

4 Experiment

4.1 Datasets

This section describes our tests of the performance of the proposed tracking algorithm on multiple databases. The MOT challenge [11] has collected extensive live video and some challenging new sequences. The video sequences are divided into two categories: static background and dynamic background. The data set is composed of 11286 frames, including 10 training and 11 testing video sequences. Some of the videos were recorded using mobile platforms and the others are from surveillance videos. Because it is composed of videos in various configurations, tracking algorithms that are particularly tuned for a specific scenario would not work well in general.

Table 1. Results on real-world datasets

Sequence	Method	MOTA↑	MOTP↑	MT↑	PT	ML↓	FP↓	FN↓	IDS↓	FG↓
PETS S2L1	MDP	80.72%	87.31%	52.63%	47.37%	0.00%	1.04%	17.91%	**16**	**14**
	RMOT	78.44%	87.46%	89.47%	10.53%	0.00%	16.18%	4.92%	21	20
	CMOT	89.57%	**87.68%**	94.74%	5.26%	0.00%	3.51%	6.48%	22	20
	Conf.Map	56.30%	79.70%	–	–	–	–	–	–	–
	Our proposed	**90.98%**	85.50%	**95.21%**	5.26%	0.00%	8.69%	7.85%	20	17
PETS S2L2	MDP	47.37%	76.08%	16.28%	69.77%	13.95%	**4.90%**	45.95%	**184**	**177**
	RMOT	38.26%	78.14%	46.51%	46.51%	**6.98%**	26.38%	**31.45%**	414	390
	CMOT	49.31%	**79.01%**	51.16%	37.21%	11.63%	12.53%	35.42%	286	268
	Our proposed	**51.73%**	77.34%	**54.78%**	42.60%	10.72%	15.33%	37.23%	420	270
ETH	MDP	60.20%	82.70%	47.58%	42.74%	9.67%	**10.06%**	28.75%	106	91
	RMOT	**66.18%**	80.24%	62.90%	31.45%	**5.65%**	14.60%	**18.41%**	88	**55**
	CMOT	60.30%	81.30%	58.87%	33.87%	7.26%	12.75%	25.95%	108	72
	Our proposed	65.57%	**84.23%**	**68.34%**	32.54%	8.47%	11.86%	24.35%	**83**	86
KITTI	MDP	67.93%	93.39%	37.01%	35.83%	27.17%	**0.54%**	30.34%	**83**	**64**
	RMOT	71.01%	86.57%	**56.30%**	41.34%	2.36%	7.59%	**14.31%**	111	96
	CMOT	59.23%	89.55%	41.09%	51.98%	6.93%	7.80%	17.41%	196	174
	Our proposed	**72.35%**	**93.62%**	55.60%	36.25%	2.36%	7.34%	18.52%	124	74

4.2 Performance

We used multiple metrics to evaluate the multiple object tracking performance as suggested by the MOT Benchmark. Multiple Object Tracking Accuracy (MOTA) measures three kinds of errors. Multiple Object Tracking Precision (MOTP) measures the overlap ratio between the bounding box of the tracker output and the ground truth, whose threshold is generally set as 0.5, which reflects the precision of determining the object locations. False positives (FPs), false negatives (FNs) and identity switch (IDS) reflect the precision of determining the object locations. Additionally, we adopt the Trajectory-Based measures (TBM), including Mostly Tracked (MT), Mostly Lost (ML), Partially Tracked (PT), and Fragment (FRG).

After analysis of the validation set, we performed training with all the training sequences, and tested the trained trackers on the test set. Table 1 shows the experiment results compared with three online methods: MDP, RMOT, and the baseline CMOT. MDP formulates the online MOT problem as decision making in a Markov decision process. Complete target tracking is achieved by assigning a life cycle to the target being tracked. RMOT uses a motion contest from multiple moving objects that are mostly insensitive to unexpected camera motions.

In addition, CMOT has an advantage over pedestrian tracking with a wealth of appearance information that is different from the ACT data set. Although these methods perform well in some scenarios, they still have limitations. As indicated by the chart, our methods have more advantages on the overall tracking performance compared with MOTA and MOTP, which shows the robustness and stability of our method. By combining structural information and salient features, the proposed method can better deal with relative motion between the target and the camera, and thus produce better results.

5 Conclusion

In this paper, We have proposed a novel online multi-object tracking framework based on conditional random field and a distinctive feature based on metric learning. In the target tracking phase, the data association utilizes the structural constraints of the trajectory. In the trajectory evaluation phase, a conditional random field model is constructed for the trajectory of multiple targets. Experimental results on a large number of datasets demonstrate the effectiveness of the proposed algorithm.

Acknowledgement. This work was supported by the National Natural Science Foundation of China (NO.61806168), Fundamental Research Funds for the Central Universities (NO. SWU117059), and Venture & Innovation Support Program for Chongqing Overseas Returnees (NO. CX2018075).

References

1. Ren, S., He, K., Girshick, R., Sun, J.: Faster R-CNN: towards real-time object detection with region proposal networks. In: International Conference on Neural Information Processing Systems, pp. 91–99 (2015)
2. Girshick, R., Donahue, J., Darrell, T., Malik, J.: Rich feature hierarchies for accurate object detection and semantic segmentation, pp. 580–587 (2013)
3. Yu, F., Li, W., Li, Q., Liu, Y., Shi, X., Yan, J.: POI: multiple object tracking with high performance detection and appearance feature. In: Hua, G., Jégou, H. (eds.) ECCV 2016. LNCS, vol. 9914, pp. 36–42. Springer, Cham (2016). https://doi.org/10.1007/978-3-319-48881-3_3
4. Zhang, L., Li, Y., Nevatia, R.: Global data association for multi-object tracking using network flows. In: IEEE Conference on Computer Vision & Pattern Recognition (2008)
5. Yang, M., Wu, Y., Jia, Y.: A hybrid data association framework for robust online multi-object tracking. IEEE Trans. Image Process. **PP**(99), 1 (2017)
6. Li, S., Wei, Y., Hoseinnezhad, R., Wang, B., Kong, L.J.: Multi-object tracking for generic observation model using labeled random finite sets. IEEE Trans. Signal Process. **66**(2), 368–383 (2017)
7. Shafiee, M.J., Azimifar, Z., Wong, A.: A deep-structured conditional random field model for object silhouette tracking. Plos One **10**(8), e0133036 (2015)
8. Shafiee, M.J., Azimifar, Z., Fieguth, P.: Model-based tracking: temporal conditional random fields (2010)

9. Zhang, B., Perina, A., Li, Z., Murino, V., Liu, J., Ji, R.: Bounding multiple Gaussians uncertainty with application to object tracking. Int. J. Comput. Vision **118**(3), 364–379 (2016)
10. Chatfield, K., Simonyan, K., Vedaldi, A., Zisserman, A.: Return of the devil in the details: delving deep into convolutional nets. arXiv preprint arXiv:1405.3531 (2014)
11. Multiple object tracking benchmark. http://motchallenge.net/

Intra-Modality Feature Interaction Using Self-attention for Visual Question Answering

Huan Shao[1], Yunlong Xu[2($\boxtimes$)], Yi Ji[1], Jianyu Yang[3], and Chunping Liu[1($\boxtimes$)]

[1] School of Computer Science and Technology, Soochow University,
Suzhou 215006, People's Republic of China
hshao@stu.suda.edu.cn, {jiyi,cpliu}@suda.edu.cn
[2] Applied Technical School of Soochow University,
Suzhou 215325, People's Republic of China
ylxu@suda.edu.cn
[3] School of Rail Transportation, Soochow University,
Suzhou 215006, People's Republic of China
jyyang@suda.edu.cn

Abstract. Better capturing the interactions of different modality is a hot research topic in visual question answering (VQA) recently. Inspired by human vision information processing, a method of VQA based on intra-modality features interactive with self-attention mechanism (IMFI-SA) is proposed. We adopted object-level features with bottom-up attention instead of feature mapping to extract the fine-grained information in images. Moreover, the interactions of intra-modality in the question and the image modality is also extracted by proposed IMFI-SA model respectively. Finally, we combined the enhanced object-level features interaction using top-down cross-attention and the question features interaction to predict the answer given a question and image. Experimental results on the VQA2.0 dataset show that the proposed method is superior to the existing method in the reasoning answer generating, especially in counting problems.

Keywords: Cross-modality · Object interaction · Visual Question Answering · Self-attention

1 Introduction

Visual question answering (VQA) [2] is a computer vision task that focuses on solving cross-modality fusion between a natural language question and a given image. With the development of annual VQA Challenge, the research of VQA has been a hot topic in the field of visual understanding. Most improvement methods are centered on three aspects: feature representation [1,18], attention [12,13] and the strategy of cross-modality fusion [6,23]. For the extraction of visual feature, most extracting methods of feature maps are based on VGG network [16] and

© Springer Nature Switzerland AG 2019
T. Gedeon et al. (Eds.): ICONIP 2019, CCIS 1143, pp. 215–222, 2019.
https://doi.org/10.1007/978-3-030-36802-9_24

ResNet network [9]. Recently Anderson et al. [1] proposed bottom-up feature with the object-level feature to replace the feature map. Besides, visual attention representation in the VQA model is a mainstream trendency [13]. The attention mechanisms in existing VQA models include hard attention [21], soft attention [21], self-attention [17] and co-attention [12]. In order to further explore the deep correlation between questions and images, the current VQA research focuses on the cross-modality feature fusion methods such as simple multiplication [2], bilinear multiplication [6] and so on.

It is very robust that human beings understand image content to answer the question, especially for some complex and reasoning questions. Therefore, imitating human understanding process, inspired by the self-attention mechanism in natural language processing (NLP) research, based on the multi-head attention mechanism [19], we propose intra-modality feature interaction using self-attention method to mine the interaction information.

2   Related Works

2.1   Bottom-Up Feature

Extracting feature maps is the basis of VQA task. The feature map contains global information, but the important information related to the answer is often just a part of the image. To reduce the dimension of visual feature and highlight the detailed information of the image, Anderson proposed the object detection tool [1] with Fastser-RCNN pre-trained on the ImageNet [15] and Visual Genome dataset [20] instead of CNN. They choose ResNet-101 as the backbone network and exploit the entire Res-5 block as the feature of the second-stage region classifier. Based on the 7×7 average pooling operation of the pre-training model, the image regions can be represented by 2048-dimensional features.

2.2   Attention Mechanism

Human attention mechanism can quickly locate and extract the important content according to the target content. In visual attention, the importance of visual contents can be achieved by assigning different weights to different image regions. In NLP, the coding representation of input sequence is regarded as a set of Key-Value pairs (K-V) and Query (Q). Given a Query and a Key-Value pair, attention mechanism can simplify the process of mapping a Query to the correct input feature. The final output of weighted summation that is determined by the Query, Key, and Value.

In the VQA task, the original joint embedding method of image and question can only achieve limited precision [2]. The simple fusion method is completely different from humans' way of answering questions. Humans can quickly achieve visual answering based on a small number of words in the question with the help of visual attention mechanism. Attention mechanism allows the model to focus on the question-related image regions [5]. Image regions with low weights have less impact on subsequent calculations, while image regions with high weights have a greater impact on model reasoning.

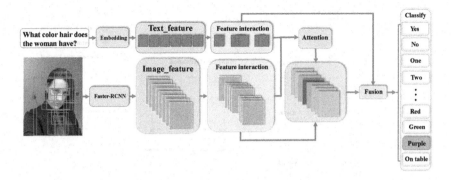

Fig. 1. Architecture of IMFI-SA. The original bottom-up feature extraction layers, highlighted in blue and the intra-modality interaction feature layers is highlighted in yellow. The cross-modality feature fusion layers, highlighted in orange, are composed of the enhanced bottom-up visual features with attention layer and the cross-model interaction feature fusion layer. (Color figure online)

2.3 Multi-head Self-attention

Self-attention, also known as intra-attention, is a mechanism of attention associated with different locations of a single sequence in order to calculate the interactive representation of the sequence. It has been proved to be very effective in many fields, such as machine reading [3] and text summary in NLP, image captioning [21] and so on. Google proposed a novel and attention-only framework for machine translation [19]. The special form of the attention model is a hybrid neural network with feedforward layer and self-attention layer. Encoding representation of input sequence in this framework is regarded as a set of key-value pairs (K, V) and query Q with K = V = Q. In other words, the self-attention Query, Key and Value are all derived from the same vector. The proposed multi-head attention [19] does not just compute attention once, but also captures the relevant information in different representational subspaces of different dimensions by counting multiple times in parallel. Independent attention output is simply connected and linearly converted into the desired dimension. Multi-head attention allows models to focus on information from different representational subspaces at different locations.

3 Self-attention Based Intra-Modality Feature Interaction Model

In this section, we describe the architecture of the intra-modality feature interaction model using self-attention (IMFI-SA). The whole structure is illustrated in Fig. 1. Compared to the model in [7], we use a similar network structure, but they focus on question feature and reinforce the impact of question type. We pay more attention to the interaction of two different modal features, and use the better interactive features to guide the model to get the correct answer.

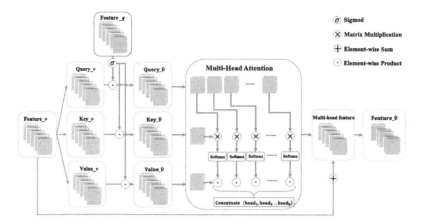

Fig. 2. Framework of the Intra-Modality Feature Interaction method (IMFI). Given feature v, vector Query, Key, and Value are achieved through the fully connected layer. Then feature q is introduced to make Query, Key, and Value with other modality information respectively. We divide the fusion of Query, Key, and Value into h by Multi-head attention, performed h times dot-product attention calculations, and then keep the dimensions unchanged after concatenate. Finally the new vector is added to the original vector.

3.1 Original Bottom-Up Features

VQA task is a joint learning task based on cross-modality information, which need to describe the content from two aspects: visual image and question text. For the questions, we discard the parts of more than 14 words in the questions, and use the zero vector-filling method to solve the questions of less than 14 words. For each sentence, we use GRU [4] and GloVe [14] to initialize the words and set the vector to 14×300. For each image, we directly feed it to Faster-RCNN to obtain $K \times 2048$-D feature representations. In the model training process, we generally set K value between 36 to 100.

3.2 Intra-Modality Feature Interaction

Given image feature v or question feature q, IMFI method can update the parameters of each feature to establish a relationship between the features. We firstly use three fully connected layers to transform the feature vector into three vectors: Query, Key and Value. This structure of can be seen in Fig. 2. Self-attention is originally used to handle textual features in single-modality tasks, but we extend this approach to multi-modal tasks. We introduce multi-modal information before making subsequent feature parameter updates. For the vector Key and Value, the same processing method as Query is used, as shown below:

$$Query_v = Linear\,(v) \quad Qu\hat{e}ry_v = Query_v \odot q \tag{1}$$

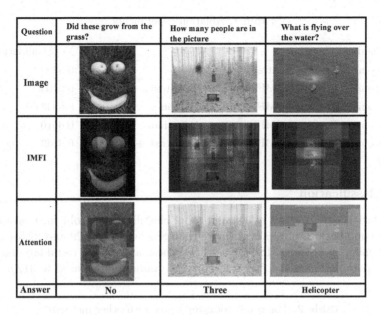

Question	Did these grow from the grass?	How many people are in the picture	What is flying over the water?
Image			
IMFI			
Attention			
Answer	No	Three	Helicopter

Fig. 3. Through our proposed IMFI-SA approach, guided by the Attention mechanism, our model can further find key objectives and ultimately reason the correct answer.

We calculate Query, Key, and Value h times respectively. In each calculation, we use dot-product for Query and Key, divide each by $\sqrt{d_k}$, and apply a softmax function to obtain the weights on the values. We set the number of heads to 8 in our experiment as follows:

$$head_i = softmax(\frac{\hat{Query}_{vi} \times \hat{Key}_{vi}^{T}}{\sqrt{d_k}}) \odot \hat{Value}_{vi} \tag{2}$$

$$MutiHead(Q,K,V) = Concat(head_1,...,head_8) \tag{3}$$

3.3   Cross-modality Feature Fusion

We use top-down attention [13] for obtaining the weights of image features. K weights are calculated by the softmax function as follows:

$$\alpha_i = softmax\left(w_a f_a\left(\hat{Feature\_v}, \hat{Feature\_q}\right)\right) \tag{4}$$

$$Att\_\hat{v} = \sum_{i=1}^{K} \alpha_i \hat{Feature\_v} \tag{5}$$

For the image vector (v) and question vector (q) entering the fusion layer, we use a simple element-wise product [24] after converting dimensions through the fully connected layer as follows:

$$Fusion\_out = Linear\left(Att\_\hat{v}\right) \odot Linear\left(\hat{Feature\_q}\right) \tag{6}$$

Table 1. Ablation experiments

Method	VQA-validation
Bottom-up Feature + Attention	0.6320
Bottom-up Feature + IMFI-image + IMFI-text	0.6103
Bottom-up Feature + IMFI-text + Attention	0.6343
Bottom-up Feature + IMFI-image + Attention	0.6510
Bottom-up Feature + IMFI-image + IMFI-text + Attention	0.6549

3.4 Classification

We regard the VQA task as a multi-label classification problem. Considering that the number of samples with more than 8 answers is 3129 and 92.5% of the total number of answers [18], we screened these answers as candidate answers, and the parameters of the fusion feature are gradually increased to 3129.

Table 2. The results of comparison with other methods

VQA2.0 test-dev					Test-standard			
Model	Yes/No	Number	Other	All	Yes/No	Number	Other	All
MCB + Attention [6]	82.20	37.70	54.80	64.20	–	–	–	63.20
MLB + Attention [10]	**84.02**	37.90	54.77	65.08	–	–	–	65.07
MFB + Attention [23]	82.50	38.30	55.20	64.60	–	–	–	–
Multi-level-attention [22]	81.80	41.20	56.70	65.30	81.3	41.9	56.5	65.2
Bottom-up + attention [1]	81.82	44.21	56.05	65.32	82.20	43.90	56.26	65.67
QGHC + Attention + Concat [7]	83.54	38.06	57.10	65.89	–	–	–	65.90
Ours (IMFI-SA + Attention)	83.16	**47.63**	**57.33**	**66.87**	**83.52**	**47.76**	**57.35**	**67.15**

4 Experiments Results and Analysis

In this section, we show that the results of ablation experiment and comparative experiments with some of the latest models on VQA 2.0 dataset [8].

4.1 Dataset

In our experiment, we choose VQA2.0 dataset proposed in 2016. It has 265,016 images from Microsoft COCO dataset [11], at least 3 questions per image, 10 ground truth answers per question and 3 plausible answers per question. In the VQA 2.0, for the yes/no question, the answer to yes and no is half-half. This makes the VQA system have to make more use of the image features.

4.2 Experiments

Table 1 shows that the results of our ablation experiments on VQA2.0-Validation dataset. We can also see the visualization results about IMFI and attention layer of the model in Fig. 3. From Table 2, our proposed method has a significant improvement in overall accuracy.

5 Conclusion

In this paper, we propose a VQA model based on self-attention to establish the intra-modality relationship. Our approach can obtain features with higher representation information and the independent modules in the model can be widely used to handle multi-modal problems. In addition, existing fusion methods do not seem to work well for our highly informative features. Based on this framework, we will further study the fusion method of multi-modal information.

Acknowledgments. Supported by National Natural Science Foundation of China (61773272, 61272258, 61301299), The Natural Science Foundation of the Jiangsu Higher Education Institutions of China (19KJA230001), the Priority Academic Program Development of Jiangsu Higher Education Institutions.

References

1. Anderson, P., et al.: Bottom-up and top-down attention for image captioning and visual question answering. In: Proceedings of the IEEE Conference on Computer Vision and Pattern Recognition, pp. 6077–6086 (2018)
2. Antol, S., et al.: VQA: visual question answering. In: Proceedings of the IEEE International Conference on Computer Vision, pp. 2425–2433 (2015)
3. Cheng, J., Dong, L., Lapata, M.: Long short-term memory-networks for machine reading. arXiv preprint arXiv:1601.06733 (2016)
4. Chung, J., Gulcehre, C., Cho, K.H., Bengio, Y.: Empirical evaluation of gated recurrent neural networks on sequence modeling. Eprint Arxiv (2014)
5. Das, A., Agrawal, H., Zitnick, L., Parikh, D., Batra, D.: Human attention in visual question answering: do humans and deep networks look at the same regions? Comput. Vis. Image Underst. **163**, 90–100 (2017)
6. Fukui, A., Park, D.H., Yang, D., Rohrbach, A., Darrell, T., Rohrbach, M.: Multimodal compact bilinear pooling for visual question answering and visual grounding. arXiv preprint arXiv:1606.01847 (2016)
7. Gao, P., et al.: Question-guided hybrid convolution for visual question answering. In: The European Conference on Computer Vision (ECCV), September 2018
8. Goyal, Y., Khot, T., Summers-Stay, D., Batra, D., Parikh, D.: Making the V in VQA matter: elevating the role of image understanding in visual question answering (2017)
9. He, K., Zhang, X., Ren, S., Sun, J.: Deep residual learning for image recognition. In: Proceedings of the IEEE Conference on Computer Vision and Pattern Recognition, pp. 770–778 (2016)
10. Kim, J.H., On, K.W., Lim, W., Kim, J., Ha, J.W., Zhang, B.T.: Hadamard product for low-rank bilinear pooling. arXiv preprint arXiv:1610.04325 (2016)

11. Lin, T.-Y., Maire, M., Belongie, S., Hays, J., Perona, P., Ramanan, D., Dollár, P., Zitnick, C.L.: Microsoft COCO: common objects in context. In: Fleet, D., Pajdla, T., Schiele, B., Tuytelaars, T. (eds.) ECCV 2014. LNCS, vol. 8693, pp. 740–755. Springer, Cham (2014). https://doi.org/10.1007/978-3-319-10602-1_48

12. Lu, J., Yang, J., Batra, D., Parikh, D.: Hierarchical question-image co-attention for visual question answering. In: Advances in Neural Information Processing Systems, pp. 289–297 (2016)

13. Patro, B., Namboodiri, V.P.: Differential attention for visual question answering. In: Proceedings of the IEEE Conference on Computer Vision and Pattern Recognition, pp. 7680–7688 (2018)

14. Pennington, J., Socher, R., Manning, C.: GloVe: global vectors for word representation. In: Proceedings of the 2014 Conference on Empirical Methods in Natural Language Processing (EMNLP), pp. 1532–1543 (2014)

15. Russakovsky, O., et al.: ImageNet large scale visual recognition challenge. Int. J. Comput. Vision **115**(3), 211–252 (2014)

16. Simonyan, K., Zisserman, A.: Very deep convolutional networks for large-scale image recognition. arXiv preprint arXiv:1409.1556 (2014)

17. Tan, Z., Wang, M., Xie, J., Chen, Y., Shi, X.: Deep semantic role labeling with self-attention. In: Thirty-Second AAAI Conference on Artificial Intelligence (2018)

18. Teney, D., Anderson, P., He, X., Hengel, A.V.D.: Tips and tricks for visual question answering: learnings from the 2017 challenge (2017)

19. Vaswani, A., et al.: Attention is all you need. In: Advances in Neural Information Processing Systems, pp. 5998–6008 (2017)

20. Vendrov, I., Kiros, R., Fidler, S., Urtasun, R.: Order-embeddings of images and language. arXiv preprint arXiv:1511.06361 (2015)

21. Xu, K., et al.: Show, attend and tell: neural image caption generation with visual attention. In: International Conference on Machine Learning, pp. 2048–2057 (2015)

22. Yu, D., Fu, J., Mei, T., Rui, Y.: Multi-level attention networks for visual question answering. In: The IEEE Conference on Computer Vision and Pattern Recognition (CVPR), July 2017

23. Yu, Z., Yu, J., Fan, J., Tao, D.: Multi-modal factorized bilinear pooling with co-attention learning for visual question answering. In: Proceedings of the IEEE International Conference on Computer Vision, pp. 1821–1830 (2017)

24. Zhou, B., Tian, Y., Sukhbaatar, S., Szlam, A., Fergus, R.: Simple baseline for visual question answering. Computer Science (2015)

Reinforcement Learning with Attention that Works: A Self-Supervised Approach

Anthony Manchin[✉], Ehsan Abbasnejad, and Anton van den Hengel

The Australian Institute for Machine Learning, The University of Adelaide,
Adelaide, Australia
{anthony.manchin,ehsan.abbasnejad,anton.vandenhengel}@adelaide.edu.au

Abstract. Attention models have had a significant positive impact on
deep learning across a range of tasks. However previous attempts at
integrating attention with reinforcement learning have failed to produce
significant improvements. Unlike the selective attention models used
in previous attempts, which constrain the attention via preconceived
notions of importance, our implementation utilises the Markovian properties inherent in the state input. We propose the first combination of
self attention and reinforcement learning that is capable of producing
significant improvements, including new state of the art results in the
Arcade Learning Environment.

Keywords: Reinforcement learning · Attention · Deep learning

1 Introduction

Research on reinforcement learning (RL) has seen accelerating advances in the
past decade. In particular, methods for deep RL have made tremendous progress
since the seminal work of Mnih *et al.* [16] on Deep Q-Networks. A number of
different approaches have progressed the state of the art on simulated tasks - in
particular in the Arcade Learning Environment [2] - to, or above that of human
players. The major focus of attention of many of these methods is the learning
and optimization of the policies. However, one thing that all of these approaches
have in common is that they process the raw input data through a convolutional
neural network (CNN).

Regardless of the chosen policy optimisation method, the underlying neural network is responsible for interpreting and encoding useful representations
of the input state. While significant efforts have focused on methods for policy
optimisation, the techniques for encoding the observations have received relatively less attention. While many methods thus use generic, off-the-shelf CNN
architectures, we instead focus on this as the main subject of study.

Taking inspiration from other areas of deep learning, we investigate the benefits of incorporating self-attention into the underlying network architecture.
Attention models were applied with remarkable success to complex visual tasks

© Springer Nature Switzerland AG 2019
T. Gedeon et al. (Eds.): ICONIP 2019, CCIS 1143, pp. 223–230, 2019.
https://doi.org/10.1007/978-3-030-36802-9_25

such as video and scene understanding [6,10,19], natural language understanding including machine translation [1,27], and generative models using generative adversarial networks (GANs) [13,24]. Although previous attempts to integrate attention with RL have been made, these attempts have largely used hand-crafted features as inputs to the attention model [25,26]. These features often do not satisfy the Markov principle. However the spatial and temporal information contained within the state input (when using stacked frames), is sufficient to satisfy the Markov principle.

For this reason we observe the work from Wang *et al.* [22], and their goal of improved attention through space, time and, space-time by combining self-attention with non-local filtering methods. The benefit of self-attention is the ability to compute representations of an input sequence by relating different positions of the input sequence. Their implementation achieves state of the art results on the Kinetics [14] and Charades [20] datasets. However, the datasets they considered are large-scale video classification problems where the changes in the input from time to time are minimal. In addition, the neural network is in a passive environment that does not require interaction. None the less, these challenges require spatial and temporal reasoning abilities that would be very useful for a reinforcement learning agent to posses. Taking inspiration from this work, we capitalise on the Markovian principle of the state input and propose a novel implementation of self-attention within the classical convolutional neural network architecture, as used by Mnih *et al.* [16]. The contributions of our paper are as follows.

- We provide a spatio-temporal self-attention mechanism for reinforcement learning and demonstrate that the network architecture has significant benefits in learning a good policy
- We present state-of-the-art results in the Arcade Learning Environment [2]. In particular, our approach significantly outperforms the baseline across a number of environments where the agent has to attend to multiple opponents and anticipate their movements in time.

2   Related Work

Rl in Video Games. Current state-of-the-art approaches for RL in the Arcade Learning Environment are built on top of the original network architecture proposed by Mnih *et al.* [16]. Alterations to this underlying network architecture have often included the implementation of recurrent neural networks (RNN). Hausknecht *et al.* [11] proposed replacing the fully connected layer, following the output of the last convolutional layer of the network with an LSTM. This allowed for a single frame input to be used, as opposed to sequentially stacked frames, with the LSTM integrating temporal information. Oh *et al.* [17] also proposed using recurrent networks with their Recurrent Memory Q-Network (FRMQN). This memory-based approach used a mechanism based on soft attention to help read from memory and was evaluated with respect to solving mazes in Minecraft (a flexible 3D world).

A different approach by Fortunato *et al.* [7] proposed adding parametric noise to the networks weights to aid efficient exploration, replacing conventional exploration heuristics. This modification generally resulted in positive improvements, lending support to the idea that carefully considered improvements to the underlying network architecture can be beneficial.

In comparison to Hausknecht *et al.* and Oh *et al.* we utilise sequentially stacked frames as input, and augment the network with an attention model which is able to demonstrate improved temporal reasoning.

Attention in RL. Sorokin *et al.* [21], which studied the effects of adding both 'soft' and 'hard' attention models to the network used by Mnih *et al.* [16] . These attention models received spatial information from a single frame processed by a CNN and temporal information from a RNN. Although this approach indicated some potential performance improvements under certain conditions, experiments were limited with results showing no systematic performance increases. In contrast, our work explores a different form of self-attention, and we demonstrate significant benefits, in both performance and interpretability for the resulting policy.

Choi *et al.* [4] also proposed combining attention with reinforcement learning for navigation purposes. This approach employed a Multi-focus Attention Network which used multiple parallel attention modules. This worked by segmenting the input, with each parallel attention layer attending to a different segment. This method was evaluated in a custom, synthetic grid-world environment, in which they reported better sample efficiency, in comparison to the standard DQN.

Zhang *et al.* [26] proposed an attention-guided imitation learning framework. They trained a model to replicate human attention with supervised gaze heatmaps. The input state was then augmented with this additional information. This style of attention fundamentally differs from that used in our work as it incorporates hand crafted features as input. Gregor *et al.* [8] also investigated visual attention using a glimpse sensory approach. However this approach only investigated the integration of visual attention at the input layer, providing the network with different 'glimpses' of the full state.

More recently Yuezhang *et al.* [25] proposed a model based upon the Broadbent filter model [3]. This approach uses the optical flow calculated between two frames to construct an attention map, which was then combined with the output of the last convolutional layer in their network. This combination led to improved results when tested on a modified version of the toy problem 'Catch', originally inspired by Mnih *et al.* [15]. However the model was unable to replicate the same types of improvements in more visually complex domains.

3   Proposed Approach

We build upon the common practice of using a CNN to encode input observations into a state representation, suitable for complex decision making. Our main contribution is to incorporate a self-attention mechanism over space and time. This architecture will be shown to provide a significant benefit in learning effective policies. Previous works have attempted to provide a form of temporal attention by pairing CNNs with RNNs [11], but this resulted in limited success for temporal reasoning. There is additional evidence to suggest that agents acting on the simplified input of a single frame, may suffer difficulties in learning useful relationships over these inputs [9,11]. We hypothesize that utilising relationships over parts of the input observations – over space and time – are crucial for an agent to execute effective policies. This motivates the use of explicit attention mechanisms.

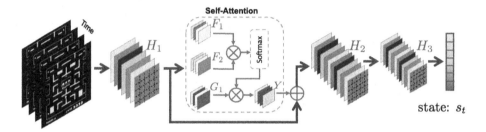

Fig. 1. Overview of the proposed architecture. We introduce a self-attention module within the CNN used to process the input observations. The resulting policy benefits significantly from the capability of selective attention over space and time.

3.1   Self-Attention Mechanism

We specifically describe the implementation of self-attention used in our approach (see Fig. 1), as originally proposed by [22]. The self-attention mechanism operates as follows. F_1, F_2, and G_1 are all 1×1 convolutions. The outputs of F_1 and F_2 are matrix multiplied together before passing through a *Softmax* activation, which is then matrix multiplied by the output from G_1. This is then passed through Y which is also a 1×1 convolution, before being added back into the original input. Our approach 'Self-Attending Double Network' (SADN), involves incorporating a block of self-attention after the convolutional layers H_1 and H_2. The incorporation of 'self-attention' in this manner allows for attention to be applied to spatial and temporal features simultaneously.

3.2   Validation Methodology

Implementation. The Arcade Learning Environment is a well established baseline which allows us to critically evaluate the effects of our proposed architecture modifications. We use Proximal Policy Optimisation [18] to train our agents over traditional DQN baselines due to its wall clock training time and improved general performance. In the interest of comparability, the open source implementation from OpenAI 'Baselines' was utilised [5]. In order to objectively identify the effects of the additional attention model, the standards set by Mnih *et al.* [16] were followed. This included preprocessing of the input image from a single 210 × 160 RGB image to a stack of four 84 × 84 grey-scale images. 'No-Op' starts were also used which prevents the agent from taking an action at the start of each game for a random number (maximum thirty) of time-steps.

4   Experiments

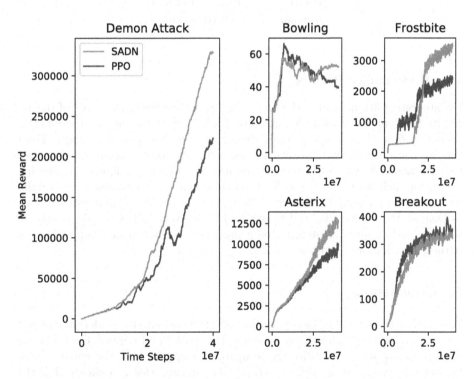

Fig. 2. Learning curves of SADN compared to the baseline PPO. Agents are trained for a total of 40M timesteps, with results averaged over three random seeds. Here we can see the clear advantage self-attention is able to provide with respect to sample efficiency.

Performance Evaluation. In order to evaluate our approach we randomly seed each different architecture for a total of three times across five different Atari games. In terms of standard training times for bench marking, [11,12,16,18] show variations between 40M to 16B+ frames. We train each model for a total of 40M time-steps, which is equivalent to 160M frames. This is inline with the evaluation methodology as presented by Fortunato *et al.* [7]. Performance is evaluated by the maximal score achieved (after averaging) during training.

Table 1. Maximal score achieved, averaged over three random seeds and trained for 40M time-steps. These results clearly demonstrate the improved performance of SADN.

	PPO	SADN
Demon Attack	222650	**329837**
Bowling	**66.33**	57.52
Frostbite	2503	**3554**
Asterix	10121	**13253**
Breakout	**398.10**	345.07

4.1 Performance Results

By integrating attention into the underlying neural network, new state of the art results for Demon Attack were achieved. Table 1 shows the maximal score after averaging over three random seeds during training for 40M time-steps. From this we can observe that integrating self-attention led to significant increases in performance in 60% of tested environments. While Fig. 2 shows the training curves for each network across all five environments. This allows us to visually see the increased sample efficiency self-attention provides in environments such as Demon Attack, Asterix, and Frostbite. Impressively SADN is able to surpass the previously highest score reported using a policy gradient method for Demon Attack, Bowling, and Frostbite [7,23].

5 Conclusions

We evaluate the benefit of incorporating self-attention into the underlying neural network architecture with direct access to the spatial and temporal information from the state input. We directly compare this approach to the classic architecture first proposed by Mnih *et al.* [16]. Our results clearly indicate that the addition of attention to the network is beneficial, and lead to significant improvements in sample efficiency across 60% of tested environments. Of particular note is the performance in the environment Demon Attack, where the addition of self-attention resulted in state-of-the-art results, far exceeding the previous reported benchmark.

Future work will seek to further investigate why attention was more beneficial in some environments compared to others, along with further testing of the proposed architecture with different optimisation techniques, including DQN methods.

Acknowledgments. We would like to thank Michele Sasdelli for his helpful discussions, and Damien Teney for his feed-back and advice on writing this paper.

References

1. Bahdanau, D., Cho, K., Bengio, Y.: Neural machine translation by jointly learning to align and translate (2014). CoRR abs/1409.0473. http://arxiv.org/abs/1409.0473

2. Bellemare, M.G., Naddaf, Y., Veness, J., Bowling, M.: The arcade learning environment: an evaluation platform for general agents. J. Artif. Intell. Res. **47**, 253–279 (2013)

3. Broadbent, D.E.: Perception and Communication (1958)

4. Choi, J., Lee, B., Zhang, B.: Multi-focus attention network for efficient deep reinforcement learning. CoRR abs/1712.04603 (2017). http://arxiv.org/abs/1712.04603

5. Dhariwal, P., et al.: Openai baselines (2017). https://github.com/openai/baselines

6. Fang, S., Xie, H., Zha, Z.J., Sun, N., Tan, J., Zhang, Y.: Attention and language ensemble for scene text recognition with convolutional sequence modeling. In: Proceedings of the 26th ACM International Conference on Multimedia, MM 2018, pp. 248–256. ACM, New York (2018). https://doi.org/10.1145/3240508.3240571

7. Fortunato, M., et al.: Noisy networks for exploration. CoRR abs/1706.10295 (2017). http://arxiv.org/abs/1706.10295

8. Gregor, M., Nemec, D., Janota, A., Pirnik, R.: A visual attention operator for playing Pac-Man, pp. 1–6, May 2018. https://doi.org/10.1109/ELEKTRO.2018.8398308

9. Greydanus, S., Koul, A., Dodge, J., Fern, A.: Visualizing and understanding Atari agents. CoRR abs/1711.00138 (2017). http://arxiv.org/abs/1711.00138

10. Han, Y.: Explore multi-step reasoning in video question answering. In: Proceedings of the 1st Workshop and Challenge on Comprehensive Video Understanding in the Wild. p. 5. CoVieW 2018. ACM, New York (2018). https://doi.org/10.1145/3265987.3265996

11. Hausknecht, M.J., Stone, P.: Deep recurrent q-learning for partially observable mdps. CoRR abs/1507.06527 (2015). http://arxiv.org/abs/1507.06527

12. Horgan, D., et al.: Distributed prioritized experience replay. CoRR abs/1803.00933 (2018). http://arxiv.org/abs/1803.00933

13. Kastaniotis, D., Ntinou, I., Tsourounis, D., Economou, G., Fotopoulos, S.: Attention-aware generative adversarial networks (ATA-GANS). CoRR abs/1802.09070 (2018). http://arxiv.org/abs/1802.09070

14. Kay, W., et al.: The kinetics human action video dataset. CoRR abs/1705.06950 (2017). http://arxiv.org/abs/1705.06950

15. Mnih, V., Heess, N., Graves, A., Kavukcuoglu, K.: Recurrent models of visual attention. CoRR abs/1406.6247 (2014). http://arxiv.org/abs/1406.6247

16. Mnih, V., et al.: Human-level control through deep reinforcement learning. Nature **518**(7540), 529–533 (2015). https://doi.org/10.1038/nature14236

17. Oh, J., Chockalingam, V., Singh, S.P., Lee, H.: Control of memory, active perception, and action in minecraft. CoRR abs/1605.09128 (2016). http://arxiv.org/abs/1605.09128

18. Schulman, J., Wolski, F., Dhariwal, P., Radford, A., Klimov, O.: Proximal policy optimization algorithms. CoRR abs/1707.06347 (2017). http://arxiv.org/abs/1707.06347

19. Shi, J., Zhang, H., Li, J.: Explainable and explicit visual reasoning over scene graphs. CoRR abs/1812.01855 (2018). http://arxiv.org/abs/1812.01855

20. Sigurdsson, G.A., Varol, G., Wang, X., Farhadi, A., Laptev, I., Gupta, A.: Hollywood in homes: crowdsourcing data collection for activity understanding. CoRR abs/1604.01753 (2016). http://arxiv.org/abs/1604.01753

21. Sorokin, I., Seleznev, A., Pavlov, M., Fedorov, A., Ignateva, A.: Deep attention recurrent q-network. CoRR abs/1512.01693 (2015). http://arxiv.org/abs/1512.01693

22. Wang, X., Girshick, R.B., Gupta, A., He, K.: Non-local neural networks. CoRR abs/1711.07971 (2017). http://arxiv.org/abs/1711.07971

23. Wu, Y., Mansimov, E., Grosse, R.B., Liao, S., Ba, J.: Scalable trust-region method for deep reinforcement learning using Kronecker-factored approximation. In: Guyon, I., et al. (eds.) Advances in Neural Information Processing Systems 30, pp. 5279–5288. Curran Associates, Inc. (2017). http://papers.nips.cc/paper/7112-scalable-trust-region-method-for-deep-reinforcement-learning-using-kronecker-factored-approximation.pdf

24. Xu, T., et al.: AttnGAN: Fine-grained text to image generation with attentional generative adversarial networks. CoRR abs/1711.10485 (2017). http://arxiv.org/abs/1711.10485

25. Yuezhang, L., Zhang, R., Ballard, D.H.: An initial attempt of combining visual selective attention with deep reinforcement learning. CoRR abs/1811.04407 (2018). http://arxiv.org/abs/1811.04407

26. Zhang, R., et al.: AGIL: learning attention from human for visuomotor tasks. CoRR abs/1806.03960 (2018). http://arxiv.org/abs/1806.03960

27. Zhao, S., Zhang, Z.: Attention-via-attention neural machine translation (2018). https://www.aaai.org/ocs/index.php/AAAI/AAAI18/paper/view/16534

Budget Cost Reduction for Label Collection with Confusability Based Exploration

Jiyi Li[1,2]([✉])

[1] University of Yamanashi, Kofu, Japan
jyli@yamanashi.ac.jp
[2] RIKEN Center for AIP, Tokyo, Japan

Abstract. Subjective judgment is an important manner of collecting labels which are used for training and evaluating models. Researchers always use a large enough and fixed number of workers to label an instance and then aggregate the labels of an instance from multiple workers into a single label. However, some easy instances only need a small number of workers and some difficult instances need more workers to reach stable aggregated labels. Using a fixed number of workers cannot efficiently use the limited budget. We thus propose an approach for reducing the cost of label collection by assigning a dynamic number of workers to each instance. We propose an Exploration-Focused Upper-Confidence Bound (EFUCB) approach which tends to explore the stable aggregated labels for all instances in the entire dataset. It iteratively selects an instance to ask the workers for one more label. To select the instances for labeling, in contrast to using the collected labels as the reward directly, it utilizes the disagreement of workers to measure the reward of collecting labels for an instance. The experiments based on real datasets verify that our approach can reduce the same budget cost with fewer influences on the aggregated labels so that the utility of the collected labels is preserved.

Keywords: Label collection · Budget cost reduction · Exploration · Confusability

1 Introduction

Subjective judgment by human beings has been an important manner for collecting labels which are used for training and evaluating models in many areas when the golden labels cannot be collected or computed automatically and need to be judged manually. Because a single worker may fail to provide the correct label, to guarantee the credibility of the label collection, people always assign an instance to multiple workers for ensuring a certain degree of redundancy. After that, they aggregate the multiple labels into a single label which can be computed by some measures like mean or label aggregation methods such as

© Springer Nature Switzerland AG 2019
T. Gedeon et al. (Eds.): ICONIP 2019, CCIS 1143, pp. 231–241, 2019.
https://doi.org/10.1007/978-3-030-36802-9_26

majority voting [9] or probabilistic models like [12]. The aggregated labels are then used as the ground truth of the data.

In this redundancy strategy, researchers always roughly use a fixed number of workers for all instances. Considering the tradeoff between the budget cost and credibility, the number is always selected as a value that is large enough to be regarded as reliable by others. For example, Snow et al. [9] proposed a dataset by using about ten workers for labeling whether one sentence can be inferred from the other one. One potential assumption in this strategy is that more workers for an instance lead to higher credibility.

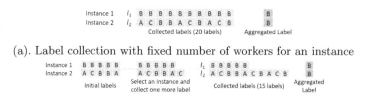

(a). Label collection with fixed number of workers for an instance

(b). Label collection with dynamic number of workers for an instance (our idea)

Fig. 1. Toy example: dynamic number of workers (labels) can decrease the total number of labels with no or minor influences on the aggregated label (by majority voting).

However, it is not necessary to use a fixed and same number of workers for all instances. Figure 1 shows a toy example of our idea. There are two instances in this example. In Fig. 1(a), a traditional way for label collection utilizes a fixed number of workers for each instance. The label sets l_1 and l_2 of two instances are different but both the aggregated labels (by majority voting) are same. As shown in Fig. 1(b), although l_2 needs ten labels to reach this majority value, it is possible for l_1 to reach the same majority value with only five labels. In other words, if we assign fewer workers to instances like instance 1, we can reduce the cost while there are no or minor influences on the collected labels.

Using a fixed number of workers for any instances cannot efficiently use the limited budget. We thus propose an approach for reducing the cost of label collection by assigning a dynamic number of workers to each instance. In the task of label collection, we have multiple targets, i.e., collecting the aggregated labels for *all instances* in the dataset (exploration); collecting more worker labels for an instance to obtain *more credible* aggregated label (exploitation); collecting *as few labels as* possible (budget cost reduction).

This exploration-exploitation scenario is similar to the multi-armed bandit (MAB) problem. Each instance can be regarded as a lever; a collected label in an iteration is the output of playing the lever. However, we cannot utilize the typical MAB directly because of a key difference between the scenario of MAB and ours. We tend to explore the stable aggregated labels of all instances in the entire dataset rather than finding one instance which has the highest expectation output, i.e., the value of the stable aggregated label of this instance.

In our scenario, to select the instances for next actions in the iterations, we cannot directly use the output of an instance as the reward like MAB, e.g., an instance with label 'C' is not more valuable than an instance with label 'A'. Therefore, instead of using the collected labels directly, we utilize the disagreement of workers (named as *confusability*) on the label set of an instance to measure the reward of an instance. We propose an Exploration-Focused Upper-Confidence Bound (EFUCB) approach for our problem.

The contributions of our paper are as follows. (1) We focus on reducing the cost of subjective label collection by assigning a dynamic number of workers to each instance so that we can utilize the budget for label collection more efficiently. (2) We propose an EFUCB approach as the instance selection algorithm which utilizes a specific reward measure based on the instance confusability. (3) The experimental results based on real datasets show that our approach can reduce the same cost with fewer influences on the aggregated labels that the baselines so that the utility of the collected labels is preserved.

2 Related Work

There are some existing works related to reducing the cost of human labeling. Some of them focused on reducing the *time cost of a single worker* in the process of creating labels [3]. Some of them focused on reducing *the budget cost by fewer instances* based on active learning in crowdsourcing context by finding the most informative and valuable instances to label fewer instances [13]. Sheng et al. [8] discussed the trade-off between selecting a new instance to label and gathering more labels for a labeled instance for supervised learning. It assumed the workers were noisy and did not model the workers' disagreement on an instance.

In contrast, our work focuses on a different aspect, i.e., reducing *the budget cost by fewer labels for an instance*. Our cost reduction approach based on the workers' disagreement on the label set of an instance can be easily used together with above-existing work on *reducing time cost* or *reducing the budget cost by fewer instances*. We assign a dynamic number of workers for each instance.

There are only a few existing works related to the budget cost of an instance [1,5]. Abraham et al. [1] tried to answer the question that how many workers are required for an instance by estimating workers performance with assessing the labeling history of pool-based workers. In our scenario, we do not access the labeling history of workers on other tasks and worker profile; the workers are stream-based. Losada et al. [5] modeled the process in the pooling-based evaluation of information retrieval systems with MAB. However, the relevance judgments are only done for the documents that were among the top retrieved documents. It thus can utilize the typical MAB and select the most relevant documents but is not proper for our topic.

3   Budget Cost Reduction Approach

3.1   Definition

For a given data collection $\mathcal{D}$, we denote $d_i \in \mathcal{D}$ as an instance and the number of instances is m. $\boldsymbol{l}_i = \{l_{i1}, l_{i2}, \ldots, l_{in_i}\}$ is the set of labels of d_i, and n_i is the number of labels in $\boldsymbol{l}_i$. $\mathcal{R}$ is the set of unique label values and $l_{ij} \in \mathcal{R}$. The labels can be discrete categories (e.g., sentiment category $\mathcal{R} = \{$negative, neutral, positive, unrelated$\}$) or continuous ratings (e.g., relevance degree $\mathcal{R} = \{1, 2, 3, 4\}$). We name the persons who provide the labels as *workers*.

When collecting labels, people usually empirically set a fixed number which can be regarded as large enough to guarantee the credibility by others. We use the labels collected in such a way as the reference for comparison. We define the label set of an instance with such a large enough label number as the final *stable* label set $\hat{\boldsymbol{l}}_i$ of this instance and define this fixed number of labels as $\hat{n}_i$.

In this paper, we propose a label collection approach which iteratively selects an instance and collects a label for this instance. We have a fixed budget which can be used to collect a limited total number of labels $\sum_i n_i$. This total number of labels is smaller than the total number of labels for the stable label sets $\sum_i \hat{n}_i$, i.e., $\sum_i n_i < \sum_i \hat{n}_i$. The average number of labels $\bar{n} = \sum_i n_i / m$ for an instance is smaller than $\hat{n}_i$. Our approach assigns proper number of labels to different instances so that the aggregated labels (computed by a label aggregation method f such as majority voting) of the collected label set $\boldsymbol{l}_i$ are as similar as possible with that of the stable label sets $\hat{\boldsymbol{l}}_i$, i.e., $f(\boldsymbol{l}_i) \approx f(\hat{\boldsymbol{l}}_i)$.

Note that $\hat{\mathcal{L}}$ is only used as the reference label sets for the evaluation of the approaches, it does NOT exist when using our approach to collect labels in real applications. Our approach is collecting labels for unlabeled datasets and is not removing the labels from the existing label sets.

We clarify some settings on the workers and instances to ensure the practical capability of our proposal. We assume that the workers are stream-based and we cannot access the workers' profile and historical performance, we do not know the exact set of candidate workers and their abilities in advance; we don't distinguish each worker and assume that they have similar behavior and incentivized with the same degree; we assume that the instances are independent of each other because we focus on the labels of a single instance.

3.2   Exploration-Focused Upper-Confidence Bound Approach

In the process of label collection, we iteratively select an instance and ask the label of this instance from a worker; we finally obtain the label sets for all instances. This process is similar to the multi-armed bandit (MAB) problem. Each instance can be regarded as a lever; a collected label in an iteration is the output of playing the lever. However, in typical MAB, people keep selecting the instances to find the instance which has the highest expected reward. One of the most important differences between our scenario and typical MAB is that we focus on exploring the entire dataset and collecting label set which has similar aggregated values with the stable labels of all instances, rather than finding one instance which has highest expectation output.

Typical MAB utilizes the outputs of the lever as its reward. However, a label of a selected instance cannot be used as the reward in our scenario because the values of labels have no sense on the rewards. For example, we cannot assume that an instance with label 'C' is more valuable than an instance with label 'A'. It results in that we cannot use the output of an instance as a reward like MAB. We thus cannot directly adopt the existing MAB technologies to our scenario. We need to find other proper information as the reward for collecting a label for an instance.

The information we utilize is the *confusability* which is defined as the disagreement of workers on the label set of an instance. It can also represent the judgment difficulty of an instance by the workers. Our idea is that if we set fewer workers to an instance with low confusability which is easy to be judged by workers, we can reduce some costs with less loss on the aggregated label relative to the stable aggregated labels.

If the confusability of an instance is high, collecting more labels is possible to help us understand the stable aggregated labels of this instance more exactly, which can be regarded as a high reward for selecting this instance. In other words, an instance with low confusability can be regarded as having a low reward on collecting the labels and thus will be selected with fewer times; an instance with high confusability can be regarded as having a high reward and thus will be selected with more times. Note that an instance with a low reward may be an actual good instance in the dataset for applications like training models because it is easy to be judged by workers; it has higher credibility and less ambiguity. The instance with high reward is just "good" in the sense of the proposed approach in the specific process of label collection.

We propose an effective instance selection approach for reducing the cost of label collection by leveraging the confusability information with the multi-armed bandit theory. We modify the Upper Confidence Bound (UCB) approach [2], which is originally proposed for action selection in MAB. The approach we propose is named as Exploration-Focused Upper-Confidence Bound (EFUCB). We define the reward based on the confusability information and formulate our algorithm as follows.

$$\mathcal{A}(t) = \arg\max_i \left(\mathcal{C}_i(t) + \sqrt{\frac{\log t}{2 n_i(t)}} \right). \tag{1}$$

t is number of current iterations; $\mathcal{A}(t)$ is the selected action (instance) at t; $\mathcal{C}_i(t)$ is the confusability of l_i at iteration t; $n_i(t)$ is number of labels n_i of instance d_i at iteration t. This formula includes two parts, i.e., the confusability degree of current label set and the confidence interval for the expected aggregated label. It can be explained as that it tends to select an instance with higher disagreement by the workers and with a fewer number of labels.

On one hand, to obtain the initial estimations of the confusability, we first collect several labels for each instance. We define this initial number of labels for an instance as $\mathcal{N}_l$. On the other hand, there may exist some instances that are very ambiguous for the workers and required too many labels. To avoid wasting too much budget on such instances, we set an upper constraint $\mathcal{N}_u$ on the number of labels for each instance. When $n_i(t) \geq \mathcal{N}_u$, the reward of this instance is 0.

To measure the confusability, we utilize the entropy on the label set of an instance, which can illustrate the disagreement in the label set and the difficulty of judgment by workers, i.e., $C_i(t) = -\sum_k p_i^k \log p_i^k$, $p_i^k = n_i^k/n_i$, where n_i^k is the number of labels that are equal to value k. There are some existing works which consider the inconsistency of human-generated labels. Because we need to measure the disagreement of a set of workers on an instance, the measures like Cohen's kappa for the inter-rater agreement of two workers on a set of instances are not proper. Other measures like Fleiss' kappa and Krippendorff's alpha coefficient can be alternatives.

There are other types of strategies in MAB for action selection such as ϵ-greedy and Thompson sampling. How to revise other types of algorithms to fit our scenario will in our future work. Furthermore, because we do not have the best instance in all instances like the typical MAB, there is no similar theoretical analysis on the regret of reward in instance selection like that in MAB.

4   Experiments

4.1   Evaluation Method

We compare our approach with the following two baselines.

Naïve: A naïve approach which uses a fixed cut-off number $\bar{n}$ of workers for any instance. It roughly reduces the cost by setting a fixed and smaller number of workers for each instance than the original number of workers $\hat{n}_i (\bar{n} < \hat{n}_i)$ in the dataset.

UCB [2]: The Upper Confidence Bound approach proposed for the typical MAB, which utilizes the expected label value as the reward, the selected action (instance) at iteration t can be formulated as follows, $\mathcal{A}(t) = \arg\max_i \left(\mu_i + (b - a)\sqrt{\frac{\log t}{2n_i(t)}}\right)$, $\mu_i = \sum_{k=1}^{n_i(t)} \frac{l_{ik}}{n_i(t)}$. UCB cannot be directly used for all datasets in our experiments. To carry out it, we map the categories into non-negative integers in range of $[a, b]$, e.g., $\{negative, positive, unrelated\} \mapsto \{0, 1, 2\}$. Note that some discrete categories (e.g.,"$unrelated \mapsto 2$") do not contain the continuous meaning with other categories.

As described in the approach section, we cannot use the regret to directly evaluate the performance of an instance selection algorithm in EFMAB. We need some indirect measures to evaluate the approaches. We thus utilize the correlation between label set $\mathcal{L} = \{l_i\}_i$ created by an approach and the reference label set (stable label set) $\hat{\mathcal{L}} = \{\hat{l}_i\}_i$ in the dataset. The evaluation metric is *similarity*, i.e., the fraction of the aggregated labels in $\mathcal{L}$ which are consistent with that in $\hat{\mathcal{L}}$. We fix the budget cost and evaluate the quality of labels.

The label aggregation method we use is *majority voting*. Although some more sophistical models (e.g., [12]) may be able to generate better aggregated labels, majority voting is still one of the most widely used ones, because of its effectiveness and simplicity for implementation in the real-world applications. Our work concentrates on reducing the cost in the stage of collecting labels and does not focus on the performance of the label aggregation method. In addition, although adding more labels may change the values of the aggregated labels, we assume that the entire labels in these datasets are the stable label sets. It is reasonable because existing works treat them as the ground truth when using these datasets for other topics.

Table 1. Statistics of the real datasets. $|\mathcal{R}|$ is the number of unique label values. lpi (label per instance) is the average number of workers for each instance.

Dataset	$|\mathcal{R}|$	#instance	#worker	#label	lpi
lp50 [4]	5	1225	83	12226	9.98
popularity [7]	2	500	143	10000	20.00
temporal [9]	2	462	76	4620	10.00
rte [9]	2	800	164	8000	10.00
weather [10]	5	300	110	6000	20.00
smile [12]	2	159	17	1950	12.26
duck [11]	2	108	39	4212	39.0
face [6]	4	584	27	5242	8.98

4.2 Experimental Settings

Table 1 lists the information of the datasets. These datasets are originally provided to verify the methods proposed for the corresponding topics in those papers. In this paper, we use them for a different topic, i.e., reducing the cost for label collection. These datasets do not only provide the data of the aggregated labels but also the non-aggregated labels of all workers so that we can use in our experiments. We select datasets with diverse factors to show the performance of our approach in different cases.

To simulate the process of labeling, when we have selected an instance based on an approach, we choose a label by a worker in the dataset. For reproducibility, we choose the labels and workers based on current orders of labels (workers) of an instance in the dataset, instead of randomly choosing one worker from the workers who label the instance.

We carry out an approach on each dataset with multiple trials. In each experimental trial, we shuffle the label orders of the instances with a specific rule. We generate the worker permutation of the instance based on the order of worker id. The first permutation is the same with the order of worker id. Next, we shift the workers with one position by moving the first worker to the end of the permutation and generate another permutation. We compute the average performance on ten permutations.

We set the fixed budget cost by controlling the average number of labels $\bar{n} = \sum_i n_i/m$. For our approach and the baselines, we set $\bar{n}$ in $\{6,8\}$. For the naïve approach, $\bar{n}$ is the unique parameter. For EFUCB and UCB, we use $\mathcal{N}_l$ at least to three to get enough redundancy for initializing the computation. The average number of additional labels collected by EFUCB (or UCB) is denoted as $\bar{n}_c = \bar{n} - \mathcal{N}_l$. We set the candidate values of average additional labels $\bar{n}_c$ as $\{1,2,3\}$. The maximum number of labels $\mathcal{N}_u$ of an instance is equal to the number of stable labels of this instance in the original dataset. The reduced budge cost can be computed by the average number of stable labels $\hat{n}$ in the dataset, i.e., $(\hat{n} - \bar{n})/\hat{n} * 100\%$.

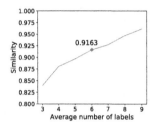 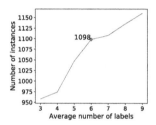

(a). Similarity of the label sets generated by different $\bar{n}$

(b). #instances which have reached stable aggregated labels at different $\bar{n}$

Fig. 2. Influence of the number of workers on the ground truth. The dataset is lp50 (1225 instances in total). It shows that different instances need a different number of workers to reach the same values with the stable aggregated label.

4.3   Experimental Results

We first investigate the influence of the worker number on the collected labels. We utilize the lp50 dataset for this investigation and using the same worker permutation with the original dataset. Figure 2(a) shows the performance of the Naïve approach with different number of labels $\bar{n}$. It shows that when $\bar{n}$ increases, the similarity increases; the aggregated labels are more approximated to the stable aggregated ones. When the $\bar{n}$ is not so high, e.g., $\bar{n} = 6$, the similarity has been already high. It shows the possibility of reducing the number of labels while we can still obtain similar aggregated labels with the stable ones. Figure 2(b) visualizes the number of instances which have reached the same values with the stable aggregated label at different $\bar{n}$. It shows that different instances need a different number of workers.

Table 2 lists the main results. First, it shows that EFUCB prominently outperforms both UCB and the Naïve approach. The aggregated labels generated by EFUCB is generally more similar to the stable aggregated labels than that

Table 2. Experimental results on the real datasets

$\hat{n}$	$\mathcal{N}_l + \bar{n}_c$	EFUCB	UCB	Naïve	Cost	$\hat{n}$	$\mathcal{N}_l + \bar{n}_c$	EFUCB	UCB	Naïve	Cost
lp50						popularity					
6	5+1	**0.9559**	0.9484	0.9163	−39.88%	6	5+1	**0.9920**	0.9754	0.9798	−70.00%
6	4+2	**0.9684**	0.9634	0.9163		6	4+2	**0.9994**	0.9664	0.9798	
6	3+3	**0.9823**	0.9634	0.9163		6	3+3	**0.9994**	0.9616	0.9798	
8	7+1	0.9849	**0.9887**	0.9469	−19.84%	8	7+1	**0.9972**	0.9828	0.9854	−60.00%
8	6+2	**0.9986**	0.9914	0.9469		8	6+2	**1.0000**	0.9798	0.9854	
8	5+3	**0.9999**	0.9914	0.9469		8	5+3	**1.0000**	0.9754	0.9854	
temporal						rte					
6	5+1	**0.9740**	0.8805	0.8595	−40.00%	6	5+1	**0.9405**	0.8850	0.8876	−40.00%
6	4+2	**0.9755**	0.8234	0.8595		6	4+2	**0.9517**	0.8349	0.8876	
6	3+3	**0.9742**	0.8799	0.8595		6	3+3	**0.9704**	0.8787	0.8876	
8	7+1	**0.9933**	0.9613	0.9686	−20.00%	8	7+1	**0.9814**	0.9249	0.9374	−20.00%
8	6+2	**1.0000**	0.9307	0.9686		8	6+2	**0.9998**	0.8938	0.9374	
8	5+3	**1.0000**	0.9777	0.9686		8	5+3	**0.9992**	0.9304	0.9374	
weather						smile					
6	5+1	**0.9233**	0.8940	0.9080	−70.00%	6	5+1	**0.8805**	0.8371	0.8189	−51.08%
6	4+2	**0.9440**	0.8937	0.9080		6	4+2	**0.8956**	0.7698	0.8189	
6	3+3	**0.9493**	0.8963	0.9080		6	3+3	**0.9302**	0.8151	0.8189	
8	7+1	**0.9460**	0.9167	0.9243	−60.00%	8	7+1	**0.9459**	0.8742	0.8742	−34.77%
8	6+2	**0.9547**	0.9217	0.9243		8	6+2	**0.9711**	0.8214	0.8742	
8	5+3	**0.9613**	0.9130	0.9243		8	5+3	**0.9755**	0.8484	0.8742	
duck						face					
6	5+1	**0.8046**	0.7528	0.7139	−84.62%	6	5+1	**0.9663**	0.9402	0.9182	−33.16%
6	4+2	**0.8083**	0.7398	0.7139		6	4+2	**0.9991**	0.9450	0.9182	
6	3+3	**0.8102**	0.7657	0.7139		6	3+3	**0.9955**	0.9604	0.9182	
8	7+1	**0.8444**	0.8093	0.7769	−79.49%	8	7+1	**1.0000**	0.9873	0.9584	−10.91%
8	6+2	**0.8380**	0.7667	0.7769		8	6+2	**1.0000**	0.9863	0.9584	
8	5+3	**0.8407**	0.8083	0.7769		8	5+3	**1.0000**	0.9943	0.9584	

generated by the baselines when using the same budget cost. In some cases (e.g., $\hat{n} = 8$, popularity, temporal and face datasets), it can decrease lots of cost while the collected labels are almost consistent with that in the stable label set.

Second, for a given $\hat{n}$, when EFUCB is used to collect more additional labels (e.g., $\bar{n}_c$ increases from 1 to 3), the performance of EFUCB can generally improve (e.g., weather and smile dataset). It shows that EFUCB can rationally select the instances which need more labels and avoid to select the instances which have enough labels. EFUCB can allocate the additional $m\bar{n}_c$ budget on all instances more effectively than just assigning $\bar{n}_c$ labels to each instance.

Third, for the Naïve approach, it shows that only using a fixed cut-off number of workers is not bad. For example, for $\hat{n} = 8$ in six of eight datasets, it can reach higher than 90% similarity by more or less cost reduction. It can be a practical solution in some cases because it is simple to carry out.

Forth, for UCB approach, although UCB can outperform the Naïve one in several cases for several datasets (e.g., all cases in lp50 dataset), it has worse performance than the Naïve one in many cases for many datasets (e.g., all cases in popularity and rte datasets). In addition, when $\bar{n}_c$ increases for a given $\hat{n}$, there is no observation that the performance of UCB can always improve (e.g., $\hat{n} = 8$ in duck dataset, UCB at $\bar{n}_c = 2$ performs worse than UCB at $\bar{n}_c = 1, 3$). It shows that utilizing the typical technologies in MAB directly cannot rationally select the instances which need more labels and cannot reach good performance in our topic. The proposed algorithm EFUCB can solve our problem.

5   Conclusion

In this paper, we proposed a solution for reducing the cost in the label collection by assigning a dynamic number of workers to each instance. We proposed an instance selection algorithm EFUCB which can reduce the same cost with fewer influences on the aggregated labels than the traditional UCB and a naïve approach using a fixed cut-off number of workers. The utility of the collected labels can be preserved. In future work, we will revise other types of action selection strategies in MAB to solve our problem and compare their performance.

Acknowledgments. This work was partially supported by JSPS KAKENHI Grant Number 19K20277.

References

1. Abraham, I., Alonso, O., Kandylas, V., Patel, R., Shelford, S., Slivkins, A.: How many workers to ask? Adaptive exploration for collecting high quality labels. In: Proceedings of the 39th International ACM SIGIR Conference on Research and Development in Information Retrieval, SIGIR 2016, pp. 473–482 (2016)
2. Auer, P., Cesa-Bianchi, N., Fischer, P.: Finite-time analysis of the multiarmed bandit problem. Mach. Learn. **47**(2–3), 235–256 (2002)
3. Krishna, R.A., et al.: Embracing error to enable rapid crowdsourcing. In: Proceedings of the 2016 CHI Conference on Human Factors in Computing Systems, CHI 2016, pp. 3167–3179 (2016)
4. Lee, M.D., Welsh, M.: An empirical evaluation of models of text document similarity. In: CogSci 2005, pp. 1254–1259. Erlbaum (2005)
5. Losada, D.E., Parapar, J., Barreiro, A.: Multi-armed bandits for adjudicating documents in pooling-based evaluation of information retrieval systems. Inf. Process. Manag. **53**(5), 1005–1025 (2017)
6. Mozafari, B., Sarkar, P., Franklin, M.J., Jordan, M.I., Madden, S.: Active learning for crowd-sourced databases. CoRR abs/1209.3686 (2012)
7. Pang, B., Lee, L.: A sentimental education: sentiment analysis using subjectivity summarization based on minimum cuts. In: Proceedings of the 42nd Annual Meeting on Association for Computational Linguistics, ACL 2004 (2004)

8. Sheng, V.S., Provost, F., Ipeirotis, P.G.: Get another label? Improving data quality and data mining using multiple, noisy labelers. In: Proceedings of the 14th ACM SIGKDD International Conference on Knowledge Discovery and Data Mining, KDD 2008, pp. 614–622 (2008)

9. Snow, R., O'Connor, B., Jurafsky, D., Ng, A.Y.: Cheap and fast–but is it good? Evaluating non-expert annotations for natural language tasks. In: Proceedings of the Conference on Empirical Methods in Natural Language Processing, EMNLP 2008, pp. 254–263 (2008)

10. Venanzi, M., Teacy, W., Rogers, A., Jennings, N.R.: Weather sentiment-Amazon Mechanical Turk dataset (2015)

11. Welinder, P., Branson, S., Belongie, S., Perona, P.: The multidimensional wisdom of crowds. In: Proceedings of the 23rd International Conference on Neural Information Processing Systems, NIPS 2010, pp. 2424–2432 (2010)

12. Whitehill, J., Ruvolo, P., Wu, T., Bergsma, J., Movellan, J.: Whose vote should count more: Optimal integration of labels from labelers of unknown expertise. In: Proceedings of the 22nd International Conference on Neural Information Processing Systems, NIPS 2009, pp. 2035–2043 (2009)

13. Zhong, J., Tang, K., Zhou, Z.H.: Active learning from crowds with unsure option. In: Proceedings of the 24th International Conference on Artificial Intelligence, IJCAI 2015, pp. 1061–1067 (2015)

Prediction of Refactoring-Prone Classes Using Ensemble Learning

Vamsi Krishna Aribandi[1(✉)], Lov Kumar[1(✉)], Lalita Bhanu Murthy Neti[3], and Aneesh Krishna[1,2]

[1] BITS Pilani, Hyderabad, India
f20160803@hyderabad.bits-pilani.ac.in, lovkumar505@gmail.com
[2] School of Electrical Engineering, Computing and Mathematical Sciences,
Curtin University, Perth, Australia
a.krishna@curtin.edu.au
[3] CSIS, BITS Pilani Hyderabad, Hyderabad, India
bhanu@hyderabad.bits-pilani.ac.in

Abstract. A considerable amount of software engineers' efforts go into maintaining code repositories, which involves identifying code whose structure can be improved. This often involves the identification of classes whose code requires refactoring. The early detection of refactoring-prone classes has the potential to reduce the costs and efforts that go into maintaining source code repositories. The purpose of this research is to develop prediction models using source code metrics for detecting patterns in object oriented source code, which are indicators of classes that are likely to be refactored in future iterations. In this study, four different sets of source code metrics have been considered as an input for refactoring prediction to evaluate the impact of these source code metrics on model performance. The impact of these source code metrics are evaluated using eleven different classification technique, and two different ensemble classes on seven different open source projects. Ensemble learning techniques have been shown to incorporate the diversity of patterns learnt by different classifiers, resulting in an augmented classifier that is more robust than any individual classifier. Our work also creates distinction between various sets of features for the task of predicting refactoring-prone classes.

Keywords: ELM · Kernels · Refactoring · Source code metrics

1 Research Motivation and Aim

Refactoring is the process of restructuring existing code without changing its external functional behavior to improve non functional attributes of software. While refactoring code has no impact on its functionality, it significantly affects the maintainability of a module. Ease of understanding, faster debugging and better re-usability are among the benefits of well-structured code, which is generally the result of code refactoring. Initial indications of classes that will likely

© Springer Nature Switzerland AG 2019
T. Gedeon et al. (Eds.): ICONIP 2019, CCIS 1143, pp. 242–250, 2019.
https://doi.org/10.1007/978-3-030-36802-9_27

be refactored in the future could significantly reduce costs and allow developers more time towards improving the functionality of their code.

One research gap in the study of refactoring-prone code is the modeling of relationships between source code metrics and the tendency to be refactored. Previous research has shown that source code metrics (like Cyclomatic Complexity) are good indicators for how intricate a segment of code is, but the application of modern ensemble learning techniques to this problem is relatively unexplored. One of the novel contributions of our work to the problem of detecting refactoring-prone classes is the application of various base learners such as logistic regression, artificial neural networks and radial basis function neural networks as constituents of ensemble classes. Another research contribution of the work presented in this paper is a study of the source code metrics deemed important to identify refactoring-prone classes. In this paper, three different Research Questions are addressed:

RQ1 *What is the capability of various feature selection techniques to predict refactoring-prone classes?*
RQ2 *What is the capability of various individual classification techniques to predict proneness to refactoring?*
RQ3 *To what extent can the application of ensemble learning classifiers improve upon the performance achieved by individual classifiers?*

2   Experimental Data Set

In order to ensure that this study remains repeatable, verifiable and refutable, we used openly available datasets from the tera-PROMISE repository [1,2]. This ensures that our work is easily replicable and makes it straightforward to compare it with other works. The seven open source systems are written in Java. We have considered 125 source code metrics to measure internal structure of class. Some examples of these metrics computed are Number of Outgoing Invocations (NOI), Documentation Lines of Code (DLOC), Number of Local Setters (NLS), Total Lines of Code (TLOC) etc. [3,4]. Table 1 shows the Number of classes (NC), Number of Refactored classes (NRC), and Percentage of Refactored classes (%RC) for each of the seven software. Table 1 reveals that the datasets are highly imbalanced with %RC values of 5.64%, 1.37%, 0.95%, 4.49%, 0.74%, 2.98% and 1.12% respectively.

Table 1. Experimental data set description

	antlr4	junit	MapDB	mcMMO	mct	oryx	titan
# NC	408	655	419	89	2028	504	1158
# NRC	23	9	4	4	15	15	13
% RC	5.64	1.37	0.95	4.49	0.74	2.98	1.12

3 Methodology

Figure 1 depicts the research framework used to develop code refactoring prediction model using source code metrics to predict refactoring at class level. The first stage of this process is to apply four different steps to find 4 feature sets of software metrics. Each step of feature selection prunes the previous feature set, starting with the initial set of all 125 source code metrics.

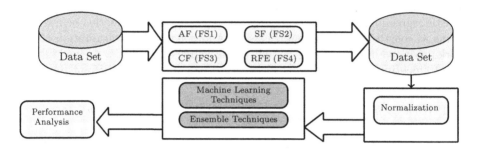

Fig. 1. Framework of proposed work

- Feature Set 1 (FS1) simply consists of all 125 metrics.
- For Feature Set 2 (FS2), upon each metric for each of the 7 datasets, Wilcoxon's rank-sum test was applied. The Wilcoxon rank-sum test tests the null hypothesis that a metric of a refactored class and the same metric of a non-refactored class were drawn from the same distribution. Here, the distributions compared are the readings of a source code metric for classes that were refactored and the readings of a source code metric for classes that were not refactored. Wilcoxon's rank-sum test was applied for each source code metric, and applied for each dataset. The features are considered as relevant if its p-value is less than 0.05. If a feature was relevant for at least 5 datasets, it was included in FS2. After feature selection in this manner, FS2 contained 21 source code metrics as shown in Table 2.
- The second class of feature selection performed was cross-correlation analysis. In any dataset, highly correlated features means there is a duplication of information, and increases the dimensionality of the data. For each dataset, a correlation matrix between the features in FS2 was constructed. If two unique features had a correlation of more than 0.7variables' correlation was calculated with the target variable REFACT. The variable with lower correlation with the target variable is then removed from consideration for that particular dataset. Cross correlation analysis was applied separately on each dataset, and the initial features considered are those of FS2.
- After cross correlation analysis, an appended dataset was temporarily constructed, consisting of data from all the 7 datasets. Since each dataset had an independently constructed FS3, the appended dataset was constructed to

construct an intermediate feature set to consider for the third feature selection technique. After running cross correlation analysis on the combined dataset similar to the construction of FS3, 12 features were obtained, and these were the features that were considered for the third feature selection technique, Recursive Feature Elimination (RFE). Given a logistic regression model, the goal of RFE is to select features by recursively considering smaller and smaller sets of features. First, the estimator is trained on the initial set of features and the importance of each feature is obtained through its coefficient (weight) in the model. Then, the least important feature is pruned from current set of features. The importance of a feature is determined by its coefficient's magnitude in the model. The procedure is recursively repeated on the pruned set until the desired number of features is eventually reached.

Table 2. Rank-sum test results

	DS1	DS2	DS3	DS4	DS5	DS6	DS7		DS1	DS2	DS3	DS4	DS5	DS6	DS7
vCC	X	X	X	X	X	X	✓	vDLOC	✓	X	✓	X	X	X	✓
vCCL	X	X	X	X	X	X	✓	vPDA	✓	X	✓	X	X	✓	✓
vCCO	X	X	X	X	X	X	✓	vPUA	✓	X	✓	X	✓	X	✓
vCI	X	X	X	X	X	X	✓	vTCD	✓	X	✓	X	✓	X	X
vCLC	X	X	X	X	X	X	✓	vTCLOC	✓	X	✓	X	✓	X	✓
vCLLC	X	X	X	X	X	X	✓	vDIT	X	X	X	X	X	X	X
vLDC	X	X	X	X	X	X	✓	vNOA	X	X	X	X	X	X	X
vLLDC	X	X	X	X	X	X	✓	vNOC	✓	X	X	X	X	X	X
vLCOM5	✓	X	X	X	✓	X	✓	vNOD	✓	X	X	X	X	X	X
vNL	✓	✓	X	✓	✓	X	✓	vNOP	X	X	X	X	X	X	X
vNLE	✓	✓	X	X	✓	X	✓	vLLOC	✓	✓	✓	X	✓	✓	✓
vWMC	✓	✓	✓	X	✓	✓	✓	vLOC	✓	✓	✓	X	✓	✓	✓
vCBO	✓	✓	X	X	✓	X	✓	vNA	X	✓	X	X	✓	✓	✓
vCBOI	✓	X	✓	✓	X	✓	X	vNG	X	X	✓	X	X	✓	✓
vNII	✓	✓	✓	✓	X	✓	X	vNLA	✓	✓	✓	X	✓	✓	✓
vNOI	✓	✓	X	X	✓	✓	X	vNLG	✓	X	✓	X	✓	X	X
vRFC	✓	✓	✓	X	✓	✓	✓	vNLM	✓	✓	✓	X	✓	✓	✓
vAD	✓	X	X	X	X	X	X	vNLPA	X	X	X	X	X	X	X
vCD	✓	X	✓	X	✓	X	X	vNLPM	✓	X	✓	X	✓	X	✓
vCLOC	✓	X	✓	X	✓	X	✓	vNLS	X	X	X	X	X	X	X

4 Experimental Results

In this section, we have summarized the results obtained using different sets of features and classification techniques. In this work, we use eleven different classification techniques such as Logistic regression (LOGR), decision tree (DST),

Table 3. AUC values

	LOGR	DST	NND	NNDM	NNDX	SVM-L	SVM-P	SVM-R	LSSVM-l	LSSVM-P	LSSVM-R	BTE	MVE
FS1 (AM)													
DS1	1.00	1.00	0.80	0.50	0.79	0.80	0.89	0.50	1.00	1.00	1.00	0.90	1.00
DS2	1.00	1.00	0.50	0.50	0.75	0.75	1.00	0.50	1.00	1.00	1.00	1.00	1.00
DS3	1.00	1.00	1.00	0.99	0.49	1.00	1.00	0.50	1.00	1.00	1.00	1.00	1.00
DS4	1.00	1.00	0.97	0.71	0.97	0.50	0.50	0.50	1.00	1.00	1.00	0.50	1.00
DS5	1.00	1.00	0.67	0.66	0.66	0.67	1.00	0.67	1.00	1.00	1.00	0.83	1.00
DS6	1.00	1.00	0.83	0.63	0.66	0.67	1.00	0.50	1.00	1.00	1.00	1.00	1.00
DS7	1.00	1.00	0.67	0.50	0.72	0.83	1.00	0.75	1.00	1.00	1.00	1.00	1.00
FS2 (SF)													
DS1	0.69	0.59	0.62	0.66	0.50	0.50	0.62	0.50	0.50	0.70	0.70	0.63	1.00
DS2	0.50	0.50	0.50	0.72	0.50	0.50	0.50	0.50	0.75	1.00	0.50	0.50	1.00
DS3	0.49	0.98	0.99	0.99	0.97	0.50	0.99	0.50	1.00	1.00	0.50	0.50	1.00
DS4	0.85	0.50	0.50	0.94	1.00	0.50	0.50	0.50	1.00	1.00	1.00	0.50	1.00
DS5	0.67	0.67	0.67	0.50	0.66	0.50	0.50	0.50	0.50	1.00	0.50	0.50	1.00
DS6	0.65	0.66	0.50	0.66	0.50	0.50	0.50	0.50	0.50	0.83	0.50	0.50	1.00
DS7	0.66	0.67	0.67	0.75	0.75	0.50	0.67	0.50	0.50	1.00	0.67	0.50	1.00
FS3 (CF)													
DS1	0.60	0.67	0.60	0.50	0.57	0.50	0.60	0.50	0.50	0.60	0.60	0.50	1.00
DS2	0.75	0.75	0.75	0.50	0.50	0.50	0.50	0.50	0.75	1.00	0.50	0.50	1.00
DS3	0.50	0.99	0.50	0.99	0.50	0.50	0.50	0.50	1.00	1.00	0.50	0.50	1.00
DS4	1.00	0.50	0.50	0.50	0.50	0.50	0.50	0.50	1.00	1.00	0.50	0.50	1.00
DS5	0.66	0.50	0.67	0.66	0.66	0.50	0.50	0.50	0.83	1.00	0.50	0.50	1.00
DS6	0.50	0.50	0.50	0.80	0.62	0.50	0.50	0.50	0.50	0.50	0.50	0.50	1.00
DS7	0.66	0.50	0.50	0.75	0.65	0.50	0.67	0.50	0.75	1.00	0.67	0.50	1.00
FS4 (RFE)													
DS1	0.50	0.69	0.63	0.50	0.50	0.50	0.50	0.50	0.50	0.80	0.50	0.50	1.00
DS2	0.50	1.00	0.50	0.50	0.50	0.50	0.50	0.50	0.50	0.50	0.50	0.50	1.00
DS3	0.50	0.99	0.50	1.00	0.50	0.50	0.50	0.50	1.00	1.00	0.50	0.50	1.00
DS4	0.50	0.50	0.50	0.50	0.50	0.50	0.50	0.50	1.00	1.00	0.50	0.50	1.00
DS5	0.67	0.50	0.50	0.50	0.66	0.50	0.50	0.50	0.50	0.50	0.50	0.50	1.00
DS6	0.50	0.50	0.50	0.50	0.50	0.50	0.50	0.50	0.50	0.50	0.50	0.50	1.00
DS7	0.50	0.50	0.50	0.50	0.67	0.50	0.50	0.50	0.50	0.50	0.50	0.50	1.00

neural network with three different training algorithm (NNDM, NNDX, NND), support vector machine with three different kernels i.e., linear kernel (SVM-L), polynomial kernel (SVM-P), and RBF kernel (SVM-R), least square support vector machine with three kernels i.e., linear kernel (LSSVM-L), polynomial kernel (LSSVM-P), RBF kernel (LSSVM-R)and two ensemble techniques such as best training ensemble and major voting ensemble (BTE and MVE) for better performance of refactoring predictions. These all techniques are validated using five fold cross validation classes on seven different data-sets. Table 3 depict the value of AUC for different classifiers on different data-sets and feature sets. From Table 3, we inferred that Refactoring prediction model developed using all metrics have relatively better AUC values as compare to feature selection classes. It can be also noticed from Table 3 that the prediction models trained using LSSVM-P have relatively better AUC values as compare to other classifiers.

5   Comparison

RQ1: What is the capability of various feature selection techniques to predict refactoring-prone classes? In this work, we have considered different feature selection techniques to prune source code metrics. The performance of these techniques are validated by examining the performance of various classification techniques across various sets of features, which are obtained after the application of three feature selection techniques.

Boxplots: Feature Selection Techniques. To understand the visual comparison showing outliers, distribution, variability of different classification techniques, we constructed the box-plot diagrams of different performance parameters such as accuracy and AUC for each feature selection technique as depicted in Fig. 2. From the box-plots diagram in Fig. 2, we notice that the model developed for refactoring prediction using all metrics has high value of performance parameters. From the box-plots diagram in Fig. 2, we also notice that FS3 (CF) has high value of performance parameters as compared to other feature selection techniques.

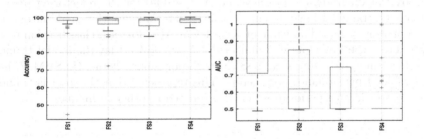

Fig. 2. Box-plots: Classification techniques

Statistical tests: Feature Selection Techniques. To understand the significance difference between the models trained using different sets of features, we conducted statistical tests on AUC value using Wilcoxon signed-rank test with Bonferroni correction as represented in Table 5. Here, the null hypothesis of this work is that *the model trained for refactoring prediction has no effect of classification techniques.* For this work, p-value initially set to 0.05 and then Bonferroni correction adjust the threshold p-value at $\dfrac{0.05}{6} = 0.0083$. Table 4 depicts the results of Wilcoxon signed-rank test with Bonferroni correction for different classification techniques using two different symbols (X for hypothesis accepted and $\sqrt{}$ for hypothesis rejected). The results reveal that the hypothesis is rejected in most of the cases i.e., models developed using different sets of features are significantly different.

RQ2: What is the capability of various feature selection techniques to predict refactoring-prone classes? Here, we analyzed box-plots, Descriptive

Table 4. Statistical tests: Classification techniques

	FS1	FS2	FS3	FS4
FS1	X	√	√	√
FS2	√	X	X	√
FS3	√	X	X	√
FS4	√	√	√	X

Statistics, and Statistical tests analysis to understand the behavior of different classification techniques for refactoring predictions. In this study, we examine the behavior of eleven different classification techniques and two ensemble techniques on seven different data sets using 5-fold cross validation technique.

Boxplots: Classification Techniques. To understand the visual comparison showing outliers, distribution, variability of different classification techniques, we constructed the box-plot diagrams of different performance parameters such as accuracy and AUC for each classifiers as depicted in Fig. 3. Figure 3 represents the box-plot diagrams of accuracy and AUC values for the refactoring prediction models trained using different classification techniques. Here, red line of each box represents the median value of accuracy or AUC for each classification technique. From the box-plots diagram in Fig. 3, we notice that the model trained for refactoring prediction using LSSVM with polynomial kernel (LSSVM-P) has high value of performance parameters. This represents that the models trained using this technique perform better as compared to other techniques.

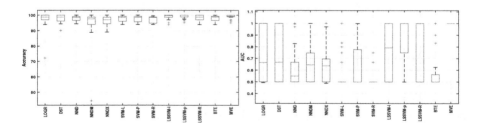

Fig. 3. Box-plots: Classification techniques

Statistical tests: Classification Techniques. To understand the significance difference between the models developed using different sets of features, we conducted statistical tests on AUC value using Wilcoxon signed-rank test with Bonferroni correction as represented in Table 5. Here, the null hypothesis of this work is that *the model trained for refactoring prediction has no effect of input set of features*. For this work, null hypothesis is rejected only if the p-value is less then $\frac{0.05}{78} = 0.00064$. Table 4 depicts the results of Wilcoxon signed-rank test with

Bonferroni correction for different sets of features using two different symbols (X for hypothesis accepted and √ for hypothesis rejected). The results reveal that the hypothesis is accepted in most of the cases i.e., models trained using different classification techniques are not significantly different. We also found from this table that models trained using major voting ensemble techniques (MVE) are significantly different.

Table 5. Statistical tests: Classification techniques

	LOGR	DST	NND	NNDM	NNDX	SVM-L	SVM-P	SVM-R	LSSVM-l	LSSVM-p	LSSVM-R	BTE	MVE
LOGR	X	X	X	X	X	X	X	√	X	X	X	X	√
DST	X	X	X	X	X	X	X	√	X	X	X	X	√
NND	X	X	X	X	X	X	X	√	X	√	X	X	√
NNDM	X	X	X	X	X	X	X	√	X	√	X	X	√
NNDX	X	X	X	X	X	X	X	√	X	√	X	X	√
SVM-L	X	X	X	X	X	X	X	X	√	√	X	X	√
SVM-P	X	X	X	X	X	X	X	X	X	√	X	X	√
SVM-R	√	√	√	√	√	X	X	X	√	√	X	X	√
LSSVM-l	X	X	X	X	X	√	X	√	X	X	X	X	√
LSSVM-p	X	X	√	√	√	√	√	√	X	X	X	√	X
LSSVM-R	X	X	X	X	X	X	X	X	X	X	X	X	√
BTE	X	X	X	X	X	X	X	X	X	√	X	X	√
MVE	√	√	√	√	√	√	√	√	√	X	√	√	X

RQ3: To what extent can the application of ensemble learning classifiers improve upon the performance achieved by individual classifiers? We have also applied two ensemble techniques such as best training ensemble and major voting ensemble (BTE and MVE) for better performance of refactoring prediction. Like individual classifiers, these techniques are evaluated based on their performance across different sets of features. The performance of these ensemble learning techniques are validated using 5-fold cross validation. From the box-plots diagram in Fig. 3, we notice that major voting ensemble technique(MVE) has high value of performance parameters as compared to all techniques. This represents that the ensemble technique perform better as compared to individual technique. From Table 4, we also notice that the refactoring prediction models developed using major voting ensemble technique(MVE) are significantly different as compared to other techniques.

6 Conclusion

Refactoring prediction is a fundamental activity in development and maintenance of software product. However, it is usually a difficult and time consuming task for software developers to identify the classes which are going for refactoring. This paper presents an in-depth investigation on different sets of source code metrics and classification techniques for refactoring prediction. In this work, we studied refactoring prediction on seven freely and publicly available data-sets at tera-PROMISE Repository. The performance different classifiers and sets of features are evaluated using descriptive statistics, box-plots and statistical tests. We found that the models developed using different sets of features are significantly different and models developed for refactoring prediction using all metrics have better performance as compared to other techniques. We also found that the models trained using different classification techniques are not significantly different and the models trained for refactoring prediction using LSSVM with polynomial kernel (LSSVM-P) have better performance as compared to other classifiers. The experimental results also confirms the superiority of ensemble techniques over individual technique.

References

1. Kádár, I., Hegedűs, P., Ferenc, R., Gyimóthy, T.: A manually validated code refactoring dataset and its assessment regarding software maintainability. In: Proceedings of the The 12th International Conference on Predictive Models and Data Analytics in Software Engineering, p. 10. ACM (2016)
2. Kádár, I., Hegedus, P., Ferenc, R., Gyimóthy, T.: A code refactoring dataset and its assessment regarding software maintainability. In: 2016 IEEE 23rd International Conference on Software Analysis, Evolution, and Reengineering (SANER), vol. 1, pp. 599–603. IEEE (2016)
3. Kumar, L., Sureka, A.: Application of LSSVM and SMOTE on seven open source projects for predicting refactoring at class level. In: 2017 24th Asia-Pacific Software Engineering Conference (APSEC), pp. 90–99. IEEE (2017)
4. Li, W., Henry, S.: Maintenance metrics for the Object-Oriented paradigm. In: Proceedings of First International Software Metrics Symposium, pp. 52–60 (1993)

Model Compression and Optimisation

Analysis on Dropout Regularization

John Sum[1(✉)] and Chi-Sing Leung[2]

[1] Institute of Technology Management, National Chung Hsing University,
Taichung 40227, Taiwan
pfsum@nchu.edu.tw
[2] Department of Electronic Engineering, City University of Hong Kong,
Kowloon Tong, KLN, Hong Kong
eeleungc@cityu.edu.hk

Abstract. Dropout, including Bernoulli dropout (equivalently random node fault) and multiplicative Gaussian noise (MGN) dropout (equivalently multiplicative node noise), has been a technique in training a neural network (NN) to achieve better performance. While simulation results have demonstrated its success, not many work has been done to explain why it works (or why it does not work). In this paper, the objective functions $\mathcal{L}(\mathbf{w})$ of the learning algorithms with Bernoulli dropout and MGN dropout are derived and thus their regularization effects are analyzed. It is found that learning with Bernoulli dropout cannot improve the generalization of a NN if its weights are not scaled down after training. If we further let $\mathcal{J}(\mathbf{w})$ be the desired measure of a NN with such inherent dropout, we clarify a misconception that $\mathcal{L}(\mathbf{w}) = \mathcal{J}(\mathbf{w})$. The model attained by learning with dropout is not the desired model that can tolerate such inherent dropout.

Keywords: Dropout · Multiplicative node noise · Random node fault

1 Introduction

Recently, dropout has been a popular technique applying in training a deep neural network (NN) to avoid overfitting [10,17,23,25]. Two dropout techniques have been introduced [25] – the Bernoulli dropout and multiplicative Gaussian noise (MGN) dropout. To be consistent with our previous researches in NNs with noise/fault, we call the Bernoulli dropout the *random node fault* (RNF) and the MGN dropout the *multiplicative node noise* (MNN). In fact, the idea of dropout could be dated back to the 90s when researchers suggested to adding input noise [4,8,9,18], weight noise [1,13–15,21,22] or random node fault [16].

Let $f(\mathbf{x}, \mathbf{z}, \mathbf{w})$ be the NN model, where $\mathbf{x} \in R^m$ is the input vector, $\mathbf{z} \in R^s$ is the hidden node vector and $\mathbf{w} \in R^n$ is the weight vector. Note that $\mathbf{z}$ is a function of $\mathbf{x}$, $\mathbf{z}$ and $\mathbf{w}$. Given a set of N samples $\mathcal{D} = \{\mathbf{x}_k, y_k\}$, the performance measure of the model is given by

$$\mathcal{V}(\mathbf{z}, \mathbf{w}) = \frac{1}{N} \sum_{k=1}^{N} \ell_k(\mathbf{z}, \mathbf{w}), \tag{1}$$

© Springer Nature Switzerland AG 2019
T. Gedeon et al. (Eds.): ICONIP 2019, CCIS 1143, pp. 253–261, 2019.
https://doi.org/10.1007/978-3-030-36802-9_28

where $\ell_k(\mathbf{z}, \mathbf{w})$ is a goodness measure of the NN on the kth sample. It could be defined as square error or cross entropy error. The stochastic gradient descent (SGD) learning is defined as follows:

$$\mathbf{w}(t) = \mathbf{w}(t-1) - \mu_t \frac{\partial \ell_t(\mathbf{z}, \mathbf{w}(t-1))}{\partial \mathbf{w}}, \tag{2}$$

where μ_t is the step size. The sample $\{\mathbf{x}_t, y_t\}$ is randomly picked from $\mathcal{D}$. With RNF, the SGD learning (2) is modified as follows:

$$\mathbf{w}(t) = \mathbf{w}(t-1) - \mu_t \frac{\partial \ell_t(\tilde{\mathbf{z}}, \mathbf{w}(t-1))}{\partial \mathbf{w}}, \tag{3}$$

where $\tilde{\mathbf{z}} = \mathbf{b} \otimes \mathbf{z}$ ($\otimes$ refers to elementwise multiplication) and $\mathbf{b}$ is a binary random vector with $P(b_j = 0) = p$ for $j = 1, \cdots, s$. We call p the fault rate. With MNN, $\tilde{\mathbf{z}} = \mathbf{z} + \mathbf{b} \otimes \mathbf{z}$ and b_j is a mean zero Gaussian random variable for $j = 1, \cdots, s$. While the ideas behind the RNF and MNN techniques are simple, theoretical analysis on their regularization effect is not so straight forward. Only a few works have been reported [2,3,5,24,35].

Next, we let $\mathcal{J}(\mathbf{w})$ be the desired performance measure $E[\mathcal{V}(\tilde{\mathbf{z}}, \mathbf{w})]$ and $\mathcal{L}(\mathbf{w})$ be the objective function of the learning algorithm (3). Without introducing too many notations, we simply use $\mathcal{J}(\mathbf{w})$ and $\mathcal{L}(\mathbf{w})$ for both cases of RNF and MNN. In the next section, learning with Bernoulli dropout will be introduced. Its objective function is derived and its regularization effect is analyzed. Learning with MGN dropout is then presented in Sect. 3 and its objective function is derived. Besides, we will clarify a misconception that the objective function of learning with MNN is the same as the desired performance measure of the neural network with MNN. A note is added in Sect. 4 to clarify a misconception that $\mathcal{L}(\mathbf{w}) = \mathcal{J}(\mathbf{w})$ for any kind of noise or fault adding in the network. Finally, the conclusion is presented in Sect. 5.

2   Random Node Fault (Bernoulli Dropout)

With RNF, the learning algorithm (3) could be rewritten as follows:

$$\mathbf{w}(t) = \mathbf{w}(t-1) - \mu_t \frac{\partial \ell_t(\mathbf{b} \otimes \mathbf{z}, \mathbf{w}(t-1))}{\partial \mathbf{w}}. \tag{4}$$

To investigate the regularization effect of RNF during training, we first show that the actual objective function $\mathcal{L}(\mathbf{w})$ for (4) is essential the same as the desired performance measure $\mathcal{J}(\mathbf{w})$. Next, we derive the desired performance measure and use it to analyze its regularization effect.

2.1   Learning Objective $\mathcal{L}(\mathbf{w})$

From (4), we can get that

$$E[\mathbf{w}(t)|\mathbf{w}(t-1)] = \mathbf{w}(t-1) - \mu_t \frac{1}{N} \sum_{k=1}^{N} \sum_{\mathbf{b}} \frac{\partial \ell_k(\mathbf{b} \otimes \mathbf{z}, \mathbf{w}(t-1))}{\partial \mathbf{w}} P(\mathbf{b})$$

$$= \mathbf{w}(t-1) - \mu_t \frac{\partial}{\partial \mathbf{w}} \sum_{\mathbf{b}} \frac{1}{N} \sum_{k=1}^{N} \ell_k(\mathbf{b} \otimes \mathbf{z}, \mathbf{w}(t-1)) P(\mathbf{b})$$

$$= \mathbf{w}(t-1) - \mu_t \frac{\partial \mathcal{J}(\mathbf{w}(t-1))}{\partial \mathbf{w}}. \tag{5}$$

The objective function is clearly identical to the desired performance measure, i.e. $\mathcal{L}(\mathbf{w}) = \mathcal{J}(\mathbf{w})$.

2.2   Regularization Effect on MLP

From (5), it is clear that the objective function $\mathcal{L}(\mathbf{w})$ is given by

$$\mathcal{L}(\mathbf{w}) = \mathcal{J}(\mathbf{w}) = \sum_{\mathbf{b}} \mathcal{V}(\mathbf{b} \otimes \mathbf{z}, \mathbf{w}) P(\mathbf{b}). \tag{6}$$

While $\mathbf{b}$ is a binary vector, it can also be referred as a network structure. Thus, the objective function of learning with random node fault, i.e. (4), is the expectation of $\mathcal{V}(\mathbf{b} \otimes \mathbf{z}, \mathbf{w})$ taking over all possible structures. To investigate the regularization effect of Bernoulli dropout, we need to find $\mathcal{L}(\mathbf{w})$. However, (6) has no simple close form solution even if the neural network is a MLP with single hidden layer and single sigmoidal output node.

For a single hidden layer MLP with s sigmoidal nodes and a linear output node, it has been shown in [33] that

$$\mathcal{L}(\mathbf{w}) = \mathcal{V}(\mathbf{w}) + p\,\mathbf{d}^T \left(\frac{1}{N} \sum_{k=1}^{N} (\mathbf{G}(\mathbf{x}_k, \mathbf{w}) - \mathbf{H}(\mathbf{x}_k, \mathbf{w})) \right) \mathbf{d}, \tag{7}$$

where $\mathbf{d}$ is the output weight vector, $\mathbf{G}(\mathbf{x}_k, \mathbf{w}) = \mathrm{diag}\{z_1^2(\mathbf{x}_k, \mathbf{w}), \cdots, z_s^2(\mathbf{x}_k, \mathbf{w})\}$ $\mathbf{H}(\mathbf{x}_k, \mathbf{w}) = \mathbf{z}(\mathbf{x}_k, \mathbf{w})\mathbf{z}(\mathbf{x}_k, \mathbf{w})^T$. The objective function (7) could also be rewritten as follows:

$$\mathcal{L}(\mathbf{w}) = \mathcal{V}(\mathbf{w}) - \frac{p}{N} \sum_{k=1}^{N} (\mathbf{d} \otimes \mathbf{z}(\mathbf{x}_k, \mathbf{w}))^T \mathbf{M} (\mathbf{d} \otimes \mathbf{z}(\mathbf{x}_k, \mathbf{w})), \tag{8}$$

where $\mathbf{M} \in R^{s \times s}$ is a special matrix with diagonal elements all zeros and off-diagonal elements all ones. It is a special matrix with the following property.

Lemma 1. *The matrix $\mathbf{M}$ has two eigenvalues, $(s-1)$ and -1. The eigenspace corresponding to the eigenvalue $(s-1)$ is a 1-D space spanned by the eigenvector $(1, 1, \cdots, 1)$. The eigenspace corresponding to the eigenvalue -1 is a $(s-1)$-D space orthogonal to $(1, 1, \cdots, 1)$.*

Therefore, regularization term in (8) has two effects, as shown in Fig. 1. The first effect is due to the eigenvalue (-1). It regulates the magnitude of $\mathbf{d} \otimes \mathbf{z}(\mathbf{x}_k, \mathbf{w})$ for $k = 1, \cdots, N$ to small value. The second effect is due to the eigenvalue $(s - 1)$. It tends to enlarge the magnitude of the projection of $\mathbf{d} \otimes \mathbf{z}(\mathbf{x}_k, \mathbf{w})$ on $(1, 1, \cdots, 1)$. In contrast to weight decay in which the effect is acting on each weight w_j, the effect of RNF is acting on each $d_j z_j$ only.

Besides, one should note that the regularization effect will be amplified if the fault rate p increases. The regularization effect due to the second eigenvalue $(s - 1)$ could be severe if the number of hidden nodes is large. If the model attained $\mathcal{L}_{min}$ is applied for prediction but random node fault does not exist anymore, it is clear from Fig. 1 that the performance of the model $\mathcal{L}_{min}$ could be terrible as it is far away from $\mathcal{V}_{min}$. To alleviate the shortcoming, one trick is to scale down d_j to $(1 - p)d_j$ after training [10, 17, 25], Fig. 1, hoping that the new model $\mathcal{H}$ could be located closer to $\mathcal{V}_{min}$.

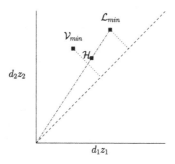

Fig. 1. Regularization effect due to Bernoulli dropout. $\mathcal{V}_{min}$ (resp. $\mathcal{L}_{min}$) is the model obtained without (resp. with) dropout. $\mathcal{H}$ is the model with the d_j in $\mathcal{L}_{min}$ being multiplied with factor $(1 - p)$.

3   Multiplicative Node Noise (MGN Dropout)

With multiplicative node noise, the learning algorithm (3) could be rewritten as follows:

$$\mathbf{w}(t) = \mathbf{w}(t - 1) - \mu_t \frac{\partial \ell_t(\mathbf{z}(t) + \Delta \mathbf{z}(t), \mathbf{w}(t - 1))}{\partial \mathbf{w}}, \tag{9}$$

where $\Delta \mathbf{z}(t) = \mathbf{b} \otimes \mathbf{z}(t)$. Consider the ith weight, we can get that

$$w_i(t) = w_i(t - 1) - \mu_t \frac{\partial \ell_t(\mathbf{z}(t) + \Delta \mathbf{z}(t), \mathbf{w}(t - 1))}{\partial w_i}. \tag{10}$$

Recall that $\mathbf{z}(t)$ depends on the data sampled at step t, $\mathbf{w}(t - 1)$ and the noise vector $\mathbf{b}$. Here, let S_M be the variance of the mean zero Gaussian noise factor b_j for all $j = 1, \cdots, s$. For small S_M, we can get that

$$\frac{\partial \ell_t(\mathbf{z}(t) + \Delta \mathbf{z}(t), \mathbf{w})}{\partial w_i} = \frac{\partial \ell_t(\mathbf{z}(t), \mathbf{w})}{\partial w_i} + \sum_j \frac{\partial^2 \ell_t(\mathbf{z}(t), \mathbf{w})}{\partial z_j \partial w_i} \Delta z_j(t)$$

$$+ \frac{1}{2} \sum_j \sum_{j'} \frac{\partial^3 \ell_t(\mathbf{z}(t), \mathbf{w})}{\partial z_j \partial z_{j'} \partial w_i} \Delta z_j(t) \Delta z_{j'}(t) + HOT. \quad (11)$$

From (11), we can derive the objective function for the learning algorithm with MNN.

3.1   Learning Objective $\mathcal{L}(\mathbf{w})$

Taking expectation of $\partial \ell_t(\mathbf{z}(t) + \Delta \mathbf{z}(t), \mathbf{w})/\partial w_i$ the space of $\mathbf{b}$ and by (10), we get that

$$E[w_i(t)|\mathbf{w}(t-1)] = w_i(t-1) - \mu_t \left\{ \frac{\partial \ell_t(\mathbf{z}(t), \mathbf{w}(t-1))}{\partial w_i} \right.$$

$$\left. + \frac{S_M}{2} \sum_j \frac{\partial^3 \ell_t(\mathbf{z}(t), \mathbf{w}(t-1))}{\partial z_j^2 \partial w_i} z_j^2(t) \right\}, \quad (12)$$

and the actual objective function of the algorithm (9) is given by

$$\mathcal{L}(\mathbf{w}) = \mathcal{V}(\mathbf{w}) + \frac{S_M}{2N} \sum_i \int \sum_j \sum_k \frac{\partial^3 \ell_k(\mathbf{z}(k), \mathbf{w})}{\partial z_j^2 \partial w_i} z_j^2(k) dw_i. \quad (13)$$

Although the learning objective has been derived, the regularization effect of the additional term $\frac{S_M}{2N} \sum_i \int \sum_j \sum_k \frac{\partial^3 \ell_k(\mathbf{z}(k), \mathbf{w})}{\partial z_j^2 \partial w_i} z_j^2(k) dw_i$ is still unknown, unlike the Bernoulli dropout.

3.2   Desired Performance Measure $\mathcal{J}(\mathbf{w})$

The desired performance measure $\mathcal{J}(\mathbf{w})$ could be derived in similar manner. Consider $\ell_k(\mathbf{z}(k), \mathbf{w})$. With MNN and S_M is small, we can get that

$$\ell_k(\tilde{\mathbf{z}}(k), \mathbf{w}) = \ell_k(\mathbf{z}(k), \mathbf{w}) + \sum_j \frac{\partial \ell_k(\mathbf{z}(k), \mathbf{w})}{\partial z_j} \Delta z_j(k)$$

$$+ \frac{1}{2} \sum_j \sum_{j'} \frac{\partial^2 \ell_k(\mathbf{z}(k), \mathbf{w})}{\partial z_j \partial z_{j'}} \Delta z_j(k) \Delta z_{j'}(k) + HOT. \quad (14)$$

Taking expectation of (14) over the noise vector $\mathbf{b}$, we can get that

$$\mathcal{J}(\mathbf{w}) = \mathcal{V}(\mathbf{w}) + \frac{S_M}{2N} \sum_j \sum_k \frac{\partial^2 \ell_k(\mathbf{z}(k), \mathbf{w})}{\partial z_j^2} z_j^2(k). \quad (15)$$

The SGD learning which minimizing (15) would be given by

$$
w_i(t) = w_i(t-1) - \mu_t \left\{ \frac{\partial \ell_t(\mathbf{z}(t), \mathbf{w}(t-1))}{\partial w_i} + \frac{S_M}{2} \sum_j \frac{\partial^3 \ell_t(\mathbf{z}(t), \mathbf{w}(t-1))}{\partial w_i \partial z_j^2} z_j^2(t) \right.
$$

$$
\left. + S_M \sum_j \frac{\partial^2 \ell_t(\mathbf{z}(t), \mathbf{w}(t-1))}{\partial z_j^2} z_j(t) \frac{\partial z_j(t)}{\partial w_i} \right\}. \tag{16}
$$

Note that $\partial^3 \ell_k(\mathbf{z}, \mathbf{w})/\partial^2 z_j \partial w_i \neq \partial^3 \ell_k(\mathbf{z}, \mathbf{w})/\partial w_i \partial^2 z_j$. Thus, $\frac{\partial^3 \ell_t(\mathbf{z}(t), \mathbf{w}(t-1))}{\partial z_j^2 \partial w_i}$ $z_j^2(t)$ in (10) is different from the factor $\frac{\partial^3 \ell_t(\mathbf{z}(t), \mathbf{w}(t-1))}{\partial w_i \partial z_j^2} z_j^2(t)$ in (16). By comparing (13) and (10) with (15) and (16), it is not difficult to see that $\mathcal{L}(\mathbf{w}) \neq \mathcal{J}(\mathbf{w})$. The SGD learning algorithms (10) and (16) are different.

4   Note on $\mathcal{L}(\mathbf{w})$ and $\mathcal{J}(\mathbf{w})$

As mentioned in [32], there are common misconceptions on the issues that adding noise during training is a regularization technique and the objective function being minimized by noise-injection training is $\mathcal{L}(\mathbf{w}) = \mathcal{J}(\mathbf{w})$. Accordingly, $\mathcal{J}(\mathbf{w})$ is used for interpreting the regularization properties of noise-injection [6,7,23] (See P.235, Section 7.5 in [7] and P.13, last paragraph of Section 7.4.3 in [6].). For instance, the authors in [23] treat $\mathcal{J}(\mathbf{w})$ as the objective function of learning with MGN dropout and use it to interpret the regularization effect of MGN dropout.

Table 1. $\mathcal{L}(\mathbf{w})$ versus $\mathcal{J}(\mathbf{w})$

Noise/fault	NN model	Learning	Equivalency	Ref.
Input noise	MLP	GD	$\mathcal{L}(\mathbf{w}) = \mathcal{J}(\mathbf{w})$	[1,4,8,9,18]
Random weight fault	MLP	GD	$\mathcal{L}(\mathbf{w}) = \mathcal{J}(\mathbf{w})$	This paper
Additive weight noise	MLP	GD	$\mathcal{L}(\mathbf{w}) = \mathcal{J}(\mathbf{w})$	[1,12,32,34]
Multiplicative weight noise	MLP	GD	$\mathcal{L}(\mathbf{w}) \neq \mathcal{J}(\mathbf{w})$	[12,32,34]
Random node fault	MLP	GD	$\mathcal{L}(\mathbf{w}) = \mathcal{J}(\mathbf{w})$	[33], this paper
Additive node noise	MLP	GD	$\mathcal{L}(\mathbf{w}) \neq \mathcal{J}(\mathbf{w})$	[27,28]
Multiplicative node noise	MLP	GD	$\mathcal{L}(\mathbf{w}) \neq \mathcal{J}(\mathbf{w})$	[29], this paper
Input Noise	RBF	GD	$\mathcal{L}(\mathbf{w}) = \mathcal{J}(\mathbf{w})$	[11]
Random weight fault	RBF	GD	$\mathcal{L}(\mathbf{w}) = \mathcal{J}(\mathbf{w})$	[27,28]
Additive weight noise	RBF	GD	$\mathcal{L}(\mathbf{w}) = \mathcal{J}(\mathbf{w})$	[11]
Multiplicative weight noise	RBF	GD	$\mathcal{L}(\mathbf{w}) \neq \mathcal{J}(\mathbf{w})$	[11]
Random node fault	RBF	GD	$\mathcal{L}(\mathbf{w}) = \mathcal{J}(\mathbf{w})$	[26,31]
Additive node noise	RBF	GD	$\mathcal{L}(\mathbf{w}) = \mathcal{J}(\mathbf{w})$	[27,28]
Multiplicative node noise	RBF	GD	$\mathcal{L}(\mathbf{w}) = \mathcal{J}(\mathbf{w})$	[27,28]
Additive weight noise	BM	BL	$\mathcal{L}(\mathbf{w}) = \mathcal{J}(\mathbf{w})$	[30]

MLP: Multilayer perceptron; RBF: Radial basis function network; BM: Boltzmann machine; GD: Gradient descent; BL: Boltzmann learning

These misconceptions could be due to earlier analysis on the connections between noise injection training and regularization effect. First, it was shown that adding input noise during training [18] is equivalent to Tikhonov regularization [4,8,9]. Second, it was shown that the objective function of injecting additive weight noise during training [19–22] is equivalent to the expected MSE of the model with additive weight noise [1]. In sequel, researchers started to confuse that the argument could be applied to other noise models. They confused that the desired performance measure $\mathcal{J}(\mathbf{w})$ of a neural network with input noise (resp. weight noise or random node fault) is the same as the objective function $\mathcal{L}(\mathbf{w})$ of the SGD learning with input noise (resp. weight noise or random node fault). However, it is not always true. The answer depends on (i) the model of neural network, (ii) the noise or fault model and (iii) the learning algorithm, see Table 1.

5   Conclusion

In this paper, we have presented a few analyses on Bernoulli dropout (equivalently the random node fault (RNF)) and MGN dropout (equivalently the multiplicative node noise (MNN)). Their actual objective functions $\mathcal{L}(\mathbf{w})$ are derived and compared with the desired performance measures $\mathcal{J}(\mathbf{w})$. It is shown that $\mathcal{L}(\mathbf{w}) = \mathcal{J}(\mathbf{w})$ for the network with RNF. For a NN with MNN, they are generally not the same, except in some special cases, like linear regressor and RBF. For a MLP with single hidden layer and a linear output node, we have further analyzed the regularization effect from the objective function $\mathcal{L}(\mathbf{w})$. By that, we provide a explanation why all the weights have to be scaled down by $(1 - p)$ after learning with Bernoulli dropout. Finally, a note has been added to clarify a misconception that $\mathcal{L}(\mathbf{w}) = \mathcal{J}(\mathbf{w})$ for learning with model noise or fault.

Acknowledgement. The works presented in this paper is supported in part by Taiwan MOST grants 105-2221-E-005-065-MY2 and 108-2221-E-005-036, and a research grant from City University of Hong Kong (7005063).

References

1. An, G.: The effects of adding noise during backpropagation training on a generalization performance. Neural Comput. **8**, 643–674 (1996)
2. Baldi, P., Sadowski, P.: The dropout learning algorithm. Artif. Intell. **210**, 78–122 (2014)
3. Baldi, P., Sadowski, P.J.: Understanding dropout. In: Advances in Neural Information Processing Systems, pp. 2814–2822 (2013)
4. Bishop, C.: Training with noise is equivalent to Tikhonov regularization. Neural Comput. **7**, 108–116 (1995)
5. Gal, Y., Ghahramani, Z.: Dropout as a Bayesian approximation: representing model uncertainty in deep learning. arXiv:1506.02142v4 (2015)
6. Ghojogh, B., Crowley, M.: The theory behind overfitting, cross validation, regularization, bagging, and boosting: tutorial. arXiv preprint arXiv:1905.12787 (2019)

7. Goodfellow, I., Bengio, Y., Courville, A.: Deep Learning. MIT Press, Cambridge (2016)
8. Grandvalet, Y., Canu, S.: Comments on "Noise injection into inputs in back propagation learning". IEEE Trans. Syst. Man Cybern. **25**(4), 678–681 (1995)
9. Grandvalet, Y., Canu, S., Boucheron, S.: Noise injection: theoretical prospects. Neural Comput. **9**(5), 1093–1108 (1997)
10. Hinton, G.E., Srivastava, N., Krizhevsky, A., Sutskever, I., Salakhutdinov, R.R.: Improving neural networks by preventing co-adaptation of feature detectors. arXiv preprint arXiv:1207.0580 (2012)
11. Ho, K., Leung, C., Sum, J.: Convergence and objective functions of some fault/noise injection-based online learning algorithms for RBF networks. IEEE Trans. Neural Netw. **21**(6), 938–947 (2010)
12. Ho, K., Leung, C., Sum, J.: Objective functions of the online weight noise injection training algorithms for MLP. IEEE Trans. Neural Netw. **22**(2), 317–323 (2011)
13. Jabri, M., Flower, B.: Weight perturbation: an optimal architecture and learning technique for analog VLSI feedforward and recurrent multilayer networks. Neural Comput. **3**(4), 546–565 (1991)
14. Jabri, M., Flower, B.: Weight perturbation: an optimal architecture and learning technique for analog VLSI feedforward and recurrent multilayer networks. IEEE Trans. Neural Netw. **3**(1), 154–157 (1992)
15. Jim, K., Giles, C., Horne, B.: An analysis of noise in recurrent neural networks: convergence and generalization. IEEE Trans. Neural Netw. **7**, 1424–1438 (1996)
16. Judd, J.S., Munro, P.W.: Nets with unreliable hidden nodes learn error-correcting codes. In: Advances in Neural Information Processing Systems 5, pp. 89–96. Morgan Kaufmann Publishers Inc., San Francisco (1993)
17. Krizhevsky, A., Sutskever, I., Hinton, G.: ImageNet classification with deep convolutional neural networks. In: Advances in Neural Information Processing Systems, pp. 1097–1105 (2012)
18. Matsuoka, K.: Noise injection into inputs in back-propagation learning. IEEE Trans. Syst. Man Cybern. **22**(3), 436–440 (1992)
19. Murray, A.: Analogue noise-enhanced learning in neural network circuits. Electron. Lett. **27**(17), 1546–1548 (1991)
20. Murray, A.: Multilayer perceptron learning optimized for on-chip implementation: a noise-robust system. Neural Comput. **4**(3), 366–381 (1992)
21. Murray, A., Edwards, P.: Synaptic weight noise during multilayer perceptron training: fault tolerance and training improvements. IEEE Trans. Neural Netw. **4**(4), 722–725 (1993)
22. Murray, A., Edwards, P.: Enhanced MLP performance and fault tolerance resulting from synaptic weight noise during training. IEEE Trans. Neural Netw. **5**(5), 792–802 (1994)
23. Nalisnick, E., Anandkumar, A., Smyth, P.: A scale mixture perspective of multiplicative noise in neural networks. arXiv preprint arXiv:1506.03208 (2015)
24. Noh, H., You, T., Mun, J., Han, B.: Regularizing deep neural networks by noise: its interpretation and optimization. In: Advances in Neural Information Processing Systems, pp. 5109–5118 (2017)
25. Srivastava, N., Hinton, G., Krizhevsky, A., Sutskever, I., Salakhutdinov, R.: Dropout: a simple way to prevent neural networks from overfitting. J. Mach. Learn. Res. **15**(1), 1929–1958 (2014)
26. Sum, J.: On a multiple nodes fault tolerant training for RBF: objective function, sensitivity analysis and relation to generalization. In: Proceedings of TAAI 2005, Tainan, Taiwan (2005)

27. Sum, J.: Misconception on the regularization effect of noise or fault injection: empirical evidence. In: Proceedings of TAAI 2019, Kaohsiung, Taiwan (2019)
28. Sum, J.: Misconception on the regularization effect of noise or fault injection: theoretical analysis. In: Proceedings of TAAI 2019, Kaohsiung, Taiwan (2019)
29. Sum, J., Leung, C.S.: Analysis on dropout regularization (2019). In submission
30. Sum, J., Leung, C.S.: Learning algorithm for Boltzmann machines with additive weight and bias noise. IEEE Trans. Neural Netw. Learn. Syst. (2019). Accepted for publication
31. Sum, J., Leung, C., Ho, K.: On node-fault-injection training of an RBF network. In: Köppen, M., Kasabov, N., Coghill, G. (eds.) ICONIP 2008. LNCS, vol. 5507, pp. 324–331. Springer, Heidelberg (2009). https://doi.org/10.1007/978-3-642-03040-6_40
32. Sum, J., Leung, C.S., Ho, K.: Convergence analyses on on-line weight noise injection-based training algorithms for MLPs. IEEE Trans. Neural Netw. Learn. Syst. $23(11)$, 1827–1840 (2012)
33. Sum, J., Leung, C.S., Ho, K.: Convergence analysis of on-line node fault injection-based training algorithms for MLP networks. IEEE Trans. Neural Netw. Learn. Syst. $23(2)$, 211–222 (2012)
34. Sum, J., Leung, C.S., Ho, K.: A limitation of gradient descent learning. IEEE Trans. Neural Netw. Learn. Syst. (2019). Accepted for publication
35. Wager, S., Wang, S., Liang, P.S.: Dropout training as adaptive regularization. In: Advances in Neural Information Processing Systems, pp. 351–359 (2013)

On a Convergence Property
of a Geometrical Algorithm
for Statistical Manifolds

Shotaro Akaho[1,4]($\boxtimes$) (iD), Hideitsu Hino[2,4](iD), and Noboru Murata[3,4](iD)

[1] National Institute of Advanced Industrial Science and Technology,
Tsukuba, Ibaraki 305-8568, Japan
s.akaho@aist.go.jp
[2] The Institute of Statistical Mathematics, Tachikawa, Tokyo 190-8562, Japan
[3] Waseda University, Shinjuku, Tokyo 169-0072, Japan
[4] RIKEN Center for Advanced Intelligence Project, Chuo, Tokyo 103-0027, Japan

Abstract. In this paper, we examine a geometrical projection algorithm for statistical inference. The algorithm is based on Pythagorean relation and it is derivative-free as well as representation-free that is useful in nonparametric cases. We derive a bound of learning rate to guarantee local convergence. In special cases of m-mixture and e-mixture estimation problems, we calculate specific forms of the bound that can be used easily in practice.

Keywords: Information geometry · Dimension reduction · Mixture model · Pythagorean theorem

1 Introduction

Information geometry is a framework to analyze statistical inference and machine learning [5]. Geometrically, statistical inference and many machine learning algorithms can be regarded as procedures to find a projection to a model subspace from a given data point. In this paper, we focus on an algorithm to find the projection.

Since the projection is given by minimizing a divergence, a common approach to finding the projection is a gradient-based method [7]. However, such an approach is not applicable in some cases. For instance, several attempts to extend the information geometrical framework to nonparametric cases [6,10,13,15], where we need to consider a function space or each data is represented as a point process. In such a case, it is difficult to compute the derivative of divergence that is necessary for gradient-based methods, and in some cases, it is difficult to deal with the coordinate explicitly.

Takano et al. [15] proposed a geometrical algorithm to find the projection for nonparametric e-mixture distribution, where the model subspace is spanned

Supported by JSPS KAKENHI Grant Number 17H01793, 19K12111.

T. Gedeon et al. (Eds.): ICONIP 2019, CCIS 1143, pp. 262–272, 2019.
https://doi.org/10.1007/978-3-030-36802-9_29

by several empirical distributions. The algorithm that is derived based on the generalized Pythagorean theorem only depends on the values of divergences. It is derivative-free as well as representation-free, and it can be applicable to many machine learning algorithms that can be regarded as finding a projection, but its convergence property has not been analyzed yet. The first contribution of this paper is to extend the algorithm to more general cases. The second contribution is to give a condition for the convergence of the algorithm, which is given as a bound of learning rate. In the case of the discrete distribution, we obtain specific forms of the bound that can be used easily in practice.

2   Projection in a Statistical Manifold

Here we briefly review the information geometry in order to explain the proposed geometrical algorithm based on generalized Pythagorean theorem [12].

Let $(S, g, \nabla, \tilde{\nabla})$ be a statistical manifold, where S is a smooth manifold with a Riemannian metric g, dual affine connections ∇ and $\tilde{\nabla}$. We consider the case that S is (dually) flat, where there exist a ∇-affine coordinate $\theta = (\theta_1, \ldots, \theta_d)$ and a $\tilde{\nabla}$-affine coordinate $\eta = (\eta_1, \ldots, \eta_d)$. For a flat manifold, there exist a pair of potential functions $\psi(\theta)$ and $\phi(\eta)$, and the two coordinates θ and η are transformed each other by Legendre transform,

$$\theta_i = \frac{\partial \phi(\eta)}{\partial \eta_i}, \quad \eta_i = \frac{\partial \psi(\theta)}{\partial \theta_i}, \quad \psi(\theta) + \phi(\eta) - \sum_{i=1}^{d} \theta_i \eta_i = 0. \tag{1}$$

A typical example of a flat manifold is an exponential family, where each member of the manifold is a distribution of a random variable x with parameter $\xi = (\xi_1, \ldots, \xi_d)$,

$$p(x; \xi) = \exp\left(\sum_{i=1}^{d} \xi_i F_i(x) - b(\xi)\right), \tag{2}$$

where $F_i(x)$ is a sufficient statistics and $\exp(-b(\xi))$ is a normalization factor. For the exponential family, there are two dual connections, called e-connection and m-connection (e: exponential, m: mixture). If we take the e-connection as the ∇-connection, ∇-affine coordinate θ is equal to ξ called e-coordinate, and $\tilde{\nabla}$-affine coordinate called m-coordinate is given by $\zeta_i = \frac{\partial b(\xi)}{\partial \xi_i} = E_\xi[F_i(x)] = \int F_i(x) p(x; \xi) dx$, where the function $b(\xi)$ becomes a potential function $\psi(\theta)$. Note that if we take the m-connection as ∇, the relation changes in a dual way, i.e., ζ becomes θ and ξ becomes η.

Here, for $p \in S$, we denote the corresponding ∇- and $\tilde{\nabla}$-coordinate by $\theta(p)$ and $\eta(p)$ respectively. Let us consider a submanifold defined by linear combinations of given K points $p_1, \ldots, p_K \in S$,

$$M = \{p \mid \theta(p) = \sum_{k=1}^{K} w_k \theta(p_k), \sum_{k=1}^{K} w_k = 1\}, \tag{3}$$

where $\mathbf{w} = (w_1, \ldots, w_K)$ is a weight vector whose sum is 1, and K is smaller than the dimensionality of θ. The submanifold M is an affine subspace and hence it is called an ∇-autoparallel (or ∇-flat) submanifold. In particular, if $K = 2$, M is a straight line of ∇-coordinate that is called ∇-geodesic.

We can also consider another submanifold in the dual coordinate,

$$\tilde{M} = \{p \mid \eta(p) = \sum_{k=1}^{K} w_k \eta(p_k), \sum_{k=1}^{K} w_k = 1\}, \tag{4}$$

which is called a $\tilde{\nabla}$-autoparallel (or $\tilde{\nabla}$-flat) submanifold. The $\tilde{\nabla}$-geodesic is defined by a straight line of $\tilde{\nabla}$-coordinate.

Now let us define a ∇-projection and a $\tilde{\nabla}$-projection from a point $q \in S$ onto a (not necessarily autoparallel) submanifold P. The ∇-projection is a point $q^* \in P$ such that ∇-geodesic connecting q and q^* is orthogonal to P at q^* with respect to the Riemannian metric $g_{ij}(\theta(q^*))$. In the statistical manifold, we take g_{ij} as

$$g_{ij}(\theta) = \frac{\partial^2 \psi(\theta)}{\partial \theta_i \partial \theta_j}, \tag{5}$$

which is equal to Fisher information for exponential family $g_{ij}(\xi) = E_\xi \left[\frac{\partial \log p(x;\xi)}{\partial \xi_i} \frac{\log p(x;\xi)}{\partial \xi_j} \right]$. In a similar way, $\tilde{\nabla}$-projection onto a submanifold P can be defined.

The following theorem gives a fundamental property of the projection (Fig. 1).

Theorem 1 (Generalized Pythagorean theorem [12]). *Let $\tilde{M}$ be a $\tilde{\nabla}$-autoparallel submanifold of a statistical manifold S, and the ∇-projection be $q^* \in \tilde{M}$ from a point $q \in S$, then for any point $p \in \tilde{M}$, the following relation holds*

$$D(p, q) = D(q^*, q) + D(p, q^*), \tag{6}$$

where D is the canonical divergence defined by

$$D(p, q) = \psi(\theta(q)) + \phi(\eta(p)) - \sum_{i=1}^{d} \eta_i(p)\theta_i(q). \tag{7}$$

By exchanging ∇ and $\tilde{\nabla}$, we have a dual relation, i.e, for a ∇-autoparallel submanifold M, the $\tilde{\nabla}$-projection $q^* \in M$ from a point $q \in S$ satisfies the relation $\tilde{D}(p, q) = \tilde{D}(q^*, q) + \tilde{D}(p, q^*)$, where $p \in M$ and $\tilde{D}$ is a dual divergence defined by $\tilde{D}(p, q) = D(q, p)$.

From this theorem, we see that a ∇-projection ($\tilde{\nabla}$-projection) onto a $\tilde{\nabla}$-autoparallel (∇-autoparallel respectively) submanifold is unique and can be found by minimizing corresponding divergence, i.e., the ∇-projection is given by

$$q^* = \arg \min_{p \in \tilde{M}} D(p, q) \tag{8}$$

and the $\tilde{\nabla}$-projection is given by $q^* = \arg \min_{p \in M} \tilde{D}(p, q)$.

3   Geometrical Algorithm for Projection

Now we explain a geometrical algorithm to find a ∇-projection (or $\tilde{\nabla}$-projection) onto a $\tilde{\nabla}$-autoparallel (and ∇-autoparallel respectively) submanifold. To avoid redundant description, we only formulate the ∇-projection onto a $\tilde{\nabla}$-autoparallel submanifold, since the dual case can be obtained by only exchanging ∇ and $\tilde{\nabla}$.

For simplicity, we assume that the projection lies in the region $\forall w_i > 0$, which can be easily generalized and discussed in Sect. 6.

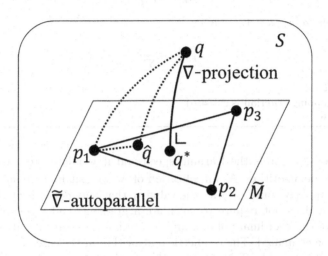

Fig. 1. The ∇-projection q^* from a point q to an $\tilde{\nabla}$-autoparallel manifold $\tilde{M}$ spanned by $\{p_k\}$, where $\hat{q}$ is a current estimate of q^*. The value γ_k defined in (9) represents the deviation from Pythagorean relation, i.e., $\gamma_k = 0$ iff $\hat{q} = q^*$, and $\gamma_k > 0$ implies $\hat{q}$ is closer to p_k while $\gamma_k < 0$ implies $\hat{q}$ is more distant from p_k.

Suppose a point $q \in S$ and a $\tilde{\nabla}$-autoparallel submanifold $\tilde{M} \subseteq S$ are given, let $\hat{q} \in \tilde{M}$ be a current estimate of the projection $q^* \in \tilde{M}$ (Fig. 1) and let us define the quantity γ_k,

$$\gamma_k = D(\hat{q}, q) + D(p_k, \hat{q}) - D(p_k, q), \tag{9}$$

which measures the deviation from the Pythagorean relation, i.e., from Eq. (6), $\gamma_k = 0$ if and only if $\hat{q} = q^*$. If $\gamma_k < 0$, that means $\hat{q}$ is closer to p_k than q^*, w_k should be decreased. On the other hand, if $\gamma_k > 0$, $\hat{q}$ is more distant from p_k than q^*, w_k should be increased. From this consideration, we can construct the Algorithm 1 to find the ∇-projection by optimizing weights $\{w_k\}_{k=1,\ldots,K}$ so that $\hat{q}$ satisfies the Pythagorean relation (6).

In the algorithm, the function $f(\gamma)$ is a positive and monotonically increasing function s.t. $f(0) = 1$, which is introduced in order to stabilize the algorithm and a typical choice of f is a sigmoid function,

$$f(\gamma) = \frac{2}{1 + \exp(-\beta\gamma)}, \quad \beta > 0. \tag{12}$$

Algorithm 1. Geometrical Algorithm A(K)

Input: $q, p_1, p_2, \ldots, p_K$
Output: w
 1: **Initialize** $\{w_k^{(0)}\}_{k=1,\ldots,K}$ s.t. $\sum_{k=1}^{K} w_k^{(0)} = 1, w_k^{(0)} > 0, t := 0$
 2: **repeat**
 3:    Calculate γ_k by (9), where $\eta(\hat{q}) = \sum_{i=1}^{K} w_k^{(t)} \eta(p_k)$, $k = 1, \ldots, K$
 4:    Update w_k, $k = 1, \ldots, K$ by

$$w_k' = w_k^{(t)} f(\gamma_k) \tag{10}$$

 5:    Normalize w_k', $k = 1, \ldots, K$ by

$$w_k^{(t+1)} = \frac{w_k'}{\sum_{k=1}^{K} w_k'} \tag{11}$$

 6:    $t := t + 1$
 7: **until** Stopping criterion is satisfied
 8: **return w**

A parameter β controls the learning speed and it is related to a convergence property of the algorithm. As initialization of **w**, we usually take $w_k^{(0)} = 1/K$.

Algorithm A(K) only requires the value γ_k that only depends on divergence. Therefore, it does not require to calculate a gradient, and the explicit representations of the coordinate of $\hat{q}$, q and $\hat{p}_k$ are not necessary if only divergence values can be estimated. The geometrical algorithm in the case of m-connection was firstly introduced by Takano et al. [15] in order to estimate a nonparametric e-mixture distribution, where the coordinate is not explicitly available. In [15], each point is represented as a point process and the divergence is estimated in a nonparametric manner [8,9]. In order to define nonparametric e-mixture without explicit coordinate representation, it uses a fact that the point on $\tilde{M}$ is characterized in terms of divergence [11],

$$\hat{q} = \arg \min_{q \in \tilde{M}} \sum_{k=1}^{K} w_k D(p_k, q). \tag{13}$$

4   Convergence for the Case of $K = 2$

4.1   A Bound for Convergence

The main purpose of this paper is to investigate the convergence property of Algorithm A(K). We first show the convergence condition for the case $K = 2$, and the result is generalized to arbitrary K.

Theorem 2. *Algorithm A(2) is locally stable when it holds*

$$\frac{df(0)}{d\gamma} < \frac{2}{w_1^*(1 - w_1^*)g(w_1^*)}, \tag{14}$$

where w_1^* is the optimal weight.

Proof. By a straight-forward calculation, we can show in general

$$\gamma_k \doteq \sum_{i=1}^{d}(\theta_i(\hat{q}) - \theta_i(q^*))(\eta_i(\hat{q}) - \eta_i(p_k)). \tag{15}$$

If the current estimate $\hat{w}_1$ is perturbed from the optimal value w_1^*, i.e., $\hat{w}_1 = w_1^* + \epsilon$ for small ϵ, then up to the first order it holds

$$\gamma_1 = (w_1^* - 1)g(w_1^*)\epsilon + o(\epsilon). \tag{16}$$

By Algorithm A(2), the weight w_1 is updated by $w_1' = \hat{w}_1 f(\gamma_1)$, then we have

$$w_1' = \hat{w}_1(1 + \frac{df(0)}{d\gamma}\gamma_1) + o(\epsilon) = w_1^* + \epsilon + w_1^*(w_1^* - 1)g(w_1^*)\frac{df(0)}{d\gamma}\epsilon + o(\epsilon), \tag{17}$$

and for w_2,

$$w_2' = \hat{w}_2 f(\gamma_2) = 1 - w_1^* - \epsilon + w_1^*(1 - w_1^*)g(w_1^*)\frac{df(0)}{d\gamma}\epsilon + o(\epsilon). \tag{18}$$

We see that $w_1' + w_2' = 1 + o(\epsilon)$, thus the normalization procedure is negligible up to the first order of ϵ. The condition that q^* is a stable point of the algorithm is given by $|w_1' - w_1^*| < |\hat{w}_1 - w_1^*| = |\epsilon|$. From Eq. (17), it is

$$|\epsilon + w_1^*(w_1^* - 1)g(w_1^*)\frac{df(0)}{d\gamma}\epsilon| < |\epsilon|, \tag{19}$$

which is equivalent to (14). □

The local stability only depends on the derivative of f at the origin. For the sigmoid function (12), $\frac{df(0)}{d\gamma} = \beta/2$. Although (14) depends on the optimal value, it is not available when the algorithm is applied in practice. To solve this problem, we have two approaches: The one is to approximate w_1^* by the current estimate $\hat{w}_1$ and use adaptively changing the derivative of f, where we need to calculate Fisher information $g(w)$. The other is to use a bound that is independent of w_1^* and g, i.e. use the following bound instead of (14)

$$\frac{df(0)}{d\gamma} < \frac{2}{\sup_w w(1 - w)g(w)}, \tag{20}$$

which is obtained explicitly in some special cases.

4.2   Special Case: Discrete Distribution

Before considering the case of general K, we give specific forms of the bound $df(0)/d\gamma$ of Eq. (20) both for the e-projection and m-projection in the case of discrete distribution. Discrete distribution can approximate even continuous distributions by sufficiently fine discretization. The nonparametric e-mixture in a continuous space [15] is also covered by this case.

The discrete distribution is formulated as

$$q(x) = \sum_{i=1}^{d} q_i \delta_i(x), \quad x \in \{1, 2, \ldots, d\}, \quad \sum_{i=1}^{d} q_i = 1, \quad q_i \geq 0. \tag{21}$$

where $\delta_i(x) = 1$ when $x = i$ and $\delta_i(x) = 0$ otherwise. The discrete distribution belongs to the exponential family,

$$q(x) = \exp\left(\sum_{i=1}^{d-1} \log \frac{q_i}{q_d} \delta_i(x) + \log q_d\right), \tag{22}$$

where we have $d-1$ independent parameters $q_1, \ldots, q_{d-1}$ and one dependent parameter q_d is given by $q_d = 1 - \sum_{i=1}^{d-1} q_i$. By taking the e-connection as the ∇-connection, $q(x)$ becomes the same form as Eq. (2) by regarding

$$F_i(x) = \delta_i(x), \quad \xi_i = \log \frac{q_i}{q_d}, \quad b(\xi) = -\log q_d, \quad i = 1, \ldots, d-1. \tag{23}$$

The dual coordinate ζ_i is given by

$$\zeta_i = \mathrm{E}_{q(x)}[F_i(x)] = q_i, \quad i = 1, \ldots, d-1. \tag{24}$$

The basis vectors in S are denoted by

$$p_k(x) = \sum_{i=1}^{d} p_{ki} \delta_i(x), \quad \sum_{i=1}^{d} p_{ki} = 1, \quad p_{ki} \geq 0. \tag{25}$$

The Case of e-Projection. First, we take the e-connection as the ∇-connection, then the ∇-projection onto the $\tilde{\nabla}$-autoparallel submanifold is the e-projection onto the m-autoparallel submanifold. A typical example of this case is the Nonnegative Matrix Factorization (NMF) whose geometrical property is examined in [4].

The m-autoparallel submanifold spanned by $p_k(x)$ is given by a set of points whose m-coordinate (24) is given by

$$\zeta_i = \sum_{k=1}^{K} w_k p_{ki}, \quad i = 1, \ldots, d-1. \tag{26}$$

Since ζ_i is the probability value, it is equivalent to the mixture distribution of $\{p_k(x)\}$.

$$p(x; \mathbf{w}) = \sum_{k=1}^{K} w_k p_k(x), \quad \sum_{k=1}^{K} w_k = 1. \tag{27}$$

Now we give a sufficient condition for convergence of the e-projection onto the m-autoparallel submanifold.

Proposition 1. *Algorithm A(2) of the e-projection onto an m-autoparallel sub-manifold for the discrete distribution is locally stable if*

$$\frac{df(0)}{d\gamma} < \frac{2}{\sum_i (\sqrt{p_{1i}} - \sqrt{p_{2i}})^2}, \tag{28}$$

where the right-hand side has a lower bound $\sqrt{2}$.

Proof. The m-autoparallel manifold spanned by $K = 2$ points can be written as $p(x; w) = w p_1(x) + (1 - w) p_2(x)$. The Riemannian metric at $p(x; w)$ is given by

$$g(w) = \mathrm{E}_w \left[\left(\frac{\partial \log p(x; w)}{\partial w} \right)^2 \right] = \sum_{x=1}^{d} \frac{1}{p(x; w)} \left(\frac{\partial p(x; w)}{\partial w} \right)^2$$

$$= \sum_{x=1}^{d} \frac{(p_1(x) - p_2(x))^2}{p(x; w)} = \sum_{i=1}^{d} \frac{(p_{1i} - p_{2i})^2}{w p_{1i} + (1 - w) p_{2i}}. \tag{29}$$

The denominator of right hand side of Eq. (20) is

$$\sup_w w(1 - w) \sum_{i=1}^{d} \frac{(p_{1i} - p_{2i})^2}{w p_{1i} + (1 - w) p_{2i}}. \tag{30}$$

The i-th term $w(1-w)\frac{(p_{1i}-p_{2i})^2}{w p_{1i}+(1-w)p_{2i}}$ has the maximum value $(\sqrt{p_{1i}}-\sqrt{p_{2i}})^2$ when $w = \sqrt{p_{2i}}/(\sqrt{p_{1i}} + \sqrt{p_{2i}})$, then Eq. (28) is bounded from upper by $\sum_i (\sqrt{p_{1i}} - \sqrt{p_{2i}})^2$, which is a Hellinger distance between $p_1(x)$ and $p_2(x)$, and we obtain the sufficient condition for local stability (28), and the right hand side has a lower bound $\sqrt{2}$.   □

The Case of m-Projection. Next, we take the m-connection as the ∇-connection, then the ∇-projection onto the $\tilde{\nabla}$-autoparallel submanifold is the m-projection onto the e-autoparallel submanifold, which is the case of e-mixture estimation [15].

The e-autoparallel submanifold spanned by $p_k(x)$ is given by a set of points whose e-coordinate (23) is given by

$$\xi_i = \sum_{k=1}^{K} w_k \log \frac{p_{ki}}{p_{kd}} = \left(\sum_{k=1}^{K} w_k \log p_{ki} \right) - \log p_{kd}, \quad i = 1, \ldots, d-1. \tag{31}$$

The corresponding probabilistic model is represented by

$$p(x; \mathbf{w}) \propto \exp \left(\sum_{k=1}^{K} w_k \log p_k(x) \right), \quad \sum_{k=1}^{K} w_k = 1, \tag{32}$$

which is a different type of mixture, log linear mixture. This type of mixture is called e-mixture, while the mixture specified by Eq. (27) is called m-mixture. Here we give a sufficient condition for convergence of the m-projection onto the e-autoparallel submanifold.

Proposition 2. *Algorithm A(2) of the m-projection onto the e-autoparallel submanifold for the discrete distribution is locally stable if*

$$\frac{df(0)}{d\gamma} \le \frac{32}{\left(\max_i \log \dfrac{p_{1i}}{p_{2i}} - \min_i \log \dfrac{p_{1i}}{p_{2i}}\right)^2}. \tag{33}$$

The right-hand side does not have a lower bound that does not depend on the coordinate p_k unlike the e-projection case, and it is left as an open problem whether there exists any such a bound.

Proof. The e-autoparallel model for $K = 2$ is written as

$$p(x; w) = \frac{1}{Z(w)} \exp(w \log p_1(x) + (1 - w) \log p_2(x)), \tag{34}$$

where w is an e-coordinate, $Z(w)$ is a normalization constant

$$Z(w) = \sum_{x=1}^{d} \exp(w \log p_1(x) + (1 - w) \log p_2(x)). \tag{35}$$

Since the discrete distribution can be written as $\log p_k(x) = \sum_{i=1}^{d} \log p_{ki} \delta_i(x)$, we have

$$p(x; w) = \frac{1}{Z(w)} \exp\left(\sum_{i=1}^{d}(w \log p_{1i} + (1 - w) \log p_{2i})\delta_i(x)\right) = \frac{1}{Z(w)} \exp\left(\sum_{i=1}^{d}(a_i w + b_i)\delta_i(x)\right), \tag{36}$$

where

$$a_i = \log(p_{1i}/p_{2i}), \quad b_i = \log p_{2i}, \quad Z(w) = \sum_{i=1}^{d} c_i(w), \quad c_i(w) = \exp(a_i w + b_i). \tag{37}$$

Note that $p(i; w) = c_i(w)/Z(w)$. The Fisher information for this model can be calculated by

$$g(w) = -\mathrm{E}_w\left[\frac{\partial^2 \log p(x; w)}{\partial w^2}\right] = \frac{1}{Z(w)}\frac{\partial^2 Z(w)}{\partial w^2} - \left(\frac{1}{Z(w)}\frac{\partial Z(w)}{\partial w}\right)^2$$

$$= \sum_{i=1}^{d} \frac{a_i^2 c_i(w)}{Z(w)} - \left(\sum_{i=1}^{d} \frac{a_i c_i(w)}{Z(w)}\right)^2 = \sum_{i=1}^{d} a_i^2 p(i; w) - \left(\sum_{i=1}^{d} a_i p(i; w)\right)^2 \tag{38}$$

The last formula represents the variance of a_i with respect to the probability weight $p(i; w)$. From Popoviciu's inequality on variances [14], $g(w)$ has an upper bound that is independent of w,

$$g(w) \le \frac{1}{4}(\max_i a_i - \min_i a_i)^2. \tag{39}$$

Since $w(1 - w) \le 1/4$, we obtain the inequality (33) of the Proposition from Eq. (20). $\qquad \square$

5   General Case

Now let us consider the general case of K. The following theorem gives a sufficient condition for local convergence.

Theorem 3. *Let w_k^*, $k = 1, \ldots, K$ be the optimal parameter. If the function f satisfies*

$$\frac{df(0)}{d\gamma} < \frac{2}{K \max_k w_k^*(1 - w_k^*)g_k(w_k^*)}, \tag{40}$$

Algorithm $A(K)$ is locally stable, where $g_k(w_k)$ denotes the Fisher information along with the $\overset{\nabla}{}$-geodesic connecting p_k and q^*.

Because the space is limited and it can be shown in a straight-forward way, we just give an outline of the proof[1]. First, we can consider a component-wise version of Algorithm $A(K)$, which updates only one w_k in each step while all other w_k's are fixed. We can show that this component-wise algorithm is equivalent to Algorithm $A(2)$ in the first order of perturbation.

By aggregating K steps of the component-wise version of Algorithm $A(K)$ for $w_1, \ldots, w_K$, the resulting algorithm can be shown to be locally equivalent to Algorithm $A(K)$ itself. Then the sufficient condition for convergence of the algorithm is obtained that the learning rate of each step is smaller than $1/K$ of the bound (14) of Algorithm $A(2)$.

6   Discussion

We have given bounds of a learning rate for the convergence of the geometrical algorithm. In this section, we will remark several issues that should be considered.

The first point is about the assumption that all the mixing weight w_k is nonnegative. In some applications such as NMF and nonparametric mixture estimation, the assumption is reasonable. However, it is easy to generalize the algorithm that accepts negative values by replacing the multiplicative update by the additive update. The framework that the weight is not necessarily non-negative was studied in the extension of principal component analysis to the statistical manifold [1,2]. The theory of the present paper can be applied to such a framework without major modifications.

In the analysis, we further assumed that the projection point lies within the region where w_k is nonnegative for simplicity. This is not always true even for applications like NMF. However, we can show that the point that achieves the minimum of divergence is given by the projection point onto a lower dimensional $\tilde{\nabla}$-autoparallel manifold that is a subset included in the boundary of $\tilde{M}$. So the theorem holds as it is without this assumption.

Next, let us consider the comparison of behavior of the proposed algorithm with a gradient descent method. By a perturbation analysis, we can show that the

---

[1] Details of the proof will appear in the full version of the paper [3].

update vector of gradient descent is locally a linear transformation of the update of the geometrical algorithm [3]. Therefore, it is supposed that the proposed algorithm behaves similarly to the gradient descent algorithm. However, a precise analysis of the comparison is open to future work.

References

1. Akaho, S.: The e-PCA and m-PCA: Dimension reduction of parameters by information geometry. In: Proceedings of the 2004 IEEE International Joint Conference on Neural Networks, 2004, vol. 1, pp. 129–134. IEEE (2004)
2. Akaho, S.: Dimension reduction for mixtures of exponential families. In: Kůrková, V., Neruda, R., Koutník, J. (eds.) ICANN 2008. LNCS, vol. 5163, pp. 1–10. Springer, Heidelberg (2008). https://doi.org/10.1007/978-3-540-87536-9_1
3. Akaho, S., Hino, H., Murata, N.: On a convergence property of a geometrical algorithm for statistical manifolds. arXiv preprint arXiv:1909.12644 [cs.LG] (2019)
4. Akaho, S., Hino, H., Nara, N., Murata, N.: Geometrical formulation of the nonnegative matrix factorization. In: Cheng, L., Leung, A.C.S., Ozawa, S. (eds.) ICONIP 2018. LNCS, vol. 11303, pp. 525–534. Springer, Cham (2018). https://doi.org/10.1007/978-3-030-04182-3_46
5. Amari, S.: Information Geometry and Its Applications, vol. 194. Springer, Tokyo (2016). https://doi.org/10.1007/978-4-431-55978-8
6. Ay, N., Jost, J., Vân Lê, H., Schwachhöfer, L.: Information geometry, vol. 64. Springer, Cham (2017). https://doi.org/10.1007/978-3-319-56478-4
7. Fujiwara, A., Amari, S.: Gradient systems in view of information geometry. Phys. D Nonlinear Phenom. **80**(3), 317–327 (1995)
8. Hino, H., Akaho, S., Murata, N.: An entropy estimator based on polynomial regression with Poisson error structure. In: Hirose, A., Ozawa, S., Doya, K., Ikeda, K., Lee, M., Liu, D. (eds.) ICONIP 2016, Part II. LNCS, vol. 9948, pp. 11–19. Springer, Cham (2016). https://doi.org/10.1007/978-3-319-46672-9_2
9. Hino, H., Koshijima, K., Murata, N.: Non-parametric entropy estimators based on simple linear regression. Comput. Stat. Data Anal. **89**, 72–84 (2015). https://doi.org/10.1016/j.csda.2015.03.011. http://www.sciencedirect.com/science/article/pii/S0167947315000791
10. Lebanon, G., et al.: Riemannian Geometry and Statistical Machine Learning. LAP LAMBERT Academic Publishing (2015)
11. Murata, N., Fujimoto, Y.: Bregman divergence and density integration. J. Math-for-Ind. (JMI) **1**(B), 97–104 (2009)
12. Nagaoka, H., Amari, S.: Differential geometry of smooth families of probability distributions. Technical report METR 82-7, University of Tokyo (1982)
13. Pistone, G.: Nonparametric information geometry. In: Nielsen, F., Barbaresco, F. (eds.) GSI 2013. LNCS, vol. 8085, pp. 5–36. Springer, Heidelberg (2013). https://doi.org/10.1007/978-3-642-40020-9_3
14. Popoviciu, T.: Sur les équations algébriques ayant toutes leurs racines réelles. Mathematica (Cluj) **9**, 129–145 (1935)
15. Takano, K., Hino, H., Akaho, S., Murata, N.: Nonparametric e-mixture estimation. Neural Comput. **28**(12), 2687–2725 (2016)

Neural Network Applications

Anomaly Detection Based on the Global-Local Anomaly Score for Trajectory Data

Chengcheng Li[1], Qing Xu[1(✉)], Cheng Peng[1], and Yuejun Guo[2]

[1] College of Intelligence and Computing, Tianjin University, Tianjin, China
qingxu@tju.edu.cn
[2] Graphics and Imaging Laboratory, Universitat de Girona, Girona, Spain

Abstract. Anomaly detection of trajectory data is important and challenging in many real applications. Many anomalous trajectory detection algorithms have been developed, however most of them cannot well handle the complex trajectories with varying local densities. Additionally, trajectory similarity measure is usually difficult for complex trajectory data. In this paper, to address these two issues, we develop a novel anomalous trajectory detection technique with two important points. First we propose a new abnormal score, Global-Local Anomaly Score (GLAS), to sensitively quantify the abnormal degree of trajectories with varying local densities. Second an effective and fast measure, Extended Power enhanced Euclidean Distance (EPED), is designed to calculate the trajectory similarity. The proposed technique is evaluated on both synthetic and real-world trajectory data, showing that the new anomaly detector outperforms both classic and state-of-the-art methods.

Keywords: Trajectory data processing · Anomaly detection

1 Introduction

With the rapid proliferation of GPS and monitoring system, massive trajectory data based on different moving objects have been generated everyday. Numerous techniques have been proposed for trajectory data analysis [1], and a wide range of applications have been generated: movement behavior analysis, destination prediction and group behavior analysis [2]. Trajectory outliers, also called anomalies, are trajectories that are significantly different from normal or expected patterns [3]. They always convey critical information of particular interest to many real scenes [4]. Anomalous trajectory detection has attracted growing attention and becomes one of the key issues for trajectory data analysis

This work has been funded by Natural Science Foundation of China under Grants No. 61471261 and No. 61771335. The author Yuejun Guo acknowledges support from Secretaria dUniversitats i Recerca del Departament dEmpresa i Coneixement de la Generalitat de Catalunya and the European Social Fund.

ⓒ Springer Nature Switzerland AG 2019
T. Gedeon et al. (Eds.): ICONIP 2019, CCIS 1143, pp. 275–285, 2019.
https://doi.org/10.1007/978-3-030-36802-9_30

[5]. Considering the trajectories with same sampling rate are very important in applications [6], in this paper we focus on this kind of trajectory data.

Anomalous trajectory detection is critically important but challenging due to the high complexity of trajectory data [2], and a plenty of algorithms have been developed [7–10]. Unfortunately most anomaly detection methods behave unsatisfactorily for complex trajectory set with largely varied local densities (the first sub-figure of Fig. 1 shows an example). This type of trajectory set is very common in the real scenes. Even the two successful state-of-the-art algorithms, Sequential Hausdorff Nearest-Neighbor Conformal Anomaly Detector (SHNN-CAD) [11] and improved Visual Assessment of Tendency plus (iVAT+) [12], somewhat suffer from this limitation. For example, SHNN-CAD uses Non-Conformity Measure (NCM) to measure the abnormal degree of a trajectory relative to the data set. However NCM based on the principle of the k-Nearest Neighbours (kNN) may result in a problem when the considered set with largely varied local densities. That is, some abnormal trajectories in dense regions may have smaller NCM values and be missed, and/or normal ones in sparse regions may have larger NCM values and be detected as anomalies.

In this paper, we take into account the local density variation of trajectories to propose a novel anomaly score, Global-Local Anomaly Score (GLAS), to properly measure the abnormal degree of trajectories. Then we propose an unsupervised Global-Local Anomaly Score based Anomalous Trajectory Detector (GLAS-ATD) to handle complex trajectory set effectively and efficiently.

Also importantly the distance/similarity measure for trajectories is a core component for anomalous trajectory detection [13,14]. Lots of trajectory distance measures have been developed, and classic ones include Euclidean Distance (ED), Hausdorff Distance (HD) and Dynamic Time Warping (DTW). All these methods have limitations [14]. ED limits the trajectories with equal length. HD ignores temporal information on trajectories [15], losing important features like motion direction [16,17]. DTW changes the temporal relationship between the corresponding sampling points of two trajectories in a non-linear way [13], negatively effecting the measurement. In this paper, we propose a new measure, called Extended Power enhanced Euclidean Distance (EPED), to measure the trajectory distance more reasonably and eliminate the limitations mentioned above. Besides, this measure can handle trajectories with both equal and unequal length.

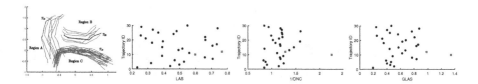

Fig. 1. The first sub-figure is an example set with varying local densities. The following three sub-figures respectively show the LAS, 1/DNC and GLAS values versus trajectory IDs of the set shown in first sub-figure. Black dots are normal trajectories and red squares (T_{10}, T_{12}) are abnormal ones. (Color figure online)

2   Anomaly Score Measure

Anomaly score, which is a measurement to evaluate how "different" a trajectory is from others in data set, transforms high-dimensional data set into a list of floating-point numbers, alleviating the difficulty of anomaly detection.

2.1   Local Anomaly Score

An abnormal trajectory is dissimilar enough from the members in its local neighborhood, while a normal trajectory is similar with its neighbors. And this is a fundamental consideration widely used in many anomaly detection methods, such as kNN [18] and SHNN-CAD [11]. Based on this consideration, we define the Local Anomaly Score (LAS) as follows

$$LAS\,(T_a) = \sum\nolimits_{T_b \in kNN(T_a)} d\,(T_a, T_b), \tag{1}$$

where $kNN\,(T_a)$ is the k-nearest neighbors of T_a, $d\,(T_a, T_b)$ is a distance measure between trajectory T_a and trajectory T_b. The more anomalous T_a is, the higher the $LAS\,(T_a)$ becomes. LAS is obtained based on the differences between the trajectory and its nearest neighbors, which means that LAS measures the local abnormal degree of a trajectory.

As for the distance measurement used in LAS, EPED, which is proposed in next section, is employed for our anomaly detector. Actually, many distance measures work well for our purpose and, EPED performs the best according to our extensive experimentation.

2.2   Global-Local Anomaly Score

LAS performs well in local regions and/or in the trajectory set with approximately uniform local densities. However most real-world sets exhibit a more complex structure and always have varying local densities and, in this case LAS behaves unsatisfactorily. Considering the example set shown in the first subfigure of Fig. 1, regions A and B are sparse and C is dense. For the trajectories T_{12}, T_{20} and T_{10}, we calculate their LAS values and obtain $LAS(T_{12}) = 0.77$, $LAS(T_{20}) = 0.66$ and $LAS(T_{10}) = 0.60$. Apparently, T_{10} would be identified as normal if only considering LAS value. But it is easy to find that T_{10} is indeed an anomaly. The issue comes from that the local densities vary in different regions in the set. On average, the distances between trajectories in dense region are smaller than those in sparse region. In this case the anomalies in dense regions are harder to be detected.

To address this issue, we enhance LAS by introducing a Density Normalization Coefficient (DNC), which is inspired by Local Outlier Factor (LOF) [19].

$$DNC\left(T_a\right) = \frac{\rho\left(T_a\right)}{\sum_{T_b \in kNN\left(T_a\right)} \rho\left(T_b\right)/k},$$ (2)

where $\rho\left(T_a\right) = k/LAS\left(T_a\right) = k/\sum_{T_b \in kNN\left(T_a\right)} d\left(T_a, T_b\right)$, here $\rho\left(T_a\right)$ is the number of trajectories contained in the distance "volume" of T_a, and $\rho\left(\cdot\right)$ can be considered as a kind of definition for local trajectory density.

Actually $DNC\left(T_a\right)$ achieves a normalization on the local density of T_a, by the average density of T_a's k-nearest neighbors. This normalization eliminates negative effects caused by the variation of local densities.

Note that, to some extent, the distance is inversely proportional to the density. That is, if T_a is abnormal compared with its neighbors, then the local density of T_a is smaller than those of T_a's k-nearest neighbors, as a result $DNC\left(T_a\right) < 1$. Therefore a novel Global-Local Anomaly Score (GLAS) is proposed, by the combination of 1/DNC and LAS, to score the trajectory abnormal degree, outperforming the use of 1/DNC and LAS alone.

$$GLAS\left(T_a\right) = \frac{1}{DNC\left(T_a\right)} * LAS\left(T_a\right)$$ (3)

The larger $GLAS\left(T_a\right)$ is, the more likely T_a is an anomaly, and vice versa. Figure 1 gives the LAS, 1/DNC and GLAS values of trajectories in the example set. Clearly, $GLAS\left(\cdot\right)$ has the best performance for distinguishing normal and abnormal trajectories, compared to both 1/DNC and LAS. For the trajectories T_{12}, T_{20} and T_{10}, their GLAS values are 0.98, 0.68, 1.34, respectively. Now T_{12} and especially T_{10} can be easily detected as anomalies, and T_{20} is normal. Obviously GLAS can appropriately and effectively measure the abnormal degree of trajectories in the data set with varying local densities.

3   The Distance Measure of Trajectories

Trajectory distance measure, also called similarity measure, is fundamental for anomalous trajectory detection. ED is the simplest distance measure and, it is one of the most commonly used in application scenarios [13,14]. Given two trajectories $T_a = (a_1, a_2, \ldots, a_n)$ and $T_b = (b_1, b_2, \ldots, b_n)$, the ED of T_a and T_b is defined as [20] $d_E(T_a, T_b) = \sqrt{\sum_{i=1}^{n} \|a_i - b_i\|^2}$, here $\|a_i - b_i\|$ is the Euclidean distance between sampling points a_i and b_i. For example, if a_i and b_i are 2D coordinates, $a_i = (a_{i,x}, a_{i,y})$ and $b_i = (b_{i,x}, b_{i,y})$, then $\|a_i - b_i\| = \sqrt{(a_{i,x} - b_{i,x})^2 + (a_{i,y} - b_{i,y})^2}$.

However, ED have some drawbacks. One issue is that ED is not sensitive enough, for identifying abnormal trajectories that are not so apparent in complex data set. Inspired by p-norm, we take advantage of the idea underlying the power function, to enlarge the differences between the corresponding sampling points of two trajectories to enhance ED, leading to a power enhanced Euclidean distance for trajectory data,

$$d_{PED}(T_a, T_b) = [\frac{1}{n} \sum_{i=1}^{n} \|a_i - b_i\|^p]^{\frac{1}{p}}, \; p > 2. \tag{4}$$

The point is that, due to the use of power p, a larger distinction between a_i and b_i is amplified in a larger degree and, a smaller distinction is increased in a smaller extent. Therefore the distinctions between abnormal and normal trajectories are enhanced, which is effective for anomaly detection.

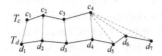

Fig. 2. An example of our extension for trajectories with unequal lengths.

Another issue is that ED requires the considered trajectories have equal length. Thus we propose an extension to generalize the power enhanced Euclidean distance to reasonably measure the distance for unequal length trajectories. We use an illustration to demonstrate how this extension works. In Fig. 2, two example trajectories T_c and T_d, with 4 and 7 sampling points respectively, are presented. Because this paper handles trajectories with same sampling rate, as mentioned in Sect. 1, our extension first keeps the temporal relationship between c_i and d_i ($i = 1, \ldots, 4$). Then the extension considers that the object producing T_c spatially stops at location c_4 but temporally keeps "moving" at the c_4 to obtain the "extended" T_c. Thus c_4 is reused three times to "correspond to" d_5, d_6 and d_7 on T_d. That is, we reuse the last sampling point of the short trajectory to make it have "equal" length with that of the long trajectory. By experimentation, this extension allows to measure the distance between trajectories with unequal lengths very well.

Given two trajectories $T_a = (a_1, a_2, \ldots, a_n)$ and $T_b = (b_1, b_2, \ldots, b_m)$, $n \leq m$. The Extended Power enhanced Euclidean Distance (EPED) is defined as

$$d_{EPED}(T_a, T_b) = [\frac{1}{m}(\sum_{i=1}^{n-1} \|a_i - b_i\|^p + \sum_{i=n}^{m} \|a_n - b_i\|^p)]^{\frac{1}{p}}. \tag{5}$$

Significantly, EPED is a truly mathematical metric to measure the similarity between trajectories with equal and/or unequal length. The complexity of EPED is $O(max(n, m))$.

4   Anomalous Trajectory Detector

Now we introduce the proposed Global-Local Anomaly Score based Anomalous Trajectory Detector (GLAS-ATD), which is unsupervised.

Given a trajectory set $T = \{T_1, T_2, \ldots, T_l\}$, we calculate the GLAS values for T_i $(i = 1, \ldots, l)$. Then the trajectory T_M, which has the maximum GLAS value, is considered as an anomaly candidate. If T_M is determined as abnormal then it is removed from T to obtain the updated set T_{update} and, the GLAS values for T_{update} is updated. Then we process T_{update} iteratively and in the same way as handling T, until the updated T_M is determined as normal, and in this case all anomalies in T have been detected. With the removal of anomalies, the complexity of data set consistently decreases and in this case the anomaly detection becomes easier.

Our criterion to determine anomalies is based on a common sense that at least a number λ percent of all the trajectories in a data set are normal. In this paper we set $\lambda = 70\%$ by experimentation. We sort all the GLAS values from small to large to obtain an ordered set of these values $\{GLAS'(i)\}$, here i is the index of the ordered value. Then we average first λ GLAS' values to obtain the mean abnormal degree of dominating normal trajectories in the set,

$$NomAS = \sum_{i=1}^{\lambda*l} GLAS'(i)/(\lambda * l), \tag{6}$$

here l is the total number of the trajectories in the data set.

We define the anomaly tolerance ε_{AT} as the ratio of the anomaly score of the candidate T_M and $NomAS$,

$$\varepsilon_{AT} = GLAS(T_M)/NomAS, \tag{7}$$

here $GLAS(T_M)$ is the GLAS value of T_M. If $\varepsilon_{AT} > \varepsilon$ then T_M is identified as an anomaly, ε is a pre-defined threshold.

5   Experiments

We evaluate the performance of GLAS-ATD and EPED, and do the comparisons between our technique with other algorithms including state-of-the-art SHNN-CAD [11] and iVAT+ [12], on synthetic and real-world trajectory sets. All the methods are implemented by C++ and run on a Windows PC with Intel Core i7 3.40 GHZ CPU and 8.0 GB RAM.

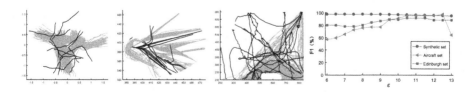

Fig. 3. The first three sub-figures show the plots of Synthetic set (only one subset: TS10), Aircraft set and Edinburgh set respectively. Gray trajectories are normal and black trajectories are abnormal. The last sub-figure shows the detection performance by GLAS-ATD under different ε values.

Data Sets: Synthetic set[1] was created by Piciarelli et al. [6], containing 1000 trajectory subsets. Each subset has 250 and 10 trajectories respectively labelled as normal and abnormal. Each trajectory has 16 sampling points. All the experimental results for this set are based on an average of those on 1000 subsets. Aircraft set[2] was extracted from the flights in the Northern California TRACON, having 450 normal and 20 abnormal trajectories, and trajectory lengths vary from 12 to 171 sampling points. Edinburgh set[3] was extracted from the Edinburgh Informatics Forum pedestrian database, with 259 normal trajectories and 18 anomalies, and trajectory lengths vary between 33 and 825 sampling points. Figure 3 show the plots of Synthetic set (one subset), Aircraft set and Edinburgh set respectively.

The Performance by GLAS-ATD: The performance of GLAS-ATD is evaluated by F_1 score [21], the higher the F_1 is, the better detection result is.

GLAS-ATD behaves very well, as shown in the last sub-figure of Fig. 3 where the best F_1 results on three sets are 98.03%, 95.20% and 91.40% respectively. Note that, when the threshold $\varepsilon \in [10, 11]$, GLAS-ATD stably achieves satisfactory results. Practically $\varepsilon = 10.5$ is used. We fix $k = \lfloor 0.020n \rfloor$ and $p = 14$, n is the total number of trajectories in data set, and this is discussed in the end of current section.

Comparison with Other Algorithms: Extensive results by GLAS-ATD significantly outperforms those by other compared methods.

For Synthetic set, we compare GLAS-ATD with SHNN-CAD, iVAT+, SVM [6] and Discords [22]. These four algorithms present results by a metric, accuracy, in their published papers. For fair comparison, GLAS-ATD is also evaluated by accuracy. Table 1 summarizes the results. Note that the accuracy values of SHNN-CAD [11], iVAT+ [12], SVM [6] and Discords [22] on Synthetic set are reported in the corresponding original papers. The accuracy by GLAS-ATD is 98.01%, outperforming those by others.

Table 1. Detection accuracy (%) of different methods on Synthetic set

	GLAS-ATD	SHNN-CAD	iVAT+	SVM	Discords
Accuracy	**98.01**	97.09	97.62	96.30	97.06

Table 2. The F_1 results (%) of different methods on two real-world data sets

	GLAS-ATD	SHNN-CAD	iVAT+
Aircraft set	**95.20**	75.00	90.0
Edinburgh set	**91.40**	83.33	83.33

---

[1] http://avires.dimi.uniud.it/papers/trclust/.

[2] https://c3.nasa.gov/dashlink/resources/132/.

[3] http://homepages.inf.ed.ac.uk/rbf/FORUMTRACKING/.

For testing our method on complex real-world Aircraft and Edinburgh sets, we implement SHNN-CAD and iVAT+ and, all the parameters used in SHNN-CAD and iVAT+ are tuned to obtain their best possible results. The F_1 results are shown in Table 2. It is clearly evident that GLAS-ATD performs the best. SHNN-CAD does not work well here, one reason is that its anomaly score cannot appropriately measure the abnormal degree of trajectories with local density variations. Another reason comes from its distance measure HD, which misses the important direction information of trajectories and does not well estimate the difference between two trajectories with large length variation. iVAT+ obtains some unsatisfactory performance mainly due to its distance measure based on DTW, which disturbs the temporal relationship between the corresponding sampling points of trajectories by a non-linear alignment.

Table 3. The F_1 results (%) of GLAS-ATD with different distance measures

	EPED	DTW	HD	ED
Synthetic set	**98.03**	97.00	97.10	97.34
Aircraft set	**95.20**	90.00	80.00	–
Edinburgh set	**91.40**	88.89	83.33	–

The Performance of GLAS-ATD with Different Distance Measures: Many measures of trajectory distance can work under the framework of GLAS-ATD. Table 3 gives the F_1 results by GLAS-ATD under EPED, ED, HD and DTW. Note that ED cannot handle trajectories with varying lengths and, for Aircraft and Edinburgh sets the corresponding results are unavailable. Obviously EPED enables GLAS-ATD to behave the best for all the data sets compared to other distance measures.

It is worth noting that, the performance of GLAS-ATD framework using other distance measures is still better than comparison works, such as SHNN-CAD and iVAT+, which indicates that in addition to EPED, our anomaly score GLAS and our detection strategy employed by GLAS-ATD also have remarkable performance. As shown in Tables 2 and 3, the F_1 values of GLAS-ATD using HD on Aircraft and Edinburgh sets are 80.00% and 83.33% respectively. SHNN-CAD is also an algorithm using HD, its F_1 results are 75.00% and 83.33% respectively. It is clear that our approach is better. Moreover, iVAT+ is performed based on DTW for Aircraft and Edinburgh sets, and the F_1 scores are 90.00% and 83.33% respectively. The results of GLAS-ATD using DTW on two real-world sets are 90.00% and 88.89% respectively. GLAS-ATD still achieves better performance.

Table 4 shows the runtime results. For the trajectories have equal length, ED is the fastest measure. But as for the real-world sets, EPED makes GLAS-ATD perform the most efficiently.

Table 4. Runtimes (Sec) of GLAS-ATD with different distance measures

	EPED	DTW	HD	ED
Synthetic set	0.042	0.223	0.171	**0.016**
Aircraft set	**0.689**	24.109	28.405	–
Edinburgh set	**0.218**	6.773	7.313	–

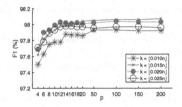 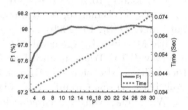

Fig. 4. Left is an exemplification of the F_1 results of GLAS-ATD under varying k and p. Right shows an illustration of the F_1 and runtime results of GLAS-ATD.

Discussion of the Parameters k and p: To better understand the effects of the k and p, we conduct experiments on GLAS-ATD under different settings of k and p. k is the number of nearest neighbors of the considered trajectory, and p is power of EPED. In this paper, for simplicity we use integers for p. We set $k \in \{\lfloor 0.010n \rfloor, \lfloor 0.015n \rfloor, \lfloor 0.020n \rfloor, \lfloor 0.025n \rfloor\}$ and $p \in [3, 200]$. Our extensive experimentation, which is exemplified by the F_1 scores on Synthetic set in left of Fig. 4, points out that $k = \lfloor 0.020n \rfloor$ performs very well for the most cases. Thus in practice we suggest $k = \lfloor 0.020n \rfloor$ is used in GLAS-ATD. In addition, when $p > 12$, with the increase of p, the accuracy performance becomes almost stable. Actually when $p = \infty$, $F_1 = 97.92\%, 98.07\%, 98.03\%$ and 97.96% respectively for different k values and, these results are the same with those for $p = 200$. Note that when p increases the runtime becomes longer (shown in right of Fig. 4), thus, considering both F_1 result and runtime, we recommend $p \in [12, 16]$.

6 Conclusion

In this paper, we have proposed a new anomalous trajectory detection technique to deal with complex trajectories, based on a novel measure GLAS for characterizing abnormal degree of trajectories. The new anomaly detector, GLAS-ATD, is established with the help of a new distance measure EPED for estimating the similarity between trajectories with both equal and unequal length. The proposed technique performs effectively and efficiently, and robustly outperforms both the classic and state-of-the-art methods.

In the near future, we will apply our technique for other kinds of complex trajectory data, such as incomplete trajectories and trajectories with varied sampling rates. We will also extend our anomaly detector for online processing.

References

1. Zheng, Y.: Trajectory data mining: an overview. ACM Trans. Intell. Syst. Technol. (TIST) **6**(3), 29 (2015)
2. Feng, Z., Zhu, Y.: A survey on trajectory data mining: techniques and applications. IEEE Access **4**, 2056–2067 (2016)
3. Hawkins, D.M.: Identification of Outliers, vol. 11. Springer, Dordrecht (1980). https://doi.org/10.1007/978-94-015-3994-4
4. Chandola, V., Banerjee, A., Kumar, V.: Anomaly detection: a survey. ACM Comput. Surv. **41**(3), 15 (2009)
5. Gupta, M., Gao, J., Aggarwal, C.C., Han, J.: Outlier detection for temporal data: a survey. IEEE Trans. Knowl. Data Eng. **26**(9), 2250–2267 (2014)
6. Piciarelli, C., Micheloni, C., Foresti, G.L.: Trajectory-based anomalous event detection. IEEE Trans. Circuits Syst. Video Technol. **18**(11), 1544–1554 (2008)
7. Li, W., Mahadevan, V., Vasconcelos, N.: Anomaly detection and localization in crowded scenes. IEEE Trans. Pattern Anal. Mach. Intell. **36**(1), 18–32 (2014)
8. Lee, J.G., Han, J., Li, X.: Trajectory outlier detection: a partition-and-detect framework. In: Proceedings IEEE International Conference on Data Engineering, pp. 140–149 (2008)
9. Zhang, D., Li, N., Zhou, Z.H., Chen, C., Sun, L., Li, S.: iBAT: detecting anomalous taxi trajectories from GPS traces. In: Proceedings International Conference on Ubiquitous Computing, pp. 99–108 (2011)
10. Ge, Y., Xiong, H., Liu, C., Zhou, Z.H.: A taxi driving fraud detection system. In: Proceedings IEEE International Conference on Data Mining, pp. 181–190 (2011)
11. Laxhammar, R., Falkman, G.: Online learning and sequential anomaly detection in trajectories. IEEE Trans. Pattern Anal. Mach. Intell. **36**(6), 1158–1173 (2014)
12. Kumar, D., Bezdek, J.C., Rajasegarar, S., Leckie, C., Palaniswami, M.: A visual-numeric approach to clustering and anomaly detection for trajectory data. Visual Comput. **33**(3), 265–281 (2017)
13. Shirkhorshidi, A.S., Aghabozorgi, S., Wah, T.Y.: A comparison study on similarity and dissimilarity measures in clustering continuous data. PLoS ONE **10**(12), e0144059 (2015)
14. Zhang, Z., Huang, K., Tan, T.: Comparison of similarity measures for trajectory clustering in outdoor surveillance scenes. In: Proceedings IEEE International Conference on Pattern Recognition, vol. 3, pp. 1135–1138 (2006)
15. Atev, S., Miller, G., Papanikolopoulos, N.P.: Clustering of vehicle trajectories. IEEE Trans. Intell. Transp. Syst. **11**(3), 647–657 (2010)
16. Guo, Y., Xu, Q., Luo, X., Wei, H., Bu, H., Sbert, M.: A group-based signal filtering approach for trajectory abstraction and restoration. Neural Comput. Appl. **12**, 1–17 (2017)
17. Guo, Y., Xu, Q., Li, P., Sbert, M., Yang, Y.: Trajectory shape analysis and anomaly detection utilizing information theory tools. Entropy **19**(7), 323 (2017)
18. Zhao, M., Saligrama, V.: Anomaly detection with score functions based on nearest neighbor graphs. In: Proceedings Advances in Neural Information Processing Systems, pp. 2250–2258 (2009)
19. Breunig, M.M., Kriegel, H.P., Ng, R.T., Sander, J.: LOF: identifying density-based local outliers. SIGMOD Rec. **29**(2), 93–104 (2000)
20. Silva, D.F., De Souza, V.M. Batista, G.E.: Time series classification using compression distance of recurrence plots. In: Proceedings IEEE International Conference Data Mining, pp. 687–696 (2014)

21. Goutte, C., Gaussier, E.: A probabilistic interpretation of precision, recall and F-score, with implication for evaluation. In: Losada, D.E., Fernández-Luna, J.M. (eds.) ECIR 2005. LNCS, vol. 3408, pp. 345–359. Springer, Heidelberg (2005). https://doi.org/10.1007/978-3-540-31865-1_25
22. Keogh, E., Lin, J., Fu, A.: HOT SAX: efficiently finding the most unusual time series subsequence. In: Proceedings IEEE International Conference Data Mining, p. 8 (2005)

Patch Selection Denoiser: An Effective Approach Defending Against One-Pixel Attacks

Dong Chen, Ruqiao Xu, and Bo Han[✉]

School of Computer Science, Wuhan University, Wuhan, China
bhan@whu.edu.cn

Abstract. A one-pixel attack applies maliciously crafted and imperceptible perturbations on just one pixel or a few pixels in an image and can mislead a target deep learning classification model. Defending against this type of attack is a relatively unexplored development in adversarial defence. In this paper, we propose a Patch Selection Denoiser (PSD) approach that removes the few potential attacking pixels in local patches without changing many pixels in a whole image. Without clean training data, it can firstly add random impulse noises to a few images to produce huge amounts of noisy images as inputs and targets in a deep residual network. Next, we can obtain a denoising model based on the Noise2Noise framework. Finally, we design a patch selection algorithm to scan a denoised image in a patch window and compare it with the corresponding part on the test image. Only the patch whose number of pixels with significant absolute difference exceeds a threshold will be detected as the local part containing potential attacking pixels. Thus, this patch will be replaced by the part in the denoised image. Evaluating our approach on a public image dataset CIFAR-10 demonstrates that it can successfully defend against one-, three-, five-pixel and JSMA attacks 98.6%, 98.0%, 97.8% and 98.9% of the time, respectively. Meanwhile, it brings almost no side effects on clean images not subject to one-pixel attacks. The state-of-the-art high defence accuracy proves the effectiveness of our approach.

Keywords: Computer vision · Neural network · One-pixel attack · Adversarial example · Defence

1 Introduction

Deep neural networks are known to be vulnerable to adversarial examples [1]. Some carefully crafted and imperceptible perturbations added to image pixels can lead deep neural networks to misclassification. These vulnerabilities can be classified into black-box attacks (UPSET [2], ANGRI [2], etc.) and white-box attacks (FGSM [3], ILCM [4], etc.). These data-driven attacks raise a great threat to some security-sensitive applications, e.g., facial recognition and self-driving vehicles.

For solving this problem, many defensive methods have been proposed [5]. Most of them are used to defend against attacks that produce adversarial examples by adding noises to the whole image, but methods for defending against "one-pixel attacks" are

© Springer Nature Switzerland AG 2019
T. Gedeon et al. (Eds.): ICONIP 2019, CCIS 1143, pp. 286–296, 2019.
https://doi.org/10.1007/978-3-030-36802-9_31

rare. In one-pixel attacks, only one, three, five, or few pixels are altered in an image for adversarial manipulation. As motivational examples, we randomly selected three images from the public image dataset CIFAR-10. Figure 1 shows their resulting images after one-pixel attack revision. However, only one pixel being changed in each of these original images has mislead the deep learning model into misclassifying objects in the pictures. Table 1 summarises our extensive experiments on CIFAR-10. We tested the One Pixel Attack [6] and Jacobian-based saliency map attack (JSMA) [7], which is similar to it, achieving a success rate as high as 72% among 500 randomly selected images and 92.3% among 400 images.

True: automobile   True: deer   True: truck
Pred: truck   Pred: airplane   Pred: dog

Fig. 1. Adversarial examples. Original images are random selected from CIFAR-10, and the attack method is One Pixel Attack. The classification network is ResNet. In each image, only one pixel has been changed, but it is still misclassified.

Table 1. One-pixel attacks results

Classification network	Attack model	Pixels	Clean images	Suc[a]	Rate (%)
ResNet	One Pixel Attack	1	500	146	29.2
		3	500	354	70.8
		5	500	363	72.6
	JSMA		400	369	92.3

[a]Suc denotes the number of images successfully attacked.

There are mainly two ways to defend against an imperceptible one-pixel attack: (i). making the deep learning classification network more robust by retraining or modifying the network structure or (ii). preprocessing images by using a denoiser. Since Moosavi-Dezfooli et al. pointed out the limitation of brute-force adversarial training [8], modifying the structure of the classification network is impractical, we aim to preprocess images by denoising them.

In this paper, we propose a Patch Select Denoiser (PSD) approach to defend against one-pixel attacks. It combines the Noise2Noise model [9] and patch selection. Though there are many methods to denoise images, Noise2Noise is a suitable choice because of its unsupervised mode and success in denoising many images mixed with different types of noise. Based on the Noise2Noise model, we improved the SRResNet [10] by adding random-valued impulse noise to a few clean images, and we used the noisy images as inputs and targets to train the classification network. In contrast to the Noise2Noise model, in which all pixels, including other clean pixels, are changed in an image, we propose a patch selection algorithm. This algorithm locates the patches that contain the attacking pixels and only clean a few patches in output images. The proposed method is simple and effective. Experiments on CIFAR-10 demonstrated a maximum defence success rate of up to 98.6%. At the same time, we proved that it has almost no side effects in terms of classifying clean images that are not subjected to a one-pixel attack.

The proposed patch selection denoiser has the following advantages. Firstly, it achieves a high defence rate against one-pixel attacks. Secondly, it just changes the attacked images locally, thus preserving the original image nature as much as possible. Meanwhile, it has no side effects on clean images not subjected to its pixels being attacked. Thirdly, the deep neural network can be trained without clean data, and we can add noise to a few images to produce millions of noisy images as inputs and targets, which is necessary to construct a deep learning model in many real-world applications.

2 Related Work

Defending against one-pixel attacks is a relatively unexplored development in adversarial defence. In the following, we will illustrate related work from two sides. One explains the current developments in one-pixel attacks. The other explains the denoisers used for image processing, which has the potential to remove the few attacking pixels as noise.

2.1 One-Pixel Attacks

Since the first attack model L-BFGS [1] was proposed, many other models have appeared. In essence, the problem of generating adversarial examples can be regarded as an optimisation problem with mathematical restrictions. Let x denote an input image, which can be represented as an n-dimensional vector, i.e., $x = (x_1, x_2, \ldots, x_n)$, and let f denote the classification neural network. For image x, its predicted probability of class t is $f_t(x)$. Let $e(x)$ denote perturbations, which can be represented as an n-dimensional vector, i.e., $e(x) = (e_1, e_2, \ldots, e_n)$, and let L denote the maximum of the perturbations, always measured by the length of $e(x)$. The goal of adversarial examples can be described as the following optimisation problem:

$$\max f_{adv}(x + e(x))$$

$$\|e(x)\| \leq L. \tag{1}$$

Among all the methods, it is most common to generate adversarial examples by restricting l_2 norms or l_∞ norms of the perturbations [11, 12]. However, Papernot et al. proposed a method (JSMA) that generates adversarial examples by restricting l_0 norms.

Instead of adding perturbations to the whole image, it only needs to change a few pixels. According to (1), the limitation can be further constrained as

$$\|e(x)\|_0 \leq d, \tag{2}$$

where d is a small number. Considering an extreme situation, just changing one pixel in the image to generate a relevant adversarial example, Su et al. [6] achieved that goal by using the concept of differential evolution [2].

2.2 Defending Against One-Pixel Attacks by Using Denoisers

The removal of a few adversarial pixels can be regarded as image denoising. Image denoising is important in computer vision, and numerous models have been proposed. Traditional methods such as NLM [13] is based on filters, GSR [14] is based on sparse coding, PGPD [15] is based on use of an effective prior, and WNNM [16] is based on low rank. With the rapid development of deep learning, some effective models based on neural networks have also appeared (e.g., UDN [17] and N3 [18]). Noise2Noise is also an important achievement based on deep learning and widely used in image denoising [19]. Noise2Noise is not a definite neural network but rather is a general framework adaptive to any neural network. It is especially suitable for situations in which it is difficult to get clean data for training. This advantage makes it excel in deep learning denoising because the deep neural network needs a large amount of training data to obtain a good structure. Noise2Noise exploits the fact that one can add zero mean noise (the mathematical expectation of the loss function) to a few clean images. With the produced millions of noisy images as inputs and targets in the training, the obtained parameters in deep neural networks are very close to the structure parameters computed by clean training images. In this way, by adding noise with different settings, one can produce huge amount of input and target images and train neural networks without knowing the clean data. This advantage is very suitable for defence against one-pixel attacks on very different images. In the Noise2Noise framework, the type of noise and its loss function should be appropriately set. The disadvantage of Noise2-Noise is that it will change the values in whole images, rather than just fix the few attacking pixels. By analysing its advantages and disadvantages, we can propose a patch selection denoiser.

3 Methods

The PSD model consists of two steps: denoising by using a deep learning model and patch selection, as Fig. 2 shows. The output denoised images from the PSD model can be sent to a deep neural network classification model as inputs for validation.

Fig. 2. Structure of the whole model.

3.1   Denoising by Using a Deep Learning Model

We selected Noise2Noise as the framework for denoising. In this framework, we only need few clean images and we added noise to them. We then used the generated millions of noisy images as inputs and targets to train the deep learning model for capturing model parameters to produce clean images.

For the purpose of defending against one-pixel attacks, the key aspects in the Noise2Noise framework are choosing the deep neural network structure, the right type of noise, and the loss function. Firstly, we chose SRResNet (shown in Fig. 3) as the deep learning structure. SRResNet is the generator network of SRGAN, which is applied to image super-resolution conversion. It mainly contains 16 residual blocks [20], and the structure of the network has no restrictions on the size of input images. The size of input images and output images is the same. The network is convenient and necessary in our work and adaptive to different data-driven defence models for many security-sensitive applications.

Secondly, by considering the theory of one-pixel attack, we added random-valued impulse noise to the original clean images and chose an annealed version of the L_0 loss function for minimisation. In comparing clean images and adversarial examples, it is a natural idea for us to think of random-valued impulse noise rather than Gaussian noise, as we will prove in the following experiments. Random-valued impulse noise retains colours of some pixels and replaces other pixels with random values rather than white or black. It gets the random value from the uniform distribution $[0, 1]^3$. Each pixel has a probability p of being replaced, and each has the probability $1 - p$ of retaining its original colour. In the training, we added such noise to all the images and divided them into two parts, inputs and targets. By considering the requirements the of Noise2Noise framework, we chose an annealed version of the L_0 loss function, which is suitable for random-valued impulse noise. It is defined as:

$$(|f_\theta(x) - y| + \mathcal{E})^\gamma, \tag{3}$$

where function f represents the network, $\varepsilon = 10^{-8}$, and γ will be annealed from 2 to 0 linearly.

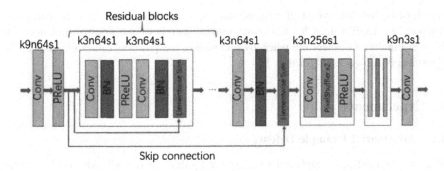

Fig. 3. SRResNet network structure.

3.2 Patch Selection

The denoising neural network can fix image noise effectively, but it will change the clean pixels in an image as well. As shown in Fig. 4, we see that the network not just denoises the frog image but also changes the pixels that should retain their value. In this way, the denoised image is much different from the original image. For solving this problem, we propose a patch selection algorithm.

Fig. 4. (Left) Original image. (Middle) Adversarial example made by the one-pixel attack method. (Right) Image denoised by the denoising neural network. The network also changes the pixels around the object while defending against attack.

The main idea of patch selection is to use a small patch window to scan a denoised image. By comparing its pixels with the corresponding patch in the original image, we can detect whether a patch contains potential attacking pixels. Based on the theory of one-pixel attack, the few attacking pixels should be much different from the original values for achieving the purpose of changing the neural network output category. Thereby, if the number of pixels with significant absolute difference in a patch window between a denoised image and its original image is higher than a threshold, we can reasonably infer with high confidence that this patch contains attacking pixels. Therefore, this patch will be replaced in the original image. Otherwise, we keep the patch in the original image. Specifically, in the implementation, we firstly divide the image into three channels (RGB) and then use a 3×3 pixel as a patch window and let it slide over an image. The window has no padding settings. The size of the patch can be adjusted, but for one-pixel attacks, that is reasonable. Next, in each patch, we compare the RGB values for the corresponding pixels of the original and denoised images. Finally, we select the patch by controlling the position values of i and j. The algorithm is shown as Algorithm 1.

4 Experiments

4.1 Adversarial Example Defence

In the experiments, we employed a widely used public dataset, CIFAR-10 [21]. In this work, we focus on defending against one-pixel attacks. Meanwhile, our approach can also defend against JSMA, which is similar to a one-pixel attack. We chose ResNet [22] to be the classification network. The network can obtain a classification accuracy in CIFAR-10 of up to 92.3% after training.

Algorithm 1 Patch Selection

Input: denoised image X_1, original image X_2, $i = $ a, $j = $ b

Output: X^*

1: $X^* \leftarrow X_2$

2: **for** each patch p in X_2 **do**

3:    $k = 0$

4:    **for** each pixel in patch p **do**

5:       **if** $\left|C_{X_1} - C_{X_2}\right| \geq a$ ($C = $ R/G/B channel) **do**

6:          $k = k+1$

7:    **end for**

8:    **if** $k \geq b$ **do**

9:       replace the patch with the same patch in X_1

10: **end for**

Firstly, we tested the dataset against one-pixel attacks. To solve the challenge by using the Noise2Noise preprocessing, we added random impulse noise with different densities to images, and we used them to train the network. Then we used it to defend against the adversarial examples. We tested different densities of noise and selected two types as representative results, as given in Table 2. The column 'Pixels' denotes the number of pixels changed in an image, the column "p" denotes the probability interval of each pixel being changed in an image, and the column "Defence Rate" denotes the rate of defence success. From the results, we see that low density random impulse noise can enable the network to achieve better performance when "Pixels" is 1, and the network gets the same results when "Pixels" is 3 or 5. This situation is intuitively understandable, because, when only one pixel is changed in an image, the network has to be more "sensitive" to the difference between pixels. Therefore, it is more likely to change the value of pixels at the edge of the object in an image.

Table 2. Defence results at different noise densities

Attack	Pixels	p	Defence rate
One-pixel attack	1	(0, 0.95)	95.2%
		(0, 0.10)	97.9%
	3	(0, 0.95)	97.2%
		(0, 0.10)	97.2%
	5	(0, 0.95)	97.0%
		(0, 0.10)	97.0%
JSMA		(0, 0.95)	96.5%
		(0, 0.10)	98.9%

Next, to further improve the defence rate, we applied patch selection to the processed images. The size of kernel was 3×3. The threshold values of i and j were set to 20 and 1, respectively, and p was kept to be (0, 0.10). The results are listed in Table 3. From the results, we find that PSD obtains higher accuracy than Noise2Noise for defending against one-pixel attacks. Meanwhile, we used the same frog image in Fig. 4 for comparing the effects of patch selection, and the results in Fig. 5 show that PSD can greatly preserve the original image nature while filtering out the attacking pixel.

Through all the results above, we have proven that our model can achieve state-of-the-art defence performance against one-pixel attacks, with defence rates of ∼98.0%. Meanwhile, because of the inherent nature of ResNet itself, it is acceptable that some processed adversarial examples cannot be classified correctly.

Table 3. Influence of patch selection

Attack	Pixels	Method	Defence rate
One pixel attack	1	Noise2Noise	97.9%
		PSD	98.6%
	3	Noise2Noise	97.2%
		PSD	98.0%
	5	Noise2Noise	97.0%
		PSD	97.8%
JSMA		Noise2Noise	98.9%
		PSD	98.9%

Fig. 5. Images after patch selection.

4.2   Test on Clean Data

To demonstrate practical applications, we tested our model on clean data. In order to eliminate the influence of classification network, we used 10000 images from the test batch of CIFAR-10, and tested 9230 images which are classified correctly. The results are given in Table 4. From the results we conclude that PSD can almost keep the original nature of the clean images and it's better than Noise2Noise method, which makes it implementable for practical applications.

Table 4.   Clean image results

Classification network	Method	Accuracy
ResNet	Noise2Noise	99.24%
	PSD	99.53%

5   Conclusion

In this study, we proposed a patch selection denoiser model to defend against one-pixel attacks. We proved that in practice it achieves state-of-the-art results by successfully defending against one-, three-, five-pixel, and JSMA attacks 98.6%, 98.0%, 97.8%, and 98.9% of the time, respectively. In contrast to other denoisers, our model produces denoised images that are very similar to the original clean images. Meanwhile, the PSD approach has almost no side effects on clean images not subject to pixel attacks. Therefore, the proposed approach has great potential for practical applications.

In future work, we aim to build a more universal defence model for other types of data-driven attacks. According to their attacking pixel distribution features, we will explore applying different kinds of noise and loss functions to the denoising neural network for defending against their attacks.

Acknowledgement. This research was supported by the National Key R&D Program of China under grant No. 2018YFB1702703, and also supported by the National Natural Science Foundation of China under grant No. U1531122, 71871170 and 61272272.

References

1. Szegedy, C., et al.: Intriguing properties of neural networks. arXiv preprint arXiv:1312.6199 (2013)
2. Das, S., Suganthan, P.N.: Differential evolution: a survey of the state-of-the-art. IEEE Trans. Evol. Comput. **15**(1), 4–31 (2011)
3. Goodfellow, I.J., Shlens, J., Szegedy, C.: Explaining and harnessing adversarial examples. arXiv preprint arXiv:1412.6572 (2014)
4. Kurakin, A., Goodfellow, I., Bengio, S.: Adversarial examples in the physical world. arXiv preprint arXiv:1607.02533 (2016)
5. Akhtar, N., Mian, A.: Threat of adversarial attacks on deep learning in computer vision: a survey. IEEE Access **6**, 14410–14430 (2018)
6. Su, J., Vargas, D.V., Sakurai, K.: One pixel attack for fooling deep neural networks. IEEE Trans. Evol. Comput. (2019)
7. Papernot, N., McDaniel, P., Jha, S., Fredrikson, M., Celik, Z.B., Swami, A.: The limitations of deep learning in adversarial settings. In: 2016 IEEE European Symposium on Security and Privacy (EuroS&P), pp. 372–387. IEEE, March 2016
8. Moosavi-Dezfooli, S.M., Fawzi, A., Fawzi, O., Frossard, P.: Universal adversarial perturbations. In: Proceedings of the IEEE Conference on Computer Vision and Pattern Recognition, pp. 1765–1773 (2017)
9. Lehtinen, J., et al.: Noise2Noise: learning image restoration without clean data. arXiv preprint arXiv:1803.04189 (2018)
10. Ledig, C., et al.: Photo-realistic single image super-resolution using a generative adversarial network. In: Proceedings of the IEEE Conference on Computer Vision and Pattern Recognition, pp. 4681–4690 (2017)
11. Moosavi-Dezfooli, S.-M., Fawzi, A., Frossard, P.: Deepfool: a simple and accurate method to fool deep neural networks. In: Proceedings of the IEEE Conference on Computer Vision and Pattern Recognition (2016)
12. Baluja, S., Fischer, I.: Adversarial transformation networks: learning to generate adversarial examples. arXiv preprint arXiv:1703.09387 (2017)
13. Kumar, B.K.S.: Image denoising based on non-local means filter and its method noise thresholding. Signal Image Video Process. **7**(6), 1211–1227 (2013)
14. Zhang, J., Zhao, D., Gao, W.: Group-based sparse representation for image restoration. IEEE Trans. Image Process. **23**(8), 3336–3351 (2014)
15. Xu, J., Zhang, L., Zuo, W., Zhang, D., Feng, X.: Patch group based nonlocal self-similarity prior learning for image denoising. In: Proceedings of the IEEE International Conference on Computer Vision, pp. 244–252 (2015)
16. Gu, S., Zhang, L., Zuo, W., Feng, X.: Weighted nuclear norm minimization with application to image denoising. In: Proceedings of the IEEE Conference on Computer Vision and Pattern Recognition, pp. 2862–2869 (2014)
17. Lefkimmiatis, S.: Universal denoising networks: a novel CNN architecture for image denoising. In: Proceedings of the IEEE Conference on Computer Vision and Pattern Recognition (2018)
18. Plötz, T., Roth, S.: Neural nearest neighbors networks. In: Advances in Neural Information Processing Systems (2018)
19. Krull, A., Tim-Oliver, B., Jug, F.: Noise2Void-Learning Denoising from Single Noisy Images. arXiv preprint arXiv:1811.10980 (2018)

20. Gross, S., Michael, W.: Training and investigating residual nets. Facebook AI Research (2016)
21. Krizhevsky, A., Hinton, G.: Learning multiple layers of features from tiny images, vol. 1. no. 4. Technical report, University of Toronto (2009)
22. He, K., Zhang, X., Ren, S., Sun, J.: Deep residual learning for image recognition. In: Proceedings of the IEEE Conference on Computer Vision and Pattern Recognition, pp. 770–778 (2016)

Demographic Prediction from Purchase Data Based on Knowledge-Aware Embedding

Yiwen Jiang[1,2,3(✉)], Wei Tang[1,2,3], Neng Gao[1,3], Ji Xiang[3], and Yijun Su[2,3]

[1] State Key Laboratory of Information Security, CAS, Beijing, China
[2] School of Cyber Security, University of Chinese Academy of Sciences, Beijing, China
[3] Institute of Information Engineering, CAS, Beijing, China
{jiangyiwen,tangwei,gaoneng,xiangji,suyijun}@iie.ac.cn

Abstract. Demographic attributes are crucial for characterizing different types of users in developing market strategy. However, in retail scenario, individual demographic information is not often available due to the difficult manual collection process. Several studies focus on inferring users' demographic attribute based on their transaction histories, but there is a common problem. Hardly work has introduced knowledge for purchase data embedding. Specifically, purchase data is informative, full of related knowledge entities and common sense. However, existing methods are unaware of such external knowledge and latent knowledge-level connections among items. To address the above problem, we propose a Knowledge-Aware Embedding (KAE) method that incorporates knowledge graph representation into demographic prediction. The KAE is a multi-channel and item-entity-aligned knowledge-aware convolutional neural network that fuses frequency-level and knowledge-level representations of purchase data. Through extensive experiments on a real world dataset, we demonstrate that KAE achieves substantial gains on state-of-the-art demographic prediction models.

Keywords: Demographic prediction · Convolutional neural networks · Knowledge graph representation

1 Introduction

Knowing users' demographic attributes is significant for many companies and retailers to make market basket analysis [1], adjust marketing strategy [6], and provide personalized recommendations [16,24]. For example, Nike's basketball shoes target mainly a relatively young (age) and male (gender) users with enough purchasing power (income). Additionally, in recommender systems, demographic information have been wildly used to improve the quality of the systems and solve the cold start problem.

© Springer Nature Switzerland AG 2019
T. Gedeon et al. (Eds.): ICONIP 2019, CCIS 1143, pp. 297–308, 2019.
https://doi.org/10.1007/978-3-030-36802-9_32

Generally, the collection of users' demographic information is difficult as users are reluctant to offer their personal information in case of data breaches. Besides, in reality, there are only partial demographic attributes known for most of users and some users have no attributes at all, as Wang et al. [20] pointed out.

There are several works that have made effort on transaction history for demographic prediction. Wang et al. [20] were the first ones to investigate the power of uses' purchase data for demographic prediction, and they proposed a Structured Neural Embedding (SNE) model which uses bag-of-item (BOI) representation as shared embedding. Resheff et al. [14] concatenated sequence embedding of transaction data with relational data to form raw user representation for demographic prediction. Kim et al. [7] proposed an Embedding Transformation Network with Attention (ETNA) model to learn the representations from purchase data with also naive shared representations at the bottom.

Although they proposed several valuable methods and made some meaningful achievements, their works overlooked a significant information for embedding. Existing methods are unaware of purchase data external knowledge and cannot fully discover latent knowledge-level connections among items. Specifically, knowledge of what user buy and what user may buy are informative and full of knowledge entities. For example, a user has a piece of transaction record with the baby clothes bought that contains two knowledge entities: "baby", "clothes". In fact, the user may also be interested in dried milk with high probability, which is strongly connected with the previous one in terms of common sense reasoning. However, previous methods are hardly able to discover latent knowledge-level connection. As a result, a user's transaction history embedding for demographic prediction will be narrowed down to a limited representation circle.

To address above problem, we propose a novel method that takes advantage of external knowledge for transaction history embedding, called Knowledge-aware embedding (KAE). Specifically, for a piece of transaction record, we first enrich its information by associating each item with a relevant entity in the knowledge graph. We also search and use the set of contextual entities of each entity (i.e., its immediate neighbors) to provide more complementary and distinguishable information. Then we fuse the item representation and knowledge-level representations of purchase data through convolutional neural networks and generate a knowledge-aware embedding. The intuitive understanding of the superiority of KAE is that it maintains the alignment of multiple representations for an purchase item and explicitly bridges different embedding spaces.

We conduct extensive experiments on a real-world benchmark dataset. The results show that KAE achieves prominent gains over the state-of-the-art methods, and prove that the usage of knowledge for transaction history embedding can bring significant improvement on all the previous models for demographic prediction. Overall, our contributions are as follows:

1. We make the first attempt to introduce the knowledge graph into demographic prediction based on purchase data.

2. We propose a novel KAE method that introduces external knowledge for transaction history embedding, and fuses item and knowledge-level representations in an aligned way through CNN.
3. We conduct extensive experiments on a real-world dataset to demonstrate the effectiveness of the proposed KAE method as compared with state-of-the-art baselines.

The rest of this paper is organized as follows. Section 2 summarizes the related works. Section 3 presents the formulation of problem. Section 4 introduces the proposed KAE in details. Section 5 talks about the experimental results. Finally in Sect. 6, we conclude our work.

2   Related Work

2.1   Demographic Prediction

Many works have been devoted to demographic prediction in different scenarios. Early work on demographic prediction are mostly based on the linguistics writing and speaking to predict demographic attributes [13,15]. Later, some works proposed to infer demographic attributes based on users' browsing history [9,17]. Then, with the fast development of online social networks and mobile computing technologies, large scale of user data are accumulated, which make it possible and also valuable to infer users' demographic attributes in network and mobile scenarios [4,12]. Also, location data have been used to predict demographic attributes [25].

Recently, some people make effort to use transaction history for demographic prediction. Wang et al. [20] first proposed to carry out demographic prediction on purchase data and constructed a SNE model to learn the shared representation based on BOI representation. Resheff et al. [14] concatenated sequence embedding of transaction data with structured relational data to form raw user representation, and obtained deep user representation through full-connected neural network. Raehyun et al. [7] proposed an ETNA model for demographic prediction which obtains task-specific representations through linear transformation based on shared BOI representation at the bottom. In this paper, on the basis of the previous work, we introduce knowledge graph for demographic prediction task.

2.2   Knowledge Graph Embedding

The goal of knowledge graph embedding is to learn a low-dimensional representation vector for each entity and relation that preserves the structural information of the original knowledge graph. At present, translation-based knowledge graph embedding methods have received great attention due to their concise models and superior performance. There are many scenarios that successfully apply knowledge graph embedding to achieve better performance, such as machine reading [22], text classification [19], and news recommendation [18]. Moreover, there are many translation-based methods included TransE [2], TransH [21], TransR [10] and TransR [10] used in this paper. Today, we make an attempt to employ knowledge graph embedding for demographic prediction via transaction data.

2.3 CNN for Representation Learning

Previous methods [7,14,20] usually represent purchasing data using an aggregation function or the BOI technique, i.e., taking goods counting statistics as the feature of purchasing history. However, these methods have insufficient representation ability for ignoring some potential information in purchasing history and are vulnerable to the sparsity problem, which leads to poor generalization performance. Inspired by the multi-channel feature of convolutional neural networks (CNN) and the successful application in the filed of computer vision [9] and sentence representation learning [3,18,23], we use CNN to fuse the multiple type of purchasing data representations.

3 Problem Formulation

In this section, we formulate the demographic prediction problem as follows. Let $D = [(x, y)]$ represents a given user, where $x = [I_1, I_2, \ldots]$ denotes a list of transaction items, each item I may be associated with an entity e in the knowledge graph. And $y = y_1, y_2, \ldots, y_K$ denotes the set of attribute label, and K is the number of attributes. According to the defined notations, we follow two types of problem used in [20]:

Partial Label Prediction: For the situation of users with partial demographic attributes. The objective is to learn a function to predict the remaining unknown attributes:

$$f : X, Y^L \to Y^U$$

where Y^L and Y^U denote the observed attributes and that to be predicted over the same set of users X respectively.

New User Prediction: For the situation of new users demographic prediction. The objective function is:

$$f : X^L, Y^L, X^N \to Y^N$$

where X^L and Y^L denote the purchase histories and attributes of labeled users, X^N and Y^N denote the purchase history and the attributes of the new users. Note that here $X^L \cap X^N = \emptyset$.

4 Knowledge-Aware Embedding

In this section, we introduce the proposed KAE method in detail. We first present the overall framework of KAE, then discuss the process of knowledge distillation from a knowledge graph, and finally talk about the representations combination through convolutional neural networks.

4.1 KAE Framework

The framework of KAE is illustrated in Fig. 1. For each user, his/her transaction history are specially processed to BOI representation, relevant entity embedding and contextual entities embedding. Then an extension of traditional CNN that allows flexibility in incorporating symbolic knowledge from a knowledge graph into BOI representation are employed to generate the final shared embedding for the next demographic prediction.

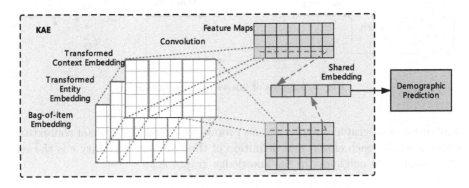

Fig. 1. Illustration of the KAE framework.

4.2 Knowledge Distillation

The process of knowledge distillation is showed in Fig. 2, which mainly consists of four steps. First, to distinguish knowledge entities in transaction history, we employ the technique of entity linking in [11,17] to disambiguate included in items by associating them with predefined entities in a knowledge graph. Based on these identified entities, we construct a sub-graph and extract all relational links among them from the original knowledge graph. There are several knowledge graphs available for academic and commercial, such as NELL[1], DBpedia[2], Google Knowledge Graph[3] and Microsoft Satori[4]. Note that we expand the knowledge sub-graph to all entities within one hop of identified ones in case of sparse relations and lacking diversity. Given the extracted knowledge graph, we apply some knowledge graph embedding methods, such as TransE [2], TransH [21], TransR [10], and TransD [5] for entity representation learning. Learned entity embedding are taken as the input for CNN.

Given that the collected raw transaction data may be insufficient and the information of learned embedding for only the identified entity in dataset is

---

[1] http://rtw.ml.cmu.edu/rtw/.

[2] https://wiki.dbpedia.org/.

[3] https://www.google.com/intl/bn/insidesearch/features/search/knowledge.html.

[4] https://searchengineland.com/library/bing/bing-satori.

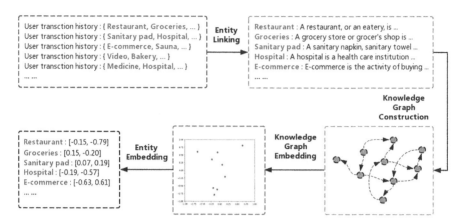

Fig. 2. Process of knowledge distillation.

limited for demographic prediction, we propose to extract additional contextual information for each entity, and definition of the "context" of entity e is the set of its immediate neighbors in the knowledge graph, ie.,

$$context(e) = \{e_i | (e, r, e_i) \in \mathcal{G} \quad or \quad (e_i, r, e) \in \mathcal{G}\}$$

where r is the relation between e and e_i, and $\mathcal{G}$ is the knowledge graph. Since the defined contextual entities closely associate with the current entity in semantics and logic, the usage of context could provide more complementary information and assist in improving the identifiability of entities. For a given entity e, the context embedding is calculated as the average of its contextual entities:

$$\bar{e} = \frac{1}{|context(e)|} \sum_{e_i \in context(e)} e_i$$

where e_i and $\bar{e}$ are the embedding of entity e_i and its context learned by knowledge graph embedding respectively.

4.3   CNN-Based Embedding Combination

As the notations given in Sect. 3, we use $x = [I_1, I_2, \ldots, I_n]$ to denote the sequence of a user's n purchased items, and use $X = [I_1 I_2 \ldots I_n] \in \mathbb{R}^{d \times n}$ to denote the item embedding matrix generated through BOI representation [20]. After the knowledge distillation introduced above, each item I_i may also be associated with an entity embedding $e_i \in \mathbb{R}^{k \times 1}$ and the corresponding context embedding $\bar{e}_i \in \mathbb{R}^{k \times 1}$, where k is the dimension of entity embedding.

   Based on the three input given above, a straightforward approach to combine the different embedding is concatenating them to the new sequence as shared embedding for the further process. However, there are some limitations for this simple concatenating strategy: (1) It breaks up the connection between items and associated entities and is unaware of their alignment. (2) Item and entity embedding are learned by different methods, which means it is not suitable to deal with them together in a single vector space. (3) The concatenating strategy implicitly forces item embedding and entity embedding to share the same dimension, which may not be optimal in practical settings since the optimal dimensions for item and entity embedding may different.

In view of the above-mentioned limitations, we propose a CNN-based method for embedding combination, which combine items frequency and knowledge information in an aligned way. As showed in Fig. 1, we use item embedding $X = [I_1 I_2 \ldots I_n]$, transformed entity embedding $g(e) = [g(e_1)g(e_2) \ldots g(e_n)]$ and transformed context embedding $g(\bar{e}) = [g(\bar{e}_1)g(\bar{e}_2) \ldots g(\bar{e}_n)]$ as input. Where e_i and $\bar{e}_i$ are set to be zero if I_i has no corresponding entity. And g is the non-linear transformation function:

$$g(e) = tanh(Me + b)$$

where $M \in \mathbb{R}^{d \times k}$ is the trainable transformation matrix and $b \in \mathbb{R}^{d \times 1}$ is the trainable bias. Since the transformation function is continuous, the entity embedding and context embedding can be mapped from the entity space to the item space as well as retaining original spatial relationship. Thus, entity embedding $g(e)$ and context embedding $g(\bar{e})$ have the same size with item embedding X and can be served as the multiple channels analogous to colored images. We therefore align and stack the three embedding matrices as follow:

$$X = [[I_1 g(e_1)g(\bar{e}_1)], [I_2 g(e_2)g(\bar{e}_2)] \ldots [I_n g(e_n)g(\bar{e}_n)]] \in \mathbb{R}^{d \times n \times 3}$$

when obtaining the multi-channel input X, we use multiple filter $h \in \mathbb{R}^{d \times l \times 3}$ with varying window sizes l to extract specific local patterns in the purchase history. The local activation of sub-matrix $X_{i:i+l-1}$ about h can be written as:

$$c_i^h = f(\mathbf{h} * X_{i:i+l-1} + b)$$

and a max-over-time pooling operation is used on the output feature to choose the largest feature:

$$\tilde{c}^h = max\{c_1^h, c_2^h, \ldots, c_{n-l+1}^h\}$$

we get the final shared embedding $e(i)$ of the i-th user by concatenating all feature $\tilde{c}^{h_i}$ together as follow:

$$e(i) = [\tilde{c}^{h_1} \tilde{c}^{h_2} \ldots \tilde{c}^{h_m}]$$

where m is the number of filters.

5   Experiments

5.1   Dataset

We use the transaction dataset came from previous work [7][5] and is collected by a Korean multi vendor loyalty program provider. The dataset consists of

---

[5] https://github.com/dmis-lab/demographic-prediction.

purchasing histories of 56,028 users with the gender, age, and marital status labels. The transaction records contain the information of user ID, company ID, type of purchased items and purchased items. The statistics of our dataset are summarized in Table 1.

Table 1. Distribution of users' attributes in dataset.

Attributes	Value	Distribution
Gender	Male	37%
	Female	63%
Age	Young	22.3%
	Adult	54.1%
	Middle age	14.3%
	Old	9.4%
Marital status	Married	19.9%
	Single	80.1%

As described in Sect. 3, we have two tasks in this paper. For the partial label prediction task, the goal is to predict the unknown attributes of users while the model is trained with the observed attributes. However, in our dataset, all demographic attributes are known for all the users. As a result, for partial prediction problem setting, we randomly set certain attributes as observed for training. For experiments on new user prediction task, we split our dataset into non overlapping sets, and the training, validation and testing split ratio are set to be 8:1:1.

5.2 Evaluation Metrics

We employ F-measure to evaluate our model. F-measure is a widely used measure method as a complement for accuracy, and it is the most popular evaluation metrics for demographic prediction. F1 score is calculated as the harmonic mean of precision and recall. The precision and recall are formulated differently depending on the following types of F1 score: micro, macro, and weighted. In our experiment, we use the weighted precision (wP), recall (wR) and F1 $(wF1)$ score as the evaluation metrics since we consider all classes to be equal important. The weighted F1 is calculated as follows:

$$wP = \sum_{y \in Y} (\frac{\sum_{i=1}^{u} I(y_i^* = \hat{y}_i \& y = \hat{y}_i)}{\sum_{i=1}^{u} I(y = \hat{y}_i)} * weight)$$

$$wR = \sum_{y \in Y} (\frac{\sum_{i=1}^{u} I(y_i^* = \hat{y}_i \& y = \hat{y}_i)}{\sum_{i=1}^{u} I(y = y_i^*)} * weight)$$

$$wF1 = 2 * \frac{wP * wR}{wP + wR}$$

where $I(\cdot)$ is an indicator function, u denotes the total number of new users, Y is the set of all label combinations to be predicted, y_i^* denotes the ground truth of attributes for the i-th new user, $\hat{y}_i$ denotes the predicted attributes, and $weight = \frac{1}{u} \sum_{i=1}^{u} I(y = y_i)$. The weighted F1 assigns a high weight to the large classes to account for label imbalance.

5.3   Baseline Models

We compare our models with state-of-the-art case studies on demographic prediction:

SNE. Structured Neural Embedding [7,20] maps users' transaction histories into a shared embedding. This embedding is processed by average pooling and then fed into a log-bilinear model for structured predictions.

FMTD. In [14], the sequences embedding of transactions and structured relational vectors are concatenated as user representation, and the user representation is fed into a full connected layer to get a deep representation for prediction.

ETNA. Embedding Transformation Network with Attention [7] model uses a shared embedding at the bottom and feeds it into an embedding transformation layer to obtain the transformed representation. Then put the transformed representation into a task-specific attention layer to take into account the importance of each element in user profile for better prediction.

5.4   Experimental Settings

We implement the baseline models as described in [7] and FMTD [14]. We search all occurred entities in the dataset as well as the ones within one hop in the Microsoft Satori knowledge graph with confidence greater than 0.8 for all edges. The dimension of both item embedding and entity embedding are set as 128. The number of filters are set as 64 for window sizes 2 and 3. We use the Adam [8] with a mini-batch size of 64 and a learning rate of $1e^{-3}$. For partial label prediction, the observed attribute ratio is set as 50%.

5.5   Comparison of Different Models

The results of comparison of different models are shown in Table 2, where the KAE contains all three embeddings with TransD. We have some findings based on the experimental results.

1. The usage of KAE for embedding could boost the performance of all the baselines for both tasks. For example, in terms of ETNA, the wF1 increases from 56.9% to 64.8% in partial label prediction task and the wF1 increases from 36.0% to 43.8% in new user prediction task.

Table 2. The experimental results of partial label prediction and new user prediction.

Model	Partial label (50%)			New user		
	wP	wR	wF1	wP	wR	wF1
SNE	52.1%	56.3%	54.2%	29.5%	35.1%	32.1%
FMTD	43.4%	45.9%	44.6%	23.7%	29.1%	26.1%
ETNA	55.4%	58.4%	56.9%	33.9 %	38.2%	36.0%
KAE+SNE	59.8%	63.2%	61.5%	35.6%	38.3%	36.9%
KAE+FMTD	50.1%	52.4%	51.2%	27.1%	33.9%	30.1%
KAE+ETNA	63.9%	65.7%	64.8%	40.5%	47.6%	43.8%

"+" denotes the combination of KAE method with previous model.

2. After combining with KAE, the ETNA achieves the maximum gains among the three previous models. This is because ETNA integrates attention mechanism, when provides more knowledge, it has the ability to select valuable ones and abandon the noise for working better.
3. Comparing the wF1 gains between two tasks, we observe that the partial label prediction task obtains a little more improvement, the reason for which might be that observed partial attributes can be a good signal for CNN learning.

5.6   Comparison Among KAE Variants

We compare the variants of KAE on two aspects to demonstrate the efficacy of the design of the KAE framework: the usage of knowledge, and the choice of knowledge graph embedding method, where we combine with ETNA model for prediction. The results are shown in Table 3, from which we can conclude that:

Table 3. Comparison among KAE variants.

Variants	Partial label (50%)			New user		
	wP	wR	wF1	wP	wR	wF1
ETNA with I	55.4%	58.4%	56.9%	33.9%	38.2%	36.0%
KAE+ETNA with I	58.2%	61.7%	59.9%	36.8%	41.7%	39.1%
KAE+ETNA with I&E	62.7%	64.5%	63.6%	39.4%	45.9%	42.4%
KAE+ETNA with I&C	61.6%	62.8%	62.2%	38.1%	44.3%	41.0%
KAE+ETNA with I&E&C	63.9%	65.7%	64.8%	40.5%	47.6%	43.8%
KAE+ETNA with TransE	62.3%	64.3%	63.3%	39.3%	46.4%	42.6%
KAE+ETNA with TransH	62.7%	64.6%	63.6%	39.6%	46.7%	42.9%
KAE+ETNA with TransR	63.1%	65.1%	64.1%	40.0%	47.1%	43.3%
KAE+ETNA with TransD	63.9%	65.7%	64.8%	40.5%	47.6%	43.8%

"I", "E" and "C" denotes item, entity and context embedding respectively.

1. The use of CNN can improve wF1 by about 3% for both tasks. Moreover, the usage of entity embedding and contextual embedding can improve wF1 by 3.7% and 2.3% for partial label prediction as well as 3.3% and 1.9% for new user prediction. And we can attain better performance by combining them together.
2. TransD method used in KAE works better. Probably because TransD is the most complicated model among the four embedding methods, which is able to better capture non-linear relationships among the knowledge graph for demographic prediction.

6   Conclusion

In this paper, we propose a knowledge-aware embedding method that takes advantage of knowledge graph representation in demographic prediction. KAE breaks the traditional naive embedding methods on transaction history, and employs CNN to combine introduced entity and context embedding with item embedding. Experimental results demonstrate that KAE can boost performance of all the baselines.

References

1. Andrews, R.L., Currim, I.S.: Identifying segments with identical choice behaviors across product categories: an intercategory logit mixture model. Int. J. Res. Mark. **19**(1), 65–79 (2002)
2. Bordes, A., Usunier, N., García-Durán, A., Weston, J., Yakhnenko, O.: Translating embeddings for modeling multi-relational data. In: 27th Annual Conference on NIPS 2013, United States, pp. 2787–2795 (2013)
3. Conneau, A., Schwenk, H., Barrault, L., LeCun, Y.: Very deep convolutional networks for natural language processing. CoRR abs/1606.01781 (2016)
4. Dong, Y., Yang, Y., Tang, J., Yang, Y., Chawla, N.V.: Inferring user demographics and social strategies in mobile social networks. In: The 20th ACM SIGKDD, KDD 2014, New York, NY, USA, pp. 15–24 (2014)
5. Ji, G., He, S., Xu, L., Liu, K., Zhao, J.: Knowledge graph embedding via dynamic mapping matrix. In: ACL 2015, Beijing, China, 26–31 July 2015, Volume 1: Long Papers, pp. 687–696 (2015)
6. Kalyanam, K., Putler, D.S.: Incorporating demographic variables in brand choice models: an indivisible alternatives framework. Mark. Sci. **16**(2), 166–181 (1997)
7. Kim, R., Kim, H., Lee, J., Kang, J.: Predicting multiple demographic attributes with task specific embedding transformation and attention network. CoRR abs/1903.10144 (2019)
8. Kingma, D.P., Ba, J.: Adam: a method for stochastic optimization. In: 3rd ICLR 2015 Conference Track Proceedings, San Diego, CA, USA (2015)
9. Krizhevsky, A., Sutskever, I., Hinton, G.E.: ImageNet classification with deep convolutional neural networks. In: 2012 26th Annual Conference on Neural Information Processing Systems, United States, pp. 1106–1114 (2012)
10. Lin, Y., Liu, Z., Sun, M., Liu, Y., Zhu, X.: Learning entity and relation embeddings for knowledge graph completion. In: Proceedings of the Twenty-Ninth AAAI, USA, pp. 2181–2187 (2015)

11. Milne, D.N., Witten, I.H.: Learning to link with Wikipedia. In: Proceedings of the 17th ACM, CIKM 2008, Napa Valley, California, USA, 26–30 October 2008, pp. 509–518 (2008)
12. Mislove, A., Viswanath, B., Gummadi, P.K., Druschel, P.: You are who you know: inferring user profiles in online social networks. In: Proceedings of the Third International Conference on WSDM
13. Otterbacher, J.: Inferring gender of movie reviewers: exploiting writing style, content and metadata. In: Proceedings of the 19th ACM Conference on Information and Knowledge Management, CIKM 2010, pp. 369–378 (2010)
14. Resheff, Y.S., Shahar, M.: Fusing multifaceted transaction data for user modeling and demographic prediction. CoRR abs/1712.07230 (2017)
15. Schler, J., Koppel, M., Argamon, S., Pennebaker, J.W.: Effects of age and gender on blogging. In: AAAI Spring Symposium, Technical Report SS-06-03, Stanford, California, USA, 27–29 March 2006, pp. 199–205 (2006)
16. Sedhain, S., Sanner, S., Braziunas, D., Xie, L., Christensen, J.: Social collaborative filtering for cold-start recommendations. In: Proceedings of the 8th ACM Conference on Recommender Systems. ACM (2014)
17. Sil, A., Yates, A.: Re-ranking for joint named-entity recognition and linking. In: 22nd ACM, CIKM 2013, San Francisco, CA, USA, 27 October–1 November 2013, pp. 2369–2374 (2013)
18. Wang, H., Zhang, F., Xie, X., Guo, M.: DKN: deep knowledge-aware network for news recommendation. In: Proceedings of the 2018 WWW Conference on World Wide Web, WWW 2018, Lyon, France, 23–27 April 2018, pp. 1835–1844 (2018)
19. Wang, J., Wang, Z., Zhang, D., Yan, J.: Combining knowledge with deep convolutional neural networks for short text classification. In: 2017 Proceedings of the Twenty-Sixth IJCAI 2017, Melbourne, Australia, pp. 2915–2921 (2017)
20. Wang, P., Guo, J., Lan, Y., Xu, J., Cheng, X.: Your cart tells you: inferring demographic attributes from purchase data. In: WSDM, pp. 173–182 (2016)
21. Wang, Z., Zhang, J., Feng, J., Chen, Z.: Knowledge graph embedding by translating on hyperplanes. In: Proceedings of the Twenty-Eighth AAAI, Canada, pp. 1112–1119 (2014)
22. Yang, B., Mitchell, T.M.: Leveraging knowledge bases in LSTMs for improving machine reading. In: ACL 2017, Vancouver, Canada, 30 July–4 August 2017, Volume 1: Long Papers, pp. 1436–1446 (2017)
23. Zhang, X., Zhao, J.J., LeCun, Y.: Character-level convolutional networks for text classification. In: 2015 Annual Conference on Neural Information Processing Systems, 7–12 December 2015, Montreal, Quebec, Canada, pp. 649–657 (2015)
24. Zhao, X.W., Guo, Y., He, Y., Jiang, H., Wu, Y: We know what you want to buy: a demographic-based system for product recommendation on microblogs. In: Proceedings of the 20th ACM SIGKDD. ACM (2014)
25. Zhong, Y., Yuan, N.J., Zhong, W., Zhang, F., Xie, X.: You are where you go: Inferring demographic attributes from location check-ins. In: 2015 Proceedings of the Eighth ACM, WSDM 2015, China, pp. 295–304 (2015)

Addressing Reward Engineering
for Deep Reinforcement Learning
on Multi-stage Task

Bin Chen[1,2] and Jianhua Su[1(✉)]

[1] The State Key Laboratory of Management and Control for Complex Systems,
Institute of Automation, Chinese Academy of Sciences, Beijing 100190, China
{chenbin2017,jianhua.su}@ia.ac.cn
[2] The school of Artificial Intelligence, University of Chinese Academy of Sciences,
Beijing, China

Abstract. In the field of robotics, it is a challenge to deal with multi-stage tasks based on Deep reinforcement learning (Deep RL). Previous researches have shown manually shaping a reward function could easily result in sub-optimal performance, hence choosing a sparse reward is a natural and sensible decision in many cases. However, it is rare for the agent to explore a non-zero reward with the increase of the horizon under the sparse reward, which makes it difficult to learn an agent to deal with multi-stage task. In this paper, we aim to develop a Deep RL based policy through fully utilizing the demonstrations to address this problem. We use the learned policy to solve some difficult multi-stage tasks, such as picking-and-place, stacking blocks, and achieve good results. A video of our experiments can be found at: https://youtu.be/6BulNjqDg3I.

Keywords: Deep reinforcement learning · Robotic manipulation · Multi-stage task

1 Introduction

In recent years, the research of robotic grasping strategies has attracted increasingly attention in the field of robotic manipulation. A key challenge is how to perform a successful grasping without a prior knowledge of the manipulated object. Analytic or model-based method can achieve excellent performance to situations that satisfy their assumption. However, the complexity and diversity of objects in a new environment have a tendency to confound these assumptions, hence learning-based methods have emerged as a powerful complement. Recently, using end-to-end approaches to handle robotic manipulation tasks has achieved great success both in the simulation environment [4] and in the real environment [9].

Using deep reinforcement learning to handle multi-stage tasks is often challenging due to the lack of professional knowledge of a special task. A way to deal

© Springer Nature Switzerland AG 2019
T. Gedeon et al. (Eds.): ICONIP 2019, CCIS 1143, pp. 309–317, 2019.
https://doi.org/10.1007/978-3-030-36802-9_33

with this situation is only to provide a sparse reward − +1 for success and 0 for failure. In many cases it is difficult for an agent to learn a feasible action at every moment in each episode for the delaying of the reward.

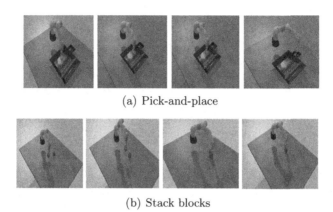

(a) Pick-and-place

(b) Stack blocks

Fig. 1. Frames from rollouts of the learned policy are shown above. (a) This is a picking-and-place task, which is a classic robotic manipulation task from industry to service robot. (b) This is a stacking blocks task, and robot must place one block to the top of another block one by one.

There are many works to deal with the grasping task with learning-based approaches, and these methods can be divided into two classes: reinforcement learning and imitation learning(similar to supervised learning or learning from demonstrations). For reinforcement learning, most of prior works [6,7] only train a Deep RL policy to grasp an object with 4-D actions (x, y, z, θ), where the last dimension specifies the anger of two gripper fingers. Hence, these policies often fail to work when grasping irregularly shaped objects. In contrast, imitation learning can learn an end-to-end policy to deal with this problem [8,11] if there is enough labeled-data to be provided for training. Although the results are impressive, it's restricted for the widespread using of imitation learning as the large amount of data are needed to be collected. In our work, we extend the learning action to the 5-D space (x, y, z, rz, θ) based on the Deep RL method, where the learned 4th dimension action is related to the posture of the object.

In reinforcement learning, learning a policy through sparse rewards has been popular with researchers for people do not need the special knowledge to design a reward function. However, learning an available policy is a huge challenge with the increase of the horizon under the sparse reward, which makes it difficult to deal with the long-horizon multi-stage task. There some related works [3,6,10] combine demonstrations with RL algorithm to deal with this problem. The work of [3,6] are essentially the same while they forcing the action generated by the policy the same as the action from the demonstration by adding an additional loss function. The method proposed in [10] is more reasonable while it gives different sampling probabilities through the importance of different transitions, which increases the using-efficiency of transitions.

In this work, we use the improved Deep reinforcement learning algorithm Deep Deterministic Policy Gradient (DDPG), combined with the our proposed method to train an agent to complete the multi-stage task with high success rate and good learning efficience. The input of the policy is only RGB image and auxiliary low-dimensional information (joint angles, gripper state, etc). Figure 1 shows the images we put into the policy and the robot manipulation tasks we are going to complete. Further more, we extend the learning action to the 5-D spaces (x, y, z, rz, θ), which ensures the agent can adjust the angle of the z-axis according to the variety of the shape of different objects and greatly improves generalization ability of the policy. In summary, the main contributions of our work are as follows:

1. In order to make full use of the demonstration data to improve learning efficiency, we artificially make the network fully utilize the state-action pairs in demo until the performance of the actions generated by the network exceeds the actions in demo.
2. Most of the work focus on learning actions limited to 4-D spaces, while we extend the learning actions to 5-D spaces, which is proved to be a significant work for improving the generalization ability of the learned policy.

2   Background

2.1   Reinforcement Learning

We consider the continuous space Markov Decision Process which can be denoted as the tuple $(S, A, P, r, \gamma, S_-)$, where S is a set of continuous states, A is a set of continuous actions, P is a state transition probability function, r is a reward function, γ is a discount factor and S_- is a set of initial continuous states. In reinforcement learning, the goal is to learn a policy $a_t = \pi(x_t)$ to maximize the expected return from the state t_i, where the return can be denoted by $R_t = \sum_{i=t}^{T} \gamma^{i-t} r_i$ and T is the horizon that the agent optimizes.

Various reinforcement learning algorithms have been proposed in order to solve the max expected return problem in RL. Most of the methods involve constructing an estimate of the expected return from a given state after taking an action:

$$Q_\pi(s_t, a_t) = E_{r_i, s_i, a_i}\left(\sum_{i=t}^{T} \gamma^{i-t} r_i | s_t, a_t\right) \qquad (1)$$

$$= E_{r_t, s_{t+1}, a_i} E[r_t + \gamma E_{a_{t+1}}(Q_\pi(s_{t+1}, a_{t+1}))] \qquad (2)$$

where the $Q_\pi(s_t, a_t)$ is called the action-value function. Equation 2 is a recursive version of Eq. 1, which is known as the Bellman equation.

In this work, we combine our method with a reinforcement learning algorithm: Deep Deterministic Policy Gradient (DDPG). DDPG is an actor-critic method which bridges the gap between policy gradient methods and value approximation method for RL, hence it composes of actor network $\pi(s)$ and critic network $Q(s, a|\theta^Q)$. However, experiments have shown if we only use single 'Q-network', it will make the learning process unstable – the parameters of the critic network for calculating the Q function are employed to calculate the gradient of the actor network while performing frequent gradient updates. We can better understand this process with reference to Eqs. (3–5). In order to stabilize learning, DDPG creates two target networks for actor network and critic network, respectively. The update of the critic network is similar to the method of supervised learning. The formula of the loss is:

$$L = \frac{1}{N} \sum_i (y_i - Q(s_i, a_i|\theta^Q))_2 \tag{3}$$

where y_i can be seen as a 'label':

$$y_i = r_i + \gamma Q'(s_{i+1}, \mu'(s_{i+1}|\theta^{\mu^i})|\theta^{Q'}) \tag{4}$$

When updating the actor network by off-policy method, the policy gradient of the strategy is as follows:

$$\nabla_{\theta^\pi} J_\beta(\pi) = E_{s\rho^\beta}[\nabla_a Q(s, a|\theta^Q)|_{a=\pi(s)} . \nabla_{\theta^\pi} \pi(s|\theta^\pi)] \tag{5}$$

Note π is an actor policy $a_t = \pi(s_t|\theta)$, and β is a behaviour policy $a_t = \pi(s_t|\theta^\pi) + N_t$ used to explore potential better policies, where N_t is a Uhlenbeck-Ornstein (UO) stochastic process as a random noise.

2.2 DDPG from Domenstrations

In our method, we use sparse reward as a feedback to an agent, in which the agent would get +1 reward when success otherwise getting 0. However, combining sparse reward with raw DDPG algorithm to deal with the multi-stage task is impossible as sparse reward can not give enough information to update parameters in right way. This problem can be partly overcome with the demonstrations as the agent can learn a suitable policy for completing the task from the demonstrations. Moreover, prioritized experience replay [10] modifies the agent to sample more important transitions from its replay buffer more frequently. The probability that each particular transition in the replay buffer is extracted as $P(i) = \frac{p_i^\alpha}{\sum_k p_k^\alpha}$, where $p(i)$ represents the priority of each transition. In the experiment, $p_i = \delta_i^2 + \lambda|\nabla_a Q(s_i, a_i|\theta^Q)|^2 + \varepsilon$, where δ_i is the last TD error calculated for this transition, the second term is the loss applied to the actor, ε is a positive constant to ensure that all transitions have a certain probability of being sampled, and λ is used to weight the contributions. DDPGfD [10] merged the above two strategies and made further improvements. It collected the demonstration before pre-training, as well as initialized the replay buffer with the demo, and pre-trained the agent with the demo transitions. Besides, they set $p_i = \delta_i^2 + \lambda|\nabla_a Q(s_i, a_i|\theta^Q)|^2 + \varepsilon + \varepsilon_D$, where ε_D is a positive constant for demo transitions to increase their probability of being sampled.

3   Methods

3.1   Behaviour Clone Loss and Reward

[6] introduced a method of adding loss function to the network, which made full use of the positive effects of the demonstration on policy training. This loss is applied only when a demonstration is sampled from the replay buffer for training. When the value function $Q(a_d)$ generated by the action a_d in the demonstration is bigger than $Q(a_t)$, it means that a_d is closer to the optimal action than a_t. At this time, they would give the policy a penalty item L_{BC}, and encouraged the policy to propose the same action as the demonstration in the given state.

$$L_{BC} = \begin{cases} |\pi(o_d) - a_d|, & if \quad Q(s_d, a_d) > Q(s_d, \pi(o_d)) \\ 0, & otherwise \end{cases} \quad (6)$$

However, this way of adding loss to the network seems still does not fully utilize the information in the demonstration. When each transition is collected from the demonstration for the reason of initializing the replay buffer and pre-training the policy for multi-stage task, the previous work always sets the reward for each state transition to 0, while only the last state transition step is set to 1, as shown in formula 7. Note that T is the total step size for an episode. However, if we set the reward of the corresponding action in most transitions to 0, the agent may consider this action to be undesirable and reject this action in training (since reward is 0, the corresponding Q_target value is smaller) even though this action may benefit completing the task. Therefore, we set the reward of all the transitions to 1 for an episode if the task is completed successfully when collecting the transition. In the early stage of training, this method ensures the policy generate the action the same as the demonstrator in the given state. In the later stage of training, we use the Q_filter method proposed in [6] to ensure that the agent can use the Q-value to determine whether to use the action in demonstration or not. It should be noted that this method is compatible with the behavior clone Loss.

$$r_t = \begin{cases} 1, & if \quad t = T \\ 0, & otherwise \end{cases} \quad (7)$$

3.2   Extend the Learning Action to 5-D Spaces

For most of the Deep RL based policy only learning a 4-D action (x, y, z, θ) to deal with the grasping task, it's often fail to work if we are going to grasp irregularly shaped objects as be shown in Fig. 2(a). Hence, in our work we extend the learning action to the 5-D space (x, y, z, rz, θ), where the learned 4th dimension action is related to the posture of the object. Figure 2(b) shows the framework of our method. The input of the actor network includes RGB image and full-low dimensional state, while the twin critic networks only receive the full-low dimensional state as input. Besides, we add the detection loss to actor loss to help the policy recognize the essential scene features quickly. The output of the actor network is a 5-D actions.

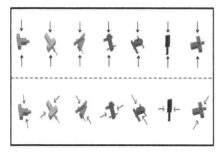 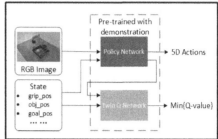

(a) Predict the possible grasping points for a given observation

(b) Framework of the proposed Method

Fig. 2. We extend the learning action to 5-D spaces. (a) If we only learn 4-D actions, the predicting points is not always leading to successfully grasping (top), while the learned 5-D actions from our method could predict the better grasping points (bottom). (b) The framework of the proposed method.

4   Experimental Results

4.1   Environmental Setup

We evaluate our algorithm on several simulated Pybullet [2] environments. In simulation environment we use the Kuka robot and a robotic grippers which proposed by Pybullet. In the collection of demonstration's data, the observation is rendered by a virtual camera and the states of environment, such as robot arm state, gripper state, goal position can be obtained by corresponding api function. In our experiments, we will perform the experiments in a 5-D space as the extra dimension representing the angle of rotation around the z-axis.

4.2   Evaluation

Comparison with Prior Work. In this section, we will demonstrate the superiority of our algorithm through comparing with several competitive methods by executing the picking-and-place and stocking blocks task. [1] proposed a 'HER' algorithm of storing experienced transitions with different goals in the replay buffer used in off-policy RL algorithms that allows to learn the policy more efficiently with sparse rewards. [5] proposed a method which incorporates several improvements to DDPG and can accomplish most of the simple tasks with excellent results, and we call this method 'Rainbow-DDPG'. [7] proposed an asymmetric actor critic algorithm for visual-based task in which the critic is trained on full states while the actor is trained on images and we call this algorithm 'AAC'.

In our experiments, we all use a fully sparse reward to train the policy, where the policy get a +1 reward if the object is at its goal position after rollouts ending.

$$r_t = \begin{cases} 1, & if \quad ||x_i - g_i|| < \delta \\ 0, & otherwise \end{cases} \tag{8}$$

where the threshold δ is 3 cm.

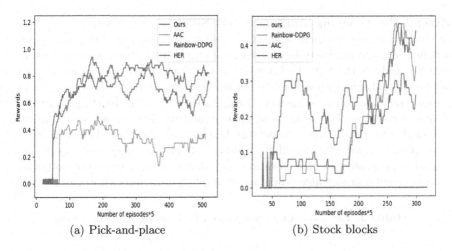

(a) Pick-and-place   (b) Stock blocks

Fig. 3. We compare our method with some existing competitive methods and as can be seen our method outperforms these methods. (a) Evaluation experiments on picking-and-placing task. (b) Evaluation experiments on stacking blocks task.

Table 1. Comparison of our method with baselines

Task	Ours	Rainbow-DDPG	AAC	HER
Pick-and-place	95%	87%	50%	0%
Stack blocks	86%	72%	38%	0%

Table 1 reports the results of our evaluation experiments. The effect of our method and 'Rainbow-DDPG' are more outstanding compared with the 'AAC' and 'HER'. The main factor is that we utilize demonstration and take some strategies to make the action generated by the policy close to the demo, which greatly improves the success rate of learning. Figure 3 shows the variation of the reward value during training. We run the each experiment 5 times in each episode, and the corresponding reward value of each time step takes the average of the past 10 episodes experimental results, so the curves report the mean of past 50 experimental results. We adapt most of the improvements proposed in 'Rainbow-DDPG', as well as our own methods, and achieve competitive results.

As can be seen from Fig. 3, HER is unable to complete this task. While Marcin Andrychowicz et al. explained in their paper that they could use HER to complete the pick-and-place task, which seems to be contrary to our experimental results. However, in their experiments, they assumed that the agent have got the knowledge of the first half rollouts action (the box is grasped) before training and the agent only needs to learn the action of rollouts in the latter half, which greatly reduces the difficulty of learning the task. In our experiments, if we don't give any extra information to the agent about the task, it's almost impossible to complete the pick-and-place task through HER.

Extend the Learning Action to 5-D Spaces. In this study, we only use the improved reinforcement learning algorithm proposed in this paper to learn a policy to grasp the irregularly shaped objects. We only compare the results of grasping objects through the learned 4-D and 5-D actions by method proposed in this paper due to the lack of existing methods for learning 5-D actions through reinforcement learning as baseline. Table 2 shows that for objects with irregular shapes, the learned 5-D actions can greatly improve the success rate of grasping. Although our results are not outstanding enough for they do not reach the level of supervised learning, the advantage of reinforcement learning does not require a large number of annotation data drives us to explore the potential of reinforcement learning algorithms to accomplish more difficult grasping tasks.

Table 2. The success rate comparison for learning different actions

Task	Learning 4-D action	Learning 5-D action
Pick-and-place for irregularly shaped objects	54 %	72 %

In order to test the ability of the agent to analyze the scene for a given image, we force the agent to predict the state of the key elements in the scene at each moment, such as object position, gripper position. In Fig. 4(b), we show the loss curves of the object-position, the robot gripper-position and the target-position, respectively. It can be seen that as the training going, the agent can analyze the location of each element accurately.

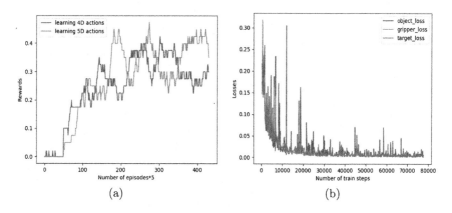

(a)   (b)

Fig. 4. (a) Evaluation experiments on grasping the irregularly shaped objects while learning different actions. (b) The loss curve of the object-location, the robot gripper-location and the target-location. It can be seen that as the training going, the agent can analyze the location of every element in the scene accurately.

5   Conclusion and Future Work

In this paper, we improve the policy based on DDPG by incorporating several improvement methods and achieve impressive results on the multi-stage task. In the experiment, compared with some baselines, the performance of the policy proposed is of the great improved. At the same time, we extend the learning action to 5-D space, which greatly improves the generalization of the policy.

In future work, we will deploy our algorithm to real word robots. For the most domain adaption technologies are achieved by enriching the diversity of elements in the scene, hence the success of the method in the real environment is depend on the diversity of environment during training. The limitations of this approach have been reflected in many of the previous works. Therefore our further research will focus on exploring a domain adaption technique that allows the algorithm to maintain comparable performance in the real world as it does in the simulation environment.

Acknowledgements. This work was supported in part by NSFC under Grant No.91848109, supported by Beijing Natural Science Foundation under Grant No.4182068 and supported by Science and Technology on Space Intelligent Control Laboratory under No. HTKJ2019KL502013.

References

1. Andrychowicz, M., et al.: Hindsight experience replay. In: Advances in Neural Information Processing Systems, pp. 5048–5058 (2017)
2. Coumans, E., Bai, Y.: Pybullet, a python module for physics simulation for games, robotics and machine learning. GitHub repository (2016)
3. Hester, T., et al.: Learning from demonstrations for real world reinforcement learning. arXiv preprint arXiv:1704.03732 (2017)
4. Levine, S., Finn, C., Darrell, T., Abbeel, P.: End-to-end training of deep visuomotor policies. J. Mach. Learn. Res. **17**(1), 1334–1373 (2016)
5. Matas, J., James, S., Davison, A.J.: Sim-to-real reinforcement learning for deformable object manipulation. arXiv preprint arXiv:1806.07851 (2018)
6. Nair, A., McGrew, B., Andrychowicz, M., Zaremba, W., Abbeel, P.: Overcoming exploration in reinforcement learning with demonstrations. In: 2018 IEEE International Conference on Robotics and Automation (ICRA), pp. 6292–6299. IEEE (2018)
7. Pinto, L., Andrychowicz, M., Welinder, P., Zaremba, W., Abbeel, P.: Asymmetric actor critic for image-based robot learning. arXiv preprint arXiv:1710.06542 (2017)
8. Pinto, L., Gupta, A.: Supersizing self-supervision: learning to grasp from 50k tries and 700 robot hours. In: 2016 IEEE International Conference on Robotics and Automation (ICRA), pp. 3406–3413. IEEE (2016)
9. Popov, I., et al.: Data-efficient deep reinforcement learning for dexterous manipulation. arXiv preprint arXiv:1704.03073 (2017)
10. Večerík, M., et al.: Leveraging demonstrations for deep reinforcement learning on robotics problems with sparse rewards. arXiv preprint arXiv:1707.08817 (2017)
11. Viereck, U., Pas, A.t., Saenko, K., Platt, R.: Learning a visuomotor controller for real world robotic grasping using simulated depth images. arXiv preprint arXiv:1706.04652 (2017)

High-Speed Synchronization
of Pulse-Coupled Phase Oscillators
on Multi-FPGA

Dinda Pramanta$^{(\boxtimes)}$ and Hakaru Tamukoh

Graduate School of Life Science and Systems Engineering, Kyushu Institute of
Technology, 2-4 Hibikino, Wakamatsu-ku, Kitakyushu, Fukuoka 808-0196, Japan
`dinda-pramanta@edu.brain.kyutech.ac.jp`, `tamukoh@brain.kyutech.ac.jp`

Abstract. This study proposes High-Speed Synchronization of Pulse-
Coupled Phase Oscillators on multiple field-programmable-gate-array
(FPGA). Winfree model is used for oscillators communication based and
two FPGAs are connected by a gigabit transceiver (GTX). In order to
verify the effect of communication delay and in-phase synchronization
phenomenon between FPGAs, we implement various number of oscillators on the multi-FPGA platform. Four-oscillator network achieved first
spike synchronization over two FPGAs within 12.47 μs and datastream
bitrate up tp 3.2 Gbps. We have successfully expanded the network from
the previous study and verified that the 10×10 pulse-coupled phase
oscillators synchronized over two FPGAs via high-speed serial communication with a 0.1 μs delay and the network reached a steady synchronization state after the spike-count reached 100 with a maximum frequency
298.014 MHz.

Keywords: GTX · Multi-FPGA · Pulse-coupled phase oscillators ·
Synchronization · Winfree model

1 Introduction

A recent advance in neuroscience [1] showed that neuron-inspired communication
architectures are giving the spotlight results. In order to model the mechanism
how the brain works for realizing brain-like very large-scale integration (VLSI),
a typical spiking communications are necessary. A mathematical model of pulse
coupled phase oscillator networks have been proposed and intensively studied
[2–7]. In these models, simply phase sensitivity function represents as an interactions between the oscillators, and the coupling between them are determined
by the timing of spike-output pulses from each oscillator. Based on this, nonlinear processing is performed with a single transition in one spike-timing, which is
only one-bit signal lines required between oscillators for communication. These
features allows for achieving hardware implementation with low-power intelligent information processing and high-speed performance, such as VLSI systems
which constructed based on spike-based computation.

© Springer Nature Switzerland AG 2019
T. Gedeon et al. (Eds.): ICONIP 2019, CCIS 1143, pp. 318–329, 2019.
https://doi.org/10.1007/978-3-030-36802-9_34

For implementing the spike-based computation and demonstrate the effectiveness of its features, proposed analog VLSI circuits [8–10] and digital circuit using field programmable gate arrays (FPGAs) [11,12]. Inside analog VLSI circuits, a region-based coupled Markov random field model using pulse-coupled phase oscillators were implemented and applied it to image region segmentation. Based on measurement result, it shows that the studied device achieved high efficiency in terms of performance per power consumption [9]. In FPGA implementation side, we established that parameterized design allow us to emulate various sizes and interconnections of pulse-coupled phase oscillator networks [12]. In a previous work [13] also, we have confirmed that spike-based computation is suitable for realizing brain-like VLSI. However, a large-scale network cannot be implemented on a single chip due the resource limitation.

Multi-FPGA is the solution for overcoming the resource limitations of a single FPGA [14] and implementing large-scale pulse-coupled phase oscillator networks. However, an issue might affect synchronization between oscillators communication delays between FPGAs, which is necessary for the function of pulse-coupled phase oscillator systems. Consequently, high-speed communication is required to implement oscillator systems over multiple FPGAs, and the synchronization state over a multi-FPGA system should be verified.

In this study, we propose an implementation of multi-FPGA communication links for pulse-coupled phase oscillators. First, we design a digital circuit of pulse-coupled phase oscillators based on the Winfree model. Secondly, building communication between two FPGAs using First-In First-Out (FIFO) interfaces, and a serial connection using a gigabit transceiver (GTX) [15]. We employed two FPGA boards for the multi-FPGA platform and connect it by using GTX serial connection. Experimental results show that a four-oscillator network reached a synchronization state, and first spike synchronization over two FPGAs took 12.47 μs, with a datastream bitrate of 3.2 Gbps. We expands and synthesize the oscillators and networks used in simulations. Orderly we evaluate 2 × 2, 3 × 3, 6 × 6, and 10 × 10 pulse-coupled phase oscillators over two FPGAs. Observation discussion phenomenon will be focus on 10 × 10 pulse-coupled phase oscillators.

2   Modeling of Pulse-Coupled Phase Oscillator

The Winfree model [3] provides an efficient way of designing pulse-coupled phase oscillator circuits. The fundamental relation for pulse-coupled phase oscillators can be expressed as follows:

$$\frac{\partial \phi_i}{\partial t} = \omega_i + Z(\phi_i) Spk(t) \tag{1}$$

Inputs from other oscillators, $Spk(t)$, are assumed here, where ϕ_i is the i-th phase variable with 2π periodicity, ω_i is the i-th natural angular frequency, $Z(\phi_i)$ is a phase sensitivity function, which gives the response of the i-th oscillator. To give the following pulse input can be expressed as follows:

$$Spk(t) = \frac{K_0}{N} \sum_{j=1}^{N} \sum_{n=1}^{\infty} \delta(t - t_{jn}) \tag{2}$$

Mathematically, δ is a Dirac delta function that represents the timing of input spikes without a pulse width. Oppositely in psychical hardware, spike pulses have a definite width Δt, during which ϕ_i is updated according to the value of $Z(\phi_i)$. Where K_0 is the coupling strength, N is the number of oscillators, and t_{jn} is the firing time.

2.1 Discretization Model

To implement the model of pulse-coupled phase oscillators in digital hardware, Eqs. (1) and (2) are discretized as follows:

$$\phi_i(t + 1) = \phi_i(t) + \omega_i + \frac{K_0}{N} \sum_{j=1}^{N} Z(\phi_i) Spk_j(t) \tag{3}$$

$$Spk_j(t) = \begin{cases} 1, & \textit{if } \phi_j(t) = \phi_{th} \\ 0, & \textit{otherwise} \end{cases} \tag{4}$$

Using the above discretization model, implementation can be simplified into logic circuits [12]. The phase value at the next time step is calculated by adding the current phase value, natural angular frequency, and the sum of products of the phase sensitivity function and input pulses. The natural angular frequency ω_i is assumed to be constant. The oscillator outputs a spike pulse when the phase variable reaches threshold value ω-th and then resets; $\omega_i = 0$.

2.2 Pulse-Coupled Phase Oscillator Dynamics

Theoretically, the oscillator operation is based on the neighbors-connections between each oscillator. This model is determined by only local interactions but can generate various phenomena in a large oscillator network based on the coupling formulas shown in Eqs. (1) and (2).

The two oscillators are coupled by the spikes Spk_i and Spk_j. Figure 1 shows a schematic of two pulse-coupled phase oscillators and a timing diagram that explains their updates. We express that the phase sensitivity function is $Z(\phi_i) = -\sin(\phi_i)$. Synchronization will occurs by updating the pulse timing, where there are two update phases for each oscillator: positive updating and negative updating. Negative updating occurs when there is spike input Spk_j to another oscillator ϕ_i and the lagging condition triggers the function $Z(\phi_i)$, making the timing of the maximum value of ϕ_i later. Positive updating occurs whenever there is spike input Spk_i to another oscillator ϕ_j and the leading condition triggers the function $Z(\phi_j)$, making the timing of the maximum value of ϕ_j earlier.

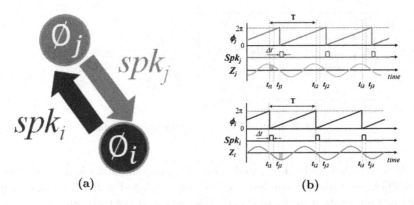

Fig. 1. Pulse-coupled phase oscillators. (a) Schematic of coupled network. (b) Phase variables (ϕ_i and ϕ_j) and pulse inputs.

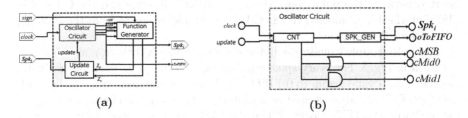

Fig. 2. Pulse-coupled phase oscillator circuit design: (a) Overall Circuit, and (b) Oscillator Circuit

3   Hardware Architecture Design

3.1   Designing Digital Pulse-Coupled Phase Oscillators

Figure 2 shows the hardware architecture of the pulse-coupled phase oscillator [11,12,16]. It includes an oscillator circuit, a function generator circuit, and an update circuit. The oscillator circuit contains an n-bit counter (CNT), a spike generator (SPK_GEN), and combinational circuits. The CNT represents a phase variable ϕ_i and counts clock inputs to implement ω_i in Eq. (3).

In each time step of t in the discretized model corresponds to a clock cycle. The signals from $cMSB$, $cMid0$, and $cMid1$ are determining the shape of the function $Z(\phi_i)$, which are also used in the function generator circuit. The function generator circuit itself combining these signals to output Zp and Zn. Output from the update circuit receives Zp, Zn, and Spk_j which also received from the other oscillator, to give update signal to the main Oscillator Circuit.

3.2   Communication System of Multi-FPGA

Block diagram of the multi-FPGA communication system is shown in Fig. 3. The Gigabit Transceiver (GTX) is a configuration system of transceiver-receiver

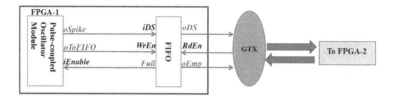

Fig. 3. Multi-FPGA communication system

(Tx/Rx), which tightly integrated with the programmable logic resources of the FPGA. In order to configure GTX, we employed a free and open high-speed communication protocol called Aurora [15]. It controls interfaces to read data (RdEn, oDS, and oEMP) and provides serial communication between the FPGAs at up to 10 Gbps [17]. Serial interface from first in-first out (FIFO) inserted between the pulse-coupled phase oscillator module and the GTX. The FIFO interface is connected to the GTX and transfers spikes from the FPGA-1 side to the FPGA-2 side.

The pulse-coupled phase oscillator circuit module has three main channels, as shown in Fig. 3. For every spike-data output (oSpike = 1) from the oscillator module, a data valid signal (oToFIFO = 1) is sent at the same time to the FIFO interface. FIFO module is used as an interface between two clock domains, i.e., the oscillator and GTX clocks. When the FIFO receives a spike (iDS = 1) with a write enable signal (WrEn = 1), the FIFO writes the spike into its internal memory. When the Full flag is high (Full = 1), the FIFO's internal memory is full and it will not accept any further data for writing (iEnable = 0).

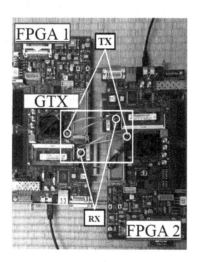

Fig. 4. Multi-FPGA platform using two Virtex6-XCVL240T ML605 boards.

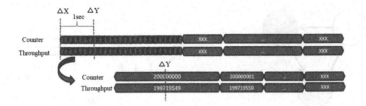

Fig. 5. Throughput measurement results using a 200 MHz clock of the Virtex6-XCVL240T ML605.

3.3 Multi-FPGA Platform over Two FPGAs

As the multi-FPGA platform, two boards of Virtex-6 XC6VLX240T ML605 Xilinx FPGA employed. Physical hardware connection between the two FPGAs are shown in Fig. 4. A SubMiniature version A (SMA) cable connection was used for the physical environment hardware between the two FPGAs' GTX channels. The ML605 boards provide a differential signal connection for each transceiver (Tx) and receiver (Rx). Therefore, two-pair of SMA cables are required to connecting each other.

4 Experimental Results

In general, serial physical connections between FPGAs incur communication delays. If it's affecting the synchronization states of the pulse-coupled phase oscillator networks, we cannot implement a large-scale network using the multi-FPGA platform. Therefore, showing the feasibility of the multi-FPGA platform is necessary, we examined basic synchronization phenomena in pulse-coupled phase oscillator networks.

4.1 Multi-FPGA Communication Performance

To evaluate the performance of the serial connection, we measured the throughput and latency of the multi-FPGA communication. We implemented a 16-bit incremental data generator (counter) in FPGA-1. FPGA-2 received the incremental data and looped them back to FPGA-1. The clock frequency of the FPGAs was set to 200 MHz. Figure 5 shows the measured throughput results for the multi-FPGA platform. "Counter" represents the number of clock cycles for FPGA-1. It counted up to 200 million every second because the clock frequency is 200 MHz. "Throughput" represents the number of valid 16-bit data values that were received in a given time. From these results, we can calculate the data rate as 16 bit * 199719549/s = 3195 Mbps ≈ 3.2 Gbps. In addition, we measured the difference in the incremental data between the Tx and Rx modules to find the latency of the multi-FPGA communication. The results show that the total latency from Tx to Rx was 20 counters or 0.1 μs.

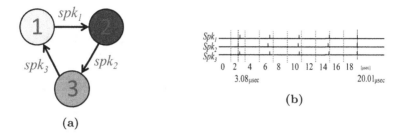

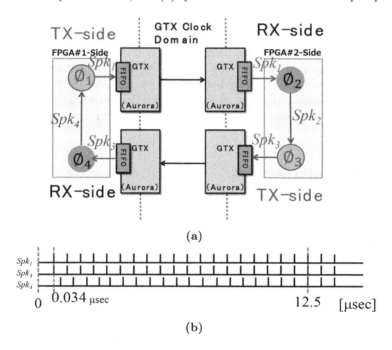

Fig. 6. Pulse-coupled spiking oscillator with three in-phase oscillators. (a) Three oscillators with rotary connections, and (b) synchronization results from ChipScope.

Fig. 7. Synchronization over two FPGAs

4.2 FPGAs Synchronization

To demonstrate the basic operation of pulse-coupled phase oscillator networks synchronization within one FPGA, we implemented a simple rotary network, as shown in Fig. 6(a). Here, we employed ChipScope, which is an in-circuit debugger provided by Xilinx [12] to observe spike pulses. Figure 6(b) shows the state of the FPGA, starting from the beginning at time zero (0 s). In this case, the first spike occurred at 3.08 μs and it reached a synchronization state at 20.01 μs. Figure 7 shows implementation of four oscillators on the multi-FPGA platform, in this case synchronization over two FPGAs. Figure 7b shows the measurement results of synchronization in FPGA-1 using ChipScope. These results show that in-phase

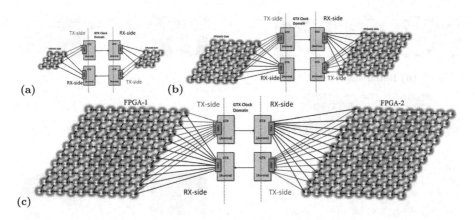

Fig. 8. The oscillator's network expansion connections over two FPGAs: (a) 3 × 3 network over two FPGAs, (b) 6 × 6 network over two FPGAs, and (c) 10 × 10 network over two FPGAs

synchronization was correctly observed at 12.47 μs. We have confirmed that our pulse-coupled phase oscillator networks were able to reach a synchronization state over the multi-FPGA platform, even though communication between the FPGAs included a 0.1 μs latency. Evaluation the hardware size of the proposed circuit, we expanded and synthesized the oscillators and networks orderly. We designed network architecture of pulse-coupled phase oscillators 2 × 2 (Fig. 7(a)), 3 × 3, 6 × 6, and 10 × 10 (Fig. 8). Observation discussion phenomenon will be focus on 10 × 10 pulse-coupled phase oscillators.

Table 1. Device utilization.

Oscillator size	2 × 2	3 × 3	6 × 6	10 × 10
LUTs	0.256%	0.35%	0.637%	1.308%
Registers	0.152%	0.171%	0.243%	0.412%
Max. freq. (MHz)	434.972	316.156	306.123	298.014

4.3   10 × 10 Oscillators Synchronization over Two FPGAs

Figure 9 shows in-phase synchronization between neighbors in the one-dimensional network composed of 10 × 10 oscillators over two FPGAs. We define the 'spike-count' as the number of spikes outputs from the oscillator located at the first row and column in the network. The spike-count roughly represents the number of update times for all oscillators in the network, and is effective for analyzing network behavior in the time domain.

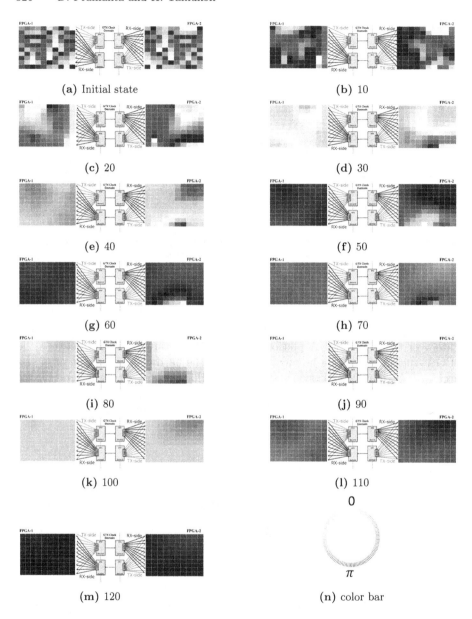

Fig. 9. Phase Synchronization of 10×10 oscillator over two FPGAs: (a) initial state, (b)–(m) network states after updating, where the number under each figure represents spike-count, (n) color bar representing the phase value [18].

In Fig. 9(a), we respectively set the initial value for each oscillator and each board to be different from all neighboring values. In this case, all oscillators were updated simultaneously by phase differences between neighbors as shown

in Fig. 9(b)–(j). In the early stage of Fig. 9(b)–(d), the phase oscillators from the
FPGA-1 and FPGA-2 updating with the difference phase. On the other hand,
in the early stage of Fig. 9(e)–(j) the phase oscillators from the FPGA-2 still
updating with the difference phase while FPGA-1 is already reached a steady
synchronization state. In the early stage of Fig. 9(k)–(m), we confirmed that the
network used in this simulation had several steady states, and the initial state
of the network affects the network steady state. The network reached a steady
synchronization state after the spike-count reached 100 of spike-count.

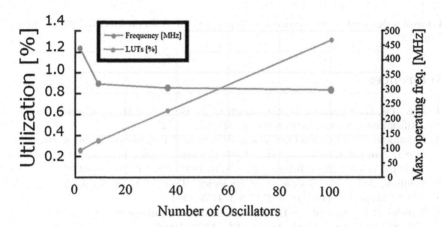

Fig. 10. Relationship between utilization, number of oscillator and max. operating
freq.

4.4   FPGA Implementation Results

Table 1 shows synthesized results for the total resources used within one FPGA.
Using ISE Design Suite software information, we can report that the proposed
model circuit was implemented with a maximum frequency of each size of oscillators. Figure 10 shows that the relationship of resource utilization and number
of oscillators will affecting the maximum operating frequency. By increasing the
resource utilization and number of oscillators, the maximum operating frequency
will slightly reduce, but still kept over 200 MHz operational frequency which is
enough for GTX communication between two FPGAs, shown in Figs. 5 and 7.
Therefore, the experimental results show that the proposed multi-FPGA system can implement and operate large pulse-coupled oscillator networks over two
FPGAs.

5   Conclusion

High-Speed Synchronization of Pulse-Coupled Phase Oscillators on Multi-FPGA
succeed. In our experiments two FPGA boards were used, four-oscillator network

achieved first spike synchronization over two FPGAs within 12.47 µs and datastream bitrate up tp 3.2 Gbps. As an extension from the previous study [13], we expanded the network and verified that the 10×10 pulse-coupled phase oscillators synchronized over two FPGAs via high-speed serial communication with a 0.1 µs delay and the network reached a steady synchronization state after the spike-count reached 100 with a maximum frequency 298.014 MHz. In future work, we will propose the different model type of network, such as Echo State Network (ESN), and then apply those model to engineering applications such as image processing for service robots.

Acknowledgment. This research is supported by JSPS KAKENHI grant number 17H01798.

References

1. Hassabis, D., Kumaran, D., Summerfield, C., Botvinick, M.: Neuroscience-inspired artificial intelligence. Neuron **95**(2), 245–258 (2017)
2. Kuramoto, Y.: Chemical Oscillations, Waves, and Turbulence, Springer, Heidelberg (1984). viii+ 156, 25 × 17 cm, 9,480 (Springer Series in Synergetics, vol. 19), 40(10):817–818 (1985). https://doi.org/10.1007/978-3-642-69689-3
3. Winfree, A.T.: The Geometry of Biological Time, vol. 12. Springer, New York (2001). https://doi.org/10.1007/978-1-4757-3484-3
4. Mirollo, R.E., Strogatz, S.H.: Synchronization of pulse-coupled biological oscillators. SIAM J. Appl. Math. **50**(6), 1645–1662 (1990)
5. Kuramoto, Y.: Collective synchronization of pulse-coupled oscillators and excitable units. Physica D Nonlinear Phenomena **50**(1), 15–30 (1991)
6. Hoppensteadt, F.C., Izhikevich, E.M.: Weakly Connected Neural Networks, vol. 126. Springer, New York (1997). https://doi.org/10.1007/978-1-4612-1828-9
7. Izhikevich, E.M.: Dynamical Systems in Neuroscience. MIT Press, Cambridge (2007)
8. Atuti, D., Kato, N., Nakada, K., Morie, T.: CMOS circuit implementation of a coupled phase oscillator system using pulse modulation approach. In: 2007 18th European Conference on Circuit Theory and Design, pp. 827–830. IEEE (2007)
9. Matsuzaka, K., Tohara, T., Nakada, K., Morie, T.: Analog CMOS circuit implementation of a pulse-coupled phase oscillator system and observation of synchronization phenomena. Nonlinear Theory Appl. IEICE **3**(2), 180–190 (2012)
10. Matsuzaka, K., Tanaka, H., Ohkubo, S., Morie, T.: VLSI implementation of coupled MRF model using pulse-coupled phase oscillators. Electron. Lett. **51**(1), 46–48 (2014)
11. Suedomi, Y., Tamukoh, H., Tanaka, M., Matsuzaka, K., Morie, T.: Parameterized digital hardware design of pulse-coupled phase oscillator model toward spike-based computing. In: Lee, M., Hirose, A., Hou, Z.-G., Kil, R.M. (eds.) ICONIP 2013. LNCS, vol. 8228, pp. 17–24. Springer, Heidelberg (2013). https://doi.org/10.1007/978-3-642-42051-1_3
12. Suedomi, Y., Tamukoh, H., Matsuzaka, K., Tanaka, M., Morie, T.: Parameterized digital hardware design of pulse-coupled phase oscillator networks. Neurocomputing **165**, 54–62 (2015)

13. Pramanta, D., Morie, T., Tamukoh, H.: Implementation of multi-FPGA communication using pulse-coupled phase oscillators. In: Proceedings of 2017 International Conference on Artificial Life And Robotics (ICAROB2017), pp. 128–131 (2017)
14. Li, J., et al.: A multidimensional configurable processor array-vocalise. IEICE Trans. Inf. Syst. **98**(2), 313–324 (2015)
15. Xilinx Inc.: Aurora 8B/10B Protocol Specification, howpublished. https://www.xilinx.com/support/documentation/ip_documentation/aurora_8b10b_protocol_spec_sp002.pdf. Accessed 28 Jun 2019
16. Pramanta, D., Morie, T., Tamukoh, H.: Synchronization of pulse-coupled phase oscillators over multi-FPGA communication links. J. Robot. Networking Artif. Life **4**(1), 91–96 (2017)
17. Athavale, A., Christensen, C.: High-speed serial i/o made simple. Xilinx inc., April 2005
18. RGB color gradient maker. http://www.perbang.dk/rgbgradient/. Accessed 28 Jun 2019

A New Approach to Automatic Heat Detection of Cattle in Video

Kitsuchart Pasupa[(✉)] and Thanawat Lodkaew

Faculty of Information Technology, King Mongkut's Institute of Technology
Ladkrabang, Bangkok 10520, Thailand
kitsuchart@it.kmitl.ac.th, lodkaew.thanawat@gmail.com

Abstract. Heat detection of cattle in video is essential for dairy farm. A
cow should be inseminated within a certain period of time in order for it
to breed successfully. After it has given birth to a calf, it produces milk.
This paper proposes the use of a set of discriminative features to detect
cattle in heat, where the features were extracted from the behaviours of
oestrus cow by a key-point analysis of locations of their body parts in a
video. We evaluated our proposed features, in terms of the algorithm's
classification accuracy of identifying cow in heat, with several machine
learning algorithms for two instances–using a global model and a number
of cattle-specific models to execute the identification. It was found that
Support Vector Machine with Radial Basis Function yielded a maximum
accuracy of 90.0% for the global model and 92.0% for the cattle-specific
models. These initial findings demonstrate that individual cows may have
different oestrus behaviours, a fact that would benefit any dairy farmers.
Our future development will be on a practical video monitoring and
detection system of cows in heat in a dairy farm.

Keywords: Behaviour analysis · Heat detection · Cattle

1 Introduction

In general, male and female cattle are reared separately in dairy farms. Techni-
cally, cow denotes female cattle. Dairy farms need female cattle more than male
cattle. Female cattle can give birth to a calf and produce milk, while male cattle
are only for meat production. Nevertheless, cows only produce milk after they
have given birth to a calf. Therefore, cows are required to be bred at the right
time because a proper breeding time comes only on one day in a month.

A conventional approach is for humans to manually detect the heat behaviour
of the cows. Heat behaviour in animals is a sexually receptive behaviour that
indicates breeding readiness. If a cow is in heat, an artificial insemination is
applied. Explicit visual signs of heat are such as mounting and sniffing events [1].
A cow may be in heat if it is mounting or sniffing another cow, and vice versa.
Using humans to detect this kind of event is laborious and costly. Long-term
continuous observation can easily lead to a detection failure due to observer's

© Springer Nature Switzerland AG 2019
T. Gedeon et al. (Eds.): ICONIP 2019, CCIS 1143, pp. 330–337, 2019.
https://doi.org/10.1007/978-3-030-36802-9_35

tiredness. If an in-heat cow is not detected and inseminated at the right time, the farm will have to wait until a month later to inseminate the same cow again. Failures to detect heat behaviour is a significant factor that leads to low dairy production and birthrate. In the same way, when a cow that is not in heat is falsely detected, the cost of time and money for semen will be wasted. Therefore, detection of cows in heat needs to be performed accurately so that the breeding process for oestrous cows can be carried out promptly and effectively.

There have been works in computer vision research that investigated cow behaviour [2–4]. Pasupa and his colleagues compared feature point matching and foreground detection methods to detect cow movement in a video [2]. They found that the foreground detection technique performed better than the other. Chowdhury et al. attempted to automatically monitor the heat of oestrus cows by utilising a deep learning technique to detect colour changes from a heat detector device [3]. The device was attached to the base of a cow's tail. If a cow is mounted by another cow, the detected colour changed. Their approach was effective with over 90% accuracy. Nevertheless, their approach did not consider other kinds of heat behaviours such as sniffing and movement. Jingqiu et al. proposed a technique to detect mounting behaviour by considering the intersection area between the minimum bounding boxes of cows' images [4]. The intersection area of the bounding boxes of two cows, one mounting the other, is different from the bounding boxes of other cows with non-oestrus behaviours. That work employed images with one-sided view to perform mounting detection, but did not taken sniffing behaviour into account as a visual sign of heat. Worse, in utilising the intersection area of bounding boxes, the approach may fail to detect this kind of heat behaviour if there is another cow behind them at a different depth in the image. Apart from works in cow behaviour research, there are some works that analyse pig's behaviour as well [5,6]. Nasirahmadi et al. proposed an automated method to detect mounting events of pigs in a video based on ellipse-fitted features [5]. They identified the head, tail, and sides of each pig by using different properties of fitted ellipse shapes. The distances from the identified parts of one pig to those of another pig were calculated to define the region of interest. If there is a mounting event, a new ellipse will form to a size of 1.3–2 pigs. Their results show that it is important to consider body parts in mounting events. Chen et al. proposed a computer vision-based method to recognise aggressive behaviour among pigs [6]. It was found that acceleration extraction between adjacent frames is useful for recognising their aggressive behaviour.

In this research, we propose a heat detection approach to identification of oestrus cows together with a new set of discriminative features of oestrus cows extracted from a video. The feature extraction principle was based on all mounting, sniffing, and motion behaviours. The obtained set of features were evaluated by a suite of machine learning algorithms.

The paper is organised as follows. Section 2 describes methods used in this work that includes a set of proposed features for identifying oestrous cow. Section 3 explains our experimental framework. Then, the results of application of the proposed features to classify oestrous cows are given and discussed in Sect. 4, followed by the conclusion in Sect. 5.

Fig. 1. The body parts that are useful for feature extraction–nose, body, and tail-head.

2   Methodology

The locations of the essential body parts of each cow in a video frame need to be defined first. After that, discriminative features that imply a certain heat behaviour of a cow can be extracted based on these locations. The details of body part extraction are described in Subsect. 2.1 and of feature extraction in Subsect. 2.2.

2.1   Body Part Extraction

There are many computer vision methods that can be used to detect body parts of humans or animals-head, body, arms, legs-in a video [7–10]. Once the body parts of a cow are detected, we can extract the proposed features. In this work, we selected nose, body, and base of tail (tail-head) as the considered parts (see Fig. 1). These parts can be used to track the motion of each cow in a corral. Moreover, cows actually make physical contact with each other with these parts, e.g., mounting, sniffing, butting heads, resting their chins on the hip of the others, hence the motions of these parts can help identify the oestrus behaviour of cows better than the motion of one part alone.

2.2   Hand-Crafted Feature Extraction

After the body parts of each cow are detected in each frame of the video, the process of feature extraction is applied. The location of each part is denoted as a single point in a Cartesian coordinate (x,y). Our proposed features can be divided into two main categories–individual-based and nearest neighbour-based features–as shown in Table 1.

Individual-Based Features. These features can, in turn, be divided into two groups, namely motion-based and coordinate-based features. Motion-based features capture motions of cows that are useful for detecting their heat behaviour. Cows in heat typically walk more because they are restless. This set of features comprises distance (F1–F3), velocity (F4–F6), and acceleration (F7–F9) of each body part of a cow. On the other hand, coordinate-based features (F10–F18) represent the appearance of each body part in a video frame. If one or more body parts, especially the body and base of tail, of a considered cow are blocked from view by body parts of another cow, they will not be detected. This means that there is a possibility of the following events: (i) a mounting event–the other cow is mounting the considered cow; (ii) sniffing event–the other cow is sniffing the considered cow; and (iii) resting event–the other cow is resting their chins on the hip or head of a considered cow. Because these two types of features had their own advantages and disadvantages, we used both to complement each other.

Nearest Neighbour-Based Features. This set of features are extracted based not only on the locations of the body parts of a considered cow, but also on the locations of the body parts of its nearest neighbouring cow. This feature may be able to capture their sniffing behaviour. If the distance between their body parts is close enough, then the considered cow is likely to sniff another cow or being sniffed by another cow.

3   Experimental Framework

3.1   Dataset

Our data were collected from Chokchai Farm in Khao Yai, Thailand. The camera was mounted at the centre of the ceiling of a corral in a top-view configuration. A fish-eye lens was used to increase the width of vision of the camera. This corral normally housed cows in heat before the insemination procedure. There were three cows in this corral. The video was recorded at 10 frames per second. A total of 2,850 images were extracted from the video for labelling purpose. Experts from Chokchai Farm labelled the body parts and recorded the in-heat or not-in-heat behaviour of each cow.

3.2   Experiment Settings

The data was split into training and test sets at 90:10 ratio. We evaluated our proposed features with various classification algorithms: Support Vector Machine (SVM) with linear and Radial Basis Function (RBF) kernels, Linear Discriminant Analysis (LDA), Multi-Layer Perceptron (MLP), and Random Forest (RF). However, most of the algorithms have hyperparameters that were required to be tuned. Therefore, we employed 10-fold cross-validation in conjunction with grid search to search for the optimal hyperparameters. The optimal models were trained by the training set with the optimal parameters and evaluated with the test set. The experiment was run 10 times with different random splits.

Table 1. The proposed set of features: individual-based features (F1–F18) and neighbour-based features (F19–F36). Features marked with * are calculated from the previous and the current frames.

#	Name	Description
F1*	distanceNoseFromPrevious	Distance between the point of nose
F2*	distanceBodyFromPrevious	Distance between the point of body
F3*	distanceTailHeadFromPrevious	Distance between the point of tail-head
F4*	velocityNose	Velocity of the point of nose
F5*	velocityBody	Velocity of the point of body
F6*	velocityTailHead	Velocity of the point of tail-head
F7*	accelerationNose	Acceleration of the point of nose
F8*	accelerationBody	Acceleration of the point of body
F9*	accelerationTailHead	Acceleration of the point of tail-head
F10	noseAppearance	Appearance of nose
F11	bodyAppearance	Appearance of body
F12	tail-headAppearance	Appearance of tail-head
F13	coordinateNoseX	x-coordinate of nose
F14	coordinateNoseY	y-coordinate of nose
F15	coordinateBodyX	x-coordinate of body
F16	coordinateBodyY	y-coordinate of body
F17	coordinateTailHeadX	x-coordinate of tail-head
F18	coordinateTailHeadY	y-coordinate of tail-head
F19	nearestDistNose2Nose	Nearest distance between nose of the considered cow to nose of another cow
F20	nearestDistNose2TailHead	Nearest distance between nose of the considered cow to tail-head of another cow
F21	nearsetDistNose2Body	Nearest distance between nose of the considered cow to body of another cow
F22	nearestDistTailHead2Nose	Nearest distance between tail-head of the considered cow to nose of another cow
F23	nearestDistTailHead2TailHead	Nearest distance between tail-head of the considered cow to tail-head of another cow
F24	nearestDistTailHead2Body	Nearest distance between tail-head of the considered cow to body of another cow
F25	nearestDistBody2Nose	Nearest distance between body of the considered cow to nose of another cow
F26	nearestDistBody2TailHead	Nearest distance between body of the considered cow to tail-head of another cow
F27	nearestDistBody2Body	Nearest distance between body of the considered cow to body of another cow
F28	nearestCowNose2Nose	A cow that its nose is nearest to nose of the considered cow
F29	nearestCowNose2TailHead	A cow that its nose is nearest to tail-head of the considered cow
F30	nearsetCowNose2Body	A cow that its nose is nearest to body of the considered cow
F31	nearestCowTailHead2Nose	A cow that its tail-head is nearest to nose of the considered cow
F32	nearestCowTailHead2TailHead	A cow that its tail-head is nearest to tail-head of the considered cow
F33	nearestCowTailHead2Body	A cow that its tail-head is nearest to body of the considered cow
F34	nearestCowBody2Nose	A cow that its body is nearest to nose of the considered cow
F35	nearestCowBody2TailHead	A cow that its body is nearest to tail-head of the considered cow
F36	nearestCowBody2Body	A cow that its body is nearest to body of the considered cow

Table 2. Performances of the global model (Accuracy and F_1-score) on detecting heat behaviour of all three cows, Best–bold.

Algorithm	Cow A		Cow B		Cow C		Average	
	Accuracy	F_1-score	Accuracy	F_1-score	Accuracy	F_1-score	Accuracy	F_1-score
SVM-RBF	**0.90**	**0.86**	**0.90**	**0.87**	**0.91**	**0.88**	**0.90**	**0.87**
SVM-Linear	0.79	0.71	0.70	0.55	0.79	0.68	0.76	0.65
RF	0.86	0.83	0.84	0.77	0.88	0.82	0.86	0.81
MLP	0.85	0.81	0.84	0.77	0.89	0.85	0.86	0.81
LDA	0.78	0.70	0.70	0.53	0.81	0.71	0.76	0.65

4   Results and Discussions

In this work, we evaluated our model in two instances: global and cow-specific models.

4.1   Global Model

In this instance, we trained the model with the obtained data from all three cows. The aim was to evaluate how well the model performed across all of the cows. Table 2 reports the performance of this model. Overall, SVM with RBF kernel was the best algorithm for detecting heat behaviour as it was able to achieve 90.33% average accuracy and 87.00% F_1-score. We also show the confusion matrix of SVM-RBF on the test set–average across all cows as shown in Table 3.

We further analysed the average weights of SVM-Linear that corresponded to our discriminative features as shown in Fig. 2. The top three discriminative features were velocityBody, accelerationBody, and accelerationTailHead. This means that velocity and acceleration extraction between adjacent frames were useful.

4.2   Cow-Specific Model

In this instance, we assumed that each cow had a specific model. Each model was trained with the training set of the considered cow and then tested with its test set. The results are shown in Table 4. Again, SVM-RBF yielded the best performance among all models. It can be clearly seen that the performances of cow-specific models were higher than the global model for all algorithms, although the number of training samples used in the cow-specific models was smaller. This indicates that each cow had its own different behaviour. To consider together all training samples from all cows, instead of from one certain cow alone, could degrade the performance.

Table 3. Confusion matrix of SVM-RBF global model on the test data of all cows, average across 10 runs.

Actual	Predicted		Accuracy
	Heat	Non-heat	(%)
Heat	91	13	87.50
Non-heat	11	170	93.92

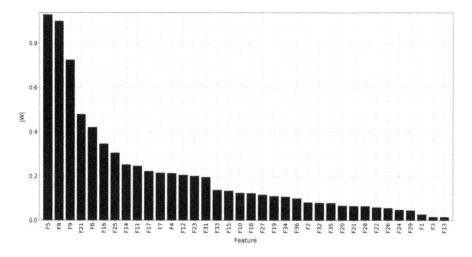

Fig. 2. The average weights of the global model by SVM-Linear across 10 runs.

Table 4. Performances of the cow-specific model (Accuracy and F_1-score) on detecting heat behaviour of all three cows, Best–bold.

Algorithm	Cow A		Cow B		Cow C		Average	
	Accuracy	F_1-score	Accuracy	F_1-score	Accuracy	F_1-score	Accuracy	F_1-score
SVM-RBF	**0.91**	**0.89**	**0.92**	**0.88**	**0.92**	**0.88**	**0.92**	**0.88**
SVM-Linear	0.81	0.75	0.79	0.68	0.84	0.76	0.81	0.73
RF	0.90	0.87	0.90	0.86	0.92	0.82	0.91	0.85
MLP	**0.91**	**0.89**	0.91	0.86	0.91	0.87	0.91	0.87
LDA	0.78	0.71	0.75	0.63	0.84	0.75	0.79	0.70

5   Conclusion

This work proposes a set of hand-crafted features to detect heat behaviour in cows in a video. To obtain these features, body part extraction was performed to detect the nose, body, and base of tail of each cow in a video frame. Then, the discriminative features were extracted. These features can be divided into two categories: individual-based and nearest neighbour-based features. We evaluated the proposed features in two instances: global and cow-specific models. It was found that the cow-specific models was able to achieve a higher performance than the global model.

References

1. O'Connor, M.L.: Heat detection and timing of insemination for cattle. Penn State Extension (2016). https://extension.psu.edu/heat-detection-and-timing-of-insemination-for-cattle
2. Pasupa, K., Pantuwong, N., Nopparit, S.: A comparative study of automatic dairy cow detection using image processing techniques. Artif. Life Robot. **20**(4), 320–326 (2015)
3. Chowdhury, S., Verma, B., Roberts, J., Corbet, N., Swain, D.: Deep learning based computer vision technique for automatic heat detection in cows. In: Proceedings of the International Conference on Digital Image Computing: Techniques and Applications (DICTA), Gold Coast, QLD, Australia, pp. 1–6 (2016)
4. Jingqiu, G., Zhihai, W., Ronghua, G., Huarui, W.: Cow behavior recognition based on image analysis and activities. Int. J. Agric. Biol. Eng. **10**(3), 165–174 (2017)
5. Nasirahmadi, A., Hensel, O., Edwards, S.A., Sturm, B.: Automatic detection of mounting behaviours among pigs using image analysis. Comput. Electron. Agric. **124**, 295–302 (2016)
6. Chen, C., Zhu, W., Ma, C., Guo, Y., Huang, W., Ruan, C.: Image motion feature extraction for recognition of aggressive behaviors among group-housed pigs. Comput. Electron. Agric. **142**, 380–387 (2017)
7. Gu, J., Lan, C., Chen, W., Han, H.: Joint pedestrian and body part detection via semantic relationship learning. Appl. Sci. **9**(4), 752 (2019)
8. Ramanathan, M., Yau, W.Y., Teoh, E.K.: Improving human body part detection using deep learning and motion consistency. In: Proceedings of the International Conference on Control, Automation, Robotics and Vision (ICARCV), Phuket, Thailand, pp. 1–5 (2016)
9. Zhang, H., et al.: SPDA-CNN: unifying semantic part detection and abstraction for fine-grained recognition. In: Proceedings of the IEEE Conference on Computer Vision and Pattern Recognition (CVPR), Las Vegas, NV, USA, pp. 1143–1152 (2016)
10. Chen, X., Mottaghi, R., Liu, X., Fidler, S., Urtasun, R., Yuille, A.L.: Detect what you can: detecting and representing objects using holistic models and body parts. CoRR abs/1406.2031 (2014)

Accurate Localization Algorithm in Wireless Sensor Networks in the Presence of Cross Technology Interference

Usman Nazir$^{(\boxtimes)}$ ⬤, Ijaz Haider Naqvi ⬤, and Murtaza Taj

Lahore University of Management Sciences, Lahore, Pakistan
{17030059,ijaznaqvi,murtaza.taj}@lums.edu.pk

Abstract. Localization of mobile nodes in wireless sensor networks (WSNs) is an active area of research. In this paper, we present a novel RSSI based localization algorithm for 802.15.4 (ZigBee) based WSNs. We propose and implement a novel range based localization algorithm to minimize cross technology interference operating in the same band. The goal is to minimize the mean square error of the localization algorithm. Hardware implementation of the algorithm is in agreement with ideal (no interference) simulation results where an accuracy of less than 0.5 m has been achieved.

Keywords: Wireless sensor network · Localization · 802.15.4 · RSSI · Kalman filter · ISM · CTI · Range-based localization · ZigBee

1 Introduction

Physical location of mobile nodes is often required for a large number of applications in wireless sensor networks (WSNs). Localization techniques [3] can be broadly classified into range free or range based, anchor free or anchor based and distributed or centralized techniques. Centralized localization techniques transfer the entire data to a centralized node, where the localization algorithm estimates the position for all of the mobile nodes. These techniques have a lot of communication overhead. In the distributed localization techniques, the mobile nodes are themselves capable of estimating their location. The location of the mobile node is transferred to the central node only in case of an event or a sink initiated query, depending on the network design, thereby reducing the communication cost significantly. Anchor free techniques do not require beacon signals from the anchor nodes but offer very limited localization accuracy whereas anchor based techniques require beacon signals from anchor nodes of known location. Range free techniques rely on attributes such as hop count, connectivity information etc. Although, these are cost effective techniques but localization results are not accurate. The Range based techniques rely on received signal strength indicator (RSSI), time of arrival (ToA), angle of arrival (AoA), time difference of arrival (TDoA) etc. ToA, AoA and TDoA, have all better localization accuracy

© Springer Nature Switzerland AG 2019
T. Gedeon et al. (Eds.): ICONIP 2019, CCIS 1143, pp. 338–346, 2019.
https://doi.org/10.1007/978-3-030-36802-9_36

Table 1. Specifications of SYNAPSE RF266 Module

Parameters	Value
Indoor Range	Up to 200 ft. at 250 kbps
Outdoor LOS Range	Up to 4000 ft. at 250 kbps
Transmit Power Output	20 dBm
RF Data Rate	250 kbps, 500 kbps, 1 Mbps, 2 Mbps
Receiver Sensitivity	−107 dBm
Supply Voltage	2.7–3.6 V
Transmit Current (Typ)	130 mA
Receive Current (Typ)	25 mA
Sleep Current (Typ)	1.18 μA (internal timer off) & 2.3 μA (internal timer on)
Topology	Mesh (SNAP)
Number of Channels	16

as compared to RSSI based techniques but require additional hardware for their implementation.

The RSSI based localization schemes are by far the most popular localization techniques employed for monitoring the location of mobile sensor nodes as it requires minimal, low cost hardware for its implementation. In RSSI based localization implementations, RF transceivers capable of measuring RSSI are used for location estimation. Multiple technologies for instance IEEE 802.15.4 (ZigBee), IEEE 802.15.1 (Bluetooth) and IEEE 802.11 (WiFi) all share the same 2.4 GHz ISM band [4]. The interference caused by these technologies causes RSSI to fluctuate rapidly, causing packet collisions and affecting the localization accuracy. Addressing cross technology interference (CTI) is therefore critical for systems operating in license free ISM bands. The highest CTI for a Zigbee based radio is caused by Wifi signals, because Wifi is often co-located with IEEE 802.15.4 networks.

In order to minimize RF interference at the receiver front end, we use an aggregation function to calculate average RSSI values. We have used SYNAPSE RF266 sensor nodes having 802.15.4 radios and specifications shown in Table 1. Detailed specifications can be found in [1]. ZigBee and WiFi radio channels are shown in Fig. 1. As shown, the bandwidth of each ZigBee and Wifi channel is 2 MHz and 22 MHz respectively, whereas separation between two adjacent ZigBee and Wifi Channels is 5 MHz [6]. In Wifi band, nonoverlapping channels are used in Europe and North America [7]. Typically, channels with high interference have higher mean RSSI values [2]. Therefore, ZigBee channel with the lowest average RSSI value is selected to acquire minimal RF interference with Wifi signals [2]. In addition to RF interference, the received signal also gets affected by White Gaussian noise at the receiver front end which has an infinite spectrum and thus infinite energy. To cater this problem, we introduce Kalman filtering at the receiver front end. We show that if RSSI after being processed for interference minimization and noise filtering is given as an input to a trilateration localization algorithm, mean square error of the localization algorithm is minimized.

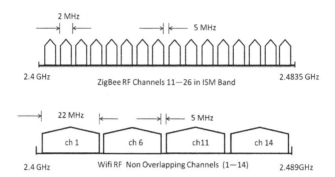

Fig. 1. ZigBee and WiFi RF channels in 2.4 GHz band. ZigBee channel is 2 MHz wide and WiFi channel is 22 MHz wide

The main contributions of this paper have been summarized as follows:

- We introduce a novel cross technology interference minimization scheme with trilateration for an improved accuracy.
- We propose the use of Kalman filter at the receiver front for noise minimization of localization algorithms.
- Hardware implementation of the localization algorithm with minimal CTI and noise reduction.

The remainder of the paper is organized as follows. Section 2 explains RSSI based localization algorithm along with CTI minimization and noise reduction techniques. Complete system design with hardware implementation has been presented in Sect. 3. Discussion on experimental results has been elaborated in Sect. 4. Node deployment strategy is discussed in Sects. 5 and 6 outlines the future directions and concludes this paper.

2   Proposed Localization Algorithm

As discussed in Sect. 1, localization algorithms can be classified into 3 categories. Each type of localization technique has its advantages or disadvantages. A detailed discussion on the merits and demerits of localization techniques is out of scope of this paper, however a summary of the state of the art for localization algorithms has been provided in previous works [3,5]. In this paper we use a distributed, range and anchor based localization algorithm to localize a target (mobile) node.

2.1 Log Distance Path Loss Model

The algorithm runs on the mobile node where RSSI of neighboring nodes is needed. In our algorithm we use an aggregation function, which averages the received RSSI over multiple acquisitions, to verify the RF interference between ZigBee and Wifi Channels. The ZigBee channel with lowest mean RSSI Value gets selected for communication. RSSI values of neighboring nodes are acquired at the mobile node over the selected frequency (channel). The mobile node then makes use of the log distance path loss model, given below in Eq. 1, to calculate its distance to the neighboring transmitting nodes.

$$RSSI(d) = RSSI(d_o) - 10n \times log(d/d_o) \tag{1}$$

Where d is the distance between transmitter and the receiver, d_0 is a reference distance which we is assumed to be 1 m. $RSSI(d_0)$ is the RSSI at reference distance taken as -45 dBm and $RSSI(d)$ is the received signal strength at distance d. Since the trilateration approach has been used to estimate the location of sensor node, RSSI value for at least three neighboring nodes is required for a valid location estimate of the mobile node. Once the location estimate from trilateration algorithm is received, a Kalman Filter is used to minimize the mean squared error. A flow chart of the localization process has been presented in the Fig. 2a.

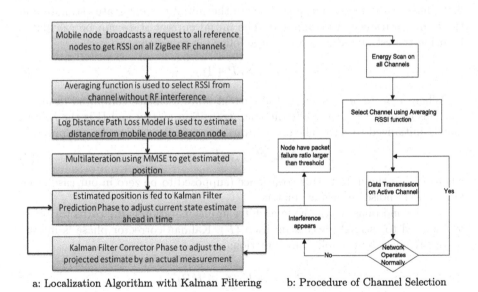

a: Localization Algorithm with Kalman Filtering   b: Procedure of Channel Selection

Fig. 2. Proposed algorithms

2.2   Trilateration

In a trilateration algorithm along with the minimum mean square estimation (MMSE), the position of the mobile node can be computed by solving the following equation:

$$\hat{P} = \begin{bmatrix} \hat{x} \\ \hat{y} \end{bmatrix} = (A^T A)^{-1} A^T B \qquad (2)$$

Where $\hat{P} = (\hat{x}, \hat{y})^T$ denotes the estimated target location, and the matrices A and B are defined as

$$A = \begin{bmatrix} 2(x_2 - x_1) \ 2(y_2 - y_1) \\ 2(x_3 - x_1) \ 2(y_3 - y_1) \end{bmatrix} \qquad (3)$$

$$B = \begin{bmatrix} d_{1,target}^2 - d_{2,target} + (x_2^2 + y_2^2 - x_1^2 - y_1^2) \\ d_{1,target}^2 - d_{3,target} + (x_3^2 + y_3^2 - x_1^2 - y_1^2) \end{bmatrix} \qquad (4)$$

Where (x_i, y_i) gives the position of the beacon nodes. Note that for the implementation of trilateration algorithm, position of at least three beacon nodes is required.

2.3   Kalman Filtering

The estimated position, $\hat{P}$, is then filtered through a Kalman filter to reduce the effects of the noise. The Kalman filter implementation has two phases. The first phase is the prediction phase where the initial current state estimate and the error covariance are computed. The initial current state can be calculated by using the estimated location by the following equation:

$$\hat{\hat{P}} = S_T \hat{P} + \Pi \qquad (5)$$

Where $\hat{\hat{P}} = (\hat{\hat{x}}, \hat{\hat{y}})^T$ is the initial current state estimate computed from the previous estimate $\hat{P}$, S_T. S_T is the state transition matrix and Π is the control matrix initialized to zero. The error covariance matrix can then be computed as

$$\hat{E} = S_T E_{int} S_T^T + Q \qquad (6)$$

where E_{int} is the initial error covariance (supposed to be zero in our case), and Q is process noise covariance matrix.

The second phase of the Kalman filter implementation is the corrector phase and is used to estimate actual position (P). Kalman corrector phase is mathematically given by the following equation:

$$P = \hat{\hat{P}} + K_k(z_k - H\hat{\hat{P}}) \qquad (7)$$

where K_k is Kalman gain expressed in Eq. (9), H is the observation matrix and z_k is the measurement vector given by following equation:

$$z_k = \begin{pmatrix} \sqrt{(\hat{x} - x_1)^2 + (\hat{y} - y_1)^2} \\ \sqrt{(\hat{x} - x_2)^2 + (\hat{y} - y_2)^2} \\ \sqrt{(\hat{x} - x_3)^2 + (\hat{y} - y_3)^2} \end{pmatrix} \qquad (8)$$

$$K_k = \frac{\hat{E}H^T}{H\hat{E}H^T + R} \tag{9}$$

In Eq. 9, R is the estimated measurement error covariance (environmental noise) and $\hat{E}$ is the predicted error covariance. The environmental noise matrix describes the noise inferred on data from the sources that lie within the path from the sensed object to the filter. Error covariance is iteratively updated using the Eq. 10.

$$\hat{E} = (1 - K_k H)\hat{E} \tag{10}$$

3   Hardware Implementation of the Proposed Algorithm

In this section, we describe the complete software and hardware implementation of localization algorithm. We make use of the wireless programming feature of the off the shelf SYNAPE sensor mote to program them Over-the-Air. The software used for programming is denoted as 'portal' and is the SYNAPSE designated software. All the nodes can be programmed simultaneously via portal. Each node is capable of receiving RSSI values from its neighboring nodes, distributed in the region of interest (ROI). Remote Procedure Calls (RPCs) are transmitted by mobile (target) node on all of ZigBee channels. The mobile node then broadcasts a request to all reference nodes which are in its range of the target node to receive their RSSI values. Thereafter, the target node tries to find out the channel with minimum RF interference from Wifi radio transmitters co-located in the environment. The built in functions of SYNAPSE motes can be used to find out the channel from which the target is receiving minimum mean energy. The RSSI value for each anchor or beacon node is measured at 100 ms intervals and then averaged out over a delay of 1 s. The aggregated RSSI value is then fed into the localization algorithm.

3.1   Channel Selection

In order to minimize CTI, the channel receiving minimum RF energy gets selected. The procedure for channel selection has been shown in Fig. 2b. First of all, the target node scans energy on all the ZigBee channels and acquires RSSI with some sampling rate. The acquired values are then averaged to minimize the effect of random fluctuations. Thereafter, the target node selects the channel with minimum interference and initializes this channel in active mode to establish communication with anchor nodes. Target node monitors this active channel continuously for any possible interference. As soon as the interference is detected on this active channel, i.e. the packet failure ratio crosses a certain set threshold, the process of channel selection gets repeated (see Fig. 2b).

3.2   Implementation of Localization Algorithm

As mentioned previously, the localization algorithm has been implemented using off the shelf SYNAPSE sensor motes [1]. In experimental setup, the target node is attached to a Laptop where SYNAPSE Portal software is running. Target node scans all ZigBee channels using built in functions and selects the channel with minimum interference, as described previously. After selection of channel with minimum RF interference, target nodes communicates to its neighbors and acquires the RSSI values of at least three neighbors (beacon) nodes. Thereafter, trilateration localization technique is used to calculate the position of target node. The mobile node can also be programmed without connecting it to the laptop using SNAPpy language through the Portal's Over-the-Air programming feature. The localization algorithm however, remains the same. The target first automatically scans the ZigBee channels and selects one channel for communication with beacon nodes. After getting average RSSI values from at least three beacons it implements trilateration algorithm to calculate its position and uses Kalman Filter for improved position accuracy.

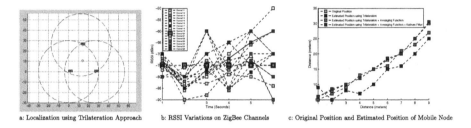

a: Localization using Trilateration Approach   b: RSSI Variations on ZigBee Channels   c: Original Position and Estimated Position of Mobile Node

Fig. 3. Experimental results

4   Experimental Results and Discussions

A snapshot of trilateration localization algorithm, implemented in the absence of Wifi interference with the SYNAPSE sensor motes, has been shown in Fig. 3a, where green is the target node and red nodes are anchor nodes. If the target node receives RSSI from more than three neighbor nodes, target node selects three neighbors with maximum RSSI. Log Distance Path Loss model (Eq. 1) is then used to find out its distances to anchor nodes and its location using trilateration algorithm. Thereafter, Kalman filter is implemented using Eqs. 5 and 7 to minimize the mean squared error of the estimated position and improve the localization accuracy. As shown in Fig. 3a, anchor nodes have positions (0,0), (30,0) and (15, 30) in cartesian coordinates respectively. Target node receives RSSI of −60 dBm from all anchor nodes where as the reference RSSI is measured to be −45 dBm at 1 m distance.

The experiments are then conducted in the presence of RF interference from Wifi transceivers. Using the built in channel analyzer tool for ZigBee motes, RSSI values, prior to any transmission, detected on all channels have been shown in Fig. 3b. RSSI values for channels 11–26 have been shown in the Fig. 3b but channel 20 has lowest variations in RSSI values w.r.t. time. Therefore, channel 20 has been selected for Zigbee transmission by the target node to communicate with the anchor nodes. Once the channel gets selected, trilateration localization algorithm is invoked in a similar manner as in the case of no interference.

Figure 3c presents the results of the trilateration algorithm with and without applying the averaging function. The curve also presents the result of localization algorithm after employing both averaging as well as Kalman filter. The deviation of the estimated position from the original position can be compared for different schemes. If we simply use Trilateration technique to estimate target position, accurate estimation of the target position is not possible. If we find estimated position of target node using Trilateration technique with the averaging function, we get a better estimate of the original position but still the deviation cannot be ignored (see Fig. 3c). Finally, if we apply Kalman filter along with the averaging function, the estimated and the original position are in good agreement with each other. Thus, with Kalman filtering, we get minimal deviation between estimated position and original positions. The result show that we achieve a localization accuracy of less than 0.5 m with proposed algorithm.

5   Beacon Node Deployment Strategy

Based on the experimental results, we proposed a nodes deployment strategy for indoor environment. With this deployment strategy, mobile node will always have connectivity with at least three beacon nodes to fulfill the requirement of trilateration based localization technique. As shown in Fig. 4, where green is the target node and red nodes are beacon nodes. If the target node receives RSSI from more than three neighbor nodes, target node selects three neighbors with maximum RSSI. Log Distance Path Loss model (Eq. 1) is then used to find out its distances to anchor nodes and its location using trilateration algorithm. Using this deployment strategy in the region of interest, mobile node will always be in the range of at least three beacon nodes.

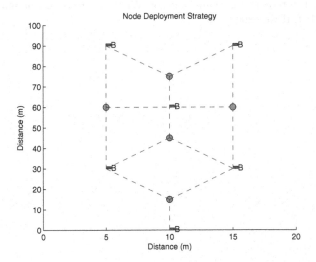

Fig. 4. Beacon node deployment strategy

6   Conclusion and Future Work

In this paper, a novel RSSI based localization technique has been introduced and implemented. The proposed scheme minimizes cross technology interference (CTI)and apply noise reduction filters to achieve very high localization accuracy. Hardware implementation of the proposed scheme has been carried out and the result suggests that the proposed scheme can easily be implemented on Zigbee based sensor motes. The combined CTI minimization and noise reduction technique achieves a localization accuracy of less than 0.5 m on commercial off the shelf sensor motes. Future work includes addressing CTI in networks using ISM band with the help of Machine Learning algorithms and Kalman Filtering.

References

1. http://www.synapse-wireless.com
2. Musaloiu-E, R., Terzis, A.: Minimising the effect of WiFi interference in 802.15.4 wireless sensor networks. Int. J. Sens. Netw. **3**(1), 43–54 (2008)
3. Nazir, U., Shahid, N., Arshad, M., Raza, S.: Classification of localization algorithms for wireless sensor network: a survey. In: 2012 International Conference on Open Source Systems and Technologies (ICOSST), pp. 1–5. IEEE (2012)
4. Noda, C., Prabh, S., Alves, M., Boano, C.A., Voigt, T.: Quantifying the channel quality for interference-aware wireless sensor networks. SIGBED Rev. **8**(4), 43–48 (2011)
5. Paul, A., Sato, T.: Localization in wireless sensor networks: a survey on algorithms, measurement techniques, applications and challenges. J. Sens. Actuator Netw. **6**(4), 24 (2017)
6. Tas, N.C., Sastry, C.R., Song, Z.: IEEE 802.15. 4 throughput analysis under IEEE 802.11 interference. In: International Symposium on Innovations and Real Time Applications of Distributed Sensor Networks (2007)
7. Yang, G., Yu, Y.: Zigbee networks performance under WLAN 802.11 b/g interference. In: 4th International Symposium on Wireless Pervasive Computing, ISWPC 2009, pp. 1–4. IEEE (2009)

On Handling Missing Values in Data Stream Mining Algorithms Based on the Restricted Boltzmann Machine

Maciej Jaworski[1]([✉]) [ID], Piotr Duda[1] [ID], Danuta Rutkowska[2] [ID],
and Leszek Rutkowski[1] [ID]

[1] Institute of Computational Intelligence,
Czestochowa University of Technology, Czestochowa, Poland
{maciej.jaworski,piotr.duda,leszek.rutkowski}@iisi.pcz.pl
[2] Information Technology Institute, University of Social Sciences, Łódź, Poland
drutkowska@san.edu.pl

Abstract. This paper addresses the issue of data stream mining using the Restricted Boltzmann Machine (RBM). Recently, it was demonstrated that the RBM can be useful as a concept drift detector in data streams with time-changing probability density. In this paper, we consider another problem which often occurs in real-life data streams, i.e. incomplete data. We propose two modifications of the RBM learning algorithms to make them able to handle missing values. The first one inserts an additional procedure before the positive phase of the Contrastive Divergence. This procedure aims at inferring the missing values in the visible layer by performing a fixed number of Gibbs steps. The second modification introduces dimension-dependent sizes of minibatches in the stochastic gradient descent method. The proposed methods are verified experimentally, demonstrating their usability for concept drift detection in data streams with incomplete data.

Keywords: Restricted Boltzmann Machine · Data stream mining · Missing values

1 Introduction

In recent years data stream mining became a very interesting and challenging branch of data mining [3,14–16]. In this paper, we define the data stream as a sequence of data elements

$$S = (\mathbf{s}_1, \mathbf{s}_2, \dots), \tag{1}$$

which potentially can be of infinite size. Each data element is a D-dimensional vector of binary values

$$\mathbf{s}_n = [s_{n,1}, \dots, s_{n,D}] \in \{0; 1\}^D \tag{2}$$

A proper data stream mining algorithm should ensure the best trade-off between the accuracy and resources consumption. In the literature, many algorithms

© Springer Nature Switzerland AG 2019
T. Gedeon et al. (Eds.): ICONIP 2019, CCIS 1143, pp. 347–354, 2019.
https://doi.org/10.1007/978-3-030-36802-9_37

based on traditional machine learning or data mining tools have been proposed, e.g. neural networks with the stochastic gradient descent method [4], decision trees [10] or ensemble methods [12].

The problem of data stream mining becomes more challenging if the underlying data distribution can change over time [17]. In this paper, we focus on the issue of applying the Restricted Boltzmann Machine (RBM) to detect possible changes in the data distribution. This idea was first proposed in [8], and extended in [9] to allow dealing with labeled data. In [11] the resource-awareness of the RBM in data stream scenario was investigated. In this paper, we continue the topic by proposing modifications of the RBM learning algorithm to handle data streams with missing values.

The RBM is a special type of a wider class of neural networks called Boltzmann Machines [7]. It consists of two layers of neurons: the visible one, consisting of D neurons $\mathbf{v} = [v_1, \ldots, v_D]$ and the hidden one, which is formed by H hidden units $\mathbf{h} = [h_1, \ldots, h_H]$. For each possible state $(\mathbf{v}, \mathbf{h})$ of the RBM an energy can be calculated, which is defined as follows

$$E(\mathbf{v}, \mathbf{h}) = -\sum_{i=1}^{D} v_i a_i - \sum_{j=1}^{H} h_j b_j - \sum_{i=1}^{D} \sum_{j=1}^{H} v_i h_j w_{ij}, \tag{3}$$

where w_{ij}, a_i and B_j are RBM weights and biases. The energy function is used to define a probability distribution of $(\mathbf{v}, \mathbf{h})$

$$P(\mathbf{v}, \mathbf{h}) = \frac{\exp\left(-E(\mathbf{v}, \mathbf{h})\right)}{Z}, \tag{4}$$

where Z is a normalization constant. Let us assume that the data stream (1) is partitioned into minibatches of size B, i.e. the t-th minibatch is given by

$$S_t = (\mathbf{s}_{Bt+1}, \ldots, \mathbf{s}_{Bt+B}), \ t = 0, 1, \ldots. \tag{5}$$

Then, the cost function for S_t is given by the following formula

$$C(S_t) = -\log P(S_t) = -\frac{1}{B} \sum_{n=1}^{B} \sum_{\mathbf{h}} \log P(\mathbf{v} = \mathbf{s}_{Bt+n}, \mathbf{h}) \tag{6}$$

and its gradient with respect to weight w_{ij} is expressed as follows (see. e.g. [2,4])

$$\frac{\partial C(S_t)}{\partial w_{ij}} = \sum_{\mathbf{v},\mathbf{h}} P(\mathbf{v}, \mathbf{h}) v_i h_j - \frac{1}{B} \sum_{n=1}^{B} \sum_{\mathbf{h}} P(\mathbf{h}|\mathbf{v} = \mathbf{s}_{Bt+n}) v_i h_j. \tag{7}$$

The first term on the right-hand side ('negative phase'), is intractable to compute and can be approximated by the Contrastive Divergence (CD) algorithm [5]. In this paper we propose some modifications of the CD algorithm, allowing the RBM to handle incomplete data.

The rest of the paper is organized as follows. In Sect. 2 the CD algorithm for learning the RBM is recalled. It is shown how it is used for approximating the gradient of the RBM cost function. In Sect. 3 two modifications are proposed which allow the RBM to handle incomplete data. Preliminary results of experimental verification of presented methods are demonstrated in Sect. 4. Conclusions are discussed in Sect. 5.

2   Contrastive Divergence Learning Algorithm

As can be seen in (7), the gradient of the cost function $\frac{\partial C}{\partial w_{ij}}$ consists of two terms. Each term is based on sampling from different probability distributions. The second term, called the 'positive phase', requires the procedure of inferring the states of the hidden units from the data element, which is presented in Algorithm 1.

Algorithm 1: Hidden layer inference based on a data element

infer($\mathbf{s}$):
$\mathbf{v} \leftarrow \mathbf{s}$;
for $j \leftarrow 1$ to H do
$\quad | \quad h_j \leftarrow P(h_j|\mathbf{v})$;
end

The first term of gradient $\frac{\partial C}{\partial w_{ij}}$, called the 'negative phase', is intractable to compute. In the CD algorithm, it is approximated by performing a Gibbs sampling algorithm [1], presented in Algorithm 2.

Algorithm 2: Gibbs sampling

GibbsSampling(K):
for $k \leftarrow 1$ to K do
$\quad$ for $i \leftarrow 1$ to D do
$\quad\quad | \quad v_i \leftarrow P(v_i|\mathbf{h})$;
$\quad$ end
$\quad$ for $j \leftarrow 1$ to H do
$\quad\quad h_j \leftarrow P(h_j|\mathbf{v})$;
$\quad\quad$ if $k < K$ then
$\quad\quad\quad h_j \leftarrow 1$ with prob.
$\quad\quad\quad h_j$, otherwise
$\quad\quad\quad h_j \leftarrow 0$;
$\quad\quad$ end
$\quad$ end
end

Algorithm 3: Gradients updating

updateGradients(sgn):
for $i \leftarrow 1$ to D do
$\quad$ for $j \leftarrow 1$ to H do
$\quad\quad \frac{\partial C}{\partial w_{ij}} \leftarrow$
$\quad\quad \frac{\partial C}{\partial w_{ij}} + sgn\frac{1}{B}v_i h_j$;
$\quad$ end
end
for $i \leftarrow 1$ to D do
$\quad | \quad \frac{\partial C}{\partial a_i} \leftarrow \frac{\partial C}{\partial a_i} + sgn\frac{1}{B}v_i$;
end
for $j \leftarrow 1$ to H do
$\quad | \quad \frac{\partial C}{\partial b_j} \leftarrow \frac{\partial C}{\partial b_j} + sgn\frac{1}{B}h_j$;
end

Algorithm 4: The Contrastive Divergence algorithm (CD)

$\text{CD}(S_t, K)$:
$\frac{\partial C}{\partial w_{ij}} = 0$, $\frac{\partial C}{\partial a_i} = 0$, $\frac{\partial C}{\partial b_{j=0}}$, $i = 1, \ldots, D$, $j = 1, \ldots, H$;
for $s \in S_t$ **do**
  infer($\mathbf{s}$);
  updateGradients(-1);
  GibbsSampling(K);
  updateGradients(1);
end

For both phases the gradient values can be updated using the procedure presented in Algorithm 3 (where $sgn = -1$ and $sgn = 1$ correspond to the positive and negative phases, respectively). Finally, the CD algorithm for minibatch S_t, consisting of all mentioned previously components, is presented in Algorithm 4.

3   RBM for Handling Incomplete Data

Algorithm 5: Missing values restoring

Restore($\mathbf{s},Q,\mathbf{m}$):
$\mathbf{v} \leftarrow s$;
for $q \leftarrow 1$ **to** Q **do**
  for $j \leftarrow 1$ **to** H **do**
    $h_j \leftarrow P(h_j|\mathbf{v})$;
    $h_j \leftarrow 1$ with prob.
    h_j, otherwise
    $h_j \leftarrow 0$;
  end
  for $i \leftarrow 1$ **to** D **do**
    if $m_i == TRUE$
    then $v_i \leftarrow P(v_i|\mathbf{h})$;
  end
end
Return $\mathbf{v}$;

Algorithm 6: Gradients updating with the masks of missing values taken into account

updateGradientsMasked(sgn, $\mathbf{m}$, $\mathbf{B}$):
for $i \leftarrow 1$ **to** D **do**
  for $j \leftarrow 1$ **to** H **do**
    if $m_i == FALSE$
    then $\frac{\partial C}{\partial w_{ij}} \leftarrow$
    $\frac{\partial C}{\partial w_{ij}} + sgn\frac{1}{B_i}v_ih_j$;
  end
end
for $i \leftarrow 1$ **to** D **do**
  if $m_i == FALSE$ **then**
  $\frac{\partial C}{\partial a_i} \leftarrow \frac{\partial C}{\partial a_i} + sgn\frac{1}{B_i}v_i$;
end
for $j \leftarrow 1$ **to** H **do**
  $\frac{\partial C}{\partial b_j} \leftarrow \frac{\partial C}{\partial b_j} + sgn\frac{1}{B}h_j$;
end

In the practical guide for training RBMs [6] several methods for inferring missing values were proposed. However, none of them seems to work fast enough to be suitable for data stream mining tasks. In the sequel, we propose two modifications of the CD algorithm to make it able to handle data streams with missing values.

For each minibatch of data elements S_t we assume that there exists a minibatch of masks $M_t = (\mathbf{m}_{Bt+1}, \ldots, \mathbf{m}_{Bt+B})$. Each mask $\mathbf{m}_n$ is a D-dimensional vector of $\{TRUE, FALSE\}$ values. If $m_{n,i}$ is $TRUE$, then the value of $s_{n,i}$ is unknown. When necessary, by default this value is assumed to be equal to 0, until it is not restored. The first modification of the basic CD algorithm is to introduce a restoring function, presented in Algorithm 5. This procedure performs Gibbs sampling, however, only unknown units of the visible layer are updated.

The second proposed modification changes the gradients updating method. In the basic CD method, updates of gradients are calculated as the arithmetic average over the whole minibatch of data (as in Algorithm 3). In our approach, we introduce variable-sized minibatches. The size of the minibatch for the i-th dimension is equal to the number of data elements, for which the mask in the i-th dimension is $FALSE$

$$B_i(M_t) = \sum_{n=1}^{B} 1_{\{m_{Bt+n,i}==FALSE\}}. \qquad (8)$$

Let $\mathbf{B} = (B_1, \ldots, B_D)$ be a D-dimensional vector of dimension-dependent mini-batch sizes. Then the method for gradients update, which takes the missing values into account, is presented in Algorithm 6.

The final form of the Contrastive Divergence algorithm for data with missing values, which we abbreviate here as CDM, is presented in Algorithm 7.

Algorithm 7: The Contrastive Divergence algorithm for data with missing values (CDM)

CDM(S_t, M_t, K, Q, $Rest$, $PosMask$, $NegMask$):

$\frac{\partial C}{\partial w_{ij}} = 0$, $\frac{\partial C}{\partial a_i} = 0$, $\frac{\partial C}{\partial b_j} = 0$, $i = 1, \ldots, D$, $j = 1, \ldots, H$;

$\mathbf{B} = 0$;

for $m \in M_t$ **do**
    for $i \leftarrow 1$ **to** D **do**
        if $m_i == FALSE$ **then** $B_i + +$;
    end
end

for $(s, m) \in (S_t, M_t)$ **do**
    if $Rest == TRUE$ **then** $s = Restore(s, Q, m)$;
    infer(s);
    if $PosMask == TRUE$ **then** $updateGradientsMasked(-1, m, B)$;
    else updateGradients(-1);
    GibbsSampling(K);
    if $NegMask == TRUE$ **then** $updateGradientsMasked(1, m, B)$;
    else updateGradients(1);
end

Comparing to the standard CD algorithm it requires several additional arguments. These are the minibatch of masks M_t, corresponding to the minibatch of data S_t, and the number of steps Q of the restoring procedure. Three last parameters, i.e. $Rest$, $PosMask$, and $NegMask$ are boolean flags, which allow to turn on or off previously discussed modifications in the CDM algorithm.

4   Experimental Results

In this section, we present some preliminary results of the experimental verification of the presented methods. The numerical simulations were carried out on the MNIST dataset [13]. It contains 60000 gray-scale images of handwritten digits of size 28×28. In experiments, we treat the dataset as a stream. The data order is mixed randomly. Then, it is processed with minibatches of size $B = 20$. For each data element, a mask of missing values was assigned. The mask was in the form of a square of size $z \times z$ pixels. The position of this square on the image was chosen randomly, with equal probability for each possible location. The parameters for learning the RBM were set as follows: $D = 784$, $H = 40$, $K = 1$, $Q = 1$, the learning rate $\eta = 0.05$. We applied standard stochastic gradient method with momentum – the friction parameter was equal to $\gamma = 0.9$.

Looking at Algorithm 7, one can see that there are many possible variants of the proposed CDM algorithm. In the simulations we focus on three of them together with the standard CD algorithm:

- CD: $Rest = FALSE$, $PosMask = FALSE$, $NegMask = FALSE$;
- CDM(TFF): $Rest = TRUE$, $PosMask = FALSE$, $NegMask = FALSE$;
- CDM(TTT): $Rest = TRUE$, $PosMask = TRUE$, $NegMask = TRUE$;

Algorithms were evaluated in the prequential manner using the reconstruction error. For the considered minibatch of data S_t a set of reconstructions $\tilde{S}_t = (\tilde{s}_{Bt+1}, \ldots, \tilde{s}_{Bt+B})$ has to be obtained first using the RBM. Then, the average reconstruction error is expressed as follows

$$R(S_t) = \frac{1}{B} \sum_{n=1}^{B} \sum_{i=1}^{D} \left(s_{Bt+n,i} - \tilde{s}_{Bt+n,i} \right)^2 . \tag{9}$$

In the first experiment, the considered algorithms were run with three various sizes of missing values masks: $z = 2$, $z = 6$ and $z = 14$. The comparison of each algorithm performance for various values of z is demonstrated in Fig. 1. As can be seen, for each algorithm the reconstruction error is positively correlated with the amount of noise in data elements. Let us now look at the results of this experiment in another configuration. In Fig. 2 the algorithms are compared for each considered value of z. Although the values of reconstruction error fluctuate significantly in each case, it is possible to notice that the algorithm with all considered previously mechanisms turned on (i.e. the CDM(TTT) algorithm) is slightly better than the two others, whereas the standard CD algorithm is always the worst. It is the most clearly seen for the case with the biggest noise (i.e. $z = 14$). Although the differences are not striking, it can be concluded that the proposed modifications improve the performance of the CD algorithm when the incomplete data have to be handled.

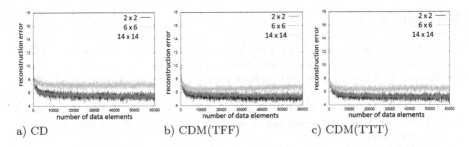

a) CD   b) CDM(TFF)   c) CDM(TTT)

Fig. 1. Reconstruction error obtained for various sizes of missing values masks for three considered algorithms.

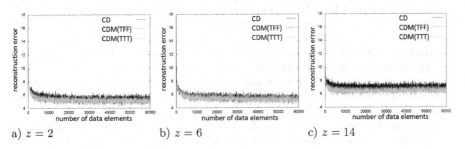

a) $z = 2$ b) $z = 6$ c) $z = 14$

Fig. 2. Reconstruction error of the CD, CDM(TFF) and CDM(TTT) algorithms for three different sizes of missing values masks.

5   Conclusions

In this paper, we considered the problem of mining stream data with missing values using the Restricted Boltzmann Machine (RBM), focusing our analysis on the Contrastive Divergence (CD) algorithm. To make it able to handle incomplete data, we proposed two modification. The first one is to introduce an additional Gibbs sampling procedure at the beginning of processing each data element. However, only those units of the visible layer are updated for which the value of the corresponding dimension in the data element is missing. In the second modification, the fixed size of minibatch is replaced by minibatches with dimension-dependent sizes. This means that not all data from the minibatch take part in updating gradients of RBM weights or visual layer biases. The proposed methods were verified experimentally, demonstrating their usability for concept drift detection in data streams with incomplete data.

Acknowledgments. The project financed under the program of the Minister of Science and Higher Education under the name "Regional Initiative of Excellence" in the years 2019–2022 project number 020/RID/2018/19, the amount of financing 12,000,000 PLN. This work was also supported by the Polish National Science Centre under grant no. 2017/27/B/ST6/02852.

References

1. Andrieu, C., de Freitas, N., Doucet, A., Jordan, M.I.: An introduction to MCMC for machine learning. Mach. Learn. **50**(1), 5–43 (2003)
2. Bengio, Y.: Learning deep architectures for AI. Found. Trends Mach. Learn. **2**(1), 1–127 (2009)
3. Devi, V.S., Meena, L.: Parallel MCNN (PMCNN) with application to prototype selection on large and streaming data. J. Artif. Intell. Soft Comput. Res. **7**(3), 155–169 (2017)
4. Goodfellow, I., Bengio, Y., Courville, A.: Deep Learning. MIT Press, Cambridge (2016). http://www.deeplearningbook.org
5. Hinton, G.E.: Training products of experts by minimizing contrastive divergence. Neural Comput. **14**(8), 1771–1800 (2002)
6. Hinton, G.E.: A practical guide to training restricted Boltzmann machines. In: Montavon, G., Orr, G.B., Müller, K.-R. (eds.) Neural Networks: Tricks of the Trade. LNCS, vol. 7700, pp. 599–619. Springer, Heidelberg (2012). https://doi.org/10.1007/978-3-642-35289-8_32
7. Hinton, G.E., Sejnowski, T.J., Ackley, D.H.: Boltzmann machines: Constraint satisfaction networks that learn. Technical report CMU-CS-84-119, Computer Science Department, Carnegie Mellon University, Pittsburgh, PA (1984)
8. Jaworski, M., Duda, P., Rutkowski, L.: On applying the restricted Boltzmann machine to active concept drift detection. In: Proceedings of the 2017 IEEE Symposium Series on Computational Intelligence, Honolulu, USA, pp. 3512–3519 (2017)
9. Jaworski, M., Duda, P., Rutkowski, L.: Concept drift detection in streams of labelled data using the restricted Boltzmann machine. In: 2018 International Joint Conference on Neural Networks (IJCNN), pp. 1–7 (2018)
10. Jaworski, M., Duda, P., Rutkowski, L.: New splitting criteria for decision trees in stationary data streams. IEEE Trans. Neural Netw. Learn. Syst. **29**(6), 2516–2529 (2018)
11. Jaworski, M., Rutkowski, L., Duda, P., Cader, A.: Resource-aware data stream mining using the restricted Boltzmann machine. In: Rutkowski, L., Scherer, R., Korytkowski, M., Pedrycz, W., Tadeusiewicz, R., Zurada, J.M. (eds.) ICAISC 2019. LNCS (LNAI), vol. 11509, pp. 384–396. Springer, Cham (2019). https://doi.org/10.1007/978-3-030-20915-5_35
12. Krawczyk, B., Cano, A.: Online ensemble learning with abstaining classifiers for drifting and noisy data streams. Appl. Soft Comput. **68**, 677–692 (2018)
13. LeCun, Y., Cortes, C.: MNIST handwritten digit database (2010). http://yann.lecun.com/exdb/mnist/
14. Lemaire, V., Salperwyck, C., Bondu, A.: A survey on supervised classification on data streams. In: Zimányi, E., Kutsche, R.-D. (eds.) eBISS 2014. LNBIP, vol. 205, pp. 88–125. Springer, Cham (2015). https://doi.org/10.1007/978-3-319-17551-5_4
15. Ramirez-Gallego, S., Krawczyk, B., García, S., Woźniak, M., Herrera, F.: A survey on data preprocessing for data stream mining: current status and future directions. Neurocomputing **239**, 39–57 (2017)
16. Rutkowski, L., Jaworski, M., Duda, P.: Stream Data Mining: Algorithms and Their Probabilistic Properties. SBD, vol. 56. Springer, Cham (2020). https://doi.org/10.1007/978-3-030-13962-9
17. Zliobaite, I., Bifet, A., Pfahringer, B., Holmes, G.: Active learning with drifting streaming data. IEEE Trans. Neural Netw. Learn. Syst. **25**(1), 27–39 (2014)

LoRa Indoor Localization Based on Improved Neural Network for Firefighting Robot

Xuechen Jin[1], Xiaoliang Xie[2(✉)], Kun An[1], Qiaoli Wang[2,3], and Jia Guo[1]

[1] School of Electrical and Control Engineering,
North University of China, Taiyuan 030051, China
tszycc@163.com, ankun@nuc.edu.cn, guojia9623@163.com
[2] The State Key Laboratory of Management and Control for Complex Systems,
Institute of Automation, Chinese Academy of Sciences, Beijing 100190, China
{xiaoliang.xie,wangqiaoli2017}@ia.ac.cn
[3] University of Chinese Academy of Sciences, Beijing 100049, China

Abstract. Trapped occupants' safety is a critical problem in the fireground and a major issue is the lack of reliable indoor localization decision-making system for firefighting. State of the art methods have failed to provide an automatic, accurate and reliable solution that can facilitate the decision-making of incident commanders. This paper aims to develop a novel smart firefighting robot to achieve this goal, by combining artificial neural network with received signal strength indication of the new wireless communication approach named Long Range (LoRa). Our solution includes a new indoor localization algorithm that contains a process for optimizing the initial weights and thresholds of BP neural networks. The solution can improve the location accuracy of trapped occupants in fire. We fully implement the algorithm in a complete indoor localization system and conduct experiments in the space of $25\,\mathrm{m} \times 25\,\mathrm{m} \times 5\,\mathrm{m}$ that involved a firefighting robot and some trapped occupants. The localization results demonstrate that our solution greatly shortens the convergence time and reduces the average and minimum location error to $0.7\,\mathrm{m}$ and $0.2\,\mathrm{m}$ respectively in a $20\,\mathrm{m} \times 15\,\mathrm{m}$ testing area.

Keywords: Firefighting robot · LoRa · Indoor localization · Mind evolution algorithm (MEA) · Received signal strength indication (RSSI)

1 Introduction

Urbanization and changes in modern cities have brought new challenges to firefighting practices. Training and research programs have been developed to meet

This work is funded by the National Key Research and Development Plan of China (2017YFE0112200) and European Commission Marie Skłodowska-Curie SMOOTH project (H2020-MSCA-RISE-2016-734875).

T. Gedeon et al. (Eds.): ICONIP 2019, CCIS 1143, pp. 355–362, 2019.
https://doi.org/10.1007/978-3-030-36802-9_38

these challenges but there are still significant losses from fires each year. In 2017, fire departments in the United States responded to more than 1319500 structure fires, which caused the losses of approximately 23 billion dollars [1]. The main reason is the lack of available information at fire scenes, so it is crucial for incident commanders (ICs) to have a accurate awareness of the evolving state of fire and people involved (victims and responders) in the fireground to make effective decisions. Such problems can be significantly reduced by developing a smart firefighting robot to perform searching and rescuing practice in the fire ground, and to facilitate the decision-making of ICs. The first task of decision-making is to determine the location of trapped occupants in the fire [2]. Therefore, How to make the correct decision and realize the indoor localization based on the information provided by the firefighting robot is a key issue of fire rescue. The communication quality is a critical issue, the communication is often lost when firefighters go to higher floors or enter basement [3]. The Worcester Polytechnic Institute proposed to use fire trucks as temporary base stations when the firefighters arrive at the site [4], but this method have suffered from long setup time. Most existing generalized localization methods only address part of the problems. For example, Global Positioning System (GPS) has the advantages of high accuracy and fast speed has been widely used in outdoor localization [5]. However, in indoor localization, there are many problems needed be faced. [6] presented a WIFI based localization system, but WiFi hotspots are greatly affected by the surrounding environment, resulting in low accuracy. [7] proposed a solution based on ultrasonic localization, and it is easily affected by multipath effect and non-line-of-sight propagation. Ultra-wideband (UWB) [8] was considered as a promising indoor localization solution, which can improved precision localization accuracy. However, it is difficult to achieve large-scale indoor coverage and the construction cost of the system is much higher than other methods. [9] presented a solution based on visible light. In recent years, LoRa has been seen as a potential alternative to existing wireless communication standards, with low power consumption and low implementation cost [10]. Besides, in localization technology, RSSI-based localization method has attracted more and more attention because of its high positioning accuracy [11]. Therefore, this paper proposes an indoor localization method which combines RSSI of LoRa with artificial neural networks (ANN).

2 Principle and Method of Non-ranging Indoor Localization Based on RSSI

2.1 The Principle of Position Fingerprint Algorithm

In the wireless sensor network, the RSSI of the node can be used to calculate its position. In order to realize the accurate location of trapped occupants in the indoor fireground, this paper proposes a method of position fingerprint database to achieve the indoor localization of LoRa environment.

The position fingerprint algorithm is derived from the pattern recognition theory [12]. Position fingerprints algorithm consists of two phases: offline phase

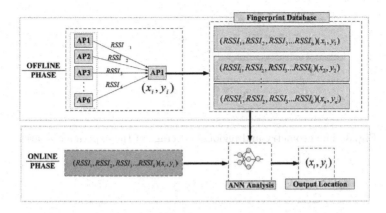

Fig. 1. RSSI-based location fingerprint algorithm flow chart.

Fig. 2. The flow chart of RSSI pre-processing.

and online phase [13]. As is shown in Fig. 1. Offline phase: the main work is to collect and store data. Since the fire field is a dynamic environment. Therefore, it is necessary to pre-process the collected data [14]. Figure 2 is the flow chart of RSSI pre-processing. Online phase: The main task of this phase is coordinate matching, matching the RSSI at the undetermined location with the RSSI in the location fingerprint database.

2.2 Indoor Localization Algorithm

The artificial neural network has strong nonlinear mapping ability, and the rules can be recorded in the weight among the neurons. In this paper, we use the artificial neural network algorithm to learn the nonlinear mapping relationship between RSSI and two-dimensional position information and calculate the coordinates of reference points. Meanwhile, an improved BP neural network through MEA is adopted to optimize the experiment. The RSSI of the node to be located is input to the trained neural network, and the output of the neural network is the coordinates of the node to be located.

3 Building the LoRa Environment

3.1 Arrangement of Experimental Environment

This paper will build the LoRa environment to achieve the indoor localization of trapped occupant in the fireground. The experiment site was selected in a building with an approximate volume of $25\,\mathrm{m} \times 25\,\mathrm{m} \times 5\,\mathrm{m}$. We evenly divided the

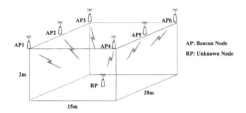

Fig. 3. The geometry and regional coverage of the experimental area.

(a) LoRa localization system (LLS)   (b) Location results displayed in LLS

Fig. 4. The localization system interface.

entire lobby into a $20\,\text{m} \times 15\,\text{m}$ grid. The selection of the sample directly affects the accuracy of the localization. In order to make the experiment more simple and accurate, we use the floor tiles of the building as the grids to establish the corresponding coordinates (the floor tiles are squares with a side length of $1\,\text{m}$), a total of 300 coordinate points. The anchor nodes were placed in each corner of the interior space designed. Similarly, in order to reduce the measurement error, we fixed the anchor nodes $2\,\text{m}$ away from the floor. The Fig. 3 shows the geometric shape and regional coverage of the experimental area.

3.2   The Design of LoRa Localization System

We use microprocessor STM32F407. ATK-LORA-01 RF module is used to connect with STM32 microcontroller, which contains LoRa RF chip SXl278 and filter circuit. In addition, a high-performance MCU is integrated into the LoRa module, so we can configure working mode and communication parameters of the module through AT instruction. In this experiment, we developed a user interface, LoRa Localization System (LLS), as shown in Fig. 4(a). Through LLS, we can configure the working mode of LoRa node. Besides, the result of localization can display on the interface with images and numbers, as is shown in Fig. 4(b). Figure 5 is the LoRa environment.

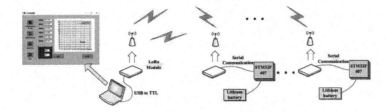

Fig. 5. The LoRa environment.

4   Data Processing and Results Analysis

4.1   Design of Network Structure and Data Collection

Through analyzing the RSSI collected at a fixed sample point, we found that the RSSI probability distribution obeys the Gaussian distribution under the indoor dynamic experimental environment, that is [15]:

$$f(RSSI) = \frac{1}{\sigma\sqrt{2\pi}} \exp\left(-\frac{(RSSI - A)^2}{2\sigma^2}\right) \tag{1}$$

Where σ is the standard deviation and A is the expectation, so it can be concluded that:

$$\overline{\sigma} = \sqrt{\frac{1}{n-1}\sum_{k=1}^{n}(RSSI - \overline{A})^2} \tag{2}$$

$$\overline{A} = \frac{1}{n}\sum_{k=1}^{n} RSSI_k \tag{3}$$

Where n is the number of RSSI collected at any reference point, and $RSSI_k$ is the RSSI measured at the kth time.

Through Gaussian filtering [15], we solved the problem that the RSSI in the indoor dynamic environment is susceptible to interference. Then the pre-processed RSSI and corresponding coordinate values are used as training samples to train the ANN.

4.2   Analysis of Indoor Localization Results

In our experiments, we designed a 3-layer BP neural network structure. The number of hidden layer nodes is given according to the empirical formula:

$$M = \sqrt{n+m} + a \tag{4}$$

Where m and n are the number of neurons in the output layer and the input layer, respectively.

By comparing the relationship between the number of hidden layer nodes and the mean square error, we select the number of hidden layer nodes as 10.

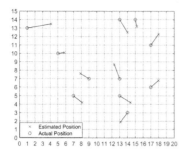 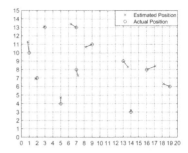

(a) Localization results without opti- (b) Localization results with optimiza-
mization algorithm   tion algorithm

Fig. 6. The localization results.

Activation functions of a hidden layer is a sigmoidal function and the output
function is a piecewise-linear function Respectively.

In the process of the experiment, six RSSI values corresponding to specified
test point are input into LLS, seen in Fig. 4(b), then the estimated position of the
test point in the experimental environment is displayed on the upper computer
interface (The * is estimated position). In our experiment, we randomly selected
11 test points to test the localization effect of the system. The localization results
are shown in the Fig. 6(a).

We quantify the magnitude of the error by calculating the Euclidean distance:

$$e_i = \sqrt{(x_i - x_r)^2 + (y_i - y_r)^2} \tag{5}$$

$$E = \frac{1}{k}\sum_{i=1}^{k} e_i \tag{6}$$

where (x_i, y_i) is the estimated position of the ith test node, (x_r, y_r) is the actual
position of the ith node, E is the average error of k test nodes, and k is the
total number of test nodes. After calculation, the maximum error of the LoRa
localization system based on BP network is 3.14 m, the minimum error is 0.80 m,
and the average error is 1.54 m. In order to improve the accuracy of RSSI-based
LoRa localization, we use the MEA to optimize the initial weights of neural
network.

This method can effectively avoid falling into local minimum values, and
can also greatly speed up the training and make the network more robust. We
also randomly selected 11 test points, using the BP neural network with MEA
to estimate the position of the test points. The test results are shown in the
Fig. 6(b). Table 1 is a comparison of the typical errors of the two localization
algorithms. From Fig. 7, we can see that the error of the localization system using
only BP network is greater than the error after processing with the optimization
algorithm. This is because the initial weights and thresholds of the network are

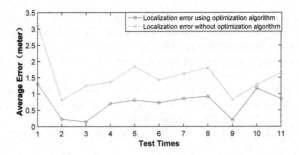

Fig. 7. Comparison of average error between the two algorithms.

randomly generated when the BP neural network is used for localization. The network will fall into the local optimal value during the training process.

Table 1. Comparison of typical localization errors between two algorithms (unit: meter).

Neural network type	Max error	Min error	Average error
BP neural network	3.1401	0.8062	1.5436
MEA-BP neural network	1.3153	0.2	0.7196

5   Conclusion

In this paper, aiming at the important requirements of fire safety, a design implementation and evaluation method of real-time positioning system based on intelligent firefighting robot decision-making is proposed. All previous work failed to provide an automatic, accurate, and reliable solution to locate the trapped occupants in harsh environment. This paper proposes a novel indoor localization platform based on firefighting robot to achieve this goal, by combining MEA-BP neutral network with RSSI of LoRa. We tested the reliability of the LoRa communication and built a LoRa environment in a realistic scenario. We fully implemented this platform, and compared our solution with the algorithm that BP neutral network is used only. Evaluation results show that our approach reduces the trapped occupants location error and outperforms the alternatives solution. In addition, The localization results from an area of $300\,\mathrm{m}^2$ demonstrate that our solution greatly shortens the convergence time and reduces the average and minimum location error to $0.7\,\mathrm{m}$ and $0.2\,\mathrm{m}$.

References

1. Evarts, B.: Fire loss in the united states during 2017. National Fire Protection Association Quincy, MA (2018)

2. Allali, S., Benchaïba, M., Ouzzani, F., Menouar, H.: No-collision grid based broadcast scheme and ant colony system with victim lifetime window for navigating robot in first aid applications. Ad Hoc Netw. **68**, 85–93 (2018)
3. Varone, C.: Firefighter safety and radio communication. Fire Eng. **156**(3), 141–164 (2003)
4. Sun, J., Cao, W., Roy, A., Liu, X.: System including base stations that provide information from which a mobile station can determine its position. US Patent Application 15/557,076, 27 September 2018
5. Kaplan, E., Hegarty, C.: Understanding GPS: Principles and Applications. Artech house, Norwood (2005)
6. Yang, C., Shao, H.R.: WiFi-based indoor positioning. IEEE Commun. Mag. **53**(3), 150–157 (2015)
7. Fan, W., Yong, P.: Improved indoor location method for ultrasonic system based on TDOA. Comput. Technol. Dev. **24**(6), 250–253 (2014)
8. Ingram, S., Harmer, D., Quinlan, M.: Ultrawideband indoor positioning systems and their use in emergencies. In: PLANS 2004. Position Location and Navigation Symposium (IEEE Cat. No. 04CH37556), pp. 706–715. IEEE (2004)
9. Song, K.U., et al.: System and method for indoor positioning using led lighting. US Patent Application 13/008,455, 21 July 2011
10. Souryal, M.R., Geissbuehler, J., Miller, L.E., Moayeri, N.: Real-time deployment of multihop relays for range extension. In: Proceedings of the 5th International Conference on Mobile Systems, Applications and Services, pp. 85–98. ACM (2007)
11. Gu, C., Jiang, L., Tan, R.: LoRa-based localization: opportunities and challenges. ArXiv Preprint ArXiv:1812.11481 (2018)
12. Le Dortz, N., Gain, F., Zetterberg, P.: WiFi fingerprint indoor positioning system using probability distribution comparison. In: 2012 IEEE International Conference on Acoustics, Speech and Signal Processing (ICASSP), pp. 2301–2304. IEEE (2012)
13. Lee-Fang Ang, J., Lee, W.K., Ooi, B.Y., Wei-Min Ooi, T., Hwang, S.O.: Pedestrian dead reckoning with correction points for indoor positioning and Wi-Fi fingerprint mapping. J. Intell. Fuzzy Syst. **35**(6), 5881–5888 (2018)
14. Hadj-Mihoub-Sidi-Moussa, H., Tedjini, S., Touhami, R.: Phase selector for RFID localization system based on RSSI filter. In: 2019 14th International Conference on Design & Technology of Integrated Systems In Nanoscale Era (DTIS), pp. 1–4. IEEE (2019)
15. Mahapatra, R.K., Shet, N.: Localization based on RSSI exploiting Gaussian and averaging filter in wireless sensor network. Arab. J. Sci. Eng. **43**(8), 4145–4159 (2018)

Neural Network Models

Stability Analysis of Multiobjective Robust Controller Employing Switching of Competitive Associative Nets

Hironobu Nakayama, Kazuya Matsuo, and Shuichi Kurogi(✉)

Kyushu Institute of Technology, Tobata, Kitakyushu, Fukuoka 804-8550, Japan
nakayama.hironobu605@mail.kyutech.jp, {matsuo,kuro}@cntl.kyutech.ac.jp
http://kurolab.cntl.kyutech.ac.jp/

Abstract. So far, we have developed a multiobjective robust controller using GPC (generalized predictive controller) and CAN2s (competitive associative neural nets) to learn and approximate Jacobian matrices of nonlinear dynamics of the plant to be controlled. Here, the CAN2 is an artificial neural net for learning efficient piecewise linear approximation of nonlinear function. In our previous studies, we have shown that the controller is capable of coping with the change of plant parameter values as well as the change of control objective by means of switching CAN2s. This paper examines the stability of the controller by means of using a linear plant to be controlled, and show the properties and the effectiveness of the present method.

Keywords: Multiobjective robust control · Switching of multiple CAN2s · Generalized predictive control · Jacobian matrix of nonlinear dynamics · Control stability

1 Introduction

So far, we have developed multiobjective robust controller using GPC (generalized predictive controller) [1] and multiple CAN2s (competitive associative nets) [2–5]. Here, a single CAN2 is an artificial neural net introduced for learning efficient piecewise linear approximation of nonlinear function by means of competitive and associative schemes [6–8]. In [2–4], we have constructed a multiobjective robust controller using multiple CAN2s to learn to approximate nonlinear dynamics of plants (specifically, overhead traveling crane models) with several parameter values and two control objectives to reduce settling time and overshoot. Our method enables the controller to cope with those objectives and the plants with different parameter values by means of switching multiple CAN2s. In [5], by means of employing a simple linear plant model similar to overhead traveling crane model, we have clarified several properties, such as the effect of enlargement of embedding dimension (or the dimension of input vector to the CAN2 for learning and prediction), how to achieve multiobjective control by the

© Springer Nature Switzerland AG 2019
T. Gedeon et al. (Eds.): ICONIP 2019, CCIS 1143, pp. 365–375, 2019.
https://doi.org/10.1007/978-3-030-36802-9_39

GPC with fixed parameter values and switching of multiple CAN2s, augmentation of CAN2s to improve the overshoot performance. However, the properties and the effectiveness of the controller have not been analyzed enough. In this paper, we analyze the stability of the controller by means of examining the poles of the transfer function from the target output to the output error. Since the poles of the transfer function change by the switching of CAN2s, we have to analyze them dynamically via numerical experiments. Here, note that a number of researches using neural networks and predictive control have been conducted in the areas of robotics, chemical engineering and so on [9,10]. The stability of GPC for linear input/output systems has been successfully addressed in a number of researches [11], while a recent research [12] for neural network-based model predictive control for nonlinear systems derives the stability conditions but the problem to prove the existence of feasible solutions has not been solved yet. On the otherhand, the present method employs piecewise linear predictive control and the stability can be analyzed by means of using piecewise linear plant models as described below, which will be applicable to nonlinear systems. There are also a number of researches on multiobjective robust predictive control [13], while the comparison to the present method has not been examined yet, which is for our future research studies.

The rest of the paper is organized as follows. In the next section, we show the multiobjective robust controller using CAN2s, and derives the transfer function to analyze the control stability. In Sect. 3, we show several results of numerical experiments, and examine the control stability and show the effectiveness of the present method followed by the conclusion in Sect. 4.

2   Multiobjective Robust Controller Using CAN2s

2.1   Learning Plant Model Using Difference Signal and CAN2

Plant Model Using Difference Signal. Suppose a plant to be controlled at a discrete time $t = 1, 2, \cdots$ has the input $u_t^{[p]}$ and the output $y_t^{[p]}$. Here, the superscript "[p]" indicates the variable related to the plant to be controlled for distinguishing the position of the load, (x, y), shown below. Furthermore, suppose that the dynamics of the plant is given by

$$y_t^{[p]} = f(x_t^{[p]}) + d_t^{[p]} \ , \tag{1}$$

where $f(\cdot)$ is a nonlinear function which may change slowly in time and $d_t^{[p]}$ represents zero-mean noise with the variance σ_d^2. The input vector $x_t^{[p]}$ consists of the input and output sequences of the plant as $x_t^{[p]} \triangleq \left(y_{t-1}^{[p]}, \cdots, \right.$ $\left. y_{t-k_y}^{[p]}, u_{t-1}^{[p]}, \cdots, u_{t-k_u}^{[p]} \right)^\top$, where k_y and k_u are the numbers of the elements, and the total embedding dimension of $x_t^{[p]}$ is given by $k = k_y + k_u$. Then, for the difference signals $\Delta y_t^{[p]} \triangleq y_t^{[p]} - y_{t-1}^{[p]}$, $\Delta u_t^{[p]} \triangleq u_t^{[p]} - u_{t-1}^{[p]}$, and

$\Delta x_t^{[p]} \triangleq x_t^{[p]} - x_{t-1}^{[p]}$, we have the relationship $\Delta y_t^{[p]} \simeq f_x \Delta x_t^{[p]}$ for small $\|\Delta x_t^{[p]}\|$, where $f_x = \partial f(x)/\partial x \big|_{x=x_{t-1}^{[p]}}$ indicates Jacobian matrix (row vector). If f_x does not change for a while after the time t, then we can predict $\Delta y_{t+l}^{[p]}$ by

$$\widehat{\Delta y}_{t+l}^{[p]} = f_x \widetilde{\Delta x}_{t+l}^{[p]} \tag{2}$$

for $l = 1, 2, \cdots$, recursively. Here, $\widetilde{\Delta x}_{t+l}^{[p]} = (\widetilde{\Delta y}_{t+l-1}^{[p]}, \cdots, \widetilde{\Delta y}_{t+l-k_y}^{[p]}, \widetilde{\Delta u}_{t+l-1}^{[p]},$ $\cdots, \widetilde{\Delta u}_{t+l-k_u}^{[p]})^\top$, and the elements are given by

$$\widetilde{\Delta y}_{t+m}^{[p]} = \begin{cases} \Delta y_{t+m}^{[p]} & \text{for } m < 1 \\ \widehat{\Delta y}_{t+m}^{[p]} & \text{for } m \geq 1 \end{cases} \quad \text{and} \quad \widetilde{\Delta u}_{t+m}^{[p]} = \begin{cases} \Delta u_{t+m}^{[p]} & \text{for } m < 0 \\ \widehat{\Delta u}_{t+m}^{[p]} & \text{for } m \geq 0. \end{cases} \tag{3}$$

Here, $\widehat{\Delta u}_{t+m}^{[p]}$ $(m \geq 0)$ is the predictive input shown in Sect. 2.2. Then, we have the prediction of the plant output from the predictive difference signals as

$$\widehat{y}_{t+l}^{[p]} = y_t^{[p]} + \sum_{m=1}^{l} \widehat{\Delta y}_{t+m}^{[p]}. \tag{4}$$

Learning Plant Model Using CAN2. A CAN2 has N units. The ith unit has a weight vector $w_i \triangleq (w_{i1}, \cdots, w_{ik})^\top \in \mathbb{R}^{k \times 1}$ and an associative matrix (row vector) $M_i \triangleq (M_{i1}, \cdots, M_{ik}) \in \mathbb{R}^{1 \times k}$ for $i \in I = \{1, 2, \cdots, N\}$. For a given dataset $D^{[n]} = \{(\Delta x_t^{[p]}, \Delta y_t^{[p]}) \mid t = 1, 2, \cdots, n\}$ obtained from the plant to be controlled, we train a CAN2 by feeding the input and output pair of the CAN2 as $(x^{[\text{can2}]}, y^{[\text{can2}]}) = (\Delta x_t^{[p]}, \Delta y_t^{[p]})$. We employ an efficient batch learning method shown in [14]. Then, for an input vector $\Delta x_t^{[p]}$, the CAN2 after the learning predicts the output $\Delta y_t^{[p]} = f_x \Delta x_t^{[p]}$ by

$$\widehat{\Delta y}_t^{[p]} = M_c \Delta x_t^{[p]}, \tag{5}$$

where c denotes the index of the unit selected by

$$c = \underset{i \in I}{\text{argmin}} \|\Delta x_t^{[p]} - w_i\|. \tag{6}$$

Since we can predict the output $\widehat{y}_{t+l}^{[p]}$ of the plant by (4) and (5), we sometimes call M_c a learning model of the plant in the following.

2.2   GPC and Control Stability

GPC Using Learning Plant Model. The GPC (Generalized Predictive Control) is an efficient method for obtaining the predictive input $\widehat{u}_t^{[p]}$ which minimizes the following control performance index [1]:

$$J = \sum_{l=1}^{N_y} \left(\widehat{y}_{t+l}^{[p]} - r_{t+l}^{[p]} \right)^2 + \lambda_u \sum_{l=1}^{N_u} \left(\widehat{\Delta u}_{t+l-1}^{[p]} \right)^2, \tag{7}$$

where $\widehat{y}^{[\mathrm{p}]}_{t+l}$ and $r^{[\mathrm{p}]}_{t+l}$ are predictive and target (desired) output, respectively. The parameters N_y, N_u and λ_u are constants to be designed for the control performance. We obtain $\widehat{u}^{[\mathrm{p}]}_t$ by means of the GPC method as follows: at a discrete time t, use the learning plant model $\boldsymbol{M}_c$ to predict $\widehat{y}^{[\mathrm{p}]}_{t+l}$ by (4) and (5). Then, owing to the linearity of these equations, the above performance index is written as $J = \|\boldsymbol{G}\Delta\boldsymbol{u}^{[\mathrm{p}]} + \overline{\boldsymbol{y}}^{[\mathrm{p}]} - \boldsymbol{r}^{[\mathrm{p}]}\|^2 + \lambda_u\|\widehat{\Delta\boldsymbol{u}}\|^2$ with $\overline{\boldsymbol{y}}^{[\mathrm{p}]} = \left(\overline{y}^{[\mathrm{p}]}_{t+1}, \cdots, \overline{y}^{[\mathrm{p}]}_{t+N_y}\right)^{\top}$, $\boldsymbol{r}^{[\mathrm{p}]} = \left(r^{[\mathrm{p}]}_{t+1}, \cdots, r^{[\mathrm{p}]}_{t+N_y}\right)^{\top}$, $\widehat{\Delta\boldsymbol{u}}^{[\mathrm{p}]} = \left(\widehat{\Delta u}^{[\mathrm{p}]}_t, \cdots, \widehat{\Delta u}^{[\mathrm{p}]}_{t+N_u-1}\right)^{\top}$. Here, $\overline{y}^{[\mathrm{p}]}_{t+l}$ denotes the natural response $\widehat{y}^{[\mathrm{p}]}_{t+l}$ of the system (1) for the null incremental input $\widehat{\Delta u}^{[\mathrm{p}]}_{t+l} = 0$ for $l \geq 0$, and we recursively obtain $\overline{y}^{[\mathrm{p}]}_{t+l} = y^{[\mathrm{p}]}_t + \sum_{m=1}^{l} \boldsymbol{M}_c\Delta\boldsymbol{x}^{[\mathrm{p}]}_{t+m}$ for $l = 1, 2, \cdots$ with updating $\Delta\boldsymbol{x}^{[\mathrm{p}]}_{t+l} = \left(\Delta y^{[\mathrm{p}]}_{t+l-1}, \cdots, \Delta y^{[\mathrm{p}]}_{t+l-k_y}, \Delta u^{[\mathrm{p}]}_{t+l-1}, \cdots, \Delta u^{[\mathrm{p}]}_{t+l-k_u}\right)^{\top}$. The ith column and the jth row of the matrix $\boldsymbol{G}$ is given by $G_{ij} = g_{i-j+N_1}$, where g_l for $l = \cdots, -2, -1, 0, 1, 2, \cdots$ is the unit step response $y^{[\mathrm{p}]}_{j+l}$ of (4) for $\widehat{y}^{[\mathrm{p}]}_{j+l} = \widehat{u}^{[\mathrm{p}]}_{j+l} = 0$ ($l < 0$) and $\widehat{u}^{[\mathrm{p}]}_{j+l} = 1(l \geq 0)$. Then, we have $\widehat{\Delta\boldsymbol{u}}^{[\mathrm{p}]}$ which minimizes J by $\widehat{\Delta\boldsymbol{u}}^{[\mathrm{p}]} = -\boldsymbol{K}\left(\overline{\boldsymbol{y}}^{[\mathrm{p}]} - \boldsymbol{r}^{[\mathrm{p}]}\right)$ and then derive the predictive input $\widehat{u}^{[\mathrm{p}]}_t = u^{[\mathrm{p}]}_{t-1} + \widehat{\Delta u}^{[\mathrm{p}]}_t$, where the feedback matrix is derived as $\boldsymbol{K} \equiv (\boldsymbol{G}^{\top}\boldsymbol{G} + \lambda_u\boldsymbol{I})^{-1}\boldsymbol{G}^{\top} \in \mathbb{R}^{N_u \times N_y}$, and $\widehat{\Delta u}^{[\mathrm{p}]}_t$ is the first element of $\widehat{\Delta\boldsymbol{u}}^{[\mathrm{p}]}$.

Transfer Function and Control Stability. We examine the stability of the control using the predictive input $u^{[\mathrm{p}]}_t = \widehat{u}^{[\mathrm{p}]}_t$ derived in the previous section as

$$u^{[\mathrm{p}]}_t = u^{[\mathrm{p}]}_{t-1} - \boldsymbol{k}_1^{\top}\left(\boldsymbol{h}_y(z)\overline{y}^{[\mathrm{p}]}_t + \boldsymbol{h}_u(z)\Delta u^{[\mathrm{p}]}_{t-1} - \boldsymbol{h}(z)r^{[\mathrm{p}]}_t\right), \qquad (8)$$

for a linear plant

$$y^{[\mathrm{p}]}_t = \sum_{i=1}^{k_y} a_i y^{[\mathrm{p}]}_{t-i} + \sum_{i=1}^{k_u} b_i u^{[\mathrm{p}]}_{t-i} = \sum_{i=1}^{k_y} a_i z^{-i} y^{[\mathrm{p}]}_t + \sum_{i=1}^{k_u} b_i z^{-i} u^{[\mathrm{p}]}_t. \qquad (9)$$

Here, $\boldsymbol{k}_1^{\top} \in \mathbb{R}^{1 \times N_y}$ denotes the first row vector of $\boldsymbol{K}$, and z^{-1} time delay operator. We denote $\overline{y}^{[\mathrm{p}]}_t = \boldsymbol{h}_y(z)\overline{y}^{[\mathrm{p}]}_t + (1 - z^{-1})\boldsymbol{h}_u(z)u^{[\mathrm{p}]}_{t-1}$ and $\boldsymbol{r}^{[\mathrm{p}]} = \boldsymbol{h}_r(z)r^{[\mathrm{p}]}_t = (z^1, \cdots, z^{N_y})^{\top}r^{[\mathrm{p}]}_t$, where note that $\overline{\boldsymbol{y}}^{[\mathrm{p}]}$ is obtained from $y^{[\mathrm{p}]}_{t-l} = z^{-l}y^{[\mathrm{p}]}_t$ and $\Delta u^{[\mathrm{p}]}_{t-1-l} = (1 - z^{-1})z^{-l}u^{[\mathrm{p}]}_{t-1}$ for $l \geq 0$ with (4) and (5). In order to obtain the transfer function from $r^{[\mathrm{p}]}_t$ to the error $e^{[\mathrm{p}]}_t = r^{[\mathrm{p}]}_t - y^{[\mathrm{p}]}_t$, we apply z-transform to (9) and (8) with initial values of $u^{[\mathrm{p}]}_t$ and $y^{[\mathrm{p}]}_t$ being 0. Then, we have

$$H^{[\mathrm{p}]}_{e/r}(z) = \frac{(1 - z^{-1})\left(1 - \sum_{i=1}^{k_y} a_i z^{-i}\right)\left(1 + z^{-1}\boldsymbol{k}_1^{\top}\boldsymbol{h}_u(z)\right) + \left(\sum_{i=1}^{k_u} b_i z^{-i}\right)\boldsymbol{k}_1^{\top}\left(\boldsymbol{h}_y(z) - \boldsymbol{h}_r(z)\right)}{(1 - z^{-1})\left(1 - \sum_{i=1}^{k_y} a_i z^{-i}\right)\left(1 + z^{-1}\boldsymbol{k}_1^{\top}\boldsymbol{h}_u(z)\right) + \left(\sum_{i=1}^{k_u} b_i z^{-i}\right)\boldsymbol{k}_1^{\top}\boldsymbol{h}_y(z)}. \qquad (10)$$

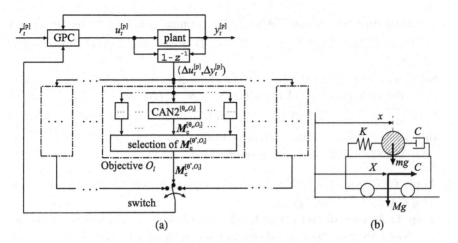

Fig. 1. Schematic diagram of (a) multiobjective controller and (b) plant model of a car and load.

Here, $H_{e/r}^{[\mathrm{p}]}(z) = E^{[\mathrm{p}]}(z)/R^{[\mathrm{p}]}(z)$, $E^{[\mathrm{p}]}(z) = \mathcal{Z}\left(e_t^{[\mathrm{p}]}\right)$, $R^{[\mathrm{p}]}(z) = \mathcal{Z}\left(r_t^{[\mathrm{p}]}\right)$, and $\mathcal{Z}\left(\cdot\right)$ denotes z-transform. From this equation, we can see that, when $e_t^{[\mathrm{p}]}$ converges, $e_\infty^{[\mathrm{p}]} = \lim_{z\to 1}(1 - z^{-1})H_{e/r}^{[\mathrm{p}]}(z)R^{[\mathrm{p}]}(z) = \lim_{z\to 1} H_{e/r}^{[\mathrm{p}]}(z)r^{[\mathrm{p}]} = 0$ for a constant target (desired) output $r_t^{[\mathrm{p}]} = r^{[\mathrm{p}]}$ or $R^{[\mathrm{p}]}(z) = r^{[\mathrm{p}]}/(1 - z^{-1})$, where we can derive $\lim_{z\to 1} h_y(z) = \lim_{z\to 1} h_r(z) = 1$ from the definition. This indicates that, without unstable pole-zero cancellation, the plant is stable when all the poles of $H_{e/r}^{[\mathrm{p}]}(z)$ has the magnitude less than 1.

2.3 Learning and Selecting CAN2s for Multiobjective Robust Control

We employ the multiobjective robust controller using CAN2s shown in [5] (see Fig. 1(a) for a schematic diagram). Here, we describe training and switching of CAN2s briefly: first, let $\theta = \theta_s$ ($\in \Theta^{[\mathrm{tst}]}; s = \{1, 2, \cdots, |\Theta^{[\mathrm{tst}]}|\}$) denote the parameter of a plant to be tested, and $O = O_l$ ($\in \Omega; l = \{1, 2, \cdots, |\Omega|\}$) a control objective. Furthermore, let CAN2$^{[\theta]}$ be a CAN2 for training the plant θ, and CAN2$^{[\Theta^{[\mathrm{tr}]}]}$ be a set of CAN2s trained for the plants with $\theta \in \Theta^{[\mathrm{tr}]}$.

1. Iterations of control and training: For each plant $\theta \in \Theta^{[\mathrm{tr}]}$, we execute iterations of the following control and training phases:

   control phase: At the first ($i = 1$) iteration, a default control sequence (see Sect. 3.2 for details) is applied to control the plant, while at the ith iteration for $i \geq 2$, control the plant by means of the GPC with CAN2$_{i-1,N}^{[\theta]}$ obtained in the previous training phase, where N denotes a number of units in a CAN2.

training phase: Obtain $CAN2_{i,N}^{[\theta]}$ by training a CAN2 with the dataset $D_i = \{(\Delta \boldsymbol{x}_t^{[p]}, \Delta y_t^{[p]}) \mid t = 1, 2, \cdots, |D_i|\}$ obtained in the above control phase.

2. Select a set of CAN2s: We execute the above iterations with $CAN2_{i,N}^{[\theta]}$ involving different number N of units for $N \in \mathcal{N} = \{N_1, N_2, \cdots, N_{|\mathcal{N}|}\}$. Among $CAN2_{i,N}^{[\theta]}$ ($i \in \mathcal{I}_{\text{it}}$, $N \in \mathcal{N}$), we obtain $CAN2^{[\theta, O_l]}$, or the best CAN2 from the point of view of the control objective $O_l \in \Omega$. Let $CAN2^{[\Theta^{[\text{tr}]}, O_l]} = \underset{\theta \in \Theta^{[\text{tr}]}}{\cup} CAN2^{[\theta, O_l]}$ denote the set of obtained CAN2s.

3. Apply GPC using $CAN2^{[\Theta^{[\text{tr}]}, O_l]}$ to a test plant $\theta_s \in \Theta^{[\text{tst}]}$: We execute GPC with $CAN2^{[\Theta^{[\text{tr}]}, O_l]}$ as follows:

   step 1: At each discrete time t, select the index c of the weight vector $\boldsymbol{w}_c$ closest to $\Delta \boldsymbol{x}_t^{[p]}$ by (5) and obtain the corresponding $\boldsymbol{M}_c^{[\theta]}$ (or $\boldsymbol{M}_c$ in (5)) for each $\theta \in \Theta^{[\text{tr}]}$.

   step 2: Select $\boldsymbol{M}_c^{[\theta^*]}$ which provides the minimum MSE (mean square prediction error) for the recent N_e predictions, or

$$\theta^* = \underset{\theta \in \Theta^{[\text{tr}]}}{\arg\min} \frac{1}{N_e} \sum_{l=0}^{N_e-1} \left\| \Delta y_{t-l}^{[p]} - \widehat{\Delta y}_{t-l}^{[p][\theta]} \right\|^2 , \tag{11}$$

   where $\widehat{\Delta y}_{t-l}^{[p][\theta]} = \boldsymbol{M}_c^{[\theta]} \Delta \boldsymbol{x}_{t-l}^{[p]}$ (see (5)) denotes the prediction by $CAN2^{[\theta, O_l]}$.

   step 3: Obtain the predictive input $\widehat{\Delta u}_t^{[p]}$ by the GPC shown in Sect. 2.2 with replacing $\boldsymbol{M}_c$ by $\boldsymbol{M}_c^{[\theta^*]}$.

3   Numerical Experiments and Analysis

3.1   Plant Model of a Car and a Load

We examine a linear plant model of a car and a load shown in [5] (see Fig. 1(b) for a schematic diagram) obtained by means of replacing the nonlinear crane examined in [4] by a load (mass) with a spring and a damper in order to analyze the present control method analytically. From the figure, we have motion equations given by

$$m\ddot{x} = -K(x - X) - C(\dot{x} - \dot{X}) \tag{12}$$

$$M\ddot{X} = F + K(x - X) \tag{13}$$

where x and X are the positions of the load and the car, respectively, m and M are the weights of the load and the car, respectively, K the spring constant, C the damping coefficient, and F is the driving force of the car. From the above equations, we have the following state-space representation for the state $\boldsymbol{x} = (x, \dot{x}, X, \dot{X})^\top$,

Table 1. Statistical summary of the performance of the controller using single $CAN2^{[20,O]}$, $CAN2^{[50,O]}$ and multiple $CAN2^{[(20,50),O]}$ with $O = ST$ and $O = OS$ for the control of test plants with $m = m_{tst} = 10, 15, 20, \cdots, 60$ [kg]. The columns of "mean", "min", "max" and "std" for "settling time" and "overshoot" indicate the minimum, maximum and standard deviation for all test plants. The boldface figures indicate the best (smallest) result in each column block, while the italicface figures show the result without corresponding to the control objective of the CAN2 shown on the leftmost column.

	Settling time t_{ST} [s]				Overshoot x_{OS} [mm]			
	mean	min	max	std	mean	min	max	std
$CAN2^{[20,ST]}$	22.12	**19.80**	28.90	3.45	*72.0*	*43.0*	*123.0*	*28.4*
$CAN2^{[50,ST]}$	24.83	21.80	30.80	2.40	*21.1*	*0.0*	*73.0*	*24.3*
$CAN2^{[(20,50),ST]}$	**21.58**	20.70	**22.50**	**0.69**	*39.0*	*25.0*	*61.0*	*10.1*
$CAN2^{[20,OS]}$	*27.70*	*26.60*	*29.10*	*0.74*	6.4	3.0	9.0	1.4
$CAN2^{[50,OS]}$	*31.29*	*24.80*	*33.60*	*2.20*	1.5	0.0	5.0	1.7
$CAN2^{[(20,50),OS]}$	*29.04*	*27.80*	*31.10*	*1.03*	**1.3**	**0.0**	**4.0**	1.4

$$\dot{x} = \begin{bmatrix} 0 & 1 & 0 & 0 \\ -\frac{K}{m} & -\frac{C}{m} & \frac{K}{m} & \frac{C}{m} \\ 0 & 0 & 0 & 1 \\ \frac{K}{M} & 0 & -\frac{K}{M} & 0 \end{bmatrix} x + \begin{bmatrix} 0 \\ 0 \\ 0 \\ \frac{1}{M} \end{bmatrix} F . \tag{14}$$

3.2 Parameter Settings

Suppose that the controller has to move the load on the car from $x = 0$ to the target position $x_d = 5$m by means of operating F. We obtain discrete signals by $u_t^{[p]} = F(tT_v)$ and $y_t^{[p]} = x(tT_v)$ with sampling period $T_S = 0.01$s. We have used $N_y = 20$, $N_u = 19$ and $\lambda_u = 0.01$ for the GPC and $N_e = 20$ duration for (11).

The parameters of the plant have been set as follows; weight of the car $M = 100$ kg, spring constant $K = 15$ kg/s^2, damping coefficient $C = 10$ kg/s, and the maximum driving force $F_{max} = 30$ N. To achieve robust control to the plant with load weight $\theta = m = m_{tst} \in \Theta^{[tst]} = \{10, 15, 20, \cdots, 60\}$ [kg], we have trained CAN2s for the plant with load weight $\theta = m = m_{tr} \in \Theta^{[tr]} = \{20, 50\}$ [kg] and obtained single $CAN2^{[20,O]}$ and $CAN2^{[50,O]}$ and the combined $CAN2^{[(20,50),O]} = CAN2^{[20,O]} \cup CAN2^{[50,O]}$ for $O = OS$ and ST, where OS and ST represent the control objectives to reduce the overshoot and settling time, respectively. For the control and training iterations shown in Sect. 2.3, we have employed $|\mathcal{I}_{it}| = 10$ iterations, and the default control sequence for the control phase at the fist iteration is as $F(t) = 0.8F_{max}$ for $0 \le t < 5$[s], 0 for $5 \le t < 10$[s] and $-0.8F_{max}$ for $10 \le t < 15$[s], which moves and stops the car at a certain position.

3.3   Results and Analysis

Statistical Result. A statistical result of settling time t_{ST} and overshoot x_{OS} obtained by the controllers using single and multiple CAN2s is shown in Table 1. We can see that multiple $CAN2^{[(20,50),ST]}$ has achieved smaller mean, max and std values than the single $CAN2^{[20,ST]}$ and $CAN2^{[50,ST]}$, while $CAN2^{[(20,50),OS]}$ has achieved smaller mean, min, max and std values than $CAN2^{[20,OS]}$ and $CAN2^{[50,OS]}$. This result shows the GPC using multiple CAN2s have achieved better control performance than using the constituent each single CAN2.

Time Course of Load Position and Poles of Transfer Function. In order to examine how the present method works, we show an example time course of the load position x of the test plant with $m = m_{tst} = 60$ kg for the control objective to minimize the overshoot in Fig. 2(a). We can see that the response for the multiple $CAN2^{[(20,50),OS]}$ seems medium of the responses for the constituent $CAN2^{[20,OS]}$ and $CAN2^{[50,OS]}$.

To analyze the behavior much more, we show selected learning models $M(m_{tr}, c, OS)$ and the transfer function $H(m_{tst}, m_{tr}, c, OS)$ in Fig. 2(b) and (c), respectively, during the control of test plant with $m = m_{tst} = 60$ kg. Here, $M(m_{tr}, c, OS)$ denotes $\boldsymbol{M}_c^{[\theta^*]}$ of $CAN2^{[m_{tr}, OS]}$ (see (11)), and $H(m_{tst}, m_{tr}, c, OS)$ denotes $H_{e/r}^{[p]}(z)$ given by (10) of the GPC using $M(m_{tr}, c, OS)$ to control the plant with $m = m_{tst}$. From (b), we can see that $M(20, 1, OS)$ and $M(50, 4, OS)$ are selected after about 25s corresponding to near the steady state of x (see (a)), while other $M(m_{tr}, c, OS)$ are selected until then. From (c), we can see that $H(60, 50, 4, OS)$ and $H(60, 20, 1, OS)$ corresponding to $M(20, 1, OS)$ and $M(50, 4, OS)$ have the poles with the magnitude less than 1 indicating stable control. On the other hand, $H(m_{tst}, m_{tr}, c, OS)$ with the magnitude of poles larger than 1 indicating unstable control, such as $H(60, 20, 8, OS)$ and $H(60, 20, 8, OS)$, are selected during transient period. Those poles are expected to contribute to not the stability but a quick response, and the selection of them before the steady state may contribute to good performance of the control objective to reduce the overshoot.

Effectiveness and Remarks. Different from the present method, usual robust control methods are designed to place the poles smaller than 1 for all the plants to be controlled through all the control duration, and then they are considered to be more conservative than the present method. On the other hand, the control stability of the present method described above is not guaranteed for all the states of the plants to be controlled. However, it is considered that the stability can be guaranteed by means of restricting the use of the learning models corresponding to unstable poles only at the transient period. For example,

(1) if $|y_t^{[p]} - r^{[p]}| \le e_\theta$, use $u_t^{[p]} = \widehat{u}_t^{[p]}$ predicted by stable learning models such as $M(20, 1, OS)$, for a threshold e_θ (> 0),
(2) otherwise, $u_t^{[p]} = \widehat{u}_t^{[p]}$.

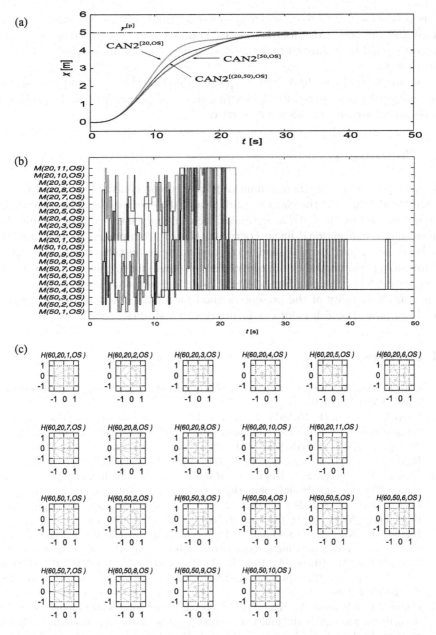

Fig. 2. (a) Time course of the load position x of the plant with $m = 60\,\mathrm{kg}$ controlled by the GPC using CAN2[20,OS] (green), CAN2[50,OS] (blue) and CAN2[(20,50),OS] (black) with the resultant overshoot $x_{OS} = 7,\ 5,\ 4$ [mm], respectively, (b) time course of selected learning models by the GPC using CAN2[20,OS] (green), CAN2[50,OS] (blue) and CAN2[(20,50),OS] (black), and (c) poles of transfer function $H^{[60,20,c,OS]}$ and $H^{[60,50,c,OS]}$. The poles are connected with lines from the poles with smaller arguments to larger arguments to see the pole distribution intuitively. (Color figure online)

This process seems to work when $y_t^{[\mathrm{p}]}$ changes gradually to $r^{[\mathrm{p}]}$. However, the proof and the conditions for the control stability and the effectiveness of this process should be examined and analyzed in detail much more, which is for our future work.

Although we do not have enough space to describe the result and analysis for the control objective to reduce the settling time, the properties and the problems to be solved are almost the same as above.

4   Conclusion

We have presented a method of multiobjective robust controller using GPC and CAN2s and examined the control stability and the effectiveness. Since the plant models learned by the CAN2s are linear, we can analyze the control stability by means of the conventional linear control methods. From numerical experiments, we have shown that learning models corresponding to unstable poles are selected in transient period and they are considered to contribute to quick response, while learning models corresponding to stable poles are selected during steady state. The control stability of the present method for all the states of the plants has not been proved, which is for our future research studies.

References

1. Clarki, D.W., Mohtadi, C.: Properties of generalized predictive control. Automatica **25**(6), 859–875 (1989)
2. Kurogi, S., Yuno, H., Nishida, T., Huang, W.: Robust control of nonlinear system using difference signals and multiple competitive associative nets. In: Lu, B.-L., Zhang, L., Kwok, J. (eds.) ICONIP 2011. LNCS, vol. 7064, pp. 9–17. Springer, Heidelberg (2011). https://doi.org/10.1007/978-3-642-24965-5_2
3. Huang, W., Kurogi, S., Nishida, T.: Robust controller for flexible specifications using difference signals and competitive associative nets. In: Huang, T., Zeng, Z., Li, C., Leung, C.S. (eds.) ICONIP 2012. LNCS, vol. 7667, pp. 50–58. Springer, Heidelberg (2012). https://doi.org/10.1007/978-3-642-34500-5_7
4. Huang, W., Kurogi, S., Nishida, T.: Performance improvement via bagging competitive associative nets for multiobjective robust controller using difference signals. In: Lee, M., Hirose, A., Hou, Z.-G., Kil, R.M. (eds.) ICONIP 2013. LNCS, vol. 8226, pp. 319–327. Springer, Heidelberg (2013). https://doi.org/10.1007/978-3-642-42054-2_40
5. Huang, W., Ishiguma, Y., Kurogi, S.: Properties of multiobjective robust controller using difference signals and multiple competitive associative nets in control of linear systems. In: Loo, C.K., Yap, K.S., Wong, K.W., Beng Jin, A.T., Huang, K. (eds.) ICONIP 2014. LNCS, vol. 8836, pp. 58–67. Springer, Cham (2014). https://doi.org/10.1007/978-3-319-12643-2_8
6. Kurogi, S., Ren, S.: Competitive associative network for function approximation and control of plants. In: Proceedings of the NOLTA 1997, pp. 775–778 (1997)
7. Kohonen, T.: Associative Memory. Communication and Cybernetics. Springer, Heidelberg (1977). https://doi.org/10.1007/978-3-642-96384-1

8. Ahalt, A.C., Krishnamurthy, A.K., Chen, P., Melton, D.E.: Competitive learning algorithms for vector quantization. Neural Netw. **3**, 277–290 (1990)
9. Kimaev, G., Ricardez-Sandoval, L.A.: Nonlinear model predictive control of a multiscale thin film deposition process using artificial neural networks. Chem. Eng. Sci. **207**(2), 1230–1245 (2019)
10. Hyatt, P., Wingate, D., Killpack, M.D.: Model-based control of soft actuators using learned non-linear discrete-time models. Front. Robot. AI **6**, 1–11 (2019)
11. Mayne, D.Q., Rawlings, J.B., Rao, C.V., Scokaert, P.O.M.: Constrained model predictive control: stability and optimality. Automatica **36**, 789–814 (2000)
12. Patan, K.: Neural network-based model predictive control: fault tolerance and stability. IEEE Trans. Control Syst. Technol. **23**, 1147–1155 (2015)
13. Li, X., Sun, J.Q.: Multi-objective optimal predictive control of signals in urban traffic network. J. Intell. Transp. Syst. Technol. Plan. Oper. **23**(4), 370–388 (2019)
14. Kurogi, S., Sawa, M., Ueno, T., Fuchikawa, Y.: A batch learning method for competitive associative net and its application to function approximation. In: Proceedings of SCI2004, vol. 5, pp. 24–28 (2004)

Interpreting Layered Neural Networks via Hierarchical Modular Representation

Chihiro Watanabe[✉]

NTT Communication Science Laboratories,
3-1, Morinosato Wakamiya, Atsugi-shi, Kanagawa Prefecture, Japan
chihiro.watanabe.xz@hco.ntt.co.jp

Abstract. Interpreting the prediction mechanism of complex models is currently one of the most important tasks in the machine learning field, especially with layered neural networks, which have achieved high predictive performance with various practical data sets. To reveal the global structure of a trained neural network in an interpretable way, a series of clustering methods have been proposed, which decompose the units into clusters according to the similarity of their inference roles. The main problems in these studies were that (1) we have no prior knowledge about the optimal resolution for the decomposition, or the appropriate number of clusters, and (2) there was no method for acquiring knowledge about whether the outputs of each cluster have a positive or negative correlation with the input and output unit values. In this paper, to solve these problems, we propose a method for obtaining a hierarchical modular representation of a layered neural network. The application of a hierarchical clustering method to a trained network reveals a tree-structured relationship among hidden layer units, based on their feature vectors defined by their correlation with the input and output unit values.

Keywords: Interpretable machine learning · Neural network · Hierarchical clustering

1 Introduction

To construct a method for interpreting the prediction mechanism of complex statistical models is currently one of the most important tasks in the machine learning field, especially with layered neural networks (or LNNs), which have achieved high predictive performance in various practical tasks. Due to their complex hierarchical structure and the nonlinear parameters that they use to process the input data, we cannot understand the function of a trained LNN as it is, and we need some kind of approximation method to convert the original function of an LNN into a simpler interpretable representation.

Recently, various methods have been proposed for interpreting the function of an LNN, and they can be roughly classified into (1) the approximation of an LNN with an interpretable model, and (2) the investigation of the roles of the

© Springer Nature Switzerland AG 2019
T. Gedeon et al. (Eds.): ICONIP 2019, CCIS 1143, pp. 376–388, 2019.
https://doi.org/10.1007/978-3-030-36802-9_40

partial structures constituting an LNN (e.g. units or layers). As for approach (1), various methods have been investigated for approximating an LNN with a linear model [10,12,14] or a decision tree [4,7,20]. For image classification tasks in particular, methods for visualizing an LNN function have been extensively studied in terms of which part of an input image affects the prediction result [2, 15,16,18,19,21]. Recent study [17] is similar to our proposed method in that it uses hierarchical clustering for interpreting an LNN, however, it differs from ours in terms of the extracted type of knowledge. The method in [17] extracts hierarchical structure of *input features* according to their relationship with the output, while our proposed method enables us to know a hierarchical structure of *hidden layer units* according to their input-output mapping functions. Approach (2) has been studied by several authors who examined the function of a given part of an LNN [1,11,13,28]. There has also been an approach designed to extract the cluster structure of a trained LNN [23,24,26] based on network analysis.

Although the above methods provide us with an interpretable representation of an LNN function with a fixed resolution (or number of clusters), there is a problem in that we do not know in advance the optimal resolution for interpreting the original network. In the methods described in the previous studies [23–27], the unit clustering results may change greatly with the cluster size setting, and there is no criterion for determining the optimal cluster size. Another problem is that the previous studies could only provide us with information about the magnitude of the relationship between a cluster and each input or output unit value, and we could not determine whether this relationship was positive or negative.

In this paper, we propose a method for extracting a hierarchical modular representation from a trained LNN, which provides us with both hierarchical clustering results with every possible number of clusters and the function of each cluster. Our proposed method mainly consists of three parts: (a) training an LNN based on an arbitrary optimization method, (b) determining the feature vectors of each hidden layer unit based on its correlation with the input and output unit values, and (c) the hierarchical clustering of the feature vectors. Unlike the clustering methods in the previous studies, the role of each cluster is computed as a centroid of the feature vectors defined by the correlations in step (b), which enables us to know the representative mapping performed by the cluster in terms of both sign and magnitude for each input or output unit.

We show experimentally the effectiveness of our proposed method in interpreting the internal mechanism of a trained LNN, by applying it to a practical data set. Based on the experimental results for the extracted hierarchical cluster structure and the role of each cluster, we discuss how the overall LNN function is structured as a collection of individual units.

2   Hierarchical Modular Representation of LNNs

2.1   Determining Feature Vectors of Hidden Layer Units

To apply hierarchical clustering to a trained LNN, we define a feature vector for each hidden layer unit. Let v_k be the feature vector of the k-th hidden layer unit in a hidden layer. Such a feature vector should reflect the role of its corresponding unit in LNN inference. Here, we propose defining such a feature vector v_k of the k-th hidden layer unit based on its correlations between each input or output unit. In previous studies [25, 27], methods have been proposed for determining the role of a unit or a unit cluster based on the square root error. However, these methods can only provide us with knowledge about the magnitude of the effect of each input unit on another unit and the effect of a unit on each output unit, not information about how a hidden layer unit is affected by each input unit and how each output unit is affected by a hidden layer unit. In other words, there is no method that can reveal whether an increase in the input unit value has a positive or negative effect on the output value of a hidden layer unit, or whether an increase in the output value of a hidden layer unit has a positive or negative effect on the output unit value. To obtain such sign information regarding the roles of each hidden layer unit, we use the following definition based on the correlation.

Definition 1 (Effect of i-th input unit on k-th hidden layer unit). *We define the effect of the i-th input unit on the k-th hidden layer unit as v_{ik}^{in}, where*

$$v_{ik}^{in} = \frac{E\left[\left(X_i^{(n)} - E[X_i^{(n)}]\right)\left(o_k^{(n)} - E[o_k^{(n)}]\right)\right]}{\sqrt{E\left[\left(X_i^{(n)} - E[X_i^{(n)}]\right)^2\right]E\left[\left(o_k^{(n)} - E[o_k^{(n)}]\right)^2\right]}}.$$

Here, $E[\cdot]$ represents the mean for all the data samples, $X_i^{(n)}$ is the i-th input unit value of the n-th data sample, and $o_k^{(n)}$ is the output of the k-th hidden layer unit for the n-th input data sample.

Definition 2 (Effect of k-th hidden layer unit on j-th output unit). *We define the effect of the k-th hidden layer unit on the j-th output unit as v_{kj}^{out}, where*

$$v_{kj}^{out} = \frac{E\left[\left(o_k^{(n)} - E[o_k^{(n)}]\right)\left(y_j^{(n)} - E[y_j^{(n)}]\right)\right]}{\sqrt{E\left[\left(o_k^{(n)} - E[o_k^{(n)}]\right)^2\right]E\left[\left(y_j^{(n)} - E[y_j^{(n)}]\right)^2\right]}}.$$

Here, $y_j^{(n)}$ is the value of the j-th output layer unit for the n-th input data sample.

We define a feature vector of each hidden layer unit based on the above definitions.

Definition 3 (Feature vector of k-th hidden layer unit). *We define the feature vector of the k-th hidden layer unit as $v_k \equiv [v_{1k}^{in}, \cdots, v_{i_0k}^{in}, v_{k1}^{out}, \cdots, v_{kj_0}^{out}]$. Here, i_0 and j_0, respectively, represent the number of the input and output units.*

By definition, all the elements of a feature vector satisfy $v_{kl} \in [-1, 1]$.

Algorithm 1. Alignment of signs of feature vectors based on cosine similarity

1: Let v_k and a_0 respectively be the feature vector for the k-th hidden layer unit and the number of iterations.
2: **for** $a = 1$ to a_0 **do**
3:    Randomly choose the k-th hidden layer unit according to the uniform distribution.
4:    **if** $\sum_{l \neq k} \frac{v_k \cdot v_l}{\sqrt{v_k \cdot v_k} \sqrt{v_l \cdot v_l}} < 0$ **then**
5:        $v_k \leftarrow -v_k$.
6:    **end if**
7: **end for**

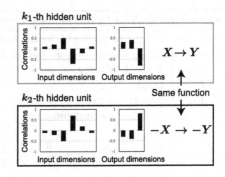

Fig. 1. An example of two hidden layer units with the same function. The corresponding feature vectors are the same, except that their signs are opposite.

Alignment of Signs of Feature Vectors Based on Cosine Similarity. The feature vectors of Definition 3 represent the roles of the hidden layer units in terms of input-output mapping. When interpreting such roles of hidden layer units, it is natural to regard the roles of any pair of units (k_1, k_2) as being the same iff they satisfy $v_{k_1} = v_{k_2}$ or $v_{k_1} = -v_{k_2}$. The latter condition corresponds to the case where the k_1-th and k_2-th units have the same correlations with input and output units except that their signs are the opposite, as depicted in Fig. 1. To regard the roles of unit pairs that satisfy one of the above conditions as the same, we propose an algorithm for aligning the signs of the feature vectors based on cosine similarity (Algorithm 1). By randomly selecting a feature vector and aligning its sign according to the sum of the cosine similarities with all the other feature vectors, the sum of the cosine similarities of all the pairs of feature vectors increases monotonically.

2.2 Hierarchical Clustering of Units in a Trained LNN

Once we have obtained the feature vectors of all the hidden layer units as described in Sect. 2.1, we can extract a hierarchical modular representation of an LNN by applying hierarchical clustering to the feature vectors. Among the several existing methods for such hierarchical clustering including single-link and complete-link, Ward's method [22] has been shown experimentally to be effective in terms of its classification sensitivity, so we employ this method in our experiments.

We start with k_0 individual hidden layer units, and sequentially combine clusters with the minimum *error sum of squares (ESS)*, which is given by

$$ESS \equiv \sum_m \left(\sum_{k:u_k \in C_m} \|v_k\|^2 - \frac{1}{|C_m|} \left\| \sum_{k:u_k \in C_m} v_k \right\|^2 \right), \tag{1}$$

Algorithm 2. Ward's hierarchical clustering method [22]

1: Let u_k and v_k, respectively, be the k-th hidden layer unit ($k = 1, \cdots, k_0$) and its corresponding feature vector, and let $\{C_m^{(t)}\}$ be the unit set assigned to the m-th cluster in the t-th iteration ($m = 1, \cdots, k_0 - t + 1$). Initially, we set $t \leftarrow 1$ and $C_m^{(1)} \leftarrow \{u_m\}$.

2: **for** $t = 2$ to $k_0 - 1$ **do**

3: $\quad (C_{m_1}^{(t-1)}, C_{m_2}^{(t-1)}) \leftarrow \arg\min_{(C_i^{(t-1)}, C_j^{(t-1)})} \Delta ESS(C_i^{(t-1)}, C_j^{(t-1)})$, where

$$\Delta ESS(C, C') \equiv \frac{|C||C'|}{|C| + |C'|} \left\| \frac{1}{|C|} \sum_{k:u_k \in C} v_k - \frac{1}{|C'|} \sum_{k:u_k \in C'} v_k \right\|^2.$$

Here, we assume $m_1 < m_2$.

4: $\quad$ Update the clusters as follows:

$$C_m^{(t)} \leftarrow \begin{cases} C_{m_1}^{(t-1)} \cup C_{m_2}^{(t-1)} & (m = m_1) \\ C_m^{(t-1)} & (1 \leq m \leq m_2 - 1, m \neq m_1) \\ C_{m+1}^{(t-1)} & (m_2 \leq m \leq k_0 - t + 1) \end{cases}.$$

5: **end for**

---

where u_k and v_k, respectively, are the k-th hidden layer unit ($k = 1, \cdots, k_0$) and its corresponding feature vector, C_m is the unit set assigned to the m-th cluster, and $|\cdot|$ represents the cluster size. From Eq. (1), the ESS is the value given by first computing the cluster size ($|C_m|$) times the variance of the feature vectors in each cluster, and then by taking the sum of all these values for all the clusters. When combining a pair of clusters (C_{m_1}, C_{m_2}) into one cluster, the ESS increases by

$$\Delta ESS = \frac{|C_{m_1}||C_{m_2}|}{|C_{m_1}| + |C_{m_2}|} \left\| \frac{1}{|C_{m_1}|} \sum_{k:u_k \in C_{m_1}} v_k - \frac{1}{|C_{m_2}|} \sum_{k:u_k \in C_{m_2}} v_k \right\|^2. \quad (2)$$

Therefore, in each iteration, we do not have to compute the error sum of squares for all the clusters, instead we simply have to compute the error increase ΔESS given by Eq. (2) for all the pairs of current clusters (C_{m_1}, C_{m_2}), find the optimal pair of clusters that achieves the minimum error increase, and combine them. We describe the whole procedure of Ward's method in Algorithm 2.

This procedure to combine a pair of clusters is repeated until all the hidden layer units are assigned to one cluster, and from the clustering result $\{C_m^{(t)}\}$ in each iteration $t = 1, \cdots, k_0 - 1$, we can obtain a hierarchical modular representation of an LNN, which connects the two extreme resolutions given by "all units are in a single cluster" and "all clusters consist of a single unit." The role of each extracted cluster can be determined from the centroid of the feature vectors of the units assigned to the cluster, which can be interpreted as a representative input-output mapping of the cluster.

3   Experiments

3.1   Experimental Settings

We applied our proposed method to the MNIST data set [8] to show its effectiveness in interpreting the mechanism of trained LNNs. Figure 2 shows training sample images for each class of digits. We trained the convolutional neural network LeNet-5 [8] by using Adam [6] to recognize 10 types of digits from input images, and set the batch size and the number of epochs at 100 and 30, respectively.

Before applying hierarchical clustering to the feature vectors, we performed the following procedure.

- We only used uniformly randomly chosen 2500 hidden layer units for computing feature vectors to save memory.
- Before computing the input and output correlations, we deleted the input and output units and hidden layer units whose values have zero standard deviations, regardless of input data.
- After computing the input and output correlations, we deleted the hidden layer units which have smaller absolute correlation coefficients than 0.5 with all the output unit values. In other words, we deleted the hidden layer units which have relatively small effect on the output.

The iteration number a_0 for the alignment of the signs of the feature vectors was set at 4000.

3.2   Experiment Using the MNIST Data Set

Figures 3 and 4 to 6, respectively, show the hierarchical cluster structure extracted from the trained LNN, and the roles or representative input-output mappings of the clusters with the three different hierarchical levels, where the number of clusters were 5, 10, and 20. From these figures, we find knowledge about the LNN structure as follows.

- At the coarsest resolution, the main function of the trained LNN is decomposed into Clusters A1 to A5 (Fig. 4). Cluster A1 captures the input information about black pixels in the shape of a "0" and white pixels in the shape of a center part of an image, and it has a positive correlation with the output unit value corresponding to "0." Cluster A2 correlates negatively with the region in the shape of a "1" in a center part and also with the upper part of an image, and positively with the other areas. It has a positive correlation with the recognition of "9," and it has a negative one with "1." Cluster A3 has a complex relationship with the input unit values, and it correlates negatively with the area with the shape of a "3." It has a positive correlation with "0," "4" and "6," and a negative correlation with "3," "7," and "9." Cluster A4 captures the 0-shaped region, however, it mainly affects the recognition result for "5," "6," and "7." Cluster A5 uses the input information of the black pixels in the upper, right, and bottom part of an image, and it correlates negatively with the digits "4" and "9."

– Cluster A2 is decomposed into three smaller clusters, B2, B3 and B4, which capture mutually different input information. On one hand, the role of cluster B4 is almost the same as that of cluster A2. On the other hand, cluster B2 is mainly affected by the region in the shape of a "0," and it affects the recognition result for various digits (for instance, "1," "3," "4," "6" and "9"). Clusters B5, B6 and B7 are both the member of cluster A3, and they have similar relationships with the input unit values (concretely, all of them correlates negatively with the C-shaped region). However, their correlation with the output are mutually different. For instance, the main recognition targets of B5 are "3," "4" and "6," while those of B7 are "6" and "7."
– In the bottom hierarchy, there are clusters with mutually similar roles, such as the pair of C13 and C14 (note that both of these clusters are the member of cluster B8). Cluster B5 consists of two smaller clusters, C8 and C9. Cluster C8 captures the input information about white pixels in the shape of a "3" and it correlates negatively with the digits "3." This cluster is also used for recognition of "0," "4," "5" and "6," while the main recognition targets of C9 are "3," "4," "6" and "7."

3.3 Comparison with Clustering Method Based on Non-negative Matrix Factorization

Here, we show the effectiveness of our proposed method by comparing it with the clustering method based on non-negative matrix factorization (or NNMF), which was proposed in a previous study [25]. In the previous study [25], the feature vectors of the hidden layer units are defined by the magnitude of the effect of each input unit value on a cluster and the effect of a cluster on each output unit value, computed by the square root error of the unit output values. By definition, the elements of such feature vectors are all non-negative, which is a necessary condition for applying NNMF to the feature vectors.

We applied the NNMF-based clustering method to the same trained network that we applied hierarchical clustering, and decomposed the network into 5, 10, and 20 clusters. Here, we set the number of iterations of the NNMF algorithm at 500. We applied the NNMF algorithm for 100 times, and used the best result in terms of the approximation error. Initial values of the two low-dimensional matrices were randomly chosen according to the normal distribution $\mathcal{N}(1, 0.5)$.

Figures 7, 8 and 9 show the representative roles of the extracted clusters. Comparing these figures with the results in Figs. 4 to 6, we can observe that the previous NNMF-based method could not capture the structures of the input and output unit values in as much detail as our proposed method, since it does not take the sign information into account. Furthermore, with the NNMF-based method, we should define the number of clusters in advance, and we cannot observe the hierarchical structure of clusters to find the optimal resolution for interpreting the roles of partial structures of an LNN (note that the clusters shown in Figs. 7, 8 and 9 do NOT have hierarchical relationship mutually).

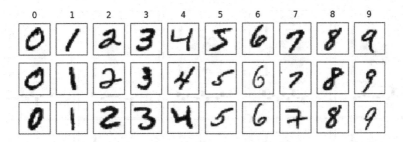

Fig. 2. Input image examples of MNIST data set.

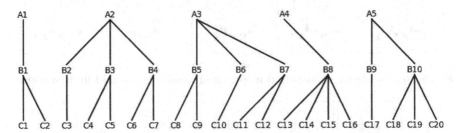

Fig. 3. Hierarchical relationship between clusters. The roles of clusters A1 to A5 in the top hierarchy are shown in Fig. 4. These clusters are decomposed into 10 smaller clusters B1 to B10, whose roles are shown in Fig. 5. Clusters B1 to B10 are decomposed into clusters C1 to C20 in the bottom hierarchy, whose relationships with the input and output unit values are shown in Fig. 6.

4   Discussion

Here, we discuss our proposed method for obtaining a hierarchical modular representation from the perspectives of statistical evaluation and visualization.

Our proposed method provides us with a series of clustering results for an arbitrary cluster size, and the resulting structure does not change if we use the same criterion (e.g. error sum of squares for Ward's method) for evaluating the similarity of the feature vectors. However, there is no way to determine which criterion yields the optimal clustering result to represent a trained LNN, due to the

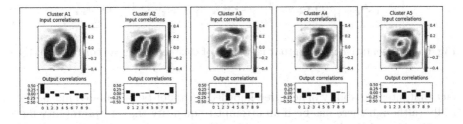

Fig. 4. Representative input-output mappings of clusters in the **top hierarchy**.

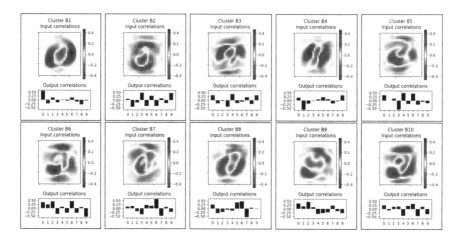

Fig. 5. Representative input-output mappings of clusters in the **middle hierarchy**.

Fig. 6. Representative input-output mappings of clusters in the **bottom hierarchy**.

fact that interpretability of acquired knowledge cannot be formulated mathematically (although there has been an attempt to quantify the interpretability for a specific task, especially image recognition [3]). This problem makes it impossible to compare different methods for interpreting LNNs quantitatively, as pointed

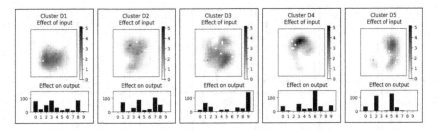

Fig. 7. Representative input-output mappings of clusters when decomposing the LNN into five clusters with **NNMF**.

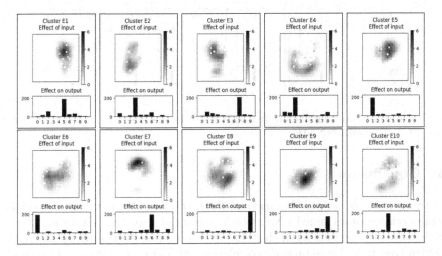

Fig. 8. Representative input-output mappings of clusters when decomposing the LNN into 10 clusters with **NNMF**.

out in the previous studies [5,9]. Therefore, the provision of a statistical evaluation method as regards both interpretability and accuracy for the resulting cluster structure constitutes important future work.

Although we can apply our proposed method to an arbitrary network structure, as long as it contains a set of units that outputs some value for a given input data sample, the visualization of the resulting hierarchical modular representations becomes more difficult with a deeper and a larger scale network structure, since a cluster may contain units in mutually distant layers. Additionally, the number of possible cluster sizes increases with the scale (or the number of units) of a network, and so it is necessary to construct a method for automatically selecting a set of representative resolutions, instead of visualizing the entire hierarchical cluster structure.

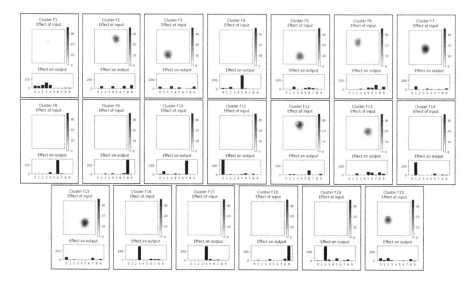

Fig. 9. Representative input-output mappings of clusters when decomposing the LNN into 20 clusters with **NNMF**.

5   Conclusion

In this paper, we proposed a method for extracting the hierarchical modular representation of a trained LNN, which consists of sequential clustering results with every possible number of clusters. By determining the feature vectors of the hidden layer units based on their correlations with input and output unit values, it also enabled us to know what range of input each cluster maps to what range of output. We showed the effectiveness of our proposed method experimentally by applying it to a practical data set and by interpreting the resulting cluster structure.

References

1. Alain, G., Bengio, Y.: Understanding intermediate layers using linear classifier probes. In: ICLR 2017 Workshop (2017)
2. Ancona, M., Ceolini, E., Öztireli, A.C., Gross, M.: Towards better understanding of gradient-based attribution methods for deep neural networks. In: International Conference on Learning Representations (2018)
3. Bau, D., Zhou, B., Khosla, A., Oliva, A., Torralba, A.: Network dissection: quantifying interpretability of deep visual representations. In: Computer Vision and Pattern Recognition (2017)
4. Craven, M., Shavlik, J.W.: Extracting tree-structured representations of trained networks. In: Advances in Neural Information Processing Systems, vol. 8, pp. 24–30 (1996)
5. Doshi-Velez, F., Kim, B.: Towards a rigorous science of interpretable machine learning. arXiv:1702.08608 (2017)

6. Kingma, D.P., Ba, J.: Adam: a method for stochastic optimization. In: International Conference on Learning Representations (2015)
7. Krishnan, R., Sivakumar, G., Bhattacharya, P.: Extracting decision trees from trained neural networks. Pattern Recogn. **32**(12), 1999–2009 (1999)
8. LeCun, Y., Bottou, L., Bengio, Y., Haffner, P.: Gradient-based learning applied to document recognition. Proc. IEEE. **86**, 2278–2324 (1998)
9. Lipton, Z.C.: The mythos of model interpretability. In: Proceedings of the 2016 ICML Workshop on Human Interpretability in Machine Learning (2016)
10. Lundberg, S.M., Lee, S.: A unified approach to interpreting model predictions. In: Advances in Neural Information Processing Systems, vol. 30, pp. 4765–4774 (2017)
11. Luo, W., Li, Y., Urtasun, R., Zemel, R.: Understanding the effective receptive field in deep convolutional neural networks. In: Advances in Neural Information Processing Systems, vol. 29, pp. 4898–4906 (2016)
12. Nagamine, T., Mesgarani, N.: Understanding the representation and computation of multilayer perceptrons: a case study in speech recognition. In: Proceedings of the 34th International Conference on Machine Learning, pp. 2564–2573 (2017)
13. Raghu, M., Gilmer, J., Yosinski, J., Sohl-Dickstein, J.: SVCCA: singular vector canonical correlation analysis for deep learning dynamics and interpretability. In: Advances in Neural Information Processing Systems, vol. 30, pp. 6076–6085 (2017)
14. Ribeiro, M.T., Singh, S., Guestrin, C.: "Why should I trust you?": explaining the predictions of any classifier. In: Proceedings of the 22nd ACM SIGKDD International Conference on Knowledge Discovery and Data Mining, pp. 1135–1144 (2016)
15. Shrikumar, A., Greenside, P., Kundaje, A.: Learning important features through propagating activation differences. In: Proceedings of the 34th International Conference on Machine Learning, pp. 3145–3153 (2017)
16. Simonyan, K., Vedaldi, A., Zisserman, A.: Deep inside convolutional networks: visualising image classification models and saliency maps. In: ICLR 2014 Workshop (2014)
17. Singh, C., Murdoch, W.J., Yu, B.: Hierarchical interpretations for neural network predictions. In: International Conference on Learning Representations (2019)
18. Springenberg, J.T., Dosovitskiy, A., Brox, T., Riedmiller, M.: Striving for simplicity: the all convolutional net. In: ICLR 2015 Workshop (2015)
19. Sundararajan, M., Taly, A., Yan, Q.: Axiomatic attribution for deep networks. In: Proceedings of the 34th International Conference on Machine Learning, pp. 3319–3328 (2017)
20. Thiagarajan, J.J., Kailkhura, B., Sattigeri, P., Ramamurthy, K.N.: Treeview: peeking into deep neural networks via feature-space partitioning. In: NIPS 2016 Workshop on Interpretable Machine Learning in Complex Systems (2016)
21. Wagner, J., Köhler, J.M., Gindele, T., Hetzel, L., Wiedemer, J.T., Behnke, S.: Interpretable and fine-grained visual explanations for convolutional neural networks. In: Computer Vision and Pattern Recognition (2019)
22. Ward, J.H.: Hierarchical grouping to optimize an objective function. J. Am. Stat. Assoc. **58**(301), 236–244 (1963)
23. Watanabe, C., Hiramatsu, K., Kashino, K.: Modular representation of autoencoder networks. In: Proceedings of 2017 IEEE Symposium on Deep Learning, 2017 IEEE Symposium Series on Computational Intelligence (2017)
24. Watanabe, C., Hiramatsu, K., Kashino, K.: Recursive extraction of modular structure from layered neural networks using variational Bayes method. In: Yamamoto, A., Kida, T., Uno, T., Kuboyama, T. (eds.) DS 2017. LNCS (LNAI), vol. 10558, pp. 207–222. Springer, Cham (2017). https://doi.org/10.1007/978-3-319-67786-6_15

25. Watanabe, C., Hiramatsu, K., Kashino, K.: Knowledge discovery from layered neural networks based on non-negative task decomposition. arXiv:1805.07137v2 (2018)
26. Watanabe, C., Hiramatsu, K., Kashino, K.: Modular representation of layered neural networks. Neural Netw. **97**, 62–73 (2018)
27. Watanabe, C., Hiramatsu, K., Kashino, K.: Understanding community structure in layered neural networks. arXiv:1804.04778 (2018)
28. Zahavy, T., Ben-Zrihem, N., Mannor, S.: Graying the black box: understanding DQNs. In: Proceedings of the 33rd International Conference on Machine Learning, pp. 1899–1908 (2016)

Application Identification of Network Traffic by Reservoir Computing

Toshiyuki Yamane[1]([⊠]), Jean Benoit Héroux[1], Hidetoshi Numata[1],
Gouhei Tanaka[2], Ryosho Nakane[2], and Akira Hirose[2]

[1] IBM Research - Tokyo, Kawasaki, Kanagawa 212-0032, Japan
{tyamane,heroux,hnumata}@jp.ibm.com
[2] Institute for Innovation in International Engineering Education, Graduate School
of Engineering, The University of Tokyo, Tokyo 113-8656, Japan
gouhei@sat.t.u-tokyo.ac.jp, nakane@cryst.t.u-tokyo.ac.jp,
ahirose@ee.t.u-tokyo.ac.jp

Abstract. We propose a method for application identification for
network traffic by reservoir computing. Different from conventional
approaches, the proposed method handles traffic flows as dynamical time
series data and enables fast and real-time identification. We apply the
proposed method to real traffic data and show that high identification
accuracy is achieved. We also discuss an implementation as physical
reservoirs based on optics and the impact of the proposed method to
5G networking.

Keywords: Application identification · Echo state network · Optical
reservoir computing · Internet of Things · 5G networking

1 Introduction

The emergence of IoT (Internet of Things) and 5G networking is making tremendous impact to IT industry. So far, the edge points of the internet generally refer
to user equipments like mobile devices. Now, with the help of massive connectivity of 5G, they include physical sensors, factory equipment, cars, robots, drones
and ultimately everything mutually interconnected and deployed all around the
world. Thus, the combination of IoT and 5G has potential to revolutionize our
society by producing new systems like cyber physical systems, self-driving cars.

From viewpoint of network management, identifying application types (e.g.
web browsing, video streaming and e-mail etc) for network traffic has been
more and more important interest to network administrators and internet service providers. It enables flexible network management such as malware detection, support of different classes of quality-of-service (QoS), service level agree-
ments (SLA), pricing, security and prioritization policy, depending on application types.

As is described in Sect. 2, machine learning approaches had great success for
application identification [9]. However, 5G network traffic generated by IoT is

© Springer Nature Switzerland AG 2019
T. Gedeon et al. (Eds.): ICONIP 2019, CCIS 1143, pp. 389–396, 2019.
https://doi.org/10.1007/978-3-030-36802-9_41

posing a new challenge to conventional machine learning approaches, such as need for hard real-time decisions and higher throughput for growing amount of traffic data. In fact, they usually operate in an off-line manner and inevitably cause significant decision delay. This is because they relies on static and summarized characteristic of entire network flows and ignore their temporal features.

Considering that communication over the network is inherently temporal, the focus of application identification in 5G IoT era should be on the real-time and on-line operation using temporal information of as few packets as possible. Use of recurrent neural networks is a natural way of such operation since they can handle temporal data and operate in on-line manner. However, there have been very few researches reported so far which use recurrent neural network for application identification [5,6]. Their architectures are based on LSTM-based recurrent neural networks and convolutional neural networks which are still hard to train and computationally expensive.

Reservoir computing (RC) can be an attractive solution for these issues of application identification. RC is a special architecture of recurrent neural network [7], which is composed of an input layer, a (hidden) reservoir layer and a readout (output) layer. The remarkable feature is that the connection weights of the input layer and the reservoir layer are initialized randomly and fixed throughout their operation and only weights in the readout layer are trained by typically well-matured simple adaptive signal processing. Therefore, RC takes both much less learning and operation cost than current large scale deep neural networks which need hard optimization of huge number of parameters. In addition, the reservoir layers can be implemented by various nonlinear physical dynamics since the reservoir layers are random network of nonlinear activation function [11]. The physical implementation of reservoirs can achieve significant performance gain such as real time, higher speed and lower power operation compared to software implementation. In this work, we apply echo state networks, which is a variant of RC, to online application identification of network traffic and show that RC can achieve high identification performance for real network traffic data. We also discuss that optics/photonics implementation of reservoir computing can be an attractive solution which is consistent with the architecture of 5G IoT edge systems.

2   Application Identification for Network Traffic

The approaches to network traffic classification can be categorized into three generations historically. The 1st generation methods relies on well-known port numbers of TCP or UDP, under the assumption that an application always uses the same port number, for example, destination port 22 for ssh. However, it hard to classify network traffic only by the destination port since many recent applications use dynamically changing port numbers for every session.

The 2nd generation methods use signature-based information by deep inspection of payload data, called deep packet inspection (DPI). The DPI can classify network traffic exactly if good signatures can be found in the payload. The disadvantage of this approach is high processing load and complexity since it need to look into the payload. If data are encrypted or specification are frequently updated, the signature-based methods can be very time- and power-consuming, or even impossible. In addition, DPI is often not allowed due to concern for violation of communication privacy.

The limitation of the 1st and 2nd generation methods prompted the 3rd generation machine learning based approaches [9]. Using conventional classifiers such as naive Bayes, K-Means, the machine learning approaches classify network traffic based on their statistical characteristics (i.e, features). Success or failure of these conventional machine learning approaches depend on the features designed by insights of domain experts.

Some recent network management tasks require that network operators make quick decisions to observed traffic and feedback to network environment as fast as possible, for example, QoS provisioning and dynamic routing. In such tasks, on-line and real-time application identification by first few packets is mandatory. However, many conventional machine learning approaches operate rather in an offline manner than in an online manner since they classify the whole traffic. That is, they need to see the entire or large portion of a traffic to calculate summarized statistics such as connection duration time, total number of packets and max, min, mean of packet sizes etc. It is necessary to see network traffic as temporal data for online classification and we will apply reservoir computing, which is a special type of recurrent neural network, described in the following section.

3 Reservoir Computing by Echo State Networks

The echo state networks (ESN) are variants of RC based on rate coding neurons in discrete time [7]. The dynamics for updating reservoir state is given by

$$\mathbf{x}(n+1) = \tanh(W_{res}\mathbf{x}(n) + W_{in}\mathbf{u}(n)), \tag{1}$$

where $\mathbf{x}(n)$ is the N-dimensional reservoir state, W_{res} is the $(N \times N)$ reservoir internal weight matrix, W_{in} is the $N \times K$ input weight matrix, $\mathbf{u}(n)$ is the K-dimensional input signal. Both the W_{res} and W_{in} are initialized randomly and fixed throughout its operation. On the other hand, the readout layer is typically defined as linear filter as follows:

$$\mathbf{y}(t) = W\mathbf{x}(t). \tag{2}$$

At training phase, the readout layer tries to minimize the difference of desired output $\{\hat{\mathbf{y}}(t), t = 1, \dots, T\}$ and the reservoir state $\{\mathbf{x}(t), t = 1, \dots, T\}$. The optimal readout weight W_{out} is given by least-mean-square principle as

$$W_{\text{out}} = \arg\min_{W} \left\{ \frac{1}{T} \sum_{t=1}^{T} \|\hat{\mathbf{y}}(t) - \mathbf{y}(t)\|^2 + \lambda \|W\| \right\}, \tag{3}$$

where the parameter $\lambda > 0$ is a regularization constant. The optimization problem can be solved analytically and W_{out} is given by Wiener solution as

$$W_{\text{out}} = UX^+, \tag{4}$$

where $X = (\mathbf{x}(1), \ldots, \mathbf{x}(T)) \in \mathbb{R}^{N \times T}$ is the reservoir state collection matrix, $U = (\mathbf{u}(1), \ldots, \mathbf{u}(T)) \in \mathbb{R}^{D \times T}$ is the teacher signal collection matrix, and $X^+ = X^T (XX^T)^{-1}$ is the Moore-Penrose pseudo inverse of X. For practical applications, it is important to control temporal memory of reservoir dynamics properly, that is, time constant of decaying correlation of present and past reservoir states. The most frequently used control parameters is spectral radius ρ of W_{res} which is often set to be close to 1, but slightly less than 1 to obtain optimal performance.

4   Experimental Results on Real Traffic Data

4.1   Datasets and Features for Experiments

In most application identification task, the target to be classified is a network traffic flow which is a sequence of packets transmitted during a communication between two host machines. Each host is specified by a combination of IP address and port number. More formally, we define *flow* as a set of successive packets sharing an unique 5-tuple of (source IP address, source port number, destination IP address, destination port number and protocol type (TCP, UDP, etc)).

To evaluate the performance of the classification, one need pre-labeled network traffic flows for ground truth. In the current work, we used packet trace dataset provided by Canadian Institute for Cybersecurity (CIC) at University of New Brunswick (UNB). We selected VPN-nonVPN dataset ISCXVPN2016 (available at https://www.unb.ca/cic/datasets/vpn.html), which is a record of real traffic observed by CIC's network environment. The advantage of this dataset is that captured data are provided for each application type such as e-mails, video streaming (youtube) and hence ground-truth labels for training are already known. Since the dataset ISCXVPN2016 include multiple flows in a fragmented way, first we need to separate and reconstruct each flow from the packet capture files. We used CICFlowmeter [4] for this flow extraction. We excluded flows with too few packets (less than 100 packets) for training. We concatenate the extracted flows and make a long dataset for each application type and two sub sequences (one for training and the other for testing) are extracted from the dataset.

Based on the prior work on application identification, we selected the following six features calculated from each packet in a flow and created six dimensional temporal data: packet inter-arrival time, packet direction (forward or backward), TCP window size in byte, packet length in byte, time-to-live (TTL).

Using the standard network traffic analysis tool tshark (https://www.wireshark.org/), we extracted these six features for each packet in a flow. The feature data are normalized to zero mean and unit standard deviation before processing by ESN.

The task in this experiment is to determine which of the applications types email, video streaming and file transfer the observed flows are generated by. In our experiment, we restricted the flows with TCP protocol since they account for the vast majority of traffic.

4.2 Application Identification Results

The first experiment is the identification of e-mail and video streaming (youtube). The entire 2,000 packet dataset for each application type is divided into two parts; first 1,000 data is used for training and the rest 1,000 data is left for testing. The parameters used in the experiment are as follows: reservoir size = 500 neurons, and spectral radius ρ = 0.9, connectivity of W_{res} = 0.9, regularization coefficient λ = 0.1. We used one-hot vector encoding for each application type, (1, 0) for e-mail and (0, 1) for video streaming. The hard decision was applied to each readout component in such a way that the maximum component is regarded as 1 while the rest are set to 0. Figure 1(a) shows the result of experiment for the classification task for e-mail and video streaming. For training phase, we obtained 0.0187 normalized root mean square error (NRMSE) for e-mail and 0.019 for video streaming. For testing phase, we obtained NRMSE = 0.3083 for e-mail and 0.3879 for video streaming. The error rates of misclassification are 0.4% (video streaming for e-mail) and 5.4% (e-mail for video streaming). To evaluate the contribution of combination of features, we performed experiments with only one of six features (Fig. 1(b)). The best two feature were tcp.len (error rate was 1.6% for e-mail, 13.3% for video streaming) and frame.len (error rate was 1.1% for e-mail, 7% for video streaming). This means that combination of six features contributes to improve the classification performance.

The second experiment is identification of three applications, e-mail, video streaming and file transfer, shown in Fig. 2(a), (b), (c). In this experiment, the one hot encoding for e-mail, video streaming and file transfer were $(1,0,0), (0,1,0)$ and $(0,0,1)$, respectively. The accuracy rates for this identification experiment were 99.4%, 99.8% and 99.9% for email, video streaming and file transfer, respectively.

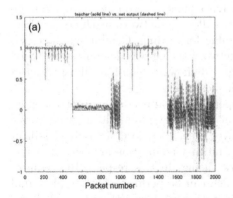
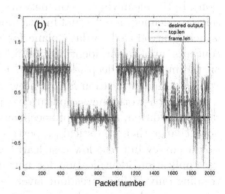

Fig. 1. (a) Comparison of desired output and readout with six features. 0 is for e-mail and 1 for video streaming. Only first component of output is shown. (b) Results for only tcp.len and only frame.len.

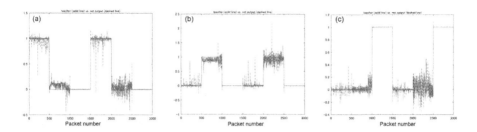

Fig. 2. (a), (b), (c): Comparison of desired output and readout for identification e-mail, video streaming and file transfer

In these experiments, e-mail and file transfer are statistically stable and can be identified with high accuracy, while video streaming has relatively high variations and its dynamics seems to be changing. Anyway, the experimental results show that RC can achieve enough identification accuracy using dynamical features of flows.

5   Reservoir Computing Built in 5G Networking Architecture

Network function virtualization (NFV) is one of the vital technologies in 5G networking since they enable flexible development, deployment and management of overall network environment and make the network dynamically programmable. In particular, NFV has enabled logical separation of physical core network resources into "network slices" on service-by-service basis for various demands of use cases, such as eMBB (enhanced mobile broadband), mMTC (massive machine type communications), URLLC (ultra-reliable low latency communication). In the meanwhile, as the network environment is becoming more and more software-defined, intellectual functionality called "in-network machine learning" is being introduced to data plane for traffic forwarding. Combining these two concepts, application identification for network slicing was demonstrated based on the in-network deep learning [8]. Figure 3 shows a high-level illustration of application identification for 5G network slicing.

In 5G IoT world, new applications appear and disappear at an unprecedented rate. In addition, updates of specification occur very often and network traffic flows change their statistical properties over time. Therefore, not only classification accuracy but also low cost learning and easy adaptation are critical metric for application identification for network slicing. In this respect, RC requires training only at linear readout layers which can be performed by simple signal processing and on-site re-learning and adaptation can be easily performed. Thus, learning, operation and re-learning cycle can be completed only in-network environments and RC-based application identification can be a cost-effective solution for 5G network slicing. In addition, traditional machine learning classifiers using entire statistics of flows can be useless when the network slicing changes their behaviors such as inter-packet arrival time, packet sizes from training phase. On the other hand, RC approaches operate in an online manner using only first few packets of flows and therefore are less sensitive to behavior changes due to network slicing.

The remarkable recent trend of RC is that they are implemented various physical dynamics [11]. From view point of application to 5G networking, implementation by optics are quite relevant; [1–3,10]. These devices can handle optical signals directly and achieve much higher speed (~multi Gbps) and lower power consumption (~several hundred mW) than software implementation of RC. Considering these advantages of physical RC, we can integrate them with edge network equipments as powerful accelerators for 5G IoT systems under severe operation constraints.

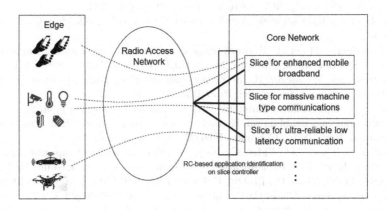

Fig. 3. Application identification for network slicing in 5G networking

6   Conclusion

We have discussed the application identification of network traffic by RC. Experiments for real traffic data showed that RC can perform application identification with enough accuracy from practical point of view. In particular, we have also proposed the application of physical implementation of reservoirs to 5G networking. The proposed architecture enables efficient and dynamic network slicing, which is a key technology for 5G IoT network, and can be very attractive solution since it enables fast and real-time classifier. Direct performance verification using optical physical reservoir computing is left to future work.

Acknowledgments. This work was supported by New Energy and Industrial Technology Development Organization (NEDO) under contract No. 18102284-0.

References

1. Héroux, J.B., Numata, H., Nakano, D.: Polymer waveguide-based reservoir computing. In: Liu, D., Xie, S., Li, Y., Zhao, D., El-Alfy, E.S. (eds.) ICONIP 2017. LNCS, vol. 10639, pp. 840–848. Springer, Cham (2017). https://doi.org/10.1007/978-3-319-70136-3_89

2. Héroux, J.B., Kanazawa, N., Nakano, D.: Delayed feedback reservoir computing with VCSEL. In: Cheng, L., Leung, A.C.S., Ozawa, S. (eds.) ICONIP 2018. LNCS, vol. 11301, pp. 594–602. Springer, Cham (2018). https://doi.org/10.1007/978-3-030-04167-0_54

3. Héroux, J.B., et al.: Optoelectronic reservoir computing with VCSEL. In: 2018 International Joint Conference on Neural Networks (IJCNN), pp. 1–6 (2018)

4. Lashkari, A.H., et al.: Characterization of Tor traffic using time based features. In: The Proceedings of the 3rd International Conference on Information System Security and Privacy. SCITEPRESS, Porto, Portugal (2017)

5. Li, R., et al.: Byte segment neural network for network traffic classification, June 2018

6. Lopez-Martin, M., et al.: Network traffic classifier with convolutional and recurrent neural networks for Internet of Things. IEEE Access 5, 18042–18050 (2017)

7. Lukoševičius, M., Jaeger, H.: Reservoir computing approaches to recurrent neural network training. Comput. Sci. Rev. 3(3), 127–149 (2009)

8. Nakao, A., Du, P.: Toward in-network deep machine learning for identifying mobile applications and enabling application specific network slicing. Inst. Electron. Inf. Commun. Eng. Trans. Commun. E101B(7), 1536–1543 (2018)

9. Perera, P., Tian, Y.C., Fidge, C., Kelly, W.: A comparison of supervised machine learning algorithms for classification of communications network traffic. In: Liu, D., Xie, S., Li, Y., Zhao, D., El-Alfy, E.S. (eds.) ICONIP 2017. LNCs, vol. 10634, pp. 445–454. Springer, Cham (2017). https://doi.org/10.1007/978-3-319-70087-8_47

10. Takeda, S., et al.: Photonic reservoir computing based on laser dynamics with external feedback. In: Hirose, A., Ozawa, S., Doya, K., Ikeda, K., Lee, M., Liu, D. (eds.) ICONIP 2016. LNCS, vol. 9947, pp. 222–230. Springer, Cham (2016). https://doi.org/10.1007/978-3-319-46687-3_24

11. Tanaka, G., et al.: Recent advances in physical reservoir computing: a review. Neural Netw. J. 115, 100–123 (2019). https://doi.org/10.1016/j.neunet.2019.03.005

A New Supervised Learning Approach: Statistical Adaptive Fourier Decomposition (SAFD)

Chunyu Tan[1(✉)], Liming Zhang[1], and Tao Qian[2]

[1] Faculty of Science and Technology, University of Macau, Macao, China
yb57416@connect.um.edu.mo, lmzhang@um.edu.mo
[2] Macau University of Science and Technology, Macao, China
tqian@must.edu.mo

Abstract. This paper proposes a new type of supervised learning approach - statistical adaptive Fourier decomposition (SAFD). SAFD uses the orthogonal rational systems, or Takenaka-Malmquist (TM) systems, to build up a learning model for the training set, based on which predictions of unknown data can be made. The approach focuses on the classification of signals or time series. AFD is a newly developed signal analysis method, which can adaptively decompose different signals into different TM systems that introduces the Fourier type but non-linear and non-negative time-frequency representation. SAFD fully integrates the learning process with the adaptability character of AFD, in which a small number of learned atoms are adequate to capture structures and features of the signals for classification. There are three advantages in SAFD. First, the features are automatically detected and extracted in the learning process. Secondly, all parameters are selected automatically by the algorithm. Finally, the learned features are mathematically represented and the characteristics can be further studied based on the induced instantaneous frequencies. The efficiency of the proposed method is verified by electrocardiography (ECG) signal classification. The experiments show promising results over other feature based learning approaches.

Keywords: Statistical adaptive Fourier decomposition · Heart beat classification · Time-frequency representation

1 Introduction

Supervised learning means that the learner observes some labeled example input–output pairs as the training set and learns a general hypothesis that maps from input to output, then makes predictions for all unseen instances using the learned hypothesis [1]. There are a number of popular supervised learning techniques in the literature, which can be divided into five categories [2]. They are Logic based algorithms, such as decision trees [3]; Perceptron-based techniques, such as neural networks [4]; Statistical learning algorithms, such as Bayesian networks [5];

© Springer Nature Switzerland AG 2019
T. Gedeon et al. (Eds.): ICONIP 2019, CCIS 1143, pp. 397–404, 2019.
https://doi.org/10.1007/978-3-030-36802-9_42

Instance-based learning, such as lazy-learning and nearest neighbor algorithms [6]; and Support Vector Machines (SVM). Each method has its own strengths and limitations. Supervised learning tasks can be further grouped into the regression and classification applications. There are a large number of classification applications related to various types of signals, including medical data, radar data, and financial data, etc. Signal processing offers effective techniques in analyzing various data, however, it has not been widely used in the supervised learning area and is not included in the above five categories.

This paper proposes a new learning technique based on adaptive Fourier decomposition (AFD), which is a newly developed signal processing technique [7]. Unlike ordinary transforms based on a pre-selected basis, AFD is based on a particular redundant dictionary leading to mono-components as composition units (atoms) of signals. AFD decomposes any given signal by suitably choosing dictionary atoms, according to some optimization principle, to form an atomic system specially for the signal to be expanded. Due to its adaptivity, different signals are represented by different systems. Combining the statistical results, signal with similar features can be represented by a common system. Based on this principle, the proposed approach can be used to classify signals. There are different variants of AFD, we choose n-best AFD in this paper that uses n-Blaschke-forms. The best n-Blaschke approximation (n-best AFD) is an alternative version of best approximation to Hardy space functions by rational functions of degree not exceeding n. The n- best approximation, due to its optimization nature, is more effective and more stable to approximate analytic signals.

The contributions of the paper are summarized as follows:

- This paper proposes a new type of supervised learning approach SAFD, in which the features are represented mathematically by the adaptively obtained well-defined orthogonal rational systems. Unlike the other transform based learning with pre-selected fixed basis, SAFD adaptively selects a parameter-determined system representing the features, and yet with fast convergence.
- Detecting and extracting the most relevant features are among very challenging tasks that often introduce manual intervention in other feature based learning models. SAFD provides a fully automatic feature detection and extraction learning process that is a superb nature in comparison with the manual feature selections. The feature detection and extraction make it to be of similarity with neural network (NN). It is, however, unlike NN, due to its automatic parameter finding and explicit mathematical representation.
- Parameter selection itself is another challenging task in learning process. SAFD offers automatic parameter selection. No parameter pre-selection is needed in SAFD.
- The commonly used feature based classification methods usually consist of two steps, including feature extraction and feature classification. The extracted features need to be put into a classifier to implement the classification. SAFD can achieve the classification by directly projecting the signal to the obtained model systems and comparing the residues without applying a separate classifier.

- The mathematical representation of the frequency features is provided in the paper, which lays foundation for feature analysis in the future.

The rest of the paper is organized as follows. The mathematical principle of the proposed SAFD is elaborated in Sect. 2. Section 3 presents the SAFD based signal classification approach in detail. In Sect. 4, the effectiveness of the method is evaluated by ECG classification. Conclusions are drawn in Sect. 5.

2   The Principle of the Proposed Statistical Adaptive Fourier Decomposition (SAFD)

The n-best Blaschke form approximation is based on the n-orthogonal rational function system, or the n-Takenaka-Malmquist system (the n-TM system in brief) [8].

Let $y(t)$ be a real-valued signal. The associated analytic signal of $y(t)$ is defined as [8]

$$y^+(t) = \frac{1}{2}\big(y(t) + iHy(t) + c_0\big), \tag{1}$$

where c_0 is the 0-th Fourier coefficient, and H is the Hilbert transformation. In the following part, we also denote y^+ as y for convenience.

For an analytic signal y, the n-best Blaschke form approximation is a function of the form

$$\tilde{y}(z) = \sum_{k=1}^{n} c_k B_k, \tag{2}$$

which best approximates y under a selection of the n-tuple of the parameters $a_1, a_2, \ldots, a_n$. Where $c_k = \langle y^+, B_k \rangle = \langle y^+, B_{a_1,\ldots,a_k} \rangle$ is the k-th coefficient of B_k.

The main learning process of SAFD is embedded in the parameter selection. An optimal selection of the parameters $a_1, a_2, \ldots, a_n$ is based on minimizing the square-distance between y and $\tilde{y}$, that is

$$\left\| y - \sum_{k=1}^{n} \langle y, B_{a_1,\ldots,a_k} \rangle B_{a_1,\ldots,a_k} \right\| = \min_{\{b_1,\ldots,b_k\} \subset \mathbb{D}} \left\| y - \sum_{k=1}^{n} \langle y, B_{b_1,\ldots,b_k} \rangle B_{b_1,\ldots,b_k} \right\|. \tag{3}$$

Cyclic AFD provided in [8] is an effective algorithm for n-best Blaschke-form approximation to solve the above optimization problems (3). It adaptively selects one more optimized parameter for each cycle.

3   SAFD Based Signal Classification

In this section, we first present the SAFD based signal classification approach, which consists of three steps, including pre-processing, learning process, and classification. Then we provide the mathematical representation of the learned features.

3.1   Pre-processing

First of all, the selected signals to be learned need to be suitably normalized as pre-processing of the training set. Then based on the statistical principle, signal averaging technique is applied, which can increase the signal-to-noise ratio. Assume signals can be divided into M classes in the training set. The i-th class is denoted as $C_i = (s_{i,1}, \ldots, s_{i,N_i})$, where N_i is the number of signals in the i-th class, $i = 1, \ldots, M$. For each class, we randomly select a certain number of signals, $(s_{i,1}, \ldots, s_{i,i_p})$, as the training set of C_i. Then the training signals of the i-th class are averaged by $\bar{s}_i = \frac{\sum_{j=1}^{i_p} s_{i,j}}{i_p}$.

3.2   The Learning Process

The learning process described in Sect. 2 is applied to $\bar{s}_i$ to select the optimal parameters $\{a_1^i, \ldots, a_{n_i}^i\}$, n_i is the number of the parameters of the i-th class. Then the corresponding weighted Blaschke products [8] are

$$B_k^i(z) = B_{a_1^i, \ldots, a_k^i} = \frac{\sqrt{1 - |a_k^i|^2}}{1 - \overline{a_k^i} z} \prod_{j=1}^{k-1} \frac{z - a_j^i}{1 - \overline{a_j^i} z}, \tag{4}$$

$\{B_k^i\}_{k=1}^{n_i}$ is the trained n-TM system of C_i.

3.3   Classification

To an unknown test signal s, it is represented as a linear combination of each trained $\{B_k^i\}_{k=1}^{n_i}$ from the i-th class, $i = 1, \ldots, M$, that is, $\tilde{s}^i = \sum_{k=1}^{n_i} c_k^i B_k^i$, where $c_k^i = \langle s, B_k^i \rangle$. The residuals of the orthogonal projections on the trained n-TM systems $\{B_k^i\}_{k=1}^{n_i}$ are

$$\left\| s - \sum_{k=1}^{n_i} c_k^i B_k^i \right\|, i = 1, \ldots, M. \tag{5}$$

Then s can be determined to the class l based on the following equation:

$$R^l(i) = \min_{i \in \{1, \ldots, M\}} R^i(s) = \min_{i \in \{1, \ldots, M\}} \| s - \tilde{s}^i \|. \tag{6}$$

3.4   Mathematical Representation of the Frequency Features

The n-best Blaschke form approximation $\tilde{y}$ of the signal y in (2) gives a non-negative time-frequency representation of y. Furthermore, the instantaneous frequency (IF) feature can be extracted by [9]

$$\theta_n'(t) = \frac{|a_n| \cos(t - \theta_{a_n}) - |a_n|^2}{1 - 2|a_n| \cos(t - \theta_{a_n}) + |a_n|^2} + \sum_{l=1}^{n-1} \frac{1 - |a_l|^2}{1 - |a_l| \cos(t - \theta_{a_l}) + |a_l|^2}, \tag{7}$$

where $\theta_n = |a_n| e^{i\theta_{a_n}}$.

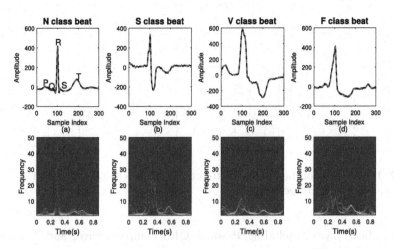

Fig. 1. Four heartbeat examples. Black lines represent the training signals and red lines represent the reconstructed signals by n-best AFD with $n = 12$. The row below is the associated time-frequency representations of the four heartbeat examples. (Color figure online)

4   Experiments

The effectiveness of the proposed learning technique is evaluated by electrocardiography (ECG) signals for heart beat classification. We evaluate our approach on well known MIT-BIH arrhythmia database[1]. The experiments are conducted on a computer with 16 GB RAM and 2.71 GHz Inter Core i5 processor and the code is implemented in MATLAB 2016a.

The annotations provided by the database are used as the labeled references for training and also used for testing the classified results. Following the Association for Advancement of Medical Instrumentation (AAMI) standard [10], the different heart beat types in MIT-BIH database are grouped into five classes: N, S, V, F and Q class. The Q class is commonly discarded according to the recommended practice and is not considered in the following heart beat classification task. The examples of the four heart beat classes are illustrated in Fig. 1.

4.1   SAFD Based ECG Classification

There are three steps in the SAFD based ECG classification approach. They are ECG signal segmentation, learning process, and classification process. In the first step, the raw ECG signals are divided into heartbeat segments following the R detection, which locates R-peak points using the provided beat locations. The training data normalization in ECG classification is as follow. We chose 300 sampling points as the segment length, 100 sampling points before the beat location and 200 sampling points after it. In this way, each segment contains

---

[1] http://www.physionet.org/physiobank/database/mitdbl.

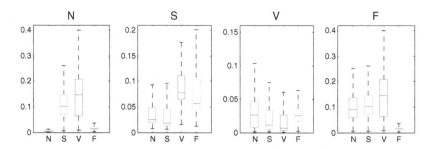

Fig. 2. The boxplots of the residual distribution for each class are projected on trained N, S, V, and F class, respectively.

a whole heart cycle, including P wave, QRS complex wave, and T wave, as illustrated in Fig. 1(a).

The proposed learning technique is trained to extract features that can represent N, S, V, and F classes, respectively. 13640 heartbeats are sampled randomly in MIT-BIH database, in which 834 heartbeats are F class, 8330 heartbeats are N class, 2072 heartbeats are S class, and 2404 heartbeats are V class. For the sake of extracting features that can represent each class and reflecting some statistical characteristics, 100 heartbeats are randomly selected in each class for training. The remaining heartbeats are used for classification validation.

In learning process, we first get the average signal in N, S, V, and F classes, respectively. In our experiment, $N = 12$ is selected for all four classes. Next, the SAFD learning process is applied to each of the labeled average N, S, V, and F classes, respectively. Four examples selected randomly from each of N, S, V, and F classes and are reconstructed by using the respective n-TM systems as illustrated in Fig. 1. It can be seen that the automatically selected features can perfectly represent the heartbeat classes they belong to. Furthermore, those respective n-TM systems possess positive IF features and there is no intersection among all IFs. The positive IFs effectively reflect the time-varying characteristics of signals, such as the morphology of heartbeats. The respective time-frequency representations of the four heartbeat examples are shown in Fig. 1.

In the classification process, tested signals are projected onto the obtained four n-TM systems, and the residuals are worked out to see which n-TM system gives rise to the minimum residual energy. The residual distributions of N, S, V, and F classes are graphically represented in Fig. 2. As shown in Fig. 2, residuals of heartbeats reach to a minimum when the heartbeats are consistent with their class labels, which suggest that the residuals have a good discriminative representation for classifying different beat types. The performance of the proposed learning technique has a 81.44% overall accuracy and other detailed results are shown in Table 1.

Table 1. Comparison the proposed method and the previous works. (# features: the number of features are extracted.)

Method	# features	N		S		V		F		Tot.
		Se	+P	Se	+P	Se	+P	Se	+P	Acc
De Chazal et al. [11]	52	86.9	99.2	75.9	38.5	77.7	81.9	89.43	0.08	81.9
Llamedo et al. [12]	39	77.55	99.47	76.46	41.34	82.94	87.97	95.36	4.23	78.0
Zhang et al. [13]	46	88.9	99.0	79.1	36.0	85.5	92.8	93.8	13.7	86.7
Herry et al. [14]	6	83.13	98.93	81.14	31.93	77.50	79.05	83.25	6.91	82.70
Proposed	–	83.58	95.45	82.25	72.48	68.53	97.23	95.78	32.38	81.44

4.2   Performance Comparisons

The experiment results of the proposed method are compared with some selected state-of-the-art feature based ECG classification methods. They are all tested and validated on the MIT-BIH arrhythmia database and follow the AAMI standard. The comparison results are illustrated in Table 1. The compared methods use a variety of features to represent the ECG signals and different types of classifiers for classification.

The performance is evaluated in terms of the sensitivity (Se) and positive predictivity ($+P$) [11]. Though the overall accuracy of the proposed method is not the best, the results are comparable. The most important is that the first three methods rely on very complicated and high dimension feature sets, which lead to three disadvantages, including manual feature selection, high computational cost, and high requirements for classifiers next step. The fourth method significantly reduces the feature number, however, the feature selection approach is still very complicated and manual selection is needed. The proposed approach classifies the ECG signal based on the selected parameters while no classifier is used. It is to be improved along with further studies on the adaptive parameter selection method.

4.3   Running Time Analysis

It takes approximately 0.373 s to complete one learning process and total $0.373 * 4 \approx 1.49$ s for the whole system training by the proposed SAFD with $N = 12$. Once the training of ECG signals is completed, the classification of ECG heartbeat is readily done. The time required for projecting a tested signal to the n-TM system space is only 0.006s. Note that this is a very short time moment, and much shorter than the time needed to finish one heart beat that the length of each heartbeat segmentation is $300/360 \approx 0.83s$. Thus, the proposed algorithm has a great potential for the real-time monitoring system. We will leave this real-time implementation as future work. Since none of other methods shown in Table 1 provides the training time, we cannot make a comparison.

5   Conclusion

This paper presents a new type of supervised learning approach SAFD, which provides fully automatic feature selection and extraction with well defined mathematical representation. It offers a new fully explainable automatic learning process for signals. The effectiveness of the proposed learning technique is demonstrated by ECG classification. This study lays foundation to further analysis of the time-frequency characteristics of the learned features.

Acknowledgment. This study is supported by the research grants: The Science and Technology Development Fund of Macao SAR FDCT 079/2016/A2, 0123/2018/A3, and MYRG 2017-00218-FST, 2018-00111-FST.

References

1. Norvig, P., Russell, S.J.: Artificial Intelligence: A Modern Approach, 3rd edn. Pearson Education Inc., New Jersey (2010)
2. Kotsiantis, S.: Supervised machine learning: a review of classification techniques. Informatica **31**, 249–268 (2007)
3. Kotsiantis, S.: Decision trees: a recent overview. Artif. Intell. Rev. **39**(4), 261–283 (2013)
4. Schmidhuber, J.: Deep learning in neural networks: an overview. Neural Netw. **61**, 85–117 (2015)
5. Cheng, J., et al.: Learning Bayesian networks from data: an information-theory based approach. Artif. Intell. **137**(1–2), 43–90 (2002)
6. Zhang, M., Zho, Z.: ML-KNN: a lazy learning approach to multi-label learning. Pattern Recognit. **40**(7), 2038–2048 (2007)
7. Qian, T., Zhang, L., Li, Z.: Algorithm of adaptive Fourier decomposition. IEEE Trans. Signal Process. **59**(12), 5899–5906 (2011)
8. Qian, T.: Cyclic AFD algorithm for the best rational approximation. Math. Methods Appl. Sci. **37**(6), 846–859 (2014)
9. Dang, P., et al.: Transient time-frequency distribution based on mono-component decompositions. Int. J. Wavelets, Multiresolution Inf. Process. **11**(3), 1350022 (2013)
10. AAMI: Testing and reporting performance results of cardiac rhythm and ST segment measurement algorithms. ANSI/AAMI EC38 (1998)
11. De Chazal, P., et al.: Automatic classification of heartbeats using ECG morphology and heartbeat interval features. IEEE Trans. Biomed. Eng. **51**(7), 1196–1206 (2004)
12. Mariano, L., Martinez, J.P.: Heartbeat classification using feature selection driven by database generalization criteria. IEEE Trans. Biomed. Eng. **58**(3), 616–625 (2011)
13. Zhang, Z., et al.: Heartbeat classification using disease specific feature selection. Comput. Biol. Med. **46**, 79–89 (2014)
14. Herry, C.L., et al.: Heart beat classification from single lead ECG using the synchrosqueezing transform. Physiol. Meas. **38**(2), 171 (2017)

Artificial Neural Network-Based Modeling for Prediction of Hardness of Austempered Ductile Iron

Ravindra V. Savangouder[1(✉)], Jagdish C. Patra[1],
and Cédric Bornand[2]

[1] Swinburne University of Technology, Melbourne, Australia
rsavangouder@swin.edu.au
[2] University of Applied Sciences HES-SO, HEIG-VD,
Yverdon-les-Bains, Switzerland

Abstract. Austempered ductile iron (ADI), because of its attractive properties, for example, high tensile strength along with good ductility is widely used in automotive industries. Such properties of ADI primarily depend on two factors: addition of a delicate proportion of several chemical compositions during the production of ductile cast iron and an isothermal heat treatment process, called austempering process. The chemical compositions, depending on the austempering temperature and its time duration, interact in a complex manner that influences the microstructure of ADI, and determines its hardness and ductility. Vickers hardness number (VHN) is commonly used as a measure of the hardness of a material. In this paper, an artificial neural network (ANN)-based modeling technique is proposed to predict the VHN of ADI by taking experimental data from literature. Extensive simulations showed that the ANN-based model can predict the VHN with a maximum mean absolute error (MAPE) of 0.22%, considering seven chemical compositions, in contrast to 0.71% reported in the recent paper considering only two chemical compositions.

Keywords: Austempered ductile iron · Artificial neural network · Modeling · Prediction · Vickers hardness number

1 Introduction

Austempered ductile iron (ADI), sparked a lot of interest among design engineers due to its high tensile strength with high fatigue strength, good wear resistance and ductility [1]. These desirable properties in ADI are achieved as a result of addition of a delicate proportion of several chemicals (alloying elements) in the ductile iron, and an isothermal heat treatment process called austempering. Since the hardness and ductility in cast-iron are inversely related, its strength and ductility, as desired for a specific application, can be widely varied by carefully selecting the proportion of chemical compositions and by varying the austempering conditions [2, 3]. Because of these characteristics, ADI is widely used in various engineering applications, e.g., manufacturing, railways, and automobiles. Some of the parts produced with ADI are gears, wheel hubs, crankshafts, connecting rods, earth moving machineries and railroads [4–6].

© Springer Nature Switzerland AG 2019
T. Gedeon et al. (Eds.): ICONIP 2019, CCIS 1143, pp. 405–413, 2019.
https://doi.org/10.1007/978-3-030-36802-9_43

ADI belongs to the family of ductile cast irons whose weight percentage of carbon (C) and silicon (Si) is relatively high as compared to other alloying elements. Ductile cast iron is obtained by magnesium (Mg) treatment immediately prior to pouring casting, which produces spheroidal or nodular graphite in ductile iron [7–9]. In order to enhance austempering heat treatment and to achieve desirable mechanical properties, elements such as copper (Cu), nickel (Ni), molybdenum (Mo), manganese (Mn), are usually added to ductile iron [2, 8]. During the austempering process, the chemicals interact in a complex and highly nonlinear manner that influences the microstructure which in turn determines the hardness of the ADI. A non-destructive hardness test of a material is commonly used in industry and research because it provides an easy, reliable and inexpensive method to determine basic properties, such as, hardness. Vickers hardness test is one of such tests in which the Vickers hardness number (VHN) of a material can be determined [10]. Due to the complexity involved in the austempering process, any analytical technique to estimate the number and the proportion of the chemicals as well as the austempering temperature and process duration is currently not available.

Recently, artificial neural networks (ANNs) have evolved as a powerful modeling technique for such complex processes. ANNs have been successfully applied to model complex processes in several fields of science and engineering [11, 12], e.g., in modeling solar cells, fuel cells and photovoltaic arrays [13–18]. There has also been a great interest in the materials science community in ANN-based modeling [19–22]. However, only a few studies of ANNs to predict hardness of ADI have been reported [23, 24]. PourAsiabi et al. [24] have proposed a multilayer perceptron (MLP)-based modeling technique to estimate VHN of ADI with eight specimens. However, they have considered only two (out of seven) chemical components, i.e., Cu and Mo, in their modeling to estimate the VHN and have reported a mean absolute error (MAPE) of 0.71%. Although a tiny proportion of other chemical elements can contribute significantly towards hardness and tensile strength of ADI [2, 8], they have neglected other elements in their model.

In the present paper, we carried out an in-depth study to predict the VHN in ADI using the experimental data reported in [24]. We proposed three MLP-based modeling schemes to predict VHN by taking two, three and seven chemical compositions in the ADI. The major contribution of this paper is as follows. In Model-1, by carefully selecting the MLP parameters, we could achieve a MAPE of 0.33% that is superior to 0.71% that reported in [24]. In Model-2 we introduced an additional important chemical composition, i.e., Mg, and shown that it can achieve a maximum MAPE of 0.27%. Next we considered seven chemical constituents in Model-3 and shown that it can achieve a MAPE of 0.22% in predicting VHN. It is noteworthy that our proposed models can predict the VHN, even beyond the range of the training parameters, accurately.

2   Austempering Process of ADI

The austempering process of the ductile iron is commonly identified as a three-step heat treatment process shown in Fig. 1 [1, 2]. In Step-1 (A-B-C), the casting is heated to and held at austenitizing temperature T_γ (815–925 °C) for approximately 2 h. This step is carried out to make the structure of the casting fully austenitic (γ). In Step-2 (C-D), rapid cooling (quenching) is performed from austenitizing temperature to austempering temperature T_A (ranging between 260–400 °C). Rapid quenching is required to avoid austenite transformation and formation of pearlite. In Step-3 (D-E), the austempering phase, the casting is held at a constant (isothermal) temperature in a salt bath over a period of 1–2 h. Two reactions take place during this isothermal heating depending on the duration of heating. In the first reaction, the austenite (γ) decomposes into acicular ferrite (α) and carbon-rich austenite (γ_{HC}). This microstructure is called ausferrite. If the casting is held at the austempering temperature for too long, a second reaction takes place, during which the carbon-rich austenite decomposes further into ferrite (α) and carbide. This second reaction makes the material brittle, which is an undesirable outcome and hence is avoided. The desirable mechanical properties (e.g., tensile strength and ductility) are obtained in ADI after the completion of the first reaction but before the onset of the second reaction. Finally, the casting is air cooled to room temperature T_{Room} (E-F) [2, 3].

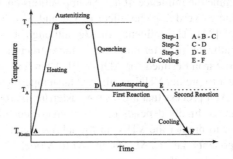

Fig. 1. The ADI austempering process.

3 The ANN and Modeling Scheme

Here, we briefly describe the ANN and the modeling scheme used to predict the VHN of ADI.

3.1 Multilayer Perceptron (MLP)

The MLP is a feed-forward ANN with an input, one or more hidden layers, and one output layer. Each layer contains one or more nonlinear processing units called 'neuron', or 'node'. Each node of a lower layer is connected to all the nodes of an upper layer through links called weights. The backpropagation (BP) algorithm is the most popular learning technique used to train the MLPs [11]. Let Ni, i = 1, 2,..., L, denote the number of nodes in layer i, in which i = 1 denotes the input layer and i = L denotes the output layer. During training process, an input vector, $X = [x_1 \ x_2 ... \ x_{N1}]$ and a target output vector, $D = [d_1 \ d_2 ... \ d_{NL}]$, are applied to the MLP. The MLP produces an output vector, $Y = [y_1 \ y_2 ... \ y_{NL}]$ depending on the weight values of different layers and the nonlinear activation function in each node.

The sum of output square error at the kth instant is given by

$$E(k) = \sum_{i=1}^{N_L} [d_i(k) - y_i(k)]^2 \tag{1}$$

The BP algorithm attempts to minimize the cost function $E(k)$ recursively by updating the weights of the network. The learning and the momentum rates used in the BP algorithm are denoted by α and β, respectively, and their values lie between 0 and 1.

3.2   The ANN-Based Modeling Scheme

Schematic diagrams of the ANN-based modeling of the ADI process are shown in Fig. 2, as a system identification process. Based on the number of chemical compositions used, we propose three MLP-based models to predict the VHN. Model-1 is proposed using *Cu* and *Mo* compositions along with austempering temperature and time in order to determine the influence of these compositions on VHN. The purpose of this model, in addition to verifying the effect of Cu and Mo, is to compare its performance with that of [24]. The influence of *Mg*, although a tiny proportion (about 0.05%), is quite significant in determining the microstructure and the mechanical properties of the ductile iron destined for ADI [7–9]. Therefore, in Model-2 we added *Mg* to Model-1 in order to predict VHN at different austempering temperature and time duration. The overall ADI process that affects the ausferrite microstructure, and in turn, its ductility and hardness, finally depends on all the major chemical compositions [2, 8]. Therefore, in order to predict the VHN, in Model-3 we considered additional four major chemical components, i.e., *C*, *Si*, *Mn* and *Ni* to Model-2.

During the training phase, the input pattern along with its corresponding desired VHN value (the measured value) is applied to the MLP-based model. The output of the model is computed after processing the input pattern in the nodes of different layers of the MLP. Thereafter, the model output is compared with the desired output to generate an error. The resulting error is used to update the weights of the model using the BP algorithm. This process is continued until the MSE_{train} settles to a small value. Thereafter, the trained model is tested for prediction of VHN.

4   Experimental Setup and the Dataset

In the current study, we consider eight specimens with different chemical compositions that have undergone the austempering process with different austempering temperatures and time durations. The experimental data are taken from the article by PourAsiabi *et al.* [24]. The chemical compositions of the eight specimens (in wt%) are given in Table 1. The specimens were austenitized and then were rapidly quenched and soaked in a salt bath at 260, 290, or 320 °C for different time durations, i.e., 30, 60, 90 or 120 min. In this way, twelve samples were created from each specimen, giving rise to a total of 96 samples. The VHN was measured for each sample under 10 kg loading. More details of the experimental setup and procedure can be found in [24].

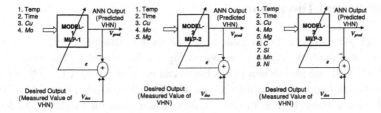

Fig. 2. The three modeling schemes.

Table 1. Chemical Composition of eight specimens in wt%. From PourAsiabi et al. [24].

Chemical composition	Specimen #								Min	Max	Avg.	SD
	1	2	3	4	5	6	7	8				
C	3.44	3.47	3.42	3.42	3.46	3.42	3.45	3.46	3.42	3.47	3.4425	0.0192
Si	2.32	2.35	2.32	2.33	2.41	2.32	2.41	2.42	2.32	2.42	2.3600	0.0424
Mn	0.24	0.24	0.24	0.23	0.25	0.25	0.26	0.26	0.23	0.26	0.2463	0.0099
Ni	1.02	1.01	1.00	1.02	1.00	1.00	1.02	1.02	1.00	1.02	1.0113	0.0093
Cu	0.50	0.50	0.51	0.50	1.00	1.01	1.01	1.00	0.50	1.01	0.7538	0.2513
Mo	0.10	0.15	0.21	0.25	0.10	0.15	0.20	0.25	0.10	0.25	0.1763	0.0566
Mg	0.051	0.052	0.048	0.049	0.053	0.053	0.049	0.054	0.048	0.054	0.0511	0.0021

5   Simulation Setup

All the input variables are normalized between 0 and 1, as per the requirement of the MLP-based modeling. Out of 96 data samples, we used 75% data (72 samples) as training set and the remaining 25% data (24 samples) are used for testing purpose. The learning parameter and momentum factor used in the BP algorithm are gradually decreased with the increase of the iteration, in order to have a faster learning. A bias unit with an output value of +1.0 is added to all layers except the output layer of the MLP. For each model, we carried out several experiments in order to achieve the best architecture and learning parameters. The hidden layer nodes were varied from 5 to 9, and the learning parameters were varied from 0.1 to 0.8. At the end, the best architecture and learning parameters were obtained. The MLP architecture of {5-7-1} in Model-1 implies that it is 3-layer MLP with 5, 7, and 1 node in the input, hidden and output layers (including the bias unit), respectively. Training continued for 10,000 iterations. The MSE at the end of training is found to be lower than -40 dB. This low value of the MLP indicates that the MLP-based models have been well-trained. The architecture of the MLP in Model-2 and Model-3 are given by {6-7-1} and {10-7-1}, respectively. Both the initial learning and momentum rates for the three models are selected as 0.5.

6 Results and Discussion

The scatter plot between the measured and the model estimated VHN values for the complete (96 samples) dataset is shown in Fig. 3a for Model-1. The values of CC, MAPE, and MSE are also shown in this Fig. It can be seen that the CC for both datasets are close to 1.0. In comparison, the CC and MAPE of PourAsiabi *et al.* [24] are 0.9912 and 0.71 respectively. Figure 3b and c shows scatter plots for Model-2 and Model-3 respectively. In Model-2, the CC values for the test set and the complete set are found to be 0.9987 and 0.9992, respectively, thus indicating the effectiveness of modeling. Low values of MAPE indicated high accuracy of estimation of VHN. In Model-3, the CC values for the test set and the complete set are found to be 0.9993 and 0.9995, respectively. The MAPE values for the two sets are found to be 0.33 and 0.22, respectively. These values are superior to the other two models. This is expected, because the VHN is dependent upon the interactions of all the alloying elements during the austempering process. Thus, Model-3 is a complete one that can capture the ADI process accurately.

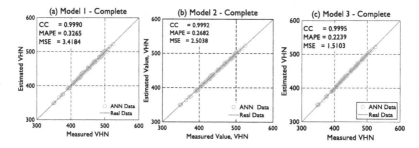

Fig. 3. Scatter plot of measured and estimated VHNs for Model-1, Model-2 and Model-3.

7 Prediction of VHN with MLP Models

Prediction performances of Model-2 at different values of temperature and at four specific austempering time durations, i.e., 30, 60, 90 and 120 min are shown in Fig. 4, for the eight specimens. The red symbols indicate the measured values and the blue symbols denote the predicted values of VHN. It can be seen that the model is able to predict VHN accurately at different values of austempering temperature ranging from 250 to 330 °C. Prediction performances of Model-3 at different values of austempering time duration at three specific austempering temperature, i.e., 260, 290, or 320 °C are shown in Fig. 5, for the eight specimens. The model is able to predict VHN accurately for austempering time duration ranging from 20 to 130 min. This aspect of the MLP-based model has not been reported in [24] or elsewhere.

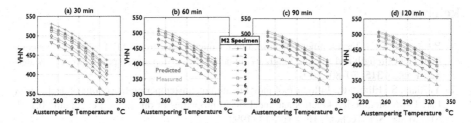

Fig. 4. Model-2 performance. Prediction of VHN at different austempering temperature duration of 30, 60, 90 and 120 min. (Color figure online)

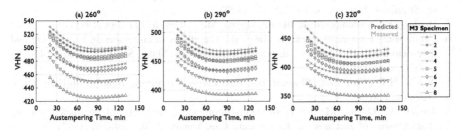

Fig. 5. Model-3 performance. Prediction of VHN at different austempering time duration for austempering temperature of 260, 290 and 320 °C. (Color figure online)

8   Conclusion

We proposed ANN-based techniques for modeling austempering process of ADI to predict VHN. These models can effectively predict VHN at a specific austempering temperature, time duration and, wt% of the seven chemical compositions. We have shown that the correlation coefficient between the measured and estimated VHN values is close to 1.0, thus, indicating high estimation accuracy. The performance of Model-1 is found to be superior to the similar model proposed in [24]. Unlike the other critical chemical constituents that are neglected in [24], the proposed Model-2 and Model-3, respectively, takes three and seven chemical constituents. The mean absolute errors are found to be as low as 0.27% and 0.22% in Model-2 and Model-3, respectively. These models can predict VHN accurately at any value (within a specified range) of austempering temperature and time duration, even beyond the range of training samples. Similar predictions were also achieved for different proportion of chemical compositions (in wt%) at a specific austempering temperature and time duration. One of the major limitations of the current work is the limited amount of available data. Here, we have considered only 96 data samples. However, in a future work, we intend increase dataset by considering other dataset available in the literature. However, these modeling techniques can help in saving time and cost of designing ADI for a specified VHN required for any application.

References

1. Putatunda, S.K.: Development of austempered ductile cast iron (ADI) with simultaneous high yield strength and fracture toughness by a novel two-step austempering process. Mater. Sci. Eng. A **315**, 70–80 (2001)
2. Yazdani, S., Elliott, R.: Influence of molybdenum on austempering behaviour of ductile iron Part 1 – Austempering kinetics and mechanical properties of ductile iron containing 0.13% Mo. Mater. Sci. Technol. **15**, 531–540 (1999)
3. Panneerselvam, S., Martis, C.J., Putatunda, S.K., Boileau, J.M.: An investigation on the stability of austenite in austempered ductile cast iron (ADI). Mater. Sci. Eng. A **626**, 237–246 (2015)
4. C.M.D. Ltd.: Austempered ductile-iron castings — advantages, production, properties and specifications. Mater. Des. **13**, 285–297 (1992)
5. Harding, R.: Austempered ductile irons-gears. Mater. Des. **6**, 177–184 (1985)
6. Cakir, M.C., Bayram, A., Isik, Y., Salar, B.: The effects of austempering temperature and time onto the machinability of austempered ductile iron. Mater. Sci. Eng. A **407**, 147–153 (2005)
7. Angus, H.T.: Cast Iron: Physical and Engineering Properties. Butterworth-Heinemann, Oxford (1978)
8. Trudel, A., Gagne, M.: Effect of composition and heat treatment parameters on the characteristics of austempered ductile irons. Can. Metall. Q. **36**, 289–298 (1997)
9. de Albuquerque Vicente, A., Moreno, J.R.S., de Abreu Santos, T.F., Espinosa, D.C.R., Tenório, J.A.S.: Nucleation and growth of graphite particles in ductile cast iron. J. Alloys Compd. **775**, 1230–1234 (2019)
10. Giannakopoulos, A.E., Larsson, P.L., Vestergaard, R.: Analysis of Vickers indentation. Int. J. Solids Struct. **31**, 2679–2708 (1994)
11. Haykin, S.: Neural Networks. Prentice Hall, Upper Saddle River (1999)
12. Patra, J.C., Kot, A.C.: Nonlinear dynamic system identification using Chebyshev functional link artificial neural networks. IEEE Trans. Syst. Man Cybern. Part B **32**, 505–511 (2002)
13. Patra, J.C.: Neural network-based model for dual-junction solar cells. Prog. Photovoltaics Res. Appl. **19**, 33–44 (2011)
14. Patra, J.C., Chakraborty, G.: e-MLP-based modeling of high-power PEM fuel cell stacks. In: 2011 IEEE International Conference on Systems, Man, and Cybernetics (SMC), Anchorage, AK, pp. 802–807 (2011)
15. Patra, J.C., Maskell, D.L.: Artificial neural network-based model for estimation of EQE of multi-junction solar cells. In: 2011 37th IEEE Photovoltaic Specialists Conference (PVSC), Seattle, WA, pp. 002279–002282 (2011)
16. Patra, J.C.: Chebyshev neural network-based model for dual-junction solar cells. IEEE Trans. Energy Convers. **26**, 132–139 (2011)
17. Jiang, L.L., Maskell, D.L., Patra, J.C.: Chebyshev functional link neural network-based modeling and experimental verification for photovoltaic arrays. In: 2012 International Joint Conference on Neural Networks (IJCNN), Brisbane, QLD, pp. 1–8 (2012)
18. Patra, J.C., Maskell, D.L.: Modeling of multi-junction solar cells for estimation of EQE under influence of charged particles using artificial neural networks. Renew. Energy. **44**, 7–16 (2012)
19. Cool, T., Bhadeshia, H.K.D.H., MacKay, D.J.C.: The yield and ultimate tensile strength of steel welds. Mater. Sci. Eng. A **223**, 186–200 (1997)

20. Malinov, S., Sha, W., McKeown, J.J.: Modeling and correlation between processing parameters and properties in titanium alloys using artificial network. Comput. Mater. Sci. **21**, 375–394 (2001)
21. Yilmaz, M., Ertunc, H.M.: The prediction of mechanical behavior for steel wires and cord materials using neural networks. Mater. Des. **28**, 599–608 (2007)
22. Reddy, N.S., Krishnaiah, J., Young, H.B., Lee, J.S.: Design of medium carbon steels by computational intelligence techniques. Comput. Mater. Sci. **101**, 120–126 (2015)
23. Yescas, M.A.: Prediction of the Vickers hardness in austempered ductile irons using neural networks. Int. J. Cast Met. Res. **15**, 513–521 (2003)
24. PourAsiabi, H., PourAsiabi, H., AmirZadeh, Z., BabaZadeh, M.: Development a multi-layer perceptron artificial neural network model to estimate the Vickers hardness of Mn–Ni–Cu–Mo austempered ductile iron. Mater. Des. **35**, 782–789 (2012)

Prediction of Hardness of Austempered Ductile Iron Using Enhanced Multilayer Perceptron Based on Chebyshev Expansion

Ravindra V. Savangouder[1(⊠)], Jagdish C. Patra[2],
and Cédric Bornand[2]

[1] Swinburne University of Technology, Melbourne, Australia
rsavangouder@swin.edu.au
[2] University of Applied Sciences HES-SO, HEIG-VD,
Yverdon-les-Bains, Switzerland

Abstract. In various industries, e.g., manufacturing, railways, and automotive, austempered ductile iron (ADI), is extensively used because of its desirable characteristics for example, high tensile strength with good ductility. The hardness and ductility of ADI can be tailor-made for a specific application by following an appropriate process. Such characteristics can be achieved by (i) adding a delicate proportion of several chemical compositions during the production of ductile cast iron and then followed by (ii) an isothermal heat treatment process, called austempering process. The chemical compositions, depending on the austempering temperature and its time duration, interact in a complex manner that influences the microstructure of ADI, and determines its hardness and ductility. Vickers hardness number (VHN) is commonly used as a measure of the hardness of a material. In this paper, we propose a computationally efficient enhanced multilayer perceptron (eMLP)-based technique to model the austempering process of ADI for prediction of VHN by taking experimental data reported in literature. By comparing the performance of the eMLP model with an MLP-based model, we have shown that the proposed model provides similar performance but with less computational complexity.

Keywords: Austempered ductile iron · Artificial neural network · Modeling · Prediction · Vickers hardness number

1 Introduction

Because of its high tensile strength with good wear resistance, high fatigue strength and good ductility, austempered ductile iron (ADI) has attracted a lot of interest among design engineers [1]. These desirable properties can be achieved by following a delicate ADI process: (i) addition of a delicate proportion of several chemicals (alloying elements) in the ductile cast iron, and (ii) an isothermal heat treatment process called austempering. The strength and ductility of ADI can be varied over a wide range by varying the chemical composition and the austempering conditions [2, 3]. Due to possibility of variation of their characteristics, ADI is widely used in various engineering applications, e.g., manufacturing, railways, and automobiles [4–6].

© Springer Nature Switzerland AG 2019
T. Gedeon et al. (Eds.): ICONIP 2019, CCIS 1143, pp. 414–422, 2019.
https://doi.org/10.1007/978-3-030-36802-9_44

Carbon (*C*) and silicon (*Si*) are relatively high in weight percent in ADI as compared to other alloying elements. Ductile cast iron is obtained by magnesium (*Mg*) treatment immediately prior to pouring casting, which produces spheroidal or nodular graphite in ductile iron [7–9]. In order to enhance austempering heat treatment and to achieve desirable mechanical properties, elements such as copper (*Cu*), nickel (*Ni*), molybdenum (*Mo*), manganese (*Mn*), are usually added to ductile iron [2, 8]. The austempering process of ADI involves complex interactions among the added chemicals during the austempering process that influences the microstructure which in turn determines the mechanical properties, e.g., hardness and ductility of the ADI. Due to its complexity, any analytical technique is not yet available to estimate the number and the proportion of the chemicals, the austempering temperature and process duration. Vickers hardness test is used to determine the VHN of a material [10].

Artificial neural networks (ANNs) have evolved as a powerful modeling technique for such complex processes. ANNs have been successfully applied to model complex processes in various fields of science and engineering [11, 12], e.g., in modeling solar cells, fuel cells, photovoltaic arrays [13–18] and material science [19–22]. Only a few studies of ANNs to predict hardness of ADI have been reported [23, 24]. PourAsiabi et al. [24] have proposed a multilayer perceptron (MLP)-based modeling technique to estimate VHN of ADI with eight specimens. They have considered only two chemical constituents, i.e., *Cu* and *Mo*, in their modeling to estimate the VHN and have reported a mean absolute error (MAE) of 0.71%. However, they have ignored other five critical chemical constituents, although these constituents play significant role in determining the hardness and ductility of ADI [2, 8].

Chebyshev functional link ANN (FLANN) is a computationally efficient ANN, as it is a single-layered structure [12, 15, 16]. Performance of these ANNs in modeling of complex systems is found to be similar to that of MLP but with much less computational load. However, performance of Chebyshev FLANN deteriorates when the dimension of the input pattern is large. In order to improve its performance, a hidden layer is added to its architecture and we named this structure as enhanced MLP (eMLP). The eMLP was first introduced successfully in our earlier work in modeling of fuel cells [18]. In the present paper, we proposed an eMLP-based scheme for modeling of the ADI process to predict VHN. With extensive simulations, we have shown that its performance in terms of prediction error is similar to that of MLP-based model.

2 Austempering Process of ADI

The austempering process of the ductile iron is commonly identified as a three-step heat treatment process shown in Fig. 1 [1, 2]. In Step-1 (A–B–C), the casting is heated to and held at austenitizing temperature T_γ (815–925 °C) for approximately 2 h. This step is carried out to make the structure of the casting fully austenitic (γ). In Step-2 (C–D), rapid cooling (quenching) is performed from austenitizing temperature to austempering temperature T_A (ranging between 260–400 °C). Quenching is required to avoid austenite transformation and formation of pearlite. In Step-3 (D–E), the austempering phase, the casting is held at a constant (isothermal) temperature in a salt bath over a period of 1–2 h. Two reactions take place during this isothermal heating depending on

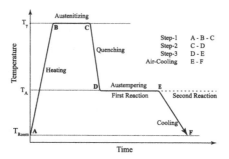

Fig. 1. The ADI austempering process.

the duration of heating. In the first reaction, the austenite (γ) decomposes into acicular ferrite (α) and carbon-rich austenite (γ_{HC}). This microstructure is called ausferrite. If the casting is held at the austempering temperature for too long, a second reaction takes place during which the carbon-rich austenite decomposes further into ferrite (α) and carbide. This second reaction makes the material brittle, which is an undesirable outcome and hence is avoided. The desirable mechanical properties (e.g., tensile strength and ductility) are obtained in ADI after the completion of the first reaction but before the onset of the second reaction. Finally, the casting is air cooled to room temperature T_{Room} (E–F) [2, 3].

3   The Modeling Scheme

3.1   Multilayer Perceptron (MLP)

The MLP is a feed-forward ANN with an input, one or more hidden layers, and one output layer. Each layer contains one or more nonlinear processing units called 'neuron', or 'node'. Each node of a lower layer is connected to all the nodes of an upper layer through links called weights. The backpropagation (BP) algorithm is the most popular learning technique used to train the MLPs [11]. Let N_i, $i = 1, 2,..., L$, denote the number of nodes in layer i, in which $i = 1$ denotes the input layer and i = L denotes the output layer. During training process, an input vector, $X = [x_1\ x_2... x_{N1}]$ and a target output vector, $D = [d_1\ d_2... d_{NL}]$, are applied to the MLP. The MLP produces an output vector, $Y = [y_1\ y_2... y_{NL}]$ depending on the weight values of different layers and the nonlinear activation function in each node.

The sum of output square error at the kth instant is given by

$$E(k) = \sum_{i=1}^{N_L} [d_i(k) - y_i(k)]^2 \qquad (1)$$

The BP algorithm attempts to minimize the cost function $E(k)$ recursively by updating the weights of the network. The learning and the momentum rates used in the BP algorithm are denoted by α and β, respectively, and their values lie between 0 and 1.

3.2   The eMLP

In the eMLP, the input patterns, before applying to the MLP, first undergo a dimensional expansion process by passing these patterns through a functional expansion (FE) block. In the FE block, each element of the input pattern is enhanced by using orthogonal Chebyshev functions. It is expected that by increasing the dimensionality, the input pattern is enhanced. These enhanced patterns will have simpler decision boundary between different classes and will facilitate to improve the prediction capability of the MLP [25]. Let the original N_1-dimensional input pattern is given by $X = [x_1 \, x_2 ... \, x_{N1}]$. Each element of this pattern x_i, $i = 1, 2, ..., N_1$ is enhanced to x_i^e as follows.

$$x_i^e = [T_1(x_i) \, T_2(x_i) \, \cdots \, T_m(x_i)], \tag{2}$$

where $T_n(\theta)$ is an nth order orthogonal Chebyshev polynomial and m is the expansion parameter. These polynomials can be recursively computed as

$$T_{n+1}(\theta) = 2\theta T_n(\theta) - T_{n-1}(\theta), \tag{3}$$

where the range of θ is between $+1.0$ and -1.0, $T0(\theta) = 1.0$, and $T1(\theta) = \theta$. Thus, the original N1-dimensional input pattern is expanded to (N1·m)-dimensional enhanced pattern. Thereafter, these enhanced patterns are applied to an MLP with only one hidden layer. We named this MLP structure as eMLP. The number of nodes in the hidden layer needs to be varied to achieve the best performance.

3.3   The MLP and eMLP-Base Modeling Schemes

Schematic diagram of an MLP-based modeling of the ADI process is shown in Fig. 2a, as a system identification process. We considered all the seven chemical components, i.e., *Cu, Mo, Mg, C, Si, Mn* and *Ni* along with austempering temperature T_A and its time duration. Figure 2b shows a schematic diagram of an eMLP-based modeling of the ADI process. The N_1-dimensional input pattern is normalized, expanded in the FE block using Chebyshev polynomials (2), (3) to enhance its features and then fed to the eMLP. Usually, an MLP requires one or two hidden layers with many nodes in each layer to achieve a satisfactory performance level. Whereas, the eMLP requires only one hidden layer with a few nodes to achieve similar level of performance. As a result, it has much less number of weights to update at each iteration of training. Therefore, it provides a computational advantage over the MLP in terms of hardware and execution time.

4   Experimental Setup and the Dataset

The experimental data were taken from the article by PourAsiabi et al. [24]. The seven chemical compositions of the eight specimens (in wt%) are given in Table 1. Eight ADI specimens were obtained by varying weight percentage of the chemical constituents and by heating them to an austenitizing temperature of 900 ± 3 °C for 90 min. Thereafter, the specimen was rapidly quenched and soaked in a salt bath at

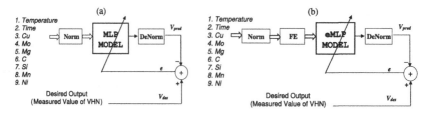

Fig. 2. The modeling schemes (a) MLP-based, (b) eMLP-based.

260, 290, or 320 °C for different time durations, i.e., 30, 60, 90 or 120 min. In this way, twelve samples were created from each specimen, giving rise to a total of 96 samples. The VHN was measured for each sample under 10 kg loading. More details of the experimental setup and procedure can be found in [24].

Table 1. Chemical composition of eight specimens in wt%. From PourAsiabi et al. [24].

Chemical composition	Specimen #								Min	Max	Avg.	SD
	1	2	3	4	5	6	7	8	Min	Max	Avg.	SD
C	3.44	3.47	3.42	3.42	3.46	3.42	3.45	3.46	3.42	3.47	3.4425	0.0192
Si	2.32	2.35	2.32	2.33	2.41	2.32	2.41	2.42	2.32	2.42	2.3600	0.0424
Mn	0.24	0.24	0.24	0.23	0.25	0.25	0.26	0.26	0.23	0.26	0.2463	0.0099
Ni	1.02	1.01	1.00	1.02	1.00	1.00	1.02	1.02	1.00	1.02	1.0113	0.0093
Cu	0.50	0.50	0.51	0.50	1.00	1.01	1.01	1.00	0.50	1.01	0.7538	0.2513
Mo	0.10	0.15	0.21	0.25	0.10	0.15	0.20	0.25	0.10	0.25	0.1763	0.0566
Mg	0.051	0.052	0.048	0.049	0.053	0.053	0.049	0.054	0.048	0.054	0.0511	0.0021

5   Simulation Setup

All the input variables are normalized between 0 and 1, as per the requirement of the MLP-based modeling. Out of 96 data samples, we used 75% data (72 samples) as training set and the remaining 25% data (24 samples) were used for test purpose. The learning parameter and momentum factor used in the BP algorithm are gradually decreased with the increase of the iteration, in order to have a faster learning. A bias unit with an output value of +1.0 was added to all layers except the output layer of the MLP. We carried out several experiments in order to achieve the best architecture and learning parameters. In case of MLP and eMLP, the hidden layer nodes were varied from 5 to 9 and from 2 to 5, respectively. The learning parameters were varied from 0.001 to 0.005. The best architecture and learning parameters obtained are shown in Table 2. All the nodes have a sigmoid activation function.

6   Results and Discussion

The scatter plot between the measured and the model estimated VHN the complete (96 samples) dataset for the MLP- and eMLP-based models are shown in Fig. 3a and b, respectively. The values of CC, MAE, and RMSE are also shown in this Fig. The CC for both datasets is close to 1.0. It can be seen in Fig. 3 that performance of the eMLP model is similar to the MLP model, but it gives advantage in terms of savings in hardware and computational load, as shown in Table 2. For the same dataset, PourAsiabi et al. [24] reported the CC and MAE as 0.9912 and 0.71%, respectively, however using only two chemical compositions, i.e., *Cu* and *Mo*. Thus, the performance of our proposed models is superior to that of [24].

Table 2. Simulation setup parameters for the MLP models

Model	Architecture	Initial value		MSE_{train} (dB)	No. of weights	Exec. time (s)
		α	β			
MLP	10-7-1	0.002	0.002	1.2	67	4.53
eMLP	19-3-1	0.002	0.002	6.8	40	4.10

7   Prediction of VHN

In this Section, we show the ability of the proposed models to predict VHN of ADI at any arbitrary austempering temperature or time (within a specific range). The predicted and measured VHNs at different values of temperature for 30 and 120 min of austempering time durations are shown in Fig. 4, for the eight specimens. The red symbols indicate the measured values and the blue symbols denote the predicted values of VHN. The predicted and measured VHNs at different values of time duration for 260 and 320 °C of austempering temperature are shown in Fig. 5, for the eight specimens. Both the MLP and eMLP-based models have similar performance and are able to predict the VHN accurately.

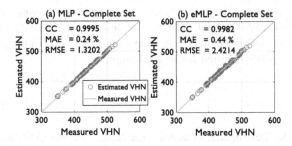

Fig. 3. Scatter plot – the measured and estimated VHNs by the MLP and eMLP models.

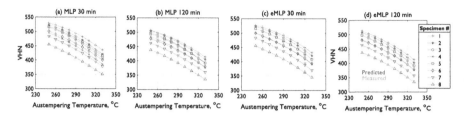

Fig. 4. Performance of VHN prediction at different austempering temperatures for the MLP and eMLP models. (Colour figure online)

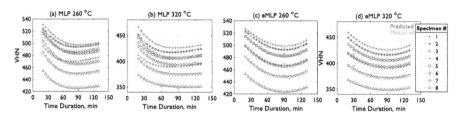

Fig. 5. Performance of VHN prediction at different austempering time durations for the MLP and eMLP models. (Colour figure online)

8 Conclusion

We proposed an enhanced MLP that is computationally efficient as it has a smaller number of weights to update than an MLP. Considering seven chemical compositions, the proposed eMLP-based scheme for modeling the ADI process to predict VHN was able to achieve similar performance to that of MLP-based scheme, in terms of prediction error. However, in the eMLP, only about 60% of the weights were used. Thus, as seen from Table 2, eMLP-based scheme provides advantages in terms of computational load and hardware requirements. The performance of the proposed models was found to be superior than the model proposed in [24], although in the latter model only two chemical compositions were used for prediction. The proposed models can predict VHN accurately at any value (within a specified range) of austempering temperature and time duration, even beyond the range of training samples. In conclusion, the proposed eMLP modeling technique can help engineers to design the ADI process to achieve a specific VHN required for an application.

References

1. Putatunda, S.K.: Development of austempered ductile cast iron (ADI) with simultaneous high yield strength and fracture toughness by a novel two-step austempering process. Mater. Sci. Eng. A **315**, 70–80 (2001)

2. Yazdani, S., Elliott, R.: Influence of molybdenum on austempering behaviour of ductile iron Part 1 – Austempering kinetics and mechanical properties of ductile iron containing 0.13% Mo. Mater. Sci. Technol. **15**, 531–540 (1999)
3. Panneerselvam, S., Martis, C.J., Putatunda, S.K., Boileau, J.M.: An investigation on the stability of austenite in austempered ductile cast iron (ADI). Mater. Sci. Eng., A **626**, 237–246 (2015)
4. Cast Metals Development Ltd.: Austempered ductile-iron castings—advantages, production, properties and specifications. Mater. Des. **13**, 285–297 (1992)
5. Harding, R.: Austempered ductile irons-gears. Mater. Des. **6**, 177–184 (1985)
6. Cakir, M.C., Bayram, A., Isik, Y., Salar, B.: The effects of austempering temperature and time onto the machinability of austempered ductile iron. Mater. Sci. Eng. A **407**, 147–153 (2005)
7. Angus, H.T.: Cast Iron: Physical and Engineering Properties. Butterworth-Heinemann, Oxford (1978)
8. Trudel, A., Gagne, M.: Effect of composition and heat treatment parameters on the characteristics of austempered ductile irons. Can. Metall. Q. **36**, 289–298 (1997)
9. de Albuquerque Vicente, A., Moreno, J.R.S., de Abreu Santos, T.F., Espinosa, D.C.R., Tenório, J.A.S.: Nucleation and growth of graphite particles in ductile cast iron. J. Alloys Compd. **775**, 1230–1234 (2019)
10. Giannakopoulos, A.E., Larsson, P.L., Vestergaard, R.: Analysis of Vickers indentation. Int. J. Solids Struct. **31**, 2679–2708 (1994)
11. Haykin, S.: Neural Networks. Prentice Hall, Upper Saddle River (1999)
12. Patra, J.C., Kot, A.C.: Nonlinear dynamic system identification using Chebyshev functional link artificial neural networks. IEEE Trans. Syst. Man Cybern. Part B **32**, 505–511 (2002)
13. Patra, J.C.: Neural network-based model for dual-junction solar cells. Prog. Photovoltaics Res. Appl. **19**, 33–44 (2011)
14. Patra, J.C., Chakraborty, G.: e-MLP-based modeling of high-power PEM fuel cell stacks. In: 2011 IEEE International Conference on Systems, Man, and Cybernetics (SMC), Anchorage, AK, pp. 802–807 (2011)
15. Patra, J.C., Maskell, D.L.: Artificial neural network-based model for estimation of EQE of multi-junction solar cells. In: 2011 37th IEEE Photovoltaic Specialists Conference (PVSC), pp. 002279–002282. IEEE, Seattle (2011)
16. Patra, J.C.: Chebyshev neural network-based model for dual-junction solar cells. IEEE Trans. Energy Convers. **26**, 132–139 (2011)
17. Jiang, L.L., Maskell, D.L., Patra, J.C.: Chebyshev functional link neural network-based modeling and experimental verification for photovoltaic arrays. In: 2012 International Joint Conference on Neural Networks (IJCNN), pp. 1–8. IEEE, Brisbane (2012)
18. Patra, J.C., Maskell, D.L.: Modeling of multi-junction solar cells for estimation of EQE under influence of charged particles using artificial neural networks. Renew. Energy **44**, 7–16 (2012)
19. Cool, T., Bhadeshia, H.K.D.H., MacKay, D.J.C.: The yield and ultimate tensile strength of steel welds. Mater. Sci. Eng., A **223**, 186–200 (1997)
20. Malinov, S., Sha, W., McKeown, J.J.: Modeling and correlation between processing parameters and properties in titanium alloys using artificial network. Comput. Mater. Sci. **21**, 375–394 (2001)
21. Yilmaz, M., Ertunc, H.M.: The prediction of mechanical behavior for steel wires and cord materials using neural networks. Mater. Des. **28**, 599–608 (2007)
22. Reddy, N.S., Krishnaiah, J., Young, H.B., Lee, J.S.: Design of medium carbon steels by computational intelligence techniques. Comput. Mater. Sci. **101**, 120–126 (2015)

23. Yescas, M.A.: Prediction of the Vickers hardness in austempered ductile irons using neural networks. Int. J. Cast Met. Res. **15**, 513–521 (2003)
24. PourAsiabi, H., PourAsiabi, H., AmirZadeh, Z., BabaZadeh, M.: Development a multi-layer perceptron artificial neural network model to estimate the Vickers hardness of Mn–Ni–Cu–Mo austempered ductile iron. Mater. Des. **35**, 782–789 (2012)
25. Pao, Y.H.: Adaptive Pattern Recognition and Neural Networks. Addison-Wesley, Boston (1989)

Improving Neural Network Classifier Using Gradient-Based Floating Centroid Method

Mazharul Islam, Shuangrong Liu, Xiaojing Zhang, and Lin Wang[(✉)]

Shandong Provincial Key Laboratory of Network Based Intelligent Computing,
University of Jinan, Jinan 250022, China
wangplanet@gmail.com

Abstract. Floating centroid method (FCM) offers an efficient way to solve a fixed-centroid problem for the neural network classifiers. However, evolutionary computation as its optimization method restrains the FCM to achieve satisfactory performance for different neural network structures, because of the high computational complexity and inefficiency. Traditional gradient-based methods have been extensively adopted to optimize the neural network classifiers. In this study, a gradient-based floating centroid (GDFC) method is introduced to address the fixed centroid problem for the neural network classifiers optimized by gradient-based methods. Furthermore, a new loss function for optimizing GDFC is introduced. The experimental results display that GDFC obtains promising classification performance than the comparison methods on the benchmark datasets.

Keywords: Neural network classifier · Classification · Loss function · Floating Centroid Method

1 Introduction

The neural network has been adopted to address considerable classification problems [3–5]. Conventionally, the classification process of the neural network is explained from the probabilistic perspective. The values of output neurons are considered as the probabilities that a sample belongs to different classes. Meanwhile, this process can also be described from a geometric perspective. The neural network is viewed as a mapping function f. Each class is coded as the unique binary string. As the input, the sample is mapped to a space by f, in which classes are represented by different fixed points, referred to centroids. The binary string of each class describes the position of centroid. The mapped sample is attached to the closest centroid (class) in according to the distance. Therefore, methods acting on the output layer of the neural network, such as one-per-class,

---

M. Islam and S. Liu—Both authors contribute equally to this article.

© Springer Nature Switzerland AG 2019
T. Gedeon et al. (Eds.): ICONIP 2019, CCIS 1143, pp. 423–431, 2019.
https://doi.org/10.1007/978-3-030-36802-9_45

softmax and error-correcting output code (ECOC) [1,2], can also be viewed as the methods of distributing the centroids.

For the one-per-class, softmax and ECOC, they are widely used in different neural network models optimized by gradient-based optimization methods, and achieve considerable successful stories [9,10]. However, for these fixed centroid methods, the fixed centroid problem (FCP) [6–8], which refers to that the locations, labels, and number of centroids are prior set before the training, restrains their performance. Because the FCP results in the reduction of the size of the set consisting of optimal neural networks, and enlarges the complexity of optimization. Although the floating centroid method (FCM) [7] affords a way to solve the fixed centroid problem, the evolutionary computation is adopted as its optimization method that impedes FCM to employ the neural networks optimized by gradient-based optimization methods.

In this study, considering the facts mentioned above, a gradient-based floating centroid method (GDFC) is proposed. The GDFC absorbs the advantages of fixed centroids methods and floating centroid method, and affords a way for the neural network classifiers optimized by gradient-based optimization method to address the fixed centroid problem. For this study, the major contributions are introduced as follows:

- The gradient-based floating centroid method is proposed to tackle fixed centroid problem for the neural network classifiers optimized by gradient-based methods.
- A new loss function, named centroid loss function, is proposed to maximize compactness of within-class and separability of between-class during training process.

The proposed GDFC method is described in Sect. 2. Experiment is reported and results on benchmark datasets are analyzed in Sect. 3. We give the conclusion, and draw the future work in Sect. 4.

2   Methodology

2.1   Gradient-Based Floating Centroid Framework

The framework of the gradient-based floating centroid is shown in Fig. 1. The training part mainly includes 4 modules: mapping by the neural network, generating centroids by K-means, coloring centroids, and calculating loss to update neural network. At first, the neural network maps the samples to the partition space. Afterwards, the centroids are generated in the partition space by using k-means algorithm. The classes' number can smaller than the number of centroids. Subsequently, these centroids are labeled by different classes. The labeling strategy is referred to coloration process. In the coloration process, if the mapped samples which is represented by one class are the majority then the corresponding centroid is colored by that class. Besides, one class can be used to label more than one centroids. After that, the neural network is iteratively updated by the proposed centroid loss function which uses the distribution information of the centroids. In the optimization process, the centroid loss function has the ability to maximize compactness of within-class and separability of between-class simultaneously; thus clear decision boundaries exist among different clusters. Finally, an optimal neural network and centroids decided by this optimal neural network are obtained.

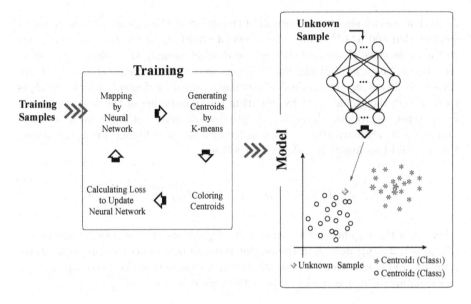

Fig. 1. Architecture of the GDFC

In the testing process, an unknown sample as input of the optimal neural network is mapped to the partition space. This unknown sample is assigned to the centroid with closest distance. For example, in Fig. 1, the unknown sample is close to the centroid$_2$, which represents class 2. Therefore, this sample categorized to class 2.

2.2   Centroid Loss Function

For the mapped samples in the partition space, the k-means algorithm is used to generate the centroids $C^{(k)}(k = 1, 2, \ldots, K)$ by clustering these mapped samples. Then, for each mapped sample, two centroids are selected from K centroids:

The first centroid is one with minimum value of the $||\cdot||_2$, and this centroid having same class with the mapped sample.

$$D_{\min}^{Self} = \arg \min_C ||\beta^{(j)} - C^{(k)}(g)||_2 \tag{1}$$

where $j = 1, 2, \ldots, m$, m is the number of samples, $k = 1, 2, .., K$, K is the number of centroids, $|| \cdot ||_2$ represents the distance. Note that we can obtain the centroid which has the same class and nearest distance to the mapped sample, denoted as C^S. The second centroid is one with minimum value of the $|| \cdot ||_2$, and the class of this centroid is different with the mapped sample.

$$D_{\min}^{Noself} = \arg \min_C ||\beta^{(j)} - C^{(k)}(g)||_2 \qquad (2)$$

Note that we can select the centroid of the different class nearest to the mapped sample, denoted as C^N. Since the target of GDFC is to put the points belongs to the same class closer and enlarge the distance among the points with different classes, thus, minimizing the $D_{\min}^{Self}$ as well as maximizing the $D_{\min}^{Noself}$ are expected. Adopting the method of stochastic gradient descent, which attempts to minimize the global error by updating the parameters of the neural network in an iterative process. Moreover, the gradient descent method is prone to over-fitting, so L_2 regularization is applied to decrease over-fitting. From the above, the centroid loss function is defined as follow,

$$E = \frac{1}{2} \sum_{q=1}^{Q} [((\beta_{qj} - C_{qj}^S)^2 - \xi \cdot (\beta_{qj} - C_{qj}^N)^2)] + \frac{\lambda}{2} \cdot \sum w^2 \qquad (3)$$

where λ is the regularization parameter. β_{qj} denotes the mapped value of the qth (q = 1, 2,.., Q) neuron in the output layer, Q represents the output neurons' number, and is also equal to the number of dimensions of the partition space. ξ is a constant, which is used to adjust the weight between $D_{\min}^{Self}$ and $D_{\min}^{Nosolf}$.

2.3 Optimization

Based on the gradient descent method, while following the back-propagation (BP) idea of training-error to iteratively update the parameters to obtain an optimal neural network. Without loss of generality, assuming that a feedforward neural network has L layers. η is a global learning rate. From the L layer to the $L - 1$ layer, the partial derivatives of the weights and biases are obtained respectively.

$$\Delta w_{qh}^{(L-1)} = -\eta \frac{\partial E}{\partial w_{qh}^{(L-1)}} = \eta \left[\delta_q^{(L)} \alpha_{hj}^{(L-1)} - \lambda w_{qh}^{(L-1)} \right] \qquad (4)$$

$$\Delta \theta_q^{(L)} = \eta \frac{\partial E}{\partial \theta_q^{(L-1)}} = \eta \frac{\partial E}{\partial \beta_{qj}} \frac{\partial \beta_{qj}}{\partial \theta_q^{(L-1)}} = -\eta \delta_q^{(L)} \qquad (5)$$

Note that,

$$\delta_q^{(L)} = -\frac{\partial E}{\partial \beta_{qj}} \frac{\partial \beta_{qj}}{\partial z_{qj}} = (C_{qj}^S - \beta_{qj}) \cdot \sigma'(z_{qj}) - \xi(C_{qj}^N - \beta_{qj}) \sigma'(z_{qj}) \qquad (6)$$

where Q is neurons' number in L layer, H represents the neurons' number in the L − 1 layer, $\Delta w_{qh}^{(L-1)}$ is the weight change value from the hth (with h = 1, 2,..., H) neuron of the L − 1 layer to the qth (q = 1, 2,.., Q) neuron of the L layer, $\Delta \theta_q^{(L)}$ is the bias change value of the qth (with q = 1, 2,.., Q) neuron in the L layer. $\sigma(\cdot)$ represents the activation function. $\beta_{qj} = \sigma(z_{qj}) = \sigma(\sum_{h=1}^{H}(w_{qh}^{(L-1)} \cdot \alpha_{hj}^{(L-1)} + \theta_q^{(L)})$, β is the activation value of the L layer neurons, α is the activation value of the L − 1 layer neurons.

Through the above derivation, we can generalize the general update formulas for the L − l layer to the L − (1 − 1) with l = 1,2,...,L − 1 layer weights and biases, can be written as below

$$\delta^{(L-l)} = \sigma'(y^{(L-l)}) \cdot \sum w^{(L-l)} \cdot \delta^{(L-l+1)} \tag{7}$$

then,

$$\Delta w^{(L-l)} = \eta\tau(x)[\delta^{(L-l)}\alpha^{(L-l+1)} - \lambda w^{(L-l+1)}] \tag{8}$$

$$\Delta\theta^{(L-l+1)} = -\eta\delta^{(L-l+1)} \tag{9}$$

Thus, the weights and biases are updated as

$$w^{(L-l)}(g+1) = w^{(L-l)}(g) + \Delta w^{(L-l)} \tag{10}$$

$$\theta^{(L-l+1)}(g+1) = \theta^{(L-l+1)}(g) + \Delta\theta^{(L-l+1)} \tag{11}$$

3 Experiments

3.1 Overview of the Datasets

Ten classification benchmark datasets is employed to evaluate models, including Pima Indians Diabetes (Diabetes), Congressional Voting Records (Vote), Risk Factors Cervical Cancer (RFCC), SPECT-heart (SPECT), Climate Model Simulation Crashes (CMSC), Website Phishing (Web), Hayes-Roth (HR), Balance Scale (Balance), Wine, Balance Scale (Balance), User Knowledge Modeling (UKM).

3.2 Comparison Methods

We choose different types of classifier in the experiment to compare the efficiency with the GDFC method. Table 1 is used to introduce these methods. For a fair comparison with the proposed method, the potentiality of these methods is explored, and their parameters are tuned by trial and error. As the evaluation metrics, Generalization Accuracy (GA) and Average F-measure (Avg.FM) [7] is adopted to evaluate the efficiency of the GDFC method. The performance of all methods is evaluated by using ten-fold cross-validation.

KNN is a classical classification method. For the KNN, the number of nearest neighbors is selected in the range {1, 30}. For the SVM, cost parameter is selected

Table 1. Comparative methods

Type	Method
Neural network-based methods	Feed-forward Neural Network (FNN)
	Nearest Neighbor Partitioning (NNP)
	Floating Centroid Method (FCM)
Other classification methods	Naïve Bayes (NB)
	Support Vector Machine (SVM)
	K-nearest Neighbor (KNN)

from the range $\{2^{-2}, 2^{-5}\}$. For the GDFC and FCM, the number of hidden layer neurons is set from range $\{1, 40\}$. The number of dimensions of the partition space is selected from $\{N, 10N\}$, and for the number of centroids is from the range $\{N, 5N\}$, where N is equal to the number of classes. The value of ξ is selected from $\{0.5, 1\}$. For NNP, hidden layer neurons number is selected from the range $\{1, 40\}$. The number of dimensions of partition space is chosen from the range $\{N, 10N\}$, and the number of centroids from the range N, 5N. The value of parameter p is fixed at 3. For FNN, the hidden layer neuron's number is selected from the range $\{1, 40\}$.

3.3 Results Analysis

We demonstrate the experimental results and the findings in this subsection. In the experiments, we compared the proposed GDFC method with six different classifiers, including SVM, NB, KNN, FNN, NNP, and FCM. Tables 2 and 3 exhibit the testing accuracy and Avg.FM of all the methods on each dataset.

Table 2. Accuracy comparison of all the methods. The unit of results is percentage.

Method	SVM	NB	KNN	FNN	NNP	FCM	GDFC
Diabetics	78.25	77.25	74.50	75.50	75.00	77.75	**79.50**
Vote	93.33	90.00	90.00	78.75	92.90	92.90	**94.17**
RFCC	92.21	81.91	88.68	88.82	87.06	90.00	**92.35**
SPECT	80.00	76.43	80.00	80.71	78.21	80.71	**82.07**
CMSC	94.25	94.91	92.55	92.55	92.73	94.00	**96.09**
Web	82.92	81.60	86.06	86.64	84.50	85.11	**88.41**
HR	57.79	70.00	76.21	71.11	79.44	77.22	**81.11**
Balance	90.63	84.60	79.68	95.08	94.60	95.87	**96.35**
Wine	**98.89**	97.22	98.33	94.44	**98.89**	**98.89**	**98.89**
UKM	94.50	87.62	83.57	89.05	95.24	95.95	**96.18**
MEAN	86.28	84.15	84.96	85.27	87.86	88.84	**90.51**

From Table 2, our proposed GDFC achieved better generalization accuracy on majority of the datasets. Only in Wine dataset, the generalization performance of SVM, NNP, and FCM is alike to the proposed method. That phenomenon clarifies that the GDFC is promising compared with all the comparative methods in terms of generalization performance.

Table 3. Avg.FM comparison of all the methods. The unit of results is percentage.

Method	SVM	NB	KNN	FNN	NNP	FCM	GDFC
Diabetics	73.27	73.73	70.1	73.07	73.25	72.21	**75.29**
Vote	93.29	89.93	89.85	82.69	92.90	92.90	**94.14**
RFCC	**87.25**	72.12	62.77	59.17	77.23	72.65	83.76
SPECT	70.11	71.53	72.24	63.16	68.12	67.46	**73.07**
CMSC	82.21	78.09	65.16	62.38	77.95	75.84	**83.79**
Web	60.32	61.69	75.29	74.83	81.40	66.55	**85.14**
Hr	59.73	71.82	76.41	69.60	80.46	77.24	**81.48**
Balance	83.76	60.23	60.11	84.87	90.23	**92.81**	89.96
Wine	98.93	97.32	98.41	93.6	**98.97**	98.88	98.93
UKM	93.5	87.45	79.26	82.13	94.88	96.11	**97.08**
MEAN	80.23	76.39	74.96	74.55	83.54	81.27	**86.26**

Table 3 shows the comparison between the proposed method with the comparative methods based on Average F-measure. For classification task, Avg.FM is a significant appraisal for the classifiers because Avg.FM summaries both precision and recall to evaluate the performance. As compared with other methods, GDFC has better performance on most of the datasets. For the RFCC, Balance and Wine dataset, the SVM, FCM and NNP has slightly better performance respectively. That demonstrates that the GDFC has a better balance between precision and recall compared to other competing methods.

Furthermore, based on mean accuracy which is the average results of each method on ten datasets. Gradient-based floating centroid method improved about **5%** (versus SVM), **8%** (versus NB), **7%** (versus KNN), **6%** (versus FNN), **3%** (versus NNP) and **2%** (versus FCM) in terms of testing accuracy. Moreover, GDFC improved about **8%** (versus SVM), **13%** (versus NB), **15%** (versus KNN), **16%** (versus FNN), **3%** (versus NNP) and **6%** (versus FCM) in terms of Avg.FM. That concludes our proposed GDFC method is superior to the other competing methods.

4 Conclusions

In this study, a novel gradient-based floating centroid method is introduced to solve the fixed-centroid problem for the gradient-based neural network classifier. Moreover, the centroid loss function is introduced for maximizing the compactness of within-class and the separability of between-class during the optimization process. Experimental results indicated that proposed gradient-based floating centroid method outperforms the comparative methods on majority of the datasets in terms of generalization accuracy and Avg.FM. In the future, different neural network structures will be investigated and employed in GDFC, considering that the mapping ability of the neural network is one of the vital factors for the classification performance of GDFC.

Acknowledgements. This work was supported by National Natural Science Foundation of China under Grant No. 61872419, No. 61573166, No. 61572230, No. 61873324, No. 81671785, No. 61672262. Shandong Provincial Natural Science Foundation No. ZR2019MF040, No. ZR2018LF005. Shandong Provincial Key R&D Program under Grant No. 2019GGX101041, No. 2018GGX101048, No. 2016ZDJS01A12, No. 2016GGX101001, No. 2017CXZC1206. Taishan Scholar Project of Shandong Province, China, under Grant No. tsqn201812077.

References

1. Bridle, J.S.: Probabilistic interpretation of feedforward classification network outputs, with relationships to statistical pattern recognition. In: Soulié, F.F., Hérault, J. (eds.) Neurocomputing. NATO ASI Series, vol. 68, pp. 227–236. Springer, Heidelberg (1990). https://doi.org/10.1007/978-3-642-76153-9_28
2. Dietterich, T.G., Bakiri, G.: Solving multiclass learning problems via error-correcting output codes. J. Artif. Int. Res. **2**(1), 263–286 (1995)
3. Jiang, G., He, H., Yan, J., Xie, P.: Multiscale convolutional neural networks for fault diagnosis of wind turbine gearbox. IEEE Trans. Ind. Electron. **PP**, 1 (2018)
4. Kamilaris, A., Prenafeta-Bold, F.X.: A review of the use of convolutional neural networks in agriculture. J. Agric. Sci. **156**(3), 312–322 (2018). https://doi.org/10.1017/S0021859618000436
5. Nazari, M., Oroojlooy, A., Snyder, L., Takac, M.: Reinforcement learning for solving the vehicle routing problem. In: Bengio, S., Wallach, H., Larochelle, H., Grauman, K., Cesa-Bianchi, N., Garnett, R. (eds.) Advances in Neural Information Processing Systems, vol. 31, pp. 9839–9849. Curran Associates, Inc. (2018)
6. Wang, L., Yang, B., Chen, Y., Zhang, X., Orchard, J.: Improving neural-network classifiers using nearest neighbor partitioning. IEEE Trans. Neural Netw. Learn. Syst. **28**(10), 2255–2267 (2017)
7. Wang, L., et al.: Improvement of neural network classifier using floating centroids. Knowl. Inf. Syst. **31**(3), 433–454 (2012)
8. Wang, L., Yang, B., Chen, Z., Abraham, A., Peng, L.: A novel improvement of neural network classification using further division of partition space. In: Mira, J., Álvarez, J.R. (eds.) IWINAC 2007. LNCS, vol. 4527, pp. 214–223. Springer, Heidelberg (2007). https://doi.org/10.1007/978-3-540-73053-8_21

9. Wibowo, A., Wiryawan, P.W., Nuqoyati, N.I.: Optimization of neural network for cancer microRNA biomarkers classification. J. Phys: Conf. Ser. **1217**, 012124 (2019)
10. Wong, Y.J., Arumugasamy, S.K., Jewaratnam, J.: Performance comparison of feedforward neural network training algorithms in modeling for synthesis of polycaprolactone via biopolymerization. Clean Technol. Environ. Policy **20**(9), 1971–1986 (2018)

MON: Multiple Output Neurons

Yasir Jan[(✉)] [iD], Ferdous Sohel [iD], Mohd Fairuz Shiratuddin [iD],
and Kok Wai Wong [iD]

Murdoch University, Perth, WA, Australia
{Y.Jan,F.Sohel,F.Shiratuddin,K.Wong}@murdoch.edu.au

Abstract. Existing basic artificial neurons merge multiple weighted inputs and generate a single activated output. This paper explores the applicability of a new structure of a neuron, which merges multiple weighted inputs like existing neurons, but instead of generating single output, it generates multiple outputs. The proposed "Multiple Output Neuron" (MON) can reduce computation in a basic XOR network. Furthermore, a MON based convolutional neural network layer (MONL) is described. Proposed MONL can backpropagate errors, thus can be used along with other CNN layers. MONL reduces the network computations, by reducing the number of filters. Reduced number of filters limits the network performance, thus MON based neuroevolution (MON-EVO) technique is also proposed. MON-EVO evolves the MONs into single output neurons for further improvement in training. Existing neuroevolution techniques do not utilize backpropagation but MONs can utilize backpropagation. Experimental networks trained using the CIFAR-10 classification dataset show that proposed MONL and MON-EVO provide a solution for reduced training computation and neuroevolution using backpropagation.

Keywords: ANN · CNN · Neuroevolution

1 Introduction

Existing basic neuron units aggregate multiple weighted inputs and apply a non linear activation to generate a single output [7]. We propose MON, a new structure of the basic neuron unit, which aggregates multiple weighted inputs but generates multiple outputs from the activated sum. MONs can be considered as the superset of the existing basic neuron units.

CNNs are designed by the weighted interconnections of basic neuron units; in which the filter weights of a layer are applied on a layer to get the next layer output [3,4]. It is assumed that if a filter/edge detector is useful in one spatial region of the layer, it should also be useful for detecting edges in other region of the same layer. Hence the filter is shared and strided across the whole layer. Other options include the locally connected layers of CNN where the weights are not shared. It is assumed that the edges in different spatial subregions are different, thus the filters of each subregion across a layer does not have to

© Springer Nature Switzerland AG 2019
T. Gedeon et al. (Eds.): ICONIP 2019, CCIS 1143, pp. 432–439, 2019.
https://doi.org/10.1007/978-3-030-36802-9_46

be the same. But what if the edges/shapes in one region of the layer are not exactly same but 'slightly' different from edges/shapes in other regions of the layer? If a weighted filter is applied on two almost similar edges in a layer, the outputs of both maybe slightly different, even though we may want them to be expressed equally for the next layer. Therefore, to rectify for the slight change in inputs, the outputs should have multiple variations. Therefore, network should have shared weights but with options of slight changes in the output values. To implement and test such a behaviour, the paper proposes MON based CNN layer (MONL). In MONL, the neurons will generate multiple linearly shifted outputs from preceeding filter outputs. MONL allows the network to reduce the filters in preceeding layer, and then generates multiple interconnections with the next layer. The effect of each interconnection, such as in dropout [11] or pruning [1], is not the scope of this paper. But this paper discusses how generating multiple linearly shifted outputs and their interconnections can assist in training.

MON network training is further improved by using MON nased neuroevolution (MON-EVO) technique. Previously, evolutionary techniques [6,12,13] suggest how the number of neurons in a layer and their weight parameters are updated in each generation. In deep neural networks, the network parameters are high in number, therefore the evolutionary algorithms applied on the network parameters become computationally extensive [8,13]. Some evolutionary techniques propose the network structure in the form of gene cells, where each gene represents a cell of operations [5]. The issue with these techniques is that they are based on the selection process, thus the efficiency of backpropagation is not utilized and the evolved network may improve/degrade abruptly. In the proposed MON-EVO, the network is trained usign backpropagation, and the evolution occurs without changing the parameters and interconnections, therefore the progression becomes gradual instead of an abrupt change like other evolutionary selection techniques. The experiments for MONL and MON-EVO are performed on CIFAR-10 dataset [2].

Section 2 discusses the background of MON. Section 3 presents the structure and training of MONL. Section 4 discusses the MON-EVO. Section 5 describes the experimental setups. Section 6 presents the results, while Sect. 7 concludes the paper.

2 Background

Artificial neuron's structure was designed to mimic neurons in human brain. The basic neuron takes in multiple inputs $x_1, x_2, ...x_n$, and generates an aggregate weighted sum $S = \sum_{i=1}^{n} x_i \times w_i$. The aggregate is activated using a nonlinear function σ to generate a single output $y = \sigma(S + b)$, where b is the bias. For ease of computations, bias b can be considered as w_0 multiplied with input $x_0 = 1$, and considered as part of S.

We would like to make few assumptions about biological neurons, which may or maynot coincide with the exisitng study. Each neuron in a human brain have more than one output connections, which may have slightly different output value than the other. Based on this idea we propose MONs in which each output of the same neuron is slightly different from the other.

Proposed MON outputs can be calculated as in Eq. (1).

$$y_n = \sigma(S) + b_n \tag{1}$$

where $n = 1, 2, 3, ...M$ for M number of outputs of MON. Each different MON output is calculated by adding a bias value before or after the activated output.

2.1   Classification

A single neuron can be a linear classifier. A less trained classifier may need further iterations of slopes and intercept, to make it a perfect classifier. However instead of numerous iterations, it is possible that by only linearly shifting i.e. changing the bias value, the classifier would give us the better result. Therefore, if classifier with multiple intercepts (bias values) could be tested in parallel, then search space could be expanded including the linear shifts. Hence, it is proposed that a neuron should be trained with multiple bias values, in parallel.

2.2   XOR Network

XOR network is an example of a non linearly separable problem. If MON is used in first layer, then two different bias values can be added to the same weighted aggregate. Figure 1 shows two structures of XOR network [9] using single output neurons (Network A and B), and a MON based structure (network C) which adds two different bias values $(-1.5, -0.5)$ to the same sum, and generate two outputs for the next layer. MON based XOR network has reduced number of operations.

Each of the networks A, B, and C have different number of multiplications (m), aggregate additions(a), bias additions(b) and threshold comparisons (t) operations. Network A has total $11(m = 2, a = 3, b = 3, t = 3)$ operations. Network B has total $8(m = 1, a = 3, b = 2, t = 2)$ operations. Network C has total $6(m = 1, a = 2, b = 2, t = 1)$ operations.

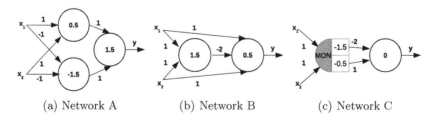

(a) Network A   (b) Network B   (c) Network C

Fig. 1. Comparison of XOR Networks, using basic single output neurons vs MON. MON based XOR network have reduced computations.

The proposed MON based XOR network is not trained using any technique, but is manually designed. The training may be done using genetic algorithm or any technique, but is not the scope of this paper. It can be observed that MONs can reduce the computations for such a simple network. Therefore it can be inferred that if MONs are used in bigger networks, the computations can be reduced for them too.

3   MONL

Proposed MONL is added after the activation layer of the CNN. Previously, if a convolutional and activation layer would generate K filter outputs, then by introducing MON layer (with M outputs) the number of convolutional filters are reduced from K to $Q = K/M$. MONL would multiply the Q outputs and bring them back to K, using bias addition. Thus the filters in CNN will be reduced but the interconnections with next layer will still be the same.

In the example, MONL takes input (Q_{ijq} filter values from previous layer, where $i \times j$ are the feature dimensions and $q = 1, 2, ...Q$). represents the filter. MON layer replicates all the Q filter values M times and to each of them adds a separate bias value m_n, where $n = 1, 2, ...M$. It generates $K = Q \times M$ filter outputs, as shown in MONL Forward Pass Fig. 2. Output K_{ijk} can be calculated using the Eq. 2.

$$K_{ijk} = \sigma_m(Q_{ijq}) + m_n \tag{2}$$

where $k = (m - 1) \times Q + q$, and σ_m is the activation function applied for each MON output. Theoretically, each σ_m could be a different function for each MON output, but for simplicity we have chosen all to be ReLU activations.

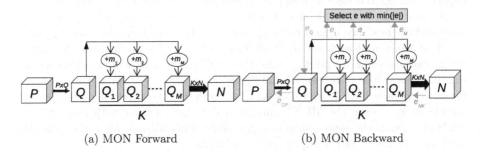

(a) MON Forward   (b) MON Backward

Fig. 2. MONL based architecture.

MONs should be able to backpropagate the errors, so they can be used with existing CNN layers. MON error propagation behaviour is inspired from the phenomenon of 'inertia', as in Newtonian Physics. MON tends to remain unchanged, therefore, MON chooses only one error with least magnitude out of the multiple errors, and propagate it to the previous layer, as shown in MONL Backward Pass in Fig. 2. Using Eq. 3, the error with minimum absolute value is selected and propagated backward.

$$e_{ijq} = argmin(e_{ijqm}) \tag{3}$$

4 MON-EVO

In CNNs, selecting number of filters in a convolutional layer is a tricky thing. Decisions of choosing the count of filters in a layer is based on previously trained networks, or by trial and error method. It would be strange to suggest that human brain has predefined number of filters and connections for each task. Human brain should behave in an evolutionary way where number of neurons are added/removed on a need-by basis. Therefore, evolutuionary CNNs should initiate with less number of filters in a layer and then gradually increase the number of filters. The issue with the gradual increase in the filters is that for each additional filter the depth of the next layer increases and thus new connections for next layer are also required. One solution is to randomly assume all the weights, for previous and next layer both. But too much randomness may generate random output, and effect the convergence abruptly. A better solution is to evolve the number of filters, but connections should already exist and does not need to be evolved. Dring evolution the existing multiple connections with the next layer can be utilized. As discussed previously, proposed MONL starts with lesser number of filters, but has multiple connections with the next layer. Therefore, MONL based networks can be ideal for such evolution.

MON evolution (MON-EVO) is inspired from the phenomenon of splitting of biological unicelllar organisms. The cells may splits and form another cells while copying the characteristics. MON-EVO evolves in a similar way by splitting each MON into multiple single output neurons, as shown in Fig. 3. Splitting can be done on a need by basis .i.e. network convergence rate. We performed experiments on self chosen iteration values .i.e. 100 and 150. After split each single output neurons will have its own weighted connections with preeceding layer and the next layer. In the example, Fig. 3, the $P \times Q$ connections are replicated M times and each set of Q filters i.e. Q_1 till Q_M would have their own input connections. For updqtign bias, the MON bias values will be added to the existing bias values of the individual filters respectively. For the output connections, we already have $K \times N$ connections. So there is no need for replication or modification. Hence, the MONL can be evolved into a layer with more filters, while preserving the input, output connections. This will not allow the network to change its behaviour abruplty after evolution.

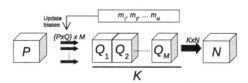

Fig. 3. MONL splitting. Input connections are replicated for each filter set, without changing weights. MON bias is added to respective filter bias. Output connections are split respectively, without changing weights.

5 Experiments

The experiments are performed on CIFAR-10 dataset [2]. The CIFAR-10 dataset is composed of 32×32 RGB natural images, each belonging to one of the 10 classes. The whole dataset has $50,000$ training images, and $10,000$ testing images.

Networks with less number of layers and connections have lesser search space, thus lesser options of improvement. Therefore, we wanted to test for improvements in smaller networks. Smaller networks are also faster to train and test. We train and test three networks for comparison. First a baseline network (*Baseline*), almost similar to Model A in [10], with 5 convolutional layers is selected. The layers in Model A are in the order of $conv1(5, 96, p = 2, s = 2, relu)$, $dropout(0.5)$, $conv2(5, 192, p = 2, s = 2, relu)$, $dropout(0.5)$, $conv3(3, 192, p = 0, s = 1)$, $conv4(1, 192, p = 0, s = 1)$, $conv5(1, 10, p = 0, s = 1)$, $avgpoo(6)$, softmax and classification. Second network (*Baseline$_3$*) has 3 times reduced number of filters in $conv2$ layer compared to (*Baseline*) network, i.e. $conv2(5, 64, p = 2, s = 2, relu)$. Third, a MONL based network (*MON$_3$*) is introduced, in which the $conv2$ layer have filters reduced 3 times i.e. $conv2(5, 64, p = 2, s = 2, relu)$, but an additional MON layer is introduced after the $conv2$ layer i.e. $MON_3(0, 0.25, 0.5)$.

$MON_3(0, 0.25, 0.5)$ means three bias values are added, having values 0, 0.25 and 0.5. There is no specific rule to select the value of bias, but since the filter size is 5×5, so considering a linear shift of 0.1 for each input, we select 0.25 and its multiple 0.5.

Baseline, *Baseline$_3$* and *MON$_3$* are all trained for 500 epochs, with learning rate 0.001. *MON$_3$* is split at two different training points i.e. at epoch 100 and 150, and the evolved networks are called $MON_3 - EVO_{100}$ and $MON_3 - EVO_{150}$ respectively. After the evolution, the networks are trained further till 500 epochs. Since before evolution networks are already trained, therefore after evolution, the learning rate is reduced to 0.0001.

6 Results

Experiments were performed using 5 different architectures i.e. *Baseline*, *Baseline$_3$*, *MON$_3$*, $MON_3 - EVO_{100}$ and $MON_3 - EVO_{150}$. Figure 4 show the training and testing curves. Accuracy curve of *MON$_3$* shows that the network does get trained by backpropagation succesfully. It also shows that the MONL based evolved networks i.e. $MON_3 - EVO_{100}$ and $MON_3 - EVO_{150}$ can be trained further after evolution. This shows that the splitting of MON can be done succesfully as well.

The training accuracies at the end of 500 epochs for the 5 networks i.e. *Baseline*, *Baseline$_3$*, *MON$_3$*, $MON_3 - EVO_{100}$ and $MON_3 - EVO_{150}$ are 0.8759, 0.7594, 0.6664, 0.8225 and 0.8263 respectively, while the testing accuracies are 0.7562, 0.7078, 0.6338, 0.7146 and 0.7153. The evolved networks perform better than the *Baseline$_3$* and *MON$_3$*.

Image input is of size 32×32, thus the filter multiplications in the original network layer are $5 \times 5 \times 192 \times 32 \times 32 = 4915200$. The (MON_3) reduces the number of filters from 192 to 64, thus reducing the multiplications to $5 \times 5 \times 64 \times 32 \times 32 = 1638400$. During each epoch $3276800\,(66\%)$ multiplications can be skipped till the evolution. Even for such a simple network with few layers filter multiplications have been reduced for each epoch. We expect ore reduction for bigger networks. Therefore, for 100 and 150 epochs the 3276800 multiplications can be skipped for each epoch i.e. a total of 3276800×100 and 3276800×150 multiplications can be skipped while training, before evolution. At the evolution, the networks get the structure similar to *Baseline* network. Further training can achieve matching accuracy compared to the *Baseline* network.

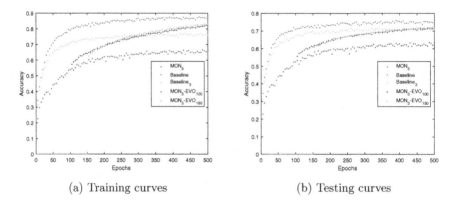

(a) Training curves   (b) Testing curves

Fig. 4. Training and testing accuracy curves of various networks at various epochs.

7 Conclusion

We conclude that during initial epochs networks should be trained using MON based architectures and backpropagation. Since they have lesser filters, they will have lesser computations. Once the network starts converging, or on need-by absis, the network can be evolved and trained further. In future, MON is suggested to be applied at multiple layers of a network. MONs may be evolved gradually in a layer by layer manner. Variation of MON shifts depending on features complexity at each depth of layer maybe applied. Thus giving more varaitions of classification. Such networks are yet to be explored.

References

1. Han, S., Pool, J., Tran, J., Dally, W.: Learning both weights and connections for efficient neural network. In: Cortes, C., Lawrence, N.D., Lee, D.D., Sugiyama, M., Garnett, R. (eds.) Advances in Neural Information Processing Systems, vol. 28, pp. 1135–1143. Curran Associates, Inc. (2015)

2. Krizhevsky, A.: Learning multiple layers of features from tiny images (2009)
3. Lecun, Y., Bottou, L., Bengio, Y., Haffner, P.: Gradient-based learning applied to document recognition. Proc. IEEE **86**(11), 2278–2324 (1998)
4. LeCun, Y., et al.: Handwritten digit recognition with a back-propagation network. In: Advances in Neural Information Processing Systems, pp. 396–404 (1990)
5. Liu, C., et al.: Progressive neural architecture search. In: Ferrari, V., Hebert, M., Sminchisescu, C., Weiss, Y. (eds.) ECCV 2018. LNCS, vol. 11205, pp. 19–35. Springer, Cham (2018). https://doi.org/10.1007/978-3-030-01246-5_2
6. Luo, R., Tian, F., Qin, T., Chen, E., Liu, T.Y.: Neural architecture optimization. In: Advances in Neural Information Processing Systems, pp. 7816–7827 (2018)
7. McCulloch, W.S., Pitts, W.: A logical calculus of the ideas immanent in nervous activity. Bull. Math. Biophys. **5**(4), 115–133 (1943). https://doi.org/10.1007/BF02478259
8. Real, E., Aggarwal, A., Huang, Y., Le, Q.V.: Regularized evolution for image classifier architecture search. Accepted for publication at AAAI 2019, AAAI Conference on Artificial Intelligence (2018). http://arxiv.org/abs/1802.01548
9. Rumelhart, D.E., Hinton, G.E., Williams, R.J.: Parallel distributed processing: explorations in the microstructure of cognition. In: Learning Internal Representations by Error Propagation, vol. 1, pp. 318–362. MIT Press, Cambridge (1986). http://dl.acm.org/citation.cfm?id=104279.104293
10. Springenberg, J., Dosovitskiy, A., Brox, T., Riedmiller, M.: Striving for simplicity: the all convolutional net. In: International Conference on Learning Representations (Workshop Track) (2015)
11. Srivastava, N., Hinton, G., Krizhevsky, A., Sutskever, I., Salakhutdinov, R.: Dropout: a simple way to prevent neural networks from overfitting. J. Mach. Learn. Res. **15**, 1929–1958 (2014). http://jmlr.org/papers/v15/srivastava14a.html
12. Suganuma, M., Shirakawa, S., Nagao, T.: A genetic programming approach to designing convolutional neural network architectures. In: Proceedings of the Genetic and Evolutionary Computation Conference, GECCO 2017, pp. 497–504. ACM, New York (2017). https://doi.org/10.1145/3071178.3071229
13. Zoph, B., Vasudevan, V., Shlens, J., Le, Q.V.: Learning transferable architectures for scalable image recognition. In: 2018 IEEE Conference on Computer Vision and Pattern Recognition, CVPR 2018, Salt Lake City, UT, USA, 18–22 June 2018, pp. 8697–8710 (2018)

Evaluation on Neural Network Models for Video-Based Stress Recognition

Alvin Kennardi[(✉)] and Jo Plested[(✉)]

Research School of Computer Science, Australian National University,
Canberra, Australia
{Alvin.Kennardi,Jo.Plested}@anu.edu.au

Abstract. We examined neural network models to perform video-based stress recognition using ANUStressDB data set [6]. Recent works on video-based stress recognition [6,11] requires feature engineering process, which is time consuming and expensive. The neural network (NN) model aims to reduce this process. In this work, we set a baseline Feed-Forward Neural Network (FFNN) and extended the model using Feature Selection Technique, namely magnitude measure technique [3] and ℓ-1 regularisation [9]. Subsequently, we performed extensive evaluation between those models with the Long Short-Term Memory (LSTM) [5] model, which are designed to store state information for time-series data. We show that feature selection technique model used significantly less parameters compare to the LSTM model with the expense of small accuracy loss. We also show that the NN models performed well in video-based stress recognition task as compare to the previous work with hand-crafted feature engineering from experts.

Keywords: Feature selection · Long Short-Term Memory · Stress recognition

1 Introduction

This paper examines several models to perform video-based stress recognition task. These include, the Feed-Forward Neural Network (FFNN) and recurrent neural network with Long Short-Term Memory (LSTM). Recent researches on video-based stress recognition system incorporates feature engineering process before making a prediction [6,11]. However, this process is expensive and time-consuming. In our work, we compared methods to improve neural network models for video-based stress recognition task to reduce the feature engineering process. We set the feed-forward neural network (FFNN) as our baseline model. We extended this model with feature selection techniques to enhance the performance. The magnitude measure technique [3] uses the absolute value of weights from a fully trained network to measure the contribution of input features towards output values. The ℓ-1 norm regularisation technique [1,8] is an embedded feature selection technique used to bring weight of irrelevant inputs to

© Springer Nature Switzerland AG 2019
T. Gedeon et al. (Eds.): ICONIP 2019, CCIS 1143, pp. 440–447, 2019.
https://doi.org/10.1007/978-3-030-36802-9_47

0, hence remove them from the model during training [9]. Finally, we compared improved NN models with a recurrent neural network model with LSTM [5] to perform video-based stress recognition task. The LSTM model worked well with time-series data set with additional parameters to store information from the previous state.

In this work, we trained and evaluated NN models with ANUStressDB [6] data set. The data set consists of statistical summary from video sample of 24 different subjects. To the best of our knowledge, extensive comparison between reduced FFNN models with the LSTM, which are designed for time-series data, has never been proposed for this data set. Our work makes two main contributions as follows:

– We evaluated the feature selection techniques, namely magnitude measure technique and ℓ-1 norm regularisation to reduce the number of parameters in FFNN. These models used significantly less parameters compare to the LSTM model with the expense of small accuracy loss.
– All the models performed well in video-based stress recognition task as compare to the model from previous work which requires hand-crafted feature engineering from the experts.

2 Method

Method section describes the data set, the models used in this experiment, and also the evaluation method used to measure the model performance.

2.1 Data Set

The ANUStressDB data set [6] consists of video samples from 24 subjects. Each video has a duration of 32 min 17 s, divided into 58110 samples (30 samples for 1 s). Each sample has 34 statistical summary features derived from the sample frames. The video has 12 clip partitions with two labels, namely stressful and not stressful. The proposed models were used to classify clip partitions into those classes. The experiment used two snippets from each class, stressful and not stressful. For every snippet, we used the samples from 20 s after snippet started until 30 s after snippet started to ensure the subjects has developed the emotion (i.e. stressful or not stressful) and reflected on the reading. For feed-forward neural network model, we considered each frame to be one data point, and for the LSTM model, the sequence consists of 15 samples or 0.5 s. For each feature, the data was standardised into zero-mean and unit-variance. We split the data set based on the snippet with 70% training set and 30% test set.

2.2 Feed-Forward Neural Network Model

Our baseline model is a feed-forward neural network with one hidden layer. The model takes an input with 34 features, denotes as $\mathbf{x} \in \mathcal{R}^{34}$. A weight matrix

$\mathbf{W}_{ih} \in \mathcal{R}^{7 \times 34}$ transforms input vector into a vector with 7-dimension. Finally, **ReLU** activation function [4] is applied to this 7-dimensional vector producing hidden vector, denotes as $\mathbf{h} \in \mathcal{R}^7$. The mapping between an input vector to a hidden unit, follows the equation below.

$$h = ReLU\,(\mathbf{W}_{ih}\mathbf{x}) \tag{1}$$

A weight matrix $\mathbf{W}_{ho}$ transforms hidden vector $\mathbf{h}$ into output vector $\mathbf{y} \in \mathcal{R}^2$. Two dimension output vector represent two classes in the data set, namely stressful and not stressful. The **softmax** layer then takes two-dimension output vector, $\mathbf{y}$ into output decision label $Y \in \{0, 1\}$ represent two classes, 0 for stressful and 1 for not stressful. Thus the decision from softmax layer follows equation below.

$$Y = softmax(\mathbf{W}_{ho}\mathbf{h}) \tag{2}$$

The error function used in the model is cross-entropy error function. The model was trained using error back-propagation [2] and optimised using Stochastic Gradient Descent (SGD) with momentum [10] with learning rate 0.1, and momentum term 0.9. The model hyper-parameters were cross-validated on the training set. We trained the baseline model over 5000 epochs.

2.3   Feature Selection Techniques for Neural Network

Feature selection techniques aim to enhance the performance of feed-forward neural network by reducing the number of input features. We presented two techniques to reduce the number of features, namely filter method (magnitude measure) [3] and embedded method (ℓ-1 norm regularisation) [1,8].

Magnitude Measure Technique. The magnitude measure technique takes into account contribution of absolute weight values that connects a hidden neuron j in the hidden layer into an output neuron k in the output layer [3]. The following equation measures the contribution from an input feature neuron i to a hidden neuron j with input vector dimension din = 34.

$$P_{ij} = \frac{|w_{ij}|}{\sum_{p=1}^{din} |w_{pj}|} \tag{3}$$

The contribution from a hidden layer j with dimension dhid = 7 to an output neuron k is measured by following equation.

$$P_{jk} = \frac{|w_{jk}|}{\sum_{r=1}^{dhid} |w_{rk}|} \tag{4}$$

The total contribution from an input feature neuron i to an output neuron k, with hidden layer dimension dhid = 7 follows the equation below.

$$Q_{ik} = \sum_{r=1}^{dhid} (P_{ir} \times P_{rk}) \tag{5}$$

We computed the Q-values for all input features in fully-trained neural network based for the data set described in Sect. 2.1, and removed two features with the lowest Q-values as suggested by [3] to produce more consistent results. The network was re-trained using reduced features. We trained the reduced model over 5000 epochs.

ℓ-1 **Norm Regularisation.** Regularisation technique is a technique to introduce additional error measure to the loss function by penalizing parameters with a big value, hence discouraging the model from over-fitting regularisation using ℓ-1 norm discourages parameters with high sum of absolute values of the parameters [9], thus creating a sparse weight (parameter) matrix solution. The capabilities of ℓ-1 norm to create this sparse solution hence bring some parameters to zero makes this technique a good candidate for feature selection. [9]. We applied ℓ-1 penalty term using following equation.

$$\text{error}_{\ell 1} = \lambda \sum_{i=0}^{\text{dhid}} \sum_{j=0}^{\text{din}} ||\text{w}_{i,j}||_1 \tag{6}$$

The hyper-parameter λ is used to control how strong the penalty applied. This error term is added in the back-propagation error function, i.e., cross entropy error. By adding ℓ-1 penalty function to the hidden layer output defined in Eq. 1, we reduced the weight parameters from $\mathbf{W}_{ih}$ into 0 and hence, removed input features contribution with 0 weight value to the hidden unit. We trained the ℓ-1 regularised model with SGD with momentum with learning rate 0.1 and momentum term 0.9. We set $\lambda = 0.0005$ and trained the model over 4500 epochs.

2.4 Long Short-Term Memory (LSTM) Model

Long Short-Term Memory (LSTM) aims to retain the information from time-series inputs by connecting the output of previous state as the input for the next state. The LSTM layers were added in front of the neural network described in Sect. 2.2. The input patterns were presented to the model in sequence of 15 samples (i.e. 0.5 s).

After sequences presented to the LSTM cell (t = 14), the LSTM cell passed feature representation $\mathbf{h}$ to the fully-connected layers. The fully-connected layers performed down-sampling of the hidden vector $\mathbf{h} \in \mathcal{R}^{34}$ to $\mathcal{R}^2$ as described in Sect. 2.2. The parameters in the LSTM cells were trained using Back-Propagation Through Time (BPTT). We trained our LSTM model using SGD with momentum with learning rate 0.01 and momentum terms 0.9. We trained the model over 1500 epochs. In order to stabilise the model performance, the model uses several modification from original fully-connected layers. Instead of **ReLU** activation function, the model uses **LeakyReLU** [7] with negative slope 0.1 to avoid dying ReLU problem due to zero gradient during back-propagation. To address over-fitting on the model, dropout layers [12] are used in the LSTM layer ($p = 0.1$) and both hidden layers ($p = 0.2$).

3   Result and Discussion

Evaluation on the model described in Sect. 2 was done by an accuracy measure metric. We used vanilla feed-forward neural network model as the baseline model. Three models using magnitude measure, ℓ-1 norm and LSTM were evaluated against the baseline. Each model is trained using training set and evaluated using the test set five times using five different random seeds.

3.1   Improving Feed-Forward Neural Network with Magnitude Measure Technique

Magnitude measure aims to remove less informative features from the model with the intention to improve the performance of the model. Table 1 summarises input removal result based on weights contribution measured using Eq. 5.

Table 1. Evaluation on the improved model using magnitude measure feature selection technique. The evaluation was done using five different random seed. The column feature pair id lists the feature ids that were removed by magnitude measure technique

Evaluation	Baseline accuracy (%)	Reduced network accuracy (%)	Feature pair id
1	73.00	85.02	14 and 29
2	83.17	80.43	14 and 17
3	77.44	82.65	14 and 26
4	78.91	81.75	14 and 21
5	80.00	82.75	14 and 20

We used the weights from the baseline model to measure the contribution from the features to the decision function. After computing the score using Eq. 5, we removed two least relevant features according to magnitude measure. We re-trained the network using 32 features to obtain reduced network model and performed an evaluation on this model. Table 1 shows that using this method we obtained better results, with the exception of second random seed evaluation. The reduced models improved the performance especially when the baseline models did not have a good accuracy (i.e. less than or equal to 80% in this task). Table 1 also shows that feature ID 14 always appears in different random seed setting, indicating that this feature was irrelevant for the task, based on magnitude measure algorithm.

3.2   Improving Feed-Forward Neural Network with ℓ-1 Norm Regularisation

The ℓ-1 norm regularisation method aims to improve the model's performance by selecting which features are relevant for the task during training time. Table 2 summarises the evaluation on regularised model against the baseline model.

Table 2. Evaluation on the improved model using ℓ-1 norm technique. The evaluation was done using five different random seeds.

Evaluation	Baseline accuracy (%)	Regularised network accuracy (%)
1	73.00	84.33
2	83.17	82.26
3	77.44	78.58
4	78.91	82.43
5	80.00	81.67

Table 2 shows the comparison between baseline model and regularised network model. Similar to the result in Sect. 3.1, the regularised model performed better, except for the second evaluation. The regularised models improved the performance especially when the baseline models did not have a good accuracy (i.e. less than or equal to 80% in this task).

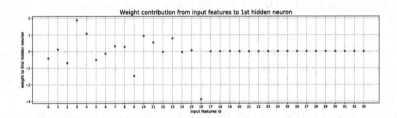

Fig. 1. Weight values that connect input features to 1st hidden unit. Several weights from input to 1st hidden unit are 0 due to ℓ-1 norm regularisation.

Figure 1 shows an example of how introducing ℓ-1 norm loss terms in the loss function forces some weights between inputs and hidden units to be 0. By forcing the weight into 0 during training, the model selects which features are relevant to the decision making in the hidden neuron and which ones are not. In the Fig. 1 input features ID 17 to 33 are not used by 1st hidden neuron to makes decisions. This may be related to the facts that input features with ID from 17 to 33 has low variance in the data set. Similarly, some feature weights connecting input feature to other hidden units are also 0, indicating that the features have less important information.

3.3   Performance Evaluation

The Table 3 summarises accuracy measurements from four models, namely the baseline, reduced network from magnitude measure, ℓ-1 regularised model and the LSTM model.

Table 3. Accuracy Measurement on the baseline model, reduced network from magnitude measure, ℓ-1 regularised model and the LSTM model.

Model	1st Evaluation	2nd Evaluation	3rd Evaluation	4th Evaluation	5th Evaluation	Average
Baseline	73.00	83.17	77.44	78.91	80.00	78.50
Magnitude measure	85.02	80.43	82.65	81.75	82.75	82.52
ℓ-1 norm	84.33	82.26	78.58	82.43	81.67	81.85
LSTM	85.94	90.10	87.15	90.45	90.10	88.75

The Table 3 shows that on average feature selection techniques improve the baseline model by around 3 to 4%. While there were cases in which reducing features impair the performance of the model, on average, the feature selection enhanced model performance. The LSTM model produced a significant improvement due to its capability to retain the information from previous input state. The LSTM model improved the accuracy by around 10% from the baseline models. From the experiment, we concluded that LSTM is the right model for a time-series data set.

The LSTM has much more trainable parameters as compare to the feed-forward neural network. The LSTM model described in this paper has 9781 trainable parameters as compared to 252 trainable parameters of the baseline model. The more trainable parameters mean the more training samples and computational resources needed to train the model. The LSTM is a data-hungry model and it may not be suitable for smaller data sets. As alternatives, we can use the FFNN model with feature selection technique, which has much less parameters with the expense of around 6–7% accuracy loss

The methods described in this paper are neural-based model. The paper in [6] used Support Vector Machine (SVM) based model to perform the task and obtain reported accuracy of 89%. The SVM model requires feature engineering to perform prediction hence it works well. The neural-based model aims to reduce the feature engineering process since this process is expensive. In this paper, we have shown that neural-based models are competitive. They are less time consuming and therefore more cost effective then other models for video-based stress recognition task.

4   Conclusion and Recommendation

In this paper, we performed extensive evaluation on the FFNN models with feature selection techniques and compare them with the LSTM model. The FFNN with feature selection techniques use much less parameters compare to the LSTM model with the expense off small accuracy loss. Our experimental results show that the NN models performed well for video-based stress recognition task, while reducing the needs of feature engineering process.

In the future, it would be interesting to use Convolutional Neural Network (CNN) to extract feature from each clip before we used RNN layer to process the video sequence. The other research direction would be to address the needs of huge training set to optimise the LSTM model using a transfer learning.

References

1. Chandrashekar, G., Sahin, F.: A survey on feature selection methods. Comput. Electr. Eng. **40**(1), 16–28 (2014). https://doi.org/10.1016/j.compeleceng.2013.11.024

2. Rumelhart, D.E., Hinton, G.E., Williams, R.J.: Learning representations by back propagating errors. Nature **323**, 533–536 (1986). https://doi.org/10.1038/323533a0

3. Gedeon, T.D.: Data mining of inputs: analysing magnitude and functional measures. Int. J. Neural Syst. **8**, 209–218 (1997). https://doi.org/10.1142/S0129065797000227

4. Glorot, X., Bordes, A., Bengio, Y.: Deep sparse rectifier neural networks. In: AISTATS (2011)

5. Hochreiter, S., Schmidhuber, J.: Long short-term memory. Neural Comput. **9**(8), 1735–1780 (1997). https://doi.org/10.1162/neco.1997.9.8.1735

6. Irani, R., Nasrollahi, K., Dhall, A., Moeslund, T.B., Gedeon, T.: Thermal superpixels for bimodal stress recognition. In: 2016 Sixth International Conference on Image Processing Theory, Tools and Applications (IPTA), pp. 1–6 (2016). https://doi.org/10.1109/IPTA.2016.7821002

7. Maas, A.L., Hannun, A.Y., Ng, A.Y.: Rectifier nonlinearities improve neural network acoustic models. In: ICML Workshop on Deep Learning for Audio, Speech and Language Processing (2013)

8. Neumann, J., Schnörr, C., Steidl, G.: Combined SVM-based feature selection and classification. Mach. Learn. **61**(1), 129–150 (2005). https://doi.org/10.1007/s10994-005-1505-9

9. Ng, A.Y.: Feature selection, L1 vs. L2 regularization, and rotational invariance. In: Proceedings of the Twenty-First International Conference on Machine Learning, ICML 2004, p. 78. ACM, New York (2004). https://doi.org/10.1145/1015330.1015435

10. Qian, N.: On the momentum term in gradient descent learning algorithms. Neural Netw. **12**(1), 145–151 (1999). https://doi.org/10.1016/S0893-6080(98)00116-6

11. Sharma, N., Dhall, A., Gedeon, T., Goecke, R.: Thermal spatio-temporal data for stress recognition. EURASIP J. Image Video Process. **2014**(1), 28 (2014). https://doi.org/10.1186/1687-5281-2014-28

12. Srivastava, N., Hinton, G., Krizhevsky, A., Sutskever, I., Salakhutdinov, R.: Dropout: a simple way to prevent neural networks from overfitting. J. Mach. Learn. Res. **15**(1), 1929–1958 (2014). http://dl.acm.org/citation.cfm?id=2627435.2670313

Pruning Convolutional Neural Network with Distinctiveness Approach

Wenrui Li[(✉)] and Jo Plested

College of Engineering and Computer Science, Australian National University,
Canberra, Australia
{wenrui.li,jo.plested}@anu.edu.au

Abstract. Convolutional Neural Network (CNN) is one of the widely used deep learning frameworks in image classification, target detection and object recognition domains. Because of the complexity of the network structures, CNNs usually contains a large number of layers and channels, which makes algorithm time-consuming. Using existing CNN architectures to solve specific problems usually leads to many redundant parameters. This paper introduces an approach based on the distinctiveness rules to prune both fully connected layers and filters of the baseline network. We experiment with different pruning means to prune multiple layer neurons and filters by the distinctiveness rules and evaluate the performance of the approach. We also discuss the influence of threshold on the selection of redundant neurons in pattern space and trade-off network scale and accuracy. The result shows using the approach will not change the structure of the baseline network and 32 filters and 161 neurons are removed. The accuracy of the pruned network reaches 80.5% compared with 87% of the original model.

Keywords: Network reduction · Distinctiveness · Pruning · Convolutional Neural Network

1 Introduction

Convolutional Neural networks (CNNs) are widely used in modern computer vision fields. Compared with traditional image processing algorithms, CNNs are more competitive to process complex images and signals in real circumstances [1]. Many well-known CNN architectures such as VGG [2], ResNet [3] and GoogleNet [4] usually consist of a considerable number of layers, which leads to the networks deeper and have more parameters to calculate. To reduce the cost of calculation, some light-weight architectures such as MobileNet [5] and ShuffleNet [6] are proposed for real-time tasks with relatively small scale of parameters to balance the calculation efficiency and the performance. However, using existing CNN architectures is not always suitable in size for specific tasks, which may introduce redundant structures and parameters. Therefore, it is necessary to simplify the network in many practical applications.

Early researches about neural network pruning such as sensitivity [7], principle components analysis [8] and singular value decomposition [9] based on neuron behaviour analysis provide observations to evaluate the redundancy of every neuron.

© Springer Nature Switzerland AG 2019
T. Gedeon et al. (Eds.): ICONIP 2019, CCIS 1143, pp. 448–455, 2019.
https://doi.org/10.1007/978-3-030-36802-9_48

However, it is still difficult to use the same means to analyse every neuron in current CNN architectures because of special structures and multiple layers. Besides, evaluating the redundancy of neurons is computationally expensive. Recent approaches for large scale CNN pruning such as increasing sparsity, quantization, compressing filters and using reinforcement learning can reduce the computational complexity of baseline networks [10].

In this paper, we propose a method to prune CNN based on the distinctiveness approach proposed by Gedeon and Harris [11]. Related experiments show that distinctiveness approach is effective for single layer pruning. We also discuss the experiment results and record some limitations when pruning the network. A conclusion and future work of the CNN pruning are mentioned at the end of the paper.

2 Method

In this section, we introduce the original the distinctiveness rules to prune the fully connected layers and filters of a baseline network and its modified version with more redundant parameters. Then we evaluate the approach performance by comparing the test accuracy and the scale of networks.

2.1 Pruning Fully Connected Layers

Based on the distinctiveness approach, we prune the single fully-connected layer and measure the redundant neurons by the angles of output vectors in pattern space. We denote the layer with k nodes as L_k, the ith output of node in this layer is $n_i \in L_k$. The vectors contain m patterns are denoted as V_i^m, the angles of two vectors is shown below.

$$\theta\left(V_i^m, V_j^m\right) = arccos\left(\frac{V_i^m \cdot V_j^m}{\left|V_j^m\right|\left|V_j^m\right|}\right) \tag{1}$$

Where V_i^m and V_j^m are vectors of output of ith and j^{th} neurons. θ is the function of calculating the angle between V_i^m and V_i^m.

To apply the network reduction, we defined two types of operators for identifying undesirable neurons based on the distinctiveness approach. When the output vectors of selected pair neurons are similar, a *remove* operator will be applied in one of the neurons, when the output vectors of selected pair neurons are complementary, the *remove* operator will be applied in both neurons. Otherwise the both neurons will be kept in the network. To measure the similarities and the complementarities of node pairs, a threshold h has been defined, which filters the angles between node pairs. the conditions are shown below.

$$operator = \begin{cases} remove\left(V_i^m\right) & \theta < h \\ keep\left(V_i^m, V_j^m\right) & h \leq \theta \leq 180 - h \\ remove\left(V_i^m, V_j^m\right) & \theta > 180 - h \end{cases} \quad (2)$$

Iterating all pairs of nodes in a layer until redundant nodes have been removed. Pruning multiple layers is more difficult than a single layer since the pruning result will influence the next layer, which will finally influence the performance of the overall network. We compare two ways of implementing the distinctiveness function to explore the influence across layers.

We denote the previous layer as L_{k-1}, assume the number of redundant nodes in L_{k-1} is m. One method is analysing the hidden neuron output layer by layer and prune all redundant neurons in one time, which means the output of all the neurons are from the original network. Another way is that we a prune single layer and refactor the network then analyse the output of next layer using the pruned network, which means the output of next layer is dependent on the pruned network by the previous step. The benefits of the first method are simple and it has little influence on the next layers. While the second method allows the distinctiveness to check the overall quality for every single layer and can have an independent threshold for every layer. In this study, we compare the robustness and the accuracy of the pruned network by these two methods and analyse the influence of the distinctiveness approach. While the difference is using the data pruned or separated it.

2.2   Pruning Filters

Different with pruning the fully connected layers, the filters pruning need to reshape the output matrix of each filter into vectors before applying the distinctiveness function. Since the filter usually generate more parameters than single neuron in fully connected layers, it is necessary to reduce the length of vectors for redundancy checking. We assume that all outputs are independent and have the same influence on the vectors, we randomly select m feature maps for vector calculation. It is important to make sure every channel selects the feature maps with the same sequence. In this paper, we calculate a filter as a complete unit, which means if a filter is redundant, then the whole channel will be removed. Relevant researches show that filters can also be pruned in sub-filter granularity to increase the sparsity of the network, which just removes some redundant weights from filters, but it has been proven that regular structures have benefits for hardware acceleration [12] (Fig. 1).

In this study, we compare two ways of implementing the distinctiveness function to explore the influence across layers. One method is analysing the hidden neuron output layer by layer and prune all redundant neurons in one time, which means the output of all the neurons are from the original network. Another way is that we prune a single layer and refactor the network, then analyse the output of next layer using the pruned network, which means the output of next layer is dependent on the pruned network by the previous step. The benefits of the first method are simple and has little influence on the next layers while the second method allows the distinctiveness to check the overall

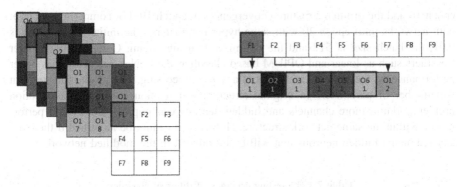

Fig. 1. Random select the output of every filter and reshape the parameter matrices to vectors. We concatenate the parameters according to the same sequence for every filter.

quality for every single layer and can have independent threshold for every layer. In this study, we compare the robustness and the accuracy of the pruned network by these two methods and analysis the influence by the distinctiveness approach.

3 Results and Discussion

To identify the overall performance of the pruning research, we use the LeNet [13] as a baseline model and set a controlling group with proportionally increase the scale of the network. The reason we use the control group is to determine the relationship between the numbers of classes and the scale of the network and observe the redundancy of the network when design a network structure or simply modify an existed structure for particular problem. The test dataset is EMNIST [14] with 47 classification problem.

Table 1. The parameters of original LeNet and Modified LeNet

Layer	Original size	Increase size	Original channels	Increase channels
Conv 1	–	–	6	28
Max Pool 1	–	–	6	28
Conv 2	–	–	16	75
Max Pool 2	–	–	16	75
Linear 1	16 * 5 * 5	75 * 5 * 5	–	–
Linear 2	120	564	–	–
Linear 3	84	395	–	–

3.1 Training Result

The comparison of different classifiers is shown in Table 2, it shows the modified network reaches the accuracy of 87.4%, which is 6.1% higher than the original structure. In the training process, the modified structure reaches the convergence since epoch 16 and the original structure convergences at epoch 19. For comparing, we train both networks until epoch 20 with same hyper parameters. The initial learning rate is 0.01, the batch size is 512 and the optimizer is using Adam. Compared with other classifiers such as linear and OPIUM based classifier, the LeNet-5 structure has better performance on the same dataset without any pre-processing. The result indicates that the number of parameters in original structure is not enough for 47 classification problem, adding more channels and hidden neurons can improve the overall performance within the same network structure. However, it cannot be identified if there are any redundant hidden neurons that will be introduced in the modified network.

Table 2. The testing accuracy of different classifier

Classifier	Accuracy	Pre-process
LeNet	81.3%	None
Linear-based [13]	50.9%	None
OPIUM-based [13]	78.0%	None
Modified-LeNet	87.4%	None

3.2 Pruning Result

We applied two implementations of the distinctiveness on both original LeNet and modified LeNet, the numbers of reduced neurons for each hidden layer (L1/L2) and the accuracy after pruning are recorded in Table 3. The threshold represents the best fit threshold for different network by measuring the predicting time and accuracy. The overall result indicates that there are no obvious differences in accuracy by using two methods while the prune all function has keep little higher accuracy compared with the prune by steps function. The best threshold of two networks is similar to 16 redundant neurons in LeNet and 161 redundant neurons in Modified LeNet.

Table 3. The result of two methods for pruning the fully connected layers

Structure	Method	Threshold (degree)	Redundant neurons (L1)	Redundant neurons (L2)	Test accuracy
LeNet	Prune all	36	15	1	76.2%
	Prune by steps	36	15	1	76.2%
Modified LeNet	Prune all	14	118	43	85.1%
	Prune by steps	14	118	43	85.0%

The Fig. 2 shows the trend of testing accuracy and the number of redundant neurons concerning the different threshold. From the result, the modified LeNet is more sensitive than the original LeNet regarding to the threshold. Most of the angles between two neuron output vectors in LeNet are near to 90°, which will not likely be influenced by the threshold. The experiment result indicates Modified LeNet contains more redundant parameters while original LeNet does not contain enough hidden neurons.

When pruning the filters, we experiment in the same condition as prune the fully-connected layers. Before the pruning, the output of each filter should be reshaped as a vector for comparison. To reduce computation complexity, we randomly select the output of each filters and ensure all data are reshaped as the same order. The result is shown in Table 4, where Prune By Steps has better test accuracy than Prune All filters together. From the experiment, it is not easy to recover the training parameters after removing a whole channel, which leads to decrease accuracy, it is expected to retrain the model after using pruning approach.

Table 4. The result of two methods for pruning the filters

Structure	Method	Threshold (degree)	Redundant filters (F1)	Redundant filters (F2)	Test accuracy
LeNet	Prune all	15	1	6	61.3%
	Prune by steps	15	1	4	70.1%
Modified LeNet	Prune all	10	9	29	73.0%
	Prune by steps	10	8	24	80.5%

3.3   Discussion

According to the experiment of this study, we find the original LeNet structure is not sufficient to solve the classification problem with 47 classes. When we use a larger scale network, which has a similar structure of the original LeNet. There is some improvement in test accuracy by increasing the scale of the network, but it has limit benefit to the network performance and will introduce more redundant neurons. The result also demonstrates multiple-layer network can be pruned by the distinctiveness approach and get the optimum pruning size by selecting a pruning threshold. Two methods of distinctiveness show the similar performance to prune the neurons of fully connected layers. While pruning the filters, it will significantly influence the output of the next layer, we are expecting to retrain the model after pruning the filters to improve the accuracy. The benefits of pruning all neurons once is simple to use while the pruning by steps is flexible to assign different threshold to the specific layers, which helps more complex network reduction process to observe the redundancy between the different layers. For determining the scale of the network, it is better to make a balance between the calculation efficiency and the performance.

Further experiment focuses on the selection of pruning threshold h, we define different thresholds to prune the network. According to the observation, if h is larger, the more neurons will be pruned and the decrease the accuracy. According to the trend of decreasing curve, we can define a best fit threshold for the network. However, there is no further heuristic for searching the optimum threshold.

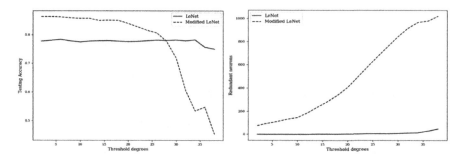

Fig. 2. The top diagram shows the relationship between the threshold and testing accuracy, the bottom diagram shows the number of redundant neurons according to the different threshold.

Overall, the experiment demonstrates the feasibility of the distinctiveness approach to prune a CNN structure, explore the relationship between the network scale and the accuracy, summarise two possible methods that implement the pruning approach and find the best threshold by observing the trend of testing accuracy.

4   Conclusion

In this paper, we apply the distinctiveness approach to prune both fully connected layers and filters of baseline network and explore the different pruning sequences when applying the distinctiveness approach to multiple layers. The result shows there is no obvious difference to prune all redundant neurons and prune them layer by layer for fully connected layers. While pruning the filters, the result of previous layer will influence the output of the current layer. Therefore, we are expecting to retrain the network after pruning to improve accuracy. The overall result illustrates that distinctiveness approach can be used in pruning hidden neurons and filters in the baseline network (LeNet) and reaches 80.5% of accuracy on EMNIST dataset with a small change of the original model. The approach will not change the structure of the network. However, for selecting threshold, there are no golden rules to determine for specific tasks, experiment is also expected. For filter pruning, it influences the output of the next layer, the pruned network should be retrained to ensure the performance. In conclusion, our approach has positive effect on reducing the CNN scale by pruning fully connected layers and filters, but there is no sure card for the optimum threshold.

References

1. Jaswal, D., Sowmya, V., Soman, K.P.: Image classification using convolutional neural networks. Int. J. Adv. Res. Technol. **3**(6), 1161–1168 (2014)
2. Simonyan, K., Zisserman, A.: Very deep convolutional networks for large-scale image recognition (2014)

3. He, K., Zhang, X., Ren, S., Sun, S.: Deep residual learning for image recognition. In: Proceedings of the IEEE Conference on Computer Vision and Pattern Recognition, pp. 770–778 (2015)
4. Szegedy, C., et al.: Going deeper with convolutions. In: Proceedings of the IEEE conference on Computer Vision and Pattern Recognition, pp. 1–9 (2015)
5. Howard, A.G., et al.: MobileNets: efficient convolutional neural networks for mobile vision applications (2017)
6. Zhang, X., Zhou, X., Lin, M., Sun, J.: ShuffleNet: an extremely efficient convolutional neural network for mobile devices. In: Computer Vision and Pattern Recognition (2017)
7. Karnin, E.: A simple procedure for pruning back-propagation trained neural networks. IEEE Trans. Neural Netw. 1(1), 239–242 (1990)
8. Kanjilal, P., Banerjee, D.: On the application of orthogonal transformation for design and analysis of feed-forward networks. IEEE Trans. Neural Netw. 6(5), 1061–1070 (1995)
9. Park, Y., Murray, T., Chen, C.: Predicting sun spots using a layered perceptron neural network. IEEE Trans. Neural Netw. 7(2), 501–505 (1996)
10. Huang, Q., Zhou, K., You, S., Neumann, U.: Learning to prune filters in convolutional neural networks. In: 2018 IEEE Winter Conference on Applications of Computer Vision (WACV), pp. 709–718. IEEE (2018)
11. Gedeon, T., Harris, D.: Network Reduction Techniques. Department of Computer Science, Brunel University (1990)
12. Mao, H., et al.: Exploring the regularity of sparse structure in convolutional neural networks. arXiv preprint arXiv:1705.08922 (2017)
13. LeCun, Y., Bottou, L., Bengio, Y., Haffner, P.: Gradient-based learning to applied document recognition. Proc. IEEE 86(11), 2278–2324 (1998)
14. Cohen, G., Afshar, S., Tapson, J., van Schaik, A.: EMNIST: an extension of MNIST to handwritten letters (2017)

Semantic and Graph Based Approaches

Semi-supervised Domain Adaptation with Representation Learning for Semantic Segmentation Across Time

Assia Benbihi[1,3]($\boxtimes$), Matthieu Geist[4], and Cédric Pradalier[1,2]

[1] UMI 2958 GeorgiaTech-CNRS, 57000 Metz, France
abenbihi@georgiatech-metz.fr
[2] GeorgiaTech Lorraine, 57000 Metz, France
[3] CentraleSupélec, Université Paris-Saclay, 57000 Paris, France
[4] Université de Lorraine - CNRS LIEC, 57000 Metz, France

Abstract. Deep learning generates state-of-the-art semantic segmentation provided that a large number of images together with pixel-wise annotations are available. To alleviate the expensive data collection process, we propose a semi-supervised domain adaptation method for the specific case of images with similar semantic content but different pixel distributions. A network trained with supervision on a past dataset is finetuned on the new dataset to conserve its features maps. The domain adaptation becomes a simple regression between feature maps and does not require annotations on the new dataset. This method reaches performances similar to classic transfer learning on the PASCAL VOC dataset with synthetic transformations.

Keywords: Domain adaptation · Semantic segmentation

1 Introduction

Recent deep learning applications such as environment monitoring [15] and autonomous driving [3] rely on semantic segmentation of datasets with redundant images. As monitoring progresses and new datasets are acquired, their data distributions may change due to light variations or camera upgrades. This can prevent a Convolutional Neural Network (CNN) trained at time t to generalize at a later time. Each time the distribution deviates from the past one, the network needs to be fine-tuned or retrained from scratch, both of which require ground-truth pixel-wise annotations. To avoid the burden of annotating new datasets, we propose a semi-supervised method to transfer the network knowledge across time. We do so by transferring the deep representations learned by the network instead of classic transductive transfer learning [14], also called finetuning. Figure 1 illustrates our method on a synthetic transformation emulating a camera downgrade: a first network H_{θ_1} is trained on a dataset D_1 with pixel-wise supervision. Then, a second network H_{θ_2} is trained to generate deep representations for a different dataset D_2 to match the deep representations of H_{θ_1} on D_1.

© Springer Nature Switzerland AG 2019
T. Gedeon et al. (Eds.): ICONIP 2019, CCIS 1143, pp. 459–466, 2019.
https://doi.org/10.1007/978-3-030-36802-9_49

This approach has been studied in [1] for classification in the Unsupervised and Transfer Learning Challenge. One of the limitations in this work was the lack of instances pairs with the same semantic content. In the segmentation applications we address, such pairs arise naturally which solves this issue.

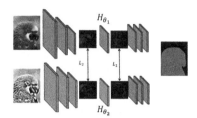

Fig. 1. Top: The trained and frozen (gray) network provides ground truth deep representations. Down: The trainable layers (blue) must learn the deep representations. (Color figure online)

One application example is environment monitoring across time [15]: a DCNN is used to segment land categories on aerial images over a period from 1955 to 2015. The 2015 dataset is made of RGB digital images whereas the 1955 data is made of black and white analog images that have been digitized. A network trained on the 2015 data can not generalize on the 1955 one because the images do not have the same color domain nor the same resolution or grain. And even though the surveys cover the same area, the images are not perfectly aligned due to small changes in land use, which prevent the use of 2015 annotations to train a network on the 1955 data. What stays coarsely invariant for one image across surveys is the high-level semantic content.

We emulate equivalent transformations on the PASCAL VOC 12 dataset [5]. Synthetic data is necessary to gather ideal baselines since they require pixel-wise annotated images over several input distribution (Sect. 3.2). To the best of our knowledge, there is no such dataset publicly available. Experiments show that our transfer method reaches the same performance as standard transfer learning and classic training. Our contribution is twofold: (i) we extend the initial work of [1] to semantic segmentation on a wider database, (ii) we provide visual insight on the CNN adaptation behavior. The next section describes the state-of-the-art in transfer learning for DCNN. Sections 3 and 4 present our method and results.

2   Related Work

Domain adaptation through feature transfer is first applied to classification: [7] uses auto-encoders to learn high-level representations of sentiments from various datasets to better classify Amazon reviews. And [1] tackles the challenge of digits classification where some classes are not represented in the training set. Both learn a data transformation that best captures each class's unique features so that a trained classifier can even process class instances from new domains. This transformation is evaluated with the classification score. One of the best methods relies on unsupervised auto-encoder training that forces the network to capture the relevant features to reconstruct the input.

We draw inspiration from these works and transfer the features maps of a segmentation CNN across data domains. Our work differs from [1,7] in the target task and the feature learning method. First, contrary to previous classification focus, we test our method on semantic segmentation on the same dataset across time. Also, we tackle a different category of domain adaptation where the output variable is the same but the input pixel distribution varies. Second, we do not follow the previous feature training methods: we use the annotations from one dataset to train a segmentation network, then utilize the features that the network autonomously learned as ground-truth feature representations. For every new segmentation dataset, we force the network to match these representations. This method is semi-supervised since it requires only to match images of the same locations, which can be done during the data collection process using odometry.

Given that [15] did not release their segmentation dataset and [8] only provides odometry datasets, we rely on three synthetic transformations of the PASCAL VOC 12 dataset [5] to test our method. Our transfer learning method reach the same performance as finetuning and standard training.

3   Method

3.1   Training

Let H_θ be a network model with weights θ, X an image and $\mathcal{F}_\theta(X)$ the set a set of feature maps of the network H_θ, with $\mathcal{F}_\theta(X) = \{F_\theta^l(X),\ l \in L\}$. H_{θ_1} is an instance of H trained with supervision on D_1. H_{θ_2} is another instance of H first initialized with the weights θ_1. H_{θ_2} is then finetuned on image pairs $\{(X_1, X_2) \subset D_1 \times D_2\}$ with the same semantic content but different pixel distributions. The feature maps $\mathcal{F}_{\theta_2}(X_2)$ must match the corresponding $\mathcal{F}_{\theta_1}(X_1)$ without ground-truth annotations on D_2. A loss is computed for each feature map and is back-propagated only in the layers lower than l: $\mathcal{L}^l(X) = w_l \|F_{\theta_1}^l - F_{\theta_2}^l\|^2, w_l \in \mathbb{R}$. In our implementation, the feature weights w_l are constant over the training. Investigations over dynamic weighting strategies have been left for future work.

3.2   Evaluation

Our transfer method is evaluated with the segmentation performance of the adapted network H_{θ_2} on the new dataset D_2. We use the standard segmentation evaluataion metrics: the mean accuracy (acc) and the mIOU. We compare against three baselines: baseline B_0 measures the performance of the first network H_{θ_1} on the new dataset D_2, i.e. how well H_{θ_1} generalizes to a dataset with the same content but a different pixel distribution. This also gives a quantitative measure of the image transformation between two datasets. In B_1, we train

H_{θ_2} with full supervision on D_2 using the pixel-wise annotations from D_1. This ideal setting provides the performance our transfer learning should aim at. The last baseline B_2 measures the performance of H_{θ_2} when it is initialized with θ_1 and then classically fine-tuned on D_2 with D_1 annotations. This provides the performance of classic supervised fine-tuning our method should reach while not using explicit annotations.

3.3 Visualization

In addition to the proposed domain adaptation techniques described above, this paper also developed a visualization technique inspired from existing approaches used in other contexts. Specifically, we design an optimization to observe the evolution of the representations induced by the feature regression during the domain adaptation. An image $X_1 \in D_1$ is fed to H_{θ_2} to generates a set of features $\mathcal{F}_{\theta_2}(X_1)$. Starting from white noise, we then generate the image $\hat{X}_1$ that minimizes $\sum_{l \in L} \|F_{\theta_2}^l(X_1) - F_{\theta_2}^l(\hat{X}_1)\|$, i.e. a version of X_1 as seen by H_{θ_2}. Visually, $\hat{X}_1$ has the same content as X_1 with the visual aspect of D_2.

The optimization is inspired from the feature map inversion method [12] to integrate style transfer into it. In addition, we adapt the style transfer method from [6,10] to generate image content and style from only one network instead of two. The final optimization is performed as follows: we initialize $\hat{X}_1$ with white noise and feed it to H_{θ_2}. For each feature map, a content loss and a style loss are computed and backpropagated into the image. $\hat{X}_1$ is optimized with Stochastic Gradient Descent (SGD). We use the losses and Gram matrix definitions from [6]. The content loss is a simple $\mathcal{L}_2$ loss between feature maps $\mathcal{L}_{content}(l) = \frac{1}{2}\|F_{\theta_2}^l(X_1) - F_{\theta_2}^l(X)\|^2$. The style loss is a normalized $\mathcal{L}_2$ loss between the Gram matrix of the feature maps. This matrix is computed from a 2D projection of each feature map $F_{\theta_2}^l$: $\mathcal{L}_{style}(l) \propto \|G(F_{\theta_2}^l(X_1)) - G(F_{\theta_2}^l(X))\|^2$.

4 Experiments

4.1 Dataset

We run tests on three synthetic transformations of the augmented version [9] of the PASCAL VOC12 dataset [5] resulting in 10 582 training images and 1449 validation images with 21 semantic classes. The regression is trained on the 10 582 original images of D_1 and the transformed ones of D_2. H_{θ_2} is then evaluated on the transformed validation set of D_2.

Three transformations T_1, T_2, T_3 with increasing perturbations are generated with GIMP[1] resulting in D_2^1, D_2^2, D_2^3 (Fig. 2). We use the 'photocopy' filter T_1 to emulate a change of color and saturation. This problem arises in long-term environment monitoring where recent datasets are numerical RGB images and older datasets are collected with the numerization of analogic pictures [15]. We address the issue of image misalignment and noise with the ripple distortion T_2.

---

[1] https://www.gimp.org/.

This is typical in natural environment monitoring such as in the dataset from [8]. Finally, we mix a change of texture and misalignment with edge noise in the cubism filter T_3.

Fig. 2. Synthetic transformations. Column 0: PASCAL. Left-Right: transformation. Photocopy (Distortion: 32.5%), Ripple (62.6%), Cubism (94.0%)

For each transformation T_i, the image distortion between D_1 and D_2^i is quantized with the normalized performance degradation of the network H_{θ_1} on D_2^i. After training H_{θ_1} on D_1 with the original DeepLab V3 [2] setup, the accuracy and mIOU reach respectively 79.92% and 69.22%. With $\mathrm{acc}(H_{\theta_1}, D_2)$ and $\mathrm{mIOU}(H_{\theta_1}, D_2)$ the performances of H_{θ_1} on D_2, the image distortion between D_1 and D_2 is: $\frac{1}{2}\left(\frac{|79.92 - \mathrm{acc}(H_{\theta_1}, D_2)|}{79.92} + \frac{|69.22 - \mathrm{mIOU}(H_{\theta_1}, D_2)|}{69.22}\right)$.

The dataset distortions (Fig. 2) comfort the visual intuition that the three transformations exhibit an increasing level of complexity that challenges the network trained only on D_1.

4.2 Experimental Setup

Supervised Training of H_{θ_1}. H follows the state-of-the-art VGG-16 architecture [17] from DeepLabV3 [2] for training time considerations. and converges within only 5 h of training on an NVIDIA 1080Ti. As in [2], we train the network for 20 000 iterations with a batch size of 10, SGD with a momentum of 0.9, a weight decay of 0.5 and the "poly" learning rate policy initialized at 2.5×10^{-4} and *power* = 0.9.

Feature Map Training of H_{θ_2}. We run the unsupervised domain adaption training on several sets of $\mathcal{F}$ to better understand the hierarchical model of network representations.

An intuition gathered from the literature [4,13,16] suggests that early layers capture low-level representations such as colors and edges, whereas higher layers embed more complex representations such as object contours and their label. According to this intuition, we could expect that regression of high layers are more relevant than lower ones when the transformation is significant. To test this assumption, we run individual regressions on one feature map and compare the performance of H_{θ_2}. We choose the feature map post max-pooling as they show the highest representations shifts in consecutive layers. When we display features of a VGG bloc, the features of successive convolutional layers look highly similar whereas we always observe significant variations after the pooling layers.

We also test the correlations between the network representation levels. We run the regression on all the post-pooling layers simultaneously with two weighting strategies. In the first one, W_{inc}, w_l increases with the layer level, i.e. we favor the deeper representations. We do the contrary with the second one, W_{dec}, and rely more on low-level features. In both cases, we use the following uniform weight distribution $[0.2, 0.4, 0.6, 0.8, 0.9]$.

4.3   Experimental Results

Despite training without supervision, our method reaches similar or higher performance than classic supervised fine-tuning (Fig. 3). Figure 4 (left) shows an example of how the regression actually makes the deep representations of H_{θ_2} on D_2 converge towards the deep representation of H_{θ_1} on D_1. Also, Fig. 4 (right) exhibits that the new deep representations embed the image style of d_2.

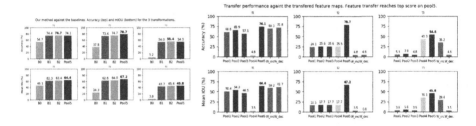

Fig. 3. Left: Transfer performance against the baselines. Right: Transfer performance with respect to the transfered features maps. Transfering on `pool5` gives the best scores.

Figure 3-left shows that the regression results on `pool5` reach similar performances to classic fine-tuning. The B_0 line recalls the performance of H_{θ_1} on the transformed dataset D_2. Our method improves significantly the metrics compared to B_0 which means that transferring deep representations is indeed a relevant transfer learning method for semantic segmentation. Classic finetuning B_2 outperforms the training from scratch B_1 on D_2: this comforts the results from [11] on the importance of weight initialization and that finetuning is relevant for boosting the performances. Our method always outperforms B_2's mIOU and reaches similar mean accuracy. This shows that transfer learning of deep representations can replace classic CNN finetuning without performance degradation.

Figure 3-right summarizes the performances of individual and parallel regressions to better understand the network representation hierarchy. The best performances are reached with the individual regression on the highest post-pooling layer `pool5`. This suggests that high-level representations are the most relevant to transfer for semantic segmentation. The metrics for each regression vary with the transformation type. The experiments do not allow to draw a general conclusion but we can gather an intuition on which representations are the most relevant to transfer. For the color and saturation transformation T_1, the transfer

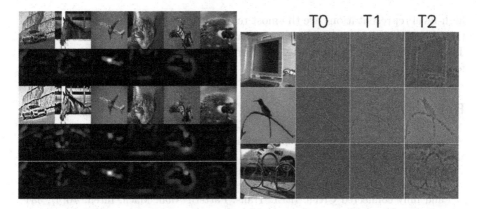

Fig. 4. Left: Deep representation evolution. Line 1: original images. Line 2: one feature of H_{θ_1} on D_1. Line 3: transformed image. Line 4: the same feature of H_{θ_1} on D_2. Line 5: the same feature of H_{θ_2} on D_2. **Right: Style reconstruction.** The finetuned network is fed with the left-most image. We reconstruct the image as seen by the three finetuned networks (right images). The generated image has the same content and the style on which the network is finetuned.

of layers up to `pool3` improves B_0 which is not the case for T_2 and T_3. One explanation can be that T_1 conserves the alignment of the image and that color processing is handled in the low-level layers. T_2 and T_3 maintain the image color domain but modify the contours of the semantic units either with regular noise as in the ripple transformation or with random one in the cubism effect. Object contours are features usually generated in deeper layers which can explain why the transfer of `pool_5` only is relevant for these datasets. These results also suggest that when low-level layers are not relevant to transfer, they can hinder the transfer of the relevant layers. For example, the transfer learning on multiple layers performs worse than the transfer on `pool5` only. For T_3, we observe that a weight distribution that favors high layers performs better than W_{dec} but worse than the transfer of `pool5`. The overall intuition we can draw consists in relying in high-level feature transfer even though low-level layers can be relevant for color domain perturbations.

5   Conclusion

We have introduced a method for unsupervised domain adaptation for semantic segmentation that relies on the transfer of a CNN deep representations. It is relevant for applications with redundant semantic content and a drift of the pixel distributions such as autonomous-driving or long-term environment monitoring for which new datasets covering a similar semantic content are acquired over time. This method shows similar performance to classic fine-tuning on three synthetic transformations of the PASCAL VOC dataset that emulates color domain variations, resolution degradation and noise. Quantitative results suggest that high-level representations are the most relevant to transfer even though low-level transfer also reach acceptable performance for color-domain transformations. These observations comfort the recurrent intuitions on the semantics of CNN features.

References

1. Bengio, Y.: Deep learning of representations for unsupervised and transfer learning. In: Proceedings of ICML Workshop on Unsupervised and Transfer Learning (2012)
2. Chen, L.C., Papandreou, G., Kokkinos, I., Murphy, K., Yuille, A.L.: DeepLab: Semantic image segmentation with deep convolutional nets, atrous convolution, and fully connected CRFs. IEEE Trans. Pattern Anal. Mach. Intell. **40**(4), 834–848 (2018)
3. Cordts, M., et al.: The cityscapes dataset for semantic urban scene understanding. In: Computer Vision and Pattern Recognition (2016)
4. Dosovitskiy, A., Brox, T.: Inverting visual representations with convolutional networks. In: Computer Vision and Pattern Recognition (2016)
5. Everingham, M., Eslami, S.A., Van Gool, L., Williams, C.K., Winn, J., Zisserman, A.: The pascal visual object classes challenge: a retrospective. Int. J. Comput. Vis. **111**(1), 98–136 (2015)
6. Gatys, L.A., Ecker, A.S., Bethge, M.: Image style transfer using convolutional neural networks. In: Computer Vision and Pattern Recognition (2016)
7. Glorot, X., Bordes, A., Bengio, Y.: Domain adaptation for large-scale sentiment classification: a deep learning approach. In: Proceedings of the 28th International Conference on Machine Learning (ICML-2011), pp. 513–520 (2011)
8. Griffith, S., Chahine, G., Pradalier, C.: Symphony lake dataset. Int. J. Robot. Res. **36**(11), 1151–1158 (2017)
9. Hariharan, B., Arbeláez, P., Bourdev, L., Maji, S., Malik, J.: Semantic contours from inverse detectors. In: International Conference on Computer Vision (2011)
10. Johnson, J., Alahi, A., Fei-Fei, L.: Perceptual losses for real-time style transfer and super-resolution. In: European Conference on Computer Vision (2016)
11. Lamblin, P., Bengio, Y.: Important gains from supervised fine-tuning of deep architectures on large labeled sets. In: NIPS* 2010 Deep Learning and Unsupervised Feature Learning Workshop, pp. 1–8 (2010)
12. Mahendran, A., Vedaldi, A.: Understanding deep image representations by inverting them. In: Computer Vision and Pattern Recognition (2015)
13. Oquab, M., Bottou, L., Laptev, I., Sivic, J.: Learning and transferring mid-level image representations using convolutional neural networks. In: Computer Vision and Pattern Recognition (2014)
14. Pan, S.J., Yang, Q.: A survey on transfer learning. IEEE Trans. Knowl. Data Eng. **22**(10), 1345–1359 (2010)
15. Richard, A., Benbihi, A., Pradalier, C., Perez, V., Van Couwenberghe, R., Durand, P.: Automated segmentation and classification of land use from overhead imagery. In: 14th International Conference on Precision Agriculture (2018)
16. Simonyan, K., Vedaldi, A., Zisserman, A.: Deep inside convolutional networks: visualising image classification models and saliency maps. arXiv:1312.6034
17. Simonyan, K., Zisserman, A.: Very deep convolutional networks for large-scale image recognition. arXiv:1409.1556

Regularizing Variational Autoencoders for Molecular Graph Generation

Xin Li[1], Xiaoqing Lyu[1(✉)], Hao Zhang[2], Keqi Hu[3], and Zhi Tang[1]

[1] Peking University, Beijing, China
{l_x,lvxiaoqing,tangzhi}@pku.edu.cn
[2] Beijing Institute of Technology, Beijing, China
gcrth@outlook.com
[3] China University of Mining and Technology, Beijing, China
keqihu@student.cumtb.edu.cn

Abstract. Deep generative models for graphs are promising for being able to sidestep expensive search procedures in the huge space of chemical compounds. However, incorporating complex and non-differentiable property metrics into a generative model remains a challenge. In this work, we formulate a differentiable objective to regularize a variational autoencoder model that we design for graphs. Experiments demonstrate that the regularization performs excellently when used for generating molecules since it can not only improve the performance of objectives optimization task but also generate molecules with high quality in terms of validity and novelty.

Keywords: Variational autoencoder · Regularization · Molecule generation · Graph generation

1 Introduction

Generating molecules with desired properties is a challenging task with important applications such as drug design. In the last few years, considerable works [1,7,8,12] using deep generative models including variational autoencoders (VAE) and generative adversarial networks (GAN) for molecule generation make use of the domain specific SMILES representation of molecules [22], a linear string notation to describe molecular structures. However, the main drawback of SMILES is that there is difficulty in capturing molecular similarity since small changes can result in drastically different structures. This shortcoming prevents some generative models from learning smooth latent variables. With recent progress in the area of deep learning on graphs [3,6,13,20,21], deep generative models for molecular graphs are attracting surging interests since the graph representation can overcome limitations of the SMILES [2,10,14,15,21].

In the task of molecular graph generation, one of the key challenges lies in the difficulty of incorporating highly complex and non-differentiable property metrics into a generative model. The two main strategies to achieve this end remain

T. Gedeon et al. (Eds.): ICONIP 2019, CCIS 1143, pp. 467–476, 2019.
https://doi.org/10.1007/978-3-030-36802-9_50

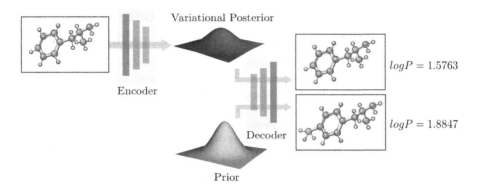

Fig. 1. Overview of the proposed OpVAE. The bottom flow depicts the regularization that encourages decoder to generate molecules with desired properties, which is detailed in Sect. 4.

reinforcement learning-based and Bayesian optimization-based approaches. The reinforcement learning-based methods [2,8] use reinforcement learning-based objective to provide a gradient to the policy towards the desired properties, which requires an extra reward network to predict the immediate reward. However, the extra network may arise the convergence difficulties. Besides, the reward network may not be able to get the correct prediction. The Bayesian optimization-based approaches [10,12] perform Bayesian optimization to navigate into regions of latent space that decode into molecules with particular properties. Specifically, such methods can be divided into two phases. The first phase will focus on training a generative model. During the second phase, the Bayesian optimization is performed in the latent space. So the performance of Bayesian optimization during the second phase depends largely on the smoothness of the latent space learned during the previous phase.

In this work, we design a variational autoencoder for matrix representations of graphs. We then formulate a regularizer to encourage the generation of molecules with desired properties, which avoids extra networks and the requirement of smoothness of latent space. Monte Carlo approximation of the regularizer is used in the training procedure. Since the approximation is differentiable, we can train the model by stochastic gradient optimization methods. We demonstrate the effectiveness of our framework with two benchmark molecule datasets to generate molecules with desired properties.

2   Related Work

Recently, there has been significant advances in molecule generation. Previous works [1,7,8,12] have explored the generative models on SMILES. Gómez-Bombarelli et al. [7] built generative models with recurrent neural networks. Kusner et al. [12] utilized a parse tree from a context-free grammar to improve the validity of generated molecules. Dai et al. [1] took a step further towards the validity by enforcing constraints on the generative model. In addition to the above VAE-based works, Guimaraes et al. [8] used GAN to address the generation. For graph representations, Simonovsky et al. [21] have explored generating molecular graphs by extending VAE. Jin et al. [10] proposed a generative model by combining a tree-structured scaffold with original graphs. Liu et al. [14] incorporated chemical constraints into generative models by specifying a generative procedure and employing masking. Ma et al. [15] reformulated the constrained objective of VAE by constructing Lagrangian function.

3   Model

3.1   Variational Autoencoder

We adopt AEVB algorithm [11] to learn a generative model $p_\theta(G|z)$ using a particular encoder $q_\phi(z|G)$, which is an approximation of actual posterior $p_\theta(z|G)$ that is intractable. The objective is to maximize the evidence lower bound (ELBO) with respect to θ and ϕ:

$$\mathcal{L}_{ELBO} = -D_{KL}(q_\phi(z|G)||p_\theta(z)) + \mathbb{E}_{q_\phi(z|G)}[\log p_\theta(G|z)]. \tag{1}$$

The first term as a regularization is the divergence of the variational posterior $q_\phi(z|G)$ from the prior $p_\theta(z)$[1], which allows for learning more general latent representations instead of simply encoding an identity mapping. The expectation term, interpreted as negative reconstruction loss, is maximized when $p(G|z)$ assigns a high probability to the observed G.

3.2   Molecules as Graphs

Each molecule can be represented by an undirected graph with a set of labeled nodes associated with the atoms and a set of labeled edges associated with bonds between atoms. We restrict the domain to a collection of molecular graphs which have at most N nodes, $T-1$ node types and $R-1$ edge types.

The decoder outputs a matrix $\widetilde{X} \in R^{N \times T}$ and a tensor $\widetilde{A} \in R^{N \times N \times R}$, which are denoted by $\widetilde{G} = (\widetilde{A}, \widetilde{X})$. The row $\widetilde{X}(i,:)$ is a categorical distribution over the type of node i, which satisfy $\sum_{t=0}^{T} \widetilde{X}(i,t) = 1$. $\widetilde{X}(i,0)$ is the probability that node i is nonexistent. Similarly, the fiber $\widetilde{A}(i,j,:)$ is a categorical distribution over the edge type between nodes i and j, which satisfy $\sum_{r=0}^{R} \widetilde{A}(i,j,r) = 1$. $\widetilde{A}(i,j,0)$ is the probability that the edge between nodes i and j is nonexistent.

We assume that the node type and the edge type are independent. The tuple $\widetilde{G}$ hence becomes a random graph model. A one-hot $G = (A, X)$ can be sampled via categorical sampling from $\widetilde{G} = (\widetilde{A}, \widetilde{X})$.

---

[1] The prior is a standard normal in this paper.

3.3   Encoder and Decoder

In this paper, the encoder is parameterized as a diagonal normal distribution $q(z|G) = \mathcal{N}(z; \mu, \psi)$ with covariance matrix $\psi = \text{diag}(\sigma^2)$, where μ and σ are outputs of the encoding graph neural networks.

Suppose that $h_i^{(l)}$ is the hidden state of the node i at layer l. We define the following layer-wise propagation rule similar to [20] for the signal $h_i^{(l+1)}$ of the node i:

$$h_i^{(l+1)} = \sigma\left(\sum_{r=1}^{R-1} \sum_{j=1}^{N} \frac{\widehat{A}(i,j,r)}{c_{i,r}} f_r^{(l)}(h_j^{(l)}, x_i)\right), \tag{2}$$

where $\widehat{A}(:,:,r) = A(:,:,r) + I$ for each edge label r, with I being the identity matrix, which means a self-connection between layers is added for each edge type. The messages from neighbors (including the node i itself) that are obtained by an edge type-specific affine function $f_r^{(l)}$ are then accumulated. Besides linear transformation, we utilize normalization constant $c_{i,r}$ to ensure that the accumulation of messages will not completely change the scale of the feature representations. In this paper, $c_{i,r} = |\mathcal{N}_{i,r}|$, where $\mathcal{N}_{i,r}$ is the set of neighbors of the node i (including i itself) and the edge label between the node i and node $j \in \mathcal{N}_{i,r} \setminus \{i\}$ is r. Finally, the normalized messages are passed through an element-wise activation function $\sigma(\cdot)$ such as ReLU [16]. Note that R and N have the same meaning as in Sect. 3.2. It remains to define the initial hidden state at the first layer:

$$h_i^{(1)} = \sigma\left(\sum_{r=1}^{R-1} \sum_{j=1}^{N} \frac{\widehat{A}(i,j,r)}{c_{i,r}} f_r^{(0)}(x_i)\right), \tag{3}$$

where $x_i = X(i,:)$.

For graph-level outputs, we consider the following aggregation method proposed by [13] after $L - 1$ layers of propagation:

$$h_G = \tanh\left(\sum_{i=1}^{N} \sigma(v(h_i^{(L)}, x_i)) \odot \tanh(u(h_i^{L}, x_i))\right), \tag{4}$$

where v and u are neural networks that take the concatenation of h_i^L and x_i as input and their output layers are both linear. $\sigma(v(h_i^{(L)}, x_i))$ is explained as a soft attention mechanism to decide how relevant node i can be to the current molecule generation task. We then perform an element-wise multiplication $\odot$ and take a sum over all weighted output vectors of the nodes to obtain the graph-level representation h_G. Finally, the μ and σ are generated from two multi-layer perceptrons (MLPs).

As mentioned in Sect. 3.1, the decoder draws latent variables z from the variational posterior $q_\sigma(z|G)$ or the prior $p_\theta(z)$ and outputs a random graph model $\widetilde{G}$ where a graph G is sampled. This is simply done by MLPs in this paper.

4 Regularization

4.1 Formularization

In this section, we design an interpretable regularizer to encourage the generation of molecule with desired properties in this paper. The differentiable regularization term can provide a gradient to the decoder towards the desired metrics. More details will be discussed after the review of several traditional approaches to calculating the chemical properties.

- **LogP:** The octanol-water partition coefficient (logP) serves as a measure of lipophilicity. Developed by [5], the atom-based method is an effective approach to calculating logP, which assigns to the individual atoms in the molecule additive contributions to molecular logP. The logP of small molecules can be calculated as the sum of the contributions of each of the atoms in the molecules. Since the high accuracies have been achieved in [23], where the 68 atomic contributions to logP have been determined, the approach is a common and standard approach to calculating logP.
- **SAS:** The synthetic accessibility score (SAS) [4] has been used as a measure to estimate ease of synthesis of molecules. It is calculated based on a combination of fragment contributions and a complexity penalty[2]. Between the two constituent parts of SAS, the fragmentScore is calculated as a sum of contributions of all fragments in the molecule divided by the number of fragments in this molecule, whereas the complexityPenalty is calculated as a combination of structural features such as rings, stereo centers, etc.

The central idea of this paper is to use regularization that encourages the decoder to generate molecules with desired properties to formulate a differentiable objective. Let S be the property to be calculated. It is justifiable that the expectation of S with respect to the distribution $p_\theta(G|z)$ is utilized as a regularizer. The expectation may then be formally written as

$$\mathbb{E}p_\theta(G|z)[S] = \sum_G p_\theta(G|z)S_G, \tag{5}$$

By analyzing the calculation schemes of logP and SAS, we can unify these methods into a single common framework. A specific substructure in a molecular graph is called a pattern. Each pattern is associated with one additive contribution. The property of a molecule G to be calculated is the summation of all the contributions for the patterns that occur in this molecule. For the calculation of logP in [5], each one of 68 atomic types can be regarded as a specific pattern with one central atom, its neighboring atoms and the bonds between the central atom and its neighbors. For the calculation of SAS in [4], patterns include the fragments, rings, etc. For any properties to be calculated, one specific pattern is associated with one contribution. Let the set of possible patterns be denoted as

---

[2] **SAS = fragmentScore − complexityPenalty.**

Q. Given a molecular graph G, the properties mentioned above can be calculated according to

$$S_G = \sum_{q \in Q} n_q c_q, \tag{6}$$

where S_G is the property of molecule G, n_q is the number of occurrences of the pattern q, and c_q is the contribution for pattern q.

By combining Eq. 5 with Eq. 6, we have

$$\mathbb{E}_{p_\theta(G|z)}[S] = \sum_G \sum_{q \in Q} p_\theta(G|z) n_q c_q, \tag{7}$$

We let $\mathbb{E}_{p_\theta(G|z)}[n_q] = \sum_G p_\theta(G|z) n_q$, which is the expectation of number n_q with respect to the distribution $p_\theta(G|z)$. $\mathbb{E}_{p_\theta(G|z)}[n_q]$ can also be interpreted as the probability of pattern q of a graph sampled from $p_\theta(G|z)$. Let $\mathbb{E}_{p_\theta(G|z)}[n_q] = p_q$ for simplicity. We obtain

$$\mathbb{E}_{p_\theta(G|z)}[S] = \sum_{q \in Q} p_q c_q, \tag{8}$$

Equation 8 indicates that the expectaion can be evaluated as long as the problem of computing p_q is solved. However, evaluating p_q is computationally expensive. In practice, we can appeal to Monte Carol approximation for evaluating the expectation of S, which is differentiable with respect to θ. More details will be discussed in Sect. 4.2.

4.2   Training

We train our model to maximize the following objective function:

$$\mathcal{L} = \mathcal{L}_{ELBO} + \mathbb{E}_{p_\theta(G|z)}[S]. \tag{9}$$

As done in [11], reparameterized trick and Monte Carlo gradient estimator are employed for evaluating $\mathcal{L}_{ELBO}$. In order to evaluate $\mathbb{E}_{p_\theta(G|z)}[S]$, we also consider Monte Carol approximation as mentioned above. Latent variables are first sampled from the prior $p_\theta(z)$. The decoder takes z as input and outputs $\widetilde{G}$. Then we sample patterns from $\widetilde{G}$. For each $q^{(m)} \sim \widetilde{G}$, we further assume that an occurrence of a pattern can be represented by a 2-tuple $q^{(m)} = (V^{(m)}, E^{(m)})$, where $V^{(m)}$ is the set of atoms in this occurrence, $E^{(m)}$ is the set of bonds in this occurrence. Under the independence assumption in Sect. 3.2, the probability $p_{q^{(m)}}$ of a pattern for one specific occurrence is given by

$$p_{q^{(m)}} = \prod_{it \in V^{(m)}} \prod_{ijr \in E^{(m)}} p_{it} p_{ijr}, \tag{10}$$

where p_{it} is an element of $\widetilde{X}$, i.e., $\widetilde{X}(i,t)$ and p_{ijr} is an element of $\widetilde{A}$, i.e., $\widetilde{A}(i,j,r)$. The spirit is that the atom it is represented by node i in $\widetilde{X}$ and its

Table 1. Comparison with baselines in terms of validity and novelty on QM9 and ZINC. The rows "% Valid" and "% Novel" are the validity and novelty in percentages, respectively. The results of baselines are copied from [21] and [15].

Method	QM9		ZINC	
	% Valid	% Novel	% Valid	% Novel
OpVAE	**100.00**	**99.97**	**100.00**	**100.00**
CVAE	10.30	90.00	0.70	**100.00**
GVAE	60.20	80.90	7.20	**100.00**
GraphVAE	55.70	76.00	57.10	71.90
SeVAE	96.60	97.50	34.90	**100.00**

label is t, which is a similar explanation to the bond ijr. Because p_{it} and p_{ijr} are differentiable with respect to θ, parameters can be updated by using stochastic gradient ascent. Note that the latent variable z for evaluating $\mathbb{E}_{p_\theta(G|z)}[S]$ is sampled from the prior $p_\theta(z)$ as shown in the bottom flow of Fig. 1.

5 Experiments

5.1 Datasets

Two benchmark datasets are used for the experiments. **QM9**[18] is a subset of the massive 166 billion organic molecules GDB-17 chemical database [19]. The dataset contains about 134 K molecular graph of up to 9 heavy atoms with 4 distinct atomic numbers (carbon, oxygen, nitrogen and fluorine) and 4 bond types. **ZINC** [9] is a curated set that contains about 250K drug molecules of up to 38 heavy atoms with 9 distinct atomic numbers and 4 bond types.

5.2 Quality of the Generated Molecules

Baselines. We compared our method OpVAE against CVAE [7], GVAE [12], GraphVAE [21] and SeVAE [15]. For evaluation, we sampled 1000 latent vectors from the prior and performed maximum-likelihood decoding for each one.

Evaluation Measures. We use the following 2 statistics to evaluate the quality of the generated molecules. **Validity** is defined as the ratio between the number of valid and all generated molecules sampled from the prior. **Novelty** is defined as the ratio between the number of valid samples that don't occur in the training set and the number of all valid samples.

Comparison with Baselines. Table 1 reports the results obtained by the proposed method and baselines. our approach OpVAE shows a significant improvement over its competitors in terms of validity and novelty since OpVAE is not designed to boost the validity or novelty percentage. For the validity, an intuitive explanation might be that the regularization invests much effort in forcing

Table 2. Optimization for different objectives. The results of baselines are copied from [21] and [15]. The Naive VAE is trained to maximize $\mathcal{L}_{ELBO}$.

Objective	OpVAE	VAE (Naive)	ORGAN	MolGAN
LogP	**0.96**	0.51	0.55	0.89
SAS	**1.00**	0.67	0.83	0.95

decoder to generate patterns with higher properties and these patterns are chemically valid, which encourages sampled molecular graphs that implicitly contain the patterns to be valid. In the same way, novel molecules are obtained when the encoder explores the possibilities for generating patterns with higher properties.

5.3   Objectives Optimization

In order to demonstrate the effectiveness of the proposed regularization, following [8] and [2], we chose to optimize the objectives LogP and SAS and compare against their works. We trained our model over full QM9 dataset for 30 epochs similarly to the experiments performed by [2] but differently from those of [8], where the model is trained on 5k QM9 subset. We normalized all scores within $[0,1]$ by using the codes[3] of [2]. As shown in Table 2, SeVAE beats both MolGAN and ORGAN in terms of objective scores, which proves that the proposed regularization is very effective.

6   Conclusion

Incorporating complex and non-differentiable property metrics into deep generative models is a challenging subject. We built graph neural networks into VAE for generating molecular graph. By resorting to a differentiable regularizer, we addressed the property optimization. We introduced several metrics to validate the quality of our proposed method. The advantages of the proposed method are reflected in experimental results.

Acknowledgement. This work was supported by the National Natural Science Foundation of China under Grant 61876003. It is a research achievement of Key Laboratory of Science, Techonology and Standard in Press Industry (Key Laboratory of Intelligent Press Media Technology).

References

1. Dai, H., Tian, Y., Dai, B., Skiena, S., Song, L.: Syntax-directed variational autoencoder for structured data. In: 6th International Conference on Learning Representations, Conference Track Proceedings, ICLR 2018, Vancouver, BC, Canada, 30 April–3 May 2018. OpenReview.net (2018). https://openreview.net/forum?id=SyqShMZRb

---

[3] Available at https://github.com/nicola-decao/MolGAN.

2. De Cao, N., Kipf, T.: MolGAN: an implicit generative model for small molecular graphs. arXiv preprint arXiv:1805.11973 (2018)
3. Duvenaud, D.K., et al.: Convolutional networks on graphs for learning molecular fingerprints. In: Cortes, C., Lawrence, N.D., Lee, D.D., Sugiyama, M., Garnett, R. (eds.) Advances in Neural Information Processing Systems 28: Annual Conference on Neural Information Processing Systems 2015, Montreal, Quebec, Canada, 7–12 December 2015, pp. 2224–2232 (2015). http://papers.nips.cc/paper/5954-convolutional-networks-on-graphs-for-learning-molecular-fingerprints
4. Ertl, P., Schuffenhauer, A.: Estimation of synthetic accessibility score of drug-like molecules based on molecular complexity and fragment contributions. J. Cheminformatics 1, 8 (2009). https://doi.org/10.1186/1758-2946-1-8
5. Ghose, A.K., Crippen, G.M.: Atomic physicochemical parameters for three-dimensional-structure-directed quantitative structure-activity relationships. 2. modeling dispersive and hydrophobic interactions. J. Chem. Inf. Comput. Sci. 27(1), 21–35 (1987)
6. Gilmer, J., Schoenholz, S.S., Riley, P.F., Vinyals, O., Dahl, G.E.: Neural message passing for quantum chemistry. In: Precup and Teh [17], pp. 1263–1272. http://proceedings.mlr.press/v70/
7. Gómez-Bombarelli, R., et al.: Automatic chemical design using a data-driven continuous representation of molecules. CoRR abs/1610.02415 (2016). http://arxiv.org/abs/1610.02415
8. Guimaraes, G.L., Sanchez-Lengeling, B., Farias, P.L.C., Aspuru-Guzik, A.: Objective-reinforced generative adversarial networks (ORGAN) for sequence generation models. CoRR abs/1705.10843 (2017). http://arxiv.org/abs/1705.10843
9. Irwin, J.J., Sterling, T., Mysinger, M.M., Bolstad, E.S., Coleman, R.G.: ZINC: a free tool to discover chemistry for biology. J. Chem. Inf. Model. 52(7), 1757–1768 (2012)
10. Jin, W., Barzilay, R., Jaakkola, T.: Junction tree variational autoencoder for molecular graph generation. arXiv preprint arXiv:1802.04364 (2018)
11. Kingma, D.P., Welling, M.: Auto-encoding variational Bayes. In: Bengio, Y., LeCun, Y. (eds.) 2nd International Conference on Learning Representations, Conference Track Proceedings, ICLR 2014, Banff, AB, Canada, 14–16 April 2014 (2014). http://arxiv.org/abs/1312.6114
12. Kusner, M.J., Paige, B., Hernández-Lobato, J.M.: Grammar variational autoencoder. In: Precup and Teh [17], pp. 1945–1954. http://proceedings.mlr.press/v70/kusner17a.html
13. Li, Y., Tarlow, D., Brockschmidt, M., Zemel, R.S.: Gated graph sequence neural networks. In: Bengio, Y., LeCun, Y. (eds.) 4th International Conference on Learning Representations, Conference Track Proceedings, ICLR 2016, San Juan, Puerto Rico, 2–4 May 2016 (2016). https://iclr.cc/archive/www/doku.php%3Fid=iclr2016:accepted-main.html
14. Liu, Q., Allamanis, M., Brockschmidt, M., Gaunt, A.: Constrained graph variational autoencoders for molecule design. In: Advances in Neural Information Processing Systems, pp. 7795–7804 (2018)
15. Ma, T., Chen, J., Xiao, C.: Constrained generation of semantically valid graphs via regularizing variational autoencoders. In: Bengio, S., Wallach, H.M., Larochelle, H., Grauman, K., Cesa-Bianchi, N., Garnett, R. (eds.) Advances in Neural Information Processing Systems 31: Annual Conference on Neural Information Processing Systems 2018, NeurIPS 2018, Montréal, Canada, 3–8 December 2018, pp. 7113–7124 (2018). http://papers.nips.cc/paper/7942-constrained-generation-of-semantically-valid-graphs-via-regularizing-variational-autoencoders

16. Nair, V., Hinton, G.E.: Rectified linear units improve restricted Boltzmann machines. In: Fürnkranz, J., Joachims, T. (eds.) Proceedings of the 27th International Conference on Machine Learning (ICML 2010), Haifa, Israel, 21–24 June 2010, pp. 807–814. Omnipress (2010). https://icml.cc/Conferences/2010/papers/432.pdf

17. Precup, D., Teh, Y.W. (eds.): Proceedings of the 34th International Conference on Machine Learning, ICML 2017, Sydney, NSW, Australia, 6–11 August 2017, Proceedings of Machine Learning Research, vol. 70. PMLR (2017). http://proceedings.mlr.press/v70/

18. Ramakrishnan, R., Dral, P.O., Rupp, M., Von Lilienfeld, O.A.: Quantum chemistry structures and properties of 134 kilo molecules. Sci. Data **1**, 140022 (2014)

19. Ruddigkeit, L., Van Deursen, R., Blum, L.C., Reymond, J.L.: Enumeration of 166 billion organic small molecules in the chemical universe database GDB-17. J. Chem. Inf. Model. **52**(11), 2864–2875 (2012)

20. Schlichtkrull, M., Kipf, T.N., Bloem, P., van den Berg, R., Titov, I., Welling, M.: Modeling relational data with graph convolutional networks. In: Gangemi, A., et al. (eds.) ESWC 2018. LNCS, vol. 10843, pp. 593–607. Springer, Cham (2018). https://doi.org/10.1007/978-3-319-93417-4_38

21. Simonovsky, M., Komodakis, N.: GraphVAE: towards generation of small graphs using variational autoencoders. In: Kůrková, V., Manolopoulos, Y., Hammer, B., Iliadis, L., Maglogiannis, I. (eds.) ICANN 2018. LNCS, vol. 11139, pp. 412–422. Springer, Cham (2018). https://doi.org/10.1007/978-3-030-01418-6_41

22. Weininger, D.: Smiles, a chemical language and information system 1 introduction to methodology and encoding rules. J. Chem. Inf. Comput. Sci. **28**(1), 31–36 (1988). https://doi.org/10.1021/ci00057a005

23. Wildman, S.A., Crippen, G.M.: Prediction of physicochemical parameters by atomic contributions. J. Chem. Inf. Comput. Sci. **39**(5), 868–873 (1999). https://doi.org/10.1021/ci990307l

Node-Edge Bilateral Attributed Network Embedding

Jingjie Mo[1,2], Neng Gao[1(✉)], Ji Xiang[1], and Daren Zha[1]

[1] State Key Laboratory of Information Security, Institute of Information
Engineering, CAS, Beijing, China
{mojingjie,gaoneng,xiangji,zhadaren}@iie.ac.cn
[2] School of Cyber Security, University of Chinese Academy of Sciences,
Beijing, China

Abstract. This paper addresses attributed network embedding which
maps the structural information and multi-modal attribute data into a
latent space. Most existing network embedding algorithms concentrate
on either node-oriented modeling or edge-oriented modeling, resulting in
unilaterally capturing information from nodes or edges. However, there
is no effective method to bilaterally extract node attributes cooperated
with edge attributes, which delineates the outline and detail of social network. To this end, we propose a novel **N**ode-**E**dge **B**ilateral **A**ttributed
Network **E**mbedding method named **NEBANE**. Regarding each edge
as a specific node, we construct a pioneering node-edge-node triangular
structure for bilateral information modeling on both nodes and edges.
Furthermore, we envisage a pairwise loss which maximizes the likelihood
of connected node pairs and of connected node-edge pairs to measure
the node-node and node-edge similarity. Empirically, experiments on two
real-world datasets, including link prediction and node classification, are
conducted in this paper. Our method achieves substantial performance
gains compared with state-of-the-art baselines (e.g., 4.21%–13.65% lift
by AUC scores for link prediction).

Keywords: Attributed network embedding · Node-edge bilateral
modeling · Triangular structure

1 Introduction

Many network analyses, such as link prediction [3], node classification [18], node
clustering [20] and anomaly detection [10], have attracted a surge of research
attention. However, due to the complex topology and heterogeneous attributes
of social network, how to automatically extract the network feature has become
a tough problem in social network mining worth extensive research. In recent
years, attributed network embedding, which maps the structural information
and multi-modal attribute data into a latent space, has developed into an effective method for network automatical feature extraction.

© Springer Nature Switzerland AG 2019
T. Gedeon et al. (Eds.): ICONIP 2019, CCIS 1143, pp. 477–488, 2019.
https://doi.org/10.1007/978-3-030-36802-9_51

Recently, a large number of structure-based network embedding methods have been proposed [1,16,17]. These algorithms are exhibited to be sufficient for plain network representation. However, structure-based approaches show limited performance due to the sparsity of social interactions. More recently, some efforts have been made for complementally capturing attribute proximity which can be divided into node-oriented and edge-oriented. Node-oriented methods introduce heterogeneous node attributes—ranging from textual information [5,15], label information [23] to user profiles [11]—into network embedding while viewing each edge as binary. Edge-oriented methods focus on edge information—varying from individual categories [2], multiple labels [19] to the effect of time [24]—whereas neglecting node attributes. Indeed, these developed methods demonstrate the effectiveness of one-side information extraction from nodes or edges for learning better network representations. Nevertheless, these algorithms rely on either node attributes or edge attributes, failing to bilaterally co-model node and edge attributes.

Node attributes delineating the outline of social networks and edge attributes probing into the detail of social networks are complementary. A typical example in academic social networks is that a scholar may collaborate with different scholars for differing reasons. Scholar A may collaborate with scholar B in *Data Mining* while collaborating with scholar C in *Databases*. Besides, the keywords of their joint studies may indicate a part of node attributes. Scholar A may take an interest in *Data Mining* if he/she has a publication collaborating with another scholar about *Network Embedding*. On the one hand, node attribute information depicts the outline of nodes, indicating the community of nodes. On the other hand, rich and variant meanings of edges make exploration of implication relation between nodes possible. Bilateral modeling on both nodes and edges is beneficial for network representation.

On top of that, it's a challenging problem to fuse node and edge attribute information. It may run into some difficulties in dealing with heterogeneous attributes. Moreover, due to the difficulty of measuring the interplay between node and edge attribute information, there is still no sufficient way to bilaterally extract information from both nodes and edges. A bilateral model needs to be designed for jointly capturing full-scale node and edge attribute information.

To address the aforementioned challenge, we propose a novel **Node-Edge Bilateral Attributed Network Embedding** method named **NEBANE**. Considering each edge as a specific node, we construct a node-edge-node triangular structure to measure the node-node and node-edge similarity. In conclusion, the main contributions are summarized as follows:

- We formally present a new problem: node-edge bilateral attributed network embedding which aims to bilaterally co-model node and edge attributes as well as the interplay between them.
- We propose a novel method **NEBANE** to cooperate node attributes with detailed content of interactions under an original node-edge-node triangular structure via maximizing the conditional probability of connected node pairs and of connected node-edge pairs.

– We empirically evaluate the effectiveness of **NEBANE** on two real-world datasets via three classical network data mining tasks. Experimental results indicate that **NEBANE** emphasizes related crossover information with bilateral modeling.

The rest of this paper is structured as follows. First and foremost, the discussion of related work is divided into two parts in Sect. 2. Then, we propose our method **NEBANE** in Sect. 3, followed by presenting experimental results in Sect. 4. Finally, Sect. 5 concludes the paper and visions the future work.

2 Related Work

In this section, we briefly review the development of network embedding methods. The discussion is carried out in two parts, i.e., structure-based network embedding and attribute-aware network embedding.

2.1 Structure-Based Network Embedding

DeepWalk [16] is one of the pioneering works for network embedding. It borrows the idea from Skip-Gram [13] to generate node sequences by imposing random walks, which aims to preserve the similarity of nodes in the context window. On top of that, node2vec [1] designs biased random walks to balance the local and global structure proximity. LINE [17] provides a much explicit loss function to capture the first-order proximity and the second-proximity with a late fusion. Besides, neural networks are introduced in several structure-based network embedding methods. DNGR [4] exploits a random surfing model to capture structural information into a matrix and leverages an SDAE model for information compression. Analogously, a deep auto-encoder model is also adopted in SDNE [21] to extract the second-order proximity with a pairwise constraint for the first-order proximity preserving. However, merely considering the structural information, these algorithms cannot handle rich user-generated content which is beneficial for learning better representations.

2.2 Attribute-Aware Network Embedding

Besides, some efforts have been devoted to exploring the influence of user-generated content on network embedding.

Node-oriented methods explore the impact of node attributes while viewing each edge as binary. TADW [5] is the first attempt to jointly embed textual information and context proximity into the same latent space by matrix factorization. TriDNR [15] preserves structural similarity and textual information through DeepWalk and doc2vec [9] separately, and concatenates them via a late fusion. These two methods are designed for node textual information modeling which is not applicable for attributed network embedding. To capture the attribute proximity, ASNE [11] is designed as a generic framework for attributed network embedding. Preserving both structure proximity and attribute proximity, it reveals the homophily effect caused by complex topologies and rich node attribute information.

Edge-oriented methods explore the influence of semantic information in social interactions whereas neglecting node attributes. Knowledge representation learning (KRL) methods such as TransE [2], TransH [22], TransR [12] put forward a concept on modeling edge information and consider each relation as an individual category. These methods cannot deal with multi-modal edge attributes. TransNet [19] formalizes the task of Social Relation Extraction to evaluate the performance of modeling multi-label relations. Inspired by knowledge graph representation methods, TransNet introduces the translation mechanism to express the meanings of edges. Furthermore, IGE [24] is specially designed for bipartite graph recommendation. It's proposed as a deep node embedding model to quantify different meanings on edges with differing time and attributes.

These proposed approaches effectively extract one-side information from nodes or edges, which achieve high performance in network embedding. Nevertheless, these aforementioned methods concentrate on either node attributes or the meanings of edges, failing to bilaterally co-model full-scale node and edge attributes.

3   Methodology

In this section, we elaborate our proposed method for node and edge attribute preserving network embedding.

3.1   Problem Definition

Given a social attributed network $\mathcal{G} = (\mathcal{E}, \mathcal{V}, \mathbf{P}, \mathbf{Q})$, where $\mathcal{E}$ denotes the set of edges with its corresponding attributes $\mathbf{P}$, whereas $\mathcal{V}$ is the set of nodes with its corresponding attributes $\mathbf{Q}$. For each edge $e_{ij} \in \mathcal{E}$, $\mathbf{p}_{ij}$, an m-dimensional feature vector, denotes the corresponding attributes of e_{ij}. As for each node $v_i \in \mathcal{V}$, $\mathbf{q}_i$, an n-dimensional feature vector, characterizes the corresponding attributes of v_i. Our goal in this paper is to embed each node $v_i \in \mathcal{V}$ into a d-dimensional latent space.

The learned node embeddings should satisfy two properties: (1) *network structure and edge meaning capturing*, i.e., node representations should reflect the impact of interactions with different social actors; (2) *node attribute homophily preserving*, i.e., the embedding of nodes with semblable attributes should be much closer.

3.2   Overall Architecture

In this paper, we propose a node-edge bilateral attributed network embedding model **NEBANE** to jointly capture node and edge attribute information.

We convert the social network to a coupled network which can be divided into two parts: node space and edge space. Each edge is considered as a specific node, and there are two new edges between the edge node and head/tail nodes. The original relationship between nodes still remains. In this way, a node-edge-node triangular structure is constructed to bilaterally measure the node-node and node-edge similarity. As shown in Fig. 1, our method **NEBANE** consists of two components: **Node Attribute Modeling** and **Edge Attribute Modeling**. The target of node attribute modeling is to preserve the node structural information and attribute homophily, whereas edge attribute modeling aims at preserving the meanings of relations between two nodes.

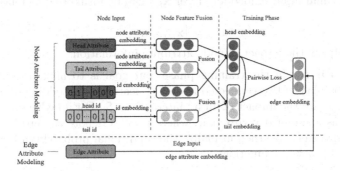

Fig. 1. The framework of **NEBANE** model.

3.3   Node Attribute Modeling

To jointly capture the strength of node structural information and node attribute homophily, we design a deep learning framework for node attribute modeling which consists of three components: **Node Input**, **Node Feature Fusion** and **Training Phase**.

Node Input. For node input, we take the head/tail id cooperated with their attributes as input. In id embedding layer, a feed-forward layer is applied to embed node id one-hot vectors to dense vectors, capturing node structural information. In attribute embedding layer, all types of attributes can be converted as attribute embedding in our work. For structured attributes, we transform their multi-hot encodings into dense vectors with a feed-forward layer and an activation function. Moreover, numerical data can be easily concatenated with multi-hot encodings. For textual information, we employ the bag-of-words to encode the content. Besides, a convolutional neural network (CNN) can be applied to encode images. The node id embedding and node attribute embedding can be denoted as follows:

$$\mathbf{x}_{i_{id}} = \mathbf{W}_{id}\mathbf{z}_i \tag{1}$$

$$\mathbf{x}_{i_{attr}} = \sigma(\mathbf{W}_{node\_attr}\mathbf{q}_i) \tag{2}$$

where $\mathbf{z}_i$ is the one-hot id encoding for node i, $\mathbf{W}_{id}$ and $\mathbf{W}_{node\_attr}$ denote the weight of id embedding layer and attribute embedding layer respectively. Besides, $\mathbf{x}_{i_{id}}$ and $\mathbf{x}_{i_{attr}}$ are the input id embedding and input attribute embedding for node i. $\sigma(\cdot)$ denotes the activation function, and we choose $tanh$ as the activation function.

Node Feature Fusion. The fusion part aims to fuse the input id embedding and input attribute embedding for better learning via the early fusion. Different fusion methods may result in different performance. Various fusion methods can be selected, such as *concatenation*, *addition*, and *element-wise product*. According to experimental results, we apply the *concatenation* fusion method in this paper. The final input embedding layer $\mathbf{x}_{i_{in}}$ can be formalized as follows:

$$\mathbf{x}_{i_{in}} = \begin{bmatrix} \mathbf{x}_{i_{id}} & \eta\mathbf{x}_{i_{attr}} \end{bmatrix} \tag{3}$$

where η adjusts the impact of node attribute information.

Training Phase. In the training phase, our goal is to maximize the conditional probability of two connected nodes. To learn the embedding of node i, we should aggregate the impact of all its interactions $\mathcal{E}_i$. For an edge $e_{ij} \in \mathcal{E}_i$, (i,j) is defined as a sampled vertex pair, where i and j denote the head and tail of edge e_{ij} separately. We then formalize the conditional probability of node i and node j as the *softmax* function:

$$P(i \mid j) = \frac{exp(\mathbf{x}_{i_{in}}^T \mathbf{x}_{j_{out}})}{\sum_{k \in \mathcal{V}} exp(\mathbf{x}_{i_{in}}^T \mathbf{x}_{k_{out}})} \tag{4}$$

where $\mathbf{x}_{i_{out}}$ is the output representation of node i. To obtain the final embedding $\mathbf{x}_i$ of node i, we fuse the input and output embedding by *addition* operation.

3.4   Edge Attribute Modeling

To preserve the meanings of relations between two nodes, we also apply a similar structure for edge attribute modeling. Considering each edge as a node, we aim to maximize the conditional probability of the edge node and the connected head/tail nodes. We process edge attributes the same way we did with node attributes. The attribute embedding of edge e_{ij} is expressed as follows:

$$\mathbf{y}_{ij} = \sigma(\mathbf{W}_{edge\_attr}\mathbf{p}_{ij}) \tag{5}$$

where $\mathbf{y}_{ij}$ denotes the attribute embedding of edge e_{ij}, and $\mathbf{W}_{edge\_attr}$ is the weight of edge attribute embedding. We also apply $tanh$ as the activation function. In this way, we also utilize the *softmax* function to measure the conditional probability of the edge node and the connected head/tail nodes, which is represented as follows:

$$P(\mathbf{y}_{ij} \mid \mathbf{x}_i) = \frac{exp(\mathbf{y}_{ij}^T \mathbf{x}_i)}{\sum_{k \in \mathcal{V}} exp(\mathbf{y}_{ij}^T \mathbf{x}_k)} \tag{6}$$

$$P(\mathbf{y}_{ij} \mid \mathbf{x}_j) = \frac{exp(\mathbf{y}_{ij}^T \mathbf{x}_j)}{\sum_{k \in \mathcal{V}} exp(\mathbf{y}_{ij}^T \mathbf{x}_k)} \tag{7}$$

3.5   Loss Functions and Optimization

Here the consolidated loss function and optimization details for the **NEBANE** model are exhibited. Our goal is to jointly preserve node similarity and the meanings of edges, which is equivalent to minimizing the following negative log-likelihood loss function $\mathcal{L}$:

$$\mathcal{L} = \alpha \mathcal{L}_{node} + \beta \mathcal{L}_{edge} + \mathcal{L}_{reg} \tag{8}$$

where α, β are two hyper-parameters to adjust the weight of each loss part. $\mathcal{L}_{node}$ denotes the node loss:

$$\mathcal{L}_{node} = -\sum_{i \in \mathcal{V}} \sum_{j \in \mathcal{N}_i} log(P(i \mid j)) \tag{9}$$

where $\mathcal{N}_i$ is the set of neighbor nodes of node i. Besides, the edge loss $\mathcal{L}_{edge}$ is formalized as:

$$\mathcal{L}_{edge} = -\sum_{e_{ij} \in \mathcal{E}} (log(P(\mathbf{y}_{ij} \mid \mathbf{x}_i) + P(\mathbf{y}_{ij} \mid \mathbf{x}_j))) \tag{10}$$

However, due to the huge cost of computing normalizers for *softmax* function, we adopt negative sampling [14] as the approximation to improve the training efficiency. For this reason, the node loss $\mathcal{L}_{node}$ and the edge loss $\mathcal{L}_{edge}$ can be approximately represented respectively as follows:

$$\mathcal{L}_{node} = -\sum_{i \in \mathcal{V}} \sum_{j \in \mathcal{N}_i} (log(\sigma(\mathbf{x}_{i_{in}}^T \mathbf{x}_{j_{out}})) \tag{11}$$
$$+ m \cdot E_{k \sim P_D}[log(\sigma(-\mathbf{x}_{i_{in}}^T \mathbf{x}_{k_{out}}))])$$

$$\mathcal{L}_{edge} = -\sum_{e_{ij} \in \mathcal{E}} (log(\sigma(\mathbf{y}_{ij}^T \mathbf{x}_i)) + log(\sigma(\mathbf{y}_{ij}^T \mathbf{x}_j)) \tag{12}$$
$$+ 2m \cdot E_{k \sim P_D}[log(\sigma(-\mathbf{y}_{ij}^T \mathbf{x}_k))])$$

where $\sigma(x) = 1/(1 + exp(-x))$ is the *sigmoid* function, and m denotes the negative sampled number. We apply the uniform distribution $P_D \sim \frac{1}{|\mathcal{V}|}$ for negative sampling, and $|\mathcal{V}|$ is the total number of nodes. $\mathcal{L}_{reg}$ indicates the regularization loss which is defined as:

$$\mathcal{L}_{reg} = \|\mathbf{W}_{id}\|_2^2 + \|\mathbf{W}_{node\_attr}\|_2^2 + \|\mathbf{W}_{edge\_attr}\|_2^2 \tag{13}$$

Adam Optimizer [8] is adopted in this paper to minimize the loss function $\mathcal{L}$, which can automatically adjust the learning rate for each parameter. We apply batch normalization [6] to reduce internal covariate shift [6]. Moreover, the dropout component is also adopted to alleviate overfitting.

4   Experiments

In this section, we systematically evaluate the performance of our method **NEBANE** on two real-world datasets compared with several cutting-edge network embedding methods. We experimentalize on link prediction and node classification.

4.1 Datasets

We conduct experiments on two types of real-world networks which are academic social network **AMiner**[1] and customer-item interaction network **Amazon**[2]. The statistics of the two datasets is summarized in Table 1.

AMiner. AMiner [7] is a large academic website for research mining services. We select authors whose published papers are more than 10 as nodes. We label the author in the domain which he/she publishes the most works. We leverage the bag-of-words of research interests and the publication titles of one author as the corresponding author's node attributes. Besides, publications are treated as edges between two co-authors, and the bag-of-words of publication titles represents edge attributes.

Amazon. We select 7 categories of items' high-score reviews (higher than 4.0) as the dataset. Customers and items are all treated as nodes, and items are labeled by categories. The bag-of-words of customers' all reviews represents node attributes. The bag-of-words of the review from a customer to an item represents attributes of the edge between them.

Table 1. Statistics of the two datasets.

Datasets	Nodes	Edges	Node attributes	Edge attributes
AMiner	4233	15384	1500	1000
Amazon	6305	61701	1000	1000

4.2 Baselines

We compare **NEBANE** with several state-of-the-art network representation learning (NRL) baselines which can be divided into two groups:

Structure-Based NRL Methods

- DeepWalk [16] exploits random walk to generate node sequences and imposes skip-gram to learn embedding.
- node2vec [1] improves random walk to balance the weight of local and global proximity.
- LINE [17] trains node embeddings via two explicit loss functions which aim at capturing the first-order and second-order proximity separately.

---

[1] https://www.aminer.cn/aminernetwork.
[2] http://jmcauley.ucsd.edu/data/amazon/.

Attribute-Aware NRL Methods

- ASNE [11] embeds attribute information to maximize the relevance of nodes and their neighbor nodes.
- TADW [5] associates textual information with structure to learn node embeddings by matrix factorization.
- TransNet [19] introduces multiple labels on edges to network embedding, and models it via the translation mechanism.

4.3 Parameter Settings

Our implementation of **NEBANE** is based on TensorFlow, and the code and datasets will be available upon acceptance. For a fair comparison, we set the embedding dimension to 256 for all methods. We optimize the model with mini-batch, and the batch size is 200. The learning rate is 0.01, and the epoch number is 20. Besides, the number of negative sampling is 5. The hyper-parameter α, β are set as 1000, 1000 respectively as default. As for baseline methods, we employ a grid search to find the best parameters.

4.4 Node Classification

For the node classification task, we learn node embeddings by training all edges. For a fair comparison, We employ an rbf-kernel SVM classifier with $\gamma = 10$ for all methods. To make a detailed quantification, we train the classifier with different ratio scales of labeled data which range from 15% to 75% by taking 10% as a step. For each train ratio scale, we conduct 5-fold cross-validation and evaluate the performance by the mean micro/macro F1-score.

Performance. Figure 2(a)–(d) visualizes the consequences of node classification task. We calculate the average of F1-scores of all training ratio scales for all methods. Notice that **NEBANE** also achieves superior performance compared with baselines, yielding 6.01%, 6.96%, 4.94% and 6.81% performance gains on micro and macro F1-score on two datasets compared with the best baseline. The reason is that **NEBANE** collects the crossover information in interactions from neighbor nodes, which reflects the feature of intragroup connection. Besides, the preponderance of **NEBANE** is more prominent given small size of training data. Given 15% of nodes as training data, **NEBANE** achieves 6.94%–18.75% gains and 8.25%–20.24% gains on micro and macro F1-score separately. It indicates that our method is effective in dealing with fewer labeled data. Furthermore, the performance of **NEBANE** is relatively stable across all training ratio scales while baselines are rather fluctuant.

Parameter Sensitivity. We also explore the parameter η, α/β ratio sensitivity on node classification. We visualize the mean F1-scores of all training ratio scales. As shown in Fig. 2(e), the performance of node classification raises to peak when the weight of attribute information η is around 1.0. Figure 2(f) suggests that when the weight of node loss α is 1–10 times to that of edge loss β we could get better results.

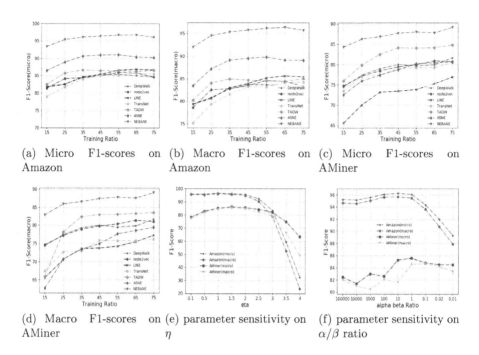

(a) Micro F1-scores on Amazon

(b) Macro F1-scores on Amazon

(c) Micro F1-scores on AMiner

(d) Macro F1-scores on AMiner

(e) parameter sensitivity on η

(f) parameter sensitivity on α/β ratio

Fig. 2. Performance and parameter sensitivity of node classification on datasets.

4.5 Link Prediction

First and foremost, we evaluate the performance of our method **NEBANE** on link prediction. We randomly hold out 60%, 40%, 20% links and generate an equal number of negative links as the test set. The remaining links are regarded as the train set. We then calculate the cosine similarity of each test node pair. To measure the ranking quality, we apply the AUC score which is widely employed in information retrieval.

Performance. We present the AUC scores in Table 2 and then calculate the average of AUC scores of all training ratio scales for all methods. Compared with the best performance of structure-based NRL methods, **NEBANE** achieves 6.63% and 9.81% improvements on two datasets respectively. It proves the utility of fusing structural and attribute information since our method sufficiently supplements the knowledge of interactions. Compared with the best performance of attribute-aware NRL methods, **NEBANE** yields 4.21% and 6.42% performance gains on two datasets respectively. Our superiority lies in the fact that **NEBANE** gains the weight of attributes related to the interaction. Relevant attributes reflect the crossover information of head and tail which is emphasized via bilateral modeling. **NEBANE** extracts more effective information with the assistance of enhancing the weight of related information.

Table 2. AUC scores for link prediction on two datasets.

Datatsets	Amazon			AMiner		
Training ratio	40%	60%	80%	40%	60%	80%
LINE	0.5349	0.5358	0.5135	0.7172	0.8085	0.8688
DeepWalk	0.7141	0.7544	0.7738	0.7526	0.8355	0.8803
node2vec	0.7224	0.7663	0.7874	0.7635	0.8452	0.9009
TransNet	0.6578	0.7351	0.78	0.7217	0.7776	0.8449
ASNE	0.7637	0.7931	0.7919	0.7217	0.7776	0.8449
TADW	0.7203	0.7295	0.747	0.8281	0.8795	0.9039
NEBANE	**0.7976**	**0.8273**	**0.8502**	**0.9191**	**0.9362**	**0.9487**

5 Conclusion and Future Work

In this paper, we propose a novel **Node-Edge Bilateral Attributed Network Embedding** method named **NEBANE**. We argue that it is essential to structure a bilateral model for capturing full-scale node and edge attributes. Considering each edge as a specific node, we construct the node-edge-node triangular structure to bilaterally measure the node-node and node-edge similarity. Focusing on both detailed content of edges and node profiles, our method captures the interplay between node pairs and between node-edge pairs via maximizing the conditional probability of connected pairs. Experiments on link prediction, node classification and node clustering demonstrate the remarkable performance of our method compared with state-of-the-art network embedding baselines. In future studies, we strive to extend the algorithm to some scenario applications, such as recommender systems and anomaly detection.

Acknowledgements. This work is supported by the National Natural Science Foundation of China (No. U163620068).

References

1. Grover, A., Leskovec, J.: node2vec: scalable feature learning for networks. In: ACM SIGKDD International Conference on Knowledge Discovery and Data Mining (2016)
2. Bordes, A., Usunier, N., Garcia-Duran, A., Weston, J., Yakhnenko, O.: Translating embeddings for modeling multi-relational data. In: Advances in Neural Information Processing Systems, pp. 2787–2795 (2013)
3. Cao, B., Liu, N.N., Yang, Q.: Transfer learning for collective link prediction in multiple heterogenous domains. In: International Conference on International Conference on Machine Learning (2010)
4. Cao, S.: Deep neural network for learning graph representations. In: Thirtieth AAAI Conference on Artificial Intelligence (2016)
5. Cheng, Y., Zhao, D., Zhao, D., Chang, E.Y., Chang, E.Y.: Network representation learning with rich text information. In: International Conference on Artificial Intelligence (2015)

6. Ioffe, S., Szegedy, C.: Batch normalization: accelerating deep network training by reducing internal covariate shift. In: International Conference on Machine Learning, pp. 448–456 (2015)
7. Jie, T., Jing, Z., Yao, L., Li, J., Zhong, S.: ArnetMiner: extraction and mining of academic social networks. In: ACM SIGKDD International Conference on Knowledge Discovery and Data Mining (2008)
8. Kingma, D., Ba, J.: Adam: a method for stochastic optimization. Computer Science (2014)
9. Le, Q., Mikolov, T.: Distributed representations of sentences and documents. In: International Conference on Machine Learning, pp. 1188–1196 (2014)
10. Li, J., Dani, H., Hu, X., Liu, H.: Radar: residual analysis for anomaly detection in attributed networks. In: IJCAI 2017 (2017)
11. Liao, L., He, X., Zhang, H., Chua, T.S.: Attributed social network embedding. IEEE Trans. Knowl. Data Eng. **30**(12), 2257–2270 (2017)
12. Lin, Y., Liu, Z., Sun, M., Liu, Y., Zhu, X.: Learning entity and relation embeddings for knowledge graph completion. In: Twenty-ninth AAAI Conference on Artificial Intelligence (2015)
13. Mikolov, T., Chen, K., Corrado, G., Dean, J.: Efficient estimation of word representations in vector space. Computer Science (2013)
14. Mikolov, T., Sutskever, I., Chen, K., Corrado, G.S., Dean, J.: Distributed representations of words and phrases and their compositionality. In: Advances in Neural Information Processing Systems, pp. 3111–3119 (2013)
15. Pan, S., Wu, J., Zhu, X., Zhang, C., Wang, Y.: Tri-party deep network representation. In: International Joint Conference on Artificial Intelligence (2016)
16. Perozzi, B., Al-Rfou, R., Skiena, S.: DeepWalk: online learning of social representations. In: ACM SIGKDD International Conference on Knowledge Discovery and Data Mining (2014)
17. Tang, J., Qu, M., Wang, M., Zhang, M., Yan, J., Mei, Q.: LINE: large-scale information network embedding. In: Proceedings of the 24th International Conference on World Wide Web, pp. 1067–1077. International World Wide Web Conferences Steering Committee (2015)
18. Tang, J., Aggarwal, C., Liu, H.: Node classification in signed social networks. In: Proceedings of the 2016 SIAM International Conference on Data Mining, pp. 54–62. SIAM (2016)
19. Tu, C., Zhang, Z., Liu, Z., Sun, M.: TransNet: translation-based network representation learning for social relation extraction. In: International Joint Conference on Artificial Intelligence (2017)
20. Vinh, N.X., Epps, J., Bailey, J.: Information theoretic measures for clusterings comparison: variants, properties, normalization and correction for chance. J. Mach. Learn. Res. **11**(1), 2837–2854 (2010)
21. Wang, D., Peng, C., Zhu, W.: Structural deep network embedding. In: ACM SIGKDD International Conference on Knowledge Discovery and Data Mining (2016)
22. Wang, Z., Zhang, J., Feng, J., Chen, Z.: Knowledge graph embedding by translating on hyperplanes. In: AAAI, vol. 14, pp. 1112–1119 (2014)
23. Xiao, H., Li, J., Xia, H.: Label informed attributed network embedding. In: Tenth ACM International Conference on Web Search and Data Mining (2017)
24. Yao, Z., Yun, X., Kong, X., Zhu, Y.: Learning node embeddings in interaction graphs. In: ACM on Conference on Information and Knowledge Management (2017)

Info-Detection: An Information-Theoretic Approach to Detect Outlier

Feng Zhao[1]($^\boxtimes$), Fei Ma[2], Yang Li[2], Shao-Lun Huang[2], and Lin Zhang[1,2]

[1] Department of Electronic Engineering, Tsinghua University, Beijing, China
zhaof17@mails.tsinghua.edu.cn, linzhang@tsinghua.edu.cn
[2] Tsinghua-Berkeley Shenzhen Institute, Tsinghua University, Beijing, China
mf17@mails.tsinghua.edu.cn,
{yangli,shaolun.huang}@sz.tsinghua.edu.cn

Abstract. Outlier detection is one of major task in unsupervised learning. We propose a cluster analysis based outlier detection method called Info-Detection. Info-Detection determines the number of outliers automatically and captures the global property of the provided data. To implement Info-Detection and overcome the global computational complexity, we use principal sequence of partition, which we improve one order of magnitude faster than the original version. Experiments show that compared with other outlier detection methods, Info-Detection achieves better accuracy with an affordable time overhead.

Keywords: Outlier detection · Clustering · Principal sequence of partition

1 Introduction

Outlier detection is an important task in data mining. It is the identification of abnormal data (called outliers) which differs from the majority of the data (called inliers) [4]. It has applications in fraud detection, loan application processing, activity monitoring etc. [5]. Many existing detection algorithms give an outlier score to each data point and the outliers are chosen as those points with highest scores. These methods need to know the number of outliers n in advance but for real-world task, we do not know how many outliers there are and mismatched n produces no good result. Also, besides distance metric, many outlier detection methods have other hyper parameters to tune, which depends heavily on dataset.

To overcome these problems, we propose Info-Detection, which do not require n in advance and is determined totally by the similarity metric used. Info-Detection is a cluster analysis based method. In this domain, density-based clustering method has been applied to outlier detection, which produces fairly good results but the density threshold still needs fine tuning [2]. Info-Detection comes from info-clustering, which is a hierarchical clustering method [3] and can choose the number of cluster for flat clustering. Info-clustering has solid theoretical foundations from information theory and can produce hierarchical tree

© Springer Nature Switzerland AG 2019
T. Gedeon et al. (Eds.): ICONIP 2019, CCIS 1143, pp. 489–496, 2019.
https://doi.org/10.1007/978-3-030-36802-9_52

for the random variables to be clustered. It can also be used to cluster data. In such case, the clustering result is the same as minimum cost clustering with $\beta = 1$ in [8]. Info-clustering can be applied to different models and directed graph is a very suitable structure. In this paper, we extend info-clustering theory on directed graph model and propose Info-Detection, which can determine the number of outlier automatically.

Info-clustering has good theoretical properties with the cost of high time complexity to implement, which limited its application. In this paper, we improve the implementation of info-clustering, which is one order of magnitude faster than the original one. We also propose a prediction scheme which can predict new observations in linear time complexity. To demonstrate our method, we compare Info-Detection with other methods on some datasets. The result shows that Info-Detection gives best result on these datasets and is suitable for different kinds of data.

The notational convention of this paper is as follows: Let x_i represents the feature vector for the i-th data point in the dataset. The directed graph is denoted by $G(V, E)$. Node index set $V = \{1, 2, \ldots, |V|\}$. Node set $Z_V = \{Z_i | i \in V\}$. Edge set E is the collection of tuple (i, j). For $B \subseteq V$, edge subset $E(B) = \{(i, j) | i, j \in B, (i, j) \in E\}$ is the edge set restricted on B. Each edge is associated with a non-negative weight w_{ij}, which is computed from x_i and x_j with a given affinity metric. $\mathcal{P}$ is a partition of V. That is, $P = \{C_1, \ldots, C_k\}, \cup_{i=1}^{k} C_i = V$ and $i \neq j \Rightarrow C_i \cap C_j = \emptyset$. $f(\cdot)$ is the graph in-cut function, defined as $f(C) = \sum_{i \neq C, j \in C, (i,j) \in E} w_{ij}$. Π is the collection of all partitions of V and $\Pi' = \Pi \backslash \{V\}$. A partial order $\mathcal{P}_1 \preceq \mathcal{P}_2$ on Π is defined as $C \in \mathcal{P}_1 \Rightarrow \exists C' \in \mathcal{P}_2 \, s.t. \, C \subseteq C'$. Finally, $f[\cdot]$ is a function defined on Π by $f[\mathcal{P}] = \sum_{C \in \mathcal{P}} f(C)$ and $\text{maximal}(F) = \{B \in F | \not\exists B' \in F \, s.t. \, B \subseteq B'\}$.

2   Formulation of Info-Detection

2.1   Overview of Info-Clustering

Info-Detection comes from info-clustering [3]. For a graph $G(V, E)$, info-clustering defines the cluster set $C_\gamma(Z_V)$ as follows:

$$I_\mathcal{P}(Z_V) := \frac{f[\mathcal{P}]}{|\mathcal{P}| - 1} \tag{1}$$

$$I(Z_V) := \min_{\mathcal{P} \in \Pi'(V)} I_\mathcal{P}(Z_V) \tag{2}$$

$$C_\gamma(Z_V) := \text{maximal}\{B \in V | |B| > 1, I(Z_B) > \gamma\} \tag{3}$$

In Eq. (1), $f[\mathcal{P}]$ can be interpreted as the inter-cluster affinity for a given partition. $f[\mathcal{P}]$ is averaged over the number of partition $|\mathcal{P}|$. For mathematical reasons, the denominator is $|\mathcal{P}| - 1$. $I(Z_B)$ is the minimum value over all averaged inter-cluster affinity and represents the shared information of Z_B. $C_\gamma(Z_V)$ collects those set whose shared information is larger than a given threshold.

It is noted that for $I_{\mathcal{P}}(Z_B)$, $\mathcal{P}$ is the partition for B and the subgraph $G(B, E(B))$ is considered when computing $f[\mathcal{P}]$.

It is shown in info-clustering theory that each set in $C_\gamma(Z_V)$ is non-intersecting and every two sets from $C(Z_V) = \bigcup_{\gamma \geq 0} C_\gamma(Z_V)$ are either disjoint or have subset relationship. Therefore, sets from $C(Z_V)$ can be put in a tree $\mathcal{T}$ with the property that A is a parent of B if $B \subset A$. To make $\mathcal{T}$ complete, $\mathcal{T}$ also includes $\{j\}$ as leaf node.

For sufficiently large γ, $C_\gamma(Z_V)$ will become empty set. The largest threshold value which makes such transition is denoted by γ_N, which has the following expression:

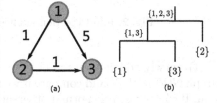

$$\gamma_N = \max_{A \subseteq V, |A| > 1} I(Z_A) \qquad (4)$$

Suppose $\gamma_N = I(Z_B)$ and B is the maximal set to achieve the maximum. This convention is used in the following sections.

Fig. 1. An illustration of info-clustering and its corresponding clustering tree

Example 1. Consider a graph $G(V, E)$ with $V = \{1, 2, 3\}$, $E = \{(1, 2), (2, 3), (1, 3)\}$. The weight values are $w_{12} = 1, w_{13} = 5, w_{23} = 1$. The graph is shown on Fig. 1(a). Let $\mathcal{P}_2 = \{\{1, 3\}, \{2\}\}$, then $f[\mathcal{P}_2] = 2$ and $I_{\mathcal{P}_2}(Z_V) = 2$ from Eq. (1). Let $\mathcal{P}_3 = \{\{1\}, \{2\}, \{3\}\}$, then $f[\mathcal{P}_3] = 7$ and $I_{\mathcal{P}_3}(Z_V) = 3.5$. Therefore $I(Z_V) = \min\{I_{\mathcal{P}_2}(Z_V), I_{\mathcal{P}_3}(Z_V), \dots\} = 2$.

$$C_\gamma(Z_V) = \begin{cases} \{\{1, 2, 3\}\} & \gamma < 2 \\ \{\{1, 3\}\} & 2 \leq \gamma < 5 \\ \emptyset & \gamma \geq \gamma_N = 5 \end{cases}$$

The clustering tree $\mathcal{T}$ for this example consists of $\{1, 2, 3\}$ as root, $\{1, 3\}$ as stem and $\{1\}, \{2\}, \{3\}$ as leaves, which are shown on Fig. 1(b).

2.2 Info-Detection Method

For our Info-Detection proposal, we use γ_N as a threshold to detect anomaly. Given $G(V, E)$, we first compute γ_N and B. Then nodes in B are inliers while nodes in $V \backslash B$ are outliers. We can score outlier Z_j for $j \in V \backslash B$ with the depth of the set $\{j\}$ in the hierarchical tree. Below is a simple example showing how Info-Detection is used.

Example 2. Consider a graph consisting of 6 nodes. Each node is corresponding to a point on Cartesian plane (Fig. 2(b)). The grid length is one unit and the edge weight $w_{ij} = \exp(-d_{ij}^2)$ where d_{ij} is the Euclidean distance between x_i and x_j on the plane. For example, $w_{45} = \exp(-5)$. The edge direction is from i to j for $i < j$. From Eq. (4), we can compute $B = \{1, 2, 3, 4\}$ and $\gamma_N = \frac{4 \exp(-1) + 2 \exp(-2)}{3} \approx 0.58$. The whole hierarchical tree $\mathcal{T}$ can also be computed, which is shown on Fig. 2(a). The outlier score for Z_6 is higher than Z_5 from the tree depth interpretation.

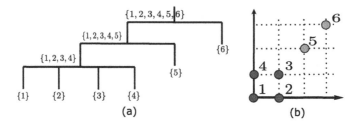

Fig. 2. Info-Detection applied to 6 points on Cartesian plane

For newly added node i'. Let $V' = B \cup \{i'\}$, we can compute γ'_N for the new graph $G(V', E(V'))$ and compare it with γ_N. From Eq. (4) $\gamma'_N \geq \gamma_N$ holds in general. If $\gamma'_N > \gamma_N$, i' is normal observation since it is more integrated with inlier nodes. Otherwise, i' is an anomaly. It can be shown that we do not need to compute γ'_N to determine whether $\gamma'_N > \gamma_N$. We summarize our main result as follows:

Proposition 1

$$\gamma'_N > \gamma_N \iff \sum_{i \in B} w_{ii'} > \gamma_N \tag{5}$$

From Proposition 1, we can see that the computational overhead to check whether new data is normal is linear to the size of existing normal data.

Info-Detection requires γ_N and B, whose computation is not a trivial task and is the main focus of the next Section.

3 Improved Principal Sequence of Partition

3.1 Existing Algorithms

It has been found that the mathematical structure of info-clustering is the same with that of principal sequence of partition (PSP) of graph. Let $\mathcal{P}_1, \ldots, \mathcal{P}_k$ denote PSP sequence where $\mathcal{P}_1 = \{V\}, \mathcal{P}_{i+1} \preceq \mathcal{P}_i$ and $\mathcal{P}_k = \{\{1\}, \ldots, \{k\}\}$. Each $\mathcal{P}_i$ is the solution to the following optimization problem:

$$h_{\mathcal{P}}(\lambda) = f[\mathcal{P}] - |\mathcal{P}|\lambda \tag{6}$$

$$h(\lambda) = \min_{\mathcal{P} \in \Pi'(V)} h_{\mathcal{P}}(\lambda) \tag{7}$$

We call λ^* a critical value for PSP if $\mathcal{P}_i, \mathcal{P}_{i+1}$ are both minimizer for $h(\lambda^*)$ in (7). The largest critical value is equal to γ_N and the largest set in $\mathcal{P}_{k-1}$ is equal to B [3].

Info-clustering is originally implemented by solving Eq. (7) for different critical values [9]. For given critical value, a procedure called Dilworth truncation (DT) can be used to get the minimum value and corresponding partition for (7) [9]. This implementation has $|V|^2\mathrm{MF}(G)$ time complexity where MF represents the time complexity of maximum flow algorithm for a graph G.

An improvement is formulated in [6] using parametric maximum flow. Although this method can achieve $|V|\mathrm{MF}(G)$ time complexity theoretically, in practice it performs not well. Besides, it also increases the space complexity and has the floating point accuracy problem.

3.2   Improved Algorithm

In this section, we give an improvement which achieves $|V|\mathrm{MF}(G)$ time complexity in general, and the space complexity is the same with [6]. Specifically, we notice that different invocation of DT is independent and does not utilize the intrinsic hierarchical structure in the original version of PSP algorithm [9]. If the hierarchical tree structure is used, we can invoke DT on smaller graph in later computation and make the total computation one order of magnitude faster than previous.

To be more specific, suppose we get $P_i = C_1, \ldots, C_t$ for the first invocation of DT. Then we can compute PSP for each $C_i(i = 1, \ldots, t)$ respectively and construct those $P_j(j > i)$ from the subgraph computation. For $P_j(j < i)$, we can contract the graph G to G^t which has t nodes by contracting each C_i to a single node. By applying DT to G^t can we get $P_j(j < i)$. We summarize our improved version in Algorithm 1.

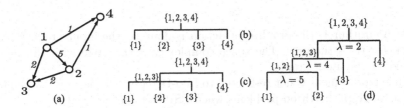

Fig. 3. Improved PSP for graph in (a), hierarchical tree evolves from (b), (c) to (d)

Example 3. We use a simple example to explain Algorithm 1. Consider a graph $G(V, E)$ with $V = \{1, 2, 3, 4\}, E = \{(1, 2), (1, 3), (2, 3), (1, 4), (2, 4)\}$. The weight values are $w_{13} = 2, w_{12} = 5, w_{23} = 2, w_{14} = 1, w_{24} = 1$. The graph is illustrated in Fig. 3(a). Initially, the hierarchical tree is shown in Fig. 3(b). Computing $\gamma' = \frac{11}{4-1}, \tilde{h} = -\frac{16}{3} < -\gamma'$ and $\mathcal{P}' = \{\{1, 2, 3\}, \{4\}\}$ from line 6 in Algorithm 1 we get the hierarchical tree shown in Fig. 3(c).

Then we run PSP for the subgraph $G[\{1, 2, 3\}], \gamma' = \frac{9}{2}, \tilde{h} = -5 < -\gamma'$ and $\mathcal{P} = \{\{1, 2\}, \{3\}\}$ we get the tree shown in Fig. 3(d). Other computation gives the edge weight of $\mathcal{T}$ shown in Fig. 3(d).

To analyze the time complexity of Algorithm 1, we suppose $|V| = n, |E| = O(n^2)$ and highest-relabel preflow algorithm is used to solve the maximum flow problem. Then DT routine has $O(n^4)$ time complexity. We use $T(n)$ to represent the time complexity of Split when the graph has n nodes. Under very general conditions, we can show that $T(n) = O(n^4)$.

Algorithm 1. An Improved Principal Sequence of Partition Algorithm

---

Require: a directed graph $G(V, E)$; edge weight map $c(e)$ for $e \in E$
Ensure: a hierarchical tree $\mathcal{T}(K, E)$ where $K \subseteq 2^V$ is node set and E is edge set
 1: initialize tree $\mathcal{T}$ with V as root node, $\{j\}(j \in V)$ as leaf node and no stem node
 2: $\mathtt{Split}(G, V)$
 3: **function** $\mathtt{Split}(\widetilde{G}, \widetilde{V})$:
 4:     w is the summation of all edge weights of $\widetilde{G}$
 5:     $\gamma' = \frac{w}{|V(\widetilde{G})| - 1}$ where $V(\widetilde{G})$ is the node set of graph $\widetilde{G}$
 6:     $(\tilde{h}, P') = \mathtt{DT}(\widetilde{G}, \gamma')$ where $\mathcal{P}'$ is minimizer of $h(\gamma')$ in Equation (7) and $\tilde{h}$ is the corresponding minimum value
 7:     **if** $\tilde{h} = -\gamma'$ **then**
 8:         add edge weight γ' in $\mathcal{T}$ starting from $\widetilde{V}$ to its children
 9:     **else**
10:         **for** S in P' and $|S| > 1$ **do**
11:             make each children of $\widetilde{V}$ in $\mathcal{T}$ have new parent S
12:             make the parent of S be $\widetilde{V}$
13:             $\mathtt{Split}(\widetilde{G}[S], S)$ where $\widetilde{G}[S]$ is the subgraph of $\widetilde{G}$ restricted on S
14:             contract S to a single node in $\widetilde{G}$
15:         **end for**
16:         $\mathtt{Split}(\widetilde{G}, \widetilde{V})$
17:     **end if**
18: **end function**

In Algorithm 1, the graph contraction is done on the input graph without creating extra structures. The space complexity is the same with the storage needed for the input graph.

Our improvement is not restricted to Info-Detection scenario. It can be used in general graph partition problems when PSP structure is needed.

4 Experimental Results

In this Section, we first illustrate the decision boundary of Info-Detection by two artificial datasets. Then we compare the running time of our implementation of PSP with previous work. Finally we conduct an experiment matrix, in which Info-Detection is compared with other commonly used outlier detection algorithms on both artificial and real-world datasets.

Info-Detection uses the boundary curve to make prediction on new observations and the boundary curve has the following form from Proposition 1:

$$\sum_{j \in B} \exp(-\gamma \|x - x_j\|^2) = \gamma_N \tag{8}$$

The weight is chosen by Gaussian kernel, which produces reasonable good result for these two artificial datasets. As shown by Fig. 4(a) and (b), the boundary curve for Info-Detection is approximately the closure of the set of inliers. We also draw the decision boundary for the elliptic envelope method in Fig. 4(c). Since the latter method assumes the inlier data is distributed in convex region, the decision boundary is an ellipse, which is not the ground truth.

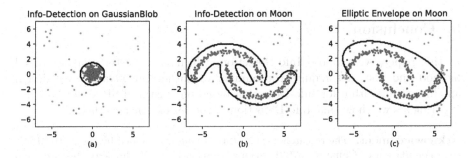

Fig. 4. Detection boundary lines of different methods on different artificial dataset

Next we compare our implementation of PSP with previous methods [6,9]. The comparison is done on Gaussian blobs dataset ($|E| = O(|V|^2)$) and a graph dataset with two hierarchical levels ($|E| = O(|V|^{3/2})$). We use the CPU times to measure the actual speed of each implementation. For each configuration, the experiment is repeated five times and the final running time for each method is averaged over each repeat. The result is shown in Fig. 5(a, b). As can be seen, our implementation is much faster than previous ones.

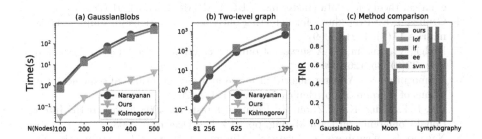

Fig. 5. Experimental results on datasets

We also compare Info-Detection with other commonly used techniques on three datasets: GaussianBlob, Moon and UCI Lymphography. The other methods include local outlier factor [1], isolation forest [7], elliptic envelope [10] and one class SVM [11]. We use two metrics to measure the overall performance of the detection: true positive rate (TPR) and true negative rate (TNR). The inlier is treated as positive sample. TPR measures the percentage of right detection in inlier set while TNR measures that in outlier set. It is difficult for a method to achieve high score on both two metrics. We manage to maximize TNR while controlling TPR $\geq 90\%$ for each method. The hyper parameters and metric are tuned for each method and the best result of TNR is shown in Fig. 5(c). We can see that Info-Detection is competitive with local outlier factor and outperforms other methods on these datasets.

5   Conclusion

In this paper, we propose Info-Detection, which is a cluster analysis based method and does not require the number of outliers in advance. We design an efficient algorithm for Info-Detection based on existing PSP algorithms. By experiments we show that Info-Detection outperforms other methods.

Acknowledgment. The research of Shao-Lun Huang was funded by the Natural Science Foundation of China 61807021, Shenzhen Science and Technology Research and Development Funds (JCYJ20170818094022586), and Innovation and entrepreneurship project for overseas high-level talents of Shenzhen (KQJSCX20180327144037831).

References

1. Breunig, M.M., Kriegel, H.P., Ng, R.T., Sander, J.: LOF: identifying density-based local outliers. SIGMOD Reco. **29**(2), 93–104 (2000)
2. Campello, R.J.G.B., Moulavi, D., Zimek, A., Sander, J.: Hierarchical density estimates for data clustering, visualization, and outlier detection. ACM Trans. Knowl. Discov. Data **10**(1), 5:1–5:51 (2015)
3. Chan, C., Al-Bashabsheh, A., Zhou, Q., Kaced, T., Liu, T.: Info-clustering: a mathematical theory for data clustering. IEEE T-MBMC **2**(1), 64–91 (2016)
4. Grubbs, F.E.: Procedures for detecting outlying observations in samples. Technometrics **11**(1), 1–21 (1969)
5. Hodge, V., Austin, J.: A survey of outlier detection methodologies. Artif. Intell. Rev. **22**(2), 85–126 (2004)
6. Kolmogorov, V.: A faster algorithm for computing the principal sequence of partitions of a graph. Algorithmica **56**(4), 394–412 (2010)
7. Liu, F.T., Ting, K.M., Zhou, Z.: Isolation forest. In: 2008 Eighth IEEE International Conference on Data Mining, pp. 413–422, December 2008
8. Nagano, K., Kawahara, Y., Iwata, S.: Minimum average cost clustering. In: NIPS 23, pp. 1759–1767. Curran Associates, Inc. (2010)
9. Narayanan, H.: The principal lattice of partitions of a submodular function. Linear Algebra Appl. **144**, 179–216 (1991)
10. Rousseeuw, P.J., Driessen, K.V.: A fast algorithm for the minimum covariance determinant estimator. Technometrics **41**(3), 212–223 (1999)
11. Schölkopf, B., Platt, J.C., Shawe-Taylor, J., Smola, A.J., Williamson, R.C.: Estimating the support of a high-dimensional distribution (2001)

Joint Grouping and Labeling
via Complete Graph Decomposition

Jinchao Ge, Zhenhua Wang$^{(\boxtimes)}$, Jiajun Meng, Jianhua Zhang,
and Shengyong Chen

Zhejiang University of Technology,
288 Liuhe Rd., Hangzhou 310023, People's Republic of China
zhhwang@zjut.edu.cn

Abstract. We introduce the complete graph decomposition approach
for joint grouping and labeling. Our framework takes into consideration
both how to group subjects and how to assign labels to them in a joint
manner, without knowing the number of groups beforehand. We model
the relations of different targets via a complete graph, which is decomposed into a set of complete subgraphs to represent distinct groups. We
implement this joint framework by fusing both deep features and rich
contextual cues with model parameters learned from data. We propose
an alternating search algorithm to solve the relevant inference problem
efficiently. We evaluate the effectiveness of the proposed approach on
human activity understanding, and show the proposed approach is competitive compared against the state-of-the-art.

Keywords: Grouping · Graph decomposition · Activity understanding

1 Introduction

Many visual applications require to assign labels to different subjects appeared
within an image, meanwhile grouping them into distinct clusters, which is called
the *joint grouping and labeling* (JGL) problem in this paper. For instance, in the
task of human activity understanding with many people [14], we are required
to recognize both each individual's action class and the interacting configuration among people, that is, which persons belong to the same interacting group.
For the tasks, the number of groups is unknown beforehand. Unfortunately,
such problems cannot be solved (or at least perfectly solved) using reflex-based
models, for example, convolutional neural networks (CNNs) or recurrent neural networks (RNNs). Essentially the reason is that CNNs or RNNs with many
layers requires regular inputs (typically grid structures) for a series of efficient
and effective convolution operations, which at last output regular results with
predesignated structures such as scalars (the classification problem), vectors (the
object detection problem) and grids (the semantic segmentation problem). However, solving JGL tasks typically requires taking as input irregular data (such as
graphs with arbitrary numbers of nodes) and making again irregular predictions.

© Springer Nature Switzerland AG 2019
T. Gedeon et al. (Eds.): ICONIP 2019, CCIS 1143, pp. 497–505, 2019.
https://doi.org/10.1007/978-3-030-36802-9_53

Fig. 1. Human activity understanding with our CGD framework. We first detect human bodies (left-most). Then we build a CGD model, where the action label of each detected target is represented by a node within a complete graph (middle-left). Each edge of the graph encodes the relation (interaction) between associated persons. We solve the JGL task via the CGD inference, which admits both person-wise labeling of actions and the grouping result (middle-right, where each shaded region represents one group, colors encode different action labels, KK means *kick* and OT denotes *others*. Here KK and OT are grouped together because the associated targets are interacting). The results are visualized (right-most) with colors of bounding boxes encoding separate groups.

It is straightforward to handle grouping and labeling tasks separately by training two different CNN models. For the labeling task, one can use a series of well-known deep models like VGGNet [13] and ResNet [3]. In terms of person-wise action recognition, Two-Stream ConvNets [12] and Temporal Segment Networks [18] are popular choices to extract both appearance and motion features of human actions. For grouping, one can implement (1) deep spectral clustering [9] with some predesignated numbers of clusters, or (2) translate the problem into a series of binary classification problems, and determine if any pair of objects belongs to the same group [19]. Unfortunately, the first approach requires a prior estimation of the number of groups, which in general is a challenging task. The second method suffers from the risk of generating conflict results. For instance, when grouping three subjects A, B and C, the results can be A and B belong to the same group, B and C are grouped together as well, but A and C belong to separate groups using, see the top row in Fig. 2 for a few examples.

This work is inspired by the subgraph decomposition method proposed in [15] and the joint graph decomposition and labeling proposed in [10]. In order to track multiple targets across time, work [15] leverages a graph representation where nodes encode body detections and edges represent if two bounding boxes belong to the identical target. Consequently, tracking translates to a minimum-cost-cut of such graphs into separate components, and each component corresponds to one track. Note this method does not require providing the number of cuts as a super-parameter. Instead it produces an appropriate number of cuts automatically by minimizing a trained cost function. Work [10] improves the graph model of [15] by introducing a new variable for each node, and modifies the associated cost function to enable the joint estimation of cuts and node labels. However, this approach still requires to take as input a sparse background graph for each instance to instantiate the cost function, which requires heuristics (*e.g* the distance between any nodes) to figure this out [7].

In this paper, we propose a novel learning framework (see Fig. 1 for an illustration) to label and group multiple targets jointly. Our approach is conceptually analogous to the method proposed in [10]. We observed that constructing sparse background graphs (as that done by [10]) are challenging for the activity understanding, we propose the complete graph decomposition (CGD) approach which takes complete graphs instead of sparse ones as input to instantiate the cost functions. In order to solve the related inference problem efficiently and effectively, we propose an alternating search algorithm. We evaluate our approach using two existing datasets. Experimental results demonstrate that our approach outperforms the state-of-the-arts.

2   Complete Graph Decomposition

Taking as input a set of images of n targets $X = \{\mathbf{x}_1, \ldots, \mathbf{x}_n\}$, our first aim for joint grouping and labeling is to decompose the set into $x \in \{1, \ldots, n\}$ mutually exclusive subsets $G = \{C_s\}_{s=1}^x$, where $C_s \subseteq X, \bigcup_s(C_s) = X$, and $C_s \bigcap C_t = \emptyset \ \forall s \neq t$ and $s, t \in \{1, \ldots, x\}$. Note x is unknown beforehand. The second aim is to predict the labels for all targets, which is represented by $\mathbf{y} = (y_1, \ldots, y_n)$, where $y_i \in \mathcal{Y} \ \forall i$.

In order to perform grouping, it is natural to consider the pairwise relation between each pair of targets. To this end, we model these pairwise relations via an undirected complete graph $\mathcal{G} = (N, E)$ with the node set $N = \{1, \ldots, n\}$ and the edge set $E = \{\{j, k\} | \{j, k\} \in N^2, j < k\}$. Then the grouping task under such graph representation is to decompose $\mathcal{G}$ into x subgraphs, each is a smaller complete graph. To model the JGL task, we propose to optimize the following energy function (parameterized by $\mathbf{w}$):

$$J_{\mathbf{w}}(G, \mathbf{y}; \mathcal{G}, X) = \frac{1}{n} \sum_{i=1}^n \langle \phi_i(y_i), \mathbf{w}_l \rangle +$$

$$\frac{1}{r} \sum_{\{j,k\} \in E} \{\langle \psi_{j,k}^1, \mathbf{w}_g^1 \rangle z_{j,k} + \langle \psi_{j,k}^0, \mathbf{w}_g^0 \rangle (1 - z_{j,k}) + \langle \eta_{j,k}(y_j, y_k), \mathbf{w}_c \rangle (1 - z_{j,k})\}, \quad (1)$$

where the binary variable $z_{j,k} \in \{0, 1\}$ is induced by the grouping result G: if the targets j and k belong to the identical group, $z_{j,k} = 0$, otherwise $z_{j,k} = 1$; $\langle \mathbf{a}, \mathbf{b} \rangle$ denotes the inner product of vectors $\mathbf{a}$ and $\mathbf{b}$; $\mathbf{w} = [\mathbf{w}_l, \mathbf{w}_g^0, \mathbf{w}_g^1, \mathbf{w}_c]$ are the model parameters to be learned from data using structured SVM [16], and $r = \frac{n(n-1)}{2}$. Functions ϕ, ψ and η output the so-called joint features in structured prediction [16]. Note the second sum in Eq. (1) enumerates over all edges of E. Hence we call the JGL task using our formulation 1 the *complete graph decomposition*, and call the complete graph $\mathcal{G}$ the *background graph*. To ease presentation, we define

$$\theta_i(y_i) := \langle \phi_i(y_i), \mathbf{w}_l \rangle, \tag{2}$$
$$\theta_{j,k}^0 := \langle \psi_{j,k}^0, \mathbf{w}_g^0 \rangle, \tag{3}$$
$$\theta_{j,k}^1 := \langle \psi_{j,k}^1, \mathbf{w}_g^1 \rangle, \tag{4}$$
$$\theta_{j,k}(y_j, y_k) := \langle \eta_{j,k}(y_j, y_k), \mathbf{w}_c \rangle. \tag{5}$$

Let $\mathbf{z} = (z_{s,t})_{s,t \in V, s < t}$. Then minimizing the cost function (1) can be translated into an equivalent form:

$$\text{argmin}_{\mathbf{z},\mathbf{y}} \frac{1}{n} \sum_i \theta_i(y_i) + \frac{1}{r} \sum_{j,k} \left\{ \theta^1_{j,k} z_{j,k} + \left(\theta_{j,k}(y_j, y_k) + \theta^0_{j,k} \right)[1 - z_{j,k}] \right\}, \quad (6a)$$

$$\text{s.t. } z_{u,v} \leq \sum_{\{a,b\} \in \mathcal{O} \setminus \{\{u,v\}\}} z_{a,b}, \ \forall \{u,v\} \in \mathcal{O}, \forall \mathcal{O} \in \text{cycles}(\mathcal{G}). \quad (6b)$$

We call the constraints within (6b) the *cycle constraints* (denoted by $\mathcal{O}$). Unfortunately, this optimization problem is NP-complete. In order to solve it efficiently and approximately, we present an alternating search algorithm in Sect. 3.

2.1 CGD for Activity Understanding with Multiple Targets

In this task we define the joint features as

$$\phi_i(y_i) = \left[p^a_i(y_i), p^m_i(y_i), \bar{p}_i(y_i), 1 \right], \quad (7)$$

$$\psi_{j,k} = \left[p^a_{j,k}, p^m_{j,k}, \bar{p}_{j,k}, \boldsymbol{\kappa}^1_{j,k}, \boldsymbol{\kappa}^2_{j,k}, \boldsymbol{\kappa}^e_{j,k} \right], \quad (8)$$

$$\eta_{j,k}(y_j, y_k) = \left[\mathbb{1}(y_j, y_k), 1 - \mathbb{1}(y_j, y_k) \right]. \quad (9)$$

Above $p^a_i(y_i)$, $p^m_i(y_i)$ represent the soft-max probabilities of assigning an action label y_i to person i, which are calculated by appearance and motion CNNs [12] taking as input the image patch occupied by the i-th person. $\bar{p}_i(y_i)$ denotes the mean of $p^a_i(y_i)$ and $p^m_i(y_i)$. Likewise, $p^a_{j,k}$ (or $p^m_{j,k}$) denotes the soft-max probabilities that person j and k belong to distinct group according to the local appearance (or motion) information. $\bar{p}_{j,k}$ denotes the mean of $p^a_{j,k}$ and $p^m_{j,k}$. $\mathbb{1}(y_j, y_k)$ outputs 1 if the associated labels are compatible, and outputs 0 otherwise. $\boldsymbol{\kappa}^1, \boldsymbol{\kappa}^2, \boldsymbol{\kappa}^e$ are defined by

$$\boldsymbol{\kappa}^1_{j,k} = [v_{j,k}, h_{j,k}, d_{j,k}, s_{j,k}], \quad (10)$$

$$\boldsymbol{\kappa}^2_{j,k} = [v^2_{j,k}, h^2_{j,k}, d^2_{j,k}, s^2_{j,k}], \quad (11)$$

$$\boldsymbol{\kappa}^e_{j,k} = [e^{-v_{j,k}}, e^{-h_{j,k}}, e^{-d_{j,k}}, e^{-s_{j,k}}], \quad (12)$$

where $v_{j,k}, h_{j,k}, d_{j,k}, s_{j,k}$ are calculated by

$$v_{j,k} = \frac{|v_j - v_k|}{\max(v_j, v_k)}, \quad h_{j,k} = \frac{|h_j - h_k|}{\max(h_j, h_k)},$$

$$d_{j,k} = \frac{\|\mathbf{c}_j - \mathbf{c}_k\|_2}{d_{\max}}, \quad s_{j,k} = \frac{|s_j - s_k|}{\max(s_j, s_k)}. \quad (13)$$

Here $v, h, s, \mathbf{c}$ represent the width, height, area and center coordinate of a bounding box respectively, $d_{\max}$ is the largest Euclidean distance between the centers of all bounding boxes.

In order to train appearance and motion CNNs which output $p^a_{j,k}$ and $p^m_{j,k}$, we need to prepare inputs for CNNs. First human bodies are detected. Next, for each pair of detected human bodies, we create a black background image which just exactly encloses the two detected bounding boxes within the image. At last patches cropped from human body areas are embedded into the background image to generate final inputs.

3 Inference with Alternating Search

Now we present our approach to solve the optimization problem (6). Fixing the grouping variable unchanged as $\mathbf{z}^*$, the optimization (6) reduces to

$$\operatorname{argmin}_{\mathbf{y}} \frac{1}{n} \sum_i \theta_i(y_i) + \frac{1}{r} \sum_{j,k,z_{j,k}^*=0} \theta_{j,k}(y_j, y_k). \tag{14}$$

Problem (14) is well-known as the discrete energy minimization [5]. Here we use loopy belief propagation to solve it to find an approximate solution.

Fixing $\mathbf{y}$ as $\mathbf{y}^*$, the inference problem (6) reduces to

$$\operatorname{argmin}_{\mathbf{z}} \sum_{j,k} \underbrace{\frac{1}{r} \left\{ \theta_{j,k}^1 - \theta_{j,k}^0 - \theta_{j,k}(y_j^*, y_k^*) \right\}}_{c_{j,k}} z_{j,k}, \tag{15a}$$

$$\text{s.t. the constraints in Equation (6b).} \tag{15b}$$

The optimization (15) is known as the weighted graph partitioning problem [2]. In order to solve it efficiently, we adapt the widely used Kernighan-Lin algorithm proposed for the problem of graph bisection [6] to our CGD problem with unknown arbitrary number of clusters. Specifically, we start from an arbitrary partition $G^0 = \{C_i^0\}_{i=1}^x$, which is feasible to (15), and iteratively improve the solution according to the differences between inner and outer costs introduced by [6]. We maintain for each node of the background graph the difference between its inner and outer costs, and maintain a flag for each partition which records if the partition has been updated (flag = *True*) or not (flag = *False*). At beginning all flags are *True*. During each iteration, for each pair of non-empty partitions which take *True* flags, we try to update the two partitions with a series of merging or moving operations (which one to choose depends on how much they decrease the objective in (15)):

- *Merging operation:* merge two partitions to generate a larger united partition.
- *Moving operation:* move a node from one partition to another.

Then we update the differences between inner and outer costs of the related nodes. Afterwards, for each partition we try to split it to generate a series of smaller partitions till the partition is not separable (*i.e.* no gain can be made by any separation). At the end of each iteration, we update all flags to mark any partition that has been modified. In addition, we calculate the total decrease of the objective earned by all updates and splits, and early stop the running if no decrease can be obtained. Finally, to approximately solve the challenging problem (6), we alternatingly search over $\mathbf{y}$ and $\mathbf{z}$ directions, which is achieved by solving (14) and (15) in turn. This local search routine guarantees to improve the solution of (6) progressively, and terminates if no progress is attained.

Table 1. Human activity understanding results on UT using CNNs and CGD. Here average precision (denoted by P), recall (denoted by R) and F1-score are presented for action labeling (P-**y**, R-**y**, F1-**y**) and grouping (P-**z**, R-**z**, F1-**z**) tasks respectively.

Method	P-**y**	R-**y**	F1-**y**	P-**z**	R-**z**	F1-**z**
Appearance net [3] (baseline)	0.516	0.493	0.504	0.879	0.836	0.857
Motion net [3] (baseline)	0.571	0.386	0.461	0.871	0.729	0.794
Two-stream net [12] (baseline)	0.641	0.548	0.591	0.889	0.801	0.843
KLj-r [10] (state-of-the-art)	0.645	**0.603**	0.624	**0.892**	0.838	0.864
CGD (ours)	**0.663**	0.600	**0.630**	0.888	**0.858**	**0.873**

Table 2. Human activity understanding results on BIT using CNNs and CGD.

Method	P-**y**	R-**y**	F1-**y**	P-**z**	R-**z**	F1-**z**
Appearance net [3] (baseline)	0.464	0.506	0.484	0.924	0.802	0.859
Motion net [3] (baseline)	0.383	0.492	0.431	0.881	0.810	0.844
Two-stream net [12] (baseline)	0.486	**0.603**	0.538	**0.927**	0.835	0.879
KLj-r [10] (state-of-the-art)	0.550	0.544	0.547	0.884	**0.855**	0.869
CGD (ours)	**0.569**	0.570	**0.569**	0.913	**0.855**	**0.883**

Fig. 2. Visualization of action labeling and people grouping. Each column represents one testing instance. Top row shows the KLj-r [10] (the state-of-the-art) results, and bottom row displays CGD results. Any persons belonging to the same group take the same color, and are connected by an edge. Correct action predictions are marked by white texts with dark green backgrounds, and incorrect ones are highlighted by red backgrounds. Note these examples contain long-range human interactions. While our approach is able to predict correct grouping and labeling results, KLj-r fails to make correct grouping predictions for all demonstrated instances.(Color figure online)

4   Experiment and Result

We evaluate our approach using two datasets including UT [11] and BIT [8]. UT is a benchmark for human interaction recognition. It includes 7 action classes including *handshake (HS)*, *hug (HG)*, *kick (KK)*, *point (PT)*, *punch (PC)*, *push (PS)* and *others (OT)*. As UT only provides frame-wise annotation of human actions, we annotate the action label and the bounding box for each person and each frame using VATIC [17]. BIT covers 9 action classes including *bend (BD)*, *box (BX)*, *handshake (HS)*, *highfive (HF)*, *hug (HG)*, *kick (KK)*, *pat (PT)*, *push (PS) and others (OT)*, where each class contains 50 videos.

Implementation Details. For UT, we choose the first 10 videos as training data and the remaining 10 videos as testing. While for BIT, we randomly select 50% videos of each class as training and the rest for testing. For testing, we detect human bodies for each frame using YOLO [4]. Since these datasets are quite small (compared with dataset like ImageNet), we augment training examples via rotating, cropping and flipping transformations. Then we sample training examples separately for each action-class to reduce the imbalance on numbers of training instances of different classes. To train the action classifier, we implemented the two-steam net [12] (a popular deep net for human action recognition) in PyTorch using ResNet-101 [3] as the architecture of appearance and motion CNNs, where appearance-CNN takes a $224 \times 224 \times 3$ RGB image as input and motion-CNN takes a $224 \times 224 \times 2$ flow data as input. We extract optical flow using FlowNet2.0 [1]. We pretrain both CNNs on UCF 101 from scratch, then finetune their weights on UT and BIT using stochastic gradient descent (SGD) with momentum (0.9). The initial learning rate is 0.001 decreased by a factor of 0.1 every 7 epochs, and the training stops in 50 epochs. Then, we train our CGD model using mini-batch gradient descent.

Results and Discussion. We compare our approach against appearance net [3], motion net [3] and two-stream net [12] which combines appearance and motion nets. We implement work [10] by ourselves and use it as the state-of-the-art (KLj-r). We provide results on action labeling and grouping separately using precision, recall and F1-score protocols, see Tables 1 and 2. Our method performs significantly better than the three popular nets on the human activity understanding tasks, and this gain indeed comes from the joint learning of action recognition and human grouping by CGD. Overall our approach performs moderately better than the state-of-the-art, mainly because our approach is able to cover long-range human interactions. Figure 2 provides grouping and labeling visualizations using a few examples. We can see that CGD gives correct labeling and coherent grouping for all examples, while KLj-r [10] fails to predict correct grouping results for all the listed examples, especially when the images contain long-range interactions.

5   Conclusion

We have presented a complete graph decomposition (CGD) approach which enables the joint learning of grouping and labeling with multiple targets. We demonstrated how to specify the cost functions of CGD for human activity understanding with multiple people. We also presented an efficient inference algorithm for CGD, which searches for grouping and labeling solutions in turn. Experimental results demonstrated that CGD outperforms the state-of-the-arts on both evaluated tasks. Though we have taken human activity understanding as instances, our CGD is applicable to other tasks which require to estimate labels and groups for many targets jointly and simultaneously, for example scene understanding, pose estimation and person re-identification.

Acknowledgement. This work is partially supported by National Natural Science Foundation of China (61802348).

References

1. Eddy, I., Nikolaus, M., Tonmoy, S., Margret, K., Alexey, D., Thomas, B.: FlowNet2.0: evolution of optical flow estimation with deep networks. In: IEEE Conference on Computer Vision and Pattern Recognition (2017)
2. Grötschel, M., Wakabayashi, Y.: A cutting plane algorithm for a clustering problem. Math. Program. **45**(1–3), 59–96 (1989)
3. He, K., Zhang, X., Ren, S., Sun, J.: Deep residual learning for image recognition. In: IEEE Conference on Computer Vision and Pattern Recognition (2016)
4. Joseph, R., Ali, F.: YOLO9000: better, faster, stronger. In: IEEE Conference on Computer Vision and Pattern Recognition (2017)
5. Kappes, J., et al.: A comparative study of modern inference techniques for discrete energy minimization problems. In: IEEE Conference on Computer Vision and Pattern Recognition (2013)
6. Kernighan, B.W., Lin, S.: An efficient heuristic procedure for partitioning graphs. Bell Syst. Tech. J. **49**(2), 291–307 (1970)
7. Keuper, M., Levinkov, E., Bonneel, N., Lavoué, G., Brox, T., Andres, B.: Efficient decomposition of image and mesh graphs by lifted multicuts. In: IEEE International Conference on Computer Vision (2015)
8. Kong, Y., Jia, Y., Fu, Y.: Learning human interaction by interactive phrases. In: Fitzgibbon, A., Lazebnik, S., Perona, P., Sato, Y., Schmid, C. (eds.) ECCV 2012. LNCS, vol. 7572, pp. 300–313. Springer, Heidelberg (2012). https://doi.org/10.1007/978-3-642-33718-5_22
9. Law, M., Urtasun, R., Zemel, R.: Deep spectral clustering learning. In: International Conference on Machine Learning (2017)
10. Levinkov, E., Tang, S., Insafutdinov, E., Andres, B.: Joint graph decomposition and node labeling by local search. In: IEEE Conference on Computer Vision and Pattern Recognition (2017)
11. Ryoo, M., Aggarwal, J.: UT-interaction dataset, ICPR contest on semantic description of human activities (SDHA). In: IEEE International Conference on Pattern Recognition Workshops (2010)
12. Simonyan, K., Zisserman, A.: Two-stream convolutional networks for action recognition in videos. In: Advances in Neural Information Processing Systems (2014)
13. Simonyan, K., Zisserman, A.: Very deep convolutional networks for large-scale image recognition. arXiv preprint arXiv:1409.1556 (2014)
14. Stergiou, A., Poppe, R.: Understanding human-human interactions: a survey. arXiv preprint arXiv:1808.00022 (2018)
15. Tang, S., Andres, B., Andriluka, M., Schiele, B.: Subgraph decomposition for multi-target tracking. In: IEEE Conference on Computer Vision and Pattern Recognition (2015)
16. Tsochantaridis, I., Joachims, T., Hofmann, T., Altun, Y.: Large margin methods for structured and interdependent output variables. J. Mach. Learn. Res. **6**(2), 1453–1484 (2006)

17. Vondrick, C., Patterson, D., Ramanan, D.: Efficiently scaling up crowdsourced video annotation. Int. J. Comput. Vis. **101**(1), 184–204 (2013)
18. Wang, L., et al.: Temporal segment networks: towards good practices for deep action recognition. In: Leibe, B., Matas, J., Sebe, N., Welling, M. (eds.) ECCV 2016. LNCS, vol. 9912, pp. 20–36. Springer, Cham (2016). https://doi.org/10.1007/978-3-319-46484-8_2
19. Wang, Z., Liu, S., Zhang, J., Chen, S., Guan, Q.: A spatio-temporal CRF for human interaction understanding. IEEE Trans. Circ. Syst. Video Technol. **27**(8), 1647–1660 (2017)

Link Prediction with Attention-Based Semantic Influence of Multiple Neighbors

Meixian Song[1], Bo Wang[1(✉)], Xindian Ma[1], Qinghua Hu[1], Xin Wang[1],
Yuexian Hou[1], and Dawei Song[2]

[1] College of Intelligence and Computing, Tianjin University, Tianjin, China
{songmeixian,bo_wang,xindianma,huqinghua,tifa_wx,yxhou}@tju.edu.cn
[2] School of Computer Science and Technology, Beijing Institute of Technology,
Beijing, China
dwsong@bit.edu.cn

Abstract. The establishment of social links is not only determined by personal interests but also by neighbors' influences, which may vary across different neighbors. However, the independent influence of each neighbor has not been separately considered on semantic level in current approaches. In this work, we predict missing social links by modeling semantic influence of each neighbor separately with an embedding approach. The semantic of influence is fine grained on each neighbor's specific interest with attention-based method. The proposed model named AIMN (**A**ttention-based semantic **I**nfluence of **M**ultiple **N**eighbors) is integrated with structure information with a uniform framework. Extensive experiments on different real-world networks demonstrate that AIMN outperforms state-of-the-art methods.

Keywords: Social networks · Link prediction · Network embedding · Semantic influence · Attention

1 Introduction

Link prediction aims to either infer the links occuring in the near future or reconstruct existing links missing in the current snapshot of a social network. In this paper, we target at the latter one. In recent studies, structural information of network and semantic attributes of vertexes are often jointly modeled [2]. Network embedding is widely used in recent studies, e.g., DeepWalk [9] performs random walks over networks. Node2vec [3] modifies the random walk in DeepWalk into biased random walks to explore the network structure. SDNE [14] captures the nonlinear network structure with auto-encoder based vertex embedding. Yang et al. [15] present TADW to improve matrix factorization. Tu et al. [12] propose context-aware embeddings toward different neighbors. With the embedding of vertexes, current studies calculate the similarity between two vertexes' embedding vectors to predict the probability of the social link between them. The idea is supported by the theory of homogeneity in sociology, which proposes two

© Springer Nature Switzerland AG 2019
T. Gedeon et al. (Eds.): ICONIP 2019, CCIS 1143, pp. 506–514, 2019.
https://doi.org/10.1007/978-3-030-36802-9_54

principles: selection and influence. Selection principle explains the similarity in social links by supposing that people befriend others who are similar to them. Influence principle supposes that people become more similar to their friends over time. In this way, the influence of one's neighbors will affect one's selection of new friends. As the example in Fig. 1, special interests of Bob and Alice can be enhanced by the neighbor influence. The enhancement makes different interests have different importance in predicting the link between Bob and Alice, e.g., 'Film' is more important than 'Music' and 'Travel'. Recent studies [4] notice the importance of influence in the task of link prediction. However, this method does not take into account the semantic influence of neighbors.

Fig. 1. Example of attention-based semantic influence of multiple neighbors. (Red fonts represent the interests enhanced (shared) by the neighbors). (Color figure online)

In this work, we introduce neighbor influence into link prediction. Each user is embedded into a single vector integrating attention-based semantic influence of each neighbor. Our main contributions are: (1) With observed neighbor relationships and textual attributes of users, we train a single joint embedding vector for each user with pair-wise semantic influence of each neighbor and structural information. (2) When modeling semantic influence, specific semantics in text are enhanced based on the attention module, which is able to select more informative part of text and denoise those noisy during training. (3) In experiments, we reveal the advantage of proposed method over state-of-the-art methods.

2   Problem Statement

Formally, let $G = (N, E, S)$ denote a social network, where $N = \{u_1, u_2, \ldots, u_n\}$ is vertexes set, E is edges set. S is textual interests set, where S_i represents the interest of vertex u_i. $V = [v_1, v_2, \ldots, v_n]$ is formed by all vertexes embedding vectors, where v_i is the embedding vector of u_i. Our research question is:

Given a graph $G = (N, E, S)$, we aim to predict the probability of existing an edge between vertexes u_i and u_j, based on the network structure and textual interest of u_i, u_j and their neighbors.

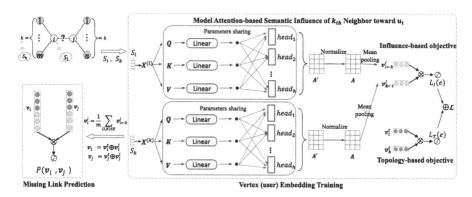

Fig. 2. The overall framework of AIMN. (The left portion predicts missing link by measuring the similarity between the connected vectors of two vertexes. The right portion train the embedding of vertexes with influence-based and structure-based vectors.)

3   AIMN Model

3.1   Overall Framework

Link Prediction. We propose a novel model named AIMN (i.e. Attention-based Semantic Influence of Multiple Neighbors). The probability of edge e_{ij} is measured by the similarity between the embedding vectors of u_i and u_j (as depicted in the left part of Fig. 2). In particular, the similarity between u_i and u_j is quantified as the proximity of two embedding vectors in euclidean space as follows:

$$P\left(\boldsymbol{v}_i, \boldsymbol{v}_j\right) = \frac{1}{1 + \exp\left(-\boldsymbol{v}_i^T \boldsymbol{v}_j\right)} \tag{1}$$

where $\boldsymbol{v}_i, \boldsymbol{v}_j \in \mathbb{R}^D$ is the embedding vector of vertex u_i and u_j, respectively. D is the dimension of the embedding vectors. Embedding vector $\boldsymbol{v}_i$ of each vertex is obtained by the concat operator $\oplus$. $\boldsymbol{v}_i = \boldsymbol{v}_i^t \oplus \boldsymbol{v}_i^I$, where $\boldsymbol{v}_i^t$ is the structure-based embedding and $\boldsymbol{v}_i^I$ is influence-based embedding.

Embedding Training. Overall, to train the embedding of each vertex in the network, we maximize the sum of probabilities of all known edges as follows:

$$\max \quad \mathcal{L} = \sum_{e \in E} L(e) \tag{2}$$

where $L(e)$ is the objective of edge e, and is a trade-off of structure-based objective ($L_T(e)$) and influence-based objective ($L_I(e)$), as shown in Eq. 3:

$$L(e) = \alpha L_T(e) + (1 - \alpha)L_I(e) \tag{3}$$

where $L_T(e_{ij}) = w_{ij} \log P(\boldsymbol{v}_i^t, \boldsymbol{v}_j^t)$ and $L_I(e_{ij}) = w_{ij} \log P(\boldsymbol{v}_i^I, \boldsymbol{v}_j^I)$, and w_{ij} is a universal weight for e_{ij}, which represents strength or polarity of relationship.

In the case of no weight, w_{ij} is 1. The structure-based embedding is regarded as parameters and is mapped into the same space with influence-based embedding. Moreover, the influence-based embedding of a vertex is modeled with the semantic of textual interests of each neighbor. Next, we introduce the detail of training influence-based embedding vectors.

3.2   Model the Influence

Given a vertex u_i, we model the influence of neighbors of u_i with a influence embedding vector v_i^I. AIMN separately models the pair-wise influence of each neighbor toward u_i, and this neighbor's interests obtained with associated textual information. (as depicted in the right part of Fig. 2) We generate the influence-based embedding v_i^I of vertex u_i by averaging all neighbors' influence-based embeddings toward u_i as follows:

$$v_i^I = \frac{1}{m} \sum_{(i,k) \in E} v_{i \leftarrow k}^I \qquad (4)$$

where m is the number of neighbors of u_i. $v_{i \leftarrow k}^I$ is the embedding vector used to model the influence of neighbor u_k toward u_i.

We assume that the influence of a neighbor can be caused by semantic of this neighbor's interests. Next, we explain how to build pair-wise influence embedding vectors $v_{i \leftarrow k}^I$ and $v_{k \leftarrow i}^I$ when given a edge e_{ij}, where vector $v_{i \leftarrow k}^I$ models the semantic influence of u_k toward u_i. We obtain $v_{i \leftarrow k}^I$ and $v_{k \leftarrow i}^I$ through four layers, i.e., embedding, multi-head attention, normalization, and output layer.

Embedding Layer. Given a word sequence $S_i = (w_1, w_2, ..., w_n)$, embedding layer transforms each word $w_i \in S_i$ into its corresponding word embedding $x_i \in \mathbb{R}^{d'}$, where x_i is word vector of the ith word in its interest information. We obtain embedding matrix as $X = [x_1, x_2, \ldots, x_n]$.

Multi-head Attention Layer. Attention Layer couples the vertexes' context vectors and produces a set of feature vectors for each word in the context. It aims to be aware of the vertex pair in an edge, and the text from a vertex can directly affect the embedding of the other vertex. Transformer [13] processes query, keys and values as matrices Q, K and V respectively. We initialize Q, K and V with embedding matrix X. The attention function can be written as follows:

$$Attention(Q, K, V) = \text{softmax}\left(\frac{QK^T}{\sqrt{d_k}}\right) V \qquad (5)$$

where d_k is the number of columns of Q and K. Based on the definition of Attention, the multi-head attention is defined as follows:

$$MultiHeadAttention(Q, K, V) = Concat\left(head_1, \ldots, head_h\right) W^O$$
$$where \quad head_i = Attention(QW_i^Q, KW_i^K, VW_i^V) \qquad (6)$$

where matrices $\boldsymbol{W}_i^Q$, $\boldsymbol{W}_i^K$ and $\boldsymbol{W}_i^V \in \mathbb{R}^{(d_{model}) \times d_k}$, and $\boldsymbol{W}^O \in \mathbb{R}^{(h \cdot d_k) \times d}$. We get the output of $MultiHeadAttention(\boldsymbol{Q}, \boldsymbol{K}, \boldsymbol{V})$ as $\boldsymbol{A}'$, where $\boldsymbol{A}' \in \mathbb{R}^{n \times d}$.

Normalization Layer. To reduce training time, we compute mean and variance which are then used to normalize $\boldsymbol{A}'$ on each training case [1].

$$\boldsymbol{\mu}_i = \frac{1}{H} \sum_{j=1}^{H} \boldsymbol{A}'_{ij}$$

$$\boldsymbol{\sigma}_i = \sqrt{\frac{1}{H} \sum_{j=1}^{H} (\boldsymbol{A}'_{ij} - \boldsymbol{\mu}_i)^2} \tag{7}$$

$$\boldsymbol{A}_{ij} = f\left(\frac{1}{\sigma_i}\left(\boldsymbol{A}'_{ij} - \boldsymbol{\mu}_i\right) + \boldsymbol{b}_i\right)$$

where H is the number of column for the matrix $\boldsymbol{A}'$, $\boldsymbol{b}_i$ is bias, and $\boldsymbol{A} \in \mathbb{R}^{n \times d}$ is the matrix after layer normalized.

Output Layer. At last, we conduct mean-pooling operations on $\boldsymbol{A}$ to generate the influence embedding vector of u_k toward u_i.

$$\boldsymbol{v}_{i \leftarrow k}^I = \frac{1}{n} \sum_{i=1}^{n} \boldsymbol{A}_{ij} \tag{8}$$

In the same way, we get the influence embedding of vertex u_i toward u_k as $\boldsymbol{v}_{k \leftarrow i}^I$.

3.3   Optimization of Model

Globally, AIMN maximize conditional probabilities on each known edge (u_i, u_k) with negative sampling, which specifies the following object function for each (u_i, u_k):

$$\log \sigma\left(\boldsymbol{v}_k^T \boldsymbol{v}_i\right) + \sum_{i=1}^{K} E_{v_n \sim P_n(v)} \left[\log \sigma\left(-\boldsymbol{v}_n^T \boldsymbol{v}_i\right)\right] \tag{9}$$

where K is the number of negative samples, $P(v) \propto d_v^{3/4}$ denotes the distribution of vertexes, d_v is the out-degree of vertex v, and σ represents the sigmoid function.

4 Experiments

4.1 Experimental Settings

Datasets. Four real-world social networks are studied. Cora [7] and HepTh [6] are citation network. We filter out papers without abstract. In Twitter [16], we select users who have tweets no less than 50. Coauthorship [11] is co-authorship network. We filter out authors without research interests. The details are shown in Table 1.

Table 1. Statistics of experimental datasets.

Dataset	Cora	HepTh	Twitter	Coauthorship
# of vertexes	2277	1038	5872	6739
# of links	5314	1990	36252	32051
Textual attributes	Abstract	Abstract	Tweets	Key words of research

Table 2. Performance evaluation on Cora.

Models	Evaluation							
	AUC				F1			
	20%	40%	60%	80%	20%	40%	60%	80%
DeepWalk	0.6522	0.8027	0.8734	0.8898	0.6028	0.7089	0.7714	0.8063
Node2vec	0.6596	0.8079	0.8684	0.8947	0.5817	0.6912	0.7338	0.7722
SDNE	0.6630	0.6799	0.7203	0.7442	0.5852	0.6285	0.6176	0.5666
TADW	0.8620	0.8979	0.9064	0.9198	0.7883	0.8190	0.8300	0.8559
CANE	0.8640	0.9130	0.9389	0.9511	0.7838	0.8109	0.8369	0.8582
AIMN	**0.8976**	**0.9201**	**0.9460**	**0.9597**	**0.8272**	**0.8420**	**0.8722**	**0.8913**

Table 3. Performance evaluation on HepTh.

Models	Evaluation							
	AUC				F1			
	20%	40%	60%	80%	20%	40%	60%	80%
DeepWalk	0.6799	0.8399	0.9130	0.9359	0.6363	0.7698	0.8361	0.8748
Node2vec	0.6976	0.8410	0.9212	0.9469	0.6347	0.7606	0.8222	0.8549
SDNE	0.6414	0.7458	0.8235	0.8608	0.5794	0.6435	0.7100	0.7163
TADW	0.8716	0.9057	0.9294	0.9308	0.8084	0.8773	0.8856	0.9050
CANE	0.8920	0.9196	0.9141	0.9520	0.8180	0.8538	0.8712	0.8958
AIMN	**0.9054**	**0.9310**	**0.9417**	**0.9717**	**0.8309**	**0.8899**	**0.8932**	**0.9187**

Methods for Comparison. We compare AIMN with following state-of-the-art base-lines. (1) Structure-based methods: DeepWalk [9], Node2vec [3], and SDNE [14]. (2) Hybrid methods (structure+semantics): TADW [15] and CANE [12].

Hyper-Parameters Settings. In the experiments, the α in Eq. 3 is 0.5. All the weight matrices are initialized by sampling from the uniform distribution $U(-0.1; 0.1)$, and all the biases are set to 0. Embedding dimension D is 200 for all methods. The h in Eq. 6 is set to 20. The number of negative samples K in Eq. 9 is set to 1. We use Adam [5] to optimize the transformed objective, and the learning rate starts with 0.0001. English word embedding are trained by word2vec [8], whose dimension is 300. Chinese word embedding are trained on Tencent AI lab embedding corpus [10], whose dimension is 200. The OOV words are randomly initialized by a uniform distribution of $(-0.25, 0.25)$.

Evaluation Methodology. We evaluate the experiments from two aspects, i.e., pair-wise accuracy and global-wise accuracy. First, we adopt AUC to test pair-wise accuracy. Each test instance is a tuple with a vertex, his/her true friend and a random negative vertex. The average accuracy is the final pair-wise accuracy. Second, the global-wise accuracy evaluates the link prediction problem as a binary classification problem. Therefore we adopt F1 score to measure the global-wise accuracy. We randomly select different subsets of tuples from dataset into training network, i.e., 20%, 40%, 60%, 80%. Then we evaluate the link prediction performance on the testing set.

4.2 AIMN Performance and Discussion

We compare AIMN with baseline methods and discuss experimental results. The results of AUC and F1 are shown in Tables 2, 3, 4 and 5.

Table 4. Performance evaluation on Twitter.

Models	Evaluation							
	AUC				F1			
	20%	40%	60%	80%	20%	40%	60%	80%
DeepWalk	0.7051	0.7739	0.8070	0.8320	0.6309	0.7099	0.7429	0.7754
Node2vec	0.7103	0.7697	0.8081	0.8376	0.6036	0.7041	0.7384	0.7698
SDNE	0.6599	0.6750	0.6880	0.7416	0.6061	0.6133	0.6225	0.5723
TADW	0.7498	0.7853	0.8104	0.8235	0.6925	0.7227	0.7397	0.7529
CANE	0.7676	0.8046	0.8252	0.8316	0.7188	0.7452	0.7565	0.7592
AIMN	**0.8480**	**0.8648**	**0.8731**	**0.8805**	**0.7698**	**0.7882**	**0.7932**	**0.8062**

Table 5. Performance evaluation on Coauthorship.

Models	Evaluation							
	AUC				F1			
	20%	40%	60%	80%	20%	40%	60%	80%
DeepWalk	0.8079	0.8925	0.9211	0.9443	0.6817	0.7806	0.8219	0.8577
Node2vec	0.7922	0.8817	0.9151	0.9403	0.6495	0.7521	0.7929	0.8232
SDNE	0.6134	0.6631	0.6727	0.6925	0.5646	0.5857	0.5748	0.5919
TADW	0.8816	0.9060	0.9191	0.9249	0.8176	0.8404	0.8561	0.8590
CANE	0.8669	0.9202	0.9346	0.9407	0.8314	0.8534	0.8718	0.8693
AIMN	**0.8883**	**0.9361**	**0.9489**	**0.9647**	**0.8340**	**0.8613**	**0.8865**	**0.9125**

Pair-Wise Accuracy. From these results of pair-wise accuracy in terms of AUC, we have following observations: (1) In most cases, hybrid baseline methods (i.e., TADW and CANE) outperform the structure-based baseline methods. However, AIMN further exceeds the latest hybrid baseline methods, indicating the general advantage of AIMN. (2) Another advantage of AIMN is working well with smaller training set. e.g., in Cora and Twitter dataset, with only 20% training set, AIMN has higher accuracy than most of the baselines with 60% training set.

Global-Wise Accuracy. We investigate the results of global-wise accuracy with F1 scores. In these results, AIMN still consistently achieves significant improvement comparing to all baselines on all datasets and training ratios. It further indicates the effectiveness of AIMN when applied to missing link prediction task, and verifies that AIMN has the capability of modeling relationships between vertexes precisely.

AIMN works better compared with state-of-the-art baselines. It verifies our assumption that social activities could have potential impacts on the neighbors, thus benefiting the relevant link prediction task.

5   Conclusion and Future Work

To predict missing social links between users with social influence, instead of measuring general external influence, we distinguish pair-wise influence of each neighbor. As main contributions, the semantic of each neighbor's pair-wise influence is fine grained with specific part of text and attention module. Each user is finally embedded into one single vector according to the semantic influence of all his neighbors. Experimental results on academic and social media networks illustrate the general advantage of the proposed method compared with state-of-the-art methods. In the future, besides the attributes of vertexes, we will also adopt the attributes of relationships to understand the semantic of interpersonal influence.

Acknowledgement. This work is supported by National Key Research and Development Program of China (2018YFC0809804), National Natural Science Foundation of China (U1736103), National Natural Science Foundation of China (Key Program, U1636203), the State Key Development Program of China (2017YFE0111900).

References

1. Ba, L.J., Kiros, R., Hinton, G.E.: Layer normalization. CoRR (2016)
2. Chang, J., Blei, D.M., et al.: Hierarchical relational models for document networks. Ann. Appl. Stat. **4**(1), 124–150 (2010)
3. Grover, A., Leskovec, J.: node2vec: scalable feature learning for networks. In: Proceedings of the SIGKDD, pp. 855–864 (2016)
4. Huo, Z., Huang, X., Hu, X.: Link prediction with personalized social influence. In: Proceedings of the AAAI, pp. 2289–2296 (2018)
5. Kingma, D., Ba, J.: Adam: a method for stochastic optimization. Computer Science (2014). https://arxiv.org/abs/1412.6980
6. Leskovec, J., Kleinberg, J.M., Faloutsos, C.: Graphs over time: densification laws, shrinking diameters and possible explanations. In: Proceedings of the SIGKDD, pp. 177–187 (2005). https://snap.stanford.edu/data/cit-HepTh.html
7. McCallum, A., Nigam, K., Rennie, J., Seymore, K.: Automating the construction of internet portals with machine learning. Inf. Retrieval **3**(2), 127–163 (2000). https://people.cs.umass.edu/mccallum/data.html
8. Mikolov, T., Sutskever, I., Kai, C., Corrado, G., Dean, J.: Distributed representations of words and phrases and their compositionality. In: Advances in Neural Information Processing Systems, vol. 26, pp. 3111–3119 (2013). https://code.google.com/p/word2vec/
9. Perozzi, B., Al-Rfou, R., Skiena, S.: Deepwalk: online learning of social representations. In: Proceedings of the SIGKDD, pp. 701–710 (2014)
10. Song, Y., Shi, S., Li, J., Zhang, H.: Directional skip-gram: explicitly distinguishing left and right context for word embeddings. In: Proceedings of the NAACL-HLT, pp. 175–180. Association for Computational Linguistics (2018). https://ai.tencent.com/ailab/nlp/
11. Tang, J., Zhang, J., Yao, L., Li, J., Zhang, L., Su, Z.: Arnetminer: extraction and mining of academic social networks. In: Proceedings of the SIGKDD, pp. 990–998 (2008). https://www.aminer.cn/aminernetwork
12. Tu, C., Liu, H., Liu, Z., Sun, M.: Cane: context-aware network embedding for relation modeling. In: Proceedings of the ACL, pp. 1722–1731 (2017)
13. Vaswani, A., et al.: Attention is all you need. In: Proceedings of the NIPS, pp. 6000–6010 (2017)
14. Wang, D., Cui, P., Zhu, W.: Structural deep network embedding. In: Proceedings of the SIGKDD, pp. 1225–1234 (2016)
15. Yang, C., Liu, Z., Zhao, D., Sun, M., Chang, E.Y.: Network representation learning with rich text information. In: Proceedings of the IJCAI, pp. 2111–2117 (2015)
16. Yuan, N.J., Zhang, F., Lian, D., Zheng, K., Yu, S., Xie, X.: We know how you live: exploring the spectrum of urban lifestyles. In: Proceedings of the COSN, pp. 3–14 (2013). https://www.microsoft.com/en-us/research/project/lifespec-learning-spectrum-urban-lifestyles-2/

U-Net with Graph Based Smoothing Regularizer for Small Vessel Segmentation on Fundus Image

Lukman Hakim[✉], Novanto Yudistira, Muthusubash Kavitha,
and Takio Kurita[✉]

Department of Information Engineering, Hiroshima University, 1-4-1 Kagamiyama,
Higashi-Hiroshima 739-8527, Japan
{lukman-hakim,cbsemaster,kavitha,tkurita}@hiroshima-u.ac.jp

Abstract. The detection of retinal blood vessels, especially the changes of small vessel condition is the most important indicator to identify the vascular network of the human body. Existing techniques focused mainly on shape of the large vessels, which is not appropriate for the disconnected small and isolated vessels. Paying attention to the low contrast small blood vessel in fundus region, first time we proposed to combine graph based smoothing regularizer with the loss function in the U-net framework. The proposed regularizer treated the image as two graphs by calculating the graph laplacians on vessel regions and the background regions on the image. The potential of the proposed graph based smoothing regularizer in reconstructing small vessel is compared over the classical U-net with or without regularizer. Numerical and visual results shows that our developed regularizer proved its effectiveness in segmenting the small vessels and reconnecting the fragmented retinal blood vessels.

Keywords: Retinal blood vessel · Graph based smoothing · Regularizer · Graph laplacians

1 Introduction

Characteristics of blood vessels in retina guide an ophthalmologist to diagnose pathologies of different eye anomalies such as age-related macular degeneration (ARMD) and diabetic retinopathy (DR) [1,2]. Additionally, it helps to identify several physiological problems, specially hypertension and some other cardiovascular diseases [3]. However, it is time consuming process to identify the disease caused blood vessels, especially the changes of states of small vessel and its characteristics. To rectify the subjective detection of retinal blood vessels, several automated system of segmentation of blood vessels were developed. However, the separation of the blood vessel is not an easy task because of the small and fragmented structure in low contrasting retinal image.

Chakraborti et al. [4] presented a self-adaptive matched filter by combining the vesselness filter with the matched filter for the detection of blood vessels on

© Springer Nature Switzerland AG 2019
T. Gedeon et al. (Eds.): ICONIP 2019, CCIS 1143, pp. 515–522, 2019.
https://doi.org/10.1007/978-3-030-36802-9_55

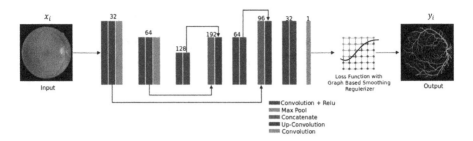

Fig. 1. U-Net framework using graph based smoothing regularizer

the retinal fundus image. Tagore et al. [5] presented a new algorithm for retinal blood vessel segmentation by using the intensity information of red and green channels of color fundus image. It helped to distinguish between vessels and its background in the phase congruency image. Recently, several researchers have been implementing convolutional neural network (CNN) for retinal blood vessel segmentation. Melinscak [6] presented retinal vessel segmentation system using ten layers of CNN. In [12], a structured prediction scheme was used to highlight the context information, while testing a comprehensive set of architectures. Fu et al. [9] combined a typical 7-layer CNN with a conditional random field and reformulated a recurrent neural network to model long-range pixel interactions. Li et al. [7] considered the vessel segmentation task as a cross-modality data transformation problem in a deep learning model. Dasgupta [8] proposed a neural network framework that iteratively classify pixels from the fundus image. Although the localization of the vessels has improved significantly with CNN, the fragmented small vessel identification is still a challenging task. Because sometimes it is located at the end of the vessel branch and failed to maintain the connectivity. Furthermore, it is difficult to detect isolated vessels in the low contrast background.

For addressing these challenges, first time we proposed a graph based smoothing regularizer that considers the image into two regions by calculating graph laplacians on vessels and its background areas. The proposed regularizer is used as a objective function in the deep CNN framework makes the network can efficiently learn the pixel connectivity of the small or isolated blood vessel structure. The effectiveness of our proposed regularization term was evaluated and compared using U-net architecture and baseline U-net. The performance of the proposed approach was also compared with the state-of-the-art networks model in reconstructing the small and isolated vessel regions.

2   Methods

In parallel to track large vessel we are interested to reconstruct the small or isolated vessels. Paying attention to segment the small vessels in the fundus region, we considered to define a regularizer that is based on the graph laplacian smoothing method. We evaluate the effectiveness of our proposed regularizer using U-Net architecture [10]. The schematic diagram in Fig. 1, describes the proposed graph based smoothing regularizer for small vessel reconstruction in the U-net framework.

2.1 Network Architecture

The network has two parts, encoder and decoder module. The encoder module of the network contains 6 convolutional layers with ReLU and 2 max-pooling layer with 32, 64 and 128 feature maps, respectively. The decoder module contains 4 convolutional layers with ReLU and one convolution layer without ReLU. The input feature maps are upsampled with the factor of 2. Skip connections are used to concatenate the feature maps after the deconvolution with the corresponding features from the encoder path. After the decoder, pixel-based probability maps and predictions are generated by a sigmoid classifier function. We used data augmentation to boost the training performance of the network. The data augmentation methods using patching were applied on the original data with the patch size of 48×48 pixels.

2.2 Graph Based Smoothing as Regularizer

Graph based smoothing regularizer is based on the graph laplacian matrix. Graph laplacian can be obtained by constructing the adjacency graph and diagonal matrix. With the help of graph laplacian, the image pixels can be interpreted as node. Every node is connected with every other node in the graph. In this study, we formally defined two graphs G_F and G_B for foreground and background, respectively as shown in Fig. 2. G_F and G_B includes a pair of (V_F, E_F), (V_B, E_B), respectively. The parameters V_F and V_B are finite set of elements called vertices, and $(E_F = \{(j_F, k_F)|j_F \in V_F, k_F \in V_F\}$ and $(E_B = \{(j_B, k_B)|j_B \in V_B, k_B \in V_B\}$ are edges.

Let us consider $(x_i, t_i)|i = 1, ..., M$, where x_i is a ith input data from the training dataset X, and t_i is a label from training dataset. The number of training samples and labels is denoted by M and N, respectively.

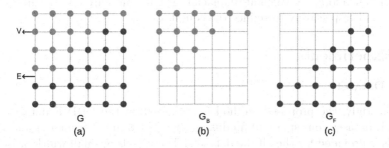

Fig. 2. Representation of set of nodes of the pixel graph to compute boundary between two regions, (a) image as an graph, (b) background region graph and (c) foreground region graph.

The proposed CNN based U-Net architecture is trained to predict the output image pixels y_i from a given input image pixels x_i. For each edge of foreground $(j_F, k_F) \in E_F$ and background $(j_B, k_B) \in E_B$ of the pixel graph, the similarity $\beta_{(j_F,k_F)}$ and $\beta_{(j_B,k_B)}$ is defined as

$$\beta_{(j_F,k_F)} = 1 - |t_{j_F} - t_{k_F}| \tag{1}$$

$$\beta_{(j_B,k_B)} = 1 - |t_{j_B} - t_{k_B}| \tag{2}$$

We introduced the regularization term for smoothing S based on foreground region F and background region B as

$$\sum_{(j_F,k_F)\in G_F} \beta_{j_F,k_F}(y_{j_F} - y_{k_F})^2 = y^T(D_F - A_F)y = y^T L_F y \tag{3}$$

$$\sum_{(j_B,k_B)\in G_B} \beta_{j_B,k_B}(y_{j_B} - y_{k_B})^2 = y^T(D_B - A_B)y = y^T L_B y \tag{4}$$

where, L_F and L_B is Laplacian graph. The adjacency and diagonal matrices is defined as following

$$\begin{pmatrix} A_F = \beta_{(j_F,k_F)}, D_F = \sum_{j_F=1}^{N} \beta_{j_F,k_F} \\ A_B = \beta_{(j_B,k_B)}, D_B = \sum_{j_B=1}^{N} \beta_{j_B,k_B} \end{pmatrix} \tag{5}$$

Smoothing S can be written as

$$S = y^T(L_F + L_B)y = y^T L_G y \tag{6}$$

The objective function O applied in this study is the summation of the binary cross entropy of each label with the regularization term using graph based smoothing, which is defined as

$$O = \sum_{i}^{M}\{t_i log(y_i) + (1 - t_i)log(1 - y_i)\} + \lambda \sum_{i}^{M} S \tag{7}$$

One of the complexity of the graph construction is depends on the size of images. Therefore, we calculate laplacian graph randomly on background and foreground respectively to reduce the complexity.

3   Experiments

3.1   Dataset

In this study, the proposed method is evaluated on the DRIVE datasets. The DRIVE dataset consists of 40 fundus images. The manual segmentations of the vessels is provided for the all the datasets. The vessels of small width or isolated pixels were defined as small vessels. The size of the image is 565×584 pixels with 8 bits per color channel. The number of training and testing data used in this study is 20 and 20, respectively.

3.2   Experimental Settings

In order to find a suitable λ value for the regularizer, we varied this parameter such as $\lambda = 0.0001, 0.00001, 0.000001$ and 0.0000001 for accurate small
vessel construction. We used Adam optimizer with learning rate of 0.001 and
100 epoch to train the model. We chose binary cross entropy loss because it
greatly improved the performance of the model. In our experiments, we compared our proposed network with graph based smoothing regularizer over the
baseline U-net without regularizer. The performance of the proposed approach
was also compared with the results of the state-of-the-art networks model. The
sensitivity (Se), specificity (Sp), accuracy (Acc) and area under the curve (AUC)
was used to evaluate the performance of the proposed approach. The proposed
approach is implemented in pytorch with Intel(R) Core(TM) i7-6700K CPU @
4.00 GHz Processor, 32 GB of RAM and Nvidia GeForce GTX 1080/PCIe/SSE2
graphic cards.

4   Results

The appropriate regularizer parameter value λ to reconstruct the width of the
small vessel retained with adequate information was found with the value of
0.000001. The proposed graph based smoothing regularizer coupling with U-
net succeeds in reconstructing both large and small vessel pixels compared over
U-net without regularizer is presented in Fig. 3. In this figure, the improvement
of the disjointed vessel connectivity is clearly observed from the patch based
fundus image.

The graph based smoothing regularizer in the U-net depends on the pixel
connectivity criterion. Hence, when we compared the AUC value of architecture
without regularizer, our approach resulting high AUC value with large number of
vessels. The qualitative and quantitative results of our approach in segmenting
most of the vessels is shown on the image patch examples (Fig. 4). Our approach achieves significantly higher performance with high AUC value (0.979) in
segmenting the small retinal blood vessels than the other state-of-the-art methods is presented in Table 1. Sensitivity of our method is moderate and it is
almost similar with other conventional methods. However, the proposed approach achieves considerably higher specificity (0.99) than all the other methods,

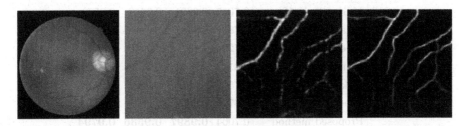

Fig. 3. Segmenting small vessels. From left to right: Fundus image, patch image, results
of network without regularizer, and with proposed graph based smoothing regularizer

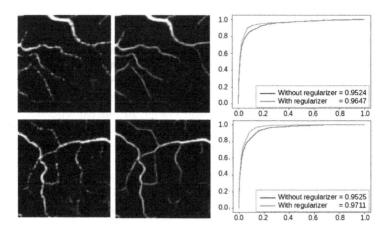

Fig. 4. Segmented small vessels. From left to right: without graph based smoothing regularizer, proposed graph based smoothing regularizer and AUC performance

which reconstructs to the segmentation of more vessels. When we considered accuracy, our approach scored higher value (0.95), and very close to [11,12] methods. Figure 5 explains examples of the analysis of vessel reconstruction, where this study focused on the small vessels. The segmentation results are colorized to demonstrate the confusion matrix: green pixels indicate the TPs and red pixels represent the FNs. Graph based network showed highly acceptable performance for small vessel reconstruction. Our method almost reconnecting all isolated vessels and it can be observed from the lager number of TP pixels. When we analyzed the network without our proposed regularizer, it produced large number of FN pixels.

Table 1. Performance comparison of our approach with state-of-the art methods interns of sensitivity, specificity, accuracy and AUC

Year	Method	Se	Sp	Acc	AUC
2016	Azzopardi et al. [13]	0.7655	0.974	0.9442	0.9614
2016	Khan et al. [14]	0.7373	0.9670	0.9501	–
2016	Zhao et al. [15]	0.7420	0.9820	0.950	0.8620
2018	Marin et al. [16]	0.7067	0.9801	0.9452	0.9588
2018	Orlando et al. [17]	0.7897	0.968	–	–
2016	Fu et al. [9]	0.7294	–	0.9470	–
2015	Wang et al. [11]	0.8173	0.9733	0.9533	0.9475
2016	Liskowski et al. [12]	0.7569	0.9816	0.9527	0.9738
–	U-net [10]	0.6707	0.9867	0.9465	0.9652
–	Proposed method	0.7064	**0.9897**	**0.9536**	**0.9794**

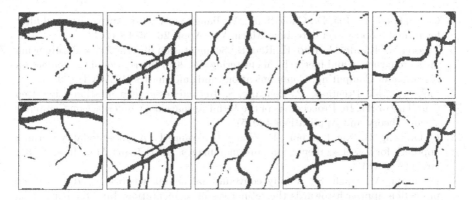

Fig. 5. Small vessel segmentation examples: Green, and red colors represented TP, and FN respectively. First row shows the segmented vessels without graph based smoothing regularizer and the second row shows the results of the proposed graph based smoothing regularizer. (Color figure online)

5   Conclusions

We newly proposed a graph based smoothing regularization term with the loss function in the U-net framework for the segmentation of small vessels in the retinal image. The proposed regularization term effectively computing the graph laplacians on both vessels and its background regions and thus it significantly reduced the segmentation errors and reconnected small fragmented vessels. Our approach can segment more number of vessels and almost reconnecting all isolated vessels than the baseline U-net without regularizer. Compared to other state-of-the-art methods, our approach demonstarted its improvement in retaining width of the small vessel and disjointed vessel connectivity through its high AUC value. Future work will focus on implementing the proposed regularization term on different retinal image datasets and different segmentation CNN architectures.

Acknowledgments. This work was partly supported by JSPS KAKENHI Grant Number 16K00239 and 18F18112.

References

1. Staal, J., Abramoff, M., Niemeijer, M., Viergever, M., van Ginneken, B.: Ridge-based vessel segmentation in color images of the retina. IEEE Trans. Med. Imaging **23**, 501–509 (2004)
2. Amin, M., Yan, H.: High speed detection of retinal blood vessels in fundus image using phase congruency. Soft. Comput. **15**, 1217–1230 (2010)
3. Sharma, A., Rani, S.: An automatic segmentation and detection of blood vessels and optic disc in retinal images. In: 2016 International Conference on Communication and Signal Processing (ICCSP) (2016)

4. Chakraborti, T., Jha, D., Chowdhury, A., Jiang, X.: A self-adaptive matched filter for retinal blood vessel detection. Mach. Vis. Appl. **26**, 55–68 (2014)

5. Tagore, M., Kande, G., Rao, E., Rao, B.: Segmentation of retinal vasculature using phase congruency and hierarchical clustering. In: 2013 International Conference on Advances in Computing, Communications and Informatics (ICACCI) (2013)

6. Melinscak, M., Prentasic, P., Loncaric, S.: Retinal vessel segmentation using deep neural networks. In: Proceedings of the 10th International Conference on Computer Vision Theory and Applications (2015)

7. Li, Q., Feng, B., Xie, L., Liang, P., Zhang, H., Wang, T.: A cross-modality learning approach for vessel segmentation in retinal images. IEEE Trans. Med. Imaging **35**, 109–118 (2016)

8. Dasgupta, A., Singh, S.: A fully convolutional neural network based structured prediction approach towards the retinal vessel segmentation. In: 2017 IEEE 14th International Symposium on Biomedical Imaging (ISBI 2017), pp. 18–21 (2017)

9. Fu, H., Xu, Y., Wong, D., Liu, J.: Retinal vessel segmentation via deep learning network and fully-connected conditional random fields. In: 2016 IEEE 13th International Symposium on Biomedical Imaging (ISBI), pp. 698–701 (2016)

10. Ronneberger, O., Fischer, P., Brox, T.: U-net: convolutional networks for biomedical image segmentation. In: Navab, N., Hornegger, J., Wells, W.M., Frangi, A.F. (eds.) MICCAI 2015. LNCS, vol. 9351, pp. 234–241. Springer, Cham (2015). https://doi.org/10.1007/978-3-319-24574-4_28

11. Wang, S., Yin, Y., Cao, G., Wei, B., Zheng, Y., Yang, G.: Hierarchical retinal blood vessel segmentation based on feature and ensemble learning. Neurocomputing **149**, 708–717 (2015)

12. Liskowski, P., Krawiec, K.: Segmenting retinal blood vessels with newline deep neural networks. IEEE Trans. Med. Imaging **35**, 2369–2380 (2016)

13. Azzopardi, G., Strisciuglio, N., Vento, M., Petkov, N.: Trainable COSFIRE filters for vessel delineation with application to retinal images. Med. Image Anal. **19**, 46–57 (2015)

14. Khan, M., Soomro, T., Khan, T., Bailey, D., Gao, J., Mir, N.: Automatic retinal vessel extraction algorithm based on contrast-sensitive schemes. In: 2016 International Conference on Image and Vision Computing New Zealand (IVCNZ), pp. 1–5 (2016)

15. Zhao, Y., Rada, L., Chen, K., Harding, S., Zheng, Y.: Automated vessel segmentation using infinite perimeter active contour model with hybrid region information with application to retinal images. IEEE Trans. Med. Imaging **34**, 1797–1807 (2015)

16. Marin, D., Aquino, A., Gegundez-Arias, M.E., Bravo, J.M.: A new supervised method for blood vessel segmentation in retinal images by using gray-level and moment invariants-based features. IEEE Trans. Med. Imaging. **30**, 146–158 (2011). https://doi.org/10.1109/TMI.2010.2064333

17. Orlando, J.I., Prokofyeva, E., Blaschko, M.B.: A discriminatively trained fully connected conditional random field model for blood vessel segmentation in fundus images. IEEE Trans. Biomed. Eng. **64**, 16–27 (2017). https://doi.org/10.1109/TBME.2016.2535311

NEOKNN: A Network Embedding Method Only Knowing Neighbor Nodes

Bencheng Yan and Chaokun Wang[✉]

School of Software, Tsinghua University, Beijing 100084, China
ybc17@mails.tsinghua.edu.cn, chaokun@tsinghua.edu.cn

Abstract. Recently, network embedding has attracted increasing research attention, due to its effectiveness in a variety of applications. Most of the existing work aims to embed varied kinds of networks, such as plain networks, dynamic networks, and attributed networks. However, all of them require sufficient information such as network structure or node attributes. In practice, we can hardly obtain the entire network structure and abundant attributes. The network information provided is often limited. For example, we may only know a few neighbors of each node. In this paper, we address an important embedding problem (i.e., embedding nodes in a limited situation), and propose a novel deep network embedding framework in a soft coupling way to infer node embedding in such a situation. Extensive experimental results demonstrate that, given limited information, our representation significantly outperforms the representations learned by state-of-the-art methods.

Keywords: Network embedding · Limited situation · Soft coupling

1 Introduction

Network embedding, which has attracted increasing research interests, represents nodes or edges by low-dimensional vectors while preserving the inherent structures of the network. The learned vectors are applied to many network related tasks, such as node classification by node embedding [5,10] and link prediction by edge embedding [8]. There is much work to be done in the field of network embedding [3–5]. All of these methods assume that the entire network structure or the sufficient node attributes should be available, and adopt a strong coupling way to learn node embeddings (i.e., all of the information is carefully considered in one loss function). For example, given the entire network structure, DeepWalk [5] generates random walk paths and directly optimizes the node embeddings. GAT [7] can infer unseen node embeddings by the patterns learned from the attributed domain.

However, in many real-world networks, such as terrorist attack networks, the entire network structure or the sufficient node attributes are very difficult or even impossible to obtain. Instead, the link information is only available within a few small local regions. For example, we may only know k neighbors of each

© Springer Nature Switzerland AG 2019
T. Gedeon et al. (Eds.): ICONIP 2019, CCIS 1143, pp. 523–531, 2019.
https://doi.org/10.1007/978-3-030-36802-9_56

node. In such a situation where the information is limited, once some of the information is unavailable, none of the existing works can be applied well.

In this paper, we propose a new network embedding problem in a limited situation where only k neighbors of each node are known. Then, we adopt a soft coupling way and propose a novel neural network to address the network embedding problem where we only know k neighbor nodes, and try to give a solution to these two kinds of cases.

The main contributions of this paper are summarized as follows: (1) We propose a novel network embedding problem where only k neighbor nodes are known. As far as we know, it is the first work which focuses on the limited information based embedding problem. (2) We adopt a soft coupling way, and propose a Network Embedding method Only Knowing k Neighbor Nodes, called NEOKNN, to address this challenging problem. (3) Extensive experimental results demonstrate that, given the limited network information, NEOKNN can effectively capture network structures than state-of-the-art methods.

2 Related Work

In this section, some related work is briefly reviewed.

Plain Network Embedding. Early work focuses on embedding static and plain networks, which only considers network structures. DeepWalk [5] and Node2vec [2] generate random walks, and learn nodes embedding by preserving the proximity of two nodes in a random walk. LINE [6] proposes the first and second order proximities and optimizes the embedding via negative sampling.

Dynamic Network Embedding. Dynamic network embedding also arouses considerable research interests. Most of these works [4,12] consider the changes of nodes or edges in different time points, and learn new coming nodes embeddings or update existing nodes embeddings. Although some of these methods [4] can infer a new coming node, these methods can only learn embeddings of new coming nodes which link to existing nodes. It is not helpful to arbitrary nodes only knowing its neighbors, which may not connect to existing nodes.

Attributed Network Embedding. Some works [1,3,11] try to make use of attributes of nodes. GCN [3] proposes a graph convolutional kernel to aggregate node attributes. DANE [1] focuses on the inconsistency problem between node attributes and network topology.

3 Preliminaries

In this section, we define related concepts and notations.

Definition 1 (*p* Hop Neighbor Nodes Set). Given a graph $G = (V, E)$, the p hop neighbor nodes set of $v \in V$ is a set $H_p(v)$ of nodes which are reachable from v within p hops.

Definition 2 (k Neighbor Nodes). The k neighbor nodes of $v \in V$ are defined as a k-size set $N_k(v)$ of nodes, and $\forall v_i \in N_k(v), v_i \in H_p(v)$, where p is a small number.

Definition 3 (k Neighbor Nodes based Center Graph). Given a graph $G = (V, E)$ and node $v \in V$, the k neighbor nodes based center graph of v is defined as $cG_{v,k} = (cV, cE)$ where $cV = N_k(v) \cup \{v\}$ and $cE = \{e_{u,v} | u, v \in cV, e_{u,v} \in E\}$, and the adjacency matrix of the graph $cG_{v,k} = (cV, cE)$ is defined as $cA_{v,k} \in R^{(k+1) \times (k+1)}$.

Definition 4 (Network Embedding Problem where we Only Know k Neighbor Nodes). Given a graph $G = (V, E)$, and a training subgraph $G' = (V', E')$, where $V' \subset V$, and $E' = \{e_{v,u} | v, u \in V', e_{u,v} \in E\}$. The network embedding problem where we only know k neighbor nodes aims to learn an embedding mapping function f by training on the graph G'. Then given a node v'' and its k neighbor nodes based center graph $cG_{v'',k} = (cV'', cE'')$, where $\forall u \in cV'', u \notin V'$, the learned function f is expected to infer the embedding of node v'' directly.

4 Proposed Model

As analyzed before, none of the existing works can deal with the limited situation. The main reason is that all of them build the model in a strong coupling way. Thus, in this work, we adopt a soft coupling way. Specifically, for each node v and the graph $cG_{v,k}$, the corresponding adjacency matrix $cA_{v,k}$ is also taken as a separate sample. The structure patterns among each node and its k neighbor nodes are preserved. In such a soft coupling way, the patterns learned from training adjacency matrices can be naturally used to infer the embedding of unseen nodes.

4.1 Neighbor Nodes Labeling

The first step of NEOKNN is to label these $k + 1$ nodes, i.e., assigning the order number to each node in a suitable way. For each sample graph $cG_{v,k}$, if all nodes in $cG_{v,k}$ are labeled in a suitable order (e.g., labeling by the degree), the model can generate a unique adjacency matrix for $cG_{v,k}$. For two isomorphic graphs $cG_{v_1,k}$ and $cG_{v_2,k}$, their adjacency matrices, i.e., $cA_{v_1,k}$ and $cA_{v_2,k}$, should be the same. However, if all nodes are randomly labeled, the model may get different adjacency matrices of $cG_{v_1,k}$ and $cG_{v_2,k}$. Then, NEOKNN may be misled into the situation where these two input graphs are different and may give different embedding vectors. For two different input graphs, labeling will assign nodes of two different input graphs to a similar relative position in the adjacency matrices if and only if their structural roles within the graphs are similar. Thus, by labeling nodes in a suitable order, NEOKNN can be easier to find the inherent relationships among the node and its k neighbor nodes across all input graphs. Actually, the adjacency matrices after labeling also preserve some patterns among nodes and its k neighbor nodes. In order to show the importance of labeling, we take NEOKNN-OL as one of our baselines (more details are described in Sect. 5.2). In this paper, we adopt the Weisfeiler-Lehman algorithm [9], a well-studied framework, for the unique assignment of node labels given a graph (The center nodes v of the graph $cG_{v,k}$ will always be labeled to number one).

4.2 Multi-scale Convolutional Kernels

The second step of NEOKNN is to effectively summarize the adjacency matrix of each graph $cG_{v,k}$, and well characterize the structure similarity [12] between nodes. Then, multi-scale convolutional kernels are proposed to capture the complex relationships and patterns in a graph $cG_{v,k}$. In addition, different kinds of convolutional kernels are simultaneously applied to the adjacency matrix. As shown in Fig. 1, different kinds of kernels can preserve different kinds of node relationships. For example, the $2 * k$ size kernel showed in Fig. 1(a) describes the common neighbor nodes of two nodes, which refers to the second-order proximity [6], and the $3 * k$ size kernel also has a similar effect with the $2 * k$ size kernel. Besides, sometimes, the $3 * 3$ size kernel showed in Fig. 1(b) can characterize the triangles in neighbors. Furthermore, to capture more triangle relationship, a dilated convolution [12] showed in Fig. 1(c) is applied to the adjacency matrix. With the help of these multi-scale convolution kernels, NEOKNN can efficiently preserve multi-kinds of the relationship among nodes.

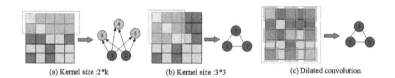

(a) Kernel size :2*k   (b) Kernel size :3*3   (c) Dilated convolution

Fig. 1. Multi-scale convolutional kernels. Different kinds of kernel can naturally preserve different relationship of nodes.

4.3 Nodes Personality Preserving

Although multi-scale convolutional kernels can well characterize k neighbor nodes information and find common patterns across all the input graphs, it is still helpful to feed the personality of each graph to NEOKNN. Generally speaking, let $e_{i,j}$ be one edge of the graph $cG_{v,k}$. If we can hardly find another edge in the same position from other adjacency matrices, the NEOKNN model may need to pay more attention to this edge $e_{i,j}$. Thus to preserve such node personality, we design two components, i.e., the diff-adjacency matrix and the diff-gate.

The Diff-Adjacency Matrix. The diff-adjacency matrix describes the difference between one center graph and another. Specifically, given a training graph

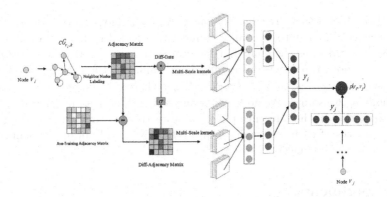

Fig. 2. The framework of NEOKNN. In the training phase, we feed the NEOKNN with a pair of nodes v_i and v_j to estimate the joint probability of these two nodes (for simplicity, we do not plot the network of node v_j, which is the same with node v_i). In the testing phase, we only feed one test node to the network each time.

$G' = (V', E')$ (Definition 4), we can get $|V'|$ number of k neighbor nodes based center graphs $cG_{v_i,k}, i = 1, 2, 3, ..., |V'|$ and the corresponding adjacency matrices $cA_i, i = 1, 2, 3, ..., |V'|$. We take the overall neighbor adjacency matrix distribution of G' as $cA_{mean} = \frac{1}{|V'|}\sum_{i=1}^{|V'|} cA_i$. Then, the diff-adjacency matrix can be expressed as $diffcA_i = cA_i - cA_{mean}$. Obviously, the low frequency edge leads to a small value in the corresponding position of the adjacency matrix. Relatively, a large value in $diffcA_i$ reflects the personality edges in $cG_{v_i,k}$. Thus, along with cA_i, we also take the $diffcA_i$ as an input of NEOKNN. In this way, NEOKNN can figure out the differences across all the nodes.

Diff-Gate. For each node and its corresponding adjacency matrix $cA_{v,k}$, to filter out useless edges and preserve important edges in the neighborhood more flexibly, we design a diff-gate as a dynamic component, controlling the flow of input to NEOKNN. Specifically, given the input adjacency matrices cA_i and $diffcA_i$ of node v_i, the diff-gate can be expressed as follows:

$$filter_{in} = \sigma(diffcA_i * W_{in} + b_{in}) \tag{1}$$

$$cA_i^* = cA_i \odot filter_{in} \tag{2}$$

where $W_{in} \in R^{k*k}$, σ is a sigmoid function, $\odot$ represents element-wise production and $filter_{in} \in R^{k*k}$ is regarded as a diff-gate to dynamically filter the input adjacency matrix.

4.4 Network Architecture

In this part, we describe the architecture of NEOKNN based on the above-mentioned components. As shown in Fig. 2, for each node v_i only knowing its k neighbor nodes, we can obtain its k neighbor nodes based center graph $cG_{v_i,k}$ and its adjacency matrix cA_i. Then, the diff-adjacency matrix $diffcA_i$ can also be obtained. Before feeding these two adjacency matrices to the neural network, a diff-gate is applied to the adjacency matrix cA_i. Next, a multi-scale kernel is applied to these two matrices. We concatenate the output of different kernels and then feed them into two fully connected layers. Finally, the output y_i of NEOKNN is taken as the embedding vector of node v_i. To train the NEOKNN model, we adopt a pair-wise proximity loss, similar to the loss proposed in [6], i.e., $Loss = log\sigma(y_i^T \cdot y_j) + \sum_{t=1}^{Ne} log\sigma(-y_t^T \cdot y_i)$ where σ is a sigmoid function. For each node v_i, we sample Ne nodes disconnected with node v_i by their degrees as negative nodes. In the inference phase, we directly feed the corresponding adjacency matrices of each test node v_i into NEOKNN, and obtain the embedding vector y_i in the last layer.

5   Experiments

5.1   Setting

Data Sets. We introduce some real data sets used in our experiment, including Wiki, DBLP, Cora, Citeseer and Pubmed. The detailed statistics of these data sets are summarized in Table 1. To satisfy the assumption where we only know k neighbors for training and test nodes, we obtain their k neighbor nodes through sampling from all of their neighbors in 2-hop.

Table 1. Data sets information

Dataset	Wiki	DBLP	Cora	Citeseer	Pubmed
#Nodes	2,405	13,184	2,708	3,327	19,717
#Edges	17,981	48,018	5,429	4,732	44,338
#Class	19	5	7	6	3

Baselines. We compared our method with DeepWalk [5], Node2vec [2] and LINE [6]. Besides, SimilarAve is considered as a baseline which first calculates the similarity of test nodes and the training nodes by the adjacency matrices of graphs $cG_{v,k}$. Then, to infer a test node embedding, we average the embedding vectors of top five similar training nodes, and take this average embedding vector as the test node embedding. Further, we propose four variants of NEOKNN. Specifically, (1) **NEOKNN-OL** only takes the adjacency matrix after labeling as the embeddings. (2) **NEOKNN-WD** does not feed the diff-adjacency matrix into the neural network. (3) **NEOKNN-WG** does not apply diff-gate in the network. (4) **NEOKNN-WDG** removes the diff-adjacency matrix and diff-gate.

Parameters Setting. In baselines, for DeepWalk, we set $t = 40$, $\gamma = 80$, and $w = 10$ as suggested in [5]. For Node2vec, the walk length, per walk and windows size are set the same as DeepWalk, and we set $p = 0.25, q = 0.25$. For our method, we set the parameters $K = 20$ and $d = 128$ for all data sets.

5.2   Classification on a Single Graph

In this task, the training subgraph $G' = (V', E')$ and the test nodes are from the same graph G. Specifically, for the dataset Wiki, we randomly sample 512 nodes from Wiki as test nodes set $TestV$. The rest nodes construct a training subgraph $G' = (V', E')$. For each test node, only k neighbor nodes are given, which are also from $TestV$ rather than V'. Then NEOKNN trains on the subgraph G', and directly infers the test node embedding. As described before, it is a challenging task, because few information of each test node is provided. Then, to evaluate the performance of the embeddings, similar to other works [2,5], a node classification is applied to these embeddings. To exhaustively evaluate the proposed model, we also report the results of LINE which is trained on the entire graph G. In this case, LINE is aware of the entire network structure, and of course gets the best results. Here, we take these results as the upper bound.

Table 2. Classification on a Single Graph Wiki

Methods	Micro-F1(%)	Macro-F1(%)
Upper Bound	69.38	47.07
DeepWalk	14.63	7.12
Node2vec	32.11	22.20
SimilarAve	25.37	7.02
NEOKNN	**60.49**	**39.37**
NEOKNN-WD	52.20	34.67
NEOKNN-WG	54.15	33.83
NEOKNN-WDG	47.32	33.54
NEOKNN-OL	42.68	26.63

The results are reported in Table 2. Some of the observations and analysis are listed as follows: (1) Compared with baselines, NEOKNN has significant improvements, and also gets comparable results with the upper bound. It indicates that NEOKNN can well deal with the situation where only k neighbor nodes of test nodes are provided. (2) NEOKNN-OL obtains better results than baselines, which demonstrates the importance of labeling process. (3) Compared with the three variants of NEOKNN, i.e., NEOKNN-WD, NEOKNN-WDG and NEOKNN-WG, NEOKNN achieves a better performance. This shows that it is effective to introduce the personality components including diff-adjacency matrix and diff-gate.

5.3   Classification Across Graphs

In this part, we try a more challenging task that infers the nodes embedding from different graphs. The only difference between this task and the classification on a single graph task is that, in this task, the test nodes and the training nodes are not from the same graph. Actually, due to the soft coupling way, NEOKNN can naturally infer the embedding of nodes across graphs. However, one problem is that the distribution of the input matrices may be different, and affects the performance of embeddings (i.e., the patterns learned from the training nodes may be not useful to the test nodes). Thus, to sufficiently understand the inference ability of NEOKNN, we design two kinds of classification tasks across graphs. First, the training graph and the testing graph are from the same domain. For instance, we train on Cora and test on Citeseer (both of them are citation graphs). Second, these two graphs are from different domains (e.g., training on Wiki and testing on Citeseer).

Table 3. Classification across Graphs

Training graph		Testing graph			
		Citeseer		Cora	
		Micro-F1(%)	Macro-F1(%)	Micro-F1(%)	Macro-F1(%)
	Upper bound	54.70	47.12	71.65	70.84
Citation networks	Citeseer	-	-	50.33	43.22
	Cora	**46.75**	**41.88**	-	-
	Pubmed	43.19	37.73	**51.37**	**43.69**
	DBLP	45.48	39.88	48.93	41.87
Others	Wiki	37.83	32.00	41.40	37.07
	Lower bound	35.00	30.42	34.54	28.87

We report the results of these two kinds of tasks in Table 3. Similar to the classification on a single graph, we also provide the results of the upper bound (i.e., LINE is directly trained on the testing graph). Besides, to have a full understanding of NEOKNN, we report the results of NEOKNN-OL which reflects the raw link information among nodes and its k neighbors. Thus, we take the results of NEOKNN-OL as the lower bound of this task. From Table 3, we can see that different training graphs have a different influence on the embedding of test nodes. Due to similar distribution, the citation training graphs (Citeseer, Cora, Pubmed and DBLP) give a better performance than other kind of graph (Wiki). As we do not feed any test nodes information to NEOKNN, NEOKNN does not obtain as good performance as the upper bound. Compared with the lower bound, it makes great progress. In practice, it is very useful. Compared with the existing works which cannot give any useful guidance in a limited situation, NEOKNN can make a great suggestion.

6   Conclusion

In this paper, we address an interesting network embedding problem where we only know k neighbor nodes, and propose a novel deep network embedding framework, called NEOKNN, to solve this problem. We conduct extensive different experiments to exhaustively evaluate the proposed model. The experimental results demonstrate that NEOKNN can capture network structures more effectively than baselines in limited network information.

Acknowledgments. This work is supported in part by the National Key Research and Development Program of China (No. 2017YFC0820402) and the National Natural Science Foundation of China (No. 61872207).

References

1. Gao, H., Huang, H.: Deep attributed network embedding. In: IJCAI, pp. 3364–3370 (2018)
2. Grover, A., Leskovec, J.: node2vec: scalable feature learning for networks. In: SIGKDD, pp. 855–864. ACM (2016)
3. Kipf, T.N., Welling, M.: Semi-supervised classification with graph convolutional networks. arXiv preprint arXiv:1609.02907 (2016)
4. Ma, J., Cui, P., Zhu, W.: Depthlgp: learning embeddings of out-of-sample nodes in dynamic networks. In: AAAI (2018)
5. Perozzi, B., Al-Rfou, R., Skiena, S.: Deepwalk: Online learning of social representations. In: SIGKDD, pp. 701–710. ACM (2014)
6. Tang, J., Qu, M., Wang, M., Zhang, M., Yan, J., Mei, Q.: Line: large-scale information network embedding. In: WWW, pp. 1067–1077 (2015)
7. Velickovic, P., Cucurull, G., Casanova, A., Romero, A., Lio, P., Bengio, Y.: Graph attention networks. arXiv preprint arXiv:1710.10903 (2017)
8. Wang, C., Wang, C., Wang, Z., Ye, X., Yu, J.X., Wang, B.: Deepdirect: learning directions of social ties with edge-based network embedding. IEEE Trans. Knowl. Data Eng. (TKDE) **31**(12), 2277–2291 (2019)
9. Weisfeiler, B., Lehman, A.: A reduction of a graph to a canonical form and an algebra arising during this reduction. Nauchno-Technicheskaya Informatsia **2**(9), 12–16 (1968)
10. Yan, B., Wang, C., Guo, G., Chen, J.: Graph neural network with rejection mechanism. In: the 7th International Conference on Big Data Applications and Services (BIGDAS2019) (2019)
11. Zhang, Z., et al.: Anrl: attributed network representation learning via deep neural networks. In: IJCAI, pp. 3155–3161 (2018)
12. Zhou, L., Yang, Y., Ren, X., Wu, F., Zhuang, Y.: Dynamic network embedding by modeling triadic closure process. In: AAAI (2018)

Hypergraph Regularized GM-pLSA: A Model to Learn the Latent Semantic Attributes

Zhengye Xiao and Yuchun Fang[✉]

School of Computer Engineering and Science,
Shanghai University, Shanghai, China
ycfang@shu.edu.cn

Abstract. Semantic attributes are proven to help improve the performance of a series of applications in the field of computer vision. The semantic attributes are usually defined by humans and labeled manually, but whether these attributes can discriminate the images or videos well should be doubted, because these attributes are not optimized. In order to solve this problem, we proposed a new model named HRGM-pLSA to learn the latent semantic attributes. Our model employs GM-pLSA to learn the latent topics in the images. To utilize the prior knowledge of predefined attributes, we use the hypergraph to present the complex correlation of the attributes in the images. We construct a regularization term of a hypergraph and integrate it into the GM-pLSA to make the learned latent semantic attributes get better performance compared to primary attributes. In this paper, we evaluate the quality of attributes in a practical application: image retrieval. The performance of our model in the application demonstrates the value of our model.

Keywords: Attribute learning · Hypergraph Regularized · Probabilistic latent semantic analysis

1 Introduction

Recently, more and more researchers have a great interest in semantic attributes. In a series of applications, semantic attributes have been proved to be effective, including image retrieval [1], face verification [2], and person re-identification [3]. Semantic attributes may be color, shape, texture or a part of an object, such as the black, oval face, chubby and hairless, and so on. They differ from a full object, such as chicken, and they are used to narrow down the gap between the low-level visual features and high-level semantic meanings.

In a dataset, the semantic attributes are usually predefined by humans previously, and people labeled the pictures according to them. However, whether these semantic attributes can discriminate the pictures well should be doubted. Firstly, most of the semantic attributes are hard to be defined, such as the mole on the left side of the mouth. We have to use a long-phrase to describe it instead

© Springer Nature Switzerland AG 2019
T. Gedeon et al. (Eds.): ICONIP 2019, CCIS 1143, pp. 532–540, 2019.
https://doi.org/10.1007/978-3-030-36802-9_57

of a brief word. This kind of semantic attributes have discrimination, but we can hardly define them. Secondly, some semantic attributes exist in many people, and they lack discrimination. Finally, in the field of recommendation system, the precision of the algorithms which use labels directly is usually lower than the algorithms which use data itself to report the users' preference [17]. Based on the three viewpoints talked above, we think that it is vital to learn latent semantic attributes which can be reflected by the data.

Many researchers have made much effort on the latent topic learning. Horster et al. [6] proposed GM-pLSA to learn the latent topic in the image. Zhong et al. [7] employed a graph to present the relationship between different video clips and proposed graph-regularized pLSA to analyze video content. In the application to latent attribute learning, we think graph-regularized pLSA is also useful, but we replace the graph with a hypergraph. The graph is too complex to show the attribute relationship between images, while the hypergraph can maintain a compact structure. Inspired by the previous works, we proposed a novel attribute learning framework named Hypergraph-Regularized pLSA with Gaussian Mixture (HRGM-pLSA). This method can help get better performance on the task of image retrieval.

As a conclusion, we summarize the contributions of this paper as follows:

1. We propose a new latent semantic attribute learning model named HRGM-pLSA which uses hypergraph as a constraint to GM-pLSA to get better performance.
2. Our model can achieve better performance in the task of image retrieval.

The structure of the rest of the paper is as follows. We start with related works in Sect. 2. Then we concretely describe the HRGM-pLSA in Sect. 3. Section 4 shows the experiment results. Our conclusion is put in Sect. 5.

2   Related Work

2.1   Attributes Learning

Traditional approaches to attributes learning tend to be supervised and the attributes are predefined by humans. Lampert et al. [4,5] and Kumar et al. [2] use SVM to train the attribute classifiers of animals and human respectively. Huang et al. [8] proposed an attribute learning framework based on the hypergraph cutting. This method uses the samples to optimize the primary attributes and get better performance. Zhang et al. [15] proposed a pose-normalized descriptors based on deformable part models to predicate attributes. Convolutional neural networks (CNNs) has become a hot topic in the field of computer vision. Also, some researchers use CNN to learn attributes to get better results, such as [16].

2.2   Image Retrieval

In this paper, the task of image retrieval is to find the same images of a specific image in an image database. Recently, many algorithms use attributes as side information to improve the precision, and the quality of attributes can influence the precision to some certain extent [2, 3, 9]. Fang et al. [10] employed attribute-hypergraph to reform the original metric and improved the precision of image retrieval. The quality of attributes can influence the performance of the algorithm and we employ it to evaluate the outcomes of attribute learning algorithms.

3 Approach

3.1 Preliminaries

Hypergraph theory was proposed by BERGE [11]. A Hypergraph is an extension of the graph. The vertex of a hypergraph is the same as the graph, and the edge of a hypergraph is called hyperedge, which is a set consist of positive numbers of vertexes. Given a hypergraph $G = (V, E)$, V is an infinite set of vertexes and E is the set of hyperedges, and the vertex and hyperedge can be defined as $v \in V, e \in E$. The incidence matrix of vertexes and hyperedges $H \in R^{|V| \times |E|}$ is defined as:

$$h(v, e) = \begin{cases} 1, & if\ v \in e \\ 0, & otherwise \end{cases}. \tag{1}$$

In the practical application, an image is donated by a vertex, and a hyperedge donates an attribute. An attribute can have many images. The hypergraph can easily present the complex correlation of attributes between images.

The pLSA with Gaussian Mixture (GM-pLSA) was proposed to solve the problem that the traditional pLSA algorithm can only deal with discrete words. However, in the application of image processing, the features of images are usually continuous variables. GM-pLSA extends the discrete variables to continuous variables so that we can use pLSA to learn the latent topic in the images. Given a set of images $D = \{d_1, d_2, \ldots, d_n\}$ and a set of features which are correspond to the images $F = \{f_1, f_2, \ldots, f_n\}$, where $f_i \in R^{d \times 1}$. We compute the likelihood function of GM-pLSA like [6], as:

$$L = \sum_{i=1}^{N} log(\sum_{m=1}^{M} \sum_{k=1}^{K} (P(d_i)P(z_m|d_i)P(g_k|z_m)N(f_i|\mu_k, \Sigma_k)), \tag{2}$$

where z_m is the latent topic, μ_k and Σ_k are the mean vector and covariance matrix under the Gaussian component g_k.

3.2 Hypergraph Regularized GM-pLSA

We propose Hypergraph Regularized GM-pLSA (HRGM-pLSA) to learn the latent semantic attributes. The main idea of HRGM-pLSA is that we use GM-pLSA with the constraint of attribute hypergraph to improve the quality of attributes compared to the primary attributes.

We use the coherence matrix to present the relation between the images in the hypergraph, which can be given by:

$$S = H \cdot H^T. \tag{3}$$

In order to embed the hypergraph into the GM-pLSA, we need to construct a regularization term of a hypergraph. The famous manifold assumption shows that there is a relation between a marginal distribution and conditional distribution. Based on this theory, if two features f_i and f_j are similar on their data, their Gaussian components are likely to have a similar conditional probability distribution. For following simplicity, the probability $P(z_m, g_k|d_i, f_i)$ is donated by $\beta^i_{m,k}$. Based on the discussion talked above, the regularization term of hypergraph should contain the similarity between $\beta^i_{m,k}$ and $\beta^j_{m,k}$. So the regularization term of hypergraph can be expressed as:

$$R = \sum_{i=1}^{N} \sum_{j=1}^{N} D(i,j)S(i,j), \tag{4}$$

where $D(i,j)$ is the similarity between $\beta^i_{m,k}$ and $\beta^j_{m,k}$, and it can be calculated by relative entropy (Kullback-Leibler divergence, KL-divergence). $D(i,j)$ can be computed by the formula:

$$
\begin{aligned}
D(i,j) = \frac{1}{2}(&\sum_{k=1}^{K} \sum_{m=1}^{M} (\beta^i_{m,k} log \frac{\beta^i_{m,k} + \beta^j_{m,k}}{2\beta^i_{m,k}} - \beta^i_{m,k} + \frac{\beta^i_{m,k} + \beta^j_{m,k}}{2}) \\
+ &\sum_{k=1}^{K} \sum_{m=1}^{M} (\beta^j_{m,k} log \frac{\beta^i_{m,k} + \beta^j_{m,k}}{2\beta^j_{m,k}} - \beta^j_{m,k} + \frac{\beta^i_{m,k} + \beta^j_{m,k}}{2})).
\end{aligned}
\tag{5}
$$

We integrate the regularization term of hypergraph into the likelihood function, and we can get a new function:

$$L = \sum_{i=1}^{N} log(\sum_{m=1}^{M} \sum_{k=1}^{K} (P(d_i)P(z_m|d_i)P(g_k|z_m)P(f_i|g_k))) - \lambda \sum_{i=1}^{N} \sum_{j=1}^{N} D(i,j)S(i,j), \tag{6}$$

where λ is a hyperparameter. Our goal is to make the expectation of L converged, and we can solve this problem by EM algorithm.

In E step, we use the Bayes rule to compute $\beta^i_{m,k}$, which can be given by:

$$\beta^i_{m,k} = \frac{P(z_m|d_i)P(g_k|z_m)P(f_i|g_k)}{\sum_{m=1}^{M} \sum_{k=1}^{K} (P(z_m|d_i)P(g_k|z_m)P(f_i|g_k))}. \tag{7}$$

And in M step, we need to compute the expectation of L and make it converged. The expectation of L can be given by:

$$E(L) = \sum_{i=1}^{N} \sum_{m=1}^{M} \sum_{k=1}^{K} \beta^i_{m,k} log((P(d_i)P(z_m|d_i)P(g_k|z_m)P(f_i|g_k))) - \lambda \sum_{i=1}^{N} \sum_{j=1}^{N} D(i,j)S(i,j). \tag{8}$$

In order to estimate the new model parameters, we can use Lagrange Multiplier to compute the parameters:

$$\sigma_k = \sum_{i=1}^{N} \sum_{m=1}^{M} \beta^i_{m,k}, P(d_i) = \frac{1}{N}, P(g_k|z_m) = \frac{\sum_{i=1}^{N} \beta^i_{m,k}}{\sum_{k=1}^{K} \sum_{i=1}^{N} \beta^i_{m,k}}, \tag{9}$$

$$P(z_m|d_i) = \frac{\sum_{k=1}^{K} \beta_{m,k}^i}{\sum_{k=1}^{K} \sum_{m=1}^{M} \beta_{m,k}^i} - \frac{\lambda}{2} \cdot \frac{\sum_{j=1}^{N} \sum_{k=1}^{K} (\beta_{m,k}^i - \beta_{m,k}^j)S(i,j)}{\sum_{k=1}^{K} \sum_{m=1}^{M} \beta_{m,k}^i}, \qquad (10)$$

$$\mu_k = \frac{\sum_{i=1}^{N} \sum_{m=1}^{M} \beta_{m,k}^i f_i}{\sigma_k} - \frac{\lambda}{2} \cdot \frac{\sum_{i=1}^{N} \sum_{j=1}^{N} \sum_{k=1}^{K} (\beta_{m,k}^i - \beta_{m,k}^j)(f_i - f_j)S(i,j)}{\sigma_k}, \qquad (11)$$

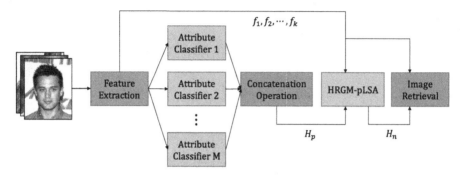

Fig. 1. The flowchart of our model named HRGM-pLSA to learn the latent attributes.

$$\Sigma_k = \frac{\sum_{i=1}^{N} \sum_{m=1}^{M} \beta_{m,k}^i (f_i - \mu_k)(f_i - \mu_k)^T}{\sigma_k} -$$
$$\frac{\lambda}{2} \cdot \frac{\sum_{i=1}^{N} \sum_{j=1}^{N} \sum_{k=1}^{K} (\beta_{m,k}^i - \beta_{m,k}^j)((f_i - \mu_k)(f_i - \mu_k)^T - (f_j - \mu_k)(f_j - \mu_k)^T)S(i,j)}{\sigma_k}.$$
$$(12)$$

Finally, $P(z_m|d_i)$ is the probability of the latent semantic attributes exist in an image that we want to learn.

Our model can be expressed by a flowchart, which is presented in Fig. 1. For a group of images, we first extract their features, then get the attributes of these images by the classifiers. By the concatenation operation, we get the original incidence matrix $H_p \in R^{n \times l}$, which can show the relation between images and attributes. Then we put H_p and the features into HRGM-pLSA to learn the new incidence matrix $H_n \in R^{n \times m}$, which is composed of the latent attributes. Finally, we can use it to finish the task of image retrieval.

4   Experiment

4.1   Experimental Setups

Datasets. We will use two datasets to confirm the value of the proposed approaches: Large-scale CelebFaces Attributes (CelebA) [12], Animals with Attributes2 (AwA2) [13]. CelebA contains more than 200 K celebrity images, each with 40 attribute annotations. We filter the attributes which may exist differences in the same people. At last, there remain 31 attributes. We select a subset of the images to be used for image retrieval testing, which contains 5000 images and 250 identities, and each one has four images. We divide another subset of images to be used for training attribute SVM classifiers, which also contains 5000 images. AwA2 contains 37322 images of 50 animal classes, and each class is annotated with 85 attributes. Following [4,5], we divide the classes into 40 classes (30337 images) to be used for training and 10 classes (6985 images) to be used for testing.

Features. We employ 3304-dimensional Uniform LBP features [14] to train the attribute classifiers of CelebA. At the stage of training our model or compared algorithms, we extract 2048-dimensional deep learning features by ResNet-50. For AwA2, we use the features extracted by ResNet-101 which are available online[1] to train the classifiers and models.

Evaluation Criteria. For image retrieval, we select the top N images as the most similar images to compute the retrieval precision:

$$precision = \frac{1}{C} \sum_{i=1}^{C} \frac{t_i}{N_i}, \tag{13}$$

where C is the number of identities or classes, t_i is the correct images, and N_i is the number of images in the ith identity or class.

Compared Methods. In the experiment, we compare three other different methods with HRGM-pLSA, including SVM, HAP, and the least-square method. We use the SVM to train the classifiers of the predefined attributes, and use them to predict the attributes of the unknown images. HAP [8] employs hypergraph learning to learn the new attribute predictors. The least-square method is used to get the attribute predictors from the labeled attributes directly.

4.2 Experiment Results

Image Retrieval. We report the performance of four different attributes in the attribute-based hypergraph learning [10], including the least-square method, primary attributes, HAP, and HRGM-pLSA (our model), where the primary attributes are predicted by SVM. We train the least-square method based on the original attributes, and we train the HAP and HRGM-pLSA based on the primary attributes. The results of the retrieval precision in the dataset CelebA based on the distance of L1, L2, SCD, Chi2 are complemented in Table 1.

Here, k is 2, the λ is set to 0.001 to prevent the probability being a negative value, and the number of latent topics H is set to 31 to be consistent with HAP and primary attributes. According to the Table 1, our model has the best performance in all four metrics. The precision of HRGM-pLSA over primary attributes is 0.12% on average. We find that the metric of L1 and L2 can more

---

[1] https://cvml.ist.ac.at/AwA/.

Table 1. The retrieval precision in CelebA

Approaches	Mean retrieval precision (%)			
	L1	L2	SCD	Chi2
Least squares	77.21	77.09	77.08	78.18
Primary attributes [2,5]	77.49	77.27	77.06	78.17
HAP [8]	77.52	77.27	77.07	78.18
HRGM-PLSA (our model)	77.70	77.45	77.10	78.23

easily present the quality of different attributes. Why is the improvement of the precision so little? We think that the number of attributes in the CelebA is small, so we can hardly learn the latent semantic attributes which can discriminate the images better. The experiments in the AwA2 confirm our view.

Table 2. The retrieval precision in AwA2

Approaches	Mean retrieval precision (%)			
	L1	L2	SCD	Chi2
Least squares	67.29	70.29	66.12	70.86
Primary attributes [2,5]	66.65	69.82	65.30	70.34
HAP [8]	66.81	69.96	65.48	70.49
HRGM-PLSA (our model)	68.75	71.37	66.76	71.66

The results of the retrieval precision in the dataset AwA2 is shown in Table 2. Here, the number of the latent topic H is set to 85 to also keep with primary attributes and HAP. From the Table 2, the precision of HRGM-pLSA over primary attributes is 1.61% on average. In AwA2, our model also gets the best precision compared to other attributes, and the L1 metric can more easily show the quality of different attributes. With the increase of the number of latent topics, our model can learn the latent semantic attributes which have discrimination more easily and get better performance. Actually, if we set the H bigger, we will get better performance, but the time cost will increase rapidly, and the improvement of accuracy may not be a lot. So how to balance the time cost and the accuracy should be considered.

Computational Cost. The computational complexity of the HRGM-pLSA for training is mainly at the M step to compute the new Σ_k. The time complexity is $O(kn^2 \cdot n^2)$, and the multiple of the vector and its transposition captures most of the time. The model is usually converged after two steps. The experimental environment is on the server, and the hardware configuration is CPU: Xeon E5-2650 2.2 GHz, GPU: Quadro P5000.

5 Conclusion

We propose an attribute learning framework named Hypergraph-Regularized pLSA with Gaussian Mixture (HRGM-pLSA) which can learn the latent semantic attributes based on the attributes that humans have defined. Our model creatively employs a hypergraph to present the complex correlation in the attributes and integrate it into GM-pLSA as a regularization term to optimize the attributes. Our model gets better performance in the task of image retrieval compared to three other kinds of attributes, and the results of the experiment demonstrate the value of our work.

In the future, we will seek attribute predictors of our model, which can map the feature space to the attribute embedding space. We can find the latent attributes of unknown images much more effectively.

Acknowledgement. The work is supported by the National Natural Science Foundation of China under Grant No.: 61976132 and the National Natural Science Foundation of Shanghai under Grant No.: 19ZR1419200.

References

1. Datta, R., Joshi, D., Li, J., et al.: Image retrieval: Ideas, influences, and trends of the new age. ACM Comput. Surv. **40**(2), 5 (2008)
2. Kumar, N., Berg, A.C., Belhumeur, P.N., et al.: Attribute and simile classifiers for face verification. In: IEEE 12th International Conference on Computer Vision, pp. 365–372 (2009)
3. Yin, Z., Zheng, W., Wu, A., et al.: Adversarial attribute-image person re-identification. In: International Joint Conference on Artificial Intelligence, pp. 1100–1106 (2018)
4. Lampert, C., Nickisch, H., Harmeling, S.: Attribute-based classification for zero-shot visual object categorization. IEEE Trans. Pattern Anal. Mach. Intell. **36**(3), 453–465 (2014)
5. Lampert, C., Nickisch, H., Harmeling, S.: Learning to detect unseen object classes by between-class attribute transfer. In: Computer Vision and Pattern Recognition, pp. 951–958 (2009)
6. Horster, E., Lienhart, R., Slaney, M., et al.: Continuous visual vocabulary models for pLSA-based scene recognition. In: Conference on Image and Video Retrieval, pp. 319–328 (2008)
7. Zhong, C., Miao, Z.: Graph regularized GM-pLSA and its applications to video content analysis. Multimedia Syst. **20**(4), 429–445 (2014)
8. Huang, S., Elhoseiny, M., Elgammal A.M., et al.: Learning hypergraph-regularized attribute predictors. In: Computer Vision and Pattern Recognition, pp. 409–417 (2015)
9. Cai, J., Zha, Z.J., Wang, M., et al.: An attribute assisted reranking model for web image search. IEEE Trans. Image Process. **24**(1), 261–272 (2014)
10. Fang, Y., Zheng, Y.: Metric learning based on attribute hypergraph. In: International Conference on Image Processing, pp. 3440–3444 (2017)
11. Berge, C.: Graphs and Hypergraphs. North-Holland, Amsterdam (1973)

12. Liu, Z., Luo, P., Wang, X., et al.: Deep learning face attributes in the wild. In: International Conference on Computer Vision, pp. 3730–3738 (2015)
13. Xian, Y., Lampert, C.H., Schiele, B., et al.: Zero-shot learning - a comprehensive evaluation of the good, the bad and the ugly. IEEE Trans. Pattern Anal. Mach. Intell. **40**(8), 1 (2018)
14. Ahonen, T., Hadid, A., Pietikainen, M., et al.: Face description with local binary patterns: application to face recognition. IEEE Trans. Pattern Anal. Mach. Intell. **28**(12), 2037–2041 (2006)
15. Zhang, N., Farrell, R., Iandola, F.N., et al.: Deformable part descriptors for fine-grained recognition and attribute prediction. In: International Conference on Computer Vision, pp. 729–736 (2013)
16. Abdulnabi, A.H., Wang, G., Lu, J., et al.: Multi-task CNN model for attribute prediction. IEEE Trans. Multimed. **17**(11), 1949–1959 (2015)
17. Koren, Y.: Factorization meets the neighborhood: a multifaceted collaborative filtering model. In: Knowledge Discovery and Data Mining, pp. 426–434 (2008)

Interactive Semantic Features Selection from Reviews for Recommendation

Tian Shi[1], Bofeng Zhang[1(✉)], Ying Lv[1], Zhuocheng Zhou[1],
and Furong Chang[1,2]

[1] School of Computer Engineering and Science,
Shanghai University, Shanghai, China
bfzhang@shu.edu.cn
[2] School of Computer Science and Technology, Kashi University,
Xinjiang, China

Abstract. In recent years, reviews information has been effectively utilized by deep learning to improve the performance of the recommendation system and alleviate the problems of sparse data and cold start. However, there is much redundant information in the reviews that has a negative effect on the performance of the recommender system, which is ignored by most existing methods. In this paper, the Interactive Semantic Features Selection (ISFS) method is proposed to more effectively select the useful information from reviews based on attention mechanisms. Specifically, each word in reviews is interactively assigned a different weight according to the value of the semantic information it contains. Experiment results on real-world datasets show that ISFS outperforms baseline recommender systems on rating prediction tasks.

Keywords: Recommender system · Deep learning · Attention mechanism · Long-Short Term Memory · Rating prediction

1 Introduction

Nowadays recommendation system is very helpful for people to get information. The traditional collaborative filtering-based recommendation method [1] has performed well in many recommendation scenarios. However, when these methods encounter the problems of cold start and sparse data, their performance will have a significant decline. Therefore, people consider using some side information such as user portraits, item descriptions, and review texts to deal with these problems. One type of method is to use the high scalability of traditional matrix factorization methods to incorporate features extracted from various side information into the model [2, 3]. In recent years, the application of deep learning has made many breakthroughs in the field of natural language processing. So, there have been models that use neural networks to jointly model user and item from reviews [4–6]. Such models not only achieve better scoring prediction accuracy, but also have better interpretability than traditional models, which will help to increase user's trust in the system. These advantages are mainly benefit from the large amount of useful information contained in the review text.

T. Gedeon et al. (Eds.): ICONIP 2019, CCIS 1143, pp. 541–549, 2019.
https://doi.org/10.1007/978-3-030-36802-9_58

However, existing methods use all user reviews and item reviews as inputs to the model, which makes it difficult for the model to obtain useful information from reviews containing a large amount of redundant information. In this paper, we think that there are two types of redundant information in the review text: (1) The words that are weakly associated with the item being evaluated. (2) The words that describe the characteristics of the item that the target user does not care about.

In order to reduce the impact of these redundant information on the accuracy of scoring prediction, we proposed the Interactive Semantic Features Selection (ISFS) method in this paper, which utilizes the attention model to interactively select useful information from the reviews. Two parallel neural networks are used to model user behaviors and item properties jointly. One is the user network, which uses the user's historical review data to model the user behaviors, and the other is the item network uses all the reviews of the item to model the item properties. Specifically, the self-attention mechanism was used to assign different weights to the hidden outputs of LSTM in the user network and add them to get the temporary representation of the user. Then the attention layer calculates the final representation of the item according to the correlation between the hidden outputs of LSTM in the item network and the temporary representation of the user. Similarly, we can get the final representation of the user. Finally, the final representations of the user and item are fed into a regression model to predict the user's rating of the item. Experimental results based on real-world dataset show that ISFS outperforms all baseline recommender systems.

2   Related Work

2.1   Traditional Recommender Systems

In traditional recommender systems, collaborative filtering according to the similarity of interest preference between users to find user's potential preferences for the items [1, 7]. However, because this method relies too much on the historical interaction information between users and items, it is vulnerable to the problems of sparse data and cold start. The content-based recommendation method finds items that are similar in content to the items in the user's interaction history, and then recommends the items with high similarity to the user. However, the traditional method is based on artificial characteristic engineering, which leads to poor effect and scalability of the model. Considering the shortcomings of the above two recommendation methods, the hybrid recommendation method combines above two methods based on the fusion of multi-source heterogeneous edge information. The challenge of this method depending on how to map multi-source heterogeneous side information into a same feature space.

2.2   Deep Learning for Recommendation

Recently, deep learning has been used not only to extract features from side information [1] but also to model the interactions between users and items [8]. However, Cheng et al. considered that deep neural networks may recommend irrelevant items when user-item interactions are sparse and high-rank, so proposed Wide & Deep

Learning to improve the generalization ability [9]. In addition, recurrent neural network was used to integrate time series information into collaborative filtering to capture dynamic changes of user interests [10, 11].

On the recommender systems that utilize reviews, deep learning was often used to capture semantic features in reviews. Wang et al. proposed Collaborative Deep Learning (CDL) based on Stack Denoising Auto-Encoder (SDAE) and Probabilistic Matrix Factorization (PMF) [12]. Kim proposed ConvMF [3], which used convolutional neural network to learn implicit representations of users and items from reviews. And then Zheng et al. proposed DeepCoNN model, which was the first model that used neural network to jointly model users and items from reviews [4]. However, these models ignored the impact of redundant information when extracting semantic features from reviews.

3   Interactive Semantic Features Selection

In this section, we introduce our proposed ISFS method in detail. First, we will present the general architecture of the ISFS and the overall workflow of the model. Then, we will introduce the representation method for words in reviews and how to learn semantic features from reviews. After that, we will show the interactive semantic feature selection method based on attention mechanism which is the main concern in this paper. Finally, we will discuss how to use the implicit representation for users and items to generates recommendations and how to train our model more effectively.

3.1   Architecture

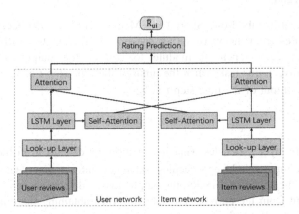

Fig. 1. The algorithm flowchart of ISFS.

The task of our proposed model, ISFS, is to predict item ratings given by users. The whole model is mainly composed of two parallel networks, one is user network Net_u and the other is item network Net_i, which have a same structure but different parameters. When we were predicting user u's rating of item i, user network's input $w^u_{1:n}$ is made up of all review texts from user u of n words except the target user's review on the target item Rev_{ui}, and item network's input $w^i_{1:m}$ is made up of all review texts from item i of m words except the Rev_{ui} (Fig. 1).

Next, the Look-up layer maps each word in the $w^u_{1:n}$ and $w^i_{1:m}$ into the vector space as the inputs of LSTM, so as to obtain the hidden layer outputs $h^u_{1:n}$ and $h^i_{1:m}$ of LSTM. Then, the first layer of attention selects useful semantic features x'_u and x'_i as temporary representation of user u and item i from $h^u_{1:n}$ and $h^i_{1:m}$ respectively.

The second layer of attention uses the user's temporary representation x'_u and the output of LSTM $h^i_{1:m}$ to interactively generate the final representation x_i for the item. Similarly, use the same method to generate the final representation x_u for the user.

Finally, a regression model is used to predict the user u's rating of item i based on x_u and x_i.

3.2   Look-up Layer

The purpose of the Look-up layer is to convert the words expressed in natural language into vectors or matrices that computers can understand. We obtain the vector representation of each word through the lookup layer while retaining the sequence information. Specifically, based on the pre-trained word vector glove6B provided by GloVe [13], we get the word vector V^k_j corresponding to the j-th word in the review text k through a search function $f\left(w^k_j\right)$, and then connect the word vector of n words to get the word vector matrix $V^k_{1:n}$.

3.3   LSTM Layer

Next layer, we utilize the Long-Short Term Memory (LSTM) architecture to learn the semantic features from the word vector sequence. Different from RNN, LSTM has three gates in each repeating module to control the output of current time step. Specifically, in this task, when the input of the time step t of LSTM is the word vector V^k_t, the output of the time step t is h_t:

$$h_t = o_t * \tanh(C_t) \tag{1}$$

The forget gate layer f_t decides what information we discard from the old cell state C_{t-1}, while the input gate layer i_t decides which new information created by $\tilde{C}_t$ will be deposited into the cell state. Finally, the output gate o_t determines what information is going to output from the current cell state C_t. After obtaining the output $h_{1:n}$ of the LSTM layer, we will use the attention mechanism to select the important information for rating prediction.

3.4   Attention Layers

In order to interactively select more useful information from the semantic features captured by the LSTM layer, this layer contains two layers of attention. The first layer of attention selects important information from the hidden layer outputs of LSTM as temporary representations of user and item. The second layer of attention uses the temporary representations of user and item to interactively guide the generation of the final representations of the item and user.

Self-attention Layer. In this layer, the temporary representations of user and item are obtained separately by adding the hidden layer outputs of LSTM in the user network and item network according to different weights.

The temporary representation x'_u of user u is calculated by:

$$x'_u = \sum_{i=1}^{n} \alpha'_i h^u_{1:n} \tag{2}$$

where the $\alpha'_{1:n}$ is attention vector generated by the attention mechanism:

$$\alpha'_{1:n} = softmax\left(\tanh\left(h^u_{1:n} \cdot w'_\alpha + b'_\alpha\right)\right) \tag{3}$$

where the w'_α and b'_α are weight matrix and bias respectively, tanh is a non-linear function and *softmax* is a normalized exponential function. And temporary representation x'_i of item i can be calculated similarly.

Attention Layer. This layer uses the temporary representations of user and item to interactively guide the generation of the final representations of the item and user. Specifically, for the user, the attention vector $\alpha_{1:n}$ is calculated by:

$$\alpha_{1:n} = softmax\left(\tanh\left(h^u_{1:n} \cdot w_\alpha \cdot x'_i + b_\alpha\right)\right) \tag{4}$$

where the x'_i is the item's temporary representation, $h^u_{1:n}$ is the LSTM's hidden layer output of user network, and the w_α and b_α are weight matrix and bias respectively.

Then the final representation x_u of the user u is calculated by:

$$x_u = \sum_{i=1}^{n} \alpha_i h^u_{1:n} \tag{5}$$

For the item, the attention vector $\beta_{1:n}$ can be calculated similarly.

3.5   Rating Prediction Layer

At the output layer, we enter the final representations of the user and item into a regression model based on Factorization Machine [6] for rating prediction. The regression model uses Eq. 6 to calculate the predicted rating $\hat{r}_{ui}$.

$$\hat{r}_{ui} = w_o + \sum_{i=1}^{|z|} w_i z_i + x^T_u \tag{6}$$

where $w_o \in R$ is the global bias, $w_i \in R^{2n}$ is the weight of dense layer, $x_u^T x_i$ is the dot product of the output from the second attention layers of Net_u and Net_i, and z_i is obtained by concatenating the final representations of the user and item.

3.6   Network Training

In ISFS, the parameters we optimized during the training process are from LSTM layer: $[W_o, W_C, W_i, W_f, b_o, b_C, b_i, b_f]$, the Attention layer: $[w, b, w', b']$, and the Rating prediction layer: $[w_o, w_i]$. The loss function and optimizer used in our training process are Mean Square Error and Adam optimizer respectively. In addition, dropout strategy and L2 constraint were added to improve the generalization ability of model.

4   Experiment

4.1   Datasets and Evaluation Metric

In our experiments, the performance of the ISFS was evaluated on five public datasets. Four of them (Musical, Instant video, Automotive, Cell phone) are review datasets from Amazon, including reviews spanning from May 1996 to July 2014. And the data from 2010 to 2011 from the open source dataset Yelp was selected for experiments. Some statistics of these datasets are shown in Table 1.

Table 1. The statistics of the datasets.

Dataset	#Users	#Items	#Reviews
Musical	1429	900	10261
Instant video	5130	1685	37126
Automotive	2928	1835	20473
Cell phone	27879	10429	194439
Yelp	44182	37739	175177

In order to more intuitively compare with the other works, the same evaluation metric, the Mean Square Error (MSE), was used to evaluate the validity of our model. The MSE is calculated as:

$$\text{MSE} = \frac{\sum_{i=1}^{N} (r_{ui} - \hat{r}_{ui})^2}{N} \tag{7}$$

where N indicates the number of users reviews on items in test set.

4.2   Experimental Settings

In our experiments, Words that not in the vocabulary were initialized with a vector of all zeros in the look up layer. The keep_prob of dropout layer, input text length and hidden layer size of the LSTM layer were 0.5, 260 and 64 respectively. In addition, all the weights in the experiments were initialized to values sampling from uniform distribution $U(-0.1, 0.1)$, and the biases were initialized with zeros. Finally, the L2 constraint during training was set to 0.001, and the training was terminated when the MSE does not decrease on the verification set for 3 epochs.

During training, we randomly selected 10% of data as the test set, 10% as the verification set, and the remaining 80% as the training set. About the choice of word embedding dimension, GloVe provides four dimensions of 50, 100, 200 and 300 for selection. It is found that ISFS can obtain best result when the word embedding dimension was 100 through multiple experiments. The specific experiment results are shown in Table 2.

Table 2. Comparing MSE results for different word embedding dimensions.

Embedding dimension	Musical	Instant video	Automotive	Cell phone	Yelp
50	0.7732	1.0792	0.9051	1.4034	1.1720
100	**0.7593**	**1.0599**	**0.8981**	**1.3776**	**1.0948**
200	0.7635	1.0612	0.9581	1.3832	1.2219
300	0.8091	1.0697	0.9627	1.3837	1.2010

4.3 Performance Evaluation

To verify the validity of our model, we compare ISFS against the following current state-of-the-art models: Matrix Factorization (MF), Probabilistic Matrix Factorization (PMF), Convolutional Matrix Factorization (ConvMF) and Deep Cooperative Neural Networks (DeepCoNN) mentioned above. The experiment results of ISFS and baselines are reported in terms of MSE in Table 3. Each result is the average of the experiments repeated 3 times on each dataset.

Table 3. MSE comparison with baselines.

Dataset	MF	PMF	ConvMF	DeepCoNN	ISFS
Musical	0.9965	0.9194	0.9069	0.8180	**0.7593**
Instant video	1.2328	1.2267	1.1243	1.0913	**1.0599**
Automotive	1.1002	1.1120	1.0602	1.0460	**0.8981**
Cell phone	1.6756	1.5111	1.4931	1.4368	**1.3776**
Yelp	1.5669	1.4484	1.3137	1.1660	**1.0948**

From the results, we can draw the following conclusions:

First, it is obvious that traditional models (MF and PMF) have shown a significant decline on Yelp dataset. Because they only use rating data and the Yelp dataset is sparse. In addition, the methods considering reviews (ConvMF, DeepCoNN and ISFS) perform better on all datasets than collaborative filtering methods (MF and PMF) considering only rating data. So, it shows that our model can get useful information from reviews to help improve performance and reduce the impact of data sparse.

Second, although the reviews are taken into account, our model and DeepCoNN 's score prediction accuracy is higher than ConvMF which based on matrix factorization. This is mainly because neural networks can model user behaviors and item properties in a nonlinear way. The nonlinear activation function was used in many places in our model and the dot product was used in the rating prediction layer, which are helpful for dealing with nonlinear problems.

Third, our model ISFS performs better than DeepCoNN on all datasets tested. Although they both use neural networks to model user behaviors and item properties from review information, our approach performs better because it differs from Deep-CoNN in the use of review information. In our model, the attention mechanism has been added to select the review information, which lead to a smaller MSE.

To illustrate that our model can select useful information from review's semantic features, a review text was selected from the test set as the input of trained model. As is shown in Fig. 2, the second layer attention weights of 260 words in the item network was visualized in a heat map. The brighter the colors in the picture, the heavier the weight of words.

Fig. 2. Visualization of attention weights.

Figure 3 shows the corresponding text fragment in Fig. 2. The highlighted words in Fig. 3 are the words with heavier attention weights. From this we can see that words with obvious emotional polarity are given heavier weights, which are very helpful for rating prediction. Such as best, great, good and etc. In addition, the back-end all black vector in Fig. 2 shows that if all zeros are used to fill short review texts, our model can also solve the problem of variable length input in recurrent neural networks.

> it beats watching a blank screen however i just dont seem to be in tune with to comedy of today this
> is the best of the best comedy stand up the fact that i was able to just watch continuously one
> comedian after another was great i had the best laughter i have had in a long time not bad didnt
> know any of the comedians but first time viewing put a smile on my face ill check out the next season
> soon funny interesting a great way to pass time i usually enjoy standup comedy and is this is a good
> show for me

Fig. 3. Words with heavier attention weights are highlighted.

5   Conclusion

In this paper, we proposed the Interactive Semantic Features Selection Networks (ISFS), which utilizes the attention model to select useful information from the reviews. We use two parallel neural networks to model user behaviors and item properties jointly. In particular, self-attention and norm-attention are used to make semantic feature selection to eliminate the negative effects of two types of redundant information. Finally, the final representations of the user and item are fed into a regression model to predict the user's rating of the item. Experimental results based on real-world datasets show that ISFS outperforms all baseline recommender systems. In addition, from the case study, our model can indeed select words with obvious emotional polarity semantic features from the reviews. Although this ability will degenerate as the length of the review text increases, it is better than the average pooling and maximum pooling in the rating prediction task.

Acknowledgment. This work was partially supported by the National Key R&D Program of China grant (No. 2017YFC0907505) and the Xinjiang Natural Science Foundation (No. 2016D01B010).

References

1. Koren, Y.: Factorization meets the neighborhood: a multifaceted collaborative filtering model. In: The 14th ACM SIGKDD International Conference on Knowledge Discovery and Data Mining, pp. 426–434. ACM (2008)
2. Wang, C., Blei, D.M.: Collaborative topic modeling for recommending scientific articles. In: The 17th ACM SIGKDD International Conference on Knowledge Discovery and Data Mining, pp. 448–456. ACM (2011)
3. Kim, D., Park, C., et al.: Convolutional matrix factorization for document context-aware recommendation. In: The 10th ACM Conference on Recommender Systems, pp. 233–240. ACM (2016)
4. Zheng, L., Noroozi, V., et al.: Joint deep modeling of users and items using reviews for recommendation. In: The Tenth ACM International Conference on Web Search and Data Mining, pp. 425–434. ACM (2017)
5. Chen, C., Zhang, M., et al.: Neural attentional rating regression with review-level explanations. In: The 2018 World Wide Web Conference, pp. 1583–1592 (2018)
6. Catherine, R., Cohen, W.: TransNets: learning to transform for recommendation. In: The Eleventh ACM Conference on Recommender Systems, pp. 288–296. ACM (2017)
7. Salakhutdinov, R., Mnih, A.: Probabilistic matrix factorization. In: The 20th International Conference on Neural Information Processing Systems, pp. 1257–1264. Curran Associates Inc. (2007)
8. Sedhain, S., Menon, A.K., et al.: AutoRec: autoencoders meet collaborative filtering. In: International Conference on World Wide Web, pp. 111–112. ACM (2015)
9. Cheng, H.-T., Koc, L., et al.: Wide & deep learning for recommender systems. In: The 1st Workshop on Deep Learning for Recommender Systems, pp. 7–10. ACM (2016)
10. Dai, H., Wang, Y., et al.: Recurrent coevolutionary latent feature processes for continuous-time recommendation. In: The 1st Workshop on Deep Learning for Recommender Systems, pp. 29–34. ACM (2016)
11. Wu, C., Ahmed, A., et al.: Recurrent recommender networks. In: the Tenth ACM International Conference on Web Search and Data Mining, pp. 495–503. ACM (2017)
12. Wang, H., Wang, N., et al.: Collaborative Deep Learning for Recommender Systems. Computer Science (2014)
13. Pennington, J., Socher, R., et al.: GloVe: global vectors for word representation (2014). https://nlp.stanford.edu/projects/glove/

DXNet: An Encoder-Decoder Architecture with XSPP for Semantic Image Segmentation in Street Scenes

Yexin Shang[1], Shan Zhong[2], Shengrong Gong[1,2(✉)], Lifan Zhou[2], and Wenhao Ying[2]

[1] School of Computer Science and Technology, Soochow University, Suzhou 215006, China
`20175227094@stu.suda.edu.cn`
[2] School of Computer Science and Engineering, Changshu Institute of Technology, Suzhou 215500, China
{`sunshine620,shrgong,zhoulifan`}`@cslg.edu.cn, cslgywh@163.com`

Abstract. Semantic image segmentation plays a crucial role in scene understanding tasks. In autonomous driving, the driving of the vehicle causes the scale changes of objects in the street scene. Although multi-scale features can be learned through concatenating multiple different atrous-convolved features, it is difficult to accurately segment pedestrians with only partial feature information due to factors such as occlusion. Therefore, we propose a Xiphoid Spatial Pyramid Pooling method integrated with detailed information. This method, while connecting the features of multiple atrous-convolved, retains the image-level features of target boundary information. Based on the above methods, we design an encoder-decoder architecture called DXNet. The encoder is composed of a deep convolution neural network and two XSPP modules, and the decoder decodes the advanced features through up-sampling operation and skips connection to gradually restore the target boundary. We evaluate the effectiveness of our approach on the Cityscapes dataset. Experimental results show that our method performs better in the case of occlusion, and the mean intersection-over-union score of our model outperforms some representative works.

Keywords: Semantic segmentation · Encoder decoder · Spatial pyramid pooling · Convolutional neural network

1 Introduction

In recent years, autonomous driving attracts more and more interest. Acquiring a comprehensive understanding of the street scene is vital for autonomous driving. Semantic image segmentation [1,2] is a crucial tool for understanding the complex relationships of semantic entities in street scenes, such as road, pedestrians, cars, and sidewalks.

© Springer Nature Switzerland AG 2019
T. Gedeon et al. (Eds.): ICONIP 2019, CCIS 1143, pp. 550–557, 2019.
https://doi.org/10.1007/978-3-030-36802-9_59

In autonomous driving, the same category such as pedestrians varies largely in scale caused by distance to the camera. Besides, under the influence of lighting and other factors, it is difficult to accurately segment pedestrians blocked by obstacles such as vehicles or trees in the street scene.

State-of-the-art semantic segmentation frameworks are mostly based on the fully convolutional network (FCN) [3]. Towards accurate semantic segmentation for street scene, knowledge graph not only depends on the prior information of scene context but also depends on the boundary location information of the target. Experiences [4–6] show that multi-scale information may help resolve ambiguous examples and lead to more robust classification. Atrous Spatial Pyramid Pooling (ASPP) [4,7] was aimed to concatenate atrous-convolved feature maps with different atrous rates so that the output feature maps contain multiple receptive field sizes. However, many pixels are misclassified because this strategy may isolate the pixels from the global scene context. We found that the possible problem for current ASPP based models is the lack of boundary information to accurately segment the boundary part of the objects.

Different from these methods, to incorporate boundary features, we propose a double XSPP network (DXNet). In addition to the ASPP module for pixel prediction, we extend the multi-scale feature to the boundary information. The global and local clues together make the final prediction more reliable. We evaluate our model on the Cityscapes dataset [8] and achieve better performance in the case of occlusion. To summarize, our work makes two contributions as follows:

(1) We introduced an SPP-based method called XSPP, which can generate features that contain multi-scale information to remedy challenging scale variations in street scenes.
(2) We propose a new encoder-decoder structure called DXNet by combining XSPP. Except for other encoder-decoders, the encoder can capture very large scale range information by employing two XSPP modules. Furthermore, the corresponding decoder can gradually recover the boundaries of objects by using skip connections and up-sampling operations.

2   Related Work

Spatial Pyramid Pooling: Spatial pyramid pooling [9] was employed to capture context information at several ranges. ParseNet [10] exploited the image-level features by performing spatial pyramid pooling to capture global context information. DeepLabV2 [4] applied several parallel atrous convolutions with different rates (called Atrous Spatial Pyramid Pooling, or ASPP) to capture multi-scale information. Pyramid Scene Parsing Net (PSPNet) [6] implemented spatial pooling at several grid scales and demonstrated excellent performance on several semantic segmentation benchmarks. These models that exploited the multi-scale information have shown promising results on several semantic segmentation benchmarks.

Encoder-Decoder: Models using encoder-decoder architecture have been successfully applied to object detection [11] and semantic segmentation [12]. Typically, this architecture consists of an encoder module and a corresponding decoder module. The encoder module gradually reduced the spatial dimension of feature maps to capture higher semantic information for classification. The corresponding decoder module gradually recovers the spatial dimension and object details for sharp object boundaries of images. For example, SegNet [13] reused the pooling indices computed in the max-pooling step of the encoder and learned extra convolutional layers to produce dense feature maps. U-Net [14] added skip connections from the encoder features to the corresponding decoder activations. These works have demonstrated the effectiveness of encoder-decoder models on several semantic segmentation benchmarks.

However, the repeated combinations of max-pooling and convolution operations of these networks reduce the spatial resolution of the resulting feature maps significantly, which leads to the loss of location information for the objects. In typical street scenes, the boundary information of the objects helps to better understand the scenes with occlusion. Hence it is important to retain the location information in the extracted image feature maps. The encoder-decoder networks can generate sharper object boundaries by gradually recovering the spatial information. But the normal encoder module can not encode multi-scale contextual information. Combining the above elements, we design a spatial pyramid pooling module composed of atrous convolutions as an encoder to capture multi-scale context information and construct an effective decoder module to obtain more refined boundaries.

3   DXNet

3.1   Overall Framework

Our proposed model is an encoder-decoder architecture, illustrated in Fig. 1.

Proposed Encoder: Firstly, we employ atrous convolution to extract multi-scale features computed by deep convolutional neural networks. Additionally, the proposed XSPP module augments the atrous spatial pyramid pooling module with the image-level features by image pooling operation. In this work, motivated by DenseNets [15], we further applying multiple XSPP modules. The output feature of former XSPP will be the input of the next XSPP, which more informative than DeepLabV3+.

Proposed Decoder: Through the processing of the encoder, we can obtain high-level semantic information of multi-scale objects. Then we employ skip connection and up-sample operator to recover object segmentation details. We apply convolution on the encoder features to reduce the number of channels. To obtain the same spatial resolution, the output features from the $XSPP_2$ are first bilinear up-sampled by a factor of 2 and then concatenated with the features from $XSPP_1$ module. After first concatenation, the concatenated features are bilinear up-sampled by a factor of 2 and then concatenated with the output features from

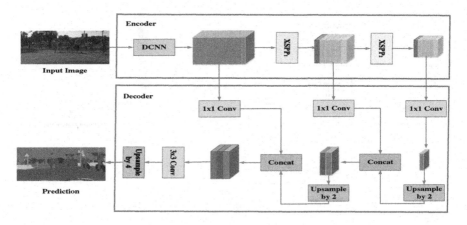

Fig. 1. Our proposed model DXNet is an encoder-decoder architecture. The encoder module can better present high-level feature with multi-scale information, while the decoder module gradually recover the boundaries of objects by skip connection and up-sample operations.

the deep convolution neural network. After this operation, we employ convolution to refine the features. In the end, we apply another simple bilinear up-sampling by a factor of 4 to obtain the final prediction. The experiment shows that using output stride $s = 16$ for the encoder module can achieve the best results for the task of semantic segmentation.

3.2 Xiphoid Spatial Pyramid Pooling

Atrous convolution [4,16] is a powerful tool that allows us to control the resolution of features computed within deep convolutional neural networks explicitly. In the case of one dimension, for each location i on the output feature map y, we formulate atrous convolution over the input feature map x as follows:

$$y[i] = \sum_{k=1}^{K} x[i + r \cdot k] \cdot w[k] \tag{1}$$

where r is the atrous rate, K denotes the filter size, and $w[k]$ is the k-th parameter of filter. The atrous rate r corresponds to the stride with which we sample the input signal. To capture multi-scaled feature maps, we can enlarge the receptive field by increase the atrous rate.

In order to simplify notations, we use $S_{K,r}(x)$ to denote an atrous convolution. XSPP in Fig. 2 can be formulated as follows:

$$y = S_{1,1}(x) + S_{3,3}(x) + S_{3,6}(x) + S_{3,12}(x) + S_{3,18}(x) + P \tag{2}$$

where P means the average pooling. XSPP is an effective module for obtaining different scale ranges, as shown in Fig. 3. Compared to ASPP [4], XSPP is able

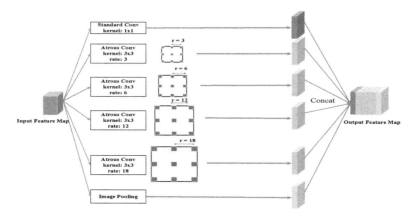

Fig. 2. The XSPP module. It consists of a 1×1 convolution, ASPP and an image pooling operation. We utilize ASPP [4] module to capture multi-scale context information.

to acquire feature maps with better scale diversity, and more pixels are involved in convolution. Based on ASPP, the key design of XSPP is adding a standard 1×1 convolution and an image pooling. This makes the entire filter like a sword, which is why we call it the xiphoid spatial pyramid pooling. The main reason for choosing a 1×1 convolution is that the output feature maps retaining much of the original objects' position information. Moreover, the purpose of adding image pooling is to obtain image-level features that contain not only a higher level of semantic information but also a portion of the location information. Since the receptive field in ASPP is pyramid-shaped, each feature map contains different scales of information, but finally the size of the feature map output by each layer of the pyramid is the same.

4 Experiments

4.1 Implementation Details

We implemented our proposed methods using Tensorflow, the popular deep learning framework. We employ the powerful aligned Xception [17] which has been modified by [7] as our network backbone. We follow the same training protocol as in [7]. The aligned Xception network with a similar training protocol in [17] is pre-trained on the ImageNet-1k dataset. During training, we use the Nesterov momentum optimizer with momentum $m = 0.9$. We adopt this with an initial learning rate of 0.05 and a weight decay of $4e - 5$. And the rate decay is set to 0.94 every 2 epoch. We apply the mean square error as the loss function in our experiment to minimize the intersection over the union. We trained with 4 GPUs for all the experiments. In practice, larger crop size consumes larger GPU memory which leads to training with smaller batch size for Batch Normalization, even though larger crop size can yield better performance for semantic image segmentation. Taking into account the memory of our GPUs, we train our model with each GPU has batch size 2 with an image size 512×512.

4.2 Ablation Study

In order to analyze the influence of different components in our framework, we design different runs in Cityscapes and report the results in Table 1. As shown in Table 1, the first line is the segmentation result obtained on the base model without the XSPP module and the decoder module. The second line shows that the result of the segmentation has been refined after the decoder module has been added. In the third experiment, we considered adding our proposed XSPP module to the model but without coding and decoding. The experimental results show that the model's segmentation ability is significantly improved by extracting multi-scale semantic information. Finally, we combine the two modules to get the encoder-decoder model we introduced in the previous section, and get good results.

Table 1. Results on Cityscapes database with different segmentation branch.

Backbone	XSPP	Decoder	mIoU (%)
Xception65 (baseline)			65.87
Xception65+Decoder		√	67.76
Xception65+XSPP	√		71.23
Xception65+XSPP+Decoder	√	√	72.68

4.3 Quantitative Evaluation

We compare our method with the state-of-art methods on Cityscapes [8], and the results are illustrated in Table 2. Noted that we use the results reported in the original paper instead of the Cityscapes leader board. In [7], after pre-trained on both ImageNet-1k and JFT-300M dataset, DeepLabV3+ was asynchronous trained with 50 GPUs with performance of 82.66% $mIoU$. However, we compare our model with the DeepLabV3+ model under the same environment, iterating over the same epochs. The experimental results show that our model is more accurate in the case of small scale and occlusion. These results further demonstrate the effectiveness of our approach.

Some examples of semantic image segmentation results on Cityscapes can be seen in Fig. 3.

5 Conclusion

In this paper, we propose DXNet to alleviate the challenging problem of semantic segmentation in a street scenes that objects vary largely in scale. Although our

Table 2. Experimental results on Cityscapes test set.

Method	mIoU (%)
FCN 8s [3]	65.3
Dilation10 [16]	67.1
DeepLabV2-CRF [4]	70.4
FRRN [1]	71.8
DeepLabV3+ [7]	70.89 (local)
Our approach	72.68

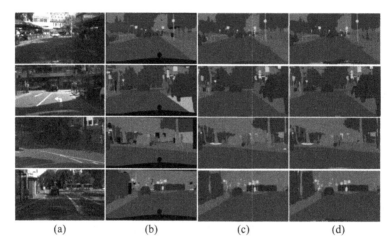

    (a)   (b)   (c)   (d)

Fig. 3. Qualitative results on the Cityscapes validation set. (a) original image, (b) ground truth annotation, (c) DeeplabV3+ [7], (d) our results.

method is not the best semantic segmentation method on the street view data set, our method performs better than some representative works, especially on some objects with only small pixels. We hope this article will contribute to the development of semantic segmentation and advance related techniques.

Acknowledgments. The authors would like to thank the anonymous reviewers for their helpful and constructive comments. This work was partially supported by the National Natural Science Foundation of China (NSFC Grant No. 61972059, 61702055, 61773272, 61272059) Natural Science Foundation of Jiangsu Province under Grant (BK20191474, BK20161268). Research and Innovation Fund of the Science and Technology Development Center of the Ministry of Education (2018A01007), and Ministry of Education Science and Technology Development Center Industry-University Research Innovation Fund (2018A02003), and Humanities and Social Sciences Foundation of the Ministry of Education under Grant 18YJCZH229.

References

1. Pohlen, T., Hermans, A., Mathias, M., Leibe, B.: Full-resolution residual networks for semantic segmentation in street scenes. In: 2017 IEEE Conference on Computer Vision and Pattern Recognition (CVPR), pp. 3309–3318 (2017)
2. Yang, M., Yu, K., Zhang, C., Li, Z., Yang, K.: DenseASPP for semantic segmentation in street scenes. In: Proceedings of the IEEE Conference on Computer Vision and Pattern Recognition, pp. 3684–3692 (2018)
3. Shelhamer, E., Long, J., Darrell, T.: Fully convolutional networks for semantic segmentation. In: 2015 IEEE Conference on Computer Vision and Pattern Recognition (CVPR), pp. 3431–3440 (2015)
4. Chen, L.-C., Papandreou, G., Kokkinos, I., Murphy, K., Yuille, A.L.: DeepLab: semantic image segmentation with deep convolutional nets, atrous convolution, and fully connected CRFs. IEEE Trans. Pattern Anal. Mach. Intell. 40, 834–848 (2018)
5. Chen, L.-C., Papandreou, G., Schroff, F., Adam, H.: Rethinking atrous convolution for semantic image segmentation. CoRR, abs/1706.05587 (2017)
6. Zhao, H., Shi, J., Qi, X., Wang, X., Jia, J.: Pyramid scene parsing network. In: 2017 IEEE Conference on Computer Vision and Pattern Recognition (CVPR), pp. 6230–6239 (2017)
7. Chen, L.-C., Zhu, Y., Papandreou, G., Schroff, F., Adam, H.: Encoder-decoder with atrous separable convolution for semantic image segmentation. In: ECCV (2018)
8. Cordts, M., et al.: The cityscapes dataset for semantic urban scene understanding. In: 2016 IEEE Conference on Computer Vision and Pattern Recognition (CVPR), pp. 3213–3223 (2016)
9. He, K., Zhang, X., Ren, S., Sun, J.: Spatial pyramid pooling in deep convolutional networks for visual recognition. IEEE Trans. Pattern Anal. Mach. Intell. 37, 1904–1916 (2014)
10. Liu, W., Rabinovich, A., Berg, A.C.: ParseNet: looking wider to see better. CoRR, abs/1506.04579 (2015)
11. Fu, C.-Y., Liu, W., Ranga, A., Tyagi, A., Berg, A.C.: Dssd: deconvolutional single shot detector. CoRR, abs/1701.06659 (2017)
12. Lin, G., Milan, A., Shen, C., Reid, I.D.: RefineNet: multi-path refinement networks for high-resolution semantic segmentation. In: 2017 IEEE Conference on Computer Vision and Pattern Recognition (CVPR), pp. 5168–5177 (2017)
13. Badrinarayanan, V., Kendall, A., Cipolla, R.: SegNet: a deep convolutional encoder-decoder architecture for image segmentation. IEEE Trans. Pattern Anal. Mach. Intell. 39, 2481–2495 (2016)
14. Ronneberger, O., Fischer, P., Brox, T.: Convolutional networks for biomedical image segmentation. In: MICCAI, U-net (2015)
15. Huang, G., Liu, Z., Van Der Maaten, L., Weinberger, K.Q.: Densely connected convolutional networks. In: Proceedings of the IEEE Conference on Computer Vision and Pattern Recognition, pp. 4700–4708 (2017)
16. Yu, F., Koltun, V.: Multi-scale context aggregation by dilated convolutions. CoRR, abs/1511.07122 (2016)
17. Chollet, F.: Xception: deep learning with depthwise separable convolutions. In: 2017 IEEE Conference on Computer Vision and Pattern Recognition (CVPR), pp. 1800–1807 (2017)

Social Network Computing

Anchor User Oriented Accordant Embedding for User Identity Linkage

Xiang Li[1,2,3], Yijun Su[2,3], Neng Gao[1,3], Wei Tang[1,2,3], Ji Xiang[3(✉)],
and Yuewu Wang[1,2,3]

[1] State Key Laboratory of Information Security,
Institute of Information Engineering, CAS, Beijing, China
{lixiang9015,gaoneng,tangwei,wangyuewu}@iie.ac.cn
[2] School of Cyber Security, University of Chinese Academy of Sciences,
Beijing, China
[3] Institute of Information Engineering, Chinese Academy of Sciences, Beijing, China
{suyijun,xiangji}@iie.ac.cn

Abstract. User Identity Linkage is to find users belonging to the same real person in different social networks. Besides, anchor users refer to matching users known in advance. However, how to match users only based on network information is still very difficult and existing embedding methods suffer from the challenge of **error propagation**. Error propagation means the error occurring in learning some users' embeddings may be propagated and amplified to other users along with edges in the network. In this paper, we propose the Anchor UseR ORiented Accordant Embedding (AURORAE) method to learn the vector representation for each user in each social network by capturing useful network information and avoiding error propagation. Specifically, AURORAE learns the potential relations between anchor users and all users, which means each user is directly connected to all anchor users and the error cannot be propagated without paths. Then, AURORAE captures the useful local structure information into final embeddings under the constraint of accordant vector representations between anchor users. Experimental results on real-world datasets demonstrate that our method significantly outperforms other state-of-the-art methods.

1 Introduction

Nowadays, the fast development of communication technology makes people easy to enjoy "second life" in Internet. To meet different needs, people have been used to appear on multiple social networks. User Identity Linkage (UIL) aims to find users belonging to the same person in different social networks [12]. UIL is a significant task due to its ability of completing user portrait and fusing diverse information based on matching users. Hence, much more useful information can be delivered to many sequent applications, such as cross-network recommendation [2,11,14,15], link prediction [1,17,18] and topic analysis [6].

However, how to solve the UIL problem purely based on network information is still very hard. Existing methods apply the network embedding technique

© Springer Nature Switzerland AG 2019
T. Gedeon et al. (Eds.): ICONIP 2019, CCIS 1143, pp. 561–572, 2019.
https://doi.org/10.1007/978-3-030-36802-9_60

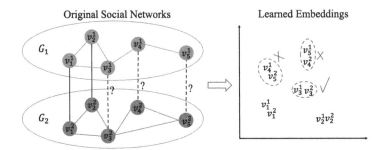

Fig. 1. A tiny example of error propagation. Given two social networks G_1 and G_2, the user space is formed by probable learned vector representations. v_i^j refers to i-th user in G_j and we already know matching user pairs (v_1^1, v_1^2) and (v_2^1, v_2^2). UIL aims to predict true matching user pairs (v_3^1, v_3^2), (v_4^1, v_4^2) and (v_5^1, v_5^2).

to learn the vector representation for each user with enough local structure information preserved [7–9,16,22]. These methods most pay much attention to the way of preseving enough useful information while ignore the main reason of misleading the learning process - **Error Propagation**. The challenge of error propagation means the error occurring in learning some users' embedding may be propagated and amplified along with the paths in networks. When the learned embeddings of two true matching users cannot correctly demonstrate they are a same real person, we think some errors happen during the process of learning embeddings. As shown in Fig. 1, from the user space, we cannot predict v_4^1 and v_4^2 are a same person. As a result, it's hard to correctly predict v_5^1 and v_5^2 are a same person due to error propagation. When user is far away from anchor users, the error would be amplified and we would not be able to predict correctly.

In this paper, to address the challenge of error propagation, we propose the **A**nchor **U**se**R** **OR**iented **A**ccordant **E**mbedding (AURORAE) method to learn the vector representation for each user in each social network by capturing useful network information and avoiding error propagation. Anchor user refers to user whose matching user has been given in advance. AURORAE learns potential graphs oriented at anchor users, which means each user is connected to all anchor users directly and not connected to users not being anchor users. By this way, due to full confidence of information from anchor users, error occurring during learning process has no direct path to be propagated and amplified. Then, when learning the user embeddings, AURORAE tends to capture useful local structure information and learn accordant embeddings to make users belonging to a same person much closer. Our contributions can be summarized as follows:

- We propose a new Anchor User Oriented Accordant Embedding method to solve the challenge of error propagation. To the best of our knowledge, we are the first to explicitly study the effect of error propagation and provide an effective solution purely based on network information.
- When handling error propagation, we give a direct but effective idea, which is to cut off the path propagating the error. Besides, a carefully devised

optimization algorithm is proposed to solve the final optimization problem formed by combining our idea with traditional embedding methods.

- Experiments on real-world Data sets demonstrate the effectiveness of our proposed method. Extension experiments show our method is superior to state-of-the-art methods under different conditions.

The rest of this paper is organized as follows: We review related work in Sect. 2. Section 3 presents our AURORAE approach and the devised optimization algorithm is proposed in Sect. 4. Experimental evaluation and comparison are shown in Sect. 5. Finally, Sect. 6 concludes the paper with a brief discussion.

2   Related Work

In this section, we review two main kinds of methods only using network information. Firstly, we introduce the traditional propagation methods. Then, we discuss recent work of embedding methods.

Propagation methods discover unknown user pairs in an iterative way from anchor users [12]. Formally, they firstly find probable matching user pairs directly connected to anchor users. Then, they view these user pairs as anchor users and predict probable matching user pairs directly connected to them. Finally, when all users have been scanned, these methods will be terminated [3,23]. Another kind of propagation methods is dividing all users into several subsets and then matching user pairs among same subset [4]. The core of propagation methods is to design reasonable similarity measures centering around anchor users [4,10]. For example, the similarity of one user pair can be computed using the number of shared identified friends [4,23]. Other metrics such as common neighbors, Jaccard's coefficient and Adamic/Adar score are extended to measure the neighborhood similarities as well [3]. However, we can find the error propagation occurring in the prediction of each propagation can explicitly influence the next prediction, which can be avoided by our AURORAE.

Recently, embedding methods have attracted much attention due to its ability of preserving network structure information. Spectral embedding has been studied at early [16]. PALE preserves neighbor links in users' representations and learns the linear/non-linear mapping among anchor users [9]. IONE models the followee/follower relationship and learn multiple representations for each user [7]. ABNE utilizes attention mechanism to distinguish different effects of neighbors [8]. DeepLink introduces the deep neural network based on the learned users' representations by random walk [22]. Existing embedding methods focus on preserving useful information and cannot avoid error propagation during learning, which is the main problem solved by our proposed AURORAE.

3   Proposed Method

We use $G_i = (V_i, A_i)$ to represent i-th social network. $A_i \in R^{n_i \times n_i}$ is the adjacency matrix, where 1 represents two users is connected. n_i means the total

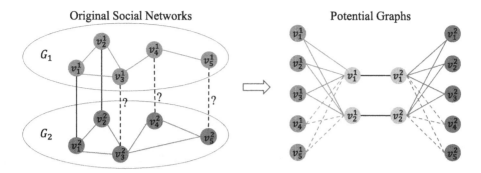

Fig. 2. Illustration of learning potential graphs oriented at anchor users. In the right, blue dash line and green dash line means edges to be learned. Blue solid line and green solid line means edges known. Besides, each edge owns a weight value to be learned, which is not shown in the figure. (Color figure online)

number of users in G_i. V_i refers to the set of all users and v_i^j means the i-th user in j-th social network. Besides, we use $H_i \in R^{n_i \times l}$ to represent the connection matrix between all users V_i and all anchor users. l is the number of all anchor users. m is the dimension of learned vector representation. In the sequel, we firstly present the way of learning potential graphs oriented at anchor users. Then, we discuss the way of learning accordant embeddings.

3.1 Learning Potential Graph Oriented at Anchor Users

In this section, we re-think the challenge of error propagation and solve it from an intuitive view. When learning the user representations, existing studies often think the information from anchor users are fully trustworthy. Besides, paths through questionable users is the essential reason of propagating and amplifying error. Hence, if questionable users cannot influence users connected to them, the error would have no path to be propagated and amplified. Naturally, as shown in Fig. 2, if all users are only connected to trustworthy anchor users and not directly connected to each other, the challenge of error propagation can be effectively avoided.

Formally, we can get the vector representations for the i-th social network by recovering original network:

$$\min_{U_i, Q_i} \frac{1}{2} ||H_i - U_i Q_i^T||_F^2 + \frac{\beta}{2} ||U_i||_F^2 , \tag{1}$$

where $U_i \in R^{n_i \times m}$ represents users' vector representations and $Q_i \in R^{l \times m}$ represents vector representations of all anchor users. β is to control the complexity of U_i and $|| \cdot ||_F$ stands for Frobenius norm.

Noting that objective (1) tries to recover connection matrix H_i exactly by user embeddings, which means only solid lines in Fig. 2 can be recovered. However, users connected to anchor users are so few that H_i is sparse and many users may be not connected to any anchor users such as users connected by dash lines in Fig. 2. In our experiment settings, we select 30% matching users as anchor users. Then, the ratio of nonzero values in H_i is 0.73%, 0.34%, 0.18% and 0.24% for social network dataset Twitter, BlogCatalog, Douban Online and Douban Offline respectively. Hence, the final user embeddings may contain many zero values and the information learned in the vector representations is not sufficient.

To solve the problem of sparse connections, we need to complete unknown connections. As shown in Fig. 2, we want to learn the weight of each dash line. Formally, we view the zero in H_i (e.g., dash line in Fig. 2) as missing edge rather than no connection. Then, we use collective matrix factorization technique [13] to complete missing edges and restrict the value in $[0, 1]$. Finally, the optimization problem for the i-th social network can be written as:

$$\min_{0 \leq U_i \leq 1, 0 \leq Q_i \leq 1} \frac{1}{2} ||H_i \odot (H_i - U_i Q_i^T)||_F^2 + \frac{\beta}{2} ||U_i||_F^2 , \tag{2}$$

where $\odot$ is the Hadamard (element-wise) product. To provide more information for learning the potential graph by (2), we need to preserve some local structure information:

$$\min_{0 \leq U_i \leq 1, 0 \leq Q_i \leq 1} \frac{1}{2} ||H_i \odot (H_i - U_i Q_i^T)||_F^2 + \frac{\alpha_i}{2} ||A_i - U_i U_i^T||_F^2 + \frac{\beta}{2} ||U_i||_F^2 , \tag{3}$$

where α_i is to control how much information to be preserved. Objective (3) can make the reconstruction loss of observed edges minimized and complete missing edges with the help of local structure information.

3.2   Learning Accordant Embedding

Optimization problem (3) can learn potential graph oriented at anchor users for each social network. As shown in Fig. 2, we can view it as two potential bipartite graphs connected by anchor users. Hence, we cannot independently learn the user embeddings for different social networks. Specifically, we should learn the accordant embedding under the constraint of anchor users.

We have two kinds of constraints of anchor users from two different aspects. On one hand, when we consider the interaction between two different social networks, we should restrict the distance between embeddings of users belonging to a same person as closer as possible. On the other hand, among each bipartite graph, we can find the the vector representations U_i also contain the embeddings of anchor users. Then, we should make the embeddings of anchor users in U_i is similar to Q_i. Finally, the optimization problem can be rewritten as:

$$\min_{0 \leq U_i \leq 1, 0 \leq Q_i \leq 1} ||Q_1 - Q_2||_F^2 + ||T_1 U_1 - Q_1||_F^2 + ||T_2 U_2 - Q_2||_F^2 , \tag{4}$$

where $T_i \in R^{l \times n_i}$ is the indicator matrix. $T_i(p, q) = 1$ if the q-th user belongs to the p-th real person. l is the number of anchor users and all anchor users are re-numbered from 1 to l.

In conclusion, the final optimization problem can be formulated as:

$$\min_{0 \leq U_i \leq 1, 0 \leq Q_i \leq 1} \sum_i \frac{1}{2} ||H_i \odot (H_i - U_i Q_i^T)||_F^2 + \frac{\alpha_i}{2} ||A_i - U_i U_i^T||_F^2 + \frac{\beta}{2} ||U_i||_F^2$$
$$\frac{\gamma}{2} (||Q_1 - Q_2||_F^2 + ||T_1 U_1 - Q_1||_F^2 + ||T_2 U_2 - Q_2||_F^2) , \qquad (5)$$

where γ is penalty parameter and we set it to a large value such as 10 compared to the small value of other parameters.

4 Optimization

Due to the nonconvexity of (5), we cannot get the optimal solution. Therefore, we utilize an alternative way to update U_i, Q_i by stochastic gradient method with multiplicative updating rules, which can ensure the nonnegativity of U_i and Q_i. After each update step, we apply the projection technique [5,19] to project elements greater than 1 in U_i and Q_i to 1. The whole algorithm is shown in Algorithm 1.

Optimize U_1, U_2: The partial derivatives of objective (5) w.r.t.$\{U_i\}$ are

$$\frac{\partial L}{\partial U_1} = H_1 \odot (U_1 Q_1^T - H_1) Q_1 + \alpha_1 (U_1 U_1^T - A_1) U_1 + \beta U_1 + \gamma T_1^T (T_1 U_1 - Q_1)$$
$$\frac{\partial L}{\partial U_2} = H_2 \odot (U_2 Q_2^T - H_2) Q_2 + \alpha_2 (U_2 U_2^T - A_2) U_2 + \beta U_2 + \gamma T_2^T (T_2 U_2 - Q_2) . \qquad (6)$$

Similar to classic nonnegative matrix factorization, we use following updating rules:

$$U_1 = U_1 \odot \sqrt{\frac{H_1 Q_1 + \alpha_1 A_1 U_1 + \gamma T_1^T Q_1}{(H_1 \odot (U_1 Q_1^T)) Q_1 + \alpha_1 U_1 U_1^T U_1 + \beta U_1 + \gamma T_1^T T_1 U_1}} \qquad (7)$$

$$U_2 = U_2 \odot \sqrt{\frac{H_2 Q_2 + \alpha_2 A_2 U_2 + \gamma T_2^T Q_2}{(H_2 \odot (U_2 Q_2^T)) Q_2 + \alpha_2 U_2 U_2^T U_2 + \beta U_2 + \gamma T_2^T T_2 U_2}} . \qquad (8)$$

We can find that it is easy to update U_1 and U_2 parallelly.

Optimize Q_1, Q_2: The partial derivates of objective (5) w.r.t.$\{Q_i\}$ are

$$\frac{\partial L}{\partial Q_1} = H_1^T \odot (Q_1 U_1^T - H_1^T) U_1 + \gamma(Q_1 - Q_2) + \gamma(Q_1 - T_1 U_1)$$
$$\frac{\partial L}{\partial Q_2} = H_2^T \odot (Q_2 U_2^T - H_2^T) U_2 + \gamma(Q_2 - Q_1) + \gamma(Q_2 - T_2 U_2) . \qquad (9)$$

Similar to U_1, U_2, we update Q_1, Q_2 by

$$Q_1 = Q_1 \odot \sqrt{\frac{H_1^T U_1 + \gamma T_1 U_1 + \gamma Q_2}{(H_1^T \odot (Q_1 U_1^T)) U_1 + 2\gamma Q_1}} \qquad (10)$$

$$Q_2 = Q_2 \odot \sqrt{\frac{H_2^T U_2 + \gamma T_2 U_2 + \gamma Q_1}{(H_2^T \odot (Q_2 U_2^T))U_2 + 2\gamma Q_2}} \ . \tag{11}$$

Algorithm 1. Anchor User Oriented Accordant Embedding (AURORAE)

Input: $G_1 = (V_1, A_1), G_2 = (V_2, A_2)$, anchor users, parameters α_i, β, γ, maximal number of iterations *maxiter*
Output: U_1, U_2 and Q_1, Q_2
1: Initialize U_1, U_2, Q_1, Q_2 with $(0,1)$ uniform distribution
2: **for** t=1:*maxiter* **do**
3:     Update U_1 by (7)
4:     Update U_2 by (8)
5:     Update Q_1 by (10)
6:     Update Q_2 by (11)
7:     **if** objective (5) converge **then** break

---

5   Experiment Study

In this section, we study the performance of our proposed AURORAE compared to several state-of-the-art methods. Besides, we evaluate the effect of different ratios of anchor users. Finally, experiment results are reported under different testing settings.

5.1   Experimental Settings

Compared Methods. To evaluate the performance of AURORAE, we select three state-of-the-art methods:

Table 1. Statistical information of datasets

Dataset	Type	#Users	#Edges	#Matching users
Douban online	Undirected	3,906	8,164	1,118
Douban offline	Undirected	1,118	1,511	
Twitter	Directed	5,120	164,919	1,609
Foursquare	Directed	5,313	76,972	

- IONE [7] models the followee and follower relationships and learns multiple representations for each user.
- ABNE [8] distinguishes the different effects of neighbors of each user and represents all users in a same vector space.
- DeepLink [22] gets the initial representations by random walk and feeds it to a deep neural network.
- AURORAE is proposed to solve the challenge of error propagation and learn accordant embeddings for users in each social network.

Datasets. We use following two pairs of datasets to evaluate different methods: (1) *Douban Online-Offline*, which is provided by [20]. Online users mean users who surf and interact with others in online Douban website and Offline users mean users who attend the offline activities. (2) *Twitter-Foursquare*, which is provided by [7,8]. Different from the first dataset, Twitter and Foursquare used in the experiments are directed networks, which means the adjacency matrix is not symmetric due to the emergence of followee/follower relationships. The statistical information of datasets is shown in Table 1.

Performance Metric. To evaluate the performance of comparison methods, *Accuracy* and *Hit Precision@k* are used to evaluate the exact prediction and top-k prediction [21]. Specially, *Hit Precision@k* allocates different weights for different rank k:

$$h(x) = \frac{k - (hit(x) - 1)}{k} \ ,$$

where $hit(x)$ is the position of correct linked user in the returned top-k users. Then, *Hit Precision@k* can be computed on N test users by $\frac{\sum_i^N h(x_i)}{N}$.

Experiment Setups. All compared methods have provided their source codes. For reproducibility, we will also provide full codes publicly. The final dimension of user representation in compared methods is set according to original papers. For our proposed AURORAE method, we set $\beta = 0.4$ and $\gamma = 10$ for all datasets. For each α_i, we tune it in $\{0.1, 0.2, 0.3, 0.4, 0.5\}$. Due to the core idea of our method, we denote r_o as the ratio of anchor users among all matching users. When testing, we set $k = 5$ for metric *Hit Precision@k*. Besides, we test each pair of datasets by different orders and report the average performance.

5.2   Experimental Results

Table 2. Overall prediction performance on different datasets with $r_o = 30\%$

Metric	Method	Online-Offline	Twitter-Foursquare
Accuracy	IONE	3.96	5.28
	ABNE	6.52	8.39
	DeepLink	26.02	1.02
	AURORAE	**37.21**	**10.66**
Hit Precision@5	IONE	8.07	10.37
	ABNE	12.26	14.86
	DeepLink	41.43	1.89
	AURORAE	**51.79**	**18.15**

Overall Prediction Performance. We evaluate the overall prediction performance for compared methods. The ratio of anchor users is 30%. As shown in Table 2, our proposed AURORAE method always behaves much better than other methods. IONE and ABNE shows much worse performance than DeepLink on Online-Offline. By contrary, DeepLink shows much worse than IONE and ABNE on Twitter-Foursquare. Hence, we can find IONE and ABNE are good at modeling directed networks while DeepLink is good at handling undirected networks. It is reasonable because IONE and ABNE are designed to handle the directed networks and DeepLink is not specially designed to handle the directed relationship. However, no matter whether networks are directed, our proposed AURORAE can achieve best performance. Compared to other methods not specially using graphs oriented at anchor users, the good performance of AURORAE demonstrates the effectiveness and generalization of learning accordant embedding based on the potential graphs oriented at anchor users.

Effect of Anchor Users. Because the core idea of our method to avoid error propagation is based on anchor users, we should study the effect of different ratios of anchor users. Hence, we report the accuracy and hit precision@5 for compared methods on dataset Douban Online-Offline with the r_o varying in $\{0.1, 0.2, 0.3, 0.4, 0.5\}$. As shown in Fig. 3, with the increasement of r_o, the performances of all compared methods rise explicitly. When we only know a few anchor users, our proposed AURORAE still shows better performance than other methods. When we know more anchor users, the performance of AURORAE increases quickly. Though ABNE and DeepLink behaves better when the number of anchor users grows, our proposed AURORAE exceeds DeepLink about 9%–13% on Hit Precision@5 and 7%–14% on Accuracy when the ratio of anchor users varying. Therefore, using potential graphs oriented at anchor users is a good way to solve the challenge of error propagation even the number of anchor users is not large.

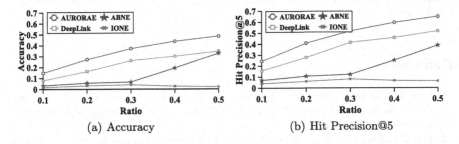

(a) Accuracy   (b) Hit Precision@5

Fig. 3. Effect of different ratios of anchor users on Douban Online-Offline.

Effect of Rank k. When testing the performance, we focus on the top-k prediction performance. In practical scenario, algorithms can provide top-k list for experts to economize time and match users effectively. Hence, we evaluate the

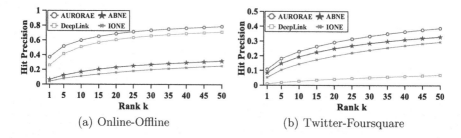

Fig. 4. Performance of different rank k on different datasets with $r_o = 30\%$.

performance for different rank k with $r_o = 30\%$. From Fig. 4, our AURORAE shows better performance than other methods. With the increasement of rank k, the performance of AURORAE can rise a lot on both two datasets, which means AURORAE can provide more accurate candidate users for sequent artificial identification.

6   Conclusion

Error Propagation is a great challenge for the task of User Identity Linkage. However, existing methods mainly seek to capture more useful information into user embeddings and lack the study of the essential reason for error propagation. In this paper, we study the reason leading to this challenge. Then, we propose the Anchor UseR ORiented Accordant Embedding (AURORAE) method to avoid error propagation by connecting each user to all anchor users and learn the accordant embedding with consistence constraint. Future work is to combine our method with more information such as label and content information.

Acknowledgments. This work is supported by the National Key Research and Development Program of China, and National Natural Science Foundation of China (No. U163620068).

References

1. Cao, X., Chen, H., Wang, X., Zhang, W., Yu, Y.: Neural link prediction over aligned networks. In: Proceedings of the 32th AAAI Conference on Artificial Intelligence, pp. 249–256 (2018)
2. Hu, G., Zhang, Y., Yang, Q.: CoNet: collaborative cross networks for cross-domain recommendation. In: Proceedings of the 27th ACM International Conference on Information and Knowledge Management, pp. 667–676 (2018)
3. Kong, X., Zhang, J., Yu, P.S.: Inferring anchor links across multiple heterogeneous social networks. In: Proceedings of the 22nd ACM International Conference on Information and Knowledge Management, pp. 179–188 (2013)

4. Korula, N., Lattanzi, S.: An efficient reconciliation algorithm for social networks. In: PVLDB, vol. 7, no. 5, pp. 377–388 (2014)

5. Koutra, D., Tong, H., Lubensky, D.: BIG-ALIGN: fast bipartite graph alignment. In: Proceedings of the 13th IEEE International Conference on Data Mining, pp. 389–398 (2013)

6. Lee, R.K.W., Hoang, T.A., Lim, E.P.: On analyzing user topic-specific platform preferences across multiple social media sites. In: Proceedings of the 26th International Conference on World Wide Web, pp. 1351–1359 (2017)

7. Liu, L., Cheung, W.K., Li, X., Liao, L.: Aligning users across social networks using network embedding. In: Proceedings of the 25th International Joint Conference on Artificial Intelligence, pp. 1774–1780 (2016)

8. Liu, L., Zhang, Y., Fu, S., Zhong, F., Hu, J., Zhang, P.: ABNE: an attention-based network embedding for user alignment across social networks. IEEE Access **7**, 23595–23605 (2019)

9. Man, T., Shen, H., Liu, S., Jin, X., Cheng, X.: Predict anchor links across social networks via an embedding approach. In: Proceedings of the 25th International Joint Conference on Artificial Intelligence, pp. 1823–1829 (2016)

10. Narayanan, A., Shmatikov, V.: De-anonymizing social networks. In: Proceedings of the 30th IEEE Symposium on Security and Privacy, pp. 173–187 (2009)

11. Perera, D., Zimmermann, R.: LSTM networks for online cross-network recommendations. In: Proceedings of the 27th International Joint Conference on Artificial Intelligence, pp. 3825–3833 (2018)

12. Shu, K., Wang, S., Tang, J., Zafarani, R., Liu, H.: User identity linkage across online social networks: a review. SIGKDD Explor. **18**(2), 5–17 (2016)

13. Singh, A.P., Gordon, G.J.: Relational learning via collective matrix factorization. In: Proceedings of the 14th ACM SIGKDD International Conference on Knowledge Discovery and Data Mining, pp. 650–658 (2008)

14. Yan, M., Sang, J., Xu, C.: Mining cross-network association for YouTube video promotion. In: Proceedings of the 22nd ACM International Conference on Multimedia, pp. 557–566 (2014)

15. Yan, M., Sang, J., Xu, C., Hossain, M.S.: A unified video recommendation by cross-network user modeling. ACM Trans. Multimedia Comput. Commun. Appl. **12**, 53:1–53:24 (2016)

16. Zafarani, R., Tang, L., Liu, H.: User identification across social media. ACM Trans. Knowl. Discov. Data **10**(2), 16:1–16:30 (2015)

17. Zhang, J., Chen, J., Zhi, S., Chang, Y., Yu, P.S., Han, J.: Link prediction across aligned networks with sparse and low rank matrix estimation. In: Proceedings of the 33rd IEEE International Conference on Data Engineering, pp. 971–982 (2017)

18. Zhang, J., Kong, X., Yu, P.S.: Predicting social links for new users across aligned heterogeneous social networks. In: Proceedings of the 13th IEEE International Conference on Data Mining, pp. 1289–1294 (2013)

19. Zhang, J., Yu, P.S.: Multiple anonymized social networks alignment. In: Proceedings of the 15th IEEE International Conference on Data Mining, pp. 599–608 (2015)

20. Zhang, S., Tong, H.: Final: fast attributed network alignment. In: Proceedings of the 22nd ACM SIGKDD International Conference on Knowledge Discovery and Data Mining, pp. 1345–1354 (2016)

21. Zhao, W., et al.: Learning to map social network users by unified manifold alignment on hypergraph. IEEE Trans. Neural Netw. Learn. Syst. **29**, 5834–5846 (2018)

22. Zhou, F., Liu, L., Zhang, K., Trajcevski, G., Wu, J., Zhong, T.: Deeplink: a deep learning approach for user identity linkage. In: Proceedings of the 37th IEEE Conference on Computer Communications, pp. 1313–1321 (2018)
23. Zhou, X., Liang, X., Zhang, H., Ma, Y.: Cross-platform identification of anonymous identical users in multiple social media networks. IEEE Trans. Knowl. Data Eng. **28**(2), 411–424 (2016)

Event Prediction in Complex Social Graphs via Feature Learning of Vertex Embeddings

Bonaventure C. Molokwu$^{(\boxtimes)}$ and Ziad Kobti

School of Computer Science, University of Windsor, 401 Sunset Avenue,
Windsor, ON N9B-3P4, Canada
{molokwub,kobti}@uwindsor.ca

Abstract. Social network graphs and structures possess implicit and latent knowledge about their respective vertices/actors and edges/links which may be exploited, using effective and efficient techniques, for predicting events within the social graphs. Understanding the intrinsic relationship patterns among spatial social actors as well as their respective properties are very crucial factors to be taken into consideration with regard to event prediction in social network graphs. Thus, in this paper, we model an event prediction task as a classification problem; and we propose a unique edge sampling approach for predicting events in social graphs by learning the context of each actor via neighboring actors/nodes with the goal of generating vector-space embeddings per actor. Successively, these relatively low-dimensional node embeddings are fed as input features to a downstream classifier for event prediction about the reference social graph. Training and evaluation of our approach was done on popular and real-world datasets, viz: Cora and Citeseer.

Keywords: Feature learning · Dimensionality reduction · Node embeddings · Feature extraction · Social graph

1 Introduction

Social (network) graphs are arbitrary structures which can be very challenging to study analytically using machine learning (ML) as well as deep learning (DL) models. Basically, a social network consists of finite set(s) of actors, and the relationship(s) defined between these actors [8]. Analyzing and learning intrinsic knowledge from communities comprising social units using some sets of standard still remains a significant research problem in social network analysis (SNA). Owing to the complex size and random nature of social network graphs; the process of developing machine learning models, and training them to predict events over a given graph data with respect to its constituent vertices (or actors)

Supported by International Business Machines (IBM).

and edges (or relationships) can be very difficult. In this regard, we propose a DL technique to address graph problems of this nature.

Primarily, the core of our approach is the Skip-gram layer which acts as an input sub-layer with respect to our proposed system architecture. The Skip-gram layer utilizes the structure of the graph in order to extract latent attributes with respect to every actor/node that constitutes the relationships/edges within the social graph. The proposed Skip-gram layer, which is essentially a dimensionality-reduction layer, is sandwiched between the downstream classifier and the input layer of the proposed system. Our proposed edge-sampling model is trained over the graph's edge list; and as a classification problem, every record of tie/relationship in the social graph is mapped to a corresponding output-event label. Furthermore, the Skip-gram sub-layer acts as an unsupervised representation (or feature) learning layer where underlying knowledge and viable facts (in the form of node embeddings) are automatically extracted from the graph data. In turn, this feature representation (edge list embeddings) serve as input for training over the downstream classifier via a supervised learning approach.

Our proposition is based on a neural architecture assembled using deep layers of stacked neural units [5]. The goal is to develop and train a neural network that is capable of learning the nonlinear distributed representation enmeshed in the graph structure [3]. We evaluated our approach/methodology against an array of state-of-the-art methodologies which serve as our baselines. The baselines (Deep-Walk [7], LINE [10], Node2Vec [2], SDNE [11]) used herein for benchmarking are based on representation learning (RL) methodology over graphs.

This paper is organized into four sections. Firstly, we have the introduction and overview of related literature. In the second section, we explained the data preprocessing techniques employed herein, the proposed Skip-gram layer, and the architecture of our proposed system. The third section discusses our experimentation results with reference to the evaluation of 4 baselines; and the last section summarizes our research findings, and in the near future work.

2   Methodology and Framework

2.1   Problem Definition

Definition 1. *Social Network: A social network, SN, can formally be expressed via Eq. 1 such that SN is a tuple comprising a set of vertices: V; a set of edges: E; a metadata function: f_V which extends the definition of the vertices' set by mapping it to a given set of attributes: V'; and a metadata function: f_E which extends the definition of the edges' set by mapping it to a given set of attributes: E'.*

$$
\begin{aligned}
SN &= (V, E, f_V, f_E) \\
G &: V, E && \text{graph function} \\
f_V &: V \rightarrow V' && \text{vertices' metadata function} \\
f_E &: E \rightarrow E' && \text{edges' metadata function}
\end{aligned}
\tag{1}
$$

Definition 2. *Node Embeddings: With reference to the social network, SN; the vector-space embeddings, X, generated by the Skip-gram layer is a mapping function, f, as expressed via Eq. 2. This function projects the representation of the graph's actors/nodes to a k-dimensional real space, $\mathbb{R}^k$, such that the existent tie(s)/edge(s) between any given pair of actors/nodes, (u_i, v_j), remain preserved via the homomorphism from V to X.*

$$f : V \to X \in \mathbb{R}^k \qquad (2)$$

Definition 3. *Event Prediction: This is essentially the conditional probability that an activity, B, will be true with reference to prior knowledge harnessed from a past event, A, which has already occurred. Formally: $Pr(B|A)$.*

2.2 Datasets

See Table 1.

Table 1. Training datasets

Dataset	Nodes	Edges	Dictionary	Classes
Cora	2,708	5,429	1,433	[C1, C2, C3, C4, C5, C6, C7]
Citeseer	3,312	4,732	3,703	[C1, C2, C3, C4, C5, C6]

Data Preprocessing Cora dataset comprises nodes and edges already encoded in numerical format. However, Citeseer dataset is made up of nodes and edges encoded in categorical (non-numerical) and numerical formats. Thus, preprocessing is necessary to transcode these categorical entities to their respective numerical representation without semantic loss. Thereafter, the numeric representation of both datasets are normalized prior to training on ML/DL models.

2.3 Proposed Skip-Gram Layer

Consider a words' vocabulary, $W : \forall w_m \in W$ where $M : m \in M$ is the number of unique words in the vocabulary. Given a target_word, w_t, in a text corpus; the "context" of w_t is defined as the words surrounding it in a given size-L window within the text corpus (comprising $N : n \in N$ words extracted from W).

The goal of the Skip-gram model is to maximize the average logarithmic probability of the context_words, w_{t+L}, being predicted as neighboring texts for the target_word, w_t, with respect to all training pairs, $\forall (w_t, w_{t+l}) \in D$. Formally, the objective function is defined as:

$$\sum_{(w_t, w_{t+l}) \in D} log Pr(w_{t+l}|w_t) = \frac{1}{N} \sum_{n=1}^{N} (\sum_{-L \leq l \leq L : l \neq 0} log Pr(w_{t+l}|w_t)) \qquad (3)$$

Consequently, in order to compute $Pr(w_{t+l}|w_t)$, we have to quantify the proximity of each target_word, w_t, with respect to its context_word, w_{t+l}. The Skip-gram model measures this adjacency/proximity as the cosine distance (or similarity) between w_t and its corresponding w_{t+l}. To accomplish this goal, every word comprising the text corpus with respect to the vocabulary, W, is encoded over a real number space, $\mathbb{R}$, using a function, f, such that $\forall\, w_t, w_{t+l}$:

$$
\begin{aligned}
&f_1 : w_t \rightarrow v_t & &v_t \in \mathbb{R} : v_t \text{ is the vector representation of target\_word} \\
&f_2 : w_{t+l} \rightarrow u_c & &u_c \in \mathbb{R} : u_c \text{ is the vector representation of context\_word} \\
&f_3 : w_m \rightarrow u_m & &u_m \in \mathbb{R} : u_m \text{ is the vector output of the } m_{th} \text{ word in } W
\end{aligned}
\tag{4}
$$

Thus, the cosine distance is calculated as the dot product between the vector representation of the target_word and the context_word. Mathematically, $Pr(w_{t+l}|w_t)$ is computed via a softmax function as defined below:

$$
Pr(w_{t+l}|w_t) = Pr(u_c|v_t) = \frac{exp(u_c \cdot v_t)}{\sum_{m=1}^{M} exp(u_m \cdot v_t)}
\tag{5}
$$

Furthermore, extending this NLP methodology to graph theory, given a social network, SN, as defined via expression 1 above; the edge list, $E[i,j] \subset G$, which is a sequence of tuples is defined via Eq. 6:

$$
E[i,j] \in \frac{M!}{(M-2)!} := \{(u_0, v_0)...(u_n, v_n)\}
$$
$$
\forall\, u_i, v_j \in \{V : v_0, v_2, ..., v_{m-1}\}
\tag{6}
$$

where (u_i, v_j) denotes a link or tie from a source vertex, u_i, to a target vertex, v_j. Consequently, expression 7 below defines the functions which aim at mapping the graph domain, G, to the words' vocabulary, W.

$$
\begin{aligned}
&f_4 : G \rightarrow W \\
&f_5 : (u_i, v_j) \mapsto (u_c, v_t) \\
&f_6 : u_m \mapsto u_m
\end{aligned}
\tag{7}
$$

where M signifies the number of unique nodes in the graph's set of vertices, V, such that: $\forall\, u_m \in V$. Therefore, the objective function of our Skip-gram layer with respect to a given graph, G, is as expressed by Eq. 8, viz:

$$
\sum_{(u_i, v_j) \in E} log Pr(u_i|v_j) = \sum_{-L \leq l \leq L : l \neq 0} log \frac{exp(u_i \cdot v_j)}{\sum_{m=1}^{M} exp(u_m \cdot v_j)}
\tag{8}
$$

2.4   Proposed Downstream Layer

The use of deep-layer structures cannot be downplayed because they are capable of training themselves to recognize and learn hidden features of a representation over consecutive levels of abstraction [4]. However, as the network structure goes deeper, more levels of abstraction to learn are generated; and this results in increased training time. Concurrently, as the network structure goes wider, more hidden parameters are injected into the network for learning, and this amplifies the training time as well. Therefore, there arises the need to strike a balance between the width and depth of the network structure; such that we obtain an efficient/effective model without increasing the chances of underfitting or overfitting in the network architecture. In a bid to conceptualize the architecture for an optimal neural network model, we apply pruning of neuron(s) based on Eq. 9; and global search methods. Thus, N_s, N_i, N_o, and N_m represent the sizes of training set, input layer, output layer, and hidden layer respectively.

$$N_m = \frac{N_s}{4 * (N_i + N_o)} \qquad (9)$$

2.5 Proposed System Framework

See Fig. 1.

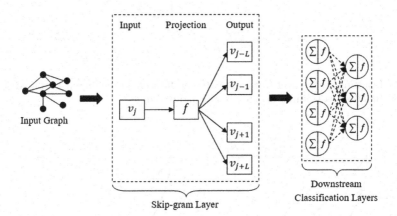

Fig. 1. Proposed system architecture

3 Results and Discussion

Our experimentation setup with regard to the Skip-gram representation learning layer were based on the hyperparameters shown in Table 2. The performance of our proposed model in comparison with 4 popular baselines (benchmark models) are as documented in Table 3. Our evaluations were done with regard to a range of objective functions. Categorical Cross Entropy was employed as the cost/loss function; while the fitness/utility was measured based on the following metrics, namely: Precision (PC), Recall (RC), F1-Score (F1), Accuracy (AC), and

Area Under the Receiver Operating Characteristic Curve (RO). Additionally, the objective functions are computed against each dataset (Cora and Citeseer respectively) with respect to the constituent classes (or categories) present in each dataset. Figures 2 and 3 graphically show the learning-progress curves of our proposed system over benchmark datasets during training. Hence, the dotted-black lines represent learning progress over the training set; and the dotted-blue lines represent learning progress over the validation set.

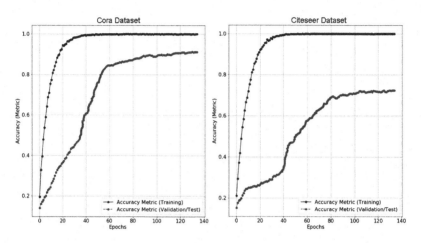

Fig. 2. Learning curves over Cora and Citeseer datasets based on our proposed system (fitness function *vs* training epochs) (Color figure online)

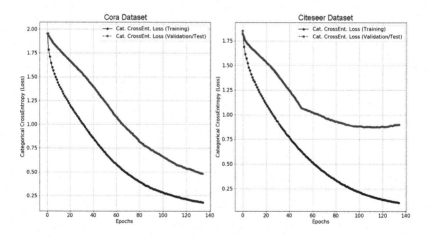

Fig. 3. Learning curves over Cora and Citeseer datasets based on our proposed system (loss function *vs* training epochs) (Color figure online)

With respect to Table 3; we have clearly tabulated the results of our multi-classification over the benchmark datasets. Thus, for each class per dataset, we have denoted the model which performed best for that classification task using a **bold font**. As can be seen from our tabular results; our proposed methodology is at the top with the highest record of fitness results followed by Node2Vec, DeepWalk, SDNE, and LINE.

Table 2. Fixed experimentation hyperparameters

Training set: 80%	Test set: 20%	Network width: 64
Batch size: 128	Optimizer: $AdaMax$	Network depth: 4
Epochs: $1.35 * 10^2$	Activation: ELU	Dropout: $4.0 * 10^{-1}$
Learning rate: $1.0 * 10^{-3}$	Learning decay: $\approx 3.3 * 10^{-5}$	Embed dimension: 64

Table 3. Experiment results over test datasets (*using Skip-gram feature learning*): model *vs* dataset for better interpretation/clarity

Model	Metric	Cora								Citeseer						
		C1	C2	C3	C4	C5	C6	C7	μ	C1	C2	C3	C4	C5	C6	μ
Skip-gram	PC	**0.90**	**1.00**	0.85	**0.96**	**0.81**	**0.97**	**1.00**	0.93	**0.58**	**0.88**	**0.68**	0.30	**0.82**	**0.58**	0.64
	RC	0.69	0.30	0.49	0.53	0.47	0.56	0.52	0.51	0.39	0.57	0.56	**0.35**	**0.46**	**0.44**	0.40
	F1	**0.78**	0.46	0.62	**0.68**	0.59	0.71	**0.68**	0.65	0.47	**0.69**	**0.61**	**0.32**	**0.59**	**0.50**	0.45
	AC	**0.91**	0.95	0.94	0.88	0.90	**0.94**	**0.94**	0.92	**0.87**	**0.85**	**0.84**	**0.93**	**0.91**	**0.86**	0.75
	RO	**0.83**	0.56	0.74	0.76	0.68	0.78	0.72	0.72	0.65	**0.77**	0.70	0.52	**0.68**	**0.69**	0.57
Node2Vec	PC	0.58	0.76	0.73	0.72	0.74	0.92	0.78	0.75	0.52	0.58	0.48	0.60	0.49	0.39	0.51
	RC	0.87	**0.53**	0.51	**0.64**	**0.61**	0.68	0.58	0.63	**0.54**	**0.61**	0.64	0.12	0.37	0.40	0.45
	F1	0.70	**0.62**	0.60	0.67	**0.67**	**0.78**	0.67	0.67	**0.53**	0.59	0.55	0.20	0.42	0.39	0.45
	AC	0.77	**0.96**	**0.95**	**0.90**	**0.92**	**0.94**	**0.94**	0.91	0.83	0.83	0.78	**0.93**	0.84	0.78	0.83
	RO	0.80	**0.76**	0.75	**0.79**	**0.79**	**0.83**	0.78	0.79	**0.72**	0.75	**0.73**	**0.56**	0.65	0.63	0.67
DeepWalk	PC	0.56	0.62	0.73	0.55	0.63	0.75	0.65	0.64	0.43	0.49	0.42	**0.67**	0.38	0.38	0.46
	RC	0.75	0.44	**0.56**	0.56	0.53	0.60	0.55	0.57	0.48	0.56	0.54	0.04	0.30	0.36	0.38
	F1	0.64	0.52	**0.63**	0.56	0.57	0.66	0.59	0.60	0.45	0.52	0.47	0.08	0.34	0.37	0.37
	AC	0.75	0.94	**0.95**	0.86	0.90	0.91	0.92	0.89	0.79	0.79	0.75	**0.93**	0.82	0.78	0.81
	RO	0.75	0.71	**0.77**	0.74	0.74	0.78	0.76	0.75	0.67	0.71	0.67	0.52	0.61	0.62	0.63
SDNE	PC	0.37	0.75	**0.88**	0.68	0.54	0.55	0.53	0.61	0.46	0.45	0.24	0.20	0.38	0.49	0.37
	RC	**0.93**	0.08	0.16	0.32	0.21	0.26	0.13	0.30	0.25	0.25	**0.78**	0.02	0.09	0.16	0.26
	F1	0.53	0.15	0.27	0.43	0.31	0.35	0.21	0.32	0.33	0.32	0.37	0.04	0.14	0.24	0.24
	AC	0.51	0.94	0.93	0.87	0.87	0.85	0.89	0.84	0.81	0.79	0.43	0.92	0.84	0.82	0.77
	RO	0.63	0.54	0.58	0.64	0.59	0.61	0.56	0.59	0.59	0.59	0.56	0.51	0.53	0.56	0.56
LINE	PC	0.35	0.67	0.58	0.67	0.68	0.62	0.67	0.61	0.27	0.28	0.28	0.38	0.33	0.24	0.30
	RC	0.91	0.11	0.16	0.36	0.19	0.18	0.17	0.30	0.22	0.48	0.42	0.06	0.14	0.15	0.25
	F1	0.51	0.19	0.25	0.47	0.29	0.28	0.27	0.32	0.24	0.35	0.34	0.10	0.19	0.19	0.24
	AC	0.47	0.94	0.92	0.87	0.88	0.86	0.90	0.83	0.75	0.65	0.65	0.92	0.82	0.76	0.76
	RO	0.60	0.55	0.58	0.67	0.59	0.58	0.58	0.59	0.54	0.58	0.57	0.53	0.54	0.52	0.55

In addition, we used a mini-batch size of 128 for training and validating as experimented by [1,9,12] because we want to ensure that sufficient patterns are extracted by the model during training before its network weights are updated. Also, our dropout regularization was implemented within the hidden-layer structure of the downstream classifier.

4   Conclusion, Future Work, and Acknowledgements

Our proposition has proven to perform well on social graphs with relatively high performance and/or fitness values. We intend to expand our experimentation scope to include additional graph-based datasets and RL-based models like GraphCN [6]. This research was conducted on a high performance IBM Power System S822LC Linux Server. Also, this work was enabled in part via support provided by SHARCNET and Compute Canada (www.computecanada.ca).

References

1. Girshick, R.B.: Fast R-CNN. In: 2015 IEEE International Conference on Computer Vision (ICCV), pp. 1440–1448 (2015)
2. Grover, A., Leskovec, J.: node2vec: scalable feature learning for networks. In: Proceedings of the International Conference on Knowledge Discovery & Data Mining, KDD, pp. 855–864 (2016)
3. Hinton, G.E.: Learning multiple layers of representation. TRENDS Cogn. Sci. **11**(10), 428–433 (2007)
4. Hinton, G.E., et al.: Deep neural networks for acoustic modeling in speech recognition. IEEE Signal Process. Mag. **29**, 82–97 (2012)
5. Goodfellow, I., Bengio, Y., Courville, A. (eds.): Deep Learning. MIT Press, Cambridge (2017)
6. Kipf, T.N., Welling, M.: Semi-supervised classification with graph convolutional networks. CoRR abs/1609.02907 (2016)
7. Perozzi, B., Al-Rfou', R., Skiena, S.: DeepWalk: online learning of social representations. ArXiv abs/1403.6652 (2014)
8. Scott, J. (ed.): Social Network Analysis. SAGE Publications Ltd., Newbury Park (2017)
9. Shin, H.C., et al.: Deep convolutional neural networks for computer-aided detection: CNN architectures, dataset characteristics and transfer learning. IEEE Trans. Med. Imaging **35**, 1285–1298 (2016)
10. Tang, J., Qu, M., Wang, M., Zhang, M., Yan, J., Mei, Q.: Line: large-scale information network embedding. In: WWW (2015)
11. Wang, D., Cui, P., Zhu, W.: Structural deep network embedding. In: KDD (2016)
12. Zhang, K., Zuo, W., Chen, Y., Meng, D., Zhang, L.: Beyond a gaussian denoiser: residual learning of deep CNN for image denoising. IEEE Trans. Image Process. **26**, 3142–3155 (2017)

A Social Recommendation Algorithm with Trust and Distrust Considering Domain Relevance

Ling Liu[(✉)], Qi Zhang, Yi Zhang, and Junhao Wen

School of Big Data and Software Engineering, Chongqing University,
Chongqing, China
{liuling, cquzhangyi, jhwen}@cqu.edu.cn,
419323619@qq.com

Abstract. Although most of the existing social recommendation algorithms can alleviate data sparsity or cold-start problems, they only measure the influence of trust relationship on recommendation precision. To accurately measure the influence of the social relationship, we propose a social recommendation algorithm based on domain relevance and integrated into the trust and distrust relationship information. The algorithm is based on the Funk-SVD algorithm in which domain relevance of users is calculated by using cluster algorithm to find the groups, the trust and distrust relationship information of user are added, and the effect of user distrust on global influence is considered. Finally, Experiments based on the well-known Epinions show that the algorithm has obvious effects in improving the recommendation quality and alleviating the cold start problem.

Keywords: Social recommender system · Distrust relationship · Domain relevance · Funk-SVD

1 Introduction

Recommendation systems can help people make better decisions in the case of information overload. Collaborative filtering recommendation is one of the mainstream recommendation algorithms [1], but they usually suffer from data sparsity or cold-start problems [2, 3]. With the rapid development of social network, in recommendation systems the applications of trust relationship of social data can alleviate these problems to a certain extent. At present, most researches on social networks focus on trust and trust transmission [4, 5], only a few researches consider the problem of distrust [6, 7]. In fact, distrust has a positive effect on social network recommendation. The distrust relationship not only can solve the sparse problem of rating data and trust relationship data, but also amend the trust relationship between users [8]. Distrust information is beneficial in rating prediction and the performance can be further improved through the use of information propagation [9]. From the data sources, Epinions has allowed users to label distrust relationships with others. Therefore, it is necessary to consider the influence of distrust relationships in social recommendation.

© Springer Nature Switzerland AG 2019
T. Gedeon et al. (Eds.): ICONIP 2019, CCIS 1143, pp. 581–588, 2019.
https://doi.org/10.1007/978-3-030-36802-9_62

Although the rating habits of users are similar with trusted friends, the trust relationship is different and multi-faceted [8]. Each user involves projects from multiple domains and has different preferences for each domain. Experts are usually only proficient in one domain. For different domains, users 'preferences and trust in friends are different. At the same time, there is a deviation between trust in the domain and the overall trust among users. Therefore, it is necessary to combine the domain information of users in order to describe more accurately the impact of social relationships among users on recommendation results.

In summary, in order to better integrate distrust into social network and consider the global influence and the domain of trust relationship, A Social Recommendation Algorithm with Trust and Distrust Based On Domain Relevance (SRecTDDR) is proposed.

2   SRecTDDR

2.1   Funk-SVD

Funk-SVD is known as Latent Factor Model (LFM), and it is an algorithm based on Singular Value Decomposition (SVD) that Simon Funk published on his blog in 2006 Netflix Prize [10].

We suppose $R = [r_{u,i}]_{N \times M}$ is a rating matrix, where N and M are the number of users and items respectively, and $r_{u,i}$ is a rating of item i from user u. The matrix R is solved into two low-rank matrix P and Q, that is $R = P^T Q$, where the matrix $P \in R^{d \times N}$ and $Q \in R^{d \times M}$ are the user and item feature matrix respectively, P^T represents its transposed matrix. Rating prediction value of item i by user u is shown in Eq. (1).

$$r_{u,i}^* = \mu + b_u + b_i + q_i^T p_u \tag{1}$$

Where μ is the global average rating, and b_u and b_i represent the user and item biases respectively. So we can learn the user and item feature matrices by minimizing the following loss function:

$$\min \ L = \frac{1}{2} \sum_{u=1}^{N} \sum_{i=1}^{M} \left(r_{u,i}^* - r_{u,i} \right)^2 \tag{2}$$

2.2   Measure the Trust and Distrust Relationship

We divides users into a triple group (u, k, g), user u trusts user k, but it does not trust user g. p_u, p_k and p_g represent the latent rating characteristics for the three users, respectively, and they represent their usual rating habits. This model uses the Euclidean distance equation to measure the difference in rating habits between users:

$$D_{u,k} = \|p_u - p_k\|_2^2 \tag{3}$$

$$D_{u,g} = \|p_u - p_g\|_2^2 \tag{4}$$

Equation (3) represents the difference in the latent rating vector between user u and its trusted friend user k, and Eq. (4) represents the difference in the latent rating vector between user u and user g that it distrusts in. The smaller the values are, the more similar their rating habits are.

Because user u trusts user k, they have certain similarities in rating preferences, then the latent rating vectors between them are similar. User u does not trust user g, so the rating habits and interest preferences between them are different. Therefore, p_u should be more similar to p_k and p_u is significantly different from p_g, then $D_{u,k}$ should be as small as possible, and $D_{u,g}$ should be as large as possible. Equation (5) measures the difference in rating habits between three users:

$$\max\left(0, D_{u,k} - D_{u,g}\right) \tag{5}$$

2.3 Computing User Domain Relevance

(1) The acquisition of user preferences. The user-item rating matrix $R = \left[r_{u,i}\right]_{N \times M}$ is decomposed into two low-rank matrices by using Fuck-SVD algorithm: $P \in R^{d \times N}$ and $Q \in R^{d \times M}$.

(2) Group discovery. According to the characteristics of user preferences, K-means clustering algorithm is used to cluster users, and cosine similarity is used to measure the similarity between users.

(3) Fusion of user preferences. We use the mean fusion strategy to fuse the preferences of users of each group, as shown in Eq. (6).

$$p_{Group} = \{avg(d_u) : \exists d_u \in \bigcup_{\forall u \in Group} p_u\} \tag{6}$$

In Eq. (6), p_{Group} represents the preference of group G, p_u represents the latent eigenvector of user u in the group; and d_u represents the value of each feature in the latent eigenvector of the user.

(4) Domain correlation calculation. The users can be divided into two types: users with trust relationship in same group and users with trust relationship in different groups. Pearson correlation coefficient is used to calculate the domain correlation between user u and user k which is its trusted friends in same group, as shown in Eq. (7):

$$f_{u,k} = Corr(u,k) = \frac{\sum_d (p_{u,d} - \bar{p}_u)(p_{Groupk,d} - \bar{p}_{Groupk})}{\sqrt{\sum_d (p_{u,d} - \bar{p}_u)^2}\sqrt{\sum_d (p_{Groupk,d} - \bar{p}_{Groupk})^2}} \tag{7}$$

The domain correlation between the users with trust relationships in different groups is weak. The group preferences can better reflect the latent preference domains among users. The method for calculating the domain correlation between users with trust relationships in different groups is shown in Eq. (8):

$$f_{u,k} = Corr(u,k) = \frac{\sum_d (p_{Groupu,d} - \bar{p}_{Groupu})(p_{Groupk,d} - \bar{p}_{Groupk})}{\sqrt{\sum_d (p_{Groupu,d} - \bar{p}_{Groupu})^2}\sqrt{\sum_d (p_{Groupk,d} - \bar{p}_{Groupk})^2}} \tag{8}$$

In Eqs. (7) and (8), p_u is the latent eigenvector of user u; p_{Groupu} and p_{Groupk} are the latent eigenvector of two groups which user u and user k are located; $\bar{p}_u$, $\bar{p}_{Groupu}$ and $\bar{p}_{Groupk}$ are the mean vectors of p_u, p_{Groupu} and p_{Groupk}, respectively. $p_{u,d}$, $p_{Groupu,d}$ and $p_{Groupk,d}$ are the eigenvalues of a feature, and d is the number of latent classes.

After adding trust and distrust relationships and considering the user domain correlation, the optimal loss function is:

$$\min L = \frac{1}{2}\sum_{u=1}^{N}\sum_{i=1}^{M}\left(r_{u,i}^* - r_{u,i}\right)^2 + \frac{\beta_1}{2}\sum_{(u,k,g)\in\Omega_S} f_{u,k}\max\left(0, D_{u,k} - D_{u,g}\right) \tag{9}$$

In Eq. (9), Parameter β_1 is the coordination factor for the users' social relationships, and ΩS is the set of triples formed by the social relations, as shown in Eq. (10). T is a trust matrix, $T_{uk}=1$ indicates that user u and K have trust relationship, and $T_{ug}=-1$ indicates that user u and g have distrust relationship.

$$\Omega_S = \{(u,k,g) \in [N] \times [N] \times [N] : T_{uk} = 1, T_{ug} = -1\} \tag{10}$$

2.4   Measure the Global Influence

According to eBay's Feedback Forum [11], the user's global influence is the result that positive evaluation number minus the negative evaluation number. If the user has more distrust relations than trust relations, his global influence is $1/N$, as shown in Eq. (11):

$$Global_u = \begin{cases} \dfrac{|Te_u| - |Dte_u|}{N}, & |Te_u| - |Dte_u| > 0 \\ \dfrac{1}{N}, & |Te_u| - |Dte_u| \le 0 \end{cases} \tag{11}$$

Where $|Te_u|$ is the number of users who trust user u, $|Dte_u|$ is the number of users who do not trust user u, and N is the total number of users.

Calculate the global influence value of all users in the social network, and rank all users according to the value. The larger the value is, the higher the ranking is, and the greater the global influence is. If $r_u \in [1,N]$, when r_u is equal to 1, it indicates that the ranking is the highest. The paper uses the Eq. (12) to map the user ranking to the interval $[0, 1]$.

$$G_u = \frac{1}{1 + \log(r_u)} \tag{12}$$

After considering the global influence of the three type of users and adding a regularization item to prevent over-fitting in the prediction process, the final optimization loss function is:

$$
\begin{aligned}
\min \ L = & \frac{1}{2} \sum_{u=1}^{N} \sum_{i=1}^{M} G_u \left(\mu + b_u + b_i + q_i^{\mathrm{T}} p_u - r_{u,i} \right)^2 + \\
& \frac{\beta_1}{2} \sum_{(u,k,g) \in \Omega_S} f_{u,k} \max \left(0, \| p_u - p_k \|_2^2 - \| p_u - p_g \|_2^2 \right) + \\
& \frac{\beta_2}{2} \sum_{(u,k,g) \in \Omega_S} M_{G_k,G_g} \| p_u - p_k \|_2^2 + \frac{\lambda}{2} (\| P \|_2^2 + \| Q \|_2^2 + b_u^2 + b_i^2)
\end{aligned} \tag{13}
$$

Where $\lambda > 0$ is regularization parameter, and $\|P\|$ and $\|Q\|$ represent Frobenius norms, and M_{G_k,G_g} represents the global influence comparison result for the user k and g, as shown in the Eq. (14). If the global influence of the distrusted user g is greater, it indicates that the social relationship of the target user is not credible, then the contribution from the social relationship to rating prediction is 0; if the global influence of trust user k is greater, the greater the influence of trusting users on the final forecast is.

$$
M_{G_k,G_g} = \begin{cases} G_k, & G_k \geq G_g \\ 0, & G_k < G_g \end{cases} \tag{14}
$$

To obtain the recommended model, the loss function Eq. (13) can be solved by the gradient descent method, and the missing values in the rating matrix R are predicted by learning the P and Q matrices. The pseudo code for the overall algorithm flow is shown in Table 1:

3 Experiments

3.1 Dataset and Evaluation Metrics

In order to evaluate the SRecTDDR model, the Epinions dataset [12] is used. In addition to rating, users can also establish trust and distrust relationships with other users where 1 indicating trust and −1 indicating distrust. Considering the time and space complexity, We use the subset of the Epinions dataset. The specific information about the dataset is shown in Table 2. These rates are divided into training sets and test sets in a 4:1 ratio in the experiments. Mean Absolute Error (MAE) [13] and Root Mean Square Error (RMSE) [13] are adopted to evaluate predictive accuracy.

Table 1. The flow of SRecTDDR algorithm

Input: Rating matrix R, trust relations T ,distrust relations Dt ,latent class number d , regulatory factors β_1 and β_2 , and the number of iteration k ,   **Output**: P and Q
1.   Cluster the users using K-means algorithm according to the rating matrix
2.   Calculate the domain correlation of users $f_{u,k}$ based on trust relations and the groups
3.   Calculate the global impact of each user based on trust and distrust relations $G\{G_{u1}, G_{u2}, \ldots\}$
4.   Initialize the matrix P and Q （with random values）
5.   Calculate $D_{u,k}$ 、 $D_{u,g}$ based on Eq.(3) and (4)
6.   Calculate the gradients based on Eq.(13)
7.   Update P and Q according to the negative gradients
8.   Continue to calculate if the number of iterations is smaller than k ; if reached, stop the calculation

Table 2. The information of epinions dataset

Number of users	Number of items	Number of ratings	Number of trust relations	Number of distrust relations	Density of ratings	Density of social relations
1510	18068	53254	1461	728	0.0019	0.0009

3.2   Experimental Results

This experiment is divided into two parts: the first part is to verify the impact of clustering number K on recommendation quality, the experimental results are shown in Fig. 1. The second part is to compare our model with traditional user-based collaborative filtering algorithm (User_CF), Funk-SVD algorithm and RecSNN [13] which captures local information and global information of social network, but ignores the influence of users who only have distrust relationship, and fails to consider the effect of distrust relationship on global influence. The experiments are compared from two different perspectives: all users and cold-start users with rating less than 5. The learning rate of gradient descent is set to 0.01 and the latent class number d is set to 10. The number of iterations is set to 100, β_1 and β_2 are set to 0.3 and 0.5, respectively. The experimental results are compared as shown in Table 3.

As shown in Fig. 1, the change of RMSE is shown when users are clustered into different categories according to their latent preferences. With the increase of K value, RMSE value decreases and reaches the minimum when the number of clusters is around 7 to 8. RMSE increases when k value gradually increases until more than 8. It shows that when users are clustered accurately according to their latent preferences, they can better measure their domain preferences, so that they can better measure the impact of user trust relationships. Therefore, in the following comparative experiments, the number of user clusters K is 8.

As shown in Table 3, SRecTDDR algorithm not only integrates the user's distrust relationship and the trust relationship, reassesses the impact of user's social relationships on rating prediction, but also uses the latent characteristics of user's rating to cluster users into domain-related groups. Each group has its own latent domain preferences, as well as the change of global influence after the addition of distrust relationship, which alleviates the problem of sparse rating data, and improves the accuracy of recommendation.

For cold-start users, RecSSN and SRecTDDR ease the sparsity of rating data to a

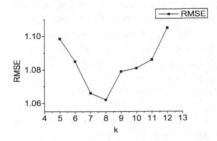

Fig. 1. The effect of the number of users cluster on recommendation results

Table 3. The comparison of experiment results for all users and cold users

Algorithm name	All users		Cold users	
	MAE	RMSE	MAE	RMSE
User_CF	0.9955	1.2005	1.0271	1.2403
Funk-SVD	0.9667	1.1780	1.0074	1.2177
RecSSN	0.8763	1.0979	0.9363	1.2013
SRecTDDR	**0.8517**	**1.0623**	**0.9157**	**1.1894**

certain extent and play a better role in improving the quality of recommendation because they use social trust and distrust relationship as supplementary data in recommendation algorithm. SRecTDDR also discovers groups similar to users 'latent preferences based on user clustering, so it can predict the latent preferences of cold-start users through user domain correlation, alleviate the problem that cold-start users cannot find similar users.

4   Conclusion

The algorithm SRecTDDR is based on Funk-SVD algorithm, which integrates trust and distrust in social relations. It calculates the domain correlation among users by clustering, uses the global influence of distrust relation to filter the trusted friends of target user, and reduces the contribution of users with low domain relevance and users with more distrust relation. The experiments show that the proposed algorithm can effectively alleviate data sparsity and cold start problems, improve the accuracy of rating prediction,and achieve better recommendation results.

Acknowledgement. This research is supported by the National Natural Science Foundation of China (61502062).

References

1. Zou, B.Y., Li, C.P., Tan, L.W., et al.: Social recommendations based on user trust and tensor factorization. Ruan Jian Xue Bao/J. Softw. **25**(12), 2852–2864 (2014)
2. Abbasi, M.A., Tang, J.L., Liu, H.: Trust-aware recommender systems. Computational trust models and machine learning: Chapman and Hall/CRC Press, pp. 11–12 (2014)
3. Wu, X.W., Liu, S.D., Zhang, Y.J., et al.: Research on social recommender systems. J. Softw. **26**(6), 1356–1372 (2015). (in Chinese)
4. Chen, T., Zhu, Q., Zhou, M.X., Wang, S.: Trust-Based recommendation algorithm in social network. Ruan Jian Xue Bao/J. Softw. **28**(3), 721–731 (2017)
5. Lei, G., Jun, M., Zhu-Min, C.: Trust strength aware social recommendation method. J. Comput. Res. Dev. **50**(9), 1805–1813 (2015)
6. Fang, H., Guo, G., Zhang, J.: Multi-faceted trust and distrust prediction for recommender systems. Decis. Support Syst. **71**, 37–47 (2015)
7. Lee, W.-P., Ma, C.-Y.: Enhancing collaborative recommendation performance by combining user preference and trust-distrust propagation in social networks. Knowl.-Based Syst. **106**, 125–134 (2016)
8. Maa, X., Lua, H., Gana, Z., Zeng, J.: An explicit trust and distrust clustering based collaborative filtering recommendation approach. Electron. Commerce Res. Appl. **25**, 29–39 (2017)
9. Lee, W.-P., Ma, C.-Y.: Enhancing collaborative recommendation performance by combining user preference and trust-distrust propagation in social networks. Knowl.-Based Syst. **106**, 125–134 (2016)
10. Funk, S.: Netflix update: try this at home (2011). http://www.sifter.org/~simon/journal/20061211.html. 01 July 2019
11. Resnick, P.: Trust among strangers in internet transactions: empirical analysis of ebay's reputation system. Econ. Internet E-Commerce **11**, 127–157 (2002)
12. Guha, R., Kumar, R., Raghavan, P., et al.: Propagation of trust and distrust. In: Proceedings of the 13th International Conference on World Wide Web, pp. 403–411. ACM, New York (2004)
13. Tang, J., Aggarwal, C., Liu, H.: Recommendations in signed social networks. In: Proceedings of the 25th International Conference on World Wide Web, pp. 31–40. ACM, New York (2016)

Spiking Neuron and Related Models

Performance Analysis of Spiking RBM with Measurement-Based Phase Change Memory Model

Masatoshi Ishii[1]([✉]), Megumi Ito[1], Wanki Kim[2], SangBum Kim[3],
Akiyo Nomura[1], Atsuya Okazaki[1], Junka Okazawa[1],
Kohji Hosokawa[1], Matt BrightSky[2], and Wilfried Haensch[2]

[1] IBM Research – Tokyo, Tokyo, Japan
ishiim@jp.ibm.com
[2] IBM Research, T.J. Watson Research Center, New York, USA
[3] Seoul National University, Seoul, South Korea

Abstract. Phase change memory (PCM) is arguably one of the most promising non-volatile memories which can be used in neuromorphic applications. When we use PCM as analog synaptic elements, non-ideality impact should be carefully taken into account. In this paper, we investigate the impact of such non-ideality items as nonlinear weight update and resistance drift on training accuracy in a spiking restricted Boltzmann machine (RBM). In addition to ideal PCM model, actual measurement-based PCM characteristics is used for this study. The resistance drift affects training accuracy, especially in a widely distributed read interval time caused by spatial and temporal sparse spike activities during training in spiking RBM. Our simulation results show that the training accuracy worsens with the increase of the weight-update nonlinearity or the resistance-drift coefficient. However, the results also suggest we can expect more than 8.95% improvement in training accuracy. This potential improvement will be possible if we use linearity-improved confined PCM cells whose median resistance drift coefficient is 0.005, compared to the resistance drift coefficient of more than 0.02 in typical existing PCM cells.

Keywords: Spiking neural network · Phase change memory · Resistance drift

1 Introduction

Artificial neural networks (ANNs) requires an overwhelming number of multiply-accumulate (MAC) operations. MAC operations are expensive computations in conventional von Neumann architecture computing systems. Various techniques and algorithms for reducing the load of MAC operations are being actively discussed in order to accelerate performance and minimize power consumption of ANNs. In addition to this software-level optimization, hardware-level acceleration is also strongly expected to achieve significant performance improvement. One of the most attractive approaches to achieve this improvement is to perform analog MAC operations massively in parallel by using memristor arrays with an inherent analogue conductance value [1–3]. There are several non-volatile memory (NVM) options for the memristor,

© Springer Nature Switzerland AG 2019
T. Gedeon et al. (Eds.): ICONIP 2019, CCIS 1143, pp. 591–599, 2019.
https://doi.org/10.1007/978-3-030-36802-9_63

such as phase change memory (PCM) [4], resistive random-access memory (ReRAM) [5], and magnetic random-access memory (MRAM) [6]. These emerging NVMs are originally developed and used as digital memories. When we utilize them as analog memories, we need to pay careful attention to their device characteristics. Several published papers reported the impact on accuracy induced by analog memories' non-ideality in neuromorphic computing [7, 8].

Such analog MAC operations can also be applied to spiking neural networks (SNNs) [9, 10]. SNNs are expected to become the next generation neural networks which will perform information processing with a sparse spike activity similar to a biological nervous system [11, 12]. Several studies have investigated the impact of non-ideality of analog memristors used in SNNs [13–15]. Nomura et al. [15] quantitatively assessed the impact of device variability on the SNN training accuracy, including device-to-device and cycle-to-cycle weight update variations. However, in addition to device variability, nonlinear weight update and resistance drift should be taken into consideration for the assessment. In this work, we investigate the impact of nonlinear weight update and resistance drift of the phase change memory on the SNN training accuracy in simulations using a spiking restricted Boltzmann machine (RBM).

2 Phase Change Memory Model

Phase change memory is arguably one of the most matured emerging NVMs. PCM exploits the ability of certain materials to rapidly change phases between two stable physical states, poly-crystalline and amorphous. In the poly-crystalline phase, the material has a regular crystalline structure and exhibits high-conductivity. In contrast, in the amorphous phase, the material has an irregular crystalline lattice and demonstrates low-conductivity. In PCM, this phase change is induced through localized joule heating caused by current injection. In NVM devices, the process of programming into high-conductivity is referred to as 'SET', and programming into low-conductivity is referred to as 'RESET'.

For neuromorphic applications, in some cases, it is sufficient for a synapse to be connected or not, so a binary response can be useful in such a synapse model. However, when we use it as analog memory for storing analog synaptic weight, continuous intermediate conductance states need to be utilized. In PCM, only the SET process can be made incremental by slowly crystallizing the amorphous state with repetitive pulses. The RESET process involves melt and quench, so it tends to be abrupt. Neuromorphic applications are sensitive to nonlinearity of conductance response, so near-linear response over most of its conductance rage is highly desirable. Also, resistance 'drift' arising from structural relaxation could be a potential impediment in on-chip learning because the read-out conductance value depends on the time elapsed from the last programming.

We implemented a flexible PCM model into our event-based hardware-aware simulator referred to in [16]. The model can express accumulative weight update behavior, resistance drift, and read noise. After initialization, the N th programming event, P_{mem} is first updated as $P_{mem} = P_{mem,N-1}e^{-1/\alpha}$ for $N = 1, 2, \ldots$. Then $G(t)$, which has seen N programming pulses, can be determined as follows.

$$\mu_{\Delta G_N} = m_1 G_{N-1}(T_0) + (c_1 + A_1 P_{mem}), \tag{1}$$

$$\sigma_{\Delta G_N} = m_2 G_{N-1}(T_0) + (c_2 + A_2 P_{mem}), \tag{2}$$

$$\Delta G_N = \mu_{\Delta G_N} + \sigma_{\Delta G_N} \chi, \tag{3}$$

$$G_N(T_0) = G_{N-1}(T_0) + \Delta G_N, \tag{4}$$

$$G(t) = G_N(T_0) \left(\frac{t - t_p}{T_0} \right)^{-\nu} + n_G, \tag{5}$$

$$\sigma_{nG} = m_3 G + c_3, \tag{6}$$

Here, χ represents a Gaussian random number of mean zero and variance 1. Another mean zero Gaussian random variable n_G captures the conductance fluctuations arising from the PCM noise, whose standard deviation is shown in Eq. (6). Please note that the conductance values predicted by the model are in μs.

3   Spiking Restricted Boltzmann Machine

A restricted Boltzmann machine is a two-layer bidirectional stochastic neural network consisting of one visible and one hidden layer. Neftci et al. [17] modified an RBM with event-driven contrastive divergence (eCD) to be constructed with stochastic leaky integrate-and-fire (LIF) neurons and to be trained with a spike-timing-dependent plasticity (STDP) weight update rule in an online and asynchronous fashion, where the spikes originate at each LIF neuron on both layers simultaneously. We slightly modified this spiking RBM model to be matched with our assumed hardware implementation.

In this work, we assume that a paired-PCM cell is used as a synaptic element with differential sensing scheme where one PCM conductance represents positive weight (G_p) and another represents negative weight (G_m) [18]. During training, the conductance of G_p and G_m is updated based on eCD algorithm, in which G_p and G_m are incremented in the data phase and the model phase respectively. However, with this one-side weight update scheme, the conductance of each PCM cell eventually reaches its maximum, and the learned weight becomes zero if both G_p and G_m reach maximum conductance. To overcome this issue, Ito et al. [19] proposed a lightweight refresh method. In their proposal, a minimized number of paired-PCM are randomly selected and checked from time to time to see if the conductance of G_p or G_m exceeds certain threshold. Then, if G_p or G_m or both exceeds the threshold conductance, the conductance value is shifted to the lower range, while keeping the conductance difference between G_p and G_m. We use the lightweight refresh method in the spiking RBM network model at a rate of 10 refresh neurons per image.

4   Simulation

To investigate the impact of the nonlinear weight update and resistance drift on the spiking RBM, we run simulations for an image recognition task by using the Modified National Institute of Standards and Technology (MNIST) hand-written image data set. Figure 1 illustrates the simulation model we used in this work. There are 832 neurons on each visible and hidden side. For training, the MNIST database images are binarized and transmitted as Poisson spike trains into 784 visible neurons in the data phase, together with four sets of ten label data sets, at a 200-Hz average spike rate. The RBM then runs without external inputs in the model phase except for bias neurons. For inference, Poisson spike trains are transmitted without the label data, then, we monitor the number of spikes on each label neuron to determine the set of label neurons that has the most spikes in total. In this work, we used randomly chosen 2,500 images for one epoch (250 images per label) from the training data set. Then, all 10,000 images are used for inference from the test data set.

4.1   Simulation with Ideal PCM Model

First, to investigate the impact of nonlinear weight update on the spiking RBM, we prepared four different ideal weight update curves shown in Fig. 2 with parameters in Table 1 used in Eqs. (1) to (6). We defined $\bar{\kappa}$ as average curvature factor by Eqs. (7) and (8), where $f(x)$ represents weight update curve, N represents number of pulses. We used 0.1 μs as a minimum conductance value by the RESET operation.

$$\kappa(x) = f''(x)/\left(1+f'(x)^2\right)^{3/2}, \tag{7}$$

$$\bar{\kappa} = \frac{1}{N}\int_0^N \kappa(x)dx, \tag{8}$$

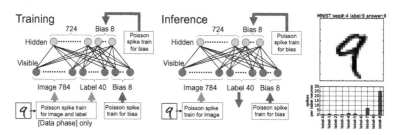

Fig. 1. Spiking RBM model for MNIST image recognition.

Table 1. PCM model parameters for four ideal weight update curves.

Model name	m_1	c_1	A_1	m_2	c_2	A_2	m_3	c_3	α	T_0	v
Ideal-A	0	0.00931	0	0	0	0	0	0	130	2	0
Ideal-B	−0.00168	0.01936	0.003	0	0	0	0	0	10	2	0
Ideal-C	−0.00168	0.01760	0.028	0	0	0	0	0	130	2	0
Ideal-D	−0.00168	0.01582	0.056	0	0	0	0	0	130	2	0

The simulation results shown in Fig. 3 indicate that the error rate degrades as the weight update linearity worsens. However, until around $\bar{\kappa}$ equals -1.84×10^{-5} (Ideal-B), the amount of degradation is limited. This indicates that there are some possibilities to minimize the impact of the weight update non-linearity as long as it is limited to some level.

We also investigate the resistance drift impact by using Ideal-A with varying parameter v, 0.02, 0.01, 0.005, and 0.0025. Figure 4 shows the simulation results. In our spiking RBM, the weight update occurs when the visible and the hidden neurons fire within a certain time window (2 ms). As spikes are sparsely generated by LIF neurons in the network, the weight update on each synapse is also sparse temporally and spatially. This results in a widely distributed read time interval from the last weight update on each PCM cell because the read and write operations from/to synapses occurs continuously. Figure 5 shows one example of distribution about the elapsed time from the last weight update measured at the end of epoch-15 with Ideal-A (drift coefficient 0.0). The time is distributed from around 1 ms to more than 1,000 s. This causes read value fluctuation in time during learning.

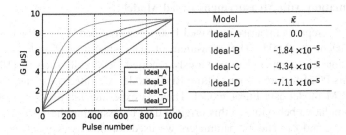

Fig. 2. Ideal weight update curves and degree of non-linearity.

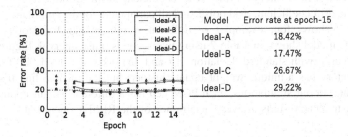

Fig. 3. MNIST simulation results with different weight update linearities.

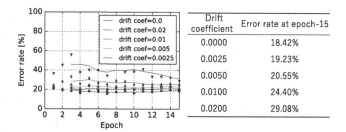

Fig. 4. MNIST simulation results using Ideal-A with different drift coefficient.

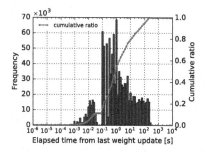

Fig. 5. Example of distribution of elapsed time after the last weight update.

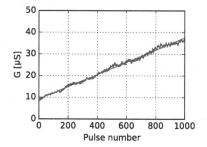

Fig. 6. Confined PCM cell measurement data plot with a fitted linear line.

4.2   Simulation with Measurement-Based Model

Kim et al. [20] report a linearity-improved 1,000-step confined PCM cell whose median drift coefficient is 0.005. The measurement data shown in Fig. 6 demonstrates improved linearity compared to previously reported PCM cells such as [10, 21]. To evaluate this PCM feature in our spiking RBM model, we fitted the measurement data into our PCM model as a linear model. Then, as experimental assessment, we added some accumulative behaviors to this original data by assigning the assumed values to m_2, c_2, A_2, m_3 and c_3. The 11 parameters we used for this assessment are shown in Table 2. In this model, 9.21 µs is used as minimum conductance which is initialized by the RESET operation. Confined-A model is a basic linear model used for comparison purpose. Confined-B includes the drift coefficient from Confined-A. Confined-C includes the drift coefficient, read noise, and all parameters for the cumulative behavior.

Figure 7 shows four examples of different weight update trajectories with the Confined-C model shown in Table-2 using two seconds pulse interval. Figure 8 shows a median curve with standard deviation (1σ and 3σ) derived from 1,000 trajectories. Each trajectory has a unique weight update history, but the median line fits well the original data. In our simulations, each PCM cell has a different weight update history, and a new trajectory starts at every occurrence of the RESET operation.

Figure 9 shows the MNIST simulation results with measurement-based models defined in Table-2. Simulation results for the Confined-B model indicate 20.11% error rate at epoch-15. It is logical these results are similar to Ideal-A with the drift coefficient of 0.005 (20.55% error rate at epoch-15). Although the assumed cumulative behaviors show a wide variation in Figs. 7 and 8, comparison results between the Confined-B and Confined-C suggest a limited accuracy loss, 2.2% at epoch-15.

If we compare the simulation results in Fig. 4, the error rate improvement from the Ideal-A with the drift coefficient of 0.02 to the Ideal-A with the drift coefficient of 0.005 is 8.53%. This indicates that we can expect a similar accuracy improvement by using in our spiking RBM the proposed PCM cell [20] whose median resistance drift coefficient is 0.005.

Table 2. PCM model parameters for four ideal weight update curves.

Model name	m_1	c_1	A_1	m_2	c_2	A_2	m_3	c_3	α	T_0	v
Confined-A	0	0.02828	0	0	0	0	0	0	65.0	2	0
Confined-B	0	0.02828	0	0	0	0	0	0	65.0	2	0.005
Confined-C	0	0.02828	0	0.008	0.006	0.021	0.03	0.18	65.0	2	0.005

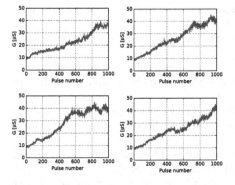

Fig. 7. Weight update trajectory examples with Confined-C model.

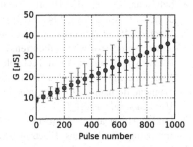

Fig. 8. Median and standard deviation derived from 1,000 trajectories of Confined-C model.

Model	Error rate at epoch-15
Confined-A	15.18%
Confined-B	20.11%
Confined-C	22.31%

Fig. 9. MNIST simulation results with Confined-A, B, and C models

5   Conclusion

We investigated the impact of the nonlinear weight update and resistance drift on the spiking RBM using a PCM ideal model and a measurement-based model with 1,000-step conductance states. The MNIST pattern recognition accuracy degrades as the weight update linearity worsens or when the resistance drift coefficient increases. Because the spiking RBM has widely distributed read interval time due to spatial and temporal sparse spike distributions, the impact of the resistance drift should be carefully considered. Our simulation results suggest that by using confined PCM cells reported in [20] whose median resistance drift coefficient is 0.005, we can potentially expect over 8.95% accuracy improvement compared to the PCM cells whose median resistance drift coefficient is more than 0.02.

References

1. Burr, G.W., et al.: Neuromorphic computing using non-volatile memory. Adv. Phys. X **2**(1), 89–124 (2017)
2. Tsai, H., et al.: Recent progress in analog memory-based accelerators for deep learning. J. Phys. D: Appl. Phys. **51** (2018). 283001
3. Ambrogio, S., et al.: Equivalent-accuracy accelerated neural-network training using analogue memory. Nature **558**, 60–67 (2018)
4. Sebastian, A., et al.: Tutorial: brain-inspired computing using phase-change memory devices. J. Appl. Phys. **124** (2018). 111101
5. Mochida, R., et al.: A 4M synapses integrated analog ReRAM based 66.5 TOPS/W neural-network processor with cell current controlled writing and flexible network architecture. In: IEEE Symposium on VLSI Technology (2018)
6. Yue, K., et al.: A brain-plausible neuromorphic on-the-fly learning system implemented with magnetic domain wall analog memristors. Sci. Adv. **5**(4) (2019). eaau8170. https://doi.org/10.1126/sciadv.aau8170
7. Lin, Y.-H., et al.: Performance impacts of analog ReRAM non-ideality on neuromorphic computing. IEEE Trans. Electron Devices **66**(3), 1289–1295 (2019)
8. Ernoult, M., et al.: Using memristors for robust local learning of hardware restricted Boltzmann machines. Sci. Rep. **9** (2019). Article number 1851
9. Kim, S., et al.: NVM neuromorphic core with 64k-cell (256-by-256) phase change memory synaptic array with on-chip neuron circuits for continuous in-situ learning. In: IEDM (2015)
10. Suri, M., et al.: Phase change memory as synapse for ultra-dense neuromorphic systems: application to complex visual pattern extraction. In: IEDM (2011)
11. Merolla, P.A., et al.: A million spiking neuron integrated circuit with a scalable communication network and interface. Science **345**(6197), 668–673 (2014)
12. Davies, M., et al.: IEEE Micro **38**(1), 82–99 (2018)
13. Boybat, I., et al.: Neuromorphic computing with multi-memristive synapses. Nature Commun. **9** (2018). Article number 2514
14. Suri, M., et al.: Impact of PCM resistance-drift in neuromorphic systems and drift-mitigation strategy. In: IEEE/ACM International Symposium on Nanoscale Architectures (2013)
15. Nomura, A., et al.: NVM weight variation impact on analog spiking neural network chip. In: Cheng, L., Leung, A.C.S., Ozawa, S. (eds.) ICONIP 2018. LNCS, vol. 11307, pp. 676–685. Springer, Cham (2018). https://doi.org/10.1007/978-3-030-04239-4_61

16. Nandakumar, S.R., et al.: A phase-change memory model for neuromorphic computing. J. Appl. Phys. **124** (2018). 152135
17. Neftci, E., et al.: Event-driven contrastive divergence for spiking neuromorphic systems. Front. Neurosci. **7**, 272 (2014)
18. Bichler, O., et al.: Visual pattern extraction using energy-efficient "2-PCM synapse" neuromorphic architecture. IEEE Trans. Electron Devices **59**(8), 2206–2214 (2012)
19. Ito, M., et al.: Lightweight refresh method for PCM-based neuromorphic circuits. In: 18th International Conference on Nanotechnology, pp. 1–4. IEEE (2018)
20. Kim, W., et al.: Confined PCM-based analog synaptic devices offering low resistance-drift and 1000 programmable states for deep learning. In: IEEE Symposium on VLSI technology (2019)
21. Boybat, I., et al.: Stochastic weight updates in phase-change memory-based synapses and their influence on artificial neural networks. In: 13th Conference on Ph.D. Research in Microelectronics and Electronics (PRIME) (2017)

Online Estimation and Control of Neuronal Nonlinear Dynamics Based on Data-Driven Statistical Approach

Shuhei Fukami and Toshiaki Omori$^{(\boxtimes)}$

Department of Electrical and Electronic Engineering, Graduate School
of Engineering, Kobe University, 1-1 Rokkodai-cho, Nada-ku, Kobe 657-8501, Japan
omori@eedept.kobe-u.ac.jp

Abstract. Due to recent development of measuring and stimulating technology, estimating and controlling neuronal dynamics become more important in neuroscience. A recent numerical study showed that the state of single neuron can not be controlled accurately by feedback control signal based on noisy observable data. Thus, it is essential to estimate hidden states in neuronal dynamics such as membrane potential and channel variable in order to realize accurate control by means of Bayesian statistical approach. For this purpose, we need to estimate the parameters of the dynamical model of neuron online used for the feedback control based on the estimated states. In this study, we propose a method for simultaneously estimating parameters and states online, and determining the control signal based on the estimated membrane potential. We use the particle filter for state estimation in order to deal with non-linear dynamics of neuron. Moreover, parameters of the neuron model are estimated online by using stochastic EM algorithm. We show that feedback control based on estimated membrane potential can be performed with online estimation of state and parameter estimation. Furthermore, we show that the controlling the state of neuron become more accurate when the control signal is determined based on our approach.

Keywords: State control of non-linear dynamics · Probabilistic time-series analysis · Online parameter estimation

1 Introduction

In recent years, experimental techniques in neuroscience regarding measurement and stimulus have been significantly developed [1]. Along with this, estimating the neural dynamics have been attempted based on measurement data. As a method of extracting the dynamics model of neuron from data, the data-driven approach of estimating the mathematical model has been studied in the framework of Bayesian statistics. For example, statistical approaches for estimating non-linear dynamics for single neurons have been proposed [2,3].

© Springer Nature Switzerland AG 2019
T. Gedeon et al. (Eds.): ICONIP 2019, CCIS 1143, pp. 600–608, 2019.
https://doi.org/10.1007/978-3-030-36802-9_64

Moreover, when applying a stimulus to a neuron, a method of determining the feedback stimulus from the estimated state of the neuron based on Bayesian statistics has been proposed [4]. In the previous study [4], the state of a neuron can be controlled more accurately by determining the control signal from the estimated value whereas that can not be controlled accurately by noisy observable data. However, in the previous methods, the state estimation is performed by assuming that the parameters of the model are known. In the case of actual operation, it is required to online estimate both parameters and state in order to determine the feedback control signal from the estimated state in the presence of unknown parameters. However, in the usual estimation of dynamics by Bayesian statistics, estimation is performed on the observed data in batch processing [5]. In order to perform feedback control based on the estimated values, it is necessary to estimate parameters online.

In this research, we propose a method for online estimation of parameters and states, and feedback control based on the estimated values simultaneously. First, we apply the state space model to the Morris Lecar model [6], which is a mathematical model of neuron. Next, a particle filter, which is a state estimation method adaptable to non-linear dynamics [7,8], is applied to this state space model in which taken into the feedback control. In addition, online parameter estimation is performed by using stochastic EM [9–12], which is one of online EM algorithm, and at the same time, a signal of feedback control is determined from the estimated value. We show that accurate feedback control can be performed by using the proposed method in the situation where only the noisy membrane potential is obtained as the observed value.

2   Estimation Algorithm

2.1   State Space Model of Neuronal Dynamics Model

According to Morris Lecar model, we assume that membrane potential V and the channel variable of the potassium ion channel n obeys the following differential equations.

$$C\frac{dV}{dt} = -g_L(V - E_L) - g_{Ca}m_\infty(V)(V - E_{Ca}) - g_K n(V - E_K) + I \quad (1)$$

$$\tau_\infty(V)\frac{dn}{dt} = \phi(n_\infty(V) - n) \quad (2)$$

Here, I represents the external input current, g_L, g_{Ca} and g_K respectively represent conductances of leak current, calcium current and potassium ion current, and E_L, E_{Ca} and E_K represent the respective equilibrium potentials, ϕ is a constant, and $\tau_\infty(V)$, $m_\infty(V)$, $n_\infty(V)$ are non-linear functions of membrane potential V.

Here we derive state space model of neuronal dynamics with feedback control. The state space model considers latent variables that can not be observed directly and observed variable generated from latent variable. In order to establish a method for estimating the non-linear neuronal dynamics from partially observable noisy data, we formulate a probabilistic model of the non-linear neuronal dynamics based on a non-linear state-space model when a feedback control is given to a neuron. We show the state space model constructed from the Morris-Lecar model shown in Eqs. (1) and (2) when performing feedback control as a system model. By discretizing Eqs. (1) and (2) with respect to time t, we obtain the difference equations for membrane potential V_t and channel variable n_t at time step t as follows:

$$V_t = V_{t-1} - \frac{\Delta t}{C}\{g_L(V_{t-1} - E_L) + g_{Ca}m_\infty(V_{t-1})(V_{t-1} - E_{Ca}) \tag{3}$$

$$+ g_K n_{t-1}(V_{t-1} - E_K) - I + I_{feedback}\} + \sqrt{\Delta t}u_t$$

$$= F_V(V_{t-1}, n_{t-1}, \boldsymbol{\theta}) + \sqrt{\Delta t}u_t$$

$$n_t = \left(1 - \Delta t\frac{\phi}{\tau_\infty(V_{t-1})}\right)n_{t-1} + \Delta t\frac{\phi}{\tau_\infty(V_{t-1})}n_\infty(V_{t-1}) + \sqrt{\Delta t}o_t \tag{4}$$

$$= F_n(n_{t-1}, V_{t-1}, \boldsymbol{\theta}) + \sqrt{\Delta t}o_t$$

where time interval is set to Δt, u_t and o_t are Gaussian noises called system noise, and feedback control signal $I_{feedback} = G \cdot \hat{V}_t$, G is a gain and $\hat{V}_t$ is the estimated value of membrane potential at time $t - 2$ by particle filter described later. We define $\boldsymbol{\theta} = \{g_K, g_{Ca}, g_L, E_K, E_{Ca}, E_L\}$ as parameters of the system model.

The system model can be expressed as the probabilistic model based on Eq. (3) as follows:

$$p(V_t|V_{t-1}, n_{t-1}, \boldsymbol{\theta}) = \mathcal{N}(V_t|F_V(V_{t-1}, n_{t-1}, \boldsymbol{\theta}), \sigma_{s_V}^2) \tag{5}$$

$$p(n_t|n_{t-1}, V_{t-1}, \boldsymbol{\theta}) = \mathcal{N}(n_t|F_n(n_{t-1}, V_{t-1}, \boldsymbol{\theta}), \sigma_{s_n}^2) \tag{6}$$

where $\sigma_{s_V}^2$, $\sigma_{s_n}^2$ are variance of system noises.

The observation model of the state space model is formulated by assuming that only noisy membrane potential y_t is observed as follows:

$$y_t = V_t + w_t \tag{7}$$

where w_t is Gaussian noise called observation noise. This equation can be expressed in the probabilistic model as follows:

$$p(y_t|V_t) = \mathcal{N}(y_t|V_t, \sigma_o^2) \tag{8}$$

Here, σ_o^2 is variance of observation noise, and y_t is observed variable.

Figure 1 (top, left) shows state space model of proposed method. The membrane potential V_t and the channel variable n_t correspond to latent variables, and noisy membrane potential y_t corresponds to an observed variable.

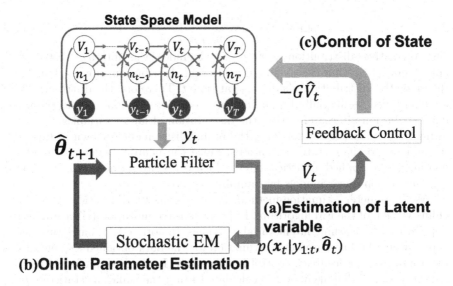

Fig. 1. Schematic of proposal method. The determination of the feedback control signal is executed through the online estimation of the state $x_t = \{V_t, n_t\}$ and parameter θ using particle filter and stochastic EM simultaneously. The observed variable y_t is obtained through the membrane potential V which is one of the latent variables based on the state space model. T is total number of time steps. The particle filter gives the filtering distribution $p(x_t|y_{1:t}, \hat{\theta}_t)$ for latent variables from y_t. The estimated parameter $\hat{\theta}_t$ is updated using this filtering distribution. Also, the control signal is determined based on the estimated membrane potential $\hat{V}_t$.

2.2 Online Estimation of Parameters and Feedback Control

In order to perform online feedback control, it is required to simultaneously perform the online estimation of parameter θ and state $x_t = \{V_t, n_t\}$ in the presence of unknown parameters. Here, we describe the method for online estimation of latent variables and parameters simultaneously using particle filter and stochastic EM (sEM) which is one of the algorithms of online expectation maximization. As shown in Fig. 1, (a) state estimation, (b) online parameter estimation, and (c) feedback control are performed in a situation where only membrane potential is obtained as the noisy observed value in the proposed method.

In the state space model, it is important to estimate latent variable $x_t = \{V_t, n_t\}$ that can not be observed directly based on observed value y_t, thereby unifying the important problems in time series analysis such as parameter estimation of the model. This estimation is called state estimation and can be classified into three cases: predictive distribution, filtering distribution, and smoothed distribution. Among these, the predictive distribution and the filtering distribution are represented by the following equations respectively.

$$p(x_t|y_{1:t-1}) = \int p(x_t|x_{t-1})p(x_{t-1}|y_{1:t-1})dx_{t-1} \qquad (9)$$

$$p(\boldsymbol{x}_t|y_{1:t}) = \frac{p(y_t|\boldsymbol{x}_t)p(\boldsymbol{x}_t|y_{1:t-1})}{p(y_t|y_{1:t-1})} \tag{10}$$

The predictive distribution $p(\boldsymbol{x}_t|y_{1:t-1})$ includes the filtering distribution $p(\boldsymbol{x}_{t-1}|y_{1:t-1})$ at time $t-1$ and the filtering distribution $p(\boldsymbol{x}_t|y_{1:t})$ includes the predictive distribution $p(\boldsymbol{x}_t|y_{1:t-1})$ at time t. Therefore, the distribution can be determined sequentially by finding the predictive distribution and the filtering distribution alternately. Also, the distribution of latent variables $\boldsymbol{x}_t$ can be estimated sequentially by applying the probabilistic models shown in Eqs. (5), (6) and (8) to these state estimations. Since the neuron's dynamics model is non-linear, we applied a particle filter, which is a state estimation method that can handle non-linear models by sampling method.

As a parameter estimation method, we use stochastic EM (sEM), which is an online version of the EM algorithm. Expectation-maximization (EM) [13,14] is a popular tool for learning latent variable models. In normal EM, batch processing is performed. Online parameter estimation for the data obtained sequentially is required to realize feedback control in this study. In the E-step of sEM at time t, expectation of log-likelihood is calculated using the sufficient statistics $\boldsymbol{\mu}_{t-1}$ at time $t-1$ as follows:

$$Q(\Theta(\boldsymbol{s})|\Theta(\boldsymbol{\mu}_{t-1})) = \langle \log p(y_{1:t}, \boldsymbol{x}_t|\Theta(\boldsymbol{s}))\rangle_{p(\boldsymbol{x}_t|y_{1:t},\Theta(\boldsymbol{\mu}_{t-1}))} \tag{11}$$

Here, estimated parameter $\hat{\boldsymbol{\theta}}_t$ is obtained by function of sufficient statistics Θ as follows:

$$\hat{\boldsymbol{\theta}}_t = \Theta(\boldsymbol{\mu}_t) \tag{12}$$

In M-step, sufficient statistics $\boldsymbol{\mu}_t$ is recalculated by maintaining an exponentially moving average $\boldsymbol{s}_t$ as an approximation of the average of sufficient statistics.

$$\boldsymbol{s}_t = \arg\max_{\boldsymbol{s}} Q(\Theta(\boldsymbol{s})|\Theta(\boldsymbol{\mu}_{t-1})) \tag{13}$$

$$\boldsymbol{\mu}_t = (1-\eta_t)\boldsymbol{\mu}_{t-1} + \eta_t\boldsymbol{s}_t \tag{14}$$

and step size η_t satisfies

$$\sum_{k=0}^{\infty}\eta_k = \infty, \sum_{k=0}^{\infty}\eta_k^2 < \infty \tag{15}$$

In sEM, estimated parameter $\hat{\boldsymbol{\theta}}_t$ are obtained by sequentially repeating E-step and M-step.

Figure 1 shows the entire framework of the proposed method. Firstly, observed data y_t is obtained according to the state space model shown in Eqs. (5), (6) and (8) at time t. Secondly, filtering distribution $p(\boldsymbol{x}_t|y_{1:t}, \hat{\boldsymbol{\theta}}_t)$ is obtained from particle filter based on parameter $\hat{\boldsymbol{\theta}}_t$. Next, parameter is updated by using sEM based on the filtering distribution. Here, function $Q(\Theta(\boldsymbol{s})|\Theta(\boldsymbol{\mu}_{t-1}))$ in Eq. (11) is calculated by using the filtering distribution. Thus, it is possible to perform simultaneous estimation of the latent variable $\boldsymbol{x}_t$ and parameter $\boldsymbol{\theta}$ sequentially. At the same time, feedback control is performed which signal is generated from estimated membrane potential $\hat{V}_t$. By repeating the above procedure, the neuronal dynamics are accurately controlled in the situation with unknown parameters.

3   Results

In this section, we verify the proposed method by using simulated data which are obtained from Eqs. (5), (6) and (8). In order to evaluate our proposed method, we consider two situations. Firstly, we consider a situation where the feedback control signal $I_{feedback}$ is generated as $I_{feedback} = G \cdot y_t$ from noisy observed data y_t directly. At $t > 15000$, control signal is given to the neuron. Secondly, we consider another situation where we sequentially estimate the model parameters $\boldsymbol{\theta}$ and the latent variable $\boldsymbol{x}_t$, and simultaneously perform feedback control $I_{feedback} = G \cdot \hat{V}_t$ based on the estimated membrane potential $\hat{V}_t$ using our method. While $0 \leq t \leq 15000$, we perform the state estimation and parameter estimation. At $t > 15000$, control signal is given to the neuron while both state estimation and parameter estimation are performed. In both of situations, we perform feedback control in order to stabilize membrane potential V_t when only a noisy membrane potential y_t within two dimensional hidden variables $\boldsymbol{x}_t$ is given.

Here we show estimated results of simultaneous estimation of $\boldsymbol{x}_t$ and parameters by means of our proposed method. Figure 2 shows that time course of the membrane potential under feedback control which is based on the noisy observed data y_t. It can be seen that the membrane potential under feedback control V_t (the green line) does not stabilize but rather becomes unstable due to the effect of noise. On the other hand, Fig. 3 shows that the membrane potential V_t (red line) become stable when the control signal is generated from the estimated membrane potential $\hat{V}_t$ (purple line). Also, we compare these two patterns of the feedback control for membrane potential in Fig. 4. The membrane potential is more stable when the feedback control is based on our method seeing the red line than the green line. Moreover, the latent variables $\boldsymbol{x}_t = \{V_t, n_t\}$ is well estimated in Figs. 3 and 5.

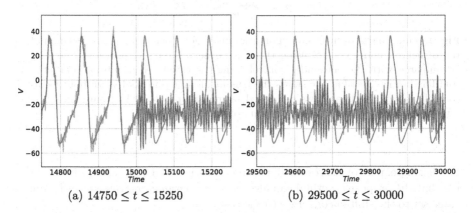

(a) $14750 \leq t \leq 15250$ (b) $29500 \leq t \leq 30000$

Fig. 2. Time course of membrane potential under feedback control which is based on the noisy observed data y_t. True membrane potential under feedback control V_t ($G \neq 0$ in Eq. 3, green line), true membrane potential without feedback control V_t ($G = 0$, blue line) and the observed data y_t (orange line) are shown. (Color figure online)

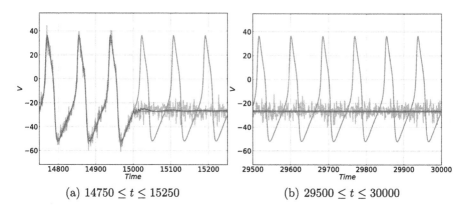

(a) $14750 \leq t \leq 15250$ (b) $29500 \leq t \leq 30000$

Fig. 3. Time course of membrane potential under feedback control which is based on the estimated membrane potential $\hat{V}_t$. True membrane potential under feedback control V_t ($G \neq 0$ in Eq. 3, red line), estimated membrane potential under feedback control $\hat{V}_t$ ($G \neq 0$, purple line), true membrane potential without feedback control V_t ($G = 0$, blue line) and the observed data y_t (orange line) are shown. (Color figure online)

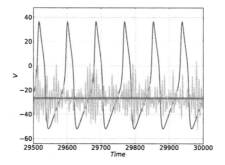

Fig. 4. Time course of membrane potential under feedback control. Membrane potential of neuron under control using the noisy observed data y_t (green line in Fig. 2) is compared with the one under control using the estimated membrane potential $\hat{V}_t$ (red line in Fig. 3). Blue line shows the true membrane potential V_t without feedback control, and purple line shows estimated membrane potential $\hat{V}_t$ from which control signal generated (in Fig. 3). (Color figure online)

Furthermore, part of the results of parameter estimation is shown in Fig. 6, showing that the proposed method can also estimate parameters well, since the initial value set to the shifted converges to the true value.

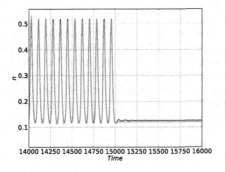

Fig. 5. Time course of channel variable n_t estimated online ($14000 \leq t \leq 16000$) when the feedback control signal is constructed by the estimated membrane potential $\hat{V}_t$ (orange line: estimated channel variable, blue line: true value). (Color figure online)

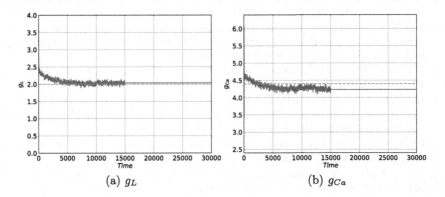

Fig. 6. Parameters estimated online (Orange line: true conductance, blue line: estimated conductance). (Color figure online)

4   Conclusion

In this paper, we have proposed the online statistical estimation method of neuronal dynamics based on Bayesian statistics, and the state control method for stabilization in a situation where only the membrane potential as the noisy observed value can be obtained. Note that the proposed method is a general framework of online estimation of control and is applicable to other nonlinear neuron models. Also, by conducting experiment using simulated data, it has been shown that model parameters and latent variables can be estimated online by using particle filter and stochastic EM. Moreover, it is shown that more accurate stabilization can be performed by feedback control using the proposed method compared with the case of directly determining the control signal from the observed value.

Acknowledgments. This work is partially supported by Grants-in-Aid for Scientific Research for Innovative Areas "Initiative for High-Dimensional Data driven Science through Deepening of Sparse Modeling" [JSPS KAKENHI Grant No. JP25120010] and for Scientific Research [JSPS KAKENHI Grant No. JP16K00330], and a Fund for the Promotion of Joint International Research (Fostering Joint International Research) [JSPS KAKENHI Grant No. JP15KK0010] from the Ministry of Education, Culture, Sports, Science and Technology of Japan, and Core Research for Evolutional Science and Technology (CREST) [Grant No. JPMJCR1914], Japan Science and Technology Agency, Japan.

References

1. Osawa, S., et al.: Optogenetically induced seizure and the longitudinal hippocampal network dynamics. PLoS One **8**(4), e60928 (2013)
2. Otsuka, S., Omori, T.: Estimation of neuronal dynamics based on sparse modeling. Neural Networks **109**, 137–146 (2019)
3. Omori, T., Hukushima, K.: Extracting nonlinear spatiotemporal dynamics in active dendrites using data-driven statistical approach. In: Journal of Physics: Conference Series, vol. 699, p. 012011. IOP Publishing (2016)
4. Ullah, G., Schiff, S.J.: Tracking and control of neuronal Hodgkin-Huxley dynamics. Phys. Rev. E **79**(4), 040901 (2009)
5. Inoue, H., Omori, T.: Bayesian estimation of neural systems using particle-Gibbs. In: Proceedings of the 2017 International Conference on Intelligent Systems, Metaheuristics and Swarm Intelligence, pp. 68–73. ACM (2017)
6. Izhikevich, E.M.: Which model to use for cortical spiking neurons? IEEE Trans. Neural Networks **15**(5), 1063–1070 (2004)
7. Kitagawa, G.: Non-Gaussian state-space modeling of nonstationary time series. J. Am. Stat. Assoc. **82**(400), 1032–1041 (1987)
8. Arulampalam, M.S., Maskell, S., Gordon, N., Clapp, T.: A tutorial on particle filters for online nonlinear/non-Gaussian Bayesian tracking. IEEE Trans. Signal Process. **50**(2), 174–188 (2002)
9. Neal, R.M., Hinton, G.E.: A view of the EM algorithm that justifies incremental, sparse, and other variants. In: Jordan, M.I. (ed.) Learning in Graphical Models, pp. 355–368. Springer, Dordrecht (1998). https://doi.org/10.1007/978-94-011-5014-9_12
10. Sato, M.A., Ishii, S.: On-line EM algorithm for the normalized Gaussian network. Neural Comput. **12**(2), 407–432 (2000)
11. Liang, P., Klein, D.: Online EM for unsupervised models. In: Proceedings of Human Language Technologies: The 2009 Annual Conference of the North American Chapter of the Association for Computational Linguistics, pp. 611–619 (2009)
12. Cappé, O., Moulines, E.: On-line expectation-maximization algorithm for latent data models. J. Roy. Stat. Soc. B (Stat. Methodol.) **71**(3), 593–613 (2009)
13. Dempster, A.P., Laird, N.M., Rubin, D.B.: Maximum likelihood from incomplete data via the EM algorithm. J. Roy. Stat. Soc.: Ser. B (Methodol.) **39**(1), 1–22 (1977)
14. Bishop, C.M.: Pattern Recognition and Machine Learning. Springer, New York (2006)

Sparse Estimation of Neuronal Network Structure with Observed Data

Ren Masahiro and Toshiaki Omori[✉]

Department of Electrical and Electronic Engineering, Graduate School of
Engineering, Kobe University, 1-1 Rokkodai-cho, Nada-ku, Kobe, Japan
181t251t@stu.kobe-u.ac.jp, omori@eedept.kobe-u.ac.jp

Abstract. Due to remarkable advances in observation technology, it
has become possible to observe membrane potentials simultaneously
from multiple neurons. Since brain functions are realized with neuronal networks in a brain, revealing structure of neuronal network is
thought to contribute to revealing brain functions. Therefore, it is necessary to address the problem of estimating neuronal network structure from observable data. In order to solve this problem, we propose
a method which combines sequential Monte Carlo method (SMC) which
estimates neuronal dynamics from observed membrane potentials and
Group LASSO which estimates neuronal network structure with estimated neuronal dynamics. We use SMC and Group LASSO because
these methods reflect non-linearity of neuronal dynamics, sparsity of neuronal networks and Dale's principle. Applying the proposed method to
simulated membrane potentials, we show that the proposed method is
effective to estimate neuronal network structure. Moreover, we show that
the proposed method is superior to conventional methods, such as linear
regression and LASSO, in estimating neuronal circuit structure.

Keywords: Neuronal network dynamics · Data-driven approach ·
Sparse modeling

1 Introduction

In recent years, remarkable advances have been made in observation technology
and these advancement have driven neuroscience field forward. Observation technology in neuroscience, such as voltage imaging, enables us to observe membrane
potentials simultaneously from multiple neurons [1,2]. While it has become possible to observe membrane potentials simultaneously from multiple neurons, we
have been faced with another problem. The problem is how we take advantage of
observable data or what we should understand to reveal brain functions. Because
neuroscience aims at elucidating brain functions, it is necessary to utilize observable data for revealing brain functions.

Incidentally, it is well-known fact that neuronal networks which are represented by synaptic connectivity are a critical substrate of brain functions.

© Springer Nature Switzerland AG 2019
T. Gedeon et al. (Eds.): ICONIP 2019, CCIS 1143, pp. 609–618, 2019.
https://doi.org/10.1007/978-3-030-36802-9_65

For instance, memory, learning, recognition and other functions are realized by synaptic plasticity and firing pattern changes [3–9]. This leads us to neuronal networks which are determined by synaptic conductances or connectivity between neurons. Namely, in order to contribute to approaching brain functions, we address neuronal network structure from observable data.

In order to realize accurate estimation of neuronal network structure, it is necessary to consider the following three neuronal properties. The first property is that neuronal states change as time goes by following non-linear dynamics. In particular, we assume that all neurons forming neuronal networks follow conductance-based neuron model as Hodgkin-Huxley (HH) equation [14] represents. The second property is that neuronal networks have sparsity in their connectivity [15]. It means that compared with full-connected networks, the number of synaptic connections is quite small in realistic networks. The third property is that neurons follow Dale's principle [9,16,17]. This property means that single neuron transmit only one type of substances, either excitatory or inhibitory chemical substance.

Estimation methods of neuronal system have been proposed for single neuron models [10–12] and for network models [13]. However neither sparse estimation nor a method taking Dale's principle into account has been proposed yet.

Considering these properties, we propose a method which estimates structure of neuronal circuit with observed membrane potentials. The proposed method combines sequential Monte Carlo method (SMC) [18,19] which is one of the time series analysis methods and Group LASSO [20] which is one of the sparse modeling methods. Applying SMC to observed membrane potentials, we estimate all neuronal states only from observed membrane potentials. Furthermore, Applying Group LASSO to estimated neuronal states, we estimate structure of neuronal circuit. Alternatively applying SMC and Group LASSO, we realize structure estimation of neuronal networks from observed membrane potentials.

In order to evaluate how valid the proposed method is for estimating neuronal networks, we use simulated membrane potentials which are generated neuronal network model based on HH equation and apply the proposed method to the simulated data. Compared the true network and the estimated network, we show that the proposed method is effective. Moreover, we show that the proposed method is superior to conventional methods. We set two methods, linear regression and least absolute shrinkage and selection operator (LASSO) [21], as conventional methods and compare the proposed method with them. The way of comparison is to use membrane potentials which are reconstructed with estimated synaptic conductances. Computing error between true membrane potentials and reconstructed membrane potentials, we show that the proposed method outperforms in estimating neuronal network structure.

2   Proposed Method

In this section, we formulate a method for estimating neuronal network structure from observed membrane potentials (Fig. 1). The proposed method combines

Dynamics estimation + Structure estimation

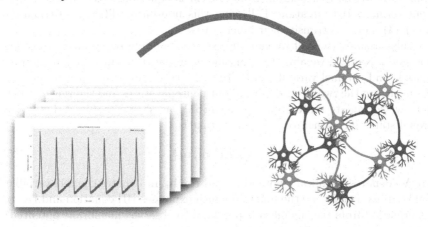

Observed membrane potentials   Estimation target of neuronal network

Fig. 1. Overview of this research. We assume that all we can observe is membrane potentials. Applying the proposed method which combines sequential Monte Carlo method (SMC) and Group LASSO to the observed membrane potentials, we realize structure estimation of neuronal network which is formed by synaptic connectivity.

sequential Monte Carlo method (SMC) and Group LASSO. SMC enables us to estimate neuronal dynamics which is non-linear state transition. Group LASSO, one of the sparse modeling methods, enables us to realize sparse estimation of neuronal network whose neurons follow Dale's principle.

2.1 Neuronal Dynamics Estimation with Sequential Monte Carlo Method

We assume that neuronal states for each neuron follow neuronal network dynamics based on Hodgkin-Huxley (HH) equation. If we consider a neuronal network which has N neurons, then the dynamics is expressed as follows:

$$C\frac{dV^i}{dt} = I_{\text{ext}}^i - I_{\text{ion}}^i - I_{\text{leak}}^i - I_{\text{syn}}^i, \tag{1}$$

$$\frac{dm^i}{dt} = \alpha_m(1 - m^i) - \beta_m m^i, \tag{2}$$

$$\frac{dh^i}{dt} = \alpha_h(1 - h^i) - \beta_h h^i, \tag{3}$$

$$\frac{dn^i}{dt} = \alpha_n(1 - n^i) - \beta_n n^i. \tag{4}$$

where C denotes membrane capacity, V^i denotes membrane potential, I_{ext}^i denotes external current, I_{ion}^i denotes ion current, I_{leak}^i denotes leak current,

I_{syn}^i denotes synaptic current, m^i, h^i and n^i denotes channel variables. These variables are of the i-th neuron. Equation (1) is equation which adds synaptic current term to HH equation. Following this non-linear differential equation, Eqs. (1)–(4), neurons transit their inner states.

We also assume that all we can observe is membrane potentials with observation noise. Namely, we aim at estimating neuronal states including membrane potentials and channel variables only from observed membrane potentials.

Considering the two assumptions about neuronal states and the way of observation, we formulate state space model (SSM) as shown in Fig. 2. Let $\mathbf{x}_t$ be neuronal states for all neurons at time t, that is

$$\mathbf{x}_t = \{\mathbf{V}_t, \mathbf{m}_t, \mathbf{h}_t, \mathbf{n}_t\}. \tag{5}$$

where $\mathbf{V}_t$ denotes membrane potentials and $\mathbf{m}_t, \mathbf{h}_t, \mathbf{n}_t$ denotes channel variables for all neurons. Let $\mathbf{f}_{\mathrm{sys}}$ be the function which expresses HH equation and $\mathbf{f}_{\mathrm{obs}}$ be transformation from true membrane potential to observed membrane potentials, we have

$$\mathbf{x}_{t+1} = \mathbf{f}_{\mathrm{sys}}(\mathbf{x}_t) + \mathbf{v}_t, \tag{6}$$

$$\mathbf{y}_{t+1} = \mathbf{f}_{\mathrm{obs}}(\mathbf{V}_{t+1}) + \mathbf{w}_{t+1}. \tag{7}$$

where,

$$\mathbf{v}_t \sim \mathcal{N}(\mu_{\mathrm{sys}}, \sigma_{\mathrm{sys}}^2), \tag{8}$$

$$\mathbf{w}_{t+1} \sim \mathcal{N}(\mu_{\mathrm{obs}}, \sigma_{\mathrm{obs}}^2). \tag{9}$$

In Eqs. (8) and (9), $\mu_{\mathrm{sys}}, \mu_{\mathrm{obs}}$ are mean of system noise and observation noise respectively, and $\sigma_{\mathrm{sys}}, \sigma_{\mathrm{obs}}$ are standard deviation of system noise and observation noise respectively. Note that the proposed method can be applied to non-Gaussian noise.

If we take probabilistic models for Eqs. (6) and (7), those are respected as $p(\mathbf{x}_{t+1} \mid \mathbf{x}_t), p(\mathbf{y}_{t+1} \mid \mathbf{x}_{t+1})$ respectively. Using these probabilistic models for all time t, we realize prediction and filtering which are key algorithms of sequential Monte Carlo method (SMC). The two algorithms, prediction and filtering, are to find following probability distributions respectively:

$$p(\mathbf{x}_{t+1} \mid \mathbf{y}_{0:t}) = \int p(\mathbf{x}_{t+1} \mid \mathbf{x}_t) p(\mathbf{x}_t \mid \mathbf{y}_{0:t}) d\mathbf{x}_t, \tag{10}$$

$$p(\mathbf{x}_{t+1} \mid \mathbf{y}_{0:t+1}) = \frac{p(\mathbf{y}_{t+1} \mid \mathbf{x}_{t+1}) p(\mathbf{x}_{t+1} \mid \mathbf{y}_{0:t})}{\int p(\mathbf{y}_{t+1} \mid \mathbf{x}_{t+1}) p(\mathbf{x}_{t+1} \mid \mathbf{y}_{0:t}) d\mathbf{x}_{t+1}}, \tag{11}$$

where $\mathbf{y}_{0:t}$ is observations from time 0 to t. By computing Eqs. (10) and (11), we realize estimation of neuronal dynamics with SMC (Fig. 2).

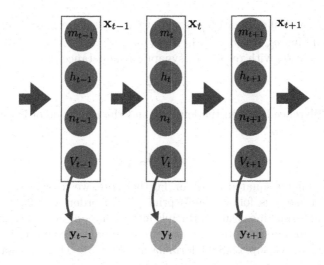

Fig. 2. State space model (SSM) for neuronal dynamics and observation. Neuronal states non-linearly transit following neuronal network model based on Hodgkin-Huxley (HH) equation. Furthermore, all we can observe is only membrane potentials V_t within higher dimensional hidden state $\mathbf{x}_t = \{V_t, m_t, h_t, n_t\}$.

2.2   Neuronal Circuit Structure Estimation with Group LASSO

It is known that synaptic current which depends on synaptic conductance has two types [22,23]. One is excitatory synaptic current and the other one is inhibitory synaptic current. Hence, I_{syn}^i in Eq. (1) is expressed in detail as follows:

$$I_{\text{syn}}^i = I_{\text{exc}}^i + I_{\text{inh}}^i. \tag{12}$$

where I_{exc}^i is excitatory synaptic current and I_{inh}^i is inhibitory synaptic current. Differences between excitatory and inhibitory synaptic currents are synaptic conductance and synaptic reversal potential [9]. With synaptic conductance and synaptic reversal potential, both synaptic currents are expressed as follows:

$$I_{\text{exc}}^i = \sum_{j=1}^{N} g_{\text{exc}}^{j \to i} s_j \left(V^i - E_{\text{exc}} \right), \tag{13}$$

$$I_{\text{inh}}^i = \sum_{j=1}^{N} g_{\text{inh}}^{j \to i} s_j \left(V^i - E_{\text{inh}} \right), \tag{14}$$

where $g_{\text{exc}}^{j \to i}$ and $g_{\text{inh}}^{j \to i}$ are excitatory and inhibitory synaptic conductances respectively, s_j is synaptic variable of pre-synaptic neuron, E_{exc} and E_{inh} are excitatory and inhibitory synaptic reversal potentials respectively. Note that if $g_{\text{exc}}^{j \to i}$ or $g_{\text{inh}}^{j \to i}$ is non-zero, there exists synaptic connectivity from the j-th to the i-th neuron.

On the other hand, if $g_{\text{exc}}^{j \to i}$ or $g_{\text{inh}}^{j \to i}$ is zero, there does not exists synaptic connectivity. Considering $g_{\text{exc}}^{j \to i}$ and $g_{\text{inh}}^{j \to i}$ in Eqs. (13) and (14), Eq. (1) can be regarded as a linear equation with respect to synaptic conductance $\mathbf{g}$:

$$\mathbf{z} = A\mathbf{g} \tag{15}$$

We define difference between the left side and the right side of Eq. (15) as cost function:

$$J := ||\mathbf{z} - A\mathbf{g}||_2^2, \tag{16}$$

where $|| \cdot ||_2^2$ denotes squared L_2-norm. Furthermore, we consider a neuroscience property that neurons follow Dale's principle. In order to solve this, we add regularization term. Since Dale's principle means that one neuron transfers either excitatory or inhibitory substance, we group synaptic connectivity from one neuron based on Group LASSO formulation (Fig. 3). Thus, we have another cost function:

$$J^* := J + R_{\text{exc}}(\mathbf{g}_{\text{exc}}) + R_{\text{inh}}(\mathbf{g}_{\text{inh}}), \tag{17}$$

where

$$R_{\text{exc}}(\mathbf{g}_{\text{exc}}) = \lambda_{\text{exc}} \sum_{j=1}^{N} \sqrt{\sum_{i=1}^{N} (g_{\text{exc}}^{j \to i})^2}, \tag{18}$$

$$R_{\text{inh}}(\mathbf{g}_{\text{inh}}) = \lambda_{\text{inh}} \sum_{j=1}^{N} \sqrt{\sum_{i=1}^{N} (g_{\text{inh}}^{j \to i})^2} \tag{19}$$

By computing synaptic conductances which minimize cost function J^* in Eq. (17), we can realize structure estimation of neuronal circuit.

3   Results

In order to investigate whether the proposed method is effective to estimate neuronal circuit, we compare true neuronal network which is estimation target in this study with the estimated network. Moreover, in order to demonstrate that the proposed method is superior to the other conventional method, linear regression and LASSO, we compare membrane potentials which are reconstructed with estimated synaptic conductances by the conventional and the proposed methods.

3.1   Comparison with Estimated Neuronal Circuits

We compared the estimated neuronal network structure with the true structure (Fig. 4). In Fig. 4, the left top and bottom graphs are the true and estimated neuronal network structure respectively. The middle top and bottom graphs show only the true and estimated excitatory neuronal network structure which

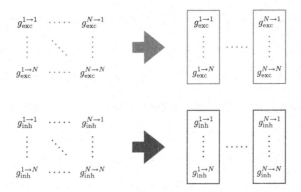

Fig. 3. Grouping pre-synaptic conductance for Group LASSO formulation. $g_{\text{exc}}^{j \to i}$ is excitatory synaptic conductances and $g_{\text{inh}}^{j \to i}$ is inhibitory synaptic conductances from the j-th $(1 \leq j \leq N)$ neuron to the i-th $(1 \leq i \leq N)$ neuron.

are extracted from the left graphs. The right top and bottom graphs show only the true and estimated inhibitory neuronal network structure which are also extracted from the left graphs.

As for excitatory neuronal network structure (middle top and bottom), both the true and the estimated neuronal network structure are equivalent. This means that sparse estimation by Group LASSO works out. As for inhibitory neuronal network structure (right top and bottom), a few misestimation can be seen in the estimated neuronal network structure (right middle), but we conclude that the sparse estimation was carried out well as a whole. The estimated neuronal network structure combines the estimated excitatory network (bottom middle) and inhibitory network (bottom right). Because of misestimation for inhibitory network, a few misestimation can be seen, such as green arrow, we conclude that the estimation result was accurate.

3.2   Comparison with Reconstructed Membrane Potentials

We reconstructed membrane potentials with the estimated synaptic conductance. In order to reproducibility, we compared the reconstructed membrane potentials and the true membrane potentials. When we reconstructed membrane potentials, we used different data set from the data set which is used to estimate synaptic conductances. We set mean absolute error (MAE) as an index to compute generalization error.

As seen in Table 1, the result of the proposed method to use Group LASSO is better than the results of the two conventional method to use linear regression or LASSO. This result indicated that the proposed method is superior to the conventional methods in estimating neuronal network structure. Moreover, the proposed method is superior in terms of generalization performance because we used different data set to compute error.

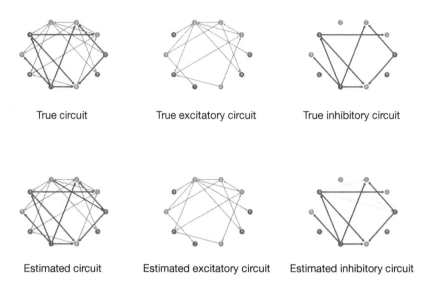

Fig. 4. The estimated neuronal network structure (bottom) and the true one (top). Red nodes and blue nodes are excitatory neurons and inhibitory neurons respectively. Red arrows and blue arrows are excitatory connectivity and inhibitory connectivity respectively. A green arrow seen in estimated circuit (bottom left) indicates the case where one neuron has both excitatory and inhibitory connectivities. (Color figure online)

Table 1. Mean absolute error (MAE) between the true and the reconstructed membrane potentials.

Linear regression	LASSO	Group LASSO
2.48	0.84	0.28

4   Conclusion

In this paper, we proposed a method which estimates neuronal network structure from observed membrane potentials. The proposed method combines sequential Monte Carlo method (SMC) and Group LASSO. SMC enables us to estimate non-linear neuronal dynamics. Group LASSO enables us to estimate neuronal network structure which has sparsity in the connectivity and follows Dale's principle.

In order to show that the proposed method is effective, we compared the estimated neuronal network structure and the true one. The result indicated effectiveness of the proposed method. Furthermore, we compared error between the true membrane potentials and the reconstructed by the proposed method and conventional methods. The result indicated superiority of the proposed method. These results show that we investigated the outstanding method to estimate neuronal network structure from observed membrane potentials.

Acknowledgement. This work is partially supported by Grants-in-Aid for Scientific Research for Innovative Areas "Initiative for High-Dimensional Data driven Science through Deepening of Sparse Modeling" [JSPS KAKENHI Grant No. JP25120010], and for Scientific Research [JSPS KAKENHI Grant No. JP16K00330], and a Fund for the Promotion of Joint International Research (Fostering Joint International Research) [JSPS KAKENHI Grant No. JP15KK0010] from the Ministry of Education, Culture, Sports, Science and Technology of Japan, and Core Research for Evolutional Science and Technology (CREST) [Grant No. JPMJCR1914], Japan Science and Technology Agency, Japan.

References

1. Kerr, J.N.D., Denk, W.: Imaging in vivo: watching the brain in action. Nat. Rev. Neurosci. **9**(3), 195–205 (2008)
2. Adam, Y., et al.: Voltage imaging and optogenetics reveal behaviour-dependent changes in hippocampal dynamics. Nature **569**, 413–417 (2019)
3. Martin, S.J., Grimwood, P.D., Morris, R.G.M.: Synaptic plasticity and memory: an evaluation of hypothesis. Ann. Rev. Neurosci. **23**, 649–711 (2000)
4. Bi, G., Poo, M.: Synaptic modification by correlated activity: Hebb's postulate revisited. Ann. Rev. Neurosci. **24**, 139–166 (2001)
5. Moser, E.I., Krobert, K.A., Moser, M., Morris, R.G.M.: Impaired Spatial learning after saturation of long-term potentiation. Science **281**(5385), 2038–2042 (1998)
6. Kayser, C., Montemurro, M.A., Logothetis, N.K., Panzeri, S.: Spike-phase coding boosts and stabilizes information carried by spatial and temporal spike patterns. Neuron **61**(4), 597–608 (2009)
7. Buzsaki, G., Draguhn, A.: Neuronal oscillation in cortical networks. Science **304**(5679), 1926–1929 (2004)
8. Mainen, Z.F., Sejnowski, T.J.: Reliability of spike timing in neocortical neurons. Science **268**(5216), 1503–1506 (1995)
9. Gerstner, W., Kistler, W.M., Naud, R., Paninski, L.: Neuronal Dynamics. Cambridge University Press, Cambridge (2014)
10. Huys, Q.J.M., Paninski, L.: Smoothing of, and parameter estimation from, noisy biophysical recordings. PLoS Comput. Biol. **5**(5), e1000379 (2009)
11. Omori, T., Hukushima, K.: Extracting nonlinear spatiotemporal dynamics in active dendrites using data-driven statistical approach. J. Phys. **669**, 012011-1-8 (2016)
12. Otsuka, S., Omori, T.: Estimation of neuronal dynamics based on sparse modeling. Neural Netw. **109**, 137–146 (2019)
13. Kataoka, S., Omori, T.: Simultaneous estimation of Hodgkin-Huxley neuronal dynamics and network connectivity based on Bayesian statistics. In: Proceedings of 16th International Sympodium on Advanced Intelligent System, pp. 812–818 (2015)
14. Hodgkin, A.L., Huxley, A.F.: A quantitative description of membrane current and its application to conduction and excitation in nerve. J. Physiol. **117**(4), 500–544 (1952)
15. Rolls, E.T., Treves, A.: Neural Networks and Brain Function. Oxford University Press, Oxford (1998)
16. Nicoll, R.A., Malenka, R.C.: A tale of two transmitters. Science **281**(5375), 360–361 (1998)
17. Squire, L.R., Berg, D., Bloom, F.E., Lac, S., Ghosh, A., Spitzer, N.C.: Fundamental Neuroscience, 4th edn. Academic Press, Cambridge (2012)

18. Gordon, N.J., Salmond, D.J., Smith, A.F.M.: Nove approach to nonlinear/non-Gaussian Bayesian state estimation. IEE Proc. F -Radar Sign. Process. **140**(2), 107–113 (1993)
19. Kitagawa, G.: Monte Carlo filter and smoother for non-Gaussian nonlinear state space models. J. Comput. Graph. Stat. **5**(1), 1–25 (1996)
20. Yuan, M., Lin, Y.: Model selection and estimation in regression with grouped variables. J. Roy. Stat. Soc. **68**(1), 49–67 (2006)
21. Tibshirani, R.: Regression shrinkage and selection via Lasso. J. Roy. Stat. Soc. **58**(1), 267–288 (1996)
22. Shepherd, G.M.: The Synaptic Organization of the Brain, 5th edn. Oxford University Press, Oxford (2004)
23. Kandel, E.R., Schwartz, J.H., Jessell, T.M., Siegelbaum, S.A., Hudspeth, A.J.: Principles of Neural Science, 5th edn. McGraw-Hill, New York (2013)

Implementation of Spiking Neural Network with Wireless Communications

Ryuya Hiraoka[1(✉)], Kazuki Matsumoto[1], Kien Nguyen[1]⬤,
Hiroyuki Torikai[2]⬤, and Hiroo Sekiya[1]⬤

[1] Chiba University, 1-33, Yayoicho, Inage Ward, Chiba-shi, Chiba 263-8522, Japan
`r.hiraoka@chiba-u.jp, sekiya@faculty.chiba-u.jp`
[2] Hosei University, 3-7-2, Kazinocho, Koganei-city, Tokyo 184-8584, Japan
`torikai@hosei.ac.jp`

Abstract. This paper proposes and implements the Spiking Neural Network (SNN) with radio-frequency wireless communications. The implemented network could obtain the XOR function through reinforcement learning. By applying the wireless communication for Internet of Things to the SNN, the SNN works with sufficient communication distance and low power consumptions for not only the line of sight environment but also the non-line of sight one. Additionally, it is unnecessary to consider communication directivity and obstacles for constructing the networks. The experimental results showed the extensibility and the scalability of the implemented system in this paper.

Keywords: Spiking Neural Network · Wireless communication · Sensor networks

1 Introduction

Recently, the brain-inspired computing has been paid attention to be open the way of the next-generation neuroscience. The Spiking Neural Network (SNN) is a neural network, whose behavior is close to natural brain activities [1–5]. It is possible to achieve intelligent signal processing such as object recognition [1], signal classification [2] and so on [3,4]. Because the SNN expresses the neuron information by spike signals, the energy consumption for information processing is much smaller than the Neuman-architecture computing.

The Wireless Sensor Network (WSN) consists of sensor nodes, which are linked by wireless communications. The WSN collects fruitful environment data continuously and automatically by sensors and transmits the data to the network server via a multi-hop route. The collected data is often used as training data of artificial intelligence. It is, however, too difficult to transmit all the environment data to the data server because of communication capacity and data explosion. Additionally, the power consumption at the wireless communication devices is also a problem.

© Springer Nature Switzerland AG 2019
T. Gedeon et al. (Eds.): ICONIP 2019, CCIS 1143, pp. 619–626, 2019.
https://doi.org/10.1007/978-3-030-36802-9_66

The concept of "Wireless Brain-Inspired Computing" (WiBIC) was originally put forward by authors [6], which is a information-processing platform based on the SNN and WSN. The WiBIC concept is that the SNN and the WSN are merged in term of 'Network'. The WiBIC is realized by connecting neurons with spike-signal wireless communications. Because of the 'wireless' characteristic, it is possible to install the WiBIC anywhere. The WiBIC has no data server in the network and can acquire intelligence in the WSN itself. Additionally, low power consumption due to the spike signal processing was a major advantage.

As the first step for embodying the WiBIC concept, the SNN, whose neurons are connecting by infrared(IR) communication, was proposed and implemented [6]. In this system, the neurons were implemented on Field Programmable Gate Arrays (FPGAs), which are linked by the IR communications. The implemented system in [6], however, had some problems. For example, the Time Division Multiple Access (TDMA) was adopted for comprehending spike-signal transmitters. It is difficult to make synchronization among many neuron nodes and many time slots should be allocated to the neuron nodes correctly and adaptively.

This paper proposes and implements the SNN with radio-frequency wireless communications. The implemented network could obtain the XOR function through reinforcement learning. By applying the wireless communication for Internet of Things (IoT), the SNN works with sufficient communication distance and low power consumptions for not only the line of sight environment but also the non-line of sight one. It is easy to comprehend the fired neurons by broadcasting Medium Access Control (MAC) header and Carrier Sense Multiple Access (CSMA) mechanism. Additionally, it is unnecessary to consider synchronization among neuron nodes, communication directivity, and obstacles for constructing the networks, which are advantages compared with the previous IR communication system. The experimental results showed the extensibility and the scalability of the implemented system in this paper.

2   Previous Works

Figure 1 shows the concept of WiBIC. The SNN neurons need to be connected wireless communication for achieving the WiBIC concept. As the first step for embodying the concept, the SNN, whose neurons are connecting by IR communication, was proposed and implemented [6]. In [6], the neurons were implemented on FPGAs and the FPGAs are linked by the infrared communications. It was confirmed that the implemented SNN with IR communications had an ability of XOR function through reinforcement learning.

For weighting the received signals, it is necessary for the signal-receiving neuron to identify the signal-transmitting neuron. In the IR communication, however, neuron nodes transmit pulse signals, which has one-bit information. Therefore, it is impossible to include the transmitter information in the pulse signal. For solving this problem, the TDMA was adopted in the IR communications.

The IR communication system implemented in [6] has, however, the following features and problems.

1. The IR communication has a strong directivity.
2. The IR communication has a short transmission distance.
3. It is possible for IR communication to transmit pulse signals in only the line of sight environment.
4. There is a problem of lack of the system extensibility because of 1–3. There is a limitation of the installation locations of the neuron nodes, which decreases in the advantage of the WiBIC concept.
5. The TDMA is also a bottleneck to extend the system. It is difficult to make a synchronization among many neuron nodes and many time slots should be allocated to the neuron nodes correctly and adaptively, which needs an advanced control strategy.

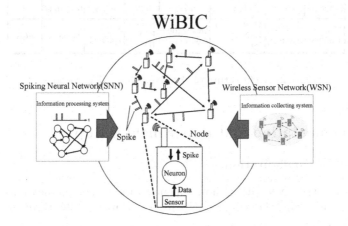

Fig. 1. Conceptual diagram of WiBIC.

3   Wireless Communication

For achieving the WiBIC concept, it is necessary to place network nodes without location consideration. Therefore, this paper proposes to apply radio-frequency wireless communication for the IoT networks to link the neuron nodes. We adopt the Lazurite 920J as radio-frequency wireless communication module [7]. By using this module, it is possible to realize 920 MHz-band wireless communications. The frequency of 920 MHz has a good diffraction characteristic, which enables non-line of sight communications. Additionally, very low power consumption is achieved in this module. We can keep 200 m communication distance. Throughput is 100 kbps, which is not high but sufficient for exchanges the spiking signal.

The wireless communication by Lazurite 920J adopts the IEEE 802.15.4e for MAC protocol, which is designed based on the CSMA. By applying CSMA, the system achieves multiple access without node synchronization. The detailed algorithm of CSMA is illustrated in [8]. In this paper, only the MAC header was transmitted. Figure 2 shows the general MAC header architecture of the IEEE 802.15.4e. It is seen from Fig. 2, the MAC header includes transmitter MAC address. There is no duplication among all the device MAC addresses. Therefore, it can be useful for identifying the transmitter. From the above, it is possible for the proposed system to have system extensibility.

Additionally, the MAC header is transmitted by broadcasting mode. In the wireless communication systems, the receiver usually ACKnowledgement packet from the receiver to the transmitter, which is unnecessary information and process in the SNN. It is specified that the receiver replies no ACK packet in the broadcasting mode. This operation is suitable for the SNN.

Octets: 2	1	0/2	0/2/8	0/2	0/2/8	0/5/6/10/14	2
Frame Control	Sequnce Number	Destination PAN ID	Destination Address	Source PAN ID	Source Address	Security	FCS

Fig. 2. Architecture of IEEE802.15.4e MAC header

4   Implementation and Experiment Setup

4.1   System Configuration

In this paper, the SNN with radio-frequency wireless communications is implemented. Additionally, the wireless SNN has an XOR function through reinforcement learning.

Figure 3(a) shows the system configuration of the implemented SNN. The SNN had two input-layer neurons, 14 hidden-layer neurons, and one output layer neuron. These neurons were connected with forwarding connections. The input neurons were linked to all hidden-layer neurons. In Fig. 3(a), $f_i(t)$, $f_h(t)$, and $f_o(t)$ express the fire states of input neuron i, hidden neuron j, and output neuron o, respectively. Additionally, w_{hi} and w_{oh} are synaptic weights between the input neuron i and hidden neuron j and those between the hidden neuron h and output neuron o, respectively. The $R(t)$ is a reward signal, which changes the synaptic weights.

4.2   Neuron Node Behavior

Figure 3(b) shows a photo of the implemented neuron device. For implementing the neuron devices, Xilinx Basys-3 Artix-7 FPGA and Lazurite 920J communication modules were used. Two input neurons were implemented on individual FPGA boards. Hidden-layer neurons, output neuron, and reward generator were installed in one FPGA board. The input neurons and hidden-layer ones are connected by the 920 MHz-band wireless communication by Lazurite 920J. The Universal Asynchronous Receiver/Transmitter (UART) unit was installed for

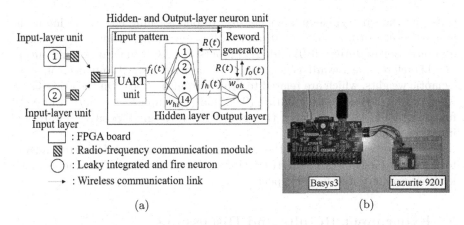

Fig. 3. (a) System configuration of the implemented system. (b) Photo of neuron unit.

communicating between FPGA and wireless communication module by serial communication.

The FPGA was designed for having the function of leaky integrated and fire neurons [5]. Additionally, hidden-node neuron h has a table of the MAC addresses of the backward neurons with the link weights w_{hi}. When a neuron is fired, the neuron-node FPGA orders to transmit a pulse signal to the communication module via UART. The communication module starts to prepare the MAC-header broadcasting immediately following the IEEE 802.15.4e.

All the communication modules receive the broadcasting frame and can comprehend which neuron is fired from the received MAC header. By matching the received-signal MAC address with the MAC addresses in the table, the spike-signal receiving neuron weights the signal and reflects the membrane potential of the neuron.

4.3   Reward Generator

In the implemented system, the reinforcement learning [9] was applied as learning mechanism. The reward generator carries out information processing for reinforcement learning. The value of $R(t)$ changes the values of w_{hi} and w_{oh}. The initial values of them are set random values between $-15.0 \leq w_{hi} < 15.0$ and $0 \leq w_{oh} < 15.0$. The detailed algorithm of reinforcement learning follows that in [6].

4.4   Learning of XOR Function

For evaluating the implemented SNN system, the XOR function, which is a classical benchmark problem of the NN, was learned and acquired in that system.

Output signals were coded by the firing rate. The firing rate for output '1' is higher than that for output '0'. Input neurons transmit 0 or 1, which are coded by the same spike-number density. In one learning epoch, we input the four patterns of $(0,0), (0,1), (1,0)$ and $(1,1)$ sequentially. When output neuron generated spikes during $(0,0)$ and $(1,1)$ inputs, the XOR function should output '0'. Therefore, the reward generator gives negative reward and decreases the synaptic weights, which are related to firing output neurons. On the other hand, when output neuron is fired during $(0,1)$ and $(1,0)$ inputs, the output of XOR function should be '1'. Namely, the reward generator gives positive reward and increases the synaptic weights.

For reinforcement learning, the input neurons transmitted the correct answer to the reward generator before transmitting the spike trains. The reinforcement learning was repeated for 150 epochs.

5   Experiment Results and Discussions

This section shows the experimental results of XOR-function learning. Two experimental results are shown in this section. One is for line of sight environment and another is for non-line of sight one.

5.1   For Line of Sight Environment

Figure 4(a) shows the firing number of the output neuron for the fixed input patterns as a function of learning epoch number in the line of sight environment. Figure 4(b) shows the photo of the experiment setup. The distances between the neuron nodes were less than 0.5 m and there was no obstacle in the network as shown in Fig. 4(b). It is seen from Fig. 4(a) that the output firing rates for all the input pattern were almost identical at the beginning of learning. As the learning epoch increases, the fire rates were separated. As shown in Fig. 4(a), the firing rate of the output neuron for the input pattern $(0, 1)$ and $(1, 0)$ is higher than that for $(0, 0)$ and $(1,1)$ posterior to learning.

From the result, it can be stated that the implemented system could learn and acquire the XOR function through reinforcement learning. This result indicates that radio-frequency wireless communications can work as links between neurons. Additionally, it can be stated that the MAC header transmission is useful for comprehending the firing neuron correctly. It is unnecessary to make synchronization among neuron nodes. The implemented system succeeded to update the link weights properly by MAC header information and reinforcement learning.

5.2   For Non-line of Sight Environment

Figure 5(a) shows the firing numbers of the output neurons for the fixed input patterns as a function of the learning epoch number in the non-line of sight environment. Figure 5(b) shows the photo of the experiment setup. The distance between input-neuron units was 1.5 m and that between the input-neuron units and the hidden- and output-layer neuron unit was 4.2 m. Additionally, there was a whiteboard between the input-neuron units and the hidden- and output-layer neuron unit as an obstacle for realizing the non-line of sight environment.

It is seen from Fig. 5(a) that the firing rates converged as the learning epoch number increases. The proposed system also could learn and acquire the XOR function even though the long-distance and non-line of sight environment.

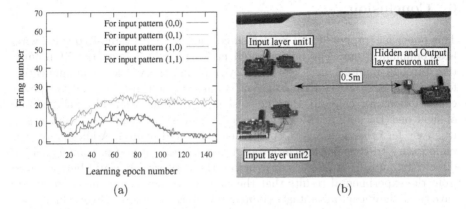

(a)

(b)

Fig. 4. Experimental results and setup for line of sight environment. (a) Firing number of the output neuron as a function of learning epoch number. (b) Experiment setup.

It can be stated from this result that the implemented neuron nodes can be placed without considering the installation locations. It is unnecessary to consider the communication directivity and the obstacles for constructing the networks. It can be shown from the experimental results that the radio-frequency

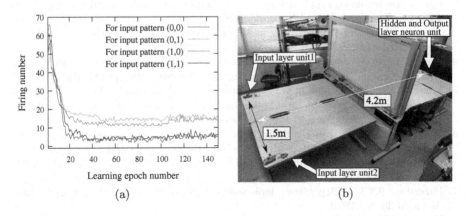

(a)

(b)

Fig. 5. Experimental results and setup for non-line of sight environment. (a) Firing number of the output neuron as a function of learning epoch number. (b) Experiment setup.

communication gives these significant advantages compared with IR communication. We confirmed that the implemented system acquired XOR function at the many harsh environments, where the SNN with IR communication cannot do. The result in Fig. 5(a) indicates the system extensibility and scalability of the proposed system.

6   Conclusion

This paper has proposed and implemented the SNN with radio-frequency wireless communications. The implemented network could acquire the XOR function through reinforcement learning. By applying the wireless communication for IoT, the SNN works with sufficient communication distance and low power consumptions for not only the line of sight environment but also the non-line of sight one. Because of the MAC header broadcasting and CSMA mechanism, it is easy for other neurons to comprehend the fired neuron. Additionally, it is unnecessary to consider synchronization among neuron nodes, communication directivity, and obstacles for constructing the networks. It could be confirmed from the experimental results that the radio-frequency communication system gave these significant advantages compared with the previous IR communication system. The obtained results indicated the system extensibility and scalability of the proposed system, which should be confirmed in the future research.

References

1. Kheradpisheh, S.R., Ganjtabesh, M., Thorpe, S.J., Masquelier, T.: STDP-based spiking deep convolutional neural networks for object recognition. Neural Netw. **99**, 56–67 (2018)
2. Dominguez-Morales, J.P., et al.: Deep spiking neural network model for time-variant signals classification: a real-time speech recognition approach. In: 2018 International Joint Conference on Neural Networks (IJCNN), pp. 1–8 (2018)
3. Mostafa, H.: Supervised learning based on temporal coding in spiking neural networks. IEEE Trans. Neural Netw. Learn. Syst. **29**(7), 3227–3235 (2018)
4. Roy, S., Basu, A.: An online unsupervised structural plasticity algorithm for spiking neural networks. IEEE Trans. Neural Netw. Learn. Syst. **28**(4), 900–910 (2017)
5. Wang, Z., Guo, L., Adjouadi, M.: A biological plausible Generalized Leaky Integrate-and-Fire neuron model. In: 2014 36th Annual International Conference of the IEEE Engineering in Medicine and Biology Society, Chicago, IL, pp. 6810–6813 (2014)
6. Matsumoto, K., Torikai, H., Sekiya, H.: XOR learning by spiking neural network with infrared communications. In: 2018 Asia-Pacific Signal and Information Processing Association Annual Summit and Conference (APSIPA ASC), Honolulu, HI, USA, pp. 1289–1292 (2018)
7. Lazurite   920J.   http://www.lapis-semi.com/lazurite-jp/products/lazurite-920j. Accessed 30 Jun 2019
8. Karl, H., Willig, A.: Protocols and Architectures for Wireless Sensor Networks, pp. 139–144. Wiley, Hoboken (2005)
9. Florian, R.V.: Reinforcement learning through modulation of spike-timing-dependent synaptic plasticity. Neural Comput. **19**(6), 1468–1502 (2007)

Text Computing Using Neural Techniques

Local Topic Detection Using Word Embedding from Spatio-Temporal Social Media

Junsha Chen[1,2], Neng Gao[1(✉)], Yifei Zhang[1,2], and Chenyang Tu[1]

[1] State Key Laboratory of Information Security,
Institute of Information Engineering, CAS, Beijing, China
{chenjunsha,gaoneng,zhangyifei,tuchenyang}@iie.ac.cn
[2] School of Cyber Security, University of Chinese Academy of Sciences,
Beijing, China

Abstract. Local topic detection from spatio-temporal social media (e.g. Twitter) plays an important role in many applications, therefore, it has attracted a surge of research attention in recent years. However, most existing studies consider time and location separately, they often assume tweets are correlated as long as their time is adjacent or their location is neighboring. But in reality, only tweets posted at both adjacent time and neighboring location tend to talk about the same local topic. To address this issue, we propose a network based embedding model to capture the correlation between time and location, and jointly model spatio-temporal information and semantic information together. This embedding model can ensure that the generated keyword embeddings are semantically coherent and spatio-temporally close. Based on the keyword embeddings, we present a novel topic model to obtain high-quality local topics. This topic model presumes that each tweet is represented by only one topic and takes the background mode of words into consideration to address the concise and noisy problem of Twitter. The experiments demonstrate that the effectiveness and efficiency of our method have been improved significantly compared to the state-of-the-art existing methods.

Keywords: Local topic detection · Word embedding · Social media

1 Introduction

With the rapid development of social media, more and more people are willing to post all kinds of valuable information on such platforms. Therefore, detecting local topics from these social media has attracted a surge of research attention in recent years. Compared to global topics, local topics are happening in a local area and during a certain period of time. Detecting such local topics is important in plenty of applications, such activity recommendation [18], local event discovery [17] and emergency warning [14]. The task of local topic detection was difficult years ago due to the lack of data sources, however, it has been made possible recently by the appearance of spatio-temporal social media like Twitter.

© Springer Nature Switzerland AG 2019
T. Gedeon et al. (Eds.): ICONIP 2019, CCIS 1143, pp. 629–641, 2019.
https://doi.org/10.1007/978-3-030-36802-9_67

Nevertheless, there exist some problems that largely limit the performance of existing local topic detection methods. (1) *Capturing the correlation between time and location.* Tweets posted at both adjacent time and neighboring location are inclined to talk about the same local topic and can be regarded as correlated, therefore, it is essential to capture the correlation between time and location. Existing methods [6,10,13,16], however, overlook such correlation, which leads to the generated local topics less spatio-temporally close. (2) *Processing the concise and noisy texts.* Tweets are quite short, which only contain a limited number of words, so it is difficult for traditional topic models [2,15] to capture the semantics from such short texts. Thus, some studies [1,4] have emerged to introduce word embedding into topic models to provide additional semantics for short texts, but these methods still suffer great limitations in processing noisy texts, like tweets.

To address above issues, we propose an effective method underpinned by two major modules to detect local topics from Twitter. The first module is a network based embedding model which can generate semantically coherent and spatio-temporally close keyword embeddings. If two keywords (e.g. *mets* and *yankees*) are semantically coherent, which are all related to baseball, and spatio-temporally close, which are all posted in tweets around Yankee Stadium and near 9pm, their representations in the latent space tend to be close. This model can not only capture the correlation between time and location, but also jointly model spatio-temporal and semantic information together. Based on the keyword embeddings, the second module is a topic model, which aims at detecting high-quality local topics by grouping the keywords of tweets into keyword clusters. Inspired by [19], in order to address the concise and noisy problem of Twitter, we assume each tweet is represented by only one topic, and the words in each tweet can either belong to the document's topic or the background mode.

Our main contributions are listed as follows:

(1) We propose a network based embedding model to capture the correlation between time and location, and fuse spatio-temporal information with semantic information. This model can generate semantically coherent and spatio-temporally close embeddings for keywords which are talking about the same local topic, and these embeddings can be directly used in following topic model.
(2) We present a novel topic model using word embeddings to generate high-quality local topics. This topic model assumes each tweet is represented by only one topic and incorporates the background mode of words into modeling in order to address the concise and noisy problem of Twitter.
(3) we collect two large-scale real-world data sets from Twitter, each containing millions of spatio-temporal tweets. And we evaluate our proposed method on both tweet data sets, the experimental results show that our proposed method outperforms the existing methods significantly.

2 Related Work

Representation Learning. Representation learning is a technique to transfer an object to a low-dimensional vector, which can solve the data sparseness problem and improve the performance of knowledge acquisition and knowledge fusion. There are considerable researches that apply representation learning in all kinds of fields. Mahdaouy et al. [11] propose a method to incorporate word embedding and semantic similarities into existing information retrieval models to deal with term mismatch. Chang et al. [3] propose an embedding model to recommend a point of interest where a user will visit next. Zhang et al. [18] propose Crossmap, which is an urban activity model based on representation learning to uncover the urban dynamics. Above methods demonstrate that representation learning is an effective way in capturing both explicit and implicit semantics, which is widely used in various tasks and can achieve good results.

Topic Detection. As the appearance of spatio-temporal social media, many studies have been done to utilize time and location for topic detection. Most studies use traditional probability graph models to detect topics. Liu et al. [9] present a model to extract global and local topics from spatio-temporal tweets. IncrAdapTL [6] proposed by Giannakopoulos et al. incorporates both time and location to find topics. Above models regard keywords as a word bag, which are independent with each other. But in reality, there exist certain correlations among keywords, for example, co-occurring keywords in the same contexts tend to discuss the same topic and share similar semantics. Therefore, a few studies have emerged to apply embedding in topic detection to maintain semantic correlation. Batmanghelich et al. [1] propose a spherical topic model (sHDP) using word embeddings, which can exploit the semantic structures of word embeddings, but this model suffer a great limitation when dealing with short texts. Zhang et al. [17] also combine embedding with topic model, and present Multi-Modal Embedding (MME) to map time, location and keywords into the same latent space, but it overlooks the correlation between time and location.

3 Preliminary

Definition 1 (Semantically coherent). *Let v_1 and v_2 be two keywords of tweets in the query window Q. If v_2 is in the context of v_1, then v_1 and v_2 are semantically close. The contexts of v_1 are generated by random walk in the network.*

Definition 2 (Spatio-temporally close). *Let v_1 and v_2 be two keywords of tweets in the query window Q, their timestamps are t_1 and t_2, and their locations are l_1 and l_2. If t_1, t_2 are adjacent and l_1, l_2 are neighboring at the same time, then v_1 and v_2 are regarded as spatio-temporally close.*

Problem Description. Let $D = \{d_1, d_2, d_3, \ldots\}$ be a spatio-temporal tweet corpus, each tweet d is represented by a tuple $<t_d, l_d, W_d>$, where t_d is its post time, l_d is its geo-coordinate and W_d is a set of keywords extracting from the tweet d. $Q = [t_s, t_e]$ is the query time window, where t_s is the start timestamp and t_e is the end timestamp. The local topic detection problem aims at finding local topics that occur in the query time window Q, and ensures that the keywords in the same local topic are semantically coherent and spatio-temporally close.

The Framework of Our Method. The appearance of a local topic often results in lots of related tweets posted around its happening place and during a certain period of time. For example, suppose *a baseball game between Mets and Bluejay* is held in NYC, participants at the stadium will post tweets during the game time, using the keywords such as *mets, bluejay, baseball* and *stadium*. These keywords discussing about the same local topic are semantically coherent and spatio-temporally close, and can form a keyword cluster as a local topic.

On the foundation of above observations, we present a local topic detection method with two modules. The first module is a network based embedding model, aiming at fusing spatio-temporal information with semantics and projecting the keywords into a low-dimensional vector space. This model can not only capture the correlation between time and location, but also maintain the semantic and spatio-temporal information of keywords, which can ensure that keywords talking about the same topic are semantically coherent and spatio-temporally close. Based on the generated keyword embeddings, the second module is a topic model, aiming to obtain high-quality local topics. Inspired by [19], consider the concise and noisy nature of tweets, we assume that each tweet is represented by only one topic and the words in tweets can either be topic words or background words. Topic words are semantically correlated, which are represented by vMF (von Mises-Fisher) distribution; background words are usually not semantically related, which are generated by multinomial distribution. The superiority of vMF distribution for modeling textual embeddings has been demonstrated in recent studies [1,7].

4   Network Based Embedding

4.1   Network Construction

We extract time, location and keywords of each tweet in the query window Q and construct a heterogeneous network as $G = (K, L, A)$, shown in Fig. 1.

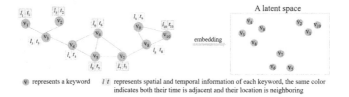

Fig. 1. Network based embedding.

- $K = \{v_i\}$ denotes the set of keywords in tweets.
- $L = \{e_{ij}\}$ denotes the set of links between two co-occurring keywords v_i and v_j in the same tweet, all the links are undirected and weighted, and the weight denotes the co-occurrence times between v_i and v_j.
- $A = \{(l_i, t_i)\}$ denotes the set of spatial and temporal information of keywords.

4.2 Embedding Details

Since keywords are natural units for embedding, we can directly use the *id* to represent them by one-hot representation. As for time and location, they are both continuous variables and there are no natural units for modeling. To address this problem, we break geographical space into equal-size regions and consider each region as a spatial unit. Similarly, we break query time interval into equal-length periods and consider each period as a basic temporal unit. Besides, in view of a keyword will occur in many regions, the multi-region information of each keyword should be included, each unit of spatial representation is a normalized occurrence frequency of this keyword in this certain spatial region. Similar to spatial information, we also use the normalized occurrence frequency for each temporal unit. After that, we adopt *early fusion* strategy to capture the correlation between time and location by concatenating spatial representation and temporal representation together, shown in Fig. 2.

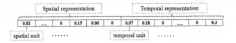

Fig. 2. Joint spatio-temporal representation for each keyword

The goal of our embedding model is to maintain the semantic similarity as well as spatio-temporal similarity among keywords. If two keywords co-occur in the same context, they tend to discuss the same topic, which indicates the semantic similarity among these two keywords; and if two keywords share adjacent time as well as neighboring location, it denotes the spatio-temporal similarity among them. It is obvious that the key to the similarity modeling depends on the estimation of pairwise similarity of keywords. Let $s(v_i, v_j)$ be the function of mapping two keywords v_i, v_j to their similarity score, and V is the vocabulary. We define the conditional probability of keyword v_j on v_i using the softmax function to measure the likelihood that v_j is connected with v_i:

$$p(v_j|v_i) = \frac{\exp(s(v_i, v_j))}{\sum_{v_{j'} \in V} \exp(s(v_i, v_{j'}))} \tag{1}$$

We further take all the neighbors N_i of v_i into account, where N_i is the set of keywords in the same contexts of v_i, which can be generated by weighted random walk. We maximize (1) over all keywords, and define the global likelihood function as:

$$likelihood = \prod_{v_i \in V} p(N_i|v_i) = \prod_{v_i \in V} \prod_{v_j \in N_i} p(v_j|v_i) \qquad (2)$$

In order to combine spatio-temporal information with semantics, we design a neural embedding model inspired by [8] and jointly model semantic information as well as spatio-temporal information in the same neural model. Therefore, the spatial and temporal information can closely interact with each other to capture the correlation between time and location, and the spatio-temporal part and the semantic part can also complement the learning of each other. In our model, the one-hot representation and the spatio-temporal representation of each keyword will be input into the neural model at the same time. After the input layer is the compression layer with two fully connected parts. One part transforms the one-hot keyword id to a dense vector u, another part encodes the spatio-temporal representation and generates a compact vector u'. Then, u and u' are connected together and fed into hidden layers, which are denoted as follows:

$$h^1 = [u, \lambda u'], \qquad h^k = \delta_k(W_k h^{k-1} + b_k), k = 1, 2, \ldots, n \qquad (3)$$

We use a strategy called *tower structure* to stack multiple hidden layers, where each successive layer has a smaller number of neurons. A trade-off parameter λ is used to adjust the importance of spatio-temporal information, and $\delta(\cdot)$ is the activation function. From the last hidden layer, we can obtain an abstractive representation h_v^n of the input keyword v. The output layer projects the output vector h_v^n into a probability vector o, which contains the predictive link probability of v to all keywords in the vocabulary V:

$$o = [p(v_1|v), p(v_2|v), \ldots, p(v_{|V|}|v)] \qquad (4)$$

By training the neural embedding model, we can get two vectors h_v^n and $\tilde{u}_v$ for each keyword v, where $\tilde{u}_v$ corresponds to a row in the weight matrix U between the last hidden layer and the output layer. And the similarity function can be defined as below:

$$s(v_i, v_j) = h_{v_i}^n \cdot \tilde{u}_{v_j} \qquad (5)$$

The similarity function can be fed into (1) to obtain the predictive link probability $p(v_j|v_i)$ in vector o. As detailed in (2), we aim at maximizing the conditional link probability over all keywords. Therefore, the whole neural network is jointly trained to maximize the likelihood about all the parameters θ.

$$\theta = \arg_\theta \max \prod_{v_i \in V} \prod_{v_j \in N_i} p(v_j|v_i) = \arg_\theta \max \sum_{v_i \in V} \sum_{v_j \in N_i} \log p(v_j|v_i)$$

$$= \arg_\theta \max \sum_{v_i \in V} \sum_{v_j \in N_i} \log \frac{\exp(h_{v_i}^n \cdot \tilde{u}_{v_j})}{\sum_{v_{j'} \in V} \exp(h_{v_i}^n \cdot \tilde{u}_{v_{j'}})} \qquad (6)$$

In order to improve the efficiency of model training, negative sampling and adaptive moment estimation are used to perform optimization. Finally, similar to [8], we use $h_v^n + \tilde{u}_v$ as the final representation for each keyword v.

5   Topic Model Using Word Embedding

On the foundation of network based embedding, we propose a topic model to divide the keywords of tweets in the query window Q into a number of clusters. Compared to traditional media, tweets are very short, which only contain a limited number of words. Thus, we assume each tweet is represented by only one topic instead of a mixture of topics. The key idea behind our model is that each keyword cluster implies a local topic around a particular location and during a certain period of time. Inspired by [19], consider the noisy nature of Twitter, we assume that words in a tweet can either belong to the document's topic or the background mode. Figure 3 shows the generative process for all the words of tweets in the query window Q, where N_d is the number of words in tweet d.

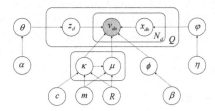

Fig. 3. Probabilistic graphical model

More formally, a tweet d is considered as a single topic, represented by z_d. The parameter that controls the latent distribution is θ, $z_d \sim Multinomial(\theta)$ and $\theta \sim Dirichlet(.|\alpha)$. Word v_{dn} can either be a topic word or a background word, and w_{dn} is the word embedding of word v_{dn}. For a topic word, it is represented by a vMF distribution, $w_{dn} \sim vMF(.|\mu, \kappa)$ and $\{\mu, \kappa\} \sim \phi(.|m, R, c)$; for a background word, $v_{dn} \sim Multinomial(.|\phi)$ and $\phi \sim Dirichlet(.|\beta)$. The fact that whether the word v_{dn} is a topic word or not is depicted by an indicator variable x_{dn}, $x_{dn} \sim Bernoulli(\psi)$ and $\psi \sim Dirichlet(.|\eta)$. When $x_{dn} = 1$, the current word v_{dn} is a topic word and it corresponds to the word embedding; otherwise, the current v_{dn} is a discrete background word. The superiority of the vMF distribution over other alternatives for modeling textual embeddings has been demonstrated in recent studies on clustering [7] and topic modeling [1].

Then, we introduce Gibbs sampling to estimate the parameters in our model, the key of Gibbs sampling is to infer the posterior distribution for z_d. Due to the space limitation, we directly give the conditional probabilities for z_d:

$$p(z_d = k|\cdot) \propto p(z_d = k|z_{\neg d}, \alpha) \cdot \prod_{v_{dn} \in d} p(v_{dn}|z_d, \phi, c, m, R, x_{dn}) \qquad (7)$$

The two quantities in (7) are given by:

$$p(z_d = k|.) \propto (n^{k, \neg d} + \alpha), \qquad p(v_{dn}|.) \propto (f(w_{dn}|\mu, \kappa))^{x_{dn}} (\phi_{v_{dn}})^{1 - x_{dn}} \qquad (8)$$

Where $n^{k,\neg d}$ denotes the number of times that topic k is sampled without counting the current tweet d, and $f(w_{dn}|\mu,\kappa)$ is the probability density of topic k's vMF distribution, where $f(w|\mu,\kappa) = C_p(\kappa)exp(\kappa\mu^T w)$, and $\kappa \geq 0$, $\|\mu\| = 1$ and the normalization constant $C_p(\kappa)$ is equal to $C_p = \frac{\kappa^{p/2-1}}{(2\pi)^{p/2}I_{p/2-1}(\kappa)}$. After the tweet topic z_d is sampled, for each word v_{dn} in tweet d, we need to sample the topic or background indicator x_{dn} according to the Bernoulli distribution:

$$p(x_{dn} = 1|.) \propto \frac{n_{x=1}^{-v_{dn}} + \eta}{n_{x=0}^{-v_{dn}} + n_{x=1}^{-v_{dn}} + 2\eta} \cdot \frac{C_p(\kappa)C_p(\|\kappa(Rm + w_{dn})\|_2)}{C_p(\|\kappa(Rm + \mathbf{w}_d^{-v_{dn}} + w_{dn})\|_2)}$$

$$p(x_{dn} = 0|.) \propto \frac{n_{x=0}^{-v_{dn}} + \eta}{n_{x=0}^{-v_{dn}} + n_{x=1}^{-v_{dn}} + 2\eta} \cdot \frac{n_{x=0}^{v_{dn}} + \beta}{\sum_{v' \in V} n_{x=0}^{v'} + |V|\beta}$$

(9)

Where $n_{x=1}^{-v_{dn}}$ and $n_{x=0}^{-v_{dn}}$ denote the number of topic words and background words respectively, without considering the current word v_{dn}. Besides, $\mathbf{w}_d^{-v_{dn}}$ is the sum of the word embeddings exclude current word, and $n_{x=0}^v$ is the number of times the current word v sampled as background word, and V is the vocabulary.

6 Experiments

6.1 Experiment Settings

Data Sets. We crawled tweets using Twitter Streaming API from 2018.03.21 to 2018.04.27 and constructed two real-world tweet data sets. The first data set, consisting of 1.7 million spatio-temporal tweets in New York, is referred to as NY. The second data set, consisting of 1.9 million spatio-temporal tweets in Los Angeles, is referred to as LA.

Baselines. We compare our method with following topic detection methods. For simplicity, we use *NBE* to represent the *Network Based Embedding* module, and represent the *Topic Model* using *Embedding* module by *TME*.

- LDA [2] is a topic model, which can find the top K keywords of each topic based on the number of topics and other prior parameters.
- BTM [15] is a topic model, which is designed to process short texts by directly modeling the generation of word co-occurrence patterns in the corpus.
- Twitter-LDA [19] is a topic model designed specifically for Twitter, which assumes a single tweet is usually about a single topic.
- IncrAdapTL [6] is a spatio-temporal topic model, which captures both spatial and temporal information in a same probabilistic graph model.

Besides, in order to demonstrate the effectiveness of NBE module, we replace it with Multi-Modal Embedding (MME) [17] proposed by Zhang et al., which can also incorporate spatial-temporal information and encode keywords into a low-dimensional space. We also replace the TME module with spherical topic model (sHDP) [1] to demonstrate the effectiveness of TME module. sHDP is also a topic model based on word embeddings. Furthermore, we compare NBE with NBE- to illustrate the importance of maintaining the correlation between time and location, the latter considers time and location separately.

6.2   Illustrative Cases

Before analyzing the quantitative results of our method, we present some illustrative cases in Fig. 4 by listing the top 3 keyword clusters and their corresponding summaries in the given query window, each keyword cluster can represent a local topic. As we can see, our method can detect high-quality local topics (e.g. sports games and entertainment parties), which can greatly benefit downstream applications like local event detection and activity recommendation.

Keyword cluster	Summary
Dodger, Angels, baseball, game, win, stadium	A game between Doger and Anger
party, night, NYU, music, bar, drink	A party held in NYU
sales, discount, shopping, food, WMT, crowd	A promotion in WMT

Keyword cluster	Summary
pillow, fight, Washington, park, square, feathers	A pillow fight in Washington park
trumptower, tower, fire, trump, hurt, escape	TrumpTower is on fire
Centralpark, sun, relax, warm, weather, weekends	Relax in central park

Keyword cluster	Summary
Coachella, festival, coachellafest, 2018, fashion, carnival	Coachella festival 2018
Mets, Brewers, player, game, fan, congrats	A game between Mets and Brewers
Yaga, gala, party, stars, 2018, entertainment	Yaga gala 2018

(a) Set query window size to 1: [2018.3.27,2018.3.27]   (b) Set query window size to 3: [2018.4.7,2018.4.9]   (c) Set query window size to 7: [2018.4.14,2018.4.20]

Fig. 4. Given the query window, detecting local topics on real-world tweet data sets.

6.3   Quantitative Results

Effectiveness Comparison. We calculate topic coherence [5, 12] for the topics generated by each method respectively. Figure 5 shows the comparison of average topic coherence in different window size on both tweet data sets, we can observe that our method outperforms the baselines on both data sets by a large margin. Then we perform above five methods to detect the topics that happened in NY from 2018.04.16 to 2018.04.18, Fig. 6 lists the top 3 topics sorted by topic coherence. As we can see, the topics generated by LDA, BTM and Twitter-LDA can not get obvious local topics, because they are all keyword-based methods, without considering time and location, so they suffer great limitations in detecting local topics happening in a local area and during a certain period of time. As for IncrAdapTL, it incorporates both time and location. Compared to above three methods, the topic coherence has been largely improved, and we can also

Data sets	LA			NY		
Query window	1	3	7	1	3	7
LDA	0.363	0.382	0.478	0.351	0.416	0.533
BTM	0.521	0.608	0.692	0.394	0.569	0.613
Twitter-LDA	0.513	0.632	0.679	0.461	0.547	0.621
IncrAdapTL	0.712	1.154	1.353	0.732	1.165	1.493
MME+TME	0.894	1.507	1.813	0.802	1.317	1.865
NBE+sHDP	0.782	1.246	1.437	0.782	1.259	1.527
NBE+TME	1.172	1.769	1.963	1.075	1.682	2.036
NBE+TME (our method)	1.512	2.493	2.897	1.577	2.396	2.742

Fig. 5. Average topic coherence of different methods in different query window size.

Methods	Topics		Coherence
	Keywords	Summary	
LDA	1. show, love, browns, traffic, baker	—	0.615
	2. place, expressway, parkway, east, place	—	0.529
	3. united, states, Get, Trump, Newyork	—	0.387
BTM	1. incident, traffic, crush, exit, clear	Traffic incident	0.872
	2. rain, fall, photo, think, birthday	—	0.669
	3. nyc, Brooklyn, stadium, block, morning	—	0.525
Twitter-LDA	1. rain, hard, heavy, fall, sudden	Heavy rain	1.064
	2. game, go, career, team, mets	—	0.972
	3. bridge, east, cloudy, flush, downtown	—	0.837
IncrAdapTL	1. party, dance, bar, dj, music	Music and dance party	1.856
	2. beer, drink, glass, dewey, game	Bear drinking game in Dewey	1.027
	3. heavyrain, rain, wet, fall, weather	Heavy rain and bad weather	0.915
our method	**1. YAGP, 2018, gala, stars, dance**	**YAGP 2018 Gala**	**2.913**
	2. subways, heavyrain, rain, swamped, drench	**Subways swamp due to heavy rain**	**2.672**
	3. yom, party, yomhaatzmaut, DJ, music	**Yom Ha'atzmaut Party**	**1.875**

Fig. 6. Top 3 topics generated by different methods in NY from 2018.4.16 to 2018.4.18.

get obvious local topics from the keyword clusters. However, the topic coherence of IncrAdapTL is still lower than our method, because our method is based on keyword embeddings, which can effectively maintain the semantic and spatio-temporal correlations among keywords, besides, our method also takes the background mode of words into account to address the concise and noisy problem of Twitter.

yanks	beachday	show
yankees	ocean	theater
mets	sun	restaurant
game	boardwalk	shopping
ballpark	sand	nightlife
newyork	sunbath	downtown
baseball	**beach**	**night**

Dogers	beachlife	nightlife
letsgodogers	yoga	date
Angels	Ocean	drink
game	Huntington	club
stadium	enjoy	restaurant
season	beachtime	bakery
baseball	**beach**	**night**

(a) Similarity queries in NY.　　　(b) Similarity queries in LA.

Fig. 7. (a) and (b) are similarity queries based on keyword embeddings generated by NBE module. Given the query keywords (*baseball, beach, and night*), we list the most similar top 6 keywords respectively by computing cosine similarity.

(a) Word embeddings generated by MME　(b) Word embeddings generated by NBE-　(c) Word embeddings generated by NBE

Fig. 8. (a), (b) and (c) are visualization of keyword embeddings generated by MME, NBE- and NBE, the embeddings are projected into two dimensions where each keyword is represented by a dot, and the keywords in the same color belong to the same topic. (Color figure online)

NBE Module Analysis. First, we perform similarity queries based on keyword embeddings in NY and LA to demonstrate whether NBE can generate semantically meaningful embeddings. As shown in Fig. 7(a) and (b), the retrieved keywords are semantically coherent with the given keyword. Then we compare NBE with MME and NBE-, Fig. 5 shows the comparison of topic coherence based on above three embedding methods. As shown, the performance of NBE based topic modeling outperforms MME and NBE- significantly. Moreover, we use t-SNE visualization to present the keyword embeddings generated by different methods, shown in Fig. 8(a), (b) and (c). Compared to MME, the other two embedding methods can get better keyword embeddings for topic modeling. Because MME embeds keywords directly on original texts, which falls short in embedding short texts, while NBE- and NBE are based on networks, which can generate rich contexts by random walk. However, NBE- considers time and location separately, leading to the generated keyword embeddings less spatio-temporally close, which demonstrate the significance of capturing the correlation between time and location. Experiments show that NBE module is capable of generating semantically coherent and spatio-temporally close keyword embeddings, which can be directly used in subsequent topic model to obtain high-quality local topics.

TME Module Analysis. As shown in Fig. 5, based on the same keyword embeddings generated from NBE module, our TME module can obtain more coherent topics than sHDP. Because sHDP assumes that each document is represented by a mixture of topics like the standard LDA, while TME presumes that each document is represented by only one topic, the latter is more suitable for modeling short texts. Besides, TME presumes that the words in tweets can either belong to the document's topic or to the background mode, which can address the noisy problem of tweets. Therefore, TME has obvious advantages compared to sHDP when dealing with concise and noisy texts, like tweets.

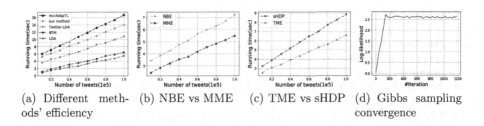

(a) Different methods' efficiency   (b) NBE vs MME   (c) TME vs sHDP   (d) Gibbs sampling convergence

Fig. 9. Efficiency study.

Efficiency Comparison. In this subsection, we analyze the efficiency of our method. Figure 9(a) shows the running time is linearly increasing as the number of tweets increases. As shown, our method is slower than LDA, BTM and twitter-LDA, because our method incorporates time and location, which brings extra time cost. Compared to IncrAdapTL, our method is more efficient, because our method utilizes neural model to get keyword embeddings quickly and only needs to sample keyword information, while IncrAdapTL needs to sample three types of information, which is time-consuming. Figure 9(b) shows the running time of NBE is little slower than MME, because NBE needs extra time to generate rich contexts by random walk to provide additional semantics. Figure 9(c) shows the running time of TME is faster than sHDP, because TME assumes that each tweet is represented by only one topic, which simplifies the sampling procedure. Figure 9(d) shows the log-likelihood variation as the number of Gibbs sampling iteration increases, we can see the log-likelihood converges very quickly, so it's sufficient to set the iterations to a relatively small number for better scalability.

7   Conclusion

In this paper, we propose a novel method to detect local topics from spatio-temporal tweet corpus. With network based embedding, we can capture the correlation between time and location, and jointly model spatio-temporal and semantic information, so that the keywords talking about the same local topic can be semantically coherent and spatio-temporally close. Based on the keyword embeddings, we present a topic model which takes background mode of keywords into consideration, and obtain high-quality local topics by clustering keywords of tweets into groups. Experiments show that our proposed method can improve the effectiveness significantly while achieving good efficiency. For future work, we will take more characteristics of tweets into account to increase the effectiveness of local topic detection and evaluate our method on more real-world data sets.

Acknowledgement. This work is supported by the National Key Research and Development Program of China, and National Natural Science Foundation of China (No. U163620068).

References

1. Batmanghelich, K., Saeedi, A., Narasimhan, K., Gershman, S.: Nonparametric spherical topic modeling with word embeddings. In: ACL (2). The Association for Computer Linguistics (2016)
2. Blei, D.M., Ng, A.Y., Jordan, M.I.: Latent dirichlet allocation. J. Mach. Learn. Res. **3**, 993–1022 (2003)
3. Chang, B., Park, Y., Park, D., Kim, S., Kang, J.: Content-aware hierarchical point-of-interest embedding model for successive POI recommendation. In: IJCAI, pp. 3301–3307. ijcai.org (2018)
4. Das, R., Zaheer, M., Dyer, C.: Gaussian LDA for topic models with word embeddings. In: ACL (1), pp. 795–804. The Association for Computer Linguistics (2015)
5. Ding, R., Nallapati, R., Xiang, B.: Coherence-aware neural topic modeling. In: EMNLP, pp. 830–836. Association for Computational Linguistics (2018)
6. Giannakopoulos, K., Chen, L.: Incremental and adaptive topic detection over social media. In: Pei, J., Manolopoulos, Y., Sadiq, S., Li, J. (eds.) DASFAA 2018. LNCS, vol. 10827, pp. 460–473. Springer, Cham (2018). https://doi.org/10.1007/978-3-319-91452-7_30

7. Gopal, S., Yang, Y.: Von mises-fisher clustering models. In: ICML. JMLR Workshop and Conference Proceedings, vol. 32, pp. 154–162. JMLR.org (2014)
8. Liao, L., He, X., Zhang, H., Chua, T.: Attributed social network embedding. IEEE Trans. Knowl. Data Eng. **30**(12), 2257–2270 (2018)
9. Liu, H., Ge, Y., Zheng, Q., Lin, R., Li, H.: Detecting global and local topics via mining Twitter data. Neurocomputing **273**, 120–132 (2018)
10. Liu, Y., Ester, M., Hu, B., Cheung, D.W.: Spatio-temporal topic models for check-in data. In: ICDM, pp. 889–894. IEEE Computer Society (2015)
11. Mahdaouy, A.E., El Alaoui, S.O., Gaussier, É.: Improving Arabic information retrieval using word embedding similarities. Int. J. Speech Technol. **21**(1), 121–136 (2018)
12. Rosner, F., Hinneburg, A., Röder, M., Nettling, M., Both, A.: Evaluating topic coherence measures. CoRR abs/1403.6397 (2014)
13. Sizov, S.: GeoFolk: latent spatial semantics in web 2.0 social media. In: WSDM, pp. 281–290. ACM (2010)
14. Xu, G., Meng, Y., Chen, Z., Qiu, X., Wang, C., Yao, H.: Research on topic detection and tracking for online news texts. IEEE Access **7**, 58407–58418 (2019)
15. Yan, X., Guo, J., Lan, Y., Cheng, X.: A biterm topic model for short texts. In: WWW, pp. 1445–1456. International World Wide Web Conferences Steering Committee/ACM (2013)
16. Yin, Z., Cao, L., Han, J., Zhai, C., Huang, T.S.: Geographical topic discovery and comparison. In: WWW, pp. 247–256. ACM (2011)
17. Zhang, C., et al.: TrioVecEvent: embedding-based online local event detection in geo-tagged tweet streams. In: KDD, pp. 595–604. ACM (2017)
18. Zhang, C., et al.: Regions, periods, activities: uncovering urban dynamics via cross-modal representation learning. In: WWW, pp. 361–370. ACM (2017)
19. Zhao, W.X., et al.: Comparing twitter and traditional media using topic models. In: Clough, P., et al. (eds.) ECIR 2011. LNCS, vol. 6611, pp. 338–349. Springer, Heidelberg (2011). https://doi.org/10.1007/978-3-642-20161-5_34

Building Mongolian TTS Front-End with Encoder-Decoder Model by Using Bridge Method and Multi-view Features

Rui Liu, Feilong Bao$^{(\boxtimes)}$, and Guanglai Gao

College of Computer Science, Inner Mongolia Key Laboratory of Mongolian
Information Processing Technology, Inner Mongolia University,
Hohhot 010021, China
liurui_imu@163.com, csfeilong@imu.edu.cn

Abstract. In the context of text-to-speech systems (TTS), a front-end
is a critical step for extracting linguistic features from given input text.
In this paper, we propose a Mongolian TTS front-end which joint training Grapheme-to-Phoneme conversion (G2P) and phrase break prediction (PB). We use a bidirectional long short-term memory (LSTM) network as the encoder side, and build two decoders for G2P and PB that
share the same encoder. Meanwhile, we put the source input features
and encoder hidden states together into the Decoder, aim to shorten the
distance between the source and target sequence and learn the alignment
information better. More importantly, to obtain a robust representation
for Mongolian words, which are agglutinative in nature and lacks sufficient training corpus, we design specific multi-view input features for it.
Our subjective and objective experiments have demonstrated the effectiveness of this proposal.

Keywords: Text-to-speech · Front-end · Phrase break ·
Grapheme-to-Phoneme · Mongolian

1 Introduction

A text-to-speech system (TTS) consists of two components. One is a front-end,
which takes a given text as its input and returns a phoneme sequence annotated with prosody information of the text. The other is a back-end, which
converts the output of a front-end into speech. For a front-end, the vital part is
Grapheme-to-Phoneme conversion (G2P) and phrase break prediction (PB), as
the intelligibility and naturalness depend on their correctness. To estimate the
correct phoneme sequence of a sentence, we need to recognize words and determine their phoneme sequences. Furthermore, to split an utterance into prosodic
units which can be easily understood by people, we need to identify the prosody
phrase boundaries of sentence.

For English and Mandarin TTS, there have been attempts at solving this
problem. G2P can be treated as a sequence prediction problem. A typical approach to G2P involves using joint sequence model [1]. Recently, there has been

© Springer Nature Switzerland AG 2019
T. Gedeon et al. (Eds.): ICONIP 2019, CCIS 1143, pp. 642–651, 2019.
https://doi.org/10.1007/978-3-030-36802-9_68

some work using long short-term memory (LSTM) networks and encoder-decoder approach for G2P problem [3,8]. PB can be treated as a sequence labeling task. Typically PB methods usually use maximum entropy Markov models (MEMMs) [9], conditional random fields (CRFs) [10], and recurrent neural networks (RNNs) [11]. But in these above works, G2P and PB are usually processed separately. However, for Mongolian TTS, the research on G2P or PB is at its initial stage. There are many works which have made great contributions [12], but the performance is less than satisfactory.

In this work, we investigate how G2P and PB can be jointly modeled while benefiting from the strong modeling capacity of the encoder-decoder models and built a Mongolian TTS front-end. We use a bidirectional recurrent neural network (RNN) as the encoder side. For decoder side, we build two decoders for G2P and PB based on unidirectional RNN separately. These two decoders share the same encoder parameter. Learning from the attention mechanism in encoder-decoder model [13], we further utilizes attention to the G2P Decoder and alignment-based PB Decoder. Such attention provides additional information to the G2P and PB. To shorten the distance between the source and target sequence and learn the alignment information better, we use bridge method inspired by the machine translation [7], in which we put the source input features and encoder hidden states together into the two Decoders.

In addition, all these methods mentioned for high resource language taking the word embeddings as input. It is hard to work with scripts of Mongolian languages, in which the necessary linguistic resources are not readily available and lack sufficient training corpus. Thus we take a multi-view approach to learning word-level representations as the encoder input for Mongolian, leverages the agglutinative property. To obtain a robust representation for Mongolian word, we first identify the sequence of morpheme (stem&suffix) automatically and encode them to a morphological representation, which captures the morphological information of the word. Then we extract acoustic features of each word. At last, the morphological vector, acoustic features and word embeddings are comprised together to a multi-view input features for each Mongolian word.

Objective experiment results show our proposed model achieves better performance than the conventional model. Subjective experiment results further show that this method is beneficial to improve the naturalness and the expression of the Mongolian synthesized speech.

2 Proposed Model

2.1 Joint Encoder-Decoder Model

Our joint Encoder-Decoder model includes one Encoder, which reads in the input Mongolian word sequence $x = (x_1, x_2, \ldots, x_T)$, and two Decoders, which generates Mongolian phoneme sequence $y = (p_1, p_2, \ldots, p_{T'})$ and the corresponding PB labels $y = (y_1, y_2, \ldots, y_T)$ simultaneously. The model structure is illustrated in Fig. 1.

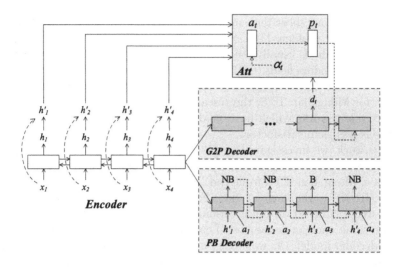

Fig. 1. Joint encoder-decoder model for G2P and PB.

Encoder Side Using Bridge Method. As shown in Fig. 1, we use bidirectional LSTM (BiLSTM) as basic component for the common encoder side on two decoders.

In conventional method, the source sequence $x = (x_1, x_2, \ldots, x_T)$ only provides source input information in generating the hidden state h_i once time, and then it is no longer used. To make the relation of source and target sequence more closely, we move source input features one step closer to the target output as illustrates in Fig. 1. After generating the final encoder hidden state h_i at each time step i, we concatenate h_i with its corresponding input features x_i as the bridge hidden vector h_i': $h_i' = [h_i, x_i]$. In this bridge method, word-level input features as part of the encoder hidden state to form the attention information and consequently have a positive effect in Decoder stage. The first and last bridge states (h_1' and h_T') carries rich information of the entire source input. We use a linear combination of h_1' and h_T', with parameters learned during training, as initial hidden state for two Decoders.

Decoder Side for G2P and PB. The Decoders are modeled as unidirectional LSTM. In PB, at decode stage, the decoder state s_i is calculated as a function of the previous decoder state s_{i-1}, the previous predicted label y_{i-1}, the aligned encoder bridge hidden state h_i', and the attention vector a_i:

$$s_i = f(s_{i-1}, y_{i-1}, h_i', a_i) \tag{1}$$

where the attention vector a_i is computed as a weight sum of the bridge encoder states $h' = (h_1', \ldots, h_T')$:

$$e_{i,k} = g(s_{i-1}, h'_k), \quad \alpha_{i,j} = \frac{\exp(e_{i,j})}{\sum_{k=1}^{T} \exp(e_{i,k})} \tag{2}$$

$$a_i = \sum_{j=1}^{T} \alpha_{i,j} h'_j \tag{3}$$

g is a feed-forward neural network. It should be note that the encoder state h_i is the explicit aligned input at each decoding step in *PB Decoder*. The attention vector a_i provides abundant context information to the *PB Decoder*.

G2P Decoder shares the same encoder hidden states with PB Decoder. The G2P Decoder predicts each phoneme p_t given the attention vector a_i and all of the previously predicted phonemes $p_1, p_2, \ldots, p_{t-1}$ in the following way:

$$s_t = g(\tilde{p}_{t-1}, s_{t-1}, a_i) \tag{4}$$

$$p(p_t | p_{<t}, x) = \text{softmax}(W_s d_t + b_s) \tag{5}$$

where s_{t-1} is the hidden state of the decoder LSTM and $\tilde{p}_{t-1}$ is the vector obtained by projecting the one hot vector corresponding to p_{t-1} using a phoneme embedding matrix E. The embedding matrix E is jointly learned with other parameters of the model. Similar to the PB Decoder, we use an attention vector a_i, a linear combination of all of the encoder bridge hidden states, at every decoder time step. But in G2P decoder, the alignment input from source word sequence is unknown. Therefore the attention vector a_i can be seen as a soft alignment between the source word sequence and target phoneme sequence. It allows the model to attend to different encoder states when decoding each output label. Motivated by [13], attention vector a_i at time step i is given by:

$$a_t = \sum_{i=1}^{T} \alpha_{i,t} h'_i \tag{6}$$

and

$$\alpha_t = \text{softmax}(u_t) \tag{7}$$

$$u_{i,t} = v^T \tanh(W_1 h_i + W_2 d_t + b_a) \tag{8}$$

where the vector v, b_a and the matrices W_1, W_2 are parameters learned jointly with the rest of the model. The score $\alpha_{i,t}$ is a weight that represents the importance of the hidden bridge encoder state h'_i in generating the phoneme p_t.

Hence, Eq. (5) can be rewritten as:

$$p(p_t | p_{<t}, x) = \text{softmax}(W_s[a_t; d_t] + b_s) \tag{9}$$

2.2   Multi-view Input Features

To obtain more robust embedding representation for Mongolian words[1], we propose two novel word-level representations, which include **Morphological vector** and **Acoustic vector**, along with **Word embedding** as input features.

Morphological Vector (m_i). We use multi-layer stacked BiLSTM network to automatically learn the mapping from the sequences of morpheme, stem and suffix, to a word-level vector.

This operation is consistent with our previous work [5,6]. Each words in a Mongolian sentences is broken down into individual smaller unit: stem and suffix, these are then mapped to a sequence of embeddings, which are passed through multi-layer (stacked) BiLSTMs to obtain the morphological vector.

Acoustic Vector. For predicting the phrase pause boundaries in speech, we can incorporate acoustic features. Here we explore three types of features widely used in TTS [14]: **Duration parameter**, **Spectrum parameter** and **Excitation parameter**. All acoustic features are extracted according to the word boundaries obtain by force alignment with respect to the reference transcriptions by using the Speech Signal Processing Toolkit (SPTK)[2].

- **Duration parameter** (t_i): Word duration are strong cues to prosodic phrase boundaries. The word duration features d_i is computed as the actual word duration divided by the mean duration of the word, clipped to a maximum value of 6. The sample mean is used for frequent words (count ≥ 15). For infrequent words we estimate the mean as the sum over the sample means for their phoneme sequences.
- **Spectrum parameter** (s_i): The spectrum vector s_i consists of Mel-generalized cepstral coefficients (MGC) vector including the zeroth coefficients.
- **Excitation parameter** (e_i): The excitation vector e_i consists of log fundamental frequency (logF0).

Word Embedding (we_i). We use Skip-Gram [15] model to train the word embedding representation we_i for Mongolian.

Therefore, our overall multi-view input feature vectors x_i designed for Mongolian is the concatenation of m_i, t_i, s_i, e_i and we_i in various combinations.

3   Experiments

3.1   Data

Mongolian Orthography dictionary as an experimental dataset in G2P. The whole dictionary, which contains about 40k items, is partitioned into training, validation and test set according to 8:1:1.

---

[1] There are two writing systems of Mongolian: Cyrillic Mongolian and traditional Mongolian. This paper only studies traditional Mongolian.

[2] http://sp-tk.sourceforge.net/.

Table 1. Ablation test with bridge method and multi-view features. **Bold** indicates the best model.

Bridge	we	m	Acoustic vector			WER (%)	F (%)
			t	s	e		
No	Yes	No	No	No	No	24.53	83.26
No	Yes	Yes	No	No	No	24.01	84.13
No	Yes	No	Yes	No	No	23.85	83.98
No	Yes	No	No	Yes	No	23.79	84.02
No	Yes	No	No	No	Yes	23.80	84.13
No	Yes	Yes	Yes	No	No	22.98	84.52
No	Yes	Yes	No	Yes	No	22.93	84.56
No	Yes	Yes	No	No	Yes	22.91	84.61
No	Yes	No	Yes	Yes	No	23.71	84.65
No	Yes	No	Yes	No	Yes	23.67	84.72
No	Yes	No	No	Yes	Yes	23.59	84.76
No	Yes	Yes	Yes	Yes	No	22.58	85.12
No	Yes	Yes	Yes	No	Yes	22.51	85.19
No	Yes	Yes	No	Yes	Yes	22.47	85.24
No	Yes	No	Yes	Yes	Yes	23.35	83.54
No	Yes	Yes	Yes	Yes	Yes	21.27	86.13
Yes	Yes	No	No	No	No	22.37	85.32
Yes	Yes	Yes	No	No	No	21.98	85.95
Yes	Yes	No	Yes	No	No	21.76	85.89
Yes	Yes	No	No	Yes	No	21.70	85.86
Yes	Yes	No	No	No	Yes	21.63	85.90
Yes	Yes	Yes	Yes	No	No	20.62	86.35
Yes	Yes	Yes	No	Yes	No	20.35	86.39
Yes	Yes	Yes	No	No	Yes	20.29	86.41
Yes	Yes	No	Yes	Yes	No	21.11	86.32
Yes	Yes	No	Yes	No	Yes	21.03	86.43
Yes	Yes	No	No	Yes	Yes	21.15	86.45
Yes	Yes	Yes	Yes	Yes	No	20.31	87.15
Yes	Yes	Yes	Yes	No	Yes	20.25	87.21
Yes	Yes	Yes	No	Yes	Yes	20.29	87.19
Yes	Yes	No	Yes	Yes	Yes	20.92	87.33
Yes	Yes	Yes	Yes	Yes	Yes	**19.46**	**88.50**

For evaluating the effectiveness of the PB model, we rely on a corpus corresponding to the Mongolian TTS database recorded by a professional native Mongolian female speaker. The corpus contains 59k sentences and more than 409k words. The speech data from the Mongolian TTS database, which was segmented according to the word boundaries obtained by forced alignment with respect to the reference transcriptions, are used to extract word-level acoustic feature.

The word embedding train data were crawled from mainstream websites in Mongolia. After cleaning web page tags and filtering longer sentences, its token size and vocabulary are about 200 million and 3 million respectively.

Table 2. Comparison with previous models and our independent model. Joint training model results on G2P and PB.

#	Model	WER (%)	F (%)
G2P	Joint sequence [1]	22.53	–
	BiLSTM [3]	22.01	–
	Encoder-decoder [8]	20.32	–
	Independent model (best model)	**19.46**	–
PB	CRF [10]	–	83.21
	LSTM [11]	–	85.16
	Independent model (best model)	–	**88.50**
G2P & PB	**Joint model**	**19.32**	**89.15**

3.2 Experiments Settings

For all experiments,we select the number of units in LSTM cell as 128. The default forget gate bias is set to 1. To prevent overfitting we use scheduled sampling with a linear decay on the decoder side. Parameters were optimised using Adam with default learning rate 1.0 and sentences were grouped into batch of size 64. We take the loss sum of two tasks as a joint loss. Performance on the training set was measured at every epoch and training was stopped if performance had not improved for 10 epoches.

In Morphological vector part, we have 3 layers stacked LSTMs, each with 512 units. We use an initial learning rate of 0.001 and reduce this learning rate by a multiplicative factor of 0.8. We use minibatch stochastic gradient descent (SGD) together with Adam using a minibatch size of 256. The embeddings for morpheme were set to 100. The network was trained 500 epochs.

Speech signals are sampled at 16 kHz, windowed by a 25-ms window shifted every 5-ms. The acoustic features vector contain 35 MGC, logF0 and word duration, totally 37 dimensions $(35 + 1 + 1 = 37)$. The word embeddings were initialised with pretrained vectors as illustrated in Sect. 2.2 and then fine-tuned during model training. For the Mongolian datasets we used 100-dimensional word embeddings.

3.3   Evaluation Metrics

In PB, results are reported on the test set in terms of the F-score (F) which is defined as the harmonic mean of the *Precision* and *Recall.* In G2P, We report *word error rate (WER)*.

3.4   Independent Model Results

Table 1 report the results on the independent Encoder-Decoder model for G2P and PB. We first using the word embedding input features $(x_i = we_i)$ to establish a strong baseline, on top of which we can add various features: m_i, t_i, s_i, e_i and use bridge method.

We note that adding any combination of features (individually or in sets) improves performance over the baseline. The proposed multi-view input features provide richer information than word embedding in Mongolian. Furthermore, morphological vector (m_i) and acoustic vector (t_i, s_i, e_i) both play a very good role in the performance for the two tasks. Specifically, in G2P, the introduction of acoustic vector contributes significantly in all metrics, compared to morphological vector. In PB, the contributions of acoustic vector and morphological vector are almost equally. We believe that the interesting phenomenon may owing to the nature of the specific tasks. For G2P, which gives every word a corresponding phoneme list. It is important to consider the pronunciation information when decoding the current word. However, the internal structure and pronunciation information is what we need to focus on in PB task for agglutinative language [2]. In Mongolian, stem and suffix serve to discriminate words based on syntactic meaning, and that these sub-word units can be used to model PB. The "$we_i + m_i + t_i + s_i + e_i$" model that uses all features has the remarkable performance over the we_i baseline.

We also notice that using bridge method, that concatenate the input features with hidden states, in conjunction with multi-view input features yields a significant improvement in F-Score. Lastly, we refer to the "*bridge method* $+ we_i + m_i + t_i + s_i + e_i$" model as our "best model".

3.5   Joint Model Results

Table 2 shows our joint model performance on G2P and PB comparing to previous methods and the independent model under same Mongolian data. As shown in this table, the joint model using encoder-decoder architecture with bridge and multi-view features achieves the best performance. This indicates that G2P and PB can be closely linked together through joint training to achieve a joint performance improvement.

3.6   Front-End Based on Joint Model

To evaluate the naturalness of the synthesized Mongolian speech from the proposed front-end component, a subjective AB preference test was conducted.

In this evaluation, a set of 20 sentences were randomly selected from test set and the synthesised speech was generated through the DNN-based Mongolian TTS system [4] based on proposed front-end and the original front-end [4]. 20 subjects were asked to choose which one was better of paired synthesis speech. Figure 2 shows the subjective evaluation results. It can be seen from the figure that the proposed front-end component, with joint model for G2P and PB, obtain the higher quality Mongolian speech.

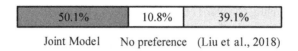

Joint Model   No preference   (Liu et al., 2018)

Fig. 2. Subjective evaluation results of the DNN-based Mongolian TTS by using proposed front-end and original front-end in [4].

4   Conclusions

In this paper, we proposed a joint Encoder-Decoder model by using bridge method and multi-view feature for joint G2P and PB and built a Mongolian TTS front-end in a unified framework. The bridge method seeks to shorten the distance between source and target word-level features from the word sequence. In view of the limitation of necessary linguistic resources are not readily available in Mongolian, the word-level multi-view features combines three parts of information (morphological, acoustic and word) to form a robust representation for Mongolian word. In addition, joint training can better utilize the close connection between G2P and PB. The proposed model achieves better performance compared to conventional front-end component in Mongolian TTS.

Acknowledgments. This research was supports by the National Natural Science Foundation of China (No.61563040, No.61773224), Natural Science Foundation of Inner Mongolian (No.2018MS06006, No.2016ZD06).

References

1. Bisani, M., Ney, H.: Joint-sequence models for grapheme-to-phoneme conversion. Speech Commun. **50**, 434–51 (2008)
2. Vadapalli, A., Bhaskararao, P., Prahallad, K.: Significance of word-terminal syllables for prediction of phrase breaks in Text-to-Speech systems for Indian languages. In: 8th ISCA Tutorial and Research Workshop on Speech Synthesis (2013)

3. Rao, K., Peng, F., Sak, H., Beaufays, F.: Grapheme-to-phoneme conversion using long short-term memory recurrent neural networks. In: 40th IEEE International Conference on Acoustics, Speech and Signal Processing, pp. 4225–4229. IEEE Press (2015)

4. Liu, R., Bao, F., Gao, G., Wang, Y.: Mongolian text-to-speech system based on deep neural network. In: Tao, J., Zheng, T.F., Bao, C., Wang, D., Li, Y. (eds.) NCMMSC 2017. CCIS, vol. 807, pp. 99–108. Springer, Singapore (2018). https://doi.org/10.1007/978-981-10-8111-8_10

5. Liu, R., Bao, F., Gao, G.: A LSTM approach with sub-word embeddings for mongolian phrase break prediction. In: 27th International Conference on Computational Linguistics, Santa Fe, New Mexico, USA, 20–26 August 2018, pp. 2448–2455 (2018)

6. Liu, R., Bao, F., Gao, G., Wang, Y.: Improving mongolian phrase break prediction by using syllable and morphological embeddings with BiLSTM model. In: 19th Annual Conference of the International Speech Communication Association, Hyderabad, India, 2–6 September 2018, pp. 57–61 (2018)

7. Kuang, S., Li, J., Branco, A., Luo, W., Xiong, D.: Attention focusing for neural machine translation by bridging source and target embeddings. In: 56th Annual Meeting of the Association for Computational Linguistics, Melbourne, Australia, 15–20 July 2018 (2018)

8. Toshniwal, S., Livescu, K.: Jointly learning to align and convert graphemes to phonemes with neural attention models. In: Spoken Language Technology Workshop. IEEE Press (2017)

9. Yu, Z., Lee, G.G., Kim, B.: Using multiple linguistic features for Mandarin phrase break prediction in maximum-entropy classification framework. In: 5th INTERSPEECH 2004 - ICSLP, International Conference on Spoken Language Processing, Jeju Island, Korea, October 2004. DBLP (2004)

10. Qian, Y., Wu, Z., Ma, X., Soong, F.: Automatic prosody prediction and detection with Conditional Random Field (CRF) models. In: 7th International Symposium on Chinese Spoken Language Processing, pp. 135–138. IEEE Press (2010)

11. Vadapalli, A., Gangashetty, S.V.: An investigation of recurrent neural network architectures using word embeddings for phrase break prediction. In: 17th INTERSPEECH, pp. 2308–2312 (2016)

12. Liu, R., Bao, F., Gao, G., Wang, W.: Mongolian prosodic phrase prediction using suffix segmentation. In: 21th International Conference on Asian Language Processing, pp. 250–253. IEEE Press (2017)

13. Bahdanau, D., Cho, K., Bengio, Y.: Neural machine translation by jointly learning to align and translate. In: Proceedings of ICLR 2015, pp. 1–15 (2014)

14. Wang, X., Lorenzo-Trueba, J., Takaki, S., Juvela, L., Yamagishi, J.: A comparison of recent waveform generation and acoustic modeling methods for neural-network-based speech synthesis. In: 43th ICASSP (2018)

15. Mikolov, T., Sutskever, I., Chen, K., Corrado, G., Dean, J.: Distributed representations of words and phrases and their compositionality. In: Advances in Neural Information Processing Systems, pp. 3111–3119 (2013)

Code Summarization with Abstract Syntax Tree

Qiuyuan Chen[1], Han Hu[2(✉)], and Zhaoyi Liu[3]

[1] College of Computer Science and Technology, Zhejiang University,
Hangzhou, China
`chenqiuyuan@zju.edu.cn`
[2] School of Software, Tsinghua University, Beijing, China
`hh17@mails.tsinghua.edu.cn`
[3] School of Shenzhen Graduate, Peking University, Shenzhen 518055, China
`1701213615@sz.pku.edu.cn`

Abstract. Code summarization, which provides a high-level description of the function implemented by code, plays a vital role in software maintenance and code retrieval. Traditional approaches focus on retrieving similar code snippets to generate summaries, and recently researchers pay increasing attention to leverage deep learning approaches, especially the encoder-decoder framework. Approaches based on encoder-decoder suffer from two drawbacks: (a) Lack of summarization in functionality level; (b) Code snippets are always too long (more than ten words), regular encoders perform poorly. In this paper, we propose a novel code representation with the help of Abstract Syntax Trees, which could describe the functionality of code snippets and shortens the length of inputs. Based on our proposed code representation, we develop Generative Task, which aims to generate summary sentences of code snippets. Experiments on large-scale real-world industrial Java projects indicate that our approaches are effective and outperform the state-of-the-art approaches in code summarization.

Keywords: Code summarization · Code clone · Code representation

1 Introduction

There is much tacit knowledge in the source code which is not consistent with human intuition [1]. To better understand the source code, code summarization is used to transform this knowledge with low cost by automatically generating functional natural language description for a code snippet.

Typical code summarization includes summarizing commit message, log text, and code comment, which is vital to comprehend the source code. Researchers leverage Information Retrieval (IR) and learning-based techniques to generate comments automatically [4,11].

Q. Chen, H. Hu and Z. Liu—Equal contribution.

© Springer Nature Switzerland AG 2019
T. Gedeon et al. (Eds.): ICONIP 2019, CCIS 1143, pp. 652–660, 2019.
https://doi.org/10.1007/978-3-030-36802-9_69

IR techniques heavily rely on whether similar the code snippets can be retrieved, how similar the code snippets are, and fail to generate comments when encountering unmatched issues. However, if no similar code snippet exists, these techniques cannot output accurate summaries [2].

Current learning-based approaches often exploit deep neural network framework to build probabilistic models of source code [2,4,7] and prevalent learning-based approaches exploit mutant encoder-decoder [2,7,15] to learn the tacit knowledge in source code and transform them into descriptive natural language.

However, current learning-based approaches suffer from two drawbacks: (a) Lack of summarization in functionality level. Vanilla deep neural networks always regard the code snippet as a sequence of plain words, only taking word-level features into account. When we meet a code snippet, we want to know more about the functionality of the code snippet, not word-level features. (b) Code snippets are always too long (more than ten words). When faced with long sequences, as we know, regular encoders perform poorly.

Faced with drawback (a), we need a new presentation of code snippets which is able to describe the functionality of code snippets. Abstract Syntax Tree (AST) is widely used in the field of Program Analysis. Figure 1(a) is a code snippet of counting characters in a string, Fig. 1 is the visualized AST of Fig. 1(a).

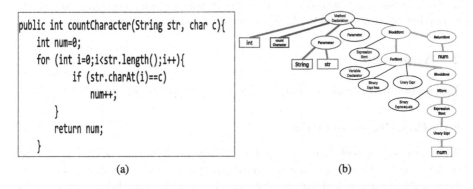

```
public int countCharacter(String str, char c){
    int num=0;
    for (int i=0;i<str.length();i++){
        if (str.charAt(i)==c)
            num++;
    }
    return num;
}
```

(a)   (b)

Fig. 1. Java method and its AST. (Color figure online)

As shown in Fig. 1, AST is a tree-structural representation of source code which describes code snippets in a specific programming language. The leaves of the tree usually refer to user-defined values which represent identifiers (such as *num, str*) or variable types (such as *String, int*) in the source code. Syntactic structures such as judgment statement (*ifStmt*) and loop (*ForStmt*) are represented as non-leaf nodes. In AST, every function module is represented as a path between two leaves or between a leaf and the root, which is called AST path. AST paths are marked red, yellow, blue or green respectively in Fig. 1.

In this paper, to overcome drawback (a), we utilize AST paths to represent code snippets. Faced with drawback (b), as mentioned early, every code snippet is represented as AST paths in our paper, so we split every long code snippet to short AST paths which could improve the performance of our encoder.

Based on our new representation of code snippets, we build an encoder-decoder model to summarize source codes in Generative Task. Generative Task aims to generate a sequence of words to describe the code snippet, the target of each code snippet is a sentence.

We collect 204,688 training pairs from 12 most popular open-source Java libraries, which are all starred more than 10,000 times on GitHub, and carry several experiments. The experiments results show that our model achieves better performance on three reliable metrics: Rouge-2, Rouge-L, and BLEU and can generate more accurate natural languages to describe the functionality of code snippets.

In summary, the contributions of our work are as follows:

- We propose a new representation of code snippets which is based on AST paths.
- We build an encoder-decoder model to summarize source codes based on our proposed code representation.
- We carry several experiments and compare the results with four baselines.

The rest of this paper is organized as follows: our approach is presented in Sect. 2. Section 3 introduces the experiments details. Section 4 introduces researches related to this work. The conclusion is shown in Sect. 5.

2   Approach

This section consists of two parts: Code Representation, and Generative Task.

2.1   Code Representation

As mentioned in Sect. 1, every input code snippet X is represented as a series of AST paths. Every AST path consists of two leaves and non-leaf nodes, so every path is seen as a sequence of its non-leaf nodes' embedding vectors and a sequence of two leaves' tokens embedding vectors[1]. Let

$$V^{nodes} = (V_1^{node}, V_2^{node}, ..., V_i^{node}) \tag{1}$$

$$V^{leaves} = (V_1^{token}, V_2^{token}, ..., V_j^{token}) \tag{2}$$

where V_{node}^i is denoted as the vector of i_{th} node. We use bi-direction LSTM to encode the V^{nodes} and V^{leaves}

$$h_1^{node}, h_2^{node}, ... h_i^{node} = BiLSTM(V_1^{node}, V_2^{node}, ..., V_i^{node}) \tag{3}$$

$$h_1^{leaves}, h_2^{leaves}, ... h_j^{leaves} = BiLSTM(V_1^{token}, V_2^{token}, ..., V_j^{token}) \tag{4}$$

---

[1] The camel-cased and underline identifiers are split into several words, for example, split *checkJavaFile* to three words: check, Java, File, so a leaf may consist of several tokens.

and concatenate the bi-direction final hidden states of $LSTM$ as the final representation of non-leaf nodes and leaf nodes.

$$E^{nodes}(V_1^{node}, V_2^{node}, ..., V_i^{node}) = [h_i^{node}; h_1^{node}] \tag{5}$$

$$E^{leaves}(V_1^{token}, V_2^{token}, ..., V_j^{token}) = [h_j^{leaves}; h_1^{leaves}] \tag{6}$$

So every AST path E^{path} is computed as

$$E^{path} = [E^{nodes}, E^{leaves}] \tag{7}$$

Suppose a snippet of code has k AST paths, so the final representation of a code snippet is

$$E^{code} = (E_1^{path}, E_2^{path}, ..., E_k^{path}) \tag{8}$$

2.2 Generative Task

Generative Task aims to generate a sequence of words to describe the code snippet. The structure is shown in Fig. 2. The task use $LSTM$ with $Attention$ as decoder to generate words one by one.

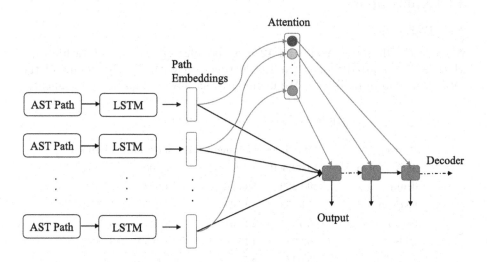

Fig. 2. The structure of generative task

Decoder. The initial state of decoder is the average of path embeddings $E^{average}$:

$$E^{average} = \frac{1}{k} \sum_{i=1}^{k} E_i^{path} \tag{9}$$

Other part is the same with the regular $LSTM$.

Attention. Similar to the regular global attention, the decoder generates outputs while attending the encoder's output vectors E^{path}. At time t, let the hidden state of $LSTM$ is h_t, the output of $LSTM$ is O_t, apply $Softmax$ to convert every dot product between h_t and E_j^{path} to a probability distribution α_t^j.

$$\alpha_t^j = softmax(E_j^{average} \cdot h_t) \tag{10}$$

where α_t^j refer to the j_{th} result in $E^{average}$ at time t. The final context vector C_t is computed as weight sum of α_t^1, α_t^2, ... α_t^j and h_t, let there are k AST paths

$$C_t = \sum_j^k \alpha_t^j \cdot h_t \tag{11}$$

Then concatenate C_t and h_t, feed it to a fully connected layer with $Softmax$ activation to the predict word at time t.

$$y_t = softmax(W_t \cdot [C_t; h_t]) \tag{12}$$

where W_t is a $length_{dictionary} \times (d_{C_t} + d_{h_t})$ weight matrix.

The loss function of this task is Multi-Class Cross Entropy Loss Function.

3   Experiments

3.1   Datasets

We collect JDK source code and 12 most popular open source Java libraries, which are all starred more than 10,000 times on GitHub. The details of the dataset are shown in Table 1. We mainly extract Java methods and comments.

Table 1. Statistics of collected projects

Project	Code lines	Comment lines	Java files	Methods
JDK source code	1,009,560	1,122,392	7,700	56,704
druid	290,239	59,086	4,023	25,836
fastjson	154,353	12,747	2,645	14,973
fresco	76,139	22,361	875	5,247
glide	76,726	14,546	649	3,712
gson	25,047	8,791	206	1,648
guava	501,702	171,651	3,170	19,940
leakcanary	5,703	1,512	81	261
RxJava-2.x	273,894	71,320	1,637	14,635
spring-boot	253,250	117,580	4,359	16,202
spring-framework	642,193	359,293	7,321	42,741
tinker	31,752	9,024	233	1,486
zxing	42,931	16,509	500	1,303
Total	3,383,489	1,986,812	33,399	204,688

3.2   Data Preprocessing

For every Java method, the method body is treated as our inputs, method names, and method comments are extracted as targets. All words are lower-cased. The camel-cased and underline identifiers are split into several words, for example, split *checkJavaFile* to three words: check, Java, File, split *get_user_name* to three words: get, user, name. All punctuation marks are removed. We add *[CLS]* at the beginning of every sentence and add *[SEP]* at the end. *[UNK]* is used to represent words outside the vocabulary. After these steps, every word is tokenized to token.

Javaparser lib[2] is used to parse Java source codes. ASTParser lib[3] is used to build AST of code. We only use the first sentence of comments since they already describe the function of methods according to Javadoc guidance[4]. Several redundant comments, such as empty comments, one-word comments, and non-English comments, are filtered. We restrict both tasks to examples where the length of code snippet is at most 20 tokens, and the length of a comment is at most ten tokens.

Finally, we collect 1,04,963 pairs of $(Code, Comment)$ for Generative Task and split them into the training set, and test set in proportion with 8:2 after shuffling the pairs.

3.3   Experiment Setting

All *LSTM* cells use 2-layers, and we set the dimensionality of the cells hidden states to 1024. In Generative Task, we set the token embeddings and hidden states to 1024. We use dropout [14] with $p = 0.2$. Adam [6] is used as our optimizer with an initial learning rate of 0.001 for optimization. Teacher forcing is used in both tasks' training phases, and the ratio of teacher forcing is 0.5. We use pytorch[5] and trained on cpus. In order to evaluate the quality of the output, following recent works in code summarization [3,5,8], we choose **Rouge-2** and **BLEU** as our metric of Generative Task, which is widely used in code summarization and machine translation.

To validate our approaches, we compare our model with four baselines in several state-of-the-art approaches:

- **CODE-NN:** CODE-NN [4] is the most famous model in code summarization.
- **Basic-Code-RNN & Code-GRU:** Basic-Code-RNN and Code-GRU are effective generation models recently proposed by Liang et al. [7].
- **Word-level Seq2Seq:** Sequence to Sequence model [15] is effective and widely used in the field of machine translation.

---

[2] https://github.com/javaparser/javaparser.
[3] http://help.eclipse.org/mars/index.jsp.
[4] https://www.oracle.com/technetwork/articles/java/index-137868.html.
[5] https://pytorch.org/tutorials/.

3.4   Experiment Results

Table 2 illustrates ROUGE-2, ROUGE-L, and BLEU results of 4 baselines and our model. The result shows that compared with other baselines, our approach outperforms all baselines in three metrics, and achieves an improvement of 0.1546 Rouge-2 points, 0.0933 Rouge-L points, and 0.0611 BLEU point compared with the best results of other approaches.

The results of Word-level Seq2Seq and Generative Task in Table 2 indicate that our AST path features are more effective than word-level features, and our proposed new code presentation could demonstrate the features of code snippets better than regarding the code snippets as a plain sequence of words.

Table 2. ROUGE-2, ROUGE-L, BLEU of baselines and our approaches

Approaches	ROUGE-2	ROUGE-L	BLEU
CODE-NN	0.0685	0.1740	0.2333
Basic CODE-RNN	0.0682	0.1749	0.0322
CODE-GRU	0.1456	0.2004	0.0312
Word-level Seq2Seq	0.3225	0.5575	0.2640
Generative task	**0.4771**	**0.6508**	**0.3251**

4   Related Work

In previous work, Information Retrieval (IR) approaches and training-based approaches are exploited to generate descriptive natural language for source code.

IR approaches are widely used in code summarization. They usually synthesize summaries by retrieving keywords from source code or searching comments from similar code snippets. Templates are utilized for several techniques [9,10,13] which select a subset of the statements and keywords from source code. Then the information is included from those statements and keywords in summary. Sridhara et al. [12] propose an automatic comment generator that identifies the content for the summary. Moreno et al. [10] proposed a template-based method to summarize Java classes rather their methods.

Recently, several studies generate natural language summaries leveraging deep learning approaches. Previous studies [11] predict comments using topic models and n-grams. Study of [1] applies a neural convolutional model with attention to summarize the source code snippets. Iyer et al. [4] propose CODE-NN that uses Long Short Term Memory (LSTM) networks with attention to produce summaries that describe C# code snippets and SQL queries. It takes source code as plain text and models the conditional distribution of the summary on two tasks, code summarization, and code retrieval. Hu et al. [2] combine the neural machine translation model and the structural information within the Java methods to generate the summaries automatically. Liang et al. [7] designed a customized recursive neural network called Code-RNN to extract features from the source code.

5   Conclusion

In this paper, we propose a novel code representation with the help of Abstract Syntax Trees, which could describe the functionality of code snippets and shortens the length of inputs. Based on our proposed code representation, we develop the Generative Task, which aims to generate summary sentences of code snippets, and models for code summarization.

The experimental results show that our approaches, which outperform state-of-the-art baselines, are significantly effective and is able to generate high-quality summary sentences. We also introduce and open our dataset, which can be revisited by other researchers in the future.

References

1. Allamanis, M., Peng, H., Sutton, C.: A convolutional attention network for extreme summarization of source code. In: International Conference on Machine Learning, pp. 2091–2100 (2016)
2. Hu, X., Li, G., Xia, X., Lo, D., Jin, Z.: Deep code comment generation. In: Proceedings of the 26th Conference on Program Comprehension, pp. 200–210. ACM (2018)
3. Hu, X., Li, G., Xia, X., Lo, D., Lu, S., Jin, Z.: Summarizing source code with transferred API knowledge. In: IJCAI, pp. 2269–2275 (2018)
4. Iyer, S., Konstas, I., Cheung, A., Zettlemoyer, L.: Summarizing source code using a neural attention model. In: Proceedings of the 54th Annual Meeting of the Association for Computational Linguistics (Volume 1: Long Papers), vol. 1, pp. 2073–2083 (2016)
5. Iyer, S., Konstas, I., Cheung, A., Zettlemoyer, L.: Mapping language to code in programmatic context (2018)
6. Kingma, D., Ba, J.: Adam: a method for stochastic optimization. Comput. Sci. (2014)
7. Liang, Y., Zhu, K.Q.: Automatic generation of text descriptive comments for code blocks. In: Thirty-Second AAAI Conference on Artificial Intelligence (2018)
8. Ling, W., et al.: Latent predictor networks for code generation (2016)
9. McBurney, P.W., McMillan, C.: Automatic documentation generation via source code summarization of method context. In: Proceedings of the 22nd International Conference on Program Comprehension, pp. 279–290. ACM (2014)
10. Moreno, L., Aponte, J., Sridhara, G., Marcus, A., Pollock, L., Vijay-Shanker, K.: Automatic generation of natural language summaries for java classes. In: 2013 21st International Conference on Program Comprehension (ICPC), pp. 23–32. IEEE (2013)

11. Movshovitz-Attias, D., Cohen, W.W.: Natural language models for predicting programming comments. In: Proceedings of the 51st Annual Meeting of the Association for Computational Linguistics (Volume 2: Short Papers), vol. 2, pp. 35–40 (2013)
12. Sridhara, G., Hill, E., Muppaneni, D., Pollock, L., Vijay-Shanker, K.: Towards automatically generating summary comments for java methods. In: Proceedings of the IEEE/ACM International Conference on Automated Software Engineering, pp. 43–52. ACM (2010)
13. Sridhara, G., Pollock, L., Vijay-Shanker, K.: Generating parameter comments and integrating with method summaries. In: 2011 IEEE 19th International Conference on Program Comprehension, pp. 71–80. IEEE (2011)
14. Srivastava, N., Hinton, G., Krizhevsky, A., Sutskever, I., Salakhutdinov, R.: Dropout: a simple way to prevent neural networks from overfitting. J. Mach. Learn. Res. **15**(1), 1929–1958 (2014)
15. Sutskever, I., Vinyals, O., Le, Q.V.: Sequence to sequence learning with neural networks. In: Advances in Neural Information Processing Systems, pp. 3104–3112 (2014)

Attention Based Shared Representation for Multi-task Stance Detection and Sentiment Analysis

Dushyant Singh Chauhan$^{(\boxtimes)}$, Rohan Kumar, and Asif Ekbal

Department of Computer Science and Engineering,
Indian Institute of Technology Patna, Bihta, India
{1821CS17,rohan.ee17,asif}@iitp.ac.in

Abstract. Stance detection and sentiment analysis are two important problems that have gained significant attention in recent time. While stance detection corresponds to detecting the attitude/position (*i.e., favor, against, and none*) of a person towards any specific event or topic, sentiment analysis deals with determining the opinion expressed by a person for a topic, event, product or a service (*i.e., positive, negative, and neutral*). We envisage these two problems to have a good correlation. For e.g., information about *favor* stance can help in the prediction of *positive* sentiment or *negative* sentiment can help in predicting the *against* stance and so on. Motivated by this, in our current work, we propose a multi-task deep neural framework to investigate whether sentiment helps in stance detection and the vice-versa. Our proposed method makes use of an attention-based shared representation for multi-task stance detection and sentiment analysis. We also deploy an attention mechanism to learn the joint-association between the words present in a tweet and a topic. We evaluate our proposed approach on the benchmark dataset of *SemEval-2016 Task 6*. The proposed multi-task approach yields higher performance compared to the state-of-the-art systems for both stance detection and sentiment analysis.

Keywords: Stance detection · Sentiment analysis · Shared representation · Attention · Natural language processing

1 Introduction

Social media has become the most usable platform nowadays. People do spend 50–70% of their time on social media platforms [1]. These platforms like YouTube, Facebook, Instagram, Twitter, etc. are very popular for its massive spreading of information via instant messages, videos, or miniature blog posts. These social media platforms have become a huge repository of news, ranging from world politics to entertainment gossip, allowing people to interact with each

---

D. S. Chauhan and R. Kumar have equal contribution and are jointly the first author.

© Springer Nature Switzerland AG 2019
T. Gedeon et al. (Eds.): ICONIP 2019, CCIS 1143, pp. 661–669, 2019.
https://doi.org/10.1007/978-3-030-36802-9_70

other with total freedom. The unique part of this platform is that it doesn't matter where people come from and irrespective of their beliefs, they convey their thoughts and discuss, argue, or complain over it and share their feelings freely.

Stance detection and sentiment analysis are two important problems that have attracted attention to the researchers in a very recent time. Stance detection corresponds to detecting the position (*i.e., against, favor, and none*) of a person towards any specific event or topic. Stance detection [2–6] is an important problem in recent time. Augenstein et al. [2] presented an auto-encoder based approach for stance detection. The author used a large set of unlabelled tweets data and train to produce a feature representation of tweet. A convolutional neural network (CNN) based system for stance detection is introduced in [4]. These works did not take into account the concept of attention for stance detection. Sun et al. [5] and Dey et al. [6] presented the attention-based framework for stance detection.

Sentiment analysis deals with determining the opinion expressed by a person for a topic, event, product, or a service (*i.e., negative, positive, and neutral*). Sentiment analysis is a well-studied topic. Sentiment analysis in twitter has been attempted in various studies [7,8]). All these systems were evaluated on SemEval-2016 Task 4[1]. These systems did not consider attention mechanism, and hence are less capable of learning the joint-association between tweet and topic words.

We formulate the problems of stance detection and sentiment analysis to have good correlation and build an effective mechanism for creating a shared-representation between these two tasks. For e.g., information about *favor* stance can help in the prediction of *positive* sentiment or *negative* sentiment can help in predicting the *against* stance and so on. Our proposed model is novel in the sense that our proposed approach applies attention over a multi-task (*i.e., stance and sentiment*) in a single step. Thus, it ensures to reveal the contributing features across tweets and topics. Further, to the best of our knowledge, this is the very first attempt at solving the two problems, namely stance detection, and sentiment analysis together.

The main contributions of our proposed work are three-fold: **(a)** *we leverage the inter-dependence of two related tasks (i.e., stance and sentiment) in improving each other's performance using an effective multi-task framework*; **(b)** *we deploy an effective attention-based mechanism that leverages contributing features by putting more attention to the important words in a tweet and topic*; and **(c)** *we present the state-of-the-art systems for stance detection and sentiment analysis on the benchmark data-set of SemEval-2016 Task 6[2].*

2   Proposed Methodology

In our proposed framework, we aim to leverage the stance detection and sentiment analysis in a multi-task learning framework. We propose an encoder-decoder based shared representation with an attention mechanism for multi-task

---

[1] http://alt.qcri.org/semeval2016/task4/.

[2] http://alt.qcri.org/semeval2016/task6/.

stance detection and sentiment analysis. We also use an attention mechanism [9] to learn the joint-association between the tweet and topic words. The objective is to emphasize on the contributing features by putting more attention to the respective word in a tweet and topic.

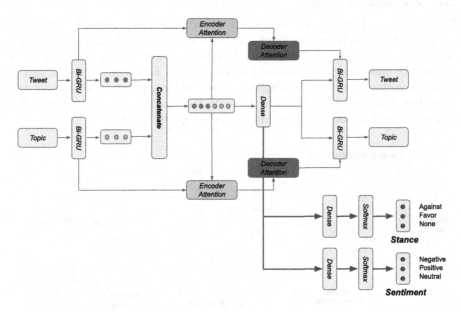

Fig. 1. Overall architecture of the proposed attention based multi-task framework.

We first map the topics and tweets to the same vector space using *GloVe* word embedding of 300 feature dimension. We then separately pass the tweet and the respective topic through a bidirectional gated recurrent unit (Bi-GRU) to obtain the contextual information at every time step. We concatenate the final hidden state of the tweet and topic in order to obtain a *shared representation* that captures the information of both tweet and topic. Subsequently, we use an attention mechanism to learn the joint-association between the words. We pass the shared representation and outputs of Bi-GRU as input for the attention layer to focus on the important words of tweet and topic.

On the other hand, we pass the obtained *shared representation* through a fully connected layer (dense) to match the size of shared representation with the hidden size of the decoder unit. Finally, the output of the attention layer and the fully-connected layer are passed to a Bi-GRU for the reconstruction of the tweet and topic. At the same time, we take the output of fully-connected layer and pass through a *softmax* activation for stance detection and sentiment classification. We illustrate and summarize the proposed methodology in Fig. 1.

We compute the final loss of our proposed multi-task model and back-propagate it to the input (*i.e., tweet and topic*). The net loss is defined as $L_{net} = L_1 + L_2 + L_3 + L_4$.

Where L_1 is the loss generated during tweet reconstruction, L_2 is the loss generated during topic reconstruction, and L_3 and L_4 are the losses generated during stance and sentiment prediction, respectively.

2.1 Encoder Attention

In our proposed framework, we deploy an attention mechanism to learn the joint-association between the tweet and topic words. We pass the shared representation and outputs of the encoder Bi-GRU as input for the attention layer to focus on the important words of tweet and topic. Extracted $attended_{vector}$ is simply a vector that contains weights for the input sequence, so the decoder and shared representation are tuned in a way to focus on the important words. We illustrate and summarize the attention mechanism in Fig. 2.

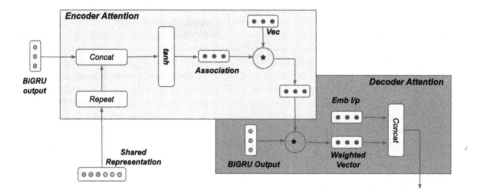

Fig. 2. Attention mechanism.

The output of Bi-GRU at each time step is a sequence of N tensors where N = tweet length, while the length of *shared representation* is a sequence of 1 tensor. Hence, we repeat the *shared representation* N times and concatenate to the output of Bi-GRU. We then pass $concat_{output}$ through *tanh* activation function that yields the *association* vector, which is defined as the correlation between Bi-GRU outputs and shared representation.

$$repeated_{SR} = repeat(SharedRepresentation, N)$$
$$concat_{output} = concat(repeated_{SR}, Bi - GRU_{Output})$$
$$association = tanh(concat_{output})$$

A new *trainable* vector '*Vec*' is used in attention layer which is initialized randomly and it learns the relationship between the tweet and topic. '*Vec*' is used to compute $Attended_{vector}$ which is defined as:

$$Attended_{vector} = softmax(Vec * Association)$$

$Attended_{vector}$ is used to compute the score for each word and is known as the final output of the encoder attention.

2.2 Decoder Attention

The decoder transforms the obtained $Attended_{vector}$ into a $weighted_{vector}$, which is the weighted sum of Bi-GRU output states.

$$Weighted_{vector} = Attended_{vector} * Bi - GRU_{Output}$$

We use separate Bi-GRU for both stance and sentiment to obtain the contextual information at every time step. We define the embedded input (*Emb I/p*) to the decoder as the input to the decoder Bi-GRU at every time step. We concatenate the embedded input with the weighted vector and use it as the input to the decoder Bi-GRU, along with h_{t-1}.

$$h_{t-1} = SharedRepresentation$$
$$x_t = concat(Input_{embedded}, Weighted_{vector})$$
$$h_t, z_t = Bi - GRU(x_t, h_{t-1})$$
$$concat_{input} = concat(Input_{embedded}, Weighted_{vector}, z_t)$$
$$Reconstructed_{Input} = fullyConnected(concat_{input})$$
$$h_{t-1} = h_t$$

Hence, we reconstruct the source sentence, and while doing so, we also improve the quality of the shared representation generated, which is then further used for stance and sentiment prediction.

3 Experimental Results and Analysis

Dataset: We evaluate our proposed approach on the benchmark dataset *SemEval-2016 Task 6*. This dataset is suitable for two tasks, *viz.* stance detection and sentiment classification. In case of stance detection, for a given tweet and a topic, the task is to predict whether the tweet is in the *against, favor*, or *none* towards the topic. Whereas for sentiment analysis, given a tweet about a topic, the task is to *negative, positive*, or *neutral* towards the topic. There is a total of six topics (i.e., Atheism, Abortion, Climate change is a real concern, Hilary Clinton, Feminist movement, and Donald Trump) in this dataset. First five topics are used for training while the topic *"Donald Trump"* is used for evaluation.

Experiments: We use PyTorch library for our experiments. We apply the grid search to find the optimal hyper-parameters. For consistency, we use the same hyper-parameters for training our proposed multi-task model. We set Bi-GRU with 200 neurons, embedding dimension with 200 neurons and dropout as 0.3. We use *Adam* optimizer with a learning rate of $1e^{-5}$, *Cross entropy* as loss function between reconstructed inputs, and *binary cross entropy* with logits loss as the criterion for stance and sentiment predictions. We evaluate our proposed approach on the benchmark dataset *SemEval-2016 Task 6*. We use the micro-average of F-scores across targets.

We report the experimental results of both single-task and multi-task learning framework in Table 1. In the single-task framework, we build separate systems for both stance detection and sentiment analysis, whereas in multi-task framework a joint-model is learned for both of these problems. For stance detection, our multi-task framework reports $F_{Against}$ of 58.47%, F_{Favor} of 78.61% and $F_{Average}$ of 68.54% compared to $F_{Against}$ of 46.25%, F_{Favor} of 78.50% and $F_{Average}$ of 62.52%, respectively, for single-task learning. Similarly for sentiment classification, our multi-task framework reports $F_{Negative}$ of 68.75%, $F_{Positive}$ of 73.33% and $F_{Average}$ of 71.04% compared to $F_{Negative}$ of 41.11%, $F_{Positive}$ of 77.22% and $F_{Average}$ of 59.16%, respectively, for single-task learning.

Table 1. Multi-task vs single-task results for Stance accuracy and sentiment analysis.

Models	Stance detection			Sentiment analysis		
	$F_{Against}$	F_{Favor}	$F_{Average}$	$F_{Negative}$	$F_{Positive}$	$F_{Average}$
Proposed Single-task learning	46.25	78.50	62.52	41.11	**77.22**	59.16
Proposed Multi-task learning	**58.47**	**78.61**	**68.54**	**68.75**	73.33	**71.04**

Comparative Analysis: We compare our proposed approach against various existing systems [2–4,10–14] that made use of the same datasets and reported the results only for stance detection. A comparative study is shown in Table 2.

We briefly outline the comparative systems as follows: Augenstein et al. [2] presented an auto-encoder based approach for stance detection. The author used a large set of unlabelled tweets data and train to produce a feature representation of tweet. Dey et al. [3] proposed an SVM based two-phase approach for stance detection from twitter. A convolutional neural network (CNN) based system for stance detection is introduced in [4]. Dias et al. [10] presented a weekly supervised approach for stance detection. Krejzl et al. [11] introduced a maximum entropy classifier-based approach for stance detection. Wojatzki et al. [12] used a sequence of the stacked classifier for stance detection in social media. In another work, a two-step pipeline-based system for automatic stance detection is proposed in [13].

The baselines [14] of SemEval-2016 Task 6 for stance detection are as follows: *Majority class*: a classifier which labels every instance with the majority class (*i.e., against or favor*) for the respective topic; *SVM-ngrams-comb:* one SVM classifier trained on 5 topics using word n-grams (1-gram, 2-gram, and 3-gram) and character n-grams (2-gram, 3-gram, 4-gram, and 5-gram) features. For stance detection, our multi-task framework reports the best F_{Favor} of 78.61% as compared to 59.72% and $F_{Average}$ of 68.54% as compared to 61.57% of the state-of-the-art. Our proposed multi-task framework outperforms the existing systems and achieve improvements of 18.89% in F_{Favor} and 6.97% in $F_{Average}$ over the state-of-the art systems.

Error Analysis: Here, we present error analysis *w.r.t.* our proposed single-task and multi-task learning frameworks. We list a few example cases in Table 3,

Table 2. Comparison of our propose multi-task framework with state of the art systems.

Module	Stance detection		
	$F_{Against}$	F_{Favor}	$F_{Average}$
USFD [2]	54.46	10.93	32.70
IIT Delhi [3]	**63.41**	59.72	61.57
Pkudblab [4]	55.17	57.39	56.28
INF-UFRGS [10]	52.09	32.56	42.32
UWB [11]	49.78	34.26	42.02
Ltl.uni-due [12]	05.71	46.56	26.14
ECNU [13]	50.20	17.96	34.08
Baselines given by SemEval 2016 Task setters [14]			
Majority class	59.44	0.00	29.72
SVM-ngrams-comb	38.45	18.42	28.43
Proposed Single-task learning	46.25	78.50	62.52
Proposed Multi-task learning	58.47	**78.61**	**68.54**

Table 3. Comparison between multi-task learning (MTL) and single-task learning (STL) frameworks.

	Utterances	Stance			Sentiment		
		Actual	STL	MTL	Actual	STL	MTL
1	*if a liberal brings up immigration hit em with the trump card trump for president trump tcot semst*	Favor	Against	Favor	Pos	Neg	Pos
2	*american was great until you embarrassed us all to death semst*	Against	Against	Against	Neg	Neg	Neg
3	*some people just want to be president others want to make america great again semst*	Favor	Against	Favor	Pos	Neg	Pos
4	*i hope every billionaire learned their lesson about talking shit about Mexicans do it in your golf clubs not on tv semst*	Against	Favor	Against	Neg	Pos	Neg
5	*you're right and i didn't vote for bonzo and not voting for hillary semst*	None	Against	Against	Neu	Neg	Neg

where the proposed multi-task framework yields better performance for both stance and sentiment classification compared to single-task learning. For example, the first utterance has gold stance label as *favor*, which was misclassified into the single-task framework. However, the multi-task framework corrects this by predicting *'favor'*. Similarly, for sentiment classification, single-task framework wrongly predicts it as *neg* while multi-task framework predicts it correctly. These analysis (c.f. Table 3) suggest that stance (*i.e., against, favor, and none*) and sentiment (*i.e., negative, positive, and neutral*) are helping each other for better prediction. Thus, we can claim that our proposed multi-task framework, indeed, captures better evidence than the single-task framework. In some cases where the proposed framework fails to capture the actual class because of highly negative words *i.e., didn't* (c.f. fifth example of Table 3).

4   Conclusion

In this paper, we have proposed a deep attentive multi-task framework for stance detection and sentiment analysis. The model can generate a shared representation from both the tasks of stance detection and sentiment analysis. We have successfully built a relationship between stance (*i.e., against, favor and none*) and sentiment (*i.e., negative, positive and neutral*). On the other hand, we have deployed an attention mechanism to learn the inter-relations between tweet and topic words. We evaluate our proposed approach on the benchmark dataset *SemEval-2016 Task 6*. The proposed multi-task approach has shown a higher performance compared to state-of-the-art systems both for stance detection and sentiment analysis.

In future work, we would like to explore our research toward multi-modal stance detection and sentiment analysis in a multi-task framework.

Acknowledgment. Asif Ekbal acknowledges the Young Faculty Research Fellowship (YFRF), supported by Visvesvaraya Ph.D. scheme of MeiTY, Government of India. The research reported here is also partially supported by Skymap Global India Private Limited".

References

1. Akakandelwa, A., Walubita, G.: Students social media use and its perceived impact on their social life: a case study of the University of Zambia (2018)
2. Augenstein, I., Vlachos, A., Bontcheva, K.: USFD at SemEval-2016 task 6: any-target stance detection on Twitter with autoencoders. In: Proceedings of the 10th International Workshop on Semantic Evaluation (SemEval-2016), pp. 389–393 (2016)
3. Dey, K., Shrivastava, R., Kaushik, S.: Twitter stance detection—a subjectivity and sentiment polarity inspired two-phase approach. In: 2017 IEEE International Conference on Data Mining Workshops (ICDMW), pp. 365–372. IEEE (2017)

4. Wei, W., Zhang, X., Liu, X., Chen, W., Wang, T.: pkudblab at SemEval-2016 task 6: a specific convolutional neural network system for effective stance detection. In: Proceedings of the 10th International Workshop on Semantic Evaluation (SemEval-2016), pp. 384–388 (2016)
5. Sun, Q., Wang, Z., Zhu, Q., Zhou, G.: Stance detection with hierarchical attention network. In: Proceedings of the 27th International Conference on Computational Linguistics, pp. 2399–2409 (2018)
6. Dey, K., Shrivastava, R., Kaushik, S.: Topical stance detection for twitter: a two-phase LSTM Model using attention. In: Pasi, G., Piwowarski, B., Azzopardi, L., Hanbury, A. (eds.) ECIR 2018. LNCS, vol. 10772, pp. 529–536. Springer, Cham (2018). https://doi.org/10.1007/978-3-319-76941-7_40
7. Nakov, P., Ritter, A., Rosenthal, S., Sebastiani, F., Stoyanov, V.: SemEval-2016 task 4: sentiment analysis in Twitter. In: Proceedings of the 10th International Workshop on Semantic Evaluation (SemEval-2016), pp. 1–18 (2016)
8. Nakov, P., Ritter, A., Rosenthal, S., Sebastiani, F., Stoyanov, V.: Evaluation measures for the SemEval-2016 task 4: sentiment analysis in Twitter (draft: Version 1.12). In: Proceedings of the 10th International Workshop on Semantic Evaluation (SemEval 2016), San Diego, California, June. Association for Computational Linguistics (2016)
9. Luong, M.T., Pham, H., Manning, C.D.: Effective approaches to attention-based neural machine translation. arXiv preprint arXiv:1508.04025 (2015)
10. Dias, M., Becker, K.: Inf-ufrgs-opinion-mining at SemEval-2016 task 6: automatic generation of a training corpus for unsupervised identification of stance in tweets. In: Proceedings of the 10th International Workshop on Semantic Evaluation (SemEval-2016), pp. 378–383 (2016)
11. Krejzl, P., Steinberger, J.: UWB at SemEval-2016 task 6: stance detection. In: Proceedings of the 10th International Workshop on Semantic Evaluation (SemEval-2016), pp. 408–412 (2016)
12. Wojatzki, M., Zesch, T.: ltl. uni-due at SemEval-2016 task 6: stance detection in social media using stacked classifiers. In: Proceedings of the 10th International Workshop on Semantic Evaluation (SemEval-2016), pp. 428–433 (2016)
13. Zhang, Z., Lan, M.: ECNU at SemEval 2016 task 6: relevant or not? Supportive or not? A two-step learning system for automatic detecting stance in tweets. In: Proceedings of the 10th International Workshop on Semantic Evaluation (SemEval-2016), pp. 451–457 (2016)
14. Mohammad, S., Kiritchenko, S., Sobhani, P., Zhu, X., Cherry, C.: SemEval-2016 task 6: detecting stance in tweets. In: Proceedings of the 10th International Workshop on Semantic Evaluation (SemEval-2016), pp. 31–41 (2016)

Multi-document Summarization Using Adaptive Composite Differential Evolution

Naveen Saini[1]([✉]), Sriparna Saha[1], Anurag Kumar[2],
and Pushpak Bhattacharyya[1]

[1] Indian Institute of Technology Patna, Bihta, Bihar, India
{naveen.pcs16,sriparna,pb}@iitp.ac.in
[2] Indian Institute of Engineering Science and Technology, Shibpur, Howrah, India
anurag122898@gmail.com

Abstract. In the current paper, a system for multi-document summarization (MLDS) is developed which simultaneously optimized different quality measures to obtain a good summary. These measures include anti-redundancy, coverage, and, readability. For optimization, multi-objective binary differential evolution (MBDE) is utilized which is an evolutionary algorithm. MBDE consists of a set of solutions and each solution represents a subset of sentences to be selected in the summary. Generally, MBDE uses a single DE variant, but, here, an ensemble of two different DE variants measuring diversity among solutions and convergence towards the global optimal solution, respectively, is employed for efficient search. Three versions of the proposed model are developed varying the syntactic/semantic similarity between sentences and DE parameters selection strategy. Results are evaluated on the standard DUC datasets using ROUGE-2 measure and significant improvements of 15.5% and 4.56% are attained by the proposed approach over the two exiting techniques.

Keywords: Multi-document summarization · Differential evolution (DE) · Word mover distance (WMD) · Multi-objective optimization (MOO)

1 Introduction

In today's world of internet, a vast and enormous amount of information is generated exponentially on almost every topic. So, there is a need for automatic summarization methods [1] to help people to get concise information on a specific topic from that large amount of data to reduce their efforts and time. Therefore, in this paper, a system for multi-document extractive summarization (MLDS) is developed in which subset of sentences are selected from the documents using different quality measures. These quality measures include readability of the summary and coverage of the central theme of the documents. Moreover, to make the summary free from redundancy, anti-redundancy measure is also considered. In the literature [2,3], lots of research works are available solving MLDS

T. Gedeon et al. (Eds.): ICONIP 2019, CCIS 1143, pp. 670–678, 2019.
https://doi.org/10.1007/978-3-030-36802-9_71

task using meta-heuristic based optimization algorithms like particle swarm optimization [3], genetic algorithm [4], differential evolution (DE) [2] etc., and, these approaches have shown great improvements over the state-of-the-art-techniques. In this paper, we have utilized binary version of differential evolution [5] as the underlying optimization strategy because of it's efficiency in solving real-life problems [1]. Brief descriptions of the existing works are reported in Table 1.

Table 1. Existing works on multi-document summarization. Here, COV: Coverage; RED: Anti-redundancy; COH: Cohesion.

Papers	Year	Opt. Technique	MOO-based	Criteria	Method
[2]	2012	SADE	No	COV, RED	Modified p-median problem
[3]	2013	PSO	No	COV, RED	Nonlinear 0-1 programming
[6]	2011	Adaptive DE	No	COV, RED	Discrete optimization
[7]	2013	PSO	No	COV, RED	Quadratic integer programming
[8]	2013	DE	No	COV, RED	Discrete optimization
[9]	2014	CHC	No	COV, RED	Binary optimization
[4]	2015	NSGA II	Yes	COV, RED	Discrete optimization

Motivation: Main drawbacks of the existing approaches are: (a) use of syntactic representations of the sentences in the documents; (b) optimization of weighted combinations of different quality measures. Therefore, to remove the first drawback, in current work, our proposed approach(s) used the recently developed word move distance (WMD) [10] to capture the semantic similarity between sentences. Note that WMD doesn't require vector representation of sentences. It makes use of word embeddings of different words obtained from word2vec [11] model trained on *Googlenews*[1] corpus which contains 3 billions words and each word vector is of 300 dimension. If two sentences are similar, then the corresponding WMD will be 0. To remove the second drawback, we have simultaneously optimized three quality measures (fitness/objective functions) discussed above using the modified multi-objective binary DE [5] (MBDE). This was done because each objective function has its own importance to obtain a good quality summary. It is important to note that MBDE consists of a set of solutions and each solution represents a subset of sentences to be selected in the summary. Moreover, in the existing MLDS systems based on DE, single DE variant is utilized to generate new solutions in each generation, but in our model, we have used two different DE variants. First one maintains diversity among solutions, while, second variant is used for convergence of solutions towards an optimal

---

[1] https://github.com/mmihaltz/word2vec-GoogleNews-vectors.

solution. In a recent survey [4], it has been shown that MLDS systems developed based on the optimization strategy, NSGA-II, are better than DE. But, in our approach, we want to illustrate that by using composition of two DE variants [12] and incorporation of semantic measures, can improve the DE.

Proposed Models: In this paper, three different versions/models of MBDE have been implemented which are enumerated below: (a) First one (Proposed-I) uses cosine distance to measure the similarity among sentences after representing sentences using well know tf-idf representation, while, the other two versions (Proposed-II and Proposed-III) make use of word mover distance (WMD) [10]; (b) MBDE parameters, namely, crossover probability (CR) and weight factor (F), play important roles in efficient search of the optimal solution and it is difficult to know the values of these parameters before implementing MBDE algorithm. Therefore, in the first and second versions, parameters are selected randomly from the available pool of values for CR and F, while, in the last one, they are selected adaptively motivated by [2].

Contributions: The major contributions of this paper are enumerated below: (a) Existing papers on MLDS discussed in Table 1 consider syntactic representations of the documents, but, in this paper, semantic representations of the documents are considered; (b) Modified differential evolution (MBDE) was never utilized as an optimizing strategy for solving MLDS task. Moreover, it's parameters, CR and F, are selected adaptively (in case of proposed approach III)

Table 2. Definition of objective functions. For proposed model-I, *dist* refers the cosine distance, while, for Model-II and III, it refers to word mover distance. All functions are of minimization type. For Model-I, sentence vectors are obtained using well known tf-idf [4] representation.

Objective function	Definition	Variables used
Anti-Redundancy	$\dfrac{|\mathcal{S}|(|\mathcal{S}|-1)}{2**\times\sum_{l,m=1,l\neq m}^{|\mathcal{S}|} dist(s_l,s_m)}$	$|\mathcal{S}|$ = Number of sentences in summary; $s_l = lth$ sentence of summary; $s_m = mth$ sentence of summary
Coverage	$\sum_{l=1}^{|\mathcal{S}|} dist(s_l,O)$	$|\mathcal{S}|$ = Number of sentences in summary; $s_l = lth$ sentence of summary; O = document vector obtained by averaging sentence vectors (used for proposed model-I)/representative sentence of the document (used for proposed model-II and III)
Readability	$\sum_{2}^{|\mathcal{S}|} dist(s_l,s_{l-1})$	$s_l = lth$ sentence of summary; s_{l-1} = previous sentence to lth sentence of the summary

motivated by the paper [2]; (c) Any multi-objective algorithm should satisfy two properties: diversity among solutions and convergence towards actual/true optimal solution. To achieve the same, composition of two DE variants (current-to-rand/1/bin and current-to-best/1/bin) are used in the current framework. The first schema incorporates diversity and the second one includes convergence.

To show the effectiveness of our proposed approach, we have used the range of topics belonging to DUC2002 standard datasets and performance is evaluated using ROUGE-2 measures. Results obtained clearly show the superiority of our proposed algorithm in comparison to two existing techniques. Results are also validated using statistical significance t-test.

2   Problem Definition

Consider a given set of documents, $T = \{Doc_1, Doc_2, \ldots, Doc_N\}$ divided into sentences $\{s_1, s_2, \ldots, s_n\}$. Our task is to select a subset of sentences $\mathcal{S} \subset T$ defined as: $S_{min} + \epsilon \leq \sum_{j=1}^{n} x_j L_{s_j} \leq S_{max} + \epsilon$, such that it simultaneously optimizes (minimization) three objective functions Ψ_{ar}, Ψ_{cov}, and, Ψ_{read}, i.e., $\min\{\Psi_{ar}(\mathcal{S}), \Psi_{cov}(\mathcal{S}), \Psi_{read}(\mathcal{S})\}$ where, x_j will be 1 if jth sentence belongs to summary $\mathcal{S}$, L_{s_j} and S_{max} are the length of jth sentence and maximum length of the summary, respectively, Ψ_{ar}, Ψ_{cov}, and, Ψ_{read} are the objectives functions or qualitative measures defining extent of anti-redundancy, coverage, and readability in the summary, respectively, ϵ is the tolerance capacity for summary length and is expressed as $\epsilon = \max_{j=1,2,\ldots,n} L_{s_j} - \min_{j=1,2,\ldots,n} L_{s_j}$. Mathematical definitions of the used objective functions are provided in Table 2.

3   Methodology

This section discusses about various steps involved in our proposed framework.

Step-1: In this step, a topic to be summarized is given as an input to our system. The topic consists of a set of documents.

Step-2: All documents are combined to form a single document and then obtained document is segmented into sentences. These sentences are passed through pre-processing steps like tokenization, removal of most frequent words like is, am are, etc., lower case conversion, and, stemming. Stemming is the process of reducing a word to common base/root form. For proposed Model-I, sentence vectors are obtained using well known tf-idf representation and then, cosine distance matrix of size $N \times N$ is computed to calculate the dissimilarity between sentences. While, for Model-II and III, dissimilarity matrix is computed using word mover distance. Note that for WMD, we are performing only lower case conversion of the text as a pre-processing step.

Step-3: This step includes initialization of solutions where each solution represents subset of sentences satisfying summary length constraint and thus, each solution presents a summary. This set of solutions is called as population P and this is initialized randomly in binary space. For example, if 10 sentences exist in the document collection and $2nd, 5th$ and $10th$ sentences should be part of the summary, then solution will be represented as $[0, 1, 0, 0, 1, 0, 0, 0, 0, 1]$.

Step-4: In this step, objective functional (Ψ_{ar}, Ψ_{cov}, and, Ψ_{read}) values for each solution are evaluated.

Step-5: For any evolutionary algorithm, in order to explore the search space efficiently or to find the global optimum solution by generating new solutions, various genetic operators are used which are mating pool generation, mutation, and crossover. Here also, new solutions/trial vectors are generated for every solution in each generation to form a new population, P'. The process followed for new solution generation is described below. Let $\boldsymbol{x}_c$ be the current solution in the population for which new solution is to be generated.

(i) Mating Pool Generation: The mating pool includes a set of solutions which can mate to generate new solutions. For the construction of the mating pool for the current solution, $(\boldsymbol{x}_c)$, a fixed number of random solutions are picked up from the population.

(ii) Mutation and Crossover: Mutation is the change in component value of the solution, while, crossover is the exchange of component values between two solutions. In our work, we have used 2 trial vector generation schemes/variants namely, current-to-rand/1/bin and current-to-best/1/bin. The generation of trial vectors using these variants are discussed in [13]. These two newly generated trial vectors using *current-to-rand/1/bin* and *current-to-best/1/bin* DE variants may violate the constraint of summary length. Therefore, these solutions are passed through constraint checking phase. If they violate the same then those are converted 'feasible' using some heuristic as discussed below:

(a) Let us denote the new solution (y''_j) as ith solution; (b) Initialize $UpdNewSol$ with zeros equal to the maximum length of the solution; (c) In case of proposed Model-I, the sentences present in the ith solution are sorted based on minimum distance with document vector. The document vector is obtained by averaging all sentence vectors of the document collection. For proposed Model-II and III, sentences are sorted based on minimum word mover distance to representative sentence in a collection of documents. The representative sentence (lets say $s_{\mathcal{R}}$) whose index in the document collection is evaluated by calculating the minimum average word mover distance of each sentence with other sentences in the document collection, i.e., $\arg\min_{\mathcal{R}} \left(\sum_{j=1, \mathcal{R} \neq j}^{n} dist_{wmd}(s_R, s_j) \right) / (n-1)$, where, $\mathcal{R} = \{1, 2 \ldots, n\}$, n is the total number of sentences, $dist_{wmd}$ is the word mover distance; (d) Fill the indices of $UpdNewSol$ with 1s until we cover 'r' indices obeying summary length constraint. Note that indices are considered in the obtained sorted order; (e) Return the $UpdNewSol$.

These new solutions (after making them feasible) form a new population P'. If population has 10 solutions, then for each solution, there will be 2 new solutions. Thus, size of new population will be 20.

Step-6: The objective functional values of generated new solutions are evaluated.

Step-7: After performing step-6, old population (P) and new population (P$'$) are merged together.

Step-8: In this step, out of merged population, only best $|P|$ solutions are selected based on dominance and non-dominance criteria. This is done using the well known non-dominating sorting algorithm and crowding distance operator [14] in the field of multi-objective optimization. More details about these operators can be found in the paper [14]. Steps 5 to 8 are executed until maximum number generations, g_{max}, is reached.

Step-9: At the end of the final generation, a set of Pareto optimal solutions are obtained and the end-user can select any solution based on his/her choice. In this paper, we have chosen that solution which has obtained good summary with respect to gold summary, made available by the annotators for the used datasets in this paper.

4   Experimental Setup and Discussion of Results

In this section, we will discuss about datasets, evaluation metric, and finally, the results obtained followed by their discussions. The results are compared among adative DE [6], NSGA-II [4] and three versions of the proposed approach (I, II, and, III). Note that out of ADE and NSGA-II, only NSGA-II is developed using the multi-objective optimization concept.

Dataset: For multi-document summarization, we have used, DUC 2002, standard dataset provided by Document Understanding Conference. For our experimentation, we have used the ten topics ranging from $d061j$ to $d070f$ from this dataset each having approx 10 documents as used in the paper [6]. The corresponding multi-document gold summaries (two in number) each of 200 words are also available for each topic.

Evaluation Measures: Results are evaluated using ROUGE-2 score which measures the overlapping 2-gram units between gold summary and our predicted summary. Moreover, we have also reported the *Range* for each topic which is the difference between the best and the worst Rouge score values in the final Pareto optimal solutions.

Parameters Used: All the experiments corresponding to three proposed models are executed for 50 (g_{max}) generations with population size of 50. For proposed model-I and II, pool of values for CR and F parameters are $[0.1, 0.2, 1.0]$ and $[0.6, 0.8, 1.0]$, respectively. Any value can be randomly selected at the time of generation of new solutions. For proposed model-III, these parameters are made adaptive similar to the paper [2]. Note that the paper [2] uses best solution and worst solution while calculating F. In current paper, the best and worst solutions will be those solutions whose average of the objective functional values are the minimum and the maximum, respectively.

Results and Discussions: In this section, we compare the results of three proposed models with the existing approaches namely, adaptive DE (in short, ADE) [6] and NSGA-II [4]. In both cases, they use ROUGE-2 score for evaluation of generated summery, therefore, we also use ROUGE-2 score as an evaluation metric for comparison purpose. Table 3 presents the average ROUGE-2 score and range of ROUGE-2 score of each topic as well as average over all topics (last row of the Table). It can be observed from this Table that, our first model based on syntactic representation of documents and utilizing cosine distance as a dissimilarity measure improves average ROUGE-2 score by 6.7% over ADE but it fails to improve average ROUGE-2 score of NSGA-II. Therefore, we explored other models considering sentences of documents in semantic space. In our second model (Proposed-II) which uses WMD improves average ROUGE-2 score of ADE and NSGA-II by 13.8% and 3.04%, respectively, and, also gives better result than our first proposed model. In our third model (Proposed-III) based on adaptive parameters and utilizing word mover distance, is able to get improvement of 15.5% and 4.56% over ADE and NSGA-II, respectively, in terms of average ROUGE-2 score. In terms of average *Range* over all topics which is the difference between best and worst Rouge-2 score values, the average range values of our proposed models are smaller than the existing works. It shows that our models are more robust than existing works. First model provides 15.6% and 23.1% less average range values compared to ADE and NSGA-II, respectively, notably, it is the smallest average range compared to all other models. Our second model provides 5.85% and 14.2% lesser values of *Range* than ADE and NSGA-II, respectively. Our third model attains the same average range value as of ADE and provides 4.5% less average range of ROUGE-2 score compared to NSGA-II. In summary, we can say that our third model, which uses composition of two DE variants with adaptive parameters and uses WMD as a dissimilarity measure between sentences, is able to generate the best result over all compared models in terms of average ROUGE-2 score. We have validated our results using statistical significant t-test at 5% significant level and it has been found that results are statistically significant.

Number of Fitness Function Evaluations: Generally, in any evolutionary based optimization strategy, the number of fitness function evaluations (NFE) [1] is reported which equals to $Max\_generations \times Population\_size \times \#Objectives$. On comparing NFEs of ADE, NSGA-II and our proposed approaches, we can conclude that both ADE and NSGA-II have 10, 0000 ($50 \times 1000 \times 2$) NFEs, while that of our proposed approach is 7500 ($50 \times 50 \times 3$) which is lesser than existing works. This indicates that our proposed approach runs faster than existing ones.

Table 3. Comparison of results obtained in terms of ROUGE-2 score values. Bold entries denote the best results. † indicates result are statistically significant at 5% significant level.

Topic	ADE		NSGA-II		Proposed-I		Proposed-II		Proposed-III	
	Rouge-2	Range	Rouge-2	Range	Rouge-2	Range	Rouge-2	Range	Rouge-2	Range
d061j	0.266	0.290	0.306	**0.263**	0.323	0.277	**0.341**	0.326	0.301	0.272
d062j	0.188	0.275	0.200	0.422	0.274	**0.211**	0.267	0.257	**0.313**	0.300
d063j	0.245	0.208	**0.275**	0.279	0.215	**0.160**	0.231	0.215	0.217	0.202
d064j	0.194	0.280	0.233	0.356	**0.307**	0.287	0.270	0.250	0.266	**0.248**
d065j	0.144	0.209	0.182	0.208	0.221	**0.192**	**0.247**	0.240	0.214	0.207
d066j	0.201	0.257	0.181	0.245	0.208	**0.180**	**0.330**	0.325	0.265	0.245
d067f	0.239	0.235	0.260	0.298	0.254	0.228	0.215	**0.205**	**0.330**	0.309
d068f	0.491	0.384	**0.496**	0.281	0.272	0.234	0.260	**0.233**	0.327	0.318
d069f	0.184	**0.166**	0.232	0.239	0.203	0.171	**0.299**	0.250	0.230	0.199
d070f	0.224	0.260	0.262	0.215	0.264	0.225	0.259	**0.190**	**0.294**	0.260
Avg.	0.238	0.256	0.263	0.281	0.254	0.216	**0.271**†	0.241	**0.275**†	0.256

5   Conclusions and Future Works

In this paper, we have proposed a model for multi-document summarization. Three objective functions are simultaneously optimized using multi-objective optimization based modified differential evolution. Instead of single variant, compositions of two different DE variants are utilized. Three models (I, II and III) of our proposed work are developed, each differs in the way of selecting values of DE parameters and similarity measure. It has been found that our third model (Proposed-III) based on adaptive parameters and utilizing word mover distance is able to get improvements of 15.5% and 4.56% over ADE and NSGA-II, respectively, in terms of average ROUGE-2 score. In future, we would like to solve MLDS task by incorporating textual entailment as one of our objective functions.

Acknowledgement. Dr. Sriparna Saha would like to acknowledge the support of Early Career Research Award of Science and Engineering Research Board (SERB) of Department of Science and Technology India to carry out this research.

References

1. Saini, N., Saha, S., Jangra, A., Bhattacharyya, P.: Extractive single document summarization using multi-objective optimization: exploring self-organized differential evolution, grey wolf optimizer and water cycle algorithm. Knowl. Based Syst. **164**, 45–67 (2019)
2. Alguliev, R.M., Aliguliyev, R.M., Isazade, N.R.: DESAMC+ DocSum: differential evolution with self-adaptive mutation and crossover parameters for multi-document summarization. Knowl. Based Syst. **36**, 21–38 (2012)

3. Alguliev, R.M., Aliguliyev, R.M., Isazade, N.R.: Formulation of document summarization as a 0–1 nonlinear programming problem. Comput. Ind. Eng. **64**(1), 94–102 (2013)

4. Saleh, H.H., Kadhim, N.J., Bara'a, A.A.: A genetic based optimization model for extractive multi-document text summarization. Iraqi J. Sci. **56**(2B), 1489–1498 (2015)

5. Wang, L., Fu, X., Menhas, M.I., Fei, M.: A modified binary differential evolution algorithm. In: Li, K., Fei, M., Jia, L., Irwin, G.W. (eds.) ICSEE/LSMS -2010. LNCS, vol. 6329, pp. 49–57. Springer, Heidelberg (2010). https://doi.org/10.1007/978-3-642-15597-0_6

6. Alguliev, R.M., Aliguliyev, R.M., Mehdiyev, C.A.: Sentence selection for generic document summarization using an adaptive differential evolution algorithm. Swarm Evol. Comput. **1**(4), 213–222 (2011)

7. Alguliev, R.M., Aliguliyev, R.M., Isazade, N.R.: CDDS: constraint-driven document summarization models. Expert Syst. Appl. **40**(2), 458–465 (2013)

8. Alguliev, R.M., Aliguliyev, R.M., Isazade, N.R.: Multiple documents summarization based on evolutionary optimization algorithm. Expert Syst. Appl. **40**(5), 1675–1689 (2013)

9. Mendoza, M., Cobos, C., León, E., Lozano, M., Rodríguez, F., Herrera-Viedma, E.: A new memetic algorithm for multi-document summarization based on CHC algorithm and Greedy search. In: Gelbukh, A., Espinoza, F.C., Galicia-Haro, S.N. (eds.) MICAI 2014. LNCS (LNAI), vol. 8856, pp. 125–138. Springer, Cham (2014). https://doi.org/10.1007/978-3-319-13647-9_14

10. Kusner, M., Sun, Y., Kolkin, N., Weinberger, K.: From word embeddings to document distances. In: International Conference on Machine Learning, pp. 957–966 (2015)

11. Mikolov, T., Chen, K., Corrado, G., Dean, J.: Efficient estimation of word representations in vector space. arXiv preprint arXiv:1301.3781 (2013)

12. Wang, Y., Cai, Z., Zhang, Q.: Differential evolution with composite trial vector generation strategies and control parameters. IEEE Trans. Evol. Comput. **15**(1), 55–66 (2011)

13. Saini, N., Saha, S., Bhattacharyya, P., Tuteja, H.: Textual entailment based figure summarization for biomedical articles. ACM Trans. Multimed. Comput. Commun. Appl. **1**(1) (2019, accepted)

14. Deb, K., Pratap, A., Agarwal, S., Meyarivan, T.A.M.T.: A fast and elitist multi-objective genetic algorithm: NSGA-II. IEEE Trans. Evol. Comput. **6**(2), 182–197 (2002)

Time-Series and Related Models

MIRU: A Novel Memory Interaction Recurrent Unit

Danyang Zheng[1,2], Zhuo Zhang[1], Hui Tian[1,2(✉)], and Ping Gong[1]

[1] School of Information and Communication Engineering,
Beijing University of Posts and Telecommunications, Beijing 100876, China
`tianhui@bupt.edu.cn`
[2] State Key Laboratory of Networking and Switching Technology,
Beijing University of Posts and Telecommunications, Beijing 100876, China

Abstract. The memory-based network is widely used in a variety of sequence modeling. Taking full use of memory is one of the challenges to build memory-based models. Existing work either as recurrent neural network, memory capacity is too small to comprehensively model information of the sequence, or as the memory network, although with an external storage structure to enhance memory, the memory is not sufficiently utilized. To address these issues, we propose a novel memory interactive recurrent unit (MIRU), which constructs a multi-dimensional memory inside the recurrent unit and employs convolution operations to interact and update memories. Finally, we test MIRU on the YELP benchmark dataset of sentiment analysis and empirical results demonstrate that MIRU significantly outperforms the advanced models.

Keywords: Recurrent neural network · Multi-dimensional memory · Memory interaction · Sentiment analysis

1 Introduction

Humans have a remarkable ability to remember the current events which are related to the past memories [1], organized into sophisticated memory systems. Owing to the memory-based network performs well in capturing long-term dependence, the memory-based network is generally suitable for sequence modeling by using memory system. Accordingly, how to construct memory-based network draws increasing research interests in the past few years.

Recurrent neural networks (RNNs) [2] have been a memory-based network successfully applied to sequential modeling in the neural network research. RNNs have the ability to distill important information from ordered sequences [2] and are good at capturing long-term dependency into their hidden states. Unfortunately, the memory of RNNs is usually too small and can not be divided into

This work is supported by the Science and Technology of the Winter Olympics under Grant 2018YFF0301201.

dimensional areas to accurately remember past events (memory is compressed into low dimensional dense vector) [3]. Existing works like memory networks with external storage structures [3,4] make it easier to learn how to interact memories across time by multi-hop mechanism, so that information can be stored and retrieved skillfully. Therefore, the method of enhanced memory (or called multi-dimensional memory) and memory interaction are profitable. Despite their success, there are two drawbacks on these works: (1) the method of constructing the multi-dimensional memory module is often not clear enough for the common sequence modeling tasks. (2) the complicated and huge memories even bring about noise [5] to some extent.

To solve these problems, we investigate the relationship of memories in the storage structure and find they have no temporal dependence. Multi-dimensional memory solely needs to consider the correlation between them. Convolutional neural networks perform well in extracting local association information. Inspired by this, we propose a novel recurrent unit, the memory interaction recurrent unit (MIRU). MIRU utilizes a multi-dimensional memory module inside the traditional RNN, but its scale and memory interaction are more fruitful than memory networks. In addition, we use multi-layer convolution operations [6] to interact memories. To verity the effect of our model, we then apply MIRU to the YELP benchmark dataset of sentiment analysis and the empirical results show MIRU outperforms the advanced methods.

2   Related Work

Recently, building a multi-dimensional network has drawn significant attention. Graves et al. [7] first proposed multi-dimensional recurrent neural networks by arranging inputs in a multi-dimensional grid. Kalchbrenner et al. [8] further enhanced multi-dimensional memory by arranging LSTM cell in a multi-dimensional grid and updated memory at both time and depth. However, their models change the topology of the overall model including inputs, causing the network to become complex. In contrast, the MRCU constructs a multi-dimensional memory module inside the recurrent unit. Therefore, we can model multi-dimensional memory simply and efficiently.

Relate studies of multi-dimensional memory interaction focus on how to efficiently fuse memory [3–5]. For example, Kumar et al. [4] proposed a dynamic memory network using episodic memory module to correlate context memories. Wu et al. [5] exploited multi-hop mechanism to select important memory and finally aggregate associated memory to interact memory. However, memory interaction of our model is different from these memory networks. MIRU adopts convolution operations to learn how to associate memories from a fixed window and updates memories dynamically.

Sentiment analysis is one of the essential tasks of natural language processing. Its main target is to assign affective labels to a document or a text. Existing works has improved the accuracy of classification [9,10]. Compared with MIRU, none of these works consider the interaction between memories, which has proven to be useful for comprehensive prediction.

3   Model

We first formulate the problem of review sentiment classification. Suppose that we have a data set $\mathcal{D} = \{s_i, y_i\}$, where s_i represents the review text $s_i = \{w_j\}_{j=1}^{T}$ and $y_i \in Y$ is the sentiment class. T indicates the length of the review text. Our goal is to learn a classification model $g(\cdot)$ with $\mathcal{D}$. For each sample $\{s_i, y_i\}$, $\widehat{y}_i = g(s_i)$, $\widehat{y}_i$ is predicted label, which is equal to or close to y_i. And then we will introduce our memory interation recurrent unit. The architecture of MIRU is illustrated in Fig. 1. Note that MIRU organizes the memory of each time steps into a multi-dimensional memory, and we exploit two-dimensional matrix to build the multi-dimensional memory $M \in \mathcal{R}^{N \times D}$, N represents the number of memory dimension, and D represents the size of memory dimension. The detailed description of MIRU is as follows.

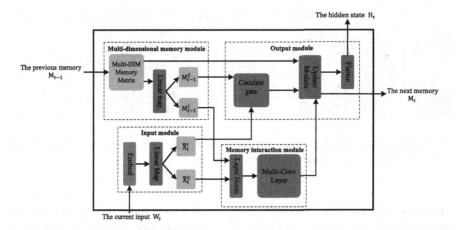

Fig. 1. The architecture of MIRU. It contains four modules, which are input module, multi-dimensional memory module, memory interaction module and output module.

Input Module. For the input review text $s = \{w_i\}_{i=1}^{T}$, the model first looks up an embedding table and represents as $X = \{X_1, X_2, \dots, X_T\}$, $X_t \in \mathcal{R}^d$, which transforms symbolic data into distributed representations, namely, word embedding [11]. In order to keep the shape of the multi-dimensional memory unchanged, the input X_t is mapped to the same shape as the 2-D memory matrix as $\tilde{X}_t^I$, $\tilde{X}_t^G \in \mathcal{R}^{N \times D}$, and then fed to memory interaction module and memory update module respectively.

$$\tilde{X}_t^I = W^c X_t \tag{1}$$

$$\tilde{X}_t^G = W^g X_t \tag{2}$$

Where W^g and W^c are trainable parameters.

Multi-dimensional Memory Module. We believe that multi-dimensional memory can help the model to organize sequence information and can provide more comprehensive information with reasoning or prediction, which is also confirmed in experiments. At each time step, the multi-dimensional memory module receives the previous step memory M_{t-1}, and then applies two linear mappings to generate the inputs of the memory interaction module and the output module.

$$M_{t-1}^I = W^I M_{t-1} + b^I \tag{3}$$

$$M_{t-1}^G = W^G M_{t-1} + b^G \tag{4}$$

where W^G, W^I, b^G, b^I are trainable parameters. In addition, each row in the 2-D memory matrix represents the different dimensions of memory, which means that the 2-D memory matrix represents different aspects of input sequence.

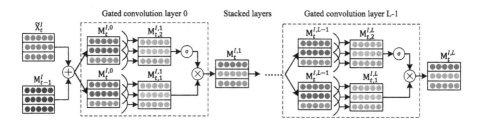

Fig. 2. The structure of memory interaction module. The module employs stacked gated convolution operations to interact memories.

Memory Interaction Module. Suppose that we have the outputs of the multi-dimensional memory module M_{t-1}^I. The memory interaction module is illustrated in Fig. 2. In the MIRU, it is necessary to interact with the previous memory M_{t-1}^I and the input of the current time step X_t. Here, we use addition method.

$$M_t^I = LN\left(\tilde{X}_t^I + M_{t-1}^I\right) \tag{5}$$

Where $LN(\cdot)$ represents the layer normalization operation. For effectively interacting memory, MIRU uses convolution operations. Following Dauphin et al. [6], we use stacked gated convolution operations to interact memories. After the interaction of the $l-1$ layer, we get the memory matrix $M_t^{I,l-1}$, and MIRU next apply two convolution operations on it respectively.

$$M_{t,1}^{I,l} = Conv(M_t^{I,l-1}) \tag{6}$$

$$M_{t,2}^{I,l} = Conv(M_t^{I,l-1}) \tag{7}$$

where $conv(\cdot)$ is a convolution operation function with K convolution kernels, each convolution kernel has a variable size of $n_w \times n_c$, where n_w is the window size and n_c is depth. Since convolution operation maintains the original matrix size, the MIRU can recursively store important information about the sequence. The output of each layer is the interactive memory matrix $M_{t,1}^{I,l}$ modulated by the gates $\sigma(M_{t,2}^{I,l})$. And the gates multiply each element of the matrix $M_{t,1}^{I,l}$:

$$M_t^{I,l} = M_{t,1}^{I,l} \cdot \sigma(M_{t,2}^{I,l}) \tag{8}$$

In the memory interaction, MIRU uses gating to clarify which associations need to be passed between layers, indicating that some associated memories may not be necessary, and we can adjust them by gating in this case. In particular, the initial stacked multi-dimensional memory interaction matrix $M_t^{I,0} = M_t^I$. The final output through the L-layer gated convolution is $M_t^{I,L}$.

Output Module. With linear mapped previous memories M_{t-1}^G from Multi-dimensional memory module, MIRU then applies gating mechanism to alleviate the problem of gradient vanishing. We define the calculation update gate z as:

$$z = \sigma\left(\tilde{X}_t^G + M_{t-1}^G\right) \tag{9}$$

Where $\sigma(\cdot)$ is the activation function and the update gate z decides to retain the size of the previous step memory. In this way, the MIRU allows information to flow unimpeded in as many time steps as possible. Then MIRU updates memory by fusing the previous memory and interactive memory with the update gate z:

$$M_t = z \cdot M_{t-1} + (1-z)M_t^{I,L} \tag{10}$$

We refer to the updated multi-dimensional memory as the unit's memory. The memory contains rich contextual information. Therefore, the MIRU can model different information about the input sequence and effectively utilizes enough information to make predictions.

4   Experiment

Datasets and Evaluation Metrics. Our model is evaluated on a large-scale YELP review dataset. The Yelp review dataset comes from the Yelp Data Challenge, which has five sentiment labels. Following Yu et al. [10] work, we extract three large-scale subsets Yelp2013, Yelp2014, Yelp2015 from YELP and employ accuracy as evaluation metric. Table 1 lists the statistics for three datasets.

Baselines. We compare MIRU with advanced models, which can be roughly divided into two categories: the first based on RNN, and the second based on CNN. *Based on RNN.* GRU: it's benchmark model of sequence modeling text classification; LSTM: it is another variant of the RNN [2]; SRNN: it is sliced RNN [10] with hierarchical structure for classification. *Based on CNN.* GCNN: it's convolution neural network with gating mechanism [6]; DPCNN: it's low complexity deep pyramid convolution network for text classification [9]; DCCNN: it's causal dilated convolution network proposed in WaveNet [12]; TCNN: it's temporal convolution network with one-dimensional causal dilated convolution and residual structure [13].

Table 1. Statistics of Yelp2013, Yelp2014 and Yelp2015 datasets.

Dataset	Class	Doc	Train set	Validation set	Test set	Average words
Yelp2015	5	897,835	718,268	89,784	89,783	108.3
Yelp2014	5	670,440	536,352	67,044	67,044	116.1
Yelp2013	5	468,608	374,887	46,861	46,860	129.2

Experimental Settings. For baseline models, if their results can be found in existing literature (e.g. SRNN, GRU, DCCNN), we copy the numbers, otherwise we implement the models following the settings reported in their literatures. The word embeddings for the datasets uses 200-dimensional GloVe embedding [11]. The number of memory dimension and the size of each memory dimension are 4 and 50. In order to thoroughly capture the relationship between memories, MIRU uses two-layer gated convolution operations. In this paper, the number of feature maps for deep convolution is 50. The parameters update for all models with Adam [14]. The minibatch size is 100 and early stopping on the validation data is adopted as a regularization strategy. We set the maximum length of the sentence to 512, because each model performs better at this length.

Table 2. Experimental results on Yelp2013, Yelp2014, Yelp2015. Evaluation metric is accuracy. The best method in each setting is in **bold**.

Model	Yelp2015		Yelp2014		Yelp2013	
	Validation	Test	Validation	Test	Validation	Test
DPCNN	0.7245	0.7249	0.6968	0.6967	0.6543	0.6545
GCNN	0.7280	0.7280	0.6980	0.6990	0.6632	0.6641
DCCNN	0.7069	0.7094	0.6846	0.6866	0.6491	0.6479
TCNN	0.7222	0.7223	0.6873	0.6909	0.6528	0.6566
SRNN	0.7309	0.7350	0.7053	0.7070	0.6718	0.6703
GRU	0.7252	0.7289	0.7037	0.7063	0.6656	0.6612
LSTM	0.7324	0.7305	0.6993	0.6980	0.6581	0.6610
MIRU	**0.7395**	**0.7403**	**0.7111**	**0.7112**	**0.6734**	**0.6776**
MIRU without MI	–	0.7368	–	–	–	–
MIRU without GM	–	0.7379	–	–	–	–

Evaluation Results. Table 2 reports the evaluation results of all models on three large-scale public datasets. As shown in Table 2, MIRU has achieved the best results on three datasets. The benchmark models of sequence modeling, GRU and LSTM, perform worse than our model, which compresses sequence knowledge into low-dimensional dense vectors. These results confirm the importance of multi-dimensional memory. Our model also performs better than SRNN.

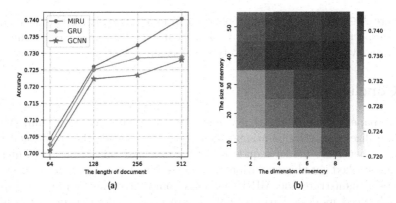

Fig. 3. (a) Performance of MIRU, GRU and GCNN across different documents length. (b) Performance in different dimensions of memory as well as the sizes of memory.

The SRNN divides the input sequence into subsequences and uses GRUs to accumulate the memory interactions by hierarchical structure. But this memory interaction is restricted and limited, and the hierarchical structure also increases the complexity of the model. Our model interacts memory directly across the entire input, ensuring memory coherence and enabling effective interaction. The RNN-based model is roughly superior to the CNN-based model, meaning that the memory-based network is naturally suitable for sequence modeling tasks.

5   Analyze

We use the YELP 2015 dataset for analyzing how MIRU works from both hyper-parameter effect as well as the ablation test. We first study the performance of the different implementations of MIRU under different length of documents. We use GRU and GCNN for comparison, where GRU is the benchmark model for sequence modeling and GCNN performs best in the CNN models. Figure 3 (a) shows how the performance of the model with the different documents length. We observe a consistent trend on all models: As the length of the document increases, performance improves. This is because the shorter the document is, the less information it contains, and the more difficult it is for the model to predict the document sentiment label. One notable point is that, with the sentence length increasing from 128 to 512, there is only a slight improvement in GCNN and GRU, but the MIRU is different. For GRU, its memory is compressed, which cannot model the comprehensive information of the sequence. For GCNN, there is no interaction between the important information. Moreover, we also investigate the effect of the dimension of memory and the size of memory on MIRU performance. As shown in Fig. 3 (b), increasing the dimension of memory can consistently improve the performance for the model has different memory sizes. We also test the effectiveness of each component of our model. We remove each component individually from MIRU, and denote the models as *MIRU without*

X, where $X \in \{MI, GM\}$ meaning memory interactions (MI), gating mechanism (GM) respectively. From Table 2, we find that each component of our model is useful, removing each one leads to the performance drop.

6 Conclusion

In this paper, we present a new memory interactive recurrent unit MIRU. MIRU utilizes a multi-dimensional memory module inside the traditional RNN to comprehensively model sequence information and takes full use of memory by using stacked gated convolution operations. Experiments on three large-scale datasets demonstrate the MIRU achieves competitive performance compared with advanced models. Various analyses are conducted to evaluate the model and verity the effectivity of multi-dimensional memory module and memory interaction.

References

1. Ciaramelli, E., Grady, C.L., Moscovitch, M.: Top-down and bottom-up attention to memory: a hypothesis (AtoM) on the role of the posterior parietal cortex in memory retrieval. Neuropsychologia **46**(7), 1828–1851 (2008)
2. Hochreiter, S., Schmidhuber, J.: Long short-term memory. Neural Comput. **9**(8), 1735–1780 (1997)
3. Weston, J., Chopra, S., Bordes, A.: Memory networks. arXiv preprint arXiv:1410.3916 (2014)
4. Kumar, A., et al.: Ask me anything: dynamic memory networks for natural language processing. In: International Conference on Machine Learning, pp. 1378–1387 (2016)
5. Wu, C.-S., Socher, R., Xiong, C.: Global-to-local memory pointer networks for task-oriented dialogue. arXiv preprint arXiv:1901.04713 (2019)
6. Dauphin, Y.N., Fan, A., Auli, M., Grangier, D.: Language modeling with gated convolutional networks. In: Proceedings of the 34th International Conference on Machine Learning, vol. 70, pp. 933–941. JMLR.org (2017)
7. Graves, A., Fernández, S., Schmidhuber, J.: Multi-dimensional recurrent neural networks. In: de Sá, J.M., Alexandre, L.A., Duch, W., Mandic, D. (eds.) ICANN 2007. LNCS, vol. 4668, pp. 549–558. Springer, Heidelberg (2007). https://doi.org/10.1007/978-3-540-74690-4_56
8. Kalchbrenner, N., Danihelka, I., Graves, A.: Grid long short-term memory. arXiv preprint arXiv:1507.01526 (2015)
9. Johnson, R., Zhang, T.: Deep pyramid convolutional neural networks for text categorization. In: Proceedings of the 55th Annual Meeting of the Association for Computational Linguistics (Volume 1: Long Papers), pp. 562–570 (2017)
10. Yu, Z., Liu, G.: Sliced recurrent neural networks. arXiv preprint arXiv:1807.02291 (2018)
11. Pennington, J., Socher, R., Manning, C.: GloVe: global vectors for word representation. In: Proceedings of the 2014 Conference on Empirical Methods in Natural Language Processing (EMNLP), pp. 1532–1543 (2014)
12. Van Den Oord, A., et a.: WaveNet: a generative model for raw audio. In: SSW, p. 125 (2016)

13. Bai, S., Kolter, J.Z., Koltun, V.: An empirical evaluation of generic convolutional and recurrent networks for sequence modeling. arXiv preprint arXiv:1803.01271 (2018)
14. Kingma, D.P., Ba, J.: Adam: a method for stochastic optimization. arXiv preprint arXiv:1412.6980 (2014)

RL-Gen: A Character-Level Text Generation Framework with Reinforcement Learning in Domain Generation Algorithm Case

Hua Cheng[(⊠)], Jing Cai, and Yiquan Fang

East China University of Science and Technology, Shanghai, China
{hcheng, fyq}@ecust.edu.cn, jcai_ecust@163.com

Abstract. Malware families often use the Domain Generation Algorithm (DGA) to communicate with the Command and Control (C&C) servers. Although machine learning and deep learning based methods have achieved good accuracy in DGA detection task, it has problems on the new DGA families with limited datasets. In this paper, RL-Gen, a Reinforcement Learning (RL) framework, is proposed to improve the performance of character-level text generation with few input samples. RL-Gen has two modules, W-Generator and Evaluator. W-Generator is an improved generation model based on WGAN-GP, which is regarded as an agent, and Evaluator acts as an environment to evaluate the generated text. Especially in DGA case, Evaluator is an effective DGA detection model (ATT-GRU). The parameters' updating of W-Generator is optimized by the reward from Evaluator, which promotes the generating abilities on speed and quality. Experiments show that the generated DGAs are sufficiently close to real DGAs, and can play an alternative role of real DGAs in detection model training. And RL-Gen gets better quality of text generation more quickly and smoothly than WGAN-GP.

Keywords: Domain Generation Algorithm · DGA detection · WGAN-GP · Reinforcement Learning · Character-level text generation

1 Introduction

A large number of malicious attacks such as Advanced Persistent Threats (APTs), botnets, etc. spread rapidly in the past two years, and continually involve to change them appearance to evade the detection. These attacks, which are remotely controlled and gather information or damage data, have become great threats to the national and commercial security. They widely utilize Domain Generation Algorithm (DGA) in communication with Command and Control (C&C) servers to avoid hardcoding the IP address or domain name in malware, which is easily detected by the signature-based anti-virus software or intrusion detection. DGA is used to generate a huge list of candidate C&C server domains, and malwares then attempt to connect to an active C&C server by querying each DNS server in turn. DGA-based malwares are highly elusive and difficult to detect using traditional defensive mechanisms and therefore have a high survivability. Nowadays, the efficient detection and prevention of DGA are the biggest challenges to face in order to guarantee the information security.

© Springer Nature Switzerland AG 2019
T. Gedeon et al. (Eds.): ICONIP 2019, CCIS 1143, pp. 690–697, 2019.
https://doi.org/10.1007/978-3-030-36802-9_73

Machine learning [1] and deep learning based [2] DGA detection methods have achieved good accuracy, and still have limitations on fewer training data that often lead to over-fit and slow update of the model. It is difficult to obtain sufficient and timely, effective DGA domain names (DGAs) in practice.

DGA domains are pseudo-random combination of characters and words in a specific dictionary. The generation of DGAs is different from the text generation task of sentences that preserves the semantic and syntactic properties of training samples. The DGA generation is a particular character-level text generation task, which keeps lexical rules learnt in DGAs. The original text generation is usually generating meaningful and coherent sentences to NLP applications.

Our contributions in this mission include:

- This paper improves the text generation task based on WGAN-GP [3], and proposes the character-level text generation model with Reinforcement Learning, which has better diversity, more stable and quicker convergence.
- In our method, W-Generator is simultaneously optimized by the GAN mechanism at each epoch and the rewards from Evaluator at each step.
- In a practical application of DGA generation, this paper proposes a DGA detection model (ATT-GRU [4]) to be the environment, which gives the evaluating results as rewards. The rewards would promote the generation ability of W-Generator.
- This paper shows that the distribution of generated DGAs is sufficiently close to the real DGAs, and can play an alternative role of real DGAs in detection model training.

2   RL-Gen Method

The character-level text generation framework with Reinforcement Learning (RL), named RL-Gen, is proposed and applied to the generation of DGAs, which provides high-quality datasets for model training in DGA detection. RL-Gen consists of two modules, which is shown in Fig. 1. The agent (W-Generator) produces samples, and the environment (Evaluator) evaluates the quality of the generated. The evaluation value is the reward at each step that will teach W-Generator to update the model parameters.

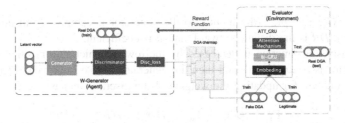

Fig. 1. RL-Gen framework

W-Generator is an improved model of WGAN-GP [3], which includes a generator and a discriminator whose parameters are updated by the mechanism of GAN.

On the DGA generation case, Evaluator is a DGA detection model (ATT-GRU) in our previous paper [4]. Here in RL-Gen, ATT-GRU is trained by the generated DGAs (instead of real DGAs) and legitimate domains, and another batch of real DGAs is a test dataset.

2.1   W-Generator

Gulrajani et al. give more specific examples to illustrate the problems caused by clipping weights and propose another way to impose Lipschitz continuity. They demonstrated problems with weight clipping in WGAN [5] and introduced WGAN-GP, where an alternative in the form of a penalty term in the discriminator loss:

$$GP|_{\hat{x}} = \mathrm{E}_{\hat{x}}[((\|\nabla_{\hat{x}}D(\hat{x})\|_2 - 1)^2] \tag{1}$$

where $\hat{x}$ is a uniform sampling along straight lines between pairs of points sampled from the data distribution P_r and the generator distribution P_g.

Although WGAN-GP has achieved success in the text generation, it has the problems of slow convergence rate, poor stability and spelling errors. It is obvious that the generated texts are not well evaluated by practical applications, and the model training is not improved by the evaluation results which could instruct the parameters updating. Therefore, a distance reward is proposed to revise the loss function of the discriminator.

Distance Reward. Distance reward (DR) is the environment's evaluation of the quality of the currently generated samples. Distance reward is expressed as following:

$$DR|_{x_g} = f(x_g) \tag{2}$$

where f is the evaluation function, x_g is the output of W-Generator.

The limited ability of the discriminator affects text generation during the adversarial training, which results the problems on slow and fluctuating convergence of generating. Since it is an underfitting issue, distance reward will be added in the loss function of the discriminator. The discriminator's minmax objective is optimized:

$$W(\mathrm{P}_r, \mathrm{P}_g) \approx \max_D \{\mathrm{E}_{x \sim \mathrm{P}_r}[D(x)] - \mathrm{E}_{\tilde{x} \sim \mathrm{P}_g}[D(\tilde{x})]\} - \lambda GP|_{\hat{x}} - \mu DR|_{x_g} \tag{3}$$

where $\tilde{x} = G(z), z \sim p(z)$ (the input z to the generator is sampled from uniform distribution p). DR is the distance reward, and μ is coefficient.

Obviously, the function is decreased by the reward which is a positive value, the discriminator is strengthened, and then the generator is also improved. Thus the new objective function for updating the weights of the discriminator is:

$$L = \mathrm{E}_{\tilde{x} \sim \mathrm{P}_g}[D(\tilde{x})] - \mathrm{E}_{x \sim \mathrm{P}_r}[D(x)] + \lambda GP|_{\hat{x}} + \mu DR|_{x_g} \tag{4}$$

For the hyper parameters, we borrow $\lambda = 10$ from [3] and use $\mu = 2$ for all our experiments no matter on which dataset.

2.2   ATT-GRU Evaluator

ATT-GRU acts as the environment to obtain real-time rewards in RL. As Fig. 2 shows that ATT-GRU is a DGA detection model based on Bi-Gated Recurrent Unit (Bi-GRU) with the attention mechanism, where Bi-GRU realizes multi-character combination encoding and character's randomness extraction, and attention mechanism focuses on the high random part in a domain name. This model can effectively identify the multiple DGA families, and has higher detection accuracy and lower false alarm rate than the traditional methods.

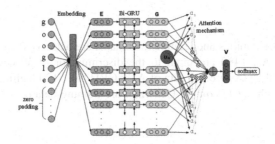

Fig. 2. DGA detection model (ATT-GRU)

2.3   Reinforcement Learning of RL-Gen

W-Generator selects action based on probabilistic and Evaluator judges the score based on W-Generator's behavior, and then W-Generator modifies the probability of action based on Evaluator's score.

And the RL-Gen framework is introduced in detail:

State. State s_i consists of the current generated samples x_i, the loss value L_i of the discriminator.

Action. An action a_i generates the next batch samples x_{i+1} from the state s_i through policy function $\pi_\theta(s_i, a_i)$.

Policy. The policy $\pi_\theta(s_i, a_i)$ determines the next action a_{i+1} of W-Generator, which means how to generate the next batch x_{i+1} under the state s_i and the reward $f_{ag}(x)$, and the parameters of generator are updated by loss and reward.

Reward. Evaluator computes a score f_{ag} based on the generated x_i of W-Generator, where $f_{ag}(x) \in [0, 1]$. The score approximating 1 indicates that the generated have better performance on training detection model and the reward is defined as follows:

$$r_{s_i}(x_i) = f_{ag}(x_i) \tag{5}$$

where f_{ag} is the detection accuracy given by ATT-GRU for the generated of each batch.

Optimization. The goal is to maximize the expected total reward, and the objective function is defined:

$$J(\theta) = E_{s_1,a_1,s_2,a_2,\ldots,s_i,a_i,s_{i+1},a_{i+1},\ldots}[\sum_{i=0}^{N} r(x_i)]$$

$$= \max V_\theta(G(\theta))$$

(6)

where V_θ is the sum of the expected future rewards that we can obtain through the policy $\pi_\theta(s_i, a_i)$ after the state s_1.

3 Experiments and Discussions

3.1 Datasets

Three typical DGA families and legitimate domains are datasets shown in the Table 1. In 2000 real DGAs, 1480 are input to the discriminator of W-Generator, and the remaining 520 real DGAs are test data of ATT-GRU [4]. 5120 legitimate domains and 5120 DGAs of each batch from W-Generator are training data of ATT-GRU.

Table 1. Datasets

Family	Label	Dataset size	Relative entropy
Legitimate	0	1280	0.861
Corebot	1	2000	0.999
Banjori	1	2000	0.930
Suppobox	1	2000	0.880

The relative entropy of characters in Corebot is as near to 1, which is fully psudo-randomly distributed. Suppobox is a dictionary-based DGA, which generate domains by words in a specific dictionary. Banjori randomly changes the first 4 characters of domains, and the latter part is fixed strings.

The relative entropy of Suppobox (0.861) is the closest to the legitimate (0.88), where legitimate domains and Suppobox have similar strategy of combination of words.

3.2 Generation Metrics

Wasserstein Distance. Wasserstein distances, which measure the distance between the distribution of real DGAs and the generated DGAs in RL-Gen and WGAN-GP [3], are illustrated in Fig. 3. The Wasserstein distance curves of RL-Gen are moving closer to 0 and promoting more stably, which means the distribution of the generated DGAs is closing to real DGAs and RL-Gen gradually achieve better generating performances.

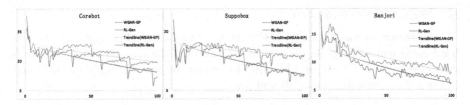

Fig. 3. Wasserstein distances of DGAs

Diversity. Diversity means that how much the generated text provides specific information, rather than "dull" and repeated information.

The same samples between the generated and real are counted, shown in Table 2. The generated DGAs of Corebot and Banjori are totally different with real DGAs. The generated Suppobox DGAs are 0.04% duplicated with real DGAs, due to the reason that Suppobox DGAs are generated from the randomly combining of words in a limited dictionary.

Table 2. Diversities of the Generated DGAs

DGA family	Generated samples	Duplicated number	Overlap ratio (%)
Corebot	5120	0	0.00
Banjori	5120	0	0.00
Suppobox	5120	2	0.04
Suppobox	10240	3	0.03

3.3 Results and Discussions

In RL-Gen, the number of total steps is 100, and each step takes place at every 100 model training epochs for the reason of efficiency.

(1) Detection accuracies with different generation models

In our experiments, the test datasets of both ATT-GRU are the same real DGAs.

The accuracies from ATT-GRU are illustrated in Fig. 4. The accuracies curves of RL-Gen are roughly divided into two stages: efficient learning phase (promotion period) and stabilization phase (fine tuning period).

From Fig. 4, the accuracies of RL-Gen are totally higher than the ones of WGAN-GP on three types of DGAs under the same generation conditions. Rewards effectively promote the generation ability quickly and smoothly, especially on Suppobox in Fig. 4, which is also reflected on Wasserstein distances in Fig. 3.

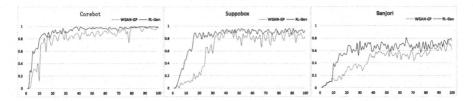

Fig. 4. Accuracies of three types of DGAs during training.

(2) Different model parameters

Two important parameters of RL-Gen influence the quality of generated DGAs, one is the coefficient μ in Eq. (4), the other is the number of the generated DGAs. Figure 5 shows the accuracies curve of ATT-GRU-RL, where the best number of the generated DGAs is 5120, and the best coefficient μ is 2.

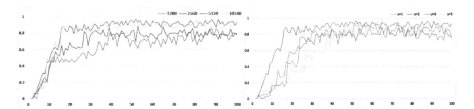

Fig. 5. Accuracies with different number of the generated DGAs and different μ

(3) The alternative role of the generated DGAs

In the following experiments, ATT-GRU trained by real DGAs is called as ATT-GRU-real, while the model trained by generated DGAs is called as ATT-GRU-gen. The test datasets of both ATT-GRU are the same real DGAs.

ATT-GRU achieve relatively high accuracies in Table 3, where accuracies of ATT-GRU-gen on Corebot and Suppobox are better than ATT-GRU-real, and the generated Banjori needs to be improved in future. Real DGAs and the generated DGAs are listed in Table 4, which show the generated DGAs have the same generating lexical rules with real DGAs, and RL-Gen has successfully imitated the algorithms of DGA.

From Tables 3 and 4, the generated DGAs can be treated as an approximation of the real samples, and can be alternatives of real DGAs in DGA detection model training.

Table 3. Detection accuracies of ATT-GRU models

DGA family	Accuracy (ATT-GRU-real)	Accuracy(ATT-GRU-gen)
Corebot	99.8%	99.9%
Suppobox	93.4%	96.3%
Banjori	98.9%	81.1%

Table 4. Real DGAs and the generated DGAs

DGA family	Real DGAs	Generated DGAs (WGAN-GP)	Generated DGAs (RL-Gen)
Corebot	sl5vwv32y2qjalu ur707vcxktw45pqr-k4q4g05	y6i0kb3xm6mhkbc ov3vipmho4y0qv-w6s01bete	ml5d52gfmv74o2s mf1lolq2ens6ib5p-efmxs47
Suppobox	increaseshort littlearrive	sufferstrong riddenbupply	riddenbecame littleairplane
Banjori	nvtpstring dpdyordf150	hnabstrinf bpydrdf15y	eulsstring emlcordf150

4 Conclusion

In this paper, a Reinforcement Learning framework (RL-Gen) based character-level text generation is proposed from the perspective of DGA detection. Experiments on typical DGAs have shown that our model can efficiently generate large and diverse DGAs with few training samples, and it has a better performance than the traditional method WGAN-GP, and the generated DGAs can replace real DGAs in detection model training.

Acknowledgement. This work has been supported by Industry-University Cooperation Project Ministry of Education (201802013016) and CERNET Next generation network Technology Innovative Project (No.: NGII20170520).

References

1. Chen, Y., Yan, S., Pang, T., Chen, R.: Detection of DGA domains based on support vector machine. In: 2018 Third International Conference on Security of Smart Cities, Industrial Control System and Communications (SSIC), Shanghai, pp. 1–4 (2018)
2. Abdelwahab, O., Elmaghraby, A.: Deep learning based vs. Markov chain based text generation for cross domain adaptation for sentiment classification. In: 2018 IEEE International Conference on Information Reuse and Integration (IRI), Salt Lake City, UT, pp. 252–255 (2018)
3. Gulrajani, I., Ahmed, F., Arjovsky, M., Dumoulin, V., Courville, A.: Improved training of wasserstein GANs. arXiv preprint arXiv:1704.00028 (2017)
4. Chen, L., Cheng, H., Fang, Y.: Detecting domain generation algorithms based on attention mechanism. J. East Chin. Univ. Sci. Technol. (Natural Science Edition). https://doi.org/10.14135/j.cnki.1006-3080.20180326002
5. Arjovsky, M., Bottou, L.: Wasserstein GAN. arXiv preprint arXiv:1701.07875 (2017)

A Conditional Random Fields Based Framework for Multiview Sequential Data Modeling

Ziang Dong, Jing Zhao$^{(\boxtimes)}$, and Shiliang Sun$^{(\boxtimes)}$

Department of Computer Science and Technology, East China Normal University,
3663 North Zhongshan Road, Shanghai 200062, People's Republic of China
jzhao2011@gmail.com, shiliangsun@gmail.com

Abstract. Multiview learning has gained much attention due to the increasing amount of multiview data which are collected from different sources or can be characterized by different types of features. How to properly handle view heterogeneity and balance information between views are the challenges in the domain of multiview learning, especially for sequential data. In this paper, we propose a multiview conditional random field model (multiview CRF) for modeling multiview sequential data, in which the model focuses on different views at a different moment with varying degrees. Especially, additional weight variables are introduced in the CRF through two different mechanisms which play a role of controlling the use of information from different views. Besides, a multiview decomposition network is designed to transform multiview observation at each moment into a unified multiview representation which is further used as an input of weight variable CRF. Variational inference with variance reduction technique is adopted for model training. We conduct experiments on two real-world datasets to compare the proposed methods with other related state-of-the-art methods. Experimental results and careful analysis show that both the weight variables and decomposition network contribute to the outstanding performance over other methods.

Keywords: Multiview learning · Conditional random field ·
Variational inference · Latent variable

1 Introduction

Recently, many multiview learning methods are well developed and applied [2,10, 12,14]. However, most of them are not suitable for sequential data as they do not specifically learn the latent dynamic in data. In order to process multiview sequential data, we mainly consider two aspects: the structured characteristics of output labels and the multiview characteristics of input features.

From the perspective of structured characteristics, probabilistic graphical models provide an effective and intuitive way to construct various structures

© Springer Nature Switzerland AG 2019
T. Gedeon et al. (Eds.): ICONIP 2019, CCIS 1143, pp. 698–706, 2019.
https://doi.org/10.1007/978-3-030-36802-9_74

with dependency. We consider designing a flexible model under the framework of probabilistic graphical models, in which some necessary conditional dependency assumptions are made on the labels of a sequence. Thus, we focus on using Conditional random field (CRF) [5] as the framework of our model to capture dependency between multiple output variables.

From the perspective of multiview characteristics, as the pairwise potentials in traditional CRFs may not faithfully deal with the heterogeneity in multiview data. Inspired by factorized latent spaces model [3], we propose a neural multiview decomposition network that learns high-level unified multiview representations to overcome the challenge of heterogeneity, where each view is correctly decomposed into common and private parts. Particularly, the proposed multiview decomposition network is regularized by making the distance between view-common representation and view-specific representation in each view as far as possible, and the distance between view-common representations as close as possible. We normalize all the distances as a regularization term added to the final objective function.

In order to better utilize the multiview representations instead of concatenating, we propose a novel multiview CRF model with additional weight variables, which is inspired by the works of attention [1,13]. Considering that the attention to different views is not consistent over time. Thus, we model the attention (weight variables) of different views at time t and present two feasible mechanisms of using latent weight variables.

The main contributions of this paper are summarized as follows:

(1) We propose a novel multiview CRF framework with a latent weight variable to control the effects of multiple views for the multiview sequential data.
(2) We propose a regularization based objective that makes the multiview representation obtained by the multiview decomposition network more flexible for the multiview CRF.
(3) We conduct comparative experiments to test each part of the model, and the results demonstrate the efficacy and superiority of our proposed model.

2   Related Work

Recently, multiview latent variable discriminant models (MVLDCRF and MVHCRF) were proposed [9], which are based on the latent variable CRF [6–8]. They learn view-common and view-specific substructures together to capture interactions between views and achieve good performance by introducing latent variables. However, due to introducing the graph of latent state variables, the training of MVLDCRF model has become more complex, so that it is difficult to run on large-scale data sets. Further, it ignores some necessary relationships between features from the same view. Unlike MVLDCRF, our proposed multiview CRF reduces computation by designing latent weight variables.

Mixture CRF was proposed to extend CRF [4], and can also be used to process multiview sequential data. Mixture CRF is to mix the potential energy functions in different views as the potential energy function of a new CRF. It is also a particular case of our latent weight CRF model when we assume that our latent weight variable has a categorical distribution and use EM for training.

3 Model

In this section, we first discuss the proposed multiview CRFs model with two alternative mechanisms, and then we present a neural multiview decomposition network to obtain multiview representations for promoting the performance of multiview CRFs. Finally, we give the learning process of multiview CRFs.

3.1 The Proposed Model

The proposed model extends the traditional CRF, which contains two parts: a neural multiview decomposition network as the bottom structure and a latent weight variable CRF model as the top structure. We show the framework of the proposed model in Fig. 1, where the two parts in our model are connected through the unified multiview representations $\boldsymbol{\Phi}(\mathbf{x}_t)$. $\boldsymbol{\Phi}(\mathbf{x}_t)$ is constructed by the multiview decomposition network with input of observations in different views, that is, $\boldsymbol{\Phi}(\mathbf{x}_t) = [\boldsymbol{\Phi}^1(\mathbf{x}_t), \dots, \boldsymbol{\Phi}^v(\mathbf{x}_t), \dots, \boldsymbol{\Phi}^V(\mathbf{x}_t)]$, $\boldsymbol{\Phi}^v(\mathbf{x}_t) = [\Phi_c^v(\mathbf{x}_t), \Phi_s^v(\mathbf{x}_t)]$. $\Phi_c^v(\mathbf{x}_t)$ denotes the view-common representation transformed from v-th view, and $\Phi_s^v(\mathbf{x}_t)$ denotes the view-specific representation transformed from v-th view.

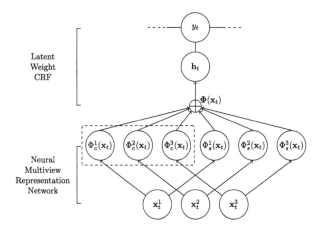

Fig. 1. The framework of multiview CRF model. Notice that the $\mathbf{h}_t$ is a view-weighted representation of multiview input $\mathbf{x}_t$. The dotted box represents the unified representation $\boldsymbol{\Phi}(\mathbf{x}_t)$, which is constructed by the neural multiview representation.

Figure 2 shows the details of the latent weight CRF model. In the latent weight CRF, $\mathbf{z}$ indicates which multiview representation (or a mixture of multiview representations) to discriminate $\mathbf{y}$. We use $\mathbf{h}$ to represent the selected multiview representation (or a mixture of multiview representations), which calculated by $\mathbf{x}$ and $\mathbf{z}$. Mainly, we explore two alternative mechanisms for the latent weight CRF to mediate multiview representation. The resulted weighted multiview representation $\mathbf{h}$ is taken into the energy function as the input. The discriminant probability can be calculated by marginalizing $\mathbf{z}$.

We jointly model the conditional probability of latent variables and outputs. The conditional probability distribution of the multiview CRF is

$$p(\mathbf{y}, \mathbf{z}|\mathbf{x}) = \frac{1}{Z(\mathbf{x})} \exp\{E(\mathbf{x}, \mathbf{y}, \mathbf{z}; \Theta)\}, \text{ and } Z(\mathbf{x}) = \sum_{\mathbf{y}, \mathbf{z}} \exp\{E(\mathbf{x}, \mathbf{y}, \mathbf{z}; \Theta)\}. \quad (1)$$

The marginal likelihood can be obtained by marginalizing out $\mathbf{z}$

$$p(\mathbf{y}|\mathbf{x}) = \sum_{\mathbf{z}} p(\mathbf{y}, \mathbf{z}|\mathbf{x}) = \sum_{\mathbf{z}} p(\mathbf{z}|\mathbf{x})p(\mathbf{y}|\mathbf{x}, \mathbf{z}). \quad (2)$$

Latent Weight CRF. We introduce the additional weight variable $\mathbf{z} = [z^1, z^2, \ldots, z^V] \in \{0,1\}^V$ into the traditional CRF. It is a one-hot vector at the slice of time t with $z^v = 1$ if the v-th view at time t is selected, where V is the number of the views. By treating view weight as an intermediate latent variable, we can assign a categorical distribution parameterized as $\{\alpha^v\}$ on z^v, i.e., $p(z^v = 1|\mathbf{x}) = \alpha^v$.

The energy function of latent weight CRF is defined as

$$E(\mathbf{x}, \mathbf{y}, \mathbf{z}) = \sum_{(m,n)\in\mathcal{F}} \theta^\mathsf{T} \mathbf{F}(\mathbf{h}, \mathbf{y}_m, \mathbf{y}_n), \quad (3)$$

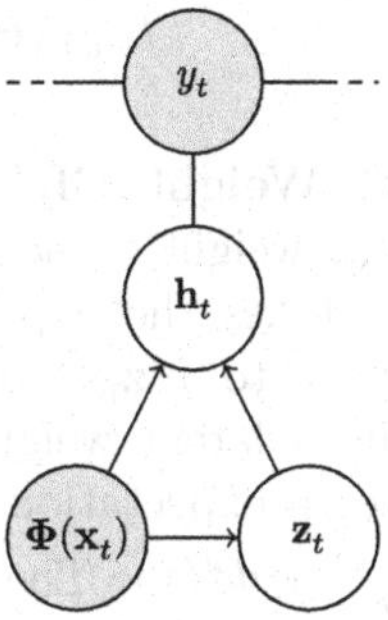

Fig. 2. A latent weight CRF. $\mathbf{h}_t$ is the input of CRF, which is constructed by selecting (or weighting) $\Phi(\mathbf{x}_t)$ with $\mathbf{z}_t$.

where $\mathbf{h} = \sum_{v=1}^{V} z^v \Phi^v(\mathbf{x})$.

Here $\Phi^v(\cdot)$ is the unified representation of $\mathbf{x}_t$ in different views, coming from the neural multiview decomposition network which will be mentioned in the following part. We can regard $\mathbf{h}$ in Eq. (3) as a more appropriate representation than $\Phi(\mathbf{x}_t) = [\Phi^1(\mathbf{x}_t), \ldots, \Phi^v(\mathbf{x}_t), \ldots, \Phi^V(\mathbf{x}_t)]$.

As the additional latent variable makes the latent weight CRF intractable, we consider using variational inference method to get a lower bound,

$$\log p(\mathbf{y}|\mathbf{x}) - \text{KL}\left[q(\mathbf{z})\|p(\mathbf{z}|\mathbf{x}, \mathbf{y})\right] = \mathbb{E}_{\mathbf{z}\sim q(\mathbf{z})}\left[\log p(\mathbf{y}|\mathbf{z}, \mathbf{x})\right] - \text{KL}\left[q(\mathbf{z})\|p(\mathbf{z}|\mathbf{x})\right] \quad (4)$$

This allows us to search over variational distributions q to improve the bound. It is tight when the variational distribution is equal to the posterior, i.e., $q(\mathbf{z}) = p(\mathbf{z}|\mathbf{x}, \mathbf{y})$. The objective in mixture CRF is a special case of the evidence lower bound with $q(\mathbf{z}) = p(\mathbf{z}|\mathbf{x})$, which is obtained by using the Jensen's inequality on the log marginal likelihood. The variational distribution $q(\mathbf{z})$ has the same distribution family as $p(\mathbf{z})$. We use a neural network to encode unified representations $\boldsymbol{\Phi}(\mathbf{x})$ and label $\mathbf{y}$ into the parameters $\boldsymbol{\theta_z}$ of the variational distribution $q(\mathbf{z}; \boldsymbol{\theta_z})$. According to the definition in Eq. (1), we have

$$p(\mathbf{y}|\mathbf{x}, \mathbf{z}) = p^v(\mathbf{y}|\mathbf{x}, z^v = 1) = \frac{1}{Z^v(\mathbf{x})} \exp\{E(\mathbf{x}, \mathbf{y}, z^v = 1)\}. \tag{5}$$

The v-th multiview representation $p(\mathbf{y}|\mathbf{x}, \mathbf{z}^v = 1)$, which is often denoted by $p^v(\mathbf{y}|\mathbf{x})$, is exactly the single-view CRF model with the parameters $\boldsymbol{\Theta}^v$. Its partition function is denoted by $Z^v(\mathbf{x})$, i.e., $Z^v(\mathbf{x}) = \exp\{E(\boldsymbol{\Phi}^v(\mathbf{x}), \mathbf{y}, z^v = 1)\}$. It immediately follows that $Z(\mathbf{x}) = \sum_{v=1}^{V} Z^v(\mathbf{x})$.

According to Eqs. (4) and (5), we can define an objective function that is evidence lower bound on the marginal log-likelihood $p(\mathbf{y}|\mathbf{x})$. The optimization goal of the latent weight CRF model is

$$\ell_{\mathrm{CRF}}(\boldsymbol{\Theta}) = \mathbb{E}_{\mathbf{z} \sim q(\mathbf{z})} \left[\log p(\mathbf{y}|\mathbf{x}, \mathbf{z}) \right] - \mathrm{KL}\left[q(\mathbf{z}) \| p(\mathbf{z}|\mathbf{x}) \right]. \tag{6}$$

Soft Weight CRF. Learning the latent weight CRF requires sampling the latent weight $\mathbf{z}$ each time. As an alternative, we proposed a soft weight CRF that takes the expectation of the weighted multiview representation $\mathbf{h}$ with respect to $p(\mathbf{z}|\mathbf{x})$ directly as the input of the model. Specially, instead of using a latent weight variable $\mathbf{z}$, we employ a deterministic network to compute an expectation over the weighted multiview representation as $\mathbb{E}_{p(\mathbf{z}|\mathbf{x})}[\mathbf{h}] = \sum_{v=1}^{V} \alpha^v \boldsymbol{\Phi}^v(\mathbf{x})$. The weight α^v for each representation $\boldsymbol{\Phi}^v(\mathbf{x})$ is computed by $\alpha^v = \frac{\exp\{e^v\}}{\sum_{k=1}^{K} \exp\{e^k\}}$, where $e^v = \psi(\boldsymbol{\Phi}^v(\mathbf{x}))$. Here $\psi(\cdot)$ is a weight model. We parametrize this weight model $\psi(\cdot)$ as a feedforward neural network.

A major benefit of the soft weight CRF is efficiency. Instead of paying a multiplicative regularization of summation in Eq. (2), the soft weight CRF can compute the expectation before computing $p(\mathbf{y}|\mathbf{x}, \mathbf{z})$ as

$$p_{\mathrm{soft}}(\mathbf{y}|\mathbf{x}) = \frac{1}{Z(\mathbf{x}, \boldsymbol{\theta})} \exp\left\{ E(\mathbb{E}_{p(\mathbf{z}|\mathbf{x})}[\mathbf{h}], \mathbf{y}; \boldsymbol{\theta}) \right\}. \tag{7}$$

Another benefit of the soft weight CRF is that it can be transferred into a general CRF to simplify the learning process. The optimization goal of soft weight CRF model is $\ell_{\mathrm{CRF}}(\boldsymbol{\Theta}) = \log p_{\mathrm{soft}}(\mathbf{y}|\mathbf{x})$.

Neural Multiview Decomposition Network. We consider building a neural network (NN) to jointly learn view-common and view-specific representations for each view, which have the ability to capture the interaction between views.

Taken a multiview sequence as an example, let $\mathbf{x} = (\mathbf{x}^1, \ldots, \mathbf{x}^V)$ be the observed sequence and $\mathbf{x}^v = \{\mathbf{x}_1^v, \ldots, \mathbf{x}_T^v\}$ represents a sequence in v-th view. We define the objective function as

$$\ell_{\mathrm{NN}}(\mathbf{W}) = -\frac{2}{V(V-1)} \sum_{u,v} \mathrm{Dist}\left(\boldsymbol{\Phi}_c^u(\mathbf{x}_t), \boldsymbol{\Phi}_c^v(\mathbf{x}_t)\right) + \frac{1}{V} \sum_{v=1}^{V} \mathrm{Dist}\left(\boldsymbol{\Phi}_c^v(\mathbf{x}_t), \boldsymbol{\Phi}_s^v(\mathbf{x}_t)\right),$$

where we normalize the distance of inter-views and intra-views to make them balanced, respectively. In our experiments, we use the cosine distance to measure the distance of inter-views and intra-views. The whole joint neural network Φ with 3-layers is defined as $\Phi(\mathbf{x}) = \sigma(\mathbf{W}_3^\mathsf{T}\sigma(\mathbf{W}_2^\mathsf{T}\sigma(\mathbf{W}_1^\mathsf{T}\mathbf{x}))))$. Each element of $\sigma(\mathbf{x})$ is the user-specified activation function $\sigma(\cdot)$. In the experiment, the decomposition network has a 3-layer fully-connected network, which consists of 300 tanh hidden units and 100 tanh output units. We have found that the three layers are the most efficient and the performance is acceptable.

3.2 Inference

We use the following objective function to estimate the parameters. This objective function consists of CRF objective and multiview constraint objective: $\ell(\Theta, \mathbf{W}) = \ell_{\mathrm{CRF}}(\Theta) + \ell_{\mathrm{NN}}(\mathbf{W})$. The parameters to be optimized are Θ which include model parameters in CRF and the variational parameters involved with $\mathbf{z}$, and $\mathbf{W}$ which is introduced in the neural multiview decomposition network. The objective function contains three parts: the variational expectation $\mathbb{E}_{\mathbf{z}\sim q(\mathbf{z})}[\log p(\mathbf{y}|\mathbf{x}, \mathbf{z})]$, $\mathrm{KL}[q(\mathbf{z})\|p(\mathbf{z}|\mathbf{x})]$, and the distance constraint for the neural network.

We parametrize the neural multiview representation network with parameter $\mathbf{W}$, which allows the gradient of the cost function to be backpropagated through. The gradient with respect to the KL portion is also easy to compute. There is an optimization issue for calculating the gradient of the term $\mathbb{E}_{\mathbf{z}\sim q(\mathbf{z})}[\log p(\mathbf{y}|\mathbf{x}, \mathbf{z})]$. Many recent methods target this issue. Among these, the approach called REINFORCE [11] along with a specialized baseline is effective. Formally, we first apply the likelihood-ratio trick to obtain an expression for the gradient with respect to the parameters Θ,

$$\frac{\partial \mathbb{E}_{\mathbf{z}}[\log p(\mathbf{y}|\mathbf{x}, \mathbf{z})]}{\partial \Theta} = \mathbb{E}_{\mathbf{z}\sim q(\mathbf{z})}\left[\frac{\partial \log p(\mathbf{y}|\mathbf{x}, \mathbf{z})}{\partial \Theta} + (\log p(\mathbf{y}|\mathbf{x}, \mathbf{z}) - B)\frac{\partial \log p(\mathbf{z}|\mathbf{x})}{\partial \Theta}\right]$$
(8)

Then, we use Monte Carlo sampling to estimate the expectation in Eq. (8). In order to reduce the variance of the stochastic approximation, we subtract the baseline term B. The ideal B would be $\mathbb{E}_{\mathbf{z}\sim q(\mathbf{z})}[\log p(\mathbf{y}|\mathbf{x}, \mathbf{z})]$, analogous to the value function in reinforcement learning. As the term $\mathbb{E}_{\mathbf{z}\sim q(\mathbf{z})}[\log p(\mathbf{y}|\mathbf{x}, \mathbf{z})]$ can hardly be computed, we use an approximation $\log p(\mathbf{y}|\mathbb{E}_{p(\mathbf{z}|\mathbf{x})}[\mathbf{h}])$ as the baseline.

According to the assumption, the view weight distribution and the variational family are categorical distributions. The view weight assumption is that $\mathbf{y}$ is discriminated from a single index of multiview representation $\Phi(\mathbf{x})$. Thus, a low-variance estimator of $\frac{\partial}{\partial \Theta}\ell_{\mathrm{CRF}}$ is easily obtained through a sample from $q(\mathbf{z})$.

4 Experiment

We evaluate our model on two real-world datasets: CoNLL-2000 and NATOPS.

Table 1. Results on the CoNLL-2000 dataset and the NATOPS dataset.

Models	CoNLL-2000	NATOPS
CRF	92.1 ± 1.0	78.09 ± 1.68
MVLDCRF	93.4 ± 0.9	84.22 ± 1.21
Mixture CRF	93.8 ± 1.0	82.62 ± 1.75
Soft weight CRF	93.8 ± 1.2	82.51 ± 1.73
Latent weight CRF	93.5 ± 0.6	82.77 ± 1.01
CRF + decomposition net	93.2 ± 1.1	80.52 ± 1.50
Soft weight CRF + Decomposition net	$\mathbf{94.7 \pm 0.6}$	84.52 ± 1.67
Latent weight CRF + Decomposition net	94.6 ± 0.4	$\mathbf{85.91 \pm 0.82}$

We compare our model with CRF, MVLDCRF [9], Mixture CRF [4] on the two dataset. Moreover, in order to analyze the role of the two components (latent weight CRF and neural decomposition network) in our model, we further make comparisons of soft weight CRF, latent weight CRF, and CRF+neural multiview decomposition network. All the results are listed in Table 1.

From the perspective of multiview, the two data sets have different properties from which we can analyze the characteristics of different models. The conll-2000 data set has sufficient information under each view. Any of these views can be used to train a single-view model to obtain good performance. In addition, the two views are highly dependent on each other. Models can be designed to maximize the consistency of high-level concepts across different views. The NATOPS data set has different characters from the first data set, whose feature only has 20 dimensions. Features in each view do not display enough information. Each view of the data may contain some knowledge that the other view does not have. Therefore, the joint representation of multiple views can provide a comprehensive and accurate description of the data. For this data set, the model is designed to maximize the complementarity of information from different views.

From the tables, we can observe that our models achieve better performance than the other methods on both two data sets with less time. Setting aside our models, MVLDCRF and Mixture CRF have better performance than CRF. In addition, MVLDCRF learns the relationship between views by constructing implicit state variable mappings, but its running time is several times more than ours. The results show that it is more stable (lower std) than CRF and Mixture CRF. Compared with MVLDCRF, our latent/soft weight CRF only introduces a latent variable with approximate inference, which makes the model more efficient and has the similar performance. And more suitable, the proposed decomposition network helps to obtain a more abstract representation that is conducive to view selection. The proposed latent weight variable for choosing multiple views and the abstract multiview representation are complementary, which make our models achieve state-of-the-art performance with better computational efficiency.

Looking into the comparison results of the two weight CRF models without using a decomposition network. They achieve similar results to Mixture CRF. It shows that when the information is sufficient, the soft weight CRF can achieve better results without approximating the posterior. In addition, we find that the latent weight CRF has a lower variance (high certainty) than the soft weight CRF, which shows that the introduced baseline of the gradient estimator helps to reduce the variance of the gradient.

The comparison results also show the effectiveness of using a multiview regularization objective. It seems that the designed multiview decomposition network makes sense. The model with multiview decomposition network gets better performance than those without using decomposition networks. Moreover, in our practice, using the decomposition network can help the latent weight CRF in our model to converge faster, because the decomposition network also contributes its learning ability to the model. Neural networks can accelerate computations with parallel computations, which makes the computational time of the decomposition network not significantly longer than CRF in one epoch. Its running time in total is equivalent to CRF, or even shorter.

5 Conclusion and Future Work

In this paper, we propose a multiview conditional random field framework for processing multiview sequential data. It is based on CRF and introduced additional latent weight variables and neural multiview decomposition networks. It can focus on the information expected in different views more accurately and maintain computational efficiency. Experiments have shown that our model is state-of-the-art, and it makes sense to add a multiview decomposition network for a high-level multiview representation. In future work, we will replace the existing neural multiview decomposition network with a better and more efficient multiview decomposition structure.

Acknowledgments. The corresponding author Shiliang Sun would like to thank supports by NSFC Project 61673179 and Shanghai Sailing Program 17YF1404600.

References

1. Bahdanau, D., Cho, K., Bengio, Y.: Neural machine translation by jointly learning to align and translate. arXiv preprint arXiv:1409.0473 (2014)
2. Baltrušaitis, T., Ahuja, C., Morency, L.P.: Multimodal machine learning: a survey and taxonomy. IEEE TPAMI **41**(2), 423–443 (2019)
3. Jia, Y., Salzmann, M., Darrell, T.: Factorized latent spaces with structured sparsity. In: NIPS, pp. 982–990 (2010)
4. Kim, M.: Mixtures of conditional random fields for improved structured output prediction. TNNLS **28**(5), 1233–1240 (2017)

5. Lafferty, J.D., McCallum, A., Pereira, F.C.: Conditional random fields: probabilistic models for segmenting and labeling sequence data. In: ICML, pp. 282–289 (2001)
6. Morency, L.P., Quattoni, A., Darrell, T.: Latent-dynamic discriminative models for continuous gesture recognition. In: CVPR, pp. 1–8 (2007)
7. Quattoni, A., Collins, M., Darrell, T.: Conditional random fields for object recognition. In: NIPS, pp. 1097–1104 (2005)
8. Quattoni, A., Wang, S., Morency, L.P., Collins, M., Darrell, T.: Hidden conditional random fields. IEEE TPAMI **10**, 1848–1852 (2007)
9. Song, Y., Morency, L.P., Davis, R.: Multi-view latent variable discriminative models for action recognition. In: CVPR, pp. 2120–2127 (2012)
10. Sun, S.: A survey of multi-view machine learning. Neural Comput. Appl. **23**(7–8), 2031–2038 (2013)
11. Williams, R.J.: Simple statistical gradient-following algorithms for connectionist reinforcement learning. Mach. Learn. **8**(3–4), 229–256 (1992)
12. Xu, C., Tao, D., Xu, C.: A survey on multi-view learning. arXiv preprint arXiv:1304.5634 (2013)
13. Xu, K., et al.: Show, attend and tell: neural image caption generation with visual attention. In: ICML, pp. 2048–2057 (2015)
14. Zhao, J., Xie, X., Xu, X., Sun, S.: Multi-view learning overview: recent progress and new challenges. Inf. Fusion **38**, 43–54 (2017)

From Signal to Image Then to Feature: Decoding Pigeon Behavior Outcomes During Goal-Directed Decision-Making Task Using Time-Frequency Textural Features

Mengmeng Li[1,2], Zhigang Shang[1,2,3(✉)], Lifang Yang[1,2],
Haofeng Wang[1,2], Kun Zhao[1], and Hong Wan[1,2,3(✉)]

[1] School of Electrical Engineering,
Zhengzhou University, Zhengzhou 450001, China
{zhigang_shang,wanhong}@zzu.edu.cn
[2] Industrial Technology Research Institute,
Zhengzhou University, Zhengzhou 450001, China
[3] Henan Key Laboratory of Brain Science and Brain-Computer Interface
Technology, Zhengzhou 450001, China

Abstract. Neural information decoding has become a hot issue in the field of brain science and the brain-computer interface. To effectively decode pigeon behavior outcomes during goal-directed decision-making task, a goal-directed experiment based on plus maze is designed and six pigeons are trained to acquire neural signals. Then, continuous wavelet transform (CWT) is employed to obtain the time-frequency images of local field potential (LFP) signals, and a series of textural features based on different texture analysis descriptors are extracted. Finally, textural features are used to decode different behavior outcomes based on random forest (RF). Decoding performances of single-channel LFP signals at different frequency bands, signals containing different numbers of channels, different types of features and classifier parameters are compared and discussed. The results show that the time-frequency textural features can be used to decode the animal behavior outcomes effectively. Specifically, the textural features of the LFP signal in the 40–60 Hz frequency band perform best in pigeon behavior decoding, and its performance is less affected by the number of channels. Different types of texture features have different decoding performances, and it seems that some local jet based gray-level statistic (LJbG) features are more suitable in this study. The performance comparison of different RF parameters shows that better decoding accuracy can be guaranteed when the number of trees is set at 6–8. These results will help us to understand the brain neuronal information processing mechanism of pigeon further.

Keywords: Decoding · Pigeon · Behavior outcome · Time-frequency image · Texture analysis · Random forest

© Springer Nature Switzerland AG 2019
T. Gedeon et al. (Eds.): ICONIP 2019, CCIS 1143, pp. 707–717, 2019.
https://doi.org/10.1007/978-3-030-36802-9_75

1 Introduction

Neural information decoding of animal behavior refers to the prediction or interpretation of external information input or intrinsic intention based on behavior detection and neural signal acquisition [1, 2], in which a series of feature extraction and pattern recognition algorithms are usually involved. It is of great significance to reveal the neural information processing mechanism of the brain and promote the development of brain-computer interface technology in the present age.

Generally, there are three stages to realize neural information decoding of animal behavior: (1) the detection of behavioral target and the acquisition of neural data, i.e., identifying the behavioral subject and obtaining corresponding neural signals; (2) feature extraction, i.e., the extraction of quantitative parameters based on neural signals that can characterize the behavior; (3) neural information decoding, i.e., realizing the recognition or classification of specific behavior. It is the most critical issue to extract effective features from neural signals in this problem. Local field potential (LFP) signals reflect the integrated electrical activity of neuron clusters in the local brain region and have good stability for long-term decoding [3]. At present, different feature extraction methods of time series signals include wavelet transform (WT) [4], independent component analysis (ICA) [5], partial least squares regression (PLSR) [6] and so on.

In recent years, texture analysis (TA) methods based on time-frequency images (TFIs) of time series signals are widely concerned. Alcin et al. [7] introduce a different multi-category EEG signal processing technique, namely TFIs representation of gray level co-occurrence matrix (GLCM) descriptors and Fisher vector encoding for automatic signal classification. Li et al. [8] employ a continuous wavelet transform (CWT) method for obtaining the time-frequency images of EEG signals to extract Gaussian mixture model (GMM) features and GLCM descriptors for epileptic seizures detection. Yuan et al. [9] use the local binary pattern (LBP) based on the TFIs to characterize the behavior of EEG activities. The application effect of neural decoding in pigeon goal-directed decision task remains to be tested.

In this paper, time-frequency textural features based on LFP signals are extracted to decode pigeon behavior outcomes during the goal-directed decision-making task. The LFP signals are acquired from pigeon goal-directed decision-making experiments. Then the CWT method is employed to obtain the TFIs. Several image descriptors based on TFIs have been employed for behavior decoding, namely gray-level histogram (GLH), GLCM, gray-level run length matrix (GRLM), gray-level gradient co-occurrence matrix (GGCM), Tamura, invariant moment (IM), local jet based gray-level statistic (LJbG) and so on. Finally, textural features are used to decode behavior outcomes, in which the comparison of different frequency bands, channels, features and classifier parameters are discussed. The results show that the time-frequency textural features can be used to decode pigeon behavior outcomes effectively.

2 Materials and Methods

2.1 Subjects and Behavioral Apparatus

Six adult pigeons (Columba livia) (450–550 g) are used in this study. All of the experiments are conducted in accordance with the Animals Act, 2006 (China), for the care and use of laboratory animals, and approved by the Life Science Ethical Review Committee of Zhengzhou University. Drug usage in the experiments complied with the Chinese Pharmacopoeia (2010 edition), approved by the Chinese Pharmacopoeia Commission. The details of experimental procedures are described in [10].

The goal-directed decision-making task and apparatus are shown in Fig. 1. Pigeons perform the task in a plus maze, which includes a waiting area, and three reward arms. If pigeons make behavior decisions (turn left, turn right, or move straight) according to the LED instruction correctly, they will receive food rewards. Rewards will not be provided if the trial is performed incorrectly. Pigeons must start from the waiting area to the reward arms and go back to the waiting area to start the next trial. Four infrared sensors are located in the arms to intercept the task-related data, which will be used to decode the behavior outcomes.

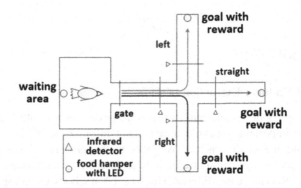

Fig. 1. The goal-directed decision-making task and apparatus.

2.2 Surgery, Data Acquisition, and Processing

Electrode implantation surgeries are carried out after pigeons perform reliably, in which behavior correct rate is more than 90% on two consecutive days. The head of the pigeon is placed in a stereotaxic holder under general anesthesia. Sixteen-channel nickel-chromium microwire array (Hong Kong Plexon Inc., Hong Kong, China) is chronically implanted in the NCL area (AP 5.5 mm; ML 7.5 mm; DV 0–3.0 mm) according to the atlas. LFP signals are recorded after 5–7 days of recovery using Cerebus^{TM} recording system (Blackrock Microsystem Inc., Salt Lake City, USA) by Butterworth low-pass filter from 0 to 250 Hz. The amplification is 4000 with 2 kHz sampling rate. The preprocessing of LFPs includes removing the baseline wander, 50 Hz power frequency interference and spatially correlated artifacts [11], and downsampling (1 kHz). In this study, we only analyze task-related LFPs data of correct trials. All data analyses are performed using MATLAB R2014a (The MathWorks Inc., Natick, USA).

Considering the ability to represent local spectral and temporal information from a non-stationary signal, WT based time-frequency representation method is used in this paper. CWT is employed to analyze non-stationary signals by choosing an arbitrary band weight and center frequency in wavelet [12], thus transient features can be captured and localized in both time and frequency information accurately [13].

For a continuous time signal $y(t)$, the wavelet transform is given as follows:

$$W(a,b) = \frac{1}{\sqrt{a}} \int_{-\infty}^{+\infty} y(t)\psi^* \left(\frac{t-b}{a}\right) dt \tag{1}$$

where a is the scaling parameter and b is its translation parameter, $\psi^*(t)$ denotes the complex conjugate of the analyzing wavelet function $\psi(t)$, in which the popular complex Morlet wavelet basis is chosen and it is defined as

$$\psi(t) = \frac{1}{\sqrt{\pi f_b}} e^{2\pi i f_c t e^{-t^2/f_b}} \tag{2}$$

where $f_b = 25$ and $f_c = 1$.

2.3 Textural Features Extraction of TFIs

Textures are a series of visual features that reflects the intrinsic attributes of images, including some important structural arrangement information of the object and the relationship with the environment. TA refers to the process of extracting textural feature parameters through certain technology to describe such attributes quantitatively or qualitatively. Recent years, different kinds of TA methods including the statistics-based method, the signal processing based method, the model-based method, and the structure-based method have been proposed. Among them, the statistics based one is most widely used. In this paper, seven methods are used to obtain 74 features from the TFIs. Details of them are described elsewhere [14, 15], and we show them in Table 1.

2.4 Random Forest Decoding Algorithm

Random forest (RF) is an ensemble learning method with the decision tree as a basic unit for classification, regression and other tasks [16]. Voting mechanisms helps to improve the performance of decision trees and overcome overfitting. It is relatively simple to implement and its operational efficiency and accuracy are relatively high, so we choose it as the decoding algorithm in this study.

Table 1. Textural features used in our experiments.

Method (index)	Extracted textural features
GLH (1–11)	Gray span, mean, variance, skewness, kurtosis, fifth center moment, sixth center moment, energy, entropy, smoothness, consistency
GLCM (12–19)	Contrast, correlation, energy, entropy, homogeneity, sum average, inertia, inverse difference moment
GRLM (20–24)	Run length, run length ratio, short run emphasis, long run emphasis, gray level distribution
GGCM (25–39)	Small gradient strengths, large gradient strengths, gray uneven representation, gradient uneven representation, energy, gray mean, gradient mean, gray mean square, gradient mean square, relevance, gray entropy, gradient entropy, mixing entropy, inertia, inverse gap
Tamura (40–43)	Coarseness, contrast, directionality, linelikeness
IM (44–50)	$M_1, M_2, M_3, M_4, M_5, M_6, M_7$
LJbG (51–74)	Gray mean, standard deviation, skewness, and kurtosis of six different jet transform (original space and the other five directions including x, y, xx, yy, xy)

3 Results

We train six pigeons (P080, P081, P085, P086, P087, and P089) to decode their behavior outcomes in this study, the number of electrodes, electrode types, number of channels, and number of trials of per pigeon are shown in Table 2. In all of 16 channels of P087 and P089, there is one confirmed to be a bad channel.

Table 2. Descriptions for the six pigeons.

Pigeon ID	Electrode type	Number of electrodes/channels	Number of trails
P080	4 × 4	16	82
P081	4 × 4	16	118
P085	4 × 4	16	44
P086	4 × 4	16	134
P087	4 × 4	15	210
P089	4 × 4	15	196

In this section, we analyze the effect of the textural features from the different frequency bands of single-channel LFP signals, different numbers of channels, different types of features and classifier parameters on the decoding performance. The results are compared and discussed below. Figure 2 shows the LFP signals and time-frequency images examples of different behaviors (left, right and straight).

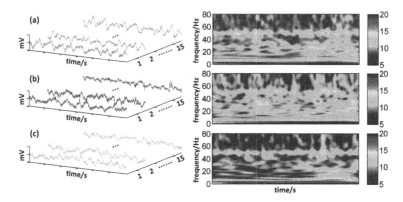

Fig. 2. LFP signals and time-frequency image examples of left (a), right (b) and (c) straight behavior.

According to the TFIs of all six pigeons, the results show that there are certain time-frequency characteristics differences in different frequency bands under different behavioral states, and texture features based on TFIs can be used to quantify these differences.

3.1 Decoding Performance of Textural Features Based on LFP Signals in the Different Frequency Bands

In order to investigate the effect of the different frequency bands of single-channel LFP signals on the decoding performance, the original LFP signals containing 0–80 Hz are segmented to eight sub-bands, i.e., 0–10 Hz, 10–20 Hz, ..., 70–80 Hz. Figure 3 depicts the decoding performance heatmap of single-channel LFP signals (in rows) and different frequency bands (in columns) using TFIs textural features. Colorbar depicts the accuracy: as the accuracy becomes higher, it appears to turn red and conversely turn blue.

It can be observed from Fig. 3 that for all six pigeons, the decoding accuracy of any channel in any band is higher than 45.00% and also higher than that of random

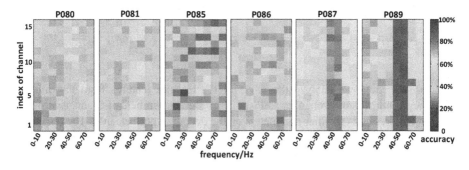

Fig. 3. Decoding performances of single-channel LFP signals in the different frequency band of six pigeons. (Color figure online)

accuracy (33.33% in this study). Even the accuracies of some channels are higher than 88.00%, such as channel 7 and channel 10 of P089. More importantly, we find that the LFP signal in the frequency band of 40–60 Hz corresponds to the best behavior decoding performance.

3.2 Effect of the Number of Channels on Decoding Performance

Figure 4 shows the relationship between decoding accuracy and the number of channels. The decoding accuracy curves in Fig. 4(a) show the mean values and standard deviations of 10 repeated experiments in which a specific number of channels are selected randomly from all of the channels. That is, for a given number of channels, we randomly select them from all channels to calculate the decoding accuracy for 10 times, and then their average and standard deviation are calculated. The results show that with the increase of the number of channels, the decoding accuracy of the LFP signal does not show a significant increasing trend. For different channels, the features of LFP signals can achieve good decoding performance, which indicates that there may be information redundancy among the LFP signals of different channels.

On this basis, the decoding accuracies of single-channel and all channels LFP signals are compared using textural features as decoding features. The results are shown in Fig. 4(b) and it can be observed that there are no significant differences between them ($p > 0.05$, the rank-sum test is used). It indicates that the effect of the number of channels to decoding performance of LFP signals is relatively slight.

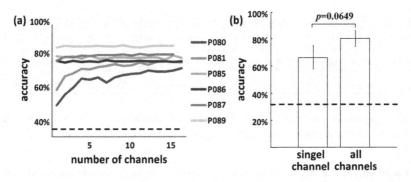

Fig. 4. Effect of the number of channels on decoding performance. (a) for the decoding accuracy curves with the number of channels; (b) for comparison of decoding accuracy between single channel and all channels. The dotted line represents the 33.33% random decoding accuracy.

In addition, as shown in Fig. 5, confusion matrices of decoding accuracies of textural features are further analyzed when all channels are used for all pigeons. The results show that the decoding performances of different behaviors of different pigeons are not consistent, and it indicates that the prediction ability of textural features for different behaviors show a certain degree of difference.

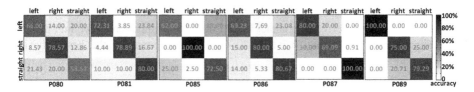

Fig. 5. Confusion matrices of decoding accuracies.

3.3 Decoding Performance of Different Types of Features

To compare the decoding performance of different types of texture features, normalized feature weights based on information gain [17] and decoding accuracy for a single feature are calculated. The results are shown in Fig. 6.

In Fig. 6, the normalized feature weights and decoding accuracy of two frequency bands are depicted by heatmap, in which colorbar depicts the accuracy: as the weight and accuracy become higher, it appears to turn red and conversely turn blue. Higher weights and accuracies correspond to the features having higher decoding performance. The numbers below the heatmap correspond to the index of the features. The results indicate that different texture features have different decoding performance, and it seems that some LJbG features (index 59 and 71) show the most prominent decoding performance. Some Tamura features (index 40), GLH feature (index 6) also show a certain degree of competitiveness.

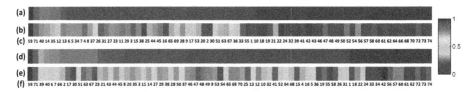

Fig. 6. Comparison of normalized feature weight (a, d), classification accuracy (b, e) and feature index (c, f) of different features belonging to band 40–50 (a, b, c) and 50–60 (d, e, f).

Furthermore, the decoding accuracy of every single feature is analyzed and the results are shown in Fig. 7.

The results show that for the features corresponding to the frequency band of 40–50 Hz and 50–60 Hz, the accuracies of most features are higher than random accuracy 33.33% (61 in 74 for 40–50 Hz, and 69 in 74 for 50–60 Hz). What is more, accuracies of some features are higher than 70% and even reach about 80.00% (index 59 and 71). It demonstrates that the effectiveness of TFIs-based texture features for decoding.

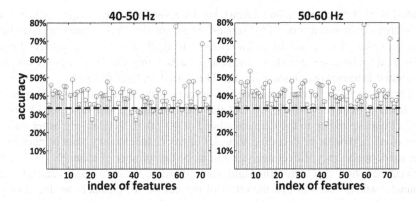

Fig. 7. Decoding performance of every single feature. The dotted lines represent the 33.33% random decoding accuracy.

3.4   Effect of Random Forest Parameter on Decoding Performance

The effect of the RF parameter, the number of trees, on the decoding performance is analyzed. Its initial value is set to 1 and the step size is set to 1. Decoding accuracy scatters distribution and polynomial fitting results are shown in Fig. 8.

Figure 8 shows that with the increase of the number of trees, the decoding accuracy increases first and then decreases. However, a significant decreasing trend is not observed. It indicates that the number of trees of RF also affects the decoding performance. In this study, it seems that the decoding performance can be guaranteed to be better when the number of trees is set between 6–8.

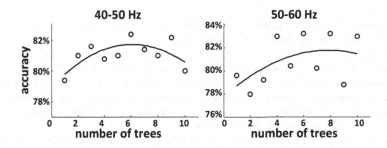

Fig. 8. Effect of the number of trees on decoding performance.

4   Conclusion

Textural features of TFIs are employed in this paper for neural information decoding. As a result, the features can realize the purpose of efficient decoding of behavior outcomes. In detail, with the pigeon as the model animal, we design a goal-directed decision-making task experiment based on the plus maze. LFP signals of six pigeons are acquired to be analyzed. To obtain the time-frequency representations, the CWT method is used by choosing an arbitrary band weight and center frequency in wavelet for analysis of non-stationary LFP signals. Then, for the TFIs, texture analysis based feature extraction is employed and a series of textural features are extracted. Finally, RF is used for behavior decoding. The results further strengthen the evidence decoding the animal behavior outcomes using the TFIs based textural features.

For the decoding performance results of textural features based on LFP signals in the different frequency bands, we find that the decoding performance of LFP signal features corresponding to the frequency band of 40–60 Hz is relatively better than any other bands, although the decoding accuracy of any channel in any band is much higher than random accuracy. Effect analysis results of the number of channels on decoding performance show that there is no significant difference between single channel and all channels, which indicates that the effect of the number of channels on decoding performance is relatively slight and LFP signals of different channels may contain much redundant information. In addition, we also analyze the decoding performance of different types of features. The results indicate that some LJbG features seem to contribute most to the decoding performance in this study. For the effect of random forest parameter on decoding performance, we investigate the relationship with the decoding performance and the number of the trees. It seems that the best decoding performance is obtained when it is set between 6–8. All of the above results will help to reveal the brain neuronal information processing mechanism of pigeon further and provide valuable references for the research of brain-computer interface.

Acknowledgment. The work is supported by the National Natural Science Foundation of China (61673353, U1304602).

References

1. Horikawa, T., Tamaki, M., Miyawaki, Y., Kamitani, Y.: Neural decoding of visual imagery during sleep. Science **340**(6132), 639–642 (2013)
2. Lu, H., Yang, S., Lin, L., Li, B., Wei, H.: Prediction of rat behavior outcomes in memory tasks using functional connections among neurons. PLoS ONE **8**(9), e74298 (2013)
3. Milekovic, T., Truccolo, W., Grun, S., Riehle, A., Brochier, T.: Local field potentials in primate motor cortex encode grasp kinetic parameters. Neuroimage **114**, 338–355 (2015)
4. Camara, C., et al.: Resting tremor classification and detection in Parkinson's disease patients. Biomed. Signal Process. **16**, 88–97 (2015)
5. Wu, Y., Zhang, L.: ECG Classification using ICA features and support vector machines. In: Lu, B.-L., Zhang, L., Kwok, J. (eds.) ICONIP 2011. LNCS, vol. 7062, pp. 146–154. Springer, Heidelberg (2011). https://doi.org/10.1007/978-3-642-24955-6_18
6. Dong, Y., Shang, Z., Li, M., Liu, X., Wan, H.: Feature reconstruction of LFP signals based on PLSR in the neural information decoding study. In: 39th Annual International Conference of the IEEE Engineering in Medicine and Biology Society (EMBC), pp. 2936–2939. IEEE, Seogwipo (2017)
7. Alcin, O.F., Siuly, S., Bajaj, V., Guo, Y., Sengur, A., Zhang, Y.: Multi-category EEG signal classification developing time-frequency texture features based Fisher vector encoding method. Neurocomputing **218**, 251–258 (2016)

8. Li, Y., Cui, W., Luo, M., Li, K., Wang, L.: Epileptic seizure detection based on time-frequency images of EEG signals using Gaussian mixture model and gray level co-occurrence matrix features. Int. J. Neur. Syst. **28**(07), 1850003 (2018)
9. Yuan, Q., Zhou, W., Xu, F., Leng, Y., Wei, D.: Epileptic EEG identification via LBP operators on wavelet coefficients. Int. J. Neur. Syst. **28**(08), 1850010 (2018)
10. Liu, X., Wan, H., Li, S., Chen, Y., Shi, L.: Adaptive common average reference for in vivo multichannel local field potentials. Biomed. Eng. Lett. **7**(1), 7–15 (2017)
11. Liu, X., Wan, H., Li, S., Shang, Z., Shi, L.: The role of nidopallium caudolaterale in the goal-directed behavior of pigeons. Behav. Brain Res. **326**, 112–120 (2017)
12. Kiymik, M.K., Guler, I., Dizibuyuk, A., Akin, M.: Comparison of STFT and wavelet transform methods in determining epileptic seizure activity in EEG signals for real-time application. Comput. Biol. Med. **35**(7), 603–616 (2005)
13. Addison, P.S., Walker, J., Guido, R.C.: Time-frequency analysis of biosignals a wavelet transform overview. IEEE Eng. Med. Biol. **28**(5), 14–29 (2009)
14. Li, M., Shang, Z., Dong, Y., Zhang, Y., Li, Y.: Application of MRI texture analysis in the study of the posterior fossa tumors growing trend in children. In: 39th Annual International Conference of the IEEE Engineering in Medicine and Biology Society (EMBC), pp. 620–623. IEEE, Seogwipo (2017)
15. Chitalia, R.D., Kontos, D.: Role of texture analysis in breast MRI as a cancer biomarker: a review. J. Magn. Reson. Imaging **49**(4), 927–938 (2019)
16. Breiman, L.: Random forests. Mach. Learn. **45**, 5–32 (2001)
17. Zdravevski, E., Lameski, P., Kulakov, A., Jakimovski, B., Filiposka, S., Trajanov, D.: Feature ranking based on information gain for large classification problems with MapReduce. In: 14th IEEE International Conference on Trust, Security and Privacy in Computing and Communications, pp. 186–191. IEEE, Helsinki (2015)

Convolutional Grid Long Short-Term Memory Recurrent Neural Network for Automatic Speech Recognition

Jiabin Xue, Tieran Zheng, and Jiqing Han(✉)

School of Computer Science and Technology,
Harbin Institute of Technology, Harbin, China
{xuejiabin,zhengtieran,jqhan}@hit.edu.cn

Abstract. The Grid Long Short-Term Memory (Grid-LSTM), which is consisted of three steps, i.e., two-dimensional grid splitting, local feature projection, and grid sequence modeling, has been widely used in Automatic Speech Recognition (ASR) tasks, since it has a strong time-frequency modeling ability. However, the network suffers from a serious problem that heavy computing time is always required. It can be found that the reason for this problem is in the last step, two cross-working LSTMs are employed to model time-frequency features in the grid via an analysis of its process. Thus, we try to speed up the Grid-LSTM by using a smaller grid and propose two enhanced Grid-LSTM models, i.e., Convolutional Grid-LSTM (ConvGrid-LSTM) and Multichannel ConvGrid-LSTM (MCConvGrid-LSTM) to reduce the grid size from the two dimensions of the Grid-LSTM respectively. In the frequency axis, we try to do this by using a large frequency stride and further to prevent performance loss by embedding a CNN in the Grid-LSTM. Moreover, in the time axis, we model several adjacent frames by the multichannel processing ability of CNN. Our method achieves 54% relative reduction of training time and 19% relative reduction of Word Error Rate (WER) for a character level End-to-End ASR task.

Keywords: Automatic Speech Recognition · Grid-LSTM · Convolutional Neural Network

1 Introduction

Recently, Grid-LSTM [5] is introduced, which models variations in the spectrogram by adopting two Long Short-Term Memories (LSTMs) [4] to unroll along two dimensions of the spectrogram, e.g., time and frequency, respectively. In general, we can divide the process of Grid-LSTM into three steps, i.e., two-dimensional grid splitting, local feature projection, and grid sequence modeling. In the first step, the two-dimensional grid splitting is achieved by using a frequency sliding window to divide the spectrum into multiple time-frequency blocks in each time frame of the spectrogram. In the second step, a projection

© Springer Nature Switzerland AG 2019
T. Gedeon et al. (Eds.): ICONIP 2019, CCIS 1143, pp. 718–726, 2019.
https://doi.org/10.1007/978-3-030-36802-9_76

matrix is employed to model local patterns for each time-frequency block within the grid. Finally, we use the grid sequence modeling to extract global time-frequency features of the spectrogram, which adopts two cross-working LSTMs in two dimensions, respectively. In the third step, the Grid-LSTM can memorize the historical time-frequency information of the processed blocks, which is helpful to extract features of the current block. As a result, the Grid-LSTM outperforms the Convolutional Neural Network (CNN) [1] on frequency modeling, and some research works on replacing the CNN with the Grid-LSTM in the Convolutional Long Short-Term Memory Deep Neural Networks (CLDNNs) [13] architecture have shown performance improvement over many ASR tasks [11].

However, it brings some good things, and some terrible problems arrive as well. One of these problems is that it is very difficult to be applied to the tasks which have to deal with massive data since it requires too much computing time. Thus, how to speed up the Grid-LSTM calculation process has become an very important research task and has attracted the attention of many researchers.

In order to cover the problem of computation speed, several works had been conducted [6,14]. In [14], a PyraMiD-LSTM method took in the LSTM state from the related time-frequency block of the previous time frame rather than from the previous time-frequency block of the current time frame. In this way, the time-frequency blocks in each time frame can be computed in parallel. Furthermore, a frequency block Grid-LSTM (fbGrid-LSTM) model had been used in [6], which split the spectrogram into multiple smaller sub-bands and handled these sub-bands in parallel.

In the above improved Grid-LSTM models, it has been shown that the fbGrid-LSTM is the best parallel speedup method and the PyraMiD-LSTM is inferior to the fbGrid-LSTM [6]. However, the performance of the Grid-LSTM may degrade since it has some shortcomings as follows. First, it is difficult to determine the divided number of sub-bands as there are no clear criteria, which has to be determined by a lot of experiments. Then, since the frame-skipping [9] is first used in this method to stack several adjacent frames into one vector, and then they split it into some sub-bands, thus it is difficult to distinguish whether each sub-band is independent of each other. Meanwhile, it also ignores the relationship between different sub-bands, which may be hurt the performance of this network.

In this paper, we try to speed up the computation process of this model while without sacrificing recognition accuracy. We find that the reason for slow computation speed is that it adopts two LSTMs to model the grid sequence between time-frequency blocks within the grid. Thus, in order to achieve the aim of speedup, we try to use a smaller grid. After referring to many methods of speedup computation [6,9,14], we focus on the following two aspects. On one hand, we reduce the size of the grid from the frequency axis and propose a new model, i.e., Convolutional Grid-LSTM (ConvGrid-LSTM), which compensates for performance loss for large stride by embedding the CNN into the Grid-LSTM. On the other hand, we further speed up our proposed model by reducing the grid size in the time dimension and propose a strengthened ConvGrid-LSTM, which is named Multichannel Convolutional Grid-LSTM (MCConvGrid-LSTM).

2   Grid-LSTM for ASR

2.1   Convolutional Grid-LSTM

The Grid-LSTM suffers from heavy computation cost in the usage process. It can be found that the first two steps in the process of the Grid-LSTM can be processed in parallel, and most of the calculations are concentrated in the third step. Therefore, we try to reduce the number of calculations for the third step. Furthermore, the number of calculations in the third step depends on the size of the grid, which is because of two LSTMs in each block are cross-work and it only enables to sequential execution. Thus we propose to achieve acceleration by reducing the grid size. According to the difference between the Grid-LSTM and LSTM, we can do this by increasing the frequency stride directly.

However, the frequency modeling ability of the Grid-LSTM will decrease as the frequency stride increasing. To overcome the problem, we first analyze the reasons for this phenomenon. In the process of the Grid-LSTM, it uses a lot of double computation for the same frequency to model local patterns on the blocks while using a small frequency stride. When we use large ones, it only computes fewer times for every time-frequency block in the grid. It is unable to model local patterns in the blocks efficiently. Furthermore, we can find that the Grid-LSTM extracts the information in each block by using a projection matrix, which is the fundamental reason for it cannot fully model and utilize the information within each block, since it is more and more difficult to model the pattern of blocks as the frequency stride increase. With that being stated, we try to reduce the performance loss with large frequency stride by improving the ability of Grid-LSTM to extract the subtle changes in the local features. Many studies have shown that CNN has strong local pattern modeling ability [2,10,12]. Thus, we try to strengthen the local frequency feature modeling capability of Grid-LSTM with a larger frequency stride by using the CNN instead of the projection operation within it and propose a new model, i.e., Convolutional Grid-LSTM (ConvGrid-LSTM). It has strong local information modeling abilities like CNN. Meanwhile, because multiple time-frequency blocks are computed in parallel, so the training and reasoning processes are sped up. Thus the proposed model reduces the amount of computation under the condition of large frequency stride.

The structure of ConvGrid-LSTM is described in Fig. 1. In the above part of this figure, we showed the process of the general Grid-LSTM, where the gt_i is the output of the ith gT-LSTM and gf_k is the output of the kth gF-LSTM, and then we display the local feature projection, which is used for ConvGrid-LSTM, in the red rectangular. It can be seen that in the local feature projection module of the ConvGrid-LSTM, we use a CNN based module instead of the projection matrix which is employed in the Grid-LSTM. And then, this CNN based module can be divided into the following three steps. First, when a time-frequency block input arrives, a convolution operation is performed for it to extract the local frequency

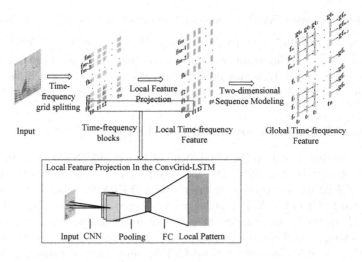

Fig. 1. The process of the ConvGrid-LSTM

feature. Next, we use a pooling operation along the frequency axis to enhance the translation invariant property in the frequency domain feature. Finally, the pooling result is changed to the specified dimension local time-frequency feature by using a fully connected layer.

In the above operations, convolution, pooling and fully connected operations can be executed in parallel within each time frame. Specifically, at each time-frequency step $s \in \{t, k\}$, the ConvGrid-LSTM will given by:

$$\mathbf{F}_{u,t,k}^{(s)} = \mathbf{x}_{t,k} \otimes \mathbf{W}_{uF}^{s}, u \in \{i, f, c, o\}, \tag{1}$$

$$\mathbf{P}_{u,t,k}^{(s)} = MeanPooling(\mathbf{F}_{u,t,k}), u \in \{i, f, c, o\}, \tag{2}$$

$$\mathbf{L}_{u,t,k}^{(s)} = \mathbf{P}_{u,t,k} * \mathbf{W}_{L}^{u} + b, u \in \{i, f, c, o\}, \tag{3}$$

$$\mathbf{q}_{u,t,k} = \mathbf{W}_{um}^{(t)} \mathbf{m}_{t-1,k}^{(t)} + \mathbf{W}_{um}^{(k)} \mathbf{m}_{t,k-1}^{(k)}, u \in \{i, f, c, o\}, \tag{4}$$

$$\mathbf{i}_{t,k}^{(s)} = \sigma(\mathbf{L}_{i,t,k}^{(s)} + \mathbf{q}_{i,t,k} + b_i^{(s)}), \tag{5}$$

$$\mathbf{f}_{t,k}^{(s)} = \sigma(\mathbf{L}_{f,t,k}^{(s)} + \mathbf{q}_{f,t,k} + b_f^{(s)}), \tag{6}$$

$$\mathbf{c}_{t,k}^{(t)} = \mathbf{f}_{t,k}^{(t)} \odot \mathbf{c}_{t-1,k}^{(t)} + \mathbf{i}_{t,k}^{(t)} \odot g(\mathbf{L}_{c,t,k}^{(t)} + \mathbf{q}_{c,t,k} + b_c^{(t)}), \tag{7}$$

$$\mathbf{c}_{t,k}^{(k)} = \mathbf{f}_{t,k}^{(k)} \odot \mathbf{c}_{t,k-1}^{(k)} + \mathbf{i}_{t,k}^{(k)} \odot g(\mathbf{L}_{c,t,k}^{(k)} + \mathbf{q}_{c,t,k} + b_c^{(k)}), \tag{8}$$

$$\mathbf{o}_{t,k}^{(s)} = \sigma(\mathbf{L}_{o,t,k}^{(s)} + \mathbf{q}_{o,t,k} + b_o^{(s)}), \tag{9}$$

$$\mathbf{m}_{t,k}^{(s)} = \mathbf{o}_{t,k}^{(s)} \odot h(\mathbf{c}_{t,k}^{(s)}), \tag{10}$$

where $\mathbf{i}_{t,k}^{(\cdot)}$, $\mathbf{f}_{t,k}^{(\cdot)}$, $\mathbf{c}_{t,k}^{(\cdot)}$ and $\mathbf{o}_{t,k}^{(\cdot)}$ denote the input, forget, memory cell and output gate activations at time-frequency step t, k respectively. $\mathbf{m}_{t,k}^{(\cdot)}$ is the output of the Grid-LSTM layer; $\mathbf{W}_{*F}^{(\cdot)}$ is the weight matrix of CNN, $\mathbf{F}_{*,t,k}^{(\cdot)}$ is the result of convolution of input time-frequency $\mathbf{x}_{t,k}$, $\mathbf{W}_{*L}^{(\cdot)}$ is the weight matrix of the linear layer, $\mathbf{L}_{*,t,k}^{(\cdot)}$ is the result of a linear layer, b is the bias of the linear layer. $\otimes$ denotes the convolution and pooling operation. The σ is a logistic sigmoid activation, and g and h are the activations of cell input and output with $\tan h$.

2.2 Multichannel Convolutional Grid-LSTM

We try to further speed up the computing of ConvGrid-LSTM by reducing the grid size along the time axis. The most common acceleration method in the training of ASR models is the Low Frame Rate (LFR) [9], which has shown that the sub-word unit in the ASR is at least several continue adjacent speech frames and speed up the ConvGrid-LSTM through down-sample the input sequence which stacking several continue adjacent speech frames into one vector. However, this method is not obvious for Grid-LSTM, since that it cannot reduce the number of time-frequency blocks within the grid.

It is well known that the adjacent speech frames have a timing correlation. Thereby, we try to fully use the correlation between adjacent frames via the capability of fusion processing multichannel information in the CNN, and propose an enhanced ConvGrid-LSTM, namely Multichannel ConvGrid-LSTM (MCConvGrid-LSTM), to further speed up the whole process of ConvGrid-LSTM by reducing the number of unrolling steps along the time axis of the grid. In order to consider the relationship between different frames while processing multiple frames in parallel, we try to explore the performance of running ConvGrid-LSTM in a smaller multichannel spectrogram. We provide a method for ConvGrid-LSTM to split the input spectrogram into a smaller multichannel spectrogram. Specifically, we split the input spectrogram $\mathbf{X}$, which is stacked via the three adjacent frames, into three smaller ones $\{\mathbf{X}_0^s, \mathbf{X}_1^s, \mathbf{X}_2^s\}$ at first step. Then, we concatenate all the spectrograms to form a multichannel spectrogram $\mathbf{X}^n$ to fully utilize the relation between different frames by using the multichannel data modeling ability of CNN. Finally, we process the $\mathbf{X}^n$ by a ConvGrid-LSTM. And at each time-frequency step $s \in \{t, k\}$, (1) is replaced by:

$$\mathbf{F}_{u,t,k}^{(s)} = \mathbf{X}_{t,k}^n \otimes \mathbf{W}_{uNF}^s, u \in \{i, f, c, o\}, \tag{11}$$

where $\mathbf{W}_{uNF}^s$ is the weight matrix of the multichannel CNN, the $\mathbf{X}_{t,k}^n$ is the multichannel time-frequency block at the tth time step and kth frequency step.

Compared with the ConvGrid-LSTM, MCConvGrid-LSTM changes the input spectrogram to a smaller multichannel spectrogram and from one input channel of ConvGrid-LSTM to three input channels. Thus, it can move fewer steps along the frequency axis and avoid the occurrence of aberrant patterns at the connection between frames.

3 Experimental Details

3.1 Experiment Data

Our experiments are conducted on the LibriSpeech corpus [7], which consists of almost 1000 h of training data. We select all 960 h training data from this corpus and use the dev-clean and the test-clean folder for development set and test set, respectively. Data preparation is implemented like Eesen [3]. The acoustic feature is 80-dimensional log-mel filter-bank energies, computed using a 25 ms window every 10 ms, which is extracted using Kaldi [8]. The language model used in this experiment is a 4-gram model that is trained using the text label of the corpus. Similar to the low frame rate (LFR) model [9], at the current frame, these features are stacked with 2 frames to the left and downsampled at a 30 ms frame rate, producing a 240-dimensional feature vector.

3.2 Results

The experimental results of different kinds of model structures involved in this paper are described in Table 1. The top part of Table 1 is for the ConvGrid-LSTMs and the lower part of this table is for the MCConvGrid-LSTMs. Where F is the frequency sliding window of Grid-LSTM, S is the frequency stride and *Share* refers to whether the weight of the CNN is shared among all time-frequency blocks. The CNN in ConvGrid-LSTM (MCConvGrid-LSTM) consists of 1 (3) input channel(s) and 32 (32) output channels and the filter size of both CNN is 1×3, which is the only convolution operation along the frequency axis. The number of linear layer units is 64 for both ConvGrid-LSTM and MCConvGrid-LSTM model.

Table 1. Experimental results for models

Model	Type of LSTM	F	S	*Share*	Time (h)	WER(test)(%)
GL_2	Grid-LSTM	16	2	Y	135.2	5.1
GL_8	Grid-LSTM	16	2	Y	71.0	5.5
GL_{16}	Grid-LSTM	16	2	Y	40.0	5.9
CGL_8	ConvGrid-LSTM	16	8	Y	72.4	5.0
CGL_{16}	ConvGrid-LSTM	16	16	Y	41.1	4.8
fGL_2	fbGrid-LSTM	3	2	Y	42.7	5.3
fGL_8	fbGrid-LSTM	3	8	Y	29.6	5.6
fGL_{16}	fbGrid-LSTM	3	16	Y	19.1	6.3
$MCCGL_8$	MCConvGrid-LSTM	3	8	Y	30.8	4.9
$MCCGL_{16}$	MCConvGrid-LSTM	3	16	Y	19.5	4.5
$NMCCGL_8$	MCConvGrid-LSTM	3	8	N	30.8	4.6
$NMCCGL_{16}$	MCConvGrid-LSTM	3	16	N	19.5	**4.3**

WER with Different Frequency Strides. We explore the performance of some kinds of Grid-LSTM variations with various frequency strides. Word Error Rate (WER) of the variations are given in Table 1.

From the top part of this table, by comparing the GL_2, GL_8 and GL_{16}, we find that the performance of Grid-LSTM is degraded greatly under the conditions of large frequency strides. By comparing the GL_2, CGL_8 and CGL_{16}, we find that the proposed ConvGrid-LSTM has almost no performance degradation under the conditions of large frequency strides. Then, we compare the WERs of all parallel processing models in the lower part of this table, which include fGL_2, fGL_8 and fGL_{16}, and find that the best result of fbGrid-LSTMs is 5.3% which is achieved by the fGL_2, while the performance decreases with the increase of the frequency stride. After that, by comparing the fGL_2, $MCCGL_8$ and $MCCGL_{16}$, we find that the performance of the MCConvGrid-LSTM has barely improved when the frequency stride S increases.

WER with/without Sharing Weights of CNN. We explore whether the sharing of CNN parameters across all time-frequency blocks affects network performance. From the lower part of Table 1 we can see that the $NfCGL$ with non-shared parameters has achieved the best performance. The best performance is obtained with $NMCCGL_{16}$. It is relatively reduced by 19% in WER comparing with fGL_2.

The results fully prove our hypothesis that the reason for the performance decrease of both Grid-LSTM and fbGrid-LSTM with a large frequency stride is that the large frequency stride hurts the local frequency feature modeling capability of the Grid-LSTM.

Training Time of Different Models. The comparison of the training time of different Grid-LSTM can be also obtained from this table. We train all models by a Tesla K80 GPU and an Intel Xeon E5-2667 CPU, the training of GL_2, CGL_8, CGL_{16}, $MCCGL_8$, $MCCGL_{16}$, $NMCCGL_8$ and $NMCCGL_{16}$ by only one GPU, the parallel training of other models by three GPUs, the other devices are same for all models.

First, we investigate the training time of the models that are spread over the entire spectrum. We compare the GL_2, CGL_8 and CGL_{16} models and find that the proposed CGL_{16} model achieves a relative 70% reduction in training time per epoch. We also investigate the training time of models that process multiple sub-bands in parallel. By comparing the fGL and $MCCGL$ models, we find that under the same performance conditions, the training time of $MCCGL_{16}$ is reduced by 54% compared with fGL_2.

4   Conclusion

This paper presented two models to accelerate computing Grid-LSTM by reducing the size of the grid in the frequency and time dimensions. First, we proposed a ConvGrid-LSTM to model local frequency feature under large frequency stride accurately. Furthermore, an enhanced version of ConvGrid-LSTM named MCConvGrid-LSTM was put forward. In this model, the ConvGrid-LSTM is run on a small multichannel spectrogram and in this manner, the computing time is fairly decreased. Finally, the experiments were conducted by using a character-level CTC-based ASR task on the LibriSpeech corpus and the results showed that the MCConvGrid-LSTM achieved 54% relative reduction of training time cost and 19% relative reduction of WER.

Acknowledgements. This research was supported by the National Key Research and Development Program of China under Grant 2017YFB1002102 and National Natural Science Foundation of China under Grant U1736210.

References

1. Abdel-Hamid, O., Mohamed, A.R., Jiang, H., Deng, L., Penn, G., Yu, D.: Convolutional neural networks for speech recognition. IEEE/ACM Trans. Audio Speech Lang. Process. **22**(10), 1533–1545 (2014)
2. Abdel-Hamid, O., Mohamed, A., Jiang, H., Penn, G.: Applying convolutional neural networks concepts to hybrid NN-HMM model for speech recognition. In: IEEE International Conference on Acoustics, Speech and Signal Processing, ICASSP, pp. 4277–4280 (2012)
3. Graves, A., Jaitly, N., Mohamed, A.: EESEN: end-to-end speech recognition using deep RNN models and WFST-based decoding. In: IEEE Workshop on Automatic Speech Recognition and Understanding, ASRU, pp. 1–4 (2015)
4. Hochreiter, S., Schmidhuber, J.: Long short-term memory. Neural Comput. **9**(8), 1735–1780 (1997)
5. Kalchbrenner, N., Danihelka, I., Graves, A.: Grid long short-term memory. In: International Conference of Learning Representation, ICLR, pp. 1–15. Open Publishing (2016)
6. Li, B., Sainath, T.N.: Reducing the computational complexity of two-dimensional LSTMs. In: INTERSPEECH, pp. 964–968 (2017)
7. Panayotov, V., Chen, G., Povey, D., Khudanpur, S.: Librispeech: an ASR corpus based on public domain audio books. In: IEEE International Conference on Acoustics, Speech and Signal Processing, ICASSP, pp. 5206–5210 (2015)
8. Povey, D., et al.: The Kaldi speech recognition toolkit. In: IEEE Workshop on Automatic Speech Recognition and Understanding, ASRU, pp. 1–4 (2011)
9. Pundak, G., Sainath, T.N.: Lower frame rate neural network acoustic models. In: INTERSPEECH, pp. 22–26 (2016)
10. Sainath, T.N., et al.: Improvements to deep convolutional neural networks for LVCSR. In: IEEE Workshop on Automatic Speech Recognition and Understanding, ASRU, pp. 315–320 (2013)
11. Sainath, T.N., Li, B.: Modeling time-frequency patterns with LSTM vs. convolutional architectures for LVCSR tasks. In: INTERSPEECH, pp. 813–817 (2016)
12. Sainath, T.N., Mohamed, A., Kingsbury, B., Ramabhadran, B.: Deep convolutional neural networks for LVCSR. In: IEEE International Conference on Acoustics, Speech and Signal Processing, ICASSP, pp. 8614–8618 (2013)

13. Sainath, T.N., Vinyals, O., Senior, A.W., Sak, H.: Convolutional, long short-term memory, fully connected deep neural networks. In: IEEE International Conference on Acoustics, Speech and Signal Processing, ICASSP, pp. 4580–4584 (2015)
14. Stollenga, M.F., Byeon, W., Liwicki, M., Schmidhuber, J.: Parallel multi-dimensional lstm, with application to fast biomedical volumetric image segmentation. In: Advances in Neural Information Processing Systems NIPS, pp. 2998–3006 (2015)

Pattern Sequence Neural Network for Solar Power Forecasting

Yang Lin[1]([✉]), Irena Koprinska[1]([✉]), Mashud Rana[2], and Alicia Troncoso[3]

[1] School of Computer Science, University of Sydney, Sydney, NSW, Australia
ylin4015@uni.sydney.edu.au, irena.koprinska@sydney.edu.au
[2] Data61, CSIRO, Sydney, Australia
mdmashud.rana@data61.csiro.au
[3] Data Science and Big Data Lab, University Pablo de Olavide, Seville, Spain
atrolor@upo.es

Abstract. We propose a new approach for time series forecasting, called PSNN, which combines pattern sequences with neural networks. It is a general approach that can be used with different pattern sequence extraction algorithms. The main idea is to build a separate prediction model for each pattern sequence type. PSNN is applicable to multiple related time series. We demonstrate its effectiveness for predicting the solar power output for the next day using Australian data from three data sources - solar power, weather and weather forecast. In our case study, we show three instantiations of PSNN by employing the pattern sequence extraction algorithms PSF, PSF1 and PSF2. The results show that PSNN achieved the most accurate results.

Keywords: Solar power forecasting · Pattern sequence similarity · Neural networks

1 Introduction

The use of solar photovoltaic (PV) systems is rapidly growing due to their improved efficiency and continued cost reduction, and the advantages of solar energy. However, it is challenging to integrate large amounts of electricity produced by PV systems into the power grid and maintain a stable electricity supply. The produced solar power is highly variable as it depends on meteorological factors such as solar irradiance, ambient temperature, clouds, dust and wind. This necessitates the development of methods for accurate PV power prediction, to ensure reliable electricity supply of grid-connected PV systems.

Different approaches for PV power forecasting have been proposed, based on statistical methods such as linear regression and autoregressive moving average [1] and machine learning methods such as Neural Networks (NNs) [1,2] and support vector regression [3,4]. Recently, the application of Pattern Sequence-based Forecasting (PSF) [5] methods has been studied for solar power forecasting in [6,7], showing promising results. PSF assigns a cluster label to each day and

© Springer Nature Switzerland AG 2019
T. Gedeon et al. (Eds.): ICONIP 2019, CCIS 1143, pp. 727–737, 2019.
https://doi.org/10.1007/978-3-030-36802-9_77

then uses a nearest neighbour approach to find similar sequences of days to the target sequence and make a prediction for the new day. One of PSF's distinct characteristics is that it predicts all values for the next day simultaneously (e.g. all half-hourly PV values for the next day), as opposed to predicting them iteratively or building a separate prediction model for each value, as the majority of the other prediction methods.

While the standard PSF algorithm uses only one time series (the time series of interest, e.g. PV data), two PSF extensions utilizing data from multiple related time series (e.g. PV, weather and weather forecast data) have been proposed in [6] and evaluated for solar power forecasting. The results showed that both extensions PSF2 and PSF1 were more accurate than the standard PSF algorithm. However, there is still an opportunity for further improvement. To make the final prediction, the PSF algorithms simply take the average of the values of the relevant days from the matched sequences. In this paper we propose to better utilise the information from the matched sequences and build a classifier to produce the final prediction. In particular, we investigate if it is possible to combine the advantages of PSF and NNs and improve the performance. The contributions of this paper are as follows:

1. We propose a novel approach combining pattern sequences with NNs, called Pattern Sequence Neural Network (PSNN). It takes as an input a sequence of cluster labels, extracts pattern sequences of different types and builds a separate NN prediction model for each of them. It combines the efficient pattern sequence extraction and similarity matching of the PSF algorithms with the advantages of NNs for modelling complex and nonlinear relationships.
2. PSNN is a general approach that can be used with different clustering and cluster sequence extraction algorithms, and can be applied to multiple related time sequences. In our case study for solar power forecasting, we show three instantiations of the PSNN approach by employing the PSF, PSF1 and PSF2 algorithms.
3. We evaluate the performance of PSNN on a Australian dataset, which includes data from three sources (PV solar, weather and weather forecast), for two years. Our results show that PSNN was the most accurate method.

2 Data

As a case study, we consider the task of simultaneously predicting the PV power output for the next day at half-hourly intervals. Given: (1) a time series of PV power output up to day d: $PV = [PV_1, \ldots, PV_d]$, where PV_i is a vector of half-hourly PV power output for day i, (2) a time series of weather vectors for the same days: $W = [W_1, \ldots, W_d]$, where W_i is a weather vector for day i, and (3) a weather forecast vector for the next day $d+1$: WF_{d+1}, our goal is to forecast PV_{d+1}, the half-hourly PV power output for day $d+1$.

2.1 Data Sources and Feature Sets

We use PV and weather data for two years - from 1 January 2015 to 31 December 2016 (731 days). Table 1 summarizes the data sources and extracted feature sets.

Table 1. Data sources and feature sets

Data source	Feature set	Attribute information
PV data	$PV \in \Re^{731 \times 20}$	Daily: half-hourly solar power between 7am and 5pm
Weather data 1	$W1 \in \Re^{731 \times 14}$	(1–6) Daily: min and max temperature, rainfall, sunshine hours, max wind gust and average solar irradiance; (7–14) At 9am and 3pm: temperature, relative humidity, cloudiness and wind speed
Weather data 2	$W2 \in \Re^{731 \times 4}$	Daily: min and max temperature, rainfall and solar irradiance. W2 is a subset of W1.
Weather forecast data	$WF \in \Re^{366 \times 4}$	Daily: min and max temperature, rainfall and average solar irradiance

Solar PV Data. This data was collected from a rooftop PV plant located at the University of Queensland in Brisbane, Australia, and is available from http://www.uq.edu.au/solarenergy/.

Weather Data. The corresponding weather data was collected from the Australian Bureau of Meteorology, http://www.bom.gov.au/climate/data/, from a weather station close to the PV plant. There are three sets of weather features: W1, W2 and WF. W1 includes the full set of 14 weather features. W2 is a subset of W1 and includes only 4 features which are frequently used in weather forecasts and available from meteorological bureaus.

The weather forecast feature set WF is obtained by adding 20% Gaussian noise to the W2 data. This is done since the weather forecasts were not available retrospectively for previous years. When making predictions for the days from the test set, the WF set replaces W2 as the weather forecast for these days.

2.2 Data Preprocessing

The raw PV data was measured every 1 min and was aggregated to 30-min intervals by taking the average value of the interval. All data was normalised.

There was a small percentage of missing values (0.82% in the PV and 0.02% in the weather data) which were replaced by using a nearest neighbour method, applied firstly to the weather data and then to the PV data as in [6].

3   Pattern Sequence Forecasting Methods

3.1   PSF

PSF [5] is a forecasting method combining clustering and sequence matching. Consider the PV power time series data $PV = (PV_1, ..., PV_n)$, where PV_i is the D dimensional vector of the PV power output for day i, $PV \in \Re^{n \times D}$. PSF firstly employs the k-means algorithm to cluster all vectors PV_i from the training data into k_1 clusters and labels them as $C_1, ..., C_{k_1}$, see Fig. 1a.

To make a prediction for a new day $d+1$, PSF extracts a sequence of w consecutive days, starting from the previous day and going backwards. This sequence is defined in terms of cluster labels and called a *target sequence*. It then matches the target sequence with the previous days to extract a set of equal sequences ES, finds the post-sequence day for each of them and obtains the final prediction by averaging the PV vectors of these post-sequence days. For example, in Fig. 1a the PV power prediction for day $d+1$ is the average of the PV vectors for days 4 and 69.

The window size w and the number of clusters k_1 are hyperparameters of the PSF algorithm and are optimised using 12-fold cross validation.

3.2   PSF1

PSF1 [6] is an extension of PSF, utilising data from more than one source. While PSF uses a single data source for clustering and sequence matching (PV data for our case study), PSF1 uses additional data source - weather forecast for our case study. It firstly clusters the training set days based on the W2 data into k_2 clusters with labels $C_1, ..., C_{k_2}$, see Fig. 1b.

To make a prediction for a new day $d+1$, PSF1 obtains the cluster label for this day by using its weather forecast vector WF, comparing it with the cluster centroids and assigning it to the cluster of the closest centroid. It then extracts a target sequence of w consecutive days from day $d+1$ backwards and including $d+1$, matches this sequence with the previous days and finds a set of equal sequences ES. The final prediction is obtained by taking the average of the PV vectors of the last days in each ES. For example, in Fig. 1b the PV power prediction for day $d+1$ is the average of the PV vectors for days 4 and 68.

3.3   PSF2

PSF2 [6] is an extension of PSF1 using two additional data sources: weather data and weather forecast data. It clusters the days from the training set in two different ways: using the weather data (k_1 clusters with labels $C_1, ..., C_{k_2}$) and weather forecast data (k_2 clusters with labels $K_1, ..., K_{k_2}$), see Fig. 1c.

The prediction for the new day $d+1$ is computed using the following steps, see Fig. 1c. First, a target sequence of w consecutive days from day d backwards and including day d is extracted based on the weather data and matched to find the set of equal sequences ES. Second, the cluster label K_x for day $d+1$

is obtained based on the weather forecast data. Third, the cluster labels of the post-sequence days for all ES are checked and if they are not the same as K_x, these sequences are excluded from ES. The final prediction for $d+1$ is formed by taking the average of the post-sequence days for each ES. For example, in Fig. 1c the PV power prediction for day $d+1$ is the average of the PV vectors for days 4 and 69. Note that day 72 is not included in the final prediction as its cluster label is K_3, which is different than the cluster label K_2 of day $d+1$.

4   Pattern Sequence Neural Network

Table 2 summarizes the proposed PSNN approach. The key idea is to identify the matched sequences as in PSF, but then instead of taking the average of the relevant days as in PSF, learn the relationship between the previous and next day for the set of relevant days using an NN classifier.

Table 2. The proposed PSNN approach

Step	Description
1	Cluster the days and generate a sequence of cluster labels
2	Find pattern sequences relevant to the task
3	Generate a training set for each type of pattern sequence
4	Aggregate training sets if the number of examples is too small
5	Build a prediction model for each pattern sequence type and use it to make a prediction for the new day

4.1   Clustering and Pattern Sequence Extraction

The first step of the proposed PSNN approach involves clustering the days and generating a sequence of cluster labels, where each day is assigned a cluster label. This can be done by employing any clustering algorithm suitable for the task, using an appropriate feature vector representation for the day. For example, in our solar power case study, we use the k-means algorithm to cluster the days based on their PV vector (in the PSNN-PSF instantiation of PSNN), W2 weather data (in PSNN-PSF1) and both the W1 and W2 data (in PSNN-PSF2).

The second step involves extracting pattern sequences that are relevant for the task. This can be done in different ways, by employing different algorithms. In our case study we use the PSF, PSF1 and PSF2 algorithms.

4.2   Generating Training Sets and Building NN Prediction Models

The next steps of PSNN involve identifying the unique pattern sequence types, creating a training set and building a separate NN prediction model for each of them. As shown in Fig. 1, the NN model takes as an input the PV and W1

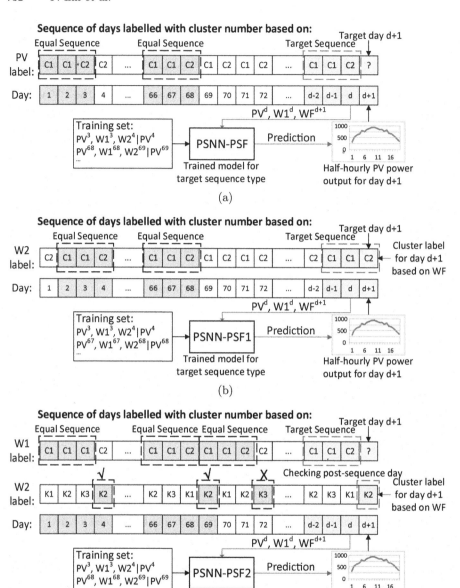

Fig. 1. The proposed PSNN approach when used in conjunction with PSF, PSF1 and PSF2 pattern sequence methods: (a) PSNN-PSF, (b) PSNN-PSF1, (c) PSNN-PSF2

features for the previous day, and the weather forecast (W2 features) for the next day, and predicts the PV data for the next day. Using information from all three data sources for the NN models was found to be beneficial in [6].

The number of unique sequences Si depends on the number of clusters and window size. For example, for PSF it is k_1^w, where k_1 is the number of clusters and w is the window size. If some pattern sequences appear less frequently, the training set for them may be too small, not allowing to build an accurate prediction model. To deal with this issue, we set a required minimum number of training examples L_{min}, expressed as a percentage of the total number of training examples, and aggregate the training sets of the pattern sequences not satisfying this condition to create a combined NN model for them.

Specifically, we apply the following aggregation rule. We identify all unique sequences Si with less than L_{min} training examples. We iteratively merge the training sets of the sequences with the same last cluster label, then the same second last cluster label and so on, until the L_{min} condition is satisfied. Thus, the aggregation rule considers the similarity between the unique pattern sequences, attempting to merge sets based on the similarity of the most recent days first.

As an example, Table 3 shows the application of the aggregation rule for PSF for $k_1 = 2$, $w = 2$ and $L_{min} = 20\%$. There are four initial sequences and two of them (S2 and S3) do not satisfy the L_{min} condition. By applying the aggregation rule, the training sets of the sequences with the same last cluster label will be aggregated, i.e. (1) S1 and S3, and (2) S2 and S4. The new training sets now satisfy the L_{min} condition, and hence two NN models will be trained - one for each of the aggregated sets.

Table 3. Aggregation of training sets - example

Initial sequences	Training set size	Aggregation rule	Final sequences
S1: C1-C1	N1: 127 (35%)	N1+N3 $\geq L_{min}\%\surd$	S1$_{new}$ = S1 or S3 = C1-C1 or C2-C1
S2: C1-C2	N2: 38 (11%) ×	N2+N4 $\geq L_{min}\%\surd$	S2$_{new}$ = S2 or S4 = C1-C2 or C2-C2
S3: C2-C1	N3: 38 (11%) ×		
S4: C2-C2	N4: 160 (44%)		

4.3 Prediction for New Days

To forecast the PV power for a new day $d + 1$, PSNN forms a target sequence depending on the pattern sequence algorithm used (PSF, PSF1 or PSF2 in our case study) and then employs the trained NN classifier for this sequence to make the prediction. For example, for PSNN-PSF1 (see Fig. 1b), the target sequence is formed by concatenating the WF cluster label of day $d+1$ with the W2 cluster labels of the previous $w-1$ days. For PSNN-PSF2 (see Fig. 1c), the target sequence is constructed by concatenating the WF cluster label of day $d+1$ with the W1 cluster labels of the previous $w-1$ days. The cluster label of $d+1$ is obtained by calculating the distance to the cluster centroids, for the clustering of the training data. Next, the pre-trained NN model for the specific target pattern sequence is used to predict the PV power for day $d+1$, taking as an input the PV and W1 vectors of day d and the WF vector of day $d+1$.

5   Experimental Setup

The PV power and corresponding weather data was split into two subsets: training and validation (the first year) and test (the second year).

Hyperparameter Tuning for PSF Models. For the PSF models and the PSF part of PSNN, the first year was used to determine the hyperparameters (number of clusters k_1 and k_2 and sequence size w) using 12-fold cross validation with grid search, consistent with the original PSF algorithm [6]. The grid search for w included values from 1 to 5. The best number of clusters was selected by varying k_1 and k_2 from 1 to 10 and evaluating three cluster quality indexes (Silhouette, Dunn and Davies-Bouldin) as described in [6].

The selected best hyperparameters were: PSF and PSF1: $k_1 = 2$, $w = 2$; PSF2: $k_1 = 2$, $k_2 = 2$, $w = 1$. The PSF parts of PSNN used the same hyperparameters as the corresponding PSF models.

Table 4. Best hyperparameters for the NN models

Model	Hidden neurons	Learning rate	L2 λ	Batch size	Epochs
NN	[25]	0.0005	0.0015	64	900
PSNN-PSF (1)	[35,25]	0.001	0.0008	64	900
PSNN-PSF (2)	[35,25]	0.001	0.001	256	500
PSNN-PSF1 (1)	[35,25]	0.003	0.0008	256	500
PSNN-PSF1 (2)	[35]	0.001	0.0015	256	900
PSNN-PSF2 (1)	[25]	0.0005	0.0008	64	500
PSNN-PSF2 (2)	[25]	0.001	0.0005	64	500
PSNN-PSF2 (3)	[30]	0.001	0.0005	256	900
PSNN-PSF2 (4)	[30]	0.0005	0.0015	64	900

Hyperparameter Tuning for NN Models. For the NN part of PSNN, the tuning of the hyperparameters was done using 5-fold cross validation with grid search. The NN training was done using the Adam optimisation algorithm. The options for the hyperparameters were: hidden layer size: 1 layer with 25, 30 and

35 neurons, 2 layers with 35 and 25, 40 and 30 neurons, respectively; learning rate: 0.0005, 0.001, 0.003, 0.005, 0.01, 0.1 and 0.3; L2 regularization parameter λ: 0.0005, 0.0008, 0.001 and 0.0015; batch size: 64 and 256, and number of epochs: 500 and 900. The activation functions were ReLu for the hidden layers and linear for the output layer, and the weight initialization mode was set to normal.

The best hyperparameters for the NN models are listed in Table 4, and were used for the evaluation on the test set. The number in brackets in the first column denotes the model built for the corresponding pattern sequence type. For example, after the aggregation, there were two pattern sequences for PSNN-PSF: PSNN-PSF (1) and PSNN-PSF (2). L_{min} was set to 20%.

Persistence Model. As a baseline used for comparison, we developed a persistence prediction model B_{per} which considers the PV power output of day d as the forecast for day $d+1$.

Evaluation Measures. To evaluate the performance on the test set, we used the Mean Absolute Error (MAE) and the Root Mean Squared Error (RMSE).

6 Results and Discussion

Table 5 shows the MAE and RMSE results of all models. We also conducted a pair-wise comparison between the prediction models for statistical significance of the differences in accuracy (MAE and RMSE, at point level) using the Wilcoxon signed-ranked test with $p < 0.05$.

Table 5. Accuracy of all methods

Model	MAE (kW)	RMSE (kW)
B_{per}	124.80	184.29
PSF	117.15	149.77
PSF1	115.55	147.72
PSF2	109.89	141.50
NN	79.46	117.65
PSNN-PSF	77.17	**106.33**
PSNN-PSF1	**77.10**	107.04
PSNN-PSF2	77.37	109.09

The main results can be summarized as follows:

- The best prediction model is PSNN-PSF1 in terms of MAE and PSNN-PSF in terms of RMSE. This shows the effectiveness of the PSNN approach in combining pattern sequences with NNs.

- All PSNN models outperform their corresponding PSF models. More specifically, PSNN-PSF outperforms PSF, PSNN-PSF1 outperforms PSF1 and PSNN-PSF2 outperforms PSF2, and the differences are statistically significant.
- All PSNN models also outperform the NN model but only the difference between PSNN-PSF and NN is statistically significant.
- Comparing the PSNN models, we can see that all three models perform similarly, with PSNN-PSF1 and PSNN-PSF performing slightly better than PSNN-PSF2. The pair differences between the three PSNN models are statistically significant except for PSNN-PSF1 vs PSNN-PSF2.
- Among the PSF models, PSF2 is the most accurate, followed by PSF1 and PSF, and the pair differences are statistically significant. This finding is consistent with [6], showing that the PSF extensions utilizing more than one data source are beneficial.
- All prediction models outperform the persistence baseline B_{per}.

We conducted additional analysis of the performance of PSNN, NN and PSF, comparing the characteristics of the days for which they performed well. We found that PSNN-PSF performs well for sunny days without rain. On the other hand, NN is the best model for rainy days and days following humid and windy days. A possible explanation is a sudden weather change, in which case the similarity of the sequence of previous days, used in PSF and PSNN but not in NN, is less important. This deserves further investigation.

7   Conclusion

In this paper, we present PSNN - a new approach combining pattern sequences with neural networks. It is a general approach that can be used with different clustering and cluster sequence extraction algorithms. It takes as an input a sequence of cluster labels, extracts pattern sequences of different types and builds a separate NN prediction model for each of them. PSNN can be applied to multiple complementary time series. In our case study for solar power forecasting, we show three instantiations of the PSNN approach by employing the pattern sequence extraction algorithms PSF, PSF1 and PSF2. We evaluate the performance of PSNN on Australian data for two years, from three sources (solar, weather and weather forecast). Our results show that PSNN was the most accurate method, with PSNN-PSF1 and PSNN-PSF obtaining the highest accuracy. All PSNN versions outperformed the PSF methods, and the differences were statistically significant. They also outperformed the NN model used for comparison but not all differences were statistically significant. Hence, we conclude that both PSNN and NN are promising approaches for solar power forecasting.

In future work we plan to investigate: (i) the use of other clustering algorithms in the PSF part, which may capture better the characteristics of the time series compared to the currently used k-means algorithm, (ii) seasonal changes and if building seasonal prediction models as in [8] can improve the accuracy, and (iii) the application of PSNN to other time series forecasting tasks.

References

1. Pedro, H.T., Coimbra, C.F.: Assessment of forecasting techniques for solar power production with no exogenous inputs. Solar Energy **86**, 2017–2028 (2012)
2. Rana, M., Koprinska, I., Agelidis, V.: Forecasting solar power generated by grid connected PV systems using ensembles of neural networks. In: International Joint Conference on Neural Networks (IJCNN) (2015)
3. Shi, J., Lee, W.J., Liu, Y., Yang, Y., Wang, P.: Forecasting power output of photovoltaic system based on weather classification and support vector machine. IEEE Trans. Ind. Appl. **48**, 1064–1069 (2012)
4. Rana, M., Koprinska, I., Agelidis, V.G.: 2D-interval forecasts for solar power production. Solar Energy **122**, 191–203 (2015)
5. Martínez-Álvarez, F., Troncoso, A., Riquelme, J.C., Aguilar Ruiz, J.S.: Energy time series forecasting based on pattern sequence similarity. IEEE Trans. Knowl. Data Eng. **23**, 1230–1243 (2011)
6. Wang, Z., Koprinska, I., Rana, M.: Solar power forecasting using pattern sequences. In: Lintas, A., Rovetta, S., Verschure, P.F.M.J., Villa, A.E.P. (eds.) ICANN 2017. LNCS, vol. 10614, pp. 486–494. Springer, Cham (2017). https://doi.org/10.1007/978-3-319-68612-7_55
7. Torres, J., Troncoso, A., Koprinska, I., Wang, Z., Martínez-Álvarez, F.: Big data solar power forecasting based on deep learning and multiple data sources. Expert Syst. **36**(4), e12394 (2019)
8. Koprinska, I., Rana, M., Agelidis, V.G.: Yearly and seasonal models for electricity load forecasting. In: International Joint Conference on Neural Networks (IJCNN), pp. 1474–1481 (2011)

Euler Recurrent Neural Network: Tracking the Input Contribution to Prediction on Sequential Data

Fengcheng Yuan[1,2], Zheng Lin[2(✉)], Weiping Wang[2], and Gang Shi[1,2]

[1] School of Cyber Security, University of Chinese Academy of Sciences,
Beijing, China
[2] Institute of Information Engineering, Chinese Academy of Sciences, Beijing, China
{yuanfengcheng,linzheng,wangweiping,shigang}@iie.ac.cn

Abstract. Recurrent neural networks (RNNs) achieve promising results on modeling sequential data. When a model produce an effective prediction, we always wonder which inputs are crucial to the specific prediction. Modern RNNs use nonlinear transformations to update their hidden states, which is hard to quantify the contributions for each input to the prediction. Inspired by the Euler Method, we propose a novel framework named Euler Recurrent Neural Network (ERNN) that uses weighted sums instead of nonlinear transformations to update its hidden states. This model can track the contribution of each input to the prediction at each time-step and achieve competitive result with fewer parameters. After quantification of their contributions to the prediction result, we can find the decisive ones among inputs and can also better understand the principle of the models in the prediction process.

Keywords: Recurrent neural network · Interpretability · Sequential data

1 Introduction

When machine learning models achieve surprising performance, we usually want to know how the model works. This answer is very significant, because by understanding the principle of the models, we can achieve better performance [13] and make more trustworthy judgments [21]. For risk control, we need further explanations for the model's applications in the fields like medical diagnosis [8] and autopilot [3].

One approach to interpreting neural networks is tracing the models' predictions back to the training data via influence functions [13]. Another work on interpreting neural networks focus on how a fixed model leads to particular predictions, such as using local explanation vectors to find the most influential features [2], locally fitting a simpler model to assert trust for a prediction or a model [21], and perturbing the test data to see how the prediction changes [1,22]. These work focus on feed-forward networks. However, recurrent neural networks may have their own characteristics.

T. Gedeon et al. (Eds.): ICONIP 2019, CCIS 1143, pp. 738–748, 2019.
https://doi.org/10.1007/978-3-030-36802-9_78

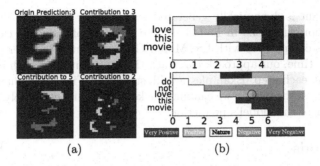

Fig. 1. (a) Explaining a handwritten digits classification made by ERNN. ERNN processes each image one pixel at a time and finally predicts the label. At the final time-step, the top 3 predicted classes are '3' ($p = 0.99$), '5' ($p = 1 \times 10^{-5}$) and '2' ($p = 5 \times 10^{-6}$). Every point has a probability distribution of classes. Points contributed to the class '3' is similar to the original image. (b) An example in sentiment analysis task. ERNN reads one word at a time. The left single column is the prediction at each time-step. Rectangles in two upper triangular matrices indicate the classification that the word contributes to at each time-step. The classification that 'love' contributes to changes from 'Positive' to 'Negative' at time 5 because of the phrase 'do not'.

In recurrent neural networks, Karpathy [11] analyzed an Long Short-Term Memory (LSTM) [9] trained on a character-based language modeling task and broke down its errors into classes, such as "rare word" errors. Murdoch [18] tracked the importance of a given input to the LSTM for a given output by telescoping sums of statistics. Sussillo [25] used nonlinear dynamical systems theory to understand RNNs by solving a set of simple but varied tasks. Hupkes [10] tried to understand the hierarchical compositionality of meaning in gated RNNs with diagnostic classification. However, it is still a challenging problem to quantify the inputs contributions to predictions in RNN models.

In this paper, we focus on the tasks of sequence classification, and further investigate these two tasks by solving one specific problem: what is the contribution of each input to the prediction? Since popular recurrent neural networks use the nonlinear transformation to update states, it is difficult for them to answer this question. To the best of our knowledge, we find that most of the previous work regarded the nonlinear transformation to update hidden states as an essential component of a recurrent neural network [5,26]. A similar work [6] used affine transformations instead of nonlinear transformations to update the hidden states. For each specific input, the model has an affine transformation depending on it. But it is difficult for the model to solve the problem with continuous values inputs.

Inspired by the Euler Method in numerical integration which uses linear additions to update states, we propose a novel framework named Euler Recurrent Neural Network (ERNN) that uses weighted sums instead of nonlinear transformations to update the hidden states. The gate units regulate the information flowing [7]. Based on this description, we view the gating values as weights. The proposed model sums up the contributions of the inputs as its hidden states, and the weight is dynamically generated by nonlinear functions similar to the gate activation functions in LSTM. In the problem of sequence classification, our model is competitive with LSTM and GRU [4], but with fewer parameters. At the same time, our approach can track the contribution of each input to the prediction at each step. Examples are illustrated in Fig. 1. By quantifying the contributions of the inputs, we can find the key inputs and effectively understand the behavior of the model's prediction.

2   Methods

2.1   Model Definition

In a sequence classification problem, we are given training points $z_1, ..., z_n$, where z_k consists of a sequence of inputs $x_1^{(k)}, x_2^{(k)}, ..., x_t^{(k)}$ and a target $y^{(k)}$. The model read an input x_i at each time-step. Our goal is to reach the target $y^{(k)}$ at the last time-step.

Inspired by the Euler Method which uses addition to update their states, our framework is defined as $\tilde{h}_t = f(x_t), h_t = \gamma_t h_{t-1} + \lambda_t \tilde{h}_t$, where λ is the forget gate, and γ is the input gate. λ and γ can be generated by a trained function, such as Restricted Boltzmann Machine (RBM) [23]. f is an input function, such as RBM or Convolutional Neural Networks (CNN) [15] etc.

Based on the proposed framework, we give two implementations. ERNN-O is a simple version, where forget gate γ_t and input gate λ_t separately depend on h_{t-1} and x_t,

$$\tilde{h}_t = \tanh(W_i x_t + b_i), \qquad h_t = \gamma_t h_{t-1} + \lambda_t \tilde{h}_t$$
$$\lambda_t = \sigma(U_\lambda h_{t-1} + W_\lambda x_t + b_\lambda), \gamma_t = \sigma(U_\gamma h_{t-1} + W_\gamma x_t + b_\gamma).$$

ERNN-X is a sophisticated version, where forget gate γ_t and input gate λ_t jointly depend on s_t,

$$\tilde{h}_t = \tanh(W_i x_t + b_i), \qquad h_t = \gamma_t h_{t-1} + \lambda_t \tilde{h}_t$$
$$\lambda_t = \sigma(U_\lambda s_t + b_\lambda), \qquad\qquad \gamma_t = \sigma(U_\gamma s_t + b_\gamma)$$
$$s_t = \tanh(U_s h_{t-1} + W_s x_t + b_s).$$

Here, $W_i, W_\lambda, W_\gamma, W_s \in \mathbb{R}^{n*m}$, $U_i, U_\lambda, U_\gamma, U_s \in \mathbb{R}^{n*n}$, $b_i, b_\lambda, b_\gamma, b_s \in \mathbb{R}^n$. σ denotes the sigmoid activation function, and $\tanh$ denotes the hyperbolic tangent activation function. In the experiments, we find that ERNN-X can effectively deal with data with longer sequence.

2.2   The Inputs Contributions to Hidden States and Predictions

Let $h_0 = 0$, the general formula of the Euler Recurrent Neural Network is

$$h_t = \sum_{k=1}^t \underbrace{\prod_{i=k+1}^t \gamma_i}_{\Gamma_{k+1}^t} \lambda_k \underbrace{f(x_k)}_{f_k} \triangleq \sum_{k=1}^t \underbrace{\Gamma_{k+1}^t \lambda_k f_k}_{C_k^t} \triangleq \sum_{k=1}^t C_k^t,$$

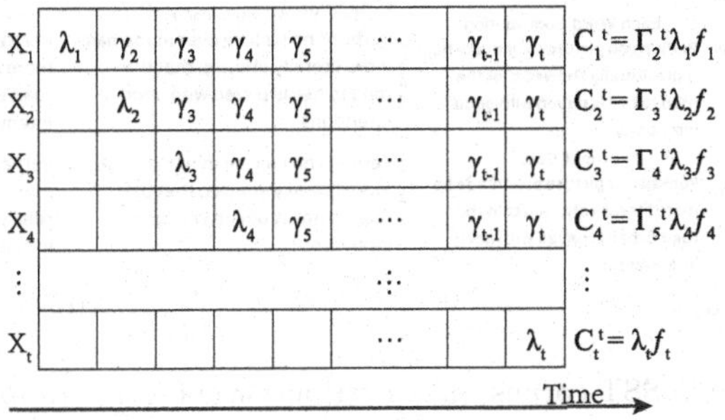

X_1	λ_1	γ_2	γ_3	γ_4	γ_5	$\cdots$	γ_{t-1}	γ_t	$C_1{}^t = \Gamma_2{}^t \lambda_1 f_1$
X_2		λ_2	γ_3	γ_4	γ_5	$\cdots$	γ_{t-1}	γ_t	$C_2{}^t = \Gamma_3{}^t \lambda_2 f_2$
X_3			λ_3	γ_4	γ_5	$\cdots$	γ_{t-1}	γ_t	$C_3{}^t = \Gamma_4{}^t \lambda_3 f_3$
X_4				λ_4	γ_5	$\cdots$	γ_{t-1}	γ_t	$C_4{}^t = \Gamma_5{}^t \lambda_4 f_4$
$\vdots$						$\ddots$		$\vdots$	$\vdots$
X_t						$\cdots$		λ_t	$C_t{}^t = \lambda_t f_t$

Time $\longrightarrow$

Fig. 2. The input contribution C_k^t to hidden state change dynamically. The vertical direction is the input in time series. Once a new input arrives, the previous inputs contributions will be changed by the forget gate.

where $C_k^t = \Gamma_{k+1}^t \lambda_k f_k$ is the contribution of the input x_k to the hidden state h_t. $\Gamma_{k+1}^t \triangleq \prod_{i=k+1}^t \gamma_i$ is the **forget factor** of the input x_k. Thus the hidden state h_t is the sum of a series of input contributions. Figure 2 illustrates that the contributions of the inputs to the hidden states vary dynamically. When a new input x_k comes, the inputs before time k must multiply the forget gate γ_k. Therefore, the contribution C_k of the input x_k varies with time.

For sequence classification problem, the probabilities are computed as

$$p_t = \mathsf{softmax}(l_t), \; l_t = W_a h_t = W_a \sum_{k=1}^t C_k^t,$$

where l_t is an unnormalized vector, which is used to track the input contribution to prediction. Specifically, the contribution of the input x_k to the prediction at time t is $W_a C_k^t$. Overall, it is the weighted sum updating method that allows us to track the contributions of inputs to predictions easily.

2.3 Relation to Long Short-Term Memory

In LSTM, the nonlinear gating units regulate the information flow into and out of the cell [7]. Our work is heavily influenced by the gating mechanism in LSTM. Obviously, the input gate and the forget gate in ERNN are similar to the ones in LSTM. All gates use nonlinear functions to compute a value. Compared with LSTM, the first difference is that ERNN only use weighted sums to update the hidden state. Another difference between the two models is that we do not use the output gate in ERNN, which makes our model more concise. The form of LSTM is as follows:

$$f_t = \sigma(W_f x_t + U_f h_{t-1} + b_f), \; i_t = \sigma(W_i x_t + U_i h_{t-1} + b_i)$$
$$o_t = \sigma(W_o x_t + U_o h_{t-1} + b_o), \; \tilde{c}_t = \tanh(W_c x_t + U_c h_{t-1} + b_c)$$
$$c_t = f_t c_{t-1} + i_t \tilde{c}_t, \qquad h_t = o_t \tanh(c_t),$$

where f_t is forget gate, i_t is input gate and o_t is output gate. To investigate the impact of the first difference, we only remove the hidden-hidden updating matrix U_c in LSTM and name it LSTM-h, $\tilde{c}_t = \tanh(W_c x_t + b_c)$.

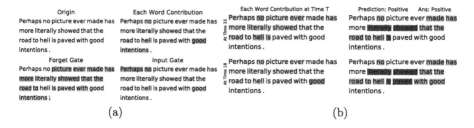

Fig. 3. An example in SST corpus. (a) Contributions of words to *the last hidden state* and gates. (b) Words contributions and predictions change dynamically.

In the experiments, we find that the LSTM-h makes a slight improvement over the LSTM on sentiment analysis task and handwritten digits classification (MNIST) task, though it has a more concise form. The impact of the second difference can be investigated by comparing ERNN with LSTM-h.

3   Experiments and Analysis

To assess the ability of ERNN on tracking the inputs contributions, we test our model on four tasks: sentiment analysis and handwritten digits classification (MNIST) task. Then we analyze how inputs influence the prediction, and discuss the relationship between two gates and inputs. We also compare our models with several popular RNNs, such as GRU and LSTM. The code for replicating our experiments is available on GitHub.

3.1   Sentiment Analysis

Task. We use three datasets, IMDB [16], Movie reviews (MR) [19] and Stanford Sentiment Treebank (SST) [24]. Data in IMDB and MR is binary labeled ('Positive' or 'Negative'). SST contains fine-grained labels ('Very Positive', 'Positive', 'Neutral', 'Negative', 'Very Negative'). In SST and MR corpus, we follow the experimental setup by Kim [12]. In IMDB corpus, we follow the standard data split rules and set aside 5,000 training samples for validation purposes.

Model Setting. All models have 200 hidden units and are trained with 300-dimension pretrained word vector[1] [17]. The pretrained word vector will not be modified during training. Models are trained for 150 epochs to minimize cross-entropy loss. Gradient clipping [20] with a threshold of 5 is applied to the loop variables. Training is performed with Adam on batches of 64. The dictionary size on IMDB and MR is 10000, and for SST we set the dictionary size as 20000. We train three models as our baseline methods: RNN, LSTM, and GRU.

---

[1] https://code.google.com/archive/p/word2vec/.

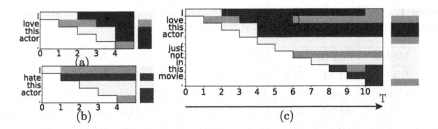

Fig. 4. Contributions of words to the *predictions*. We track every words at middle time-steps. The left single column is the prediction at time-step t. Deep red, red, white, blue and deep blue indicate 'Very positive', 'Positive', 'Nature', 'Negative' and 'Very Negative' respectively. Note images are the upper triangular matrix. (a) The word 'love' has an influence on the following words and the first words. (b) The word 'hate' affect the following words, which is obviously different from the previous sentence. (c) The contribution of the word 'love' to prediction changes when the word 'not' arrives. (Color figure online)

Case Study. We test ERNN on two sentences and explain the model in three aspects. What's the relationship between two gates and words? Fig. 3(a) illustrates that the forget gate has a small value at words 'no', 'hell', 'good' and the first word in the sentence. Then it maintains a large value at other words. The input gate has a larger value at words 'no', 'good' and the first one. Figure 3(b) illustrates that the sentiment of this sentence is negative at time-step 15. When the word 'good' appears, the value of forget gate becomes smaller (means the previous information needs to be forgotten). Thus, the contribution of the word 'no' gets smaller. Eventually, the word 'good' makes the greatest contribution. Figure 3 illustrates that two gates pick the keywords in this sentence.

How does the input gate affect the following words? To get a better understanding of the contributions of the words to the *classification*, we input another three sentences. Note that we only compare the vector l_t before softmax function. In Fig. 4, three input sentences are: 'I love this actor.', 'I hate this actor' and 'I love this actor, just not in this movie.' The only difference between the input sentences of (a) and (b) is the second word, 'love' and 'hate'. And this difference affects the contribution of the following word to the classification and the subsequent predictions. In Fig. 4(a) and (b), the contributions of the word 'this', 'actor' and '.' to the predicted classes are significantly different. The word 'this' contributes to the class 'Very Positive' instead of 'Nature' when the word 'love' appears. Figure 1(b) illustrates that the phrase 'do not' affects the contributions of the subsequent words such as 'this' and 'movie'. Affected by the word 'love','movie' contribute to the class 'Very Positive'. However, affected by phrase 'do not love', 'movie' contributes to the class 'Very Negative'. Figure 1(b) and Fig. 4(a, b) illustrate that the input gate can affect the contributions of the subsequent words.

Table 1. Classification accuracies for models on various data sets. The average and standard deviation results are reported from 10 trials.

Models	IMDB	MR	SST	MNIST
RNN	67.34 ± 10.3	66.56 ± 9.36	35.35 ± 3.10	–
LSTM	90.05 ± 0.30	77.54 ± 0.53	47.26 ± 0.72	89.33*
LSTM-h	**90.27 ± 0.28**	78.31 ± 0.63	47.93 ± 1.06	**98.49 ± 0.12**
GRU	90.14 ± 0.25	**78.34 ± 0.81**	47.51 ± 0.80	98.44 ± 0.28
ERNN-O	86.06 ± 2.82	77.95 ± 0.39	**48.27 ± 0.75**	88.63*
ERNN-X	90.00 ± 0.21	78.28 ± 0.61	47.48 ± 0.71	98.22 ± 0.14

How does the forget gate affect the previous word? A more complex example is shown in Fig. 4(c). When ERNN reads the first 4 words, it gives the same prediction as (a). However, affected by the input word 'not', the contribution of word 'love' to classification change from 'Very Positive' to 'Positive'. Figure 4(c) illustrates that the forget gate can changes the contribution of the previous word.

Analysis. Table 1 illustrates that ERNN achieves competitive results with fewer parameters on the three datasets. The average length of the reviews in IMDB is approximately 240 words, which is larger than the average length of the MR and SST datasets. We find that ERNN-O has larger standard deviation on modeling long sentences (IMDB). Meanwhile, ERNN-O is good at modeling short sentences. It achieves the best result at SST dataset. ERNN-X performs well on three datasets. Impressively, LSTM-h always achieves a better result than LSTM on all these datasets. Thus, the hidden-hidden matrix U_c may not be the key part of LSTM for the sentiment analysis task.

3.2   MNIST

Task. We evaluate our models on a sequential version of the MNIST handwritten digits classification task [15]. The model processes one pixel at a time and finally predicts the label. By flattening the 28 × 28 images into 784-d vectors, it can be reformulated as a challenging task for RNNs where long-term dependencies need to be leveraged [14].

Model Setting. We follow the standard data split rules and set aside 5,000 training samples for validation. All models are trained with 100 hidden units. After processing all pixels with an RNN, the last hidden state is fed into a softmax classifier to predict the digital class. All models are trained for 200 epochs to minimize cross-entropy loss. We train two models as our baselines, LSTM and GRU. The vanilla RNN has a very low accuracy at the test set.

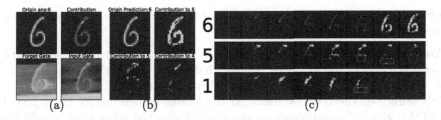

Fig. 5. (a) Contributions and gates. In the second figure, the contribution of each point to *the last hidden state* is similar to the original image. The forget gate and the input gate also show similar phenomenon and contain a large value at the bottom. (b) The contributions of inputs to *predictions*. At last time-step, The top 3 predicted classes are '6' ($p = 0.99$), '5' ($p = 2 \times 10^{-4}$) and '4' ($p = 2 \times 10^{-5}$). (c) We track 3 classes, '6', '5' and '1' at middle time-steps. The number of points that contribute to '5' decrease drastically when the point inside the red square arrives. So does the point that contributes to '1'. But the number of points that contribute to '6' increase in the last few time-steps.

Case Study. We test ERNN with one image (number '6') in test corpus and explain the model in three aspects. What's the relationship between two gates and inputs? Fig. 5(a) shows that the values of the forget gate and the input gate in the pixel of number '6' are larger than the values of other pixels in the image. After encountering the first non-zero pixel, the forget gate maintains a larger value until the last time-step. This means that ERNN-X starts to 'remember' the inputs contributions from the first non-zero pixel. Two gates also have a larger value at lines on the bottom of the image.

Which points contribute to the prediction at the last time-step? In Fig. 1(a), the top 3 predicted classes are '3', '5' and '2'. What ERNN-X picks up for the prediction is consistent with our intuition. Points in the third figure contribute to class '5'. Clearly, the number '5' usually contains a horizontal line. Figure 5(b) shows the contributions of points to predictions at the last time-step. The top 3 predicted classes are '6', '5' and '4'. Points contributed to the class '6' is similar to the original image.

How does the forget gate affect the previous point? We track 3 common predictions (class '6', '5' and '1') in Fig. 5(c). Even though the probability of class '1' at the final time-step is very low, the model has a higher probability of prediction '1' in the first 14 rows. Because points in the first 14 rows are similar to the number '1'. In the third line of Fig. 5(c), when the point inside the red square arrives, the number of points that contribute to '1' decreases sharply, since the forget gate can affect the previous inputs when the model finds that the new input is different from '1'. Thus, points that previously belong to '1' change their contribution classes. In the second line of Fig. 5(c), similar to class '1', the number of points that contribute to '5' drops sharply when the point inside the red square arrives. However, the number of points that contributes to class '6' increases drastically at the last few rows. As we can see, the points contributions to predictions can be changed by the forget gate. By tracking the input contribution to prediction at middle time-steps, we can better understand the behavior of the model's prediction.

Analysis. Table 1 illustrates that LSTM-h reaches the best result on MNIST. Given the various initial values, we find that LSTM and ERNN-O do not work as well as LSTM-h or GRU. A similar phenomenon on LSTM also appears in Le's work [14]. However, by removing the hidden-hidden matrix U_c in LSTM, LSTM-h alters this phenomenon and outperforms the other models with less parameters. When the output gate in LSTM-h is removed, the model turns out to be ERNN-O. Since ERNN-O performs unsatisfied, we think that the output gate may help leverage long-term dependencies. However, ERNN-X has a different architecture with ERNN-O, and achieves a competitive result with fewer parameters. In ERNN-X, the forget gate and the input gate jointly depending on s_t. This architecture builds strong association between the two gates, which may reduce the fluctuation of the training loss and lead to a better trainable model.

4   Conclusion

In this paper, we propose a novel model named Euler Recurrent Neural Networks which uses weighted sums instead of nonlinear transformations to update the hidden states. This model can track the contribution of each input to the prediction at each time-step. The experiment illustrates that ERNN can not only achieve competitive result with fewer parameters, but also help us better understand the principle of the models in the prediction process.

References

1. Adler, P., et al.: Auditing black-box models for indirect influence. In: 2016 IEEE 16th International Conference on Data Mining (ICDM), pp. 1–10 (2016)
2. Baehrens, D., Schroeter, T., Harmeling, S., Kawanabe, M., Hansen, K., Mãžller, K.R.: How to explain individual classification decisions. J. Mach. Learn. Res. **11**, 1803–1831 (2010)
3. Bojarski, M., et al.: End to end learning for self-driving cars. arXiv preprint arXiv:1604.07316, pp. 1–9 (2016)
4. Chung, J., Gulcehre, C., Cho, K., Bengio, Y.: Empirical evaluation of gated recurrent neural networks on sequence modeling. In: NIPS 2014 Workshop on Deep Learning, pp. 1–9 (2014)
5. Collins, J., Sohl-Dickstein, J., Sussillo, D.: Capacity and trainability in recurrent neural networks. In: Proceedings of the International Conference for Learning Representations, pp. 1–17 (2017)
6. Foerster, J.N., Gilmer, J., Sohl-Dickstein, J., Chorowski, J., Sussillo, D.: Input switched affine networks: an RNN architecture designed for interpretability. In: Proceedings of the International Conference on Machine Learning, pp. 1136–1145 (2017)
7. Greff, K., Srivastava, R.K., KoutnxEDk, J., Steunebrink, B.R., Schmidhuber, J.: LSTM: a search space odyssey. IEEE Trans. Neural Netw. Learn. Syst. **28**, 2222–2232 (2017)

8. Gulshan, V., et al.: Development and validation of a deep learning algorithm for detection of diabetic retinopathy in retinal fundus photographs. JAMA **316**(22), 2402–2410 (2016)

9. Hochreiter, S., Schmidhuber, J.: Long short-term memory. Neural Comput. **9**(8), 1735–1780 (1997)

10. Hupkes, D., Zuidema, W.: Diagnostic classification and symbolic guidance to understand and improve recurrent neural networks. In: Proceedings of Advances in Neural Information Processing Systems 2017 Workshop, pp. 1–9 (2017)

11. Karpathy, A., Johnson, J., Fei-Fei, L.: Visualizing and understanding recurrent networks. In: Proceedings of the International Conference for Learning Representations 2016 Workshop, pp. 1–12 (2016)

12. Kim, Y.: Convolutional neural networks for sentence classification. In: Proceedings of the 2014 Conference on Empirical Methods in Natural Language Processing (EMNLP), pp. 1746–1751. Association for Computational Linguistics, Doha, Qatar, October 2014

13. Koh, P.W., Liang, P.: Understanding black-box predictions via influence functions. In: Proceedings of the 34th International Conference on Machine Learning, pp. 1885–1894 (2017)

14. Le, Q.V., Jaitly, N., Hinton, G.E.: A simple way to initialize recurrent networks of rectified linear units. arXiv preprint arXiv:1504.00941, pp. 1–9 (2015)

15. LeCun, Y., Bottou, L., Bengio, Y., Haffner, P.: Gradient-based learning applied to document recognition. Proc. IEEE **86**(11), 2278–2324 (1998)

16. Maas, A.L., Daly, R.E., Pham, P.T., Huang, D., Ng, A.Y., Potts, C.: Learning word vectors for sentiment analysis. In: Proceedings of the 49th Annual Meeting of the Association for Computational Linguistics: Human Language Technologies, pp. 142–150. Association for Computational Linguistics, Portland, Oregon, USA, June 2011

17. Mikolov, T., Corrado, G., Chen, K., Dean, J.: Efficient estimation of word representations in vector space. In: Proceedings of the International Conference on Learning Representations, pp. 1–12 (2013)

18. Murdoch, W.J., Szlam, A.: Automatic rule extraction from long short term memory networks. arXiv preprint arXiv:1702.02540 (2017)

19. Pang, B., Lee, L.: Seeing stars: Exploiting class relationships for sentiment categorization with respect to rating scales. In: Proceedings of Association for Computational Linguistics, pp. 115–124 (2005)

20. Pascanu, R., Mikolov, T., Bengio, Y.: On the difficulty of training recurrent neural networks. In: Dasgupta, S., McAllester, D. (eds.) Proceedings of the 30th International Conference on Machine Learning. Proceedings of Machine Learning Research, vol. 28, pp. 1310–1318. PMLR, Atlanta, Georgia, USA, 17–19 January 2013

21. Ribeiro, M.T., Singh, S., Guestrin, C.: Why should i trust you? explaining the predictions of any classifier. In: Proceedings of the 22nd ACM SIGKDD International Conference on Knowledge Discovery and Data Mining, pp. 1135–1144. ACM (2016)

22. Simonyan, K., Vedaldi, A., Zisserman, A.: Deep inside convolutional networks: Visualising image classification models and saliency maps. CoRR abs/1312.6034 (2013)

23. Smolensky, P.: Parallel distributed processing: Explorations in the microstructure of cognition, vol. 1. chap. Information Processing in Dynamical Systems: Foundations of Harmony Theory, pp. 194–281. MIT Press, Cambridge (1986)

24. Socher, R., et al.: Recursive deep models for semantic compositionality over a sentiment treebank. In: Proceedings of the 2013 Conference on Empirical Methods in Natural Language Processing, pp. 1631–1642 (2013)
25. Sussillo, D., Barak, O.: Opening the black box: low-dimensional dynamics in high-dimensional recurrent neural networks. Neural Comput. **25**(3), 626–649 (2013)
26. Zhang, S., et al.: Architectural complexity measures of recurrent neural networks. In: Proceedings of Advances in Neural Information Processing Systems, pp. 1822–1830 (2016)

Unsupervised Neural Models

Exploiting Cluster Structure
in Probabilistic Matrix Factorization

Tao Li and Jinwen Ma[✉]

Department of Information Science, School of Mathematical Sciences
and LMAM Peking University, Beijing 100871, China
jwma@math.pku.edu.cn

Abstract. Low-rank matrix factorization is a basic model for collaborative filtering. The low-rank matrix approximation model is equivalent to represent users and items by latent factors, and rating is obtained by calculating the inner product of factors. Most low-rank matrix approximation models assume the latent factors come from a common Gaussian distribution. However, users with similar preferences or items of the same type tend to have similar factors, thus there exists cluster structure underlying user factors and item factors. In this paper, we exploit the cluster structure in the low-rank matrix factorization model to improve prediction accuracy. Experimental results on MovieLens-1m and MovieLens-10m datasets demonstrate the effectiveness of the proposed methods.

Keywords: Matrix factorization · Clustering · Markov Chain Monte Carlo

1 Introduction

Low-rank matrix factorization is one of the most popular approaches in collaborative filtering. The basic idea behind matrix factorization models is that the preferences of users and information about items can be encoded in low-dimensional latent factors, and the rating is given by the inner product of such factors. Various probabilistic matrix factorization models have been proposed, such as Probabilistic Matrix Factorization (PMF) [12], Bayesian Probabilistic Matrix Factorization (BPMF) [13] and so on. However, in these models, the prior for each user factor or movie factor is a common Gaussian distribution. We claim that this assumption ignores the cluster structures underlying users and movies. Intuitively, users with similar preferences or movies of the same type may have similar factors. Independent identically distributed Gaussian prior is not consistent with this observation.

In this paper, we adopt the idea of clustering to probabilistic matrix factorization to exploit the underlying cluster structure in data. The proposed new model is referred to as PMF_Cluster. We first derive the learning algorithm for PMF_Cluster, which is a combination of PMF and k-means. Then, we develop

© Springer Nature Switzerland AG 2019
T. Gedeon et al. (Eds.): ICONIP 2019, CCIS 1143, pp. 751–760, 2019.
https://doi.org/10.1007/978-3-030-36802-9_79

a fully Bayesian treatment of PMF_Cluster, which we refer to BPMF_MG. In BPMF_MG, we apply the Gibbs sampling method (a special case of MCMC) [3] to sample parameters from the posterior. Finally, we conduct extensive experiments on MovieLens datasets, including performance evaluation and cluster analysis. The experimental results confirm that exploiting cluster structures in data is beneficial to improving prediction accuracy.

2   Related Work

Probabilistic factor-based models date back to [5,10,11]. These models suffer from intractable exact inference. One seminal work is [12], in which probabilistic matrix factorization model was proposed. Further, a fully Bayesian treatment of PMF was developed in [13]. One distinguishing feature of [13] is the use of MCMC method for approximate inference. Recent work adopted matrix clustering techniques [16] and community detection methods [17] to address this issue.

There are several studies focusing on using the ensemble of sub-matrices for better approximation, including DFC [9], LLORMA [6,7], ACCAMS [1], WEMAREC [2], MPMA [2]. Generally speaking, these methods partition the original matrix into smaller sub-matrices, and clustering based techniques are often used for sub-matrix generation. In [15], Wang *et al.* integrated topic model and PMF. Our work is highly inspired by the studies above. However, most existing methods partition the original matrix into smaller sub-matrices and make predictions using an ensemble of local matrix factorization. The proposed methods learn cluster structures adaptively during the training procedure, and there is no need to design a weighting strategy in our methods. Although similar ideas have been investigated in [2,14], the intuitive motivation and the full Bayesian treatment of the proposed methods have never been presented.

3   Probabilistic Matrix Factorization with Clustering

3.1   From PMF to PMF_Cluster

Suppose we have M users and N movies, and the rating of user i for movie j is denoted by $\mathbf{X}_{ij}$. Let $\mathbf{U}_i \in \mathbb{R}^D, \mathbf{V}_j \in \mathbb{R}^D$ be the latent factors of user i and movie j, then the conditional distribution of $\mathbf{X}_{ij}$ given $\mathbf{U}_i, \mathbf{V}_j$ and prior distributions of factors are:

$$p(\mathbf{X}_{ij}|\mathbf{U}_i, \mathbf{V}_j, \tau) = \mathcal{N}(\mathbf{X}_{ij}|\mathbf{U}_i^T\mathbf{V}_j, \tau^{-1}) \tag{1}$$

$$p(\mathbf{U}_i|\lambda_U) = \mathcal{N}(\mathbf{U}_i|0, \lambda_U^{-1}\mathbf{I}_D), p(\mathbf{V}_j|\lambda_V) = \mathcal{N}(\mathbf{V}_j|0, \lambda_V^{-1}\mathbf{I}_D) \tag{2}$$

Parameter learning in PMF is performed by maximizing the log-posterior probability. As indicated in Eq. (2), the log-posterior is given by

$$\log p(\mathbf{U}, \mathbf{V}|\mathbf{X}, \lambda_U, \lambda_V) = \frac{\tau}{2}\sum_{i=1}^{M}\sum_{j=1}^{N}\mathbb{I}_{ij}(\mathbf{X}_{ij} - \mathbf{U}_i^T\mathbf{V}_j)^2 + \frac{\lambda_U}{2}\sum_{i=1}^{M}\|\mathbf{U}_i\|^2 + \frac{\lambda_V}{2}\sum_{j=1}^{n}\|\mathbf{V}_j\|^2. \tag{3}$$

Here, $\mathbb{I}_{ij}$ is the indicator variable which equals to 1 if $\mathbf{X}_{ij}$ is observed and 0 otherwise.

In PMF, user factors $\{\mathbf{U}_i\}_{i=1}^M$ are assumed to follow a single Gaussian distribution $\mathcal{N}(0, \lambda_U^{-1}\mathbf{I}_D)$. However, different users may have different preferences. For example, children may prefer cartoons, while teenagers may prefer science fiction. We assume that users belong to the same cluster have similar taste and thus their factors follow the same distribution. The cluster structure is unknown a prior, thus we should learn the cluster structure adaptively. Inspired by k-means algorithm, we use K vectors $\{\boldsymbol{\mu}_k\}_{k=1}^K$ to denote cluster centers, and $c(i) \in \{1, 2, \cdots, K\}$ to denote the cluster label of user i, then we can modify the regularization term of user factors: $\frac{\lambda_U}{2}\sum_{i=1}^M \|\mathbf{U}_i\|^2 \rightarrow \frac{\lambda_U}{2}\sum_{i=1}^M \|\mathbf{U}_i - \boldsymbol{\mu}_{c(i)}\|^2$.

The discussion above applies to movie factors, too. Let $\{\hat{\boldsymbol{\mu}}_j\}_{j=1}^L$ denote the cluster centers of movies factors and $\hat{c}(j)$ denote the cluster label of movie j, then the modified objective function is

$$\mathcal{L} = \frac{\tau}{2}\sum_{i=1}^M \sum_{j=1}^N \mathbb{I}_{ij}(\mathbf{X}_{ij} - \mathbf{U}_i^{\mathsf{T}}\mathbf{V}_j)^2 + \frac{\lambda_U}{2}\sum_{i=1}^M \|\mathbf{U}_i - \boldsymbol{\mu}_{c(i)}\|^2 + \frac{\lambda_V}{2}\sum_{j=1}^n \|\mathbf{V}_j - \hat{\boldsymbol{\mu}}_{\hat{c}(j)}\|^2.$$

(4)

In the rest of this paper, we refer this model as PMF_Cluster.

3.2   Learing Algorithm for PMF_Cluster

In PMF, the log-posterior probability is maximized with respect to $\{\mathbf{U}_i\}_{i=1}^N$ and $\{\mathbf{V}_j\}_{j=1}^N$ via gradient ascent. In PMF_Cluster, we need to maximize $\mathcal{L}$ with respect to not only $\mathbf{U}_i, \mathbf{V}_j$, but also $\boldsymbol{\mu}_k, c(i), \hat{\boldsymbol{\mu}}_l, \hat{c}(j)$. The introduction of cluster parameters makes the optimization problem even harder. We propose to optimize $\mathcal{L}$ in an alternative direction optimization manner. After random initialization, we first fix cluster parameters and update $\mathbf{U}_i, \mathbf{V}_j$ using gradient ascent as in PMF. Then after convergence, we keep $\mathbf{U}_i, \mathbf{V}_j$ fixed and update cluster parameters via k-means algorithm. The iteration continues until convergence. The detailed algorithm is given in Algorithm 2. When the volume of data is too large and the calculating of gradients is computationally expensive, we may use stochastic gradient ascent in the inner loop, i.e., we process the data in mini-batches.

Algorithm 1. Learning algorithm for PMF_Cluster

Input: observed entries $\mathbf{X}_{ij}$, observation indicators $\mathbb{I}_{ij}$.
Hyper parameters: prior precisions $\tau, \lambda_U, \lambda_V$, learning rate δ.
Variables: latent factors $\mathbf{U}_i, \mathbf{V}_j$, centers $\boldsymbol{\mu}_k, \hat{\boldsymbol{\mu}}_l$, and labels $c(i), \hat{c}(j)$

1: Initialize parameters by random guess
2: **while** not converged **do**
3:     **while** not converged **do**
4:         Update $\mathbf{U}_i \leftarrow \mathbf{U}_i + \delta(\tau \sum_{j=1}^M \mathbb{I}_{ij}(\mathbf{U}_i^{\mathsf{T}}\mathbf{V}_j - \mathbf{X}_{ij})\mathbf{V}_j + \lambda_U(\mathbf{U}_i - \boldsymbol{\mu}_{c(i)}))\ \forall i$
5:         Update $\mathbf{V}_j \leftarrow \mathbf{V}_j + \delta(\tau \sum_{i=1}^N \mathbb{I}_{ij}(\mathbf{U}_i^{\mathsf{T}}\mathbf{V}_j - \mathbf{X}_{ij})\mathbf{U}_i + \lambda_V(\mathbf{V}_j - \hat{\boldsymbol{\mu}}_{\hat{c}(j)}))\ \forall j$
6:     **end while**
7:     Update $\boldsymbol{\mu}_k, c(i)$ and $\hat{\boldsymbol{\mu}}_l, \hat{c}(j)$ via k-means.
8: **end while**

---

4   Bayesian Probabilistic Matrix Factorization with Mixture of Gaussian Prior

4.1   From PMF_Cluster to BPMF_MG

BPMF is the Bayesian version of PMF. Similarly, we recast PMF_cluster from the Bayesian perspective in this section. Note that in PMF_Cluster, the prior for user (or movie) factors is a mixture of K (or L) Gaussian distribution, rather than a single Gaussian distribution. The probabilistic graphical model is shown in Fig. 1. Let $\mathbf{Z}_i$ (or $\hat{\mathbf{Z}}_j$) be the latent variable indicating which component U_i (or V_j) comes from, then

$$p(\mathbf{X}_{ij}|\mathbf{U}_i, \mathbf{V}_j) = \mathcal{N}(\mathbf{X}_{ij}|\mathbf{U}_i^{\mathrm{T}}\mathbf{V}_j, \tau^{-1})$$

$$p(\mathbf{U}_i|\mathbf{Z}_i, \boldsymbol{\mu}_U, \boldsymbol{\Lambda}_U) = \prod_{k=1}^{K}[\mathcal{N}(\mathbf{U}_i|\boldsymbol{\mu}_U^k, (\boldsymbol{\Lambda}_U^k)^{-1})]^{\mathbb{I}(\mathbf{Z}_i=k)}$$

$$p(\mathbf{V}_j|\hat{\mathbf{Z}}_j, \boldsymbol{\mu}_V, \boldsymbol{\Lambda}_V) = \prod_{l=1}^{L}[\mathcal{N}(\mathbf{V}_j|\boldsymbol{\mu}_V^l, (\boldsymbol{\Lambda}_V^l)^{-1})]^{\mathbb{I}(\hat{\mathbf{Z}}_j=l)} \tag{5}$$

We place Gauss-Wishart priors on $\boldsymbol{\mu}_U, \boldsymbol{\Lambda}_U$ and $\boldsymbol{\mu}_V, \boldsymbol{\Lambda}_V$,

$$p(\boldsymbol{\mu}_U^k, \boldsymbol{\Lambda}_U^k) = \mathcal{N}(\boldsymbol{\mu}_U^k|_U, (\beta_U \boldsymbol{\Lambda}_U^k)^{-1})\mathcal{W}(\boldsymbol{\Lambda}_U^k|\nu_U, \mathbf{W}_U)$$
$$p(\boldsymbol{\mu}_V^l, \boldsymbol{\Lambda}_V^l) = \mathcal{N}(\boldsymbol{\mu}_V^l|_V, (\beta_V \boldsymbol{\Lambda}_V^l)^{-1})\mathcal{W}(\boldsymbol{\Lambda}_V^l|\nu_V, \mathbf{W}_V) \tag{6}$$

We place categorical priors for the latent variables $\mathbf{Z}_i, \hat{\mathbf{Z}}_j$ and further place Dirichlet hyper-priors for parameters $\boldsymbol{\pi}_U, \boldsymbol{\pi}_V$,

$$p(\mathbf{Z}_i|\pi_U) = \mathrm{Cat}(\mathbf{Z}_i|\pi_U) \quad , \quad p(\hat{\mathbf{Z}}_j|\pi_V) = \mathrm{Cat}(\hat{\mathbf{Z}}_j|\pi_V) \tag{7}$$

$$p(\boldsymbol{\pi}_U) = \mathrm{Dir}(\boldsymbol{\pi}_U|\boldsymbol{\alpha}_U) \quad , \quad p(\boldsymbol{\pi}_V) = \mathrm{Dir}(\boldsymbol{\pi}_V|\boldsymbol{\alpha}_V) \tag{8}$$

In comparison with BPMF, the key difference of the proposed model is the mixture of Gaussian priors for factors. Therefore, we refer this model as BPMF_MG.

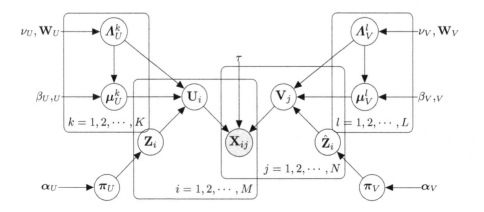

Fig. 1. Probabilistic graphical model of BPMF_MG.

4.2   Inference and Prediction

The predictive distribution of $\mathbf{X}^*_{ij}$ can be obtained by marginalizing over parameters:

$$p(\mathbf{X}^*_{ij}) = \int p(\mathbf{X}^*_{ij}|\mathbf{U}_i, \mathbf{V}_j)p(\mathbf{U}_i|\mathbf{Z}_i, \boldsymbol{\mu}_U, \boldsymbol{\Lambda}_U)p(\mathbf{V}_j|\hat{\mathbf{Z}}_j, \hat{\boldsymbol{\mu}}_V, \hat{\boldsymbol{\Lambda}}_V)$$

$$\prod_{k=1}^{K} p(\boldsymbol{\mu}^k_U, \boldsymbol{\Lambda}^k_U) \prod_{l=1}^{L} p(\hat{\boldsymbol{\mu}}^l_V, \hat{\boldsymbol{\Lambda}}^l_V)p(\mathbf{Z}_i|\boldsymbol{\pi}_U)p(\boldsymbol{\pi}_U)p(\hat{\mathbf{Z}}_j|\boldsymbol{\pi}_V)p(\boldsymbol{\pi}_V)\mathrm{d}\boldsymbol{\Theta} \tag{9}$$

Here $\boldsymbol{\Theta}$ denotes all variables except $\mathbf{X}^*_{ij}$. However, exact inference is intractable since Eq. (9) is too complicated. Instead, we develop a Markov Chain Monte Carlo procedure for parameter learning. Due to the conjugate priors in BPMF_MG, the full conditional distributions have simple form and thus Gibbs sampling is easy to implement. The full conditional distributions are summarized in the following theorem.

Theorem 1. *Let* $N_k = \sum_{i=1}^{M} \mathbf{Z}^k_i$ *and*

$$\bar{\mathbf{U}}_k = \frac{1}{N_k} \sum_{k=1}^{K} \mathbf{Z}^k_i \mathbf{U}_i, \quad \mathbf{S}_k = \sum_{i=1}^{M} \mathbf{Z}^k_i (\mathbf{U}_i - \bar{\mathbf{U}}_k)(\mathbf{U}_i - \bar{\mathbf{U}}_k)^T, \tag{10}$$

then the full conditional distributions are given by:

(i) $p(\boldsymbol{\pi}_U|\boldsymbol{\Theta}/\boldsymbol{\pi}_U) = Dir(\tilde{\boldsymbol{\alpha}})$ *where* $\tilde{\alpha}_k = \alpha_k + N_k - 1$.
(ii) $p(\mathbf{Z}_i = k|\boldsymbol{\Theta}/\mathbf{Z}_i) \propto \pi_{Uk}\mathcal{N}(\mathbf{U}_i|\boldsymbol{\mu}^k_U, (\boldsymbol{\Lambda}^k_U)^{-1})$.
(iii) *Let* $\beta = \beta_U + N_k, \nu = \nu_U + N_k$ *and* $\mathbf{m} = \frac{1}{\beta}(\beta_U \mathbf{m}_U + N_k \bar{\mathbf{U}}_k), \mathbf{W} = [\mathbf{W}^{-1}_U + \mathbf{S}_k + \frac{\beta_U N_k}{\beta_U + N_k}(\bar{\mathbf{U}}_k - \mathbf{m})(\bar{\mathbf{U}}_k - \mathbf{m})^T]^{-1}$, *then*

$$p(\boldsymbol{\mu}^k_U, \boldsymbol{\Lambda}^k_U|\boldsymbol{\Theta}/\{\boldsymbol{\mu}^k_U, \boldsymbol{\Lambda}^k_U\}) = \mathcal{N}(\boldsymbol{\mu}^k_U|\mathbf{m}, (\beta\boldsymbol{\Lambda}^k_U)^{-1})\mathcal{W}(\boldsymbol{\Lambda}^k_U|\nu, \mathbf{W}). \tag{11}$$

(iv) *Let* $k = \mathbf{Z}_i$, *then* $p(\mathbf{U}_i|\boldsymbol{\Theta}/\mathbf{U}_i) = \mathcal{N}(\mathbf{U}_i|\boldsymbol{\mu}, \boldsymbol{\Lambda}^{-1})$ *where*

$$\boldsymbol{\Lambda} = \boldsymbol{\Lambda}^k_U + \tau \sum_{j=1}^{N} \mathbb{I}_{ij}\mathbf{V}_j\mathbf{V}^T_j, \quad \boldsymbol{\mu} = \boldsymbol{\Lambda}^{-1}(\tau \sum_{j=1}^{N} \mathbb{I}_{ij}\mathbf{X}_{ij}\mathbf{V}_j + \boldsymbol{\Lambda}^k_U \boldsymbol{\mu}^k_U). \tag{12}$$

Note that we only give the full conditional distributions of the user part for simplicity, and the formulas for the movie part are very similar. We summarize the Gibbs sampling steps in Algorithm 2.

5   Experimental Results

5.1   Datasets and Experimental Settings

We use MovieLens-1m and MovieLens-10m datasets [4] in our experiments. The MovieLens 1 m dataset contains $1,000,209$ ratings of $3,706$ movies made by

6,040 MovieLens users. The MovieLens-10m dataset contains 10,000,054 ratings of 10,677 movies made by 71,567 MovieLens users. We randomly split the datasets into two parts, 80% for training and 20% for testing. For experimental purpose, we only use ratings in our experiments and ignore side-information In both datasets, movies are labeled with 18 tags according to their genres, namely, Action, Adventure, Animation, Children's, Comedy, Crime, Documentary, Drama, Fantasy, Film-Noir, Horror, Musical, Mystery, Romance, Sci-Fi, Thriller, War, Western. These tags play an important role in interpreting the cluster structure discovered by the proposed methods.

There are many hyper-parameters in PMF_Cluster and BPMF_MG. For PMF_Cluster, the parameters are set similarly as in [12]. For BPMF_MG, we set $\alpha_U = \mathbf{1}_K$ and $\alpha_V = \mathbf{1}_L$, where $\mathbf{1}_K$ denotes the length-K vector whose elements are all ones. The parameters in Gauss-Wishart prior are set as $\beta_U = \beta_V = 2, \nu_U = \nu_V = D, \mathbf{m}_U = \mathbf{m}_V = \mathbf{0}$ and $\mathbf{W}_U = \mathbf{W}_V = \mathbf{I}_D$. The parameter τ is fine-tuned to obtain the best performance. We set max_iter equals to $200 * K$ with first-half samples discarded as burn-in.

Algorithm 2. Learning algorithm for BPMF_MG

Input: observed entries $\mathbf{X}_{ij}$, observation indicators $\mathbb{I}_{ij}$.
Hyper-parameters: prior precision τ, Dirichlet parameters α_U, α_V,
Wishart parameters $_U, \nu_U, \beta_U, \mathbf{W}_U$ and $_V, \nu_V, \beta_V, \mathbf{W}_V$.
Variables: latent factors U_i, V_j, labels $\mathbf{Z}_i, \hat{\mathbf{Z}}_j$, centers μ_U, μ_V,
posterior parameters of Gaussians $\Lambda_U, \Lambda_V, \pi_U, \pi_V$.

1: Initialize parameters by random guess.
2: **for** $t = 1, 2, \cdots$,max_iter **do**
3:    **for** $i = 1, 2, \cdots, M$ **do**
4:       Sample $\mathbf{Z}_i$ according to Theorem 1.(ii) and update $N_k, \bar{\mathbf{U}}_k, \mathbf{S}_k$.
5:       Sample μ_U^k, Λ_U^k according to Theorem 1.(iii).
6:       Sample $\mathbf{U}_i$ according to Theorem 1.(iv)
7:    **end for**
8:    Sample π_U according to Theorem 1.(i)
9:    Sample $\pi_V, \hat{\mathbf{Z}}_j, \mu_V^l, \Lambda_V^l, V_j$ similarly as above.
10:   **if** $t >$num_burn_in **then**
11:      **for** each (i, j) we want to predict **do**
12:         Sample $\mathbf{X}_{ij}^t$ from $\mathcal{N}(\mathbf{U}_i^T \mathbf{V}_j, \tau^{-1})$.
13:      **end for**
14:   **end if**
15:   **for** each (i, j) we want to predict **do**
16:      $\mathbf{X}_{ij}^* = \frac{1}{\text{max\_iter} - \text{burn\_in}} \sum_{t=\text{burn\_in}+1}^{\text{max\_iter}} \mathbf{X}_{ij}^t$.
17:   **end for**
18: **end for**

5.2 Performance Evaluation

The dimension of latent factor D ranges in $\{10, 20, 30, 50\}$. For simplicity, we set $K = L$ in our experiments, and $K = 3, 5, 10$ respectively. We also report the performances of PMF and BPMF under different latent factor dimension D for comparison. The root mean square error (RMSE) is used to measure the performances of various algorithms, which can be calculated by RMSE $= \sqrt{\sum_{i,j} \mathbb{I}_{ij} (\mathbf{X}_{ij} - \hat{\mathbf{X}}_{ij})^2 / \sum_{i,j} \mathbb{I}_{ij}}$, where $\mathbb{I}_{ij}$ is the indicator and equals to 1 if and only if the rating of user i on movie j is known in the test set. The experimental results are summarized in Table 1.

From Table 1, we have several observations. First, exploiting cluster structures helps to improve the results. No matter K equals to 3, 5 or 10, the corresponding results are better than origin PMF and BPMF in most cases. However, the choice of cluster number K is a little tricky: sometimes $K = 10$ leads to better performance while sometimes $K = 5$ is better. One possible way of choosing K is cross-validation, but this is very time-consuming. Second, increasing the dimension of latent factors also helps to improve the results. Generally speaking, larger dimension leads to better results since high dimensional factors can encode more information. However, the dimension should not be too large otherwise the model will suffer from over-fitting. Besides, we note that Bayesian methods using MCMC outperform non-Bayesian methods, while the MCMC procedure requires more computing time in practice.

Table 1. Performance Evaluation (RMSE) of the proposed methods on MovieLens-1m and MovieLens-10m datasets.

Dataset	Model								
	D	PMF	PMF_Cluster			BPMF	BPMF_MG		
			$K = 3$	$K = 5$	$K = 10$		$K = 3$	$K = 5$	$K = 10$
MovieLens-1m	10	0.8596	0.8560	0.8557	0.8542	0.8474	0.8460	0.8447	0.8447
	20	0.8588	0.8538	0.8531	0.8530	0.8434	0.8415	0.8416	0.8419
	30	0.8579	0.8534	0.8522	0.8528	0.8447	0.8439	0.8430	0.8437
	50	0.8567	0.8535	0.8526	0.8518	0.8423	0.8410	0.8386	0.8410
MovieLens-10m	10	0.7988	0.7951	0.7942	0.7934	0.7881	0.7875	0.7875	0.7874
	20	0.7885	0.7897	0.7888	0.7863	0.7787	0.7777	0.7775	0.7772
	30	0.7877	0.7862	0.7854	0.7855	0.7762	0.7759	0.7761	0.7755
	50	0.7876	0.7848	0.7843	0.7846	0.7746	0.7743	0.7742	0.7738

5.3 Result Interpretation

In this part, we set $D = 10, K = L = 5$ and analyze the clusters of movies learned by PMF_Cluster on MovieLens-1m. A similar analysis also applies to other settings. As mentioned in Sect. 5.1, each movie is labeled with several category tags. For each label, we calculate the distribution over 5 clusters, and the result is illustrated in Fig. 2. Intuitively, the cluster structure underlying movies are supposed to have a strong correlation with category labels. We show the distributions of different types of movies over clusters in Fig. 2. From Fig. 2, we can see the result is consistent with our intuition. For example, the distribution of Action movies and the distribution of Adventure movies are very similar, since Animation and Adventure are closely related tags. A similar result also holds for Fantasy movies and Sci-Fi movies. Besides, we note that most Documentary movies concentrate in Cluster 3, while Cluster 3 only contains a minor proportion in other types of movies. This observation provides an interpretation for Cluster 3, *i.e.*, Cluster 3 consists of Documentary-like movies.

We point out that analyzing the relationship between cluster structure and movie tags is only for experimental purpose. One may ask: why don't we just use different Gaussian priors for each type of movie? First, each movie may have more than one genre label. Second, in this paper, we never use side-information in training phase. In fact, one primary goal of the proposed methods is to discover underlying knowledge in data without side-information. Third, cluster structure is only correlated to movie labels, but not determined by movie labels. Hopefully, the proposed methods can learn more information than tags.

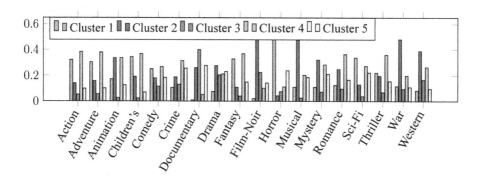

Fig. 2. Distributions of movies with different tags over 5 clusters.

5.4   Comparison with State-of-the-Art Methods

In this section, we compare PMF_Cluster and BPMF_MG with several matrix approximations based collaborative filtering algorithms. The competing methods include LLORMA [7], WEMAREC [2], MPMA [2] and MRMA [8]. The hyper-parameter of the competing methods follows the settings in the corresponding papers. As for PMF_Cluster and BPMF_MG, we choose the best result obtained in Sect. 5.2. The performances are reported in Table 2. From this table, we can see that the results of BPMF_MG are comparable or even better than the competing methods, which demonstrates the effectiveness of our proposed method. Although the results of PMF_Cluster are not very satisfying, this model has a nice intuitive interpretation, which helps us to understand BPMF_MG better. The key difference between PMF_Cluster and BPMF_MG is the inference technique. While learning parameters via gradient descent is simple and fast, this process may lead to a local minimum due to the non-convexity of objective function, or lead to the over-fitting problem. On the other hand, introducing Bayesian viewpoint complicates the model, but enables the model to integrate over uncertainty and thus obtain better results.

Table 2. Performance (RMSE) comparison on MovieLens-10m and MovieLens-10m datasets.

Dataset	Model					
	LLORMA	WEMAREC	MPMA	MRMA	PMF_Cluster	BPMF_MG
MovieLens-1m	0.8534	0.8461	0.8443	0.8391	0.8518	0.8386
MovieLens-10m	0.7912	0.7841	0.7791	0.7736	0.7843	0.7738

6 Conclusion

In this paper, we have devised two novel matrix factorization models for collaborative filtering: PMF_Cluster and BPMF_MG respectively. The basic idea is there exists cluster structure in user factors and item factors, and exploiting such cluster structure helps to improve prediction accuracy. PMF_Cluster is a modification over probabilistic matrix factorization, which introduces the idea of k-means clustering to probabilistic matrix factorization model. BPMF_MG is the Bayesian variant of PMF_Cluster, and we devised Gibbs sampling steps to make the inference. Besides, BPMF_MG can also be regarded as a modification of Bayesian probabilistic matrix factorization. This model is equivalent to replace Gaussian prior for factors with a mixture of Gaussians prior. Experimental results on real-world datasets indicate the proposed models successfully exploit cluster structure underlying data. Compared with state-of-the-art matrix approximation methods, BPMF_MG achieved comparable or better performances.

References

1. Beutel, A., Ahmed, A., Smola, A.J.: ACCAMS: additive co-clustering to approximate matrices succinctly. In: International Conference on World Wide Web (2015)
2. Chen, C., Li, D., Zhao, Y., Lv, Q., Shang, L.: WEMAREC: accurate and scalable recommendation through weighted and ensemble matrix approximation. In: ACM SIGIR International Conference on Research and Development in Information Retrieval (2015)
3. Geman, S., Geman, D.: Stochastic relaxation, gibbs distributions, and the bayesian restoration of images. IEEE Trans. Pattern Anal. Mach. Intell. **6**, 721–741 (1984)
4. Harper, F.M., Konstan, J.A.: The movielens datasets: history and context. ACM Trans. Interact. Intell. Syst. (TIIS) **5**(4), 19 (2016)

5. Hofmann, T.: Probabilistic latent semantic analysis. In: Fifteenth Conference on Uncertainty in Artificial Intelligence (1999)
6. Lee, J., Bengio, S., Kim, S., Lebanon, G., Singer, Y.: Local collaborative ranking. In: International Conference on World Wide Web (2014)
7. Lee, J., Lebanon, G., Lebanon, G., Singer, Y., Bengio, S.: LLORMA: local low-rank matrix approximation. J. Mach. Learn. Res. **17**(1), 442–465 (2016)
8. Li, D., Chen, C., Liu, W., Lu, T., Gu, N., Chu, S.: Mixture-rank matrix approximation for collaborative filtering. In: Advances in Neural Information Processing Systems (2017)
9. Mackey, L., Talwalkar, A., Jordan, M.I.: Divide-and-conquer matrix factorization. In: Advances in Neural Information Processing Systems (2011)
10. Marlin, B.: Modeling user rating profiles for collaborative filtering. In: Advances in Neural Information Processing Systems (2003)
11. Marlin, B., Zemel, R.S.: The multiple multiplicative factor model for collaborative filtering. In: International Conference on Machine Learning (2004)
12. Salakhutdinov, R., Mnih, A.: Probabilistic matrix factorization. In: Advances in Neural Information Processing Systems (2007)
13. Salakhutdinov, R., Mnih, A.: Bayesian probabilistic matrix factorization using Markov chain Monte Carlo. In: International Conference on Machine Learning (2008)
14. Shan, H., Banerjee, A.: Generalized probabilistic matrix factorizations for collaborative filtering. In: IEEE International Conference on Data Mining (2011)
15. Wang, K., Zhao, W.X., Peng, H., Wang, X.: Bayesian probabilistic multi-topic matrix factorization for rating prediction. In: International Joint Conference on Artificial Intelligence (2016)
16. Xu, B., Bu, J., Chen, C., Cai, D.: An exploration of improving collaborative recommender systems via user-item subgroups. In: International Conference on World Wide Web. ACM (2012)
17. Zhang, Y., Zhang, M., Liu, Y., Ma, S.: Improve collaborative filtering through bordered block diagonal form matrices. In: ACM SIGIR International Conference on Research and Development in Information Retrieval. ACM (2013)

Ensemble Classifier Generation Using Class-Pure Cluster Balancing

Zohaib Jan[✉] and Brijesh Verma

Center for Intelligent Systems, School of Engineering and Technology,
Central Queensland University, 160 Ann Street, Brisbane, Australia
{z.jan, b.verma}@cqu.edu.au

Abstract. Clustering based ensemble of classifiers have shown a significant improvement in classification accuracy in many real-world applications. Most of the existing clustering-based ensemble approaches generate and use predefined number of data clusters. However, datasets have different spatial structure that depends on number of characteristics for example class labels. Therefore, using a predefined set of hyperparameters to generate a clustering-based ensemble classifier is not an effective methodology. In this paper we propose a methodology to overcome this limitation by generating dataset dependent strong and balanced data clusters per class. This ensures that any spatial information that is inherent in the dataset can be exploited to train an ensemble classifier that can surpass the classification accuracy plateau. An ensemble classifier framework is proposed that benefits from this methodology and trains base classifiers on generated strong and balanced data clusters. We have evaluated the proposed approach on 8 benchmark datasets from UCI repository. Detailed experiments and results are presented in the paper, and it is evident from the results that varying the number of clusters per class does have an impact on the overall classification accuracy of the ensemble.

Keywords: Ensemble learning · Classifiers · Clustering · Diversity

1 Introduction

Ensemble classifier is a machine learning classification methodology that aims to improve the classification performance of a single classifier model by strategically combining multiple single classifiers [1]. Ensemble classifiers are very effective in classifying real-world noisy datasets because they benefit from the "perturb and combine" strategy [2] and have been widely applicable in various areas including pattern recognition and computer vision [3]. Input data is first perturbed by utilizing various strategies and multiple classifiers are trained on these perturbed training sets. All base classifiers are then strategically combined to generate an ensemble classifier.

A common strategy for perturbing the input data is known as the random subspace method [4], where random subsamples of the training data are generated typically with unique and repeating set of records. Base classifiers are trained on different perturbed subsamples and are then suitably combined. Since each base classifier is trained on a different subsample of the training data therefore, it has different learning capabilities

© Springer Nature Switzerland AG 2019
T. Gedeon et al. (Eds.): ICONIP 2019, CCIS 1143, pp. 761–769, 2019.
https://doi.org/10.1007/978-3-030-36802-9_80

and contributes to the overall ensemble diversity. Another strategy of perturbing input data is by generating random subsets of the input features rather than samples and train multiple base classifiers on generated features subsets. Examples of ensemble strategies that exploit input sample space are bagging and boosting [5, 6]. Bagging works by generating random subsamples from the input data also known as "bags" and trains base classifier on generated bags. Classification decision of all trained classifiers are fused, and ensemble is formed. Boosting on the other hand works by successively training base classifiers on data patterns which previously trained classifiers performed poorly on. Many variations of boosting [7–10] have been proposed over the years and a hallmark work is Adaboost [9] which has been very successful in classifying noisy datasets.

An ensemble strategy that exploits input feature space is Random Forest (RaF) [11]. RaF trains a multitude of Decision Trees (DT) on randomized bags and features in order to generate an ensemble of DTs with controlled variance. RaF to date is perceived as a state-of-the-art ensemble classifier and was rated as one of the best classifiers in a benchmark study of 179 classifiers on 121 benchmark datasets from UCI machine learning repository [12]. Rotation Forest (RoF) [13] is a variation of RaF that splits training data into K subsets and Principal Component Analysis (PCA) is applied to each subset. The features are rotated, and DTs are trained on the rotated feature subsets. The main reason for the robustness of both RaF and RoF is the ability to reduce the bias and variance of classifiers in an ensemble. The bias-variance decomposition theory states that a classifier's generalization error with respect to a problem can be decomposed into bias and variance [14].

Classifier accuracy and diversity are also considered to be the two competing objectives when generating ensemble classifiers. Accuracy is the ability of a classifier to generate class labels as close to the ground truth as possible, whereas, diversity is the difference in the errors of two classifiers. If classifiers are selected to generate an ensemble only based on accuracy this causes the overall ensemble accuracy to reach a plateau, whereas, if classifiers are selected based on diversity only that effects the overall classification accuracy of the ensemble negatively. Many different diversity measures have been proposed in research and they are discussed in [15]. Topic of accuracy and diversity is debatable in ensembles because some authors argued that the end objective of any ensemble classifier is to achieve higher accuracy [16]. However, others have utilized various strategies to incorporate diversity in ensemble whilst maximizing accuracy. A tradeoff must be maintained between accuracy and diversity because incorporating diversity in ensembles enables it to break through the plateau [17]. As such a rule-based accuracy and diversity comparisons-based classifier selection methodology was proposed in [18] to generate an ensemble classifier. Classifiers are selected based on increasing accuracy, if accuracy cannot be increased then diversity is measured, otherwise the classifier is discarded. Clustering-based ensemble classifiers have also been an area of focus [19–22], especially since training base classifiers on different data clusters allows for the incorporation of diversity in ensembles. Random subspace is generated by clustering input data incrementally into sparse data clusters instead of employing other subsampling strategies like bagging. Since data clusters identify dense local regions and base classifiers trained on such clusters can identify local regions more effectively thus incorporating spatial information as well. The main idea is to break a complex decision boundary into smaller local regions which can be identified by weak learners (classifiers) rather than training a classifier to learn a complex decision boundary. Moreover, according to the "no free lunch" theorem [23] a classifier performing well in one instance may not perform well in the other. Since classifiers are trained on dense local regions or simply are exposed to different aspects of the datasets, they bring with them different learning capabilities contributing to overall diversity of the ensemble as well.

While clustering-based ensemble approaches have successfully been adopted to deal with different kinds of datasets, the selection of clusters and type of clusters need a further consideration. Datasets have randomness and noise in them and generating a random subspace through clustering with a predefined upper bound of clustering for different datasets is not an ideal strategy; each dataset has different characteristics and one dataset may require more clusters as there are more dense local regions whereas others might require less. Therefore, a clear distinction of the types and number of clusters required that can generate a rich and diverse input space for training base classifiers is required. The novel contributions of this paper are as follows:

- A method for creating multiple strong class pure data clusters.
- A method for balancing strong class pure clusters by adding samples from minority classes that are closest to the cluster centroid.
- A method for fusion of strong and balanced classifiers to generate an ensemble classifier.

The remainder of this paper is organized as follows. Section 2 introduces the proposed approach. Section 3 evaluates the performance of the proposed approach using real-world benchmark datasets. Section 4 draws conclusion and give future directions.

2 Proposed Approach

2.1 Strong and Balanced Data Clusters

A data cluster is classified as a strong data cluster if it favors data samples from a specific class. To identify a cluster as a strong data cluster the standard deviation of the frequency distribution of the class labels in a cluster is calculated. For a given dataset $X = \{(x_1, y_1), (x_2, y_2), \ldots, (x_n, y_n)\}$ having feature vector $x \in \mathbb{R}^D$ each associated with a class label $y_i \in 1, .., V$ where V is the number of discrete class labels. A data cluster is a sub-sample of the input data given as $C = \{(x_1, y_1), \ldots, (x_m, y_m)\}$ where $m \ll n$. The balance measure b of a data cluster C is computed as follows:

$$b = \sqrt{\frac{\sum [f(x) - \bar{x}]^2}{m}} \tag{1}$$

where $f(x)$ is the frequency distribution of the class labels in a data cluster and is calculated as:

$$f(x) = \sum_{j=1}^{V} \sum_{i=1}^{m} I(y_j == y_i') / |m| \tag{2}$$

where y_i is a class label of feature vector x_i, y_i' is the class label in cluster C and m is number of data samples in a data cluster.

All frequency distributions are normalized in order to get values between 0 and 1. A data cluster is considered a strong data cluster if it has a skewed frequency distribution and is considered a balanced data cluster if it has a balance score of 0 or simply has a uniform class frequency distribution.

Data clusters are generated incrementally using the training dataset X until N data clusters for each class are generated. From the dataset unique class labels are identified and a row vector representing each class labels is generated as follows:

$$s = \left\{ N^1, N^2, \ldots, N^V \right\} \tag{3}$$

where V is the number of unique class labels in the dataset and N^i is the number of data clusters required for class i.

In the first iteration the dataset is clustered using $k = 2$ resulting in 2 data clusters. The clusters are identified which class they belong to and the respective N is decremented. In the next iteration k is incremented and 3 data clusters are generated. The process is repeated until there is no non-zero N in the row vector s. Clusters generated are either balanced (i.e. they have same number of samples from each class) or they are strong (i.e. they have more samples from one class than others). If a data cluster generated is balanced it is placed in the pool of data clusters and if it is strong then it will go through the process of balancing in the next step.

2.2   Balancing Strong Data Clusters

Any data cluster that does not have a uniform distribution of class labels is considered a class pure or strong data cluster. To convert a strong data cluster into a balance data cluster, data samples belonging to the minority class that are closest to the cluster centroid are added to the cluster. Considering a data cluster $C = \{(x_1, 1), (x_2, 1), (x_3, 1), (x_4, 1), (x_5, 2), (x_6, 2), (x_7, 3), (x_8, 3)\}$ having 4 data samples from class 1, 2 from class 2, and 2 from class 3. To balance the data cluster C first majority class is identified. This is done by taking the mode of the class labels which will give the majority class. The same is repeated for bimodal or trimodal data cluster using one vs all principle. A row vector of classes required to be added to the cluster is created defined as follows:

$$minority = \{0, 2, 2\} \tag{4}$$

Here each element represents the number of data samples that are required to be added to the given cluster to balance it. Distances of each minority class from the cluster centroid are computed and the samples that are closest to the centroid are added to the cluster C. In each increment if a sample is added from a given minority class into the cluster the respective element is decrement by 1. The process is repeated until all elements are 0 in the row vector *minority* or there are no more samples available. For cluster C having centroid c the distance is measured as follows:

$$argmin(d(x_i, c)) \, \forall \, x_i \in minority$$
$$where \; d(x_i, c) = \sum |x_i - c| \tag{5}$$

2.3   Ensemble Training Framework

The proposed approach starts off by generating a rich and diverse input space for training base classifiers. Instead of creating bags of random input samples we proposed a new concept of clustering input data by incrementally generating sparse data clusters identifying local regions within the data. For a training data set X having feature vector x, and class labels y, N data clusters are generated for each class by minimizing the squared Euclidean.

If the class labels are $y \in \{1, 2, 3\}$ and N is set to 3, then 3 data clusters for each class 1, 2, and 3 will be generated resulting in a total of 10 data clusters in the pool (1 data cluster for $k = 1$). The process increases the value of k incrementally until N data clusters of each class are generated. All clusters in the pool are then checked for their balance score. Any cluster that is class pure or does not have a uniform class frequency distribution goes through the process of balancing and the pool of clusters is updated. Using all clusters in the pool M', a base classifier is trained on all balanced data clusters. Each cluster is a balanced data cluster if it has the same number of data samples from each class so that a base classifier trained on such data cluster is exposed to all samples from various target classes. The benefit here is although each base classifier is exposed to all class labels however it has some inherent local expertise due to clustering and every classifier is different from other in a sense that it is trained on a different local region (on a different data cluster). All trained classifiers are then added to the base classifier pool bcp which are then utilized to generate the final ensemble predicted class labels.

3   Experiments and Analysis

To evaluate the proposed ensemble classifier 8 benchmark datasets from the UCI machine learning repository [24] were used. The details of these datasets such as number of records, number of features and class labels are given in Table 1.

Table 1. Details of benchmark datasets used in experimentation

Dataset	Number of records	Number of features	Class labels
Sonar	208	60	2
Heart	270	13	2
Bupa	345	6	2
Ionosphere	351	34	2
Breast Cancer	683	9	2
Pima Diabetes	768	8	2
Vehicle	846	18	4
Segmentation	2310	19	7

Similar to a number of recent studies [22] we have used 10-fold Cross Validation (CV) for comparative analysis. The proposed ensemble approach was implemented in MATLAB 2017 R1 [25]. For base classifier in this study we have used a multi-class version of Support Vector Machine (SVM), however any classifier can be used here as the purpose of this study was not to find the type of base classifier but to identify the effect of different clustering values on classification accuracy of the ensemble. Default implementation of k-means was used for generating data clusters. The average classification accuracy of the ensemble on with different values of clustering different datasets is summarized in Fig. 1 below.

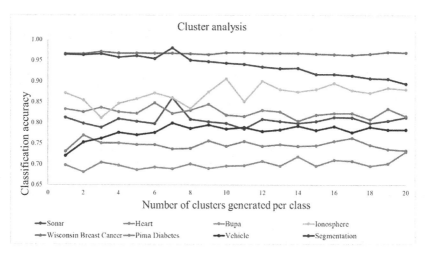

Fig. 1. Effect of different per class clusters on classification accuracy

It can be noted from Fig. 1 that varying the number of clusters per class does indeed have an impact on the classification accuracy of the ensemble. For each dataset there is an optimum value of data clusters that are needed to achieve the highest classification accuracy. This demonstrates that each dataset has different characteristics and spatial information. Therefore, by training base classifiers on different dense local regions enables an ensemble classifier to learn minuscule changes in the decision boundary enabling it to achieve higher classification accuracy.

3.1 Comparative Analysis

The classification accuracy of the proposed ensemble classifier is also compared with recent state-of-the-art ensemble classifiers. A clustering-based ensemble classifier proposed in [19], a weighted voting framework-based ensemble classifier proposed in [16], a RaF based ensemble classifier proposed in [12], and finally a RaF ensemble classifier, are used for comparative analysis. RaF is included as a benchmark ensemble classifier due to its robustness and high performance. The classification accuracies are taken directly from the respective papers with highest accuracy given in bold in Table 2. Out of 8 datasets used in experimentation the proposed ensemble classifier outperformed existing ensemble classifiers in 6. Only Sonar and Ionosphere datasets were not positively affected which means that there is little to no spatial dependency between the data and the decision boundary.

Table 2. Comparative analysis of the proposed ensemble classifier with recent state-of-the-art ensemble classifiers. Higher classification accuracies are given in bold

Dataset	Proposed approach	Fletcher 2017 [19]	Kuncheva 2014 [16]	Zhang 2015 [12]	Random Forest
Sonar	0.8613	**0.8705**	0.7620	0.8077	0.8540
Heart	**0.8481**	0.8222	0.8050	0.8237	0.8190
Bupa	**0.7329**	0.7310	0.6720	N/A	0.7273
Ionosphere	0.9060	0.9203	0.9170	**0.9470**	0.9297
Breast Cancer	**0.9714**	0.9678	0.9600	0.9667	0.9610
Pima Diabetes	**0.7713**	0.7694	0.7550	0.7526	0.7604
Vehicle	**0.7989**	0.7648	0.7250	0.7781	0.7565
Segmentation	**0.9872**	0.9675	0.9640	0.9804	0.9800

4 Conclusion

Data diversity incorporated through clustering and other means is essential when it comes to generating an ensemble classifier. However, using a predefined upper bound of clustering for every dataset is not an ideal strategy. It can be seen from the results that different datasets have different spatial dependencies and only the optimum number of balanced/strong clusters can generate an ensemble classifier that can achieve highest classification accuracy. Balancing class pure clusters enables a base classifier that is trained on such data clusters to learn the miniscule decision boundary changes, which evidently adds to the fact that instead of learning a complex decision boundary it is advantages to learn dense local regions with a simple decision boundary.

In future we will conduct further experiments on real-world and benchmark data-sets to analyze the effect of different clustering hyper-parameters on ensemble classification accuracy.

References

1. Ren, Y., Zhang, L., Suganthan, P.N.: Ensemble classification and regression-recent developments, applications and future directions. IEEE Comput. Intell. Mag. **11**(1), 41–53 (2016)
2. Breiman, L.: Bias variance, and arcing classifiers (1996)
3. Goerss, J.S.: Tropical cyclone track forecasts using an ensemble of dynamical models. Monthly Weather Rev. **128**(4), 1187–1193 (2000)
4. Barandiaran, I.: The random subspace method for constructing decision forests. IEEE Trans. Pattern Anal. Mach. Intell. **20**(8), 1–22 (1998)
5. Breiman, L.: Bagging predictors. Mach. Learn. **24**(2), 123–140 (1996)
6. Dietterich, T.G.: Ensemble methods in machine learning. Multiple Classifier Syst. **1857**, 1–15 (2000)
7. Seiffert, C., Khoshgoftaar, T.M., Van Hulse, J., Napolitano, A.: RUSBoost: a hybrid approach to alleviating class imbalance. IEEE Trans. Syst. Man Cybern. Part A Syst. Hum. **40**(1), 185–197 (2010)
8. Avidan, S.: SpatialBoost: adding spatial reasoning to AdaBoost. In: Leonardis, A., Bischof, H., Pinz, A. (eds.) ECCV 2006. LNCS, vol. 3954, pp. 386–396. Springer, Heidelberg (2006). https://doi.org/10.1007/11744085_30
9. Vezhnevets, A., Vezhnevets, V.: Modest AdaBoost-teaching AdaBoost to generalize better. Graphicon **12**(5), 987–997 (2005)
10. Domingo, C., Watanabe, O.: MadaBoost: a modification of AdaBoost. In: International Conference on Computational Learning Theory, 2000, pp. 180–189 (2000)
11. Liaw, A., Wiener, M.: Classification and regression by random forest. R News **2**(3), 18–22 (2002)
12. Zhang, L., Suganthan, P.N.: Benchmarking ensemble classifiers with novel co-trained kernal ridge regression and random vector functional link ensembles. IEEE Comput. Intell. Mag. **12** (4), 61–72 (2017)
13. Rodriguez, J.J., Kuncheva, L.I., Alonso, C.J.: Rotation forest: a new classifier ensemble method. IEEE Trans. Pattern Anal. Mach. Intell. **28**(10), 1619–1630 (2006)
14. Kohavi, R., Wolpert, D.H.: Bias plus variance decomposition for zero-one loss functions. In: International Conference on Machine Learning, 1996, vol. 96, pp. 275–283 (1996)
15. Kuncheva, L.I., Whitaker, C.J.: Measures of diversity in classifier ensembles and their relationship with the ensemble accuracy. Mach. Learn. **51**(2), 181–207 (2003)
16. Kuncheva, L.I., Rodriguez, J.J.: A weighted voting framework for classifiers ensembles. Knowl. Inf. Syst. **38**(2), 259–275 (2014)
17. Zhou, Z.-H.: Ensemble Methods: Foundations and Algorithms. CRC Press, Boca Raton (2012)
18. Asafuddoula, M., Verma, B., Zhang, M.: An incremental ensemble classifier learning by means of a rule-based accuracy and diversity comparison. In: International Joint Conference on Neural Networks, 2017, pp. 1924–1931 (2017)
19. Fletcher, S., Verma, B.: Removing bias from diverse data clusters for ensemble classification. In: International Conference on Neural Information Processing, 2017, pp. 140–149 (2017)

20. Rahman, A., Verma, B.: Ensemble classifier generation using non-uniform layered clustering and Genetic Algorithm. Knowl. Based Syst. **43**, 30–42 (2013)
21. Verma, B., Rahman, A.: Cluster-oriented ensemble classifier: Impact of multi-cluster characterization on ensemble classifier learning. IEEE Trans. Knowl. Data Eng. **24**(4), 605–618 (2012)
22. Rahman, A., Verma, B.: Novel layered clustering-based approach for generating ensemble of classifiers. IEEE Trans. Neural Networks **22**(5), 781–792 (2011)
23. Wolpert, D.H., Macready, W.G.: No free lunch theorems for optimization. IEEE Trans. Evol. Comput. **1**(1), 67–82 (1997)
24. Lichman, M.: UCI Machine Learning Repository (2013)
25. MATLAB: Statistics and Machine Learning Toolbox. The MathWorks Inc., Natick (2013)

Temporal Continuity Based Unsupervised Learning for Person Re-identification

Usman Ali[✉], Bayram Bayramli, and Hongtao Lu

Shanghai Jiao Tong University, Shanghai, China
{usmanali,bayram_bai,htlu}@sjtu.edu.cn

Abstract. Most existing person re-identification (re-id) methods generally require a large amount of, difficult to collect, identity-labeled data to act as discriminative guideline for representation learning. To overcome this problem, we propose an unsupervised center-based clustering approach capable of progressively learning and exploiting the underlying re-id discriminative information from temporal continuity within a camera. We call our framework Temporal Continuity based Unsupervised Learning (TCUL). Specifically, TCUL simultaneously does center based clustering of unlabeled (target) dataset and fine-tunes a convolutional neural network (CNN) pre-trained on irrelevant labeled (source) dataset to enhance discriminative capability of the CNN for the target dataset. Furthermore, it exploits temporally continuous nature of images within-camera jointly with spatial similarity of feature maps across-cameras to generate reliable pseudo-labels for training a re-identification model. Extensive experiments on three large-scale person re-id benchmark datasets demonstrate superiority of TCUL over existing state-of-the-art unsupervised person re-id methods.

Keywords: Unsupervised learning · Person re-identification · Clustering

1 Introduction

Person re-identification (re-id) is computer vision problem that aims to match images of a person captured by different cameras with non-overlapping views [19]. It is viewed as a search problem with a goal to retrieve the most relevant images to the top ranks [18]. Many recent works have attracted extensive research efforts to address the re-id problem using deep learning [5,10,11,17]. In spite of remarkable advancements achieved by deep learning methods, most existing techniques adopt a supervised learning paradigm to solve the re-id problem [10,17] assuming availability of sufficient manually labelled training data for each camera network. This limits the generalizing capability of a re-id model to multiple camera networks due to lack of labelled training samples under a new environment. As a result, many previous works have addressed this scarcity of labelled training data by focusing on unsupervised or semi-supervised learning

© Springer Nature Switzerland AG 2019
T. Gedeon et al. (Eds.): ICONIP 2019, CCIS 1143, pp. 770–778, 2019.
https://doi.org/10.1007/978-3-030-36802-9_81

[1,6,8]. Most of these works typically deal with relatively small datasets; without being able to exploit the potential of deep learning methods that require large-scale datasets for better performance [6,7,13]. More recently, there has been greater emphasis towards clustering and domain adaptation based deep unsupervised methods for person re-id [3,12,14]. However, performance of unsupervised learning approaches is much weaker as compared to supervised models due to difficulty in learning the discriminative representation in the absence of cross-view identity labelled matching pairs.

To improve this situation, we propose to use temporal continuity of images within a camera jointly with spatial similarity of feature maps across-cameras to generate reliable pseudo-labels for training a re-id model. The idea is that given an image in a camera, other images of the same identity would exist in the temporal vicinity of that image. However, temporal continuity is only effective within a camera; hence, it cannot be utilized to obtain cross-camera matching identity pairs. To this end, we propose a cross-camera center-based pseudo-labeling method to cluster features by their similarity to cluster-centers. An intersection between cross-camera pseudo-labels and within-camera temporal vicinity results in highly reliable pseudo-labels for fine-tuning a re-id model. In summary, we propose a self-paced progressive learning method which provides cross-view discriminative information for unsupervised person re-id task by utilizing a center-update based pseudo-labeling approach. Moreover, we exploit temporal continuity of images within a camera to sample reliable data-points. Finally, we demonstrate the effectiveness of our method on three large-scale datasets.

2 Related Work

Many works have adopted deep learning based metric learning or feature representation learning approaches to achieve state-of-the-art results in person re-id. Metric learning methods learn a similarity metric using image pairs or triplets as inputs [10,17]. Li et al. [17] used a deep siamese network to learn a similarity metric directly from image pixels and Yi et al. [10] proposed a patch matching based filter pairing network to classify similar and dissimilar images. On the other hand, feature representation learning generally uses classification models to address re-id task. A pedestrian identity-predicting model is trained to extract features from last layers as discriminative embedding during testing [19]. Hermans et al. [5] employed an efficient variant of triplet loss based on distance among features to learn identity discriminative representation. Some works like [2,20] demonstrated the effectiveness of generative adversarial networks (GANs) to generate additional samples for re-id training. In this work, we adopt a supervised model trained on source data as our baseline re-id model.

Recently, some works focus on unsupervised re-id using deep learning. Li et al. [9] used video-based person trajectory information obtained from tracking algorithms to create pseudo-labels. Wang et al. [15] proposed a network which learns to transfer attribute-semantic and identity-discriminative representation across domains. Peng et al. [14] developed a multi-task dictionary learning framework for cross-dataset representation transfer. A few works also apply GANs to

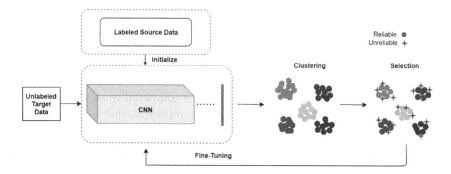

Fig. 1. Best viewed in color. Architecture for TCUL. A CNN model initialized on a labeled source dataset is used to extract features from unlabeled target dataset for similarity based clustering. Then reliable samples are selected by exploiting temporal continuity within a camera for fine-tuning the CNN model. Apart from initialization, all other steps are repeated for multiple iterations. (Color figure online)

adapt representation from one domain to another [1]. Fan et al. proposed a clustering and fine-tuning based method in [3] known as progressive unsupervised learning (PUL) which has a similar learning objective as ours. However, unlike PUL, which uses *k-means* for clustering and performs distance based sample selection, we propose a more stable center-update based clustering and utilize temporal continuity to sample reliable data points.

3   The Proposed Method

In this section, we present our proposed method TCUL illustrated in Fig. 1 and Algorithm 1.

3.1   Cluster Assignment

The first step is to assign each data point in unlabeled target dataset (T) to a cluster which is equivalent to defining a class labels. Following Ding et al. [2] we use similarity between feature representation of a person image and cluster centers as a measure to decide which center the image belongs to. Inspired by many recent domain adaptation methods for unsupervised person re-id [3,12,14], we initialize our baseline model for feature extraction, f_{θ_0}, using an irrelevant labeled dataset (S) generated with a completely different camera network and having no identity overlap with T. Cluster centers are initialized using standard *k-means* clustering on features, and distance between an image and cluster center is measured by cosine similarity:

$$sim(x_i\,,\,c_k) = \frac{f_\theta(x_i)\,.\,c_k}{|f_\theta(x_i)|\,|c_k|} \tag{1}$$

where $f_\theta(x_i)$ represents feature vector of x_i and c_k denotes kth cluster center out of K total clusters. Now, we can assign a label l to each data point as:

$$l(x_i) = \underset{k}{argmax} \; sim(x_i, c_k), \quad where \; k \in [1, K] \tag{2}$$

Fig. 2. Top and bottom rows show image fragments sorted by Frame IDs from cameraID 2 of DukeMTMC and Market-1501, respectively. Each fragment consists of images from one camera during a very short period. Multiple images of one person exist in a fragment verifying their temporal closeness within-camera.

3.2 Sample Selection

Cluster assignment process yields unreliable pseudo-labels for T due to existence of domain-specific features in S used to train baseline model f_{θ_0} [12]. One naive method to filter out unreliable data points is to use cosine similarity of features with cluster-centers [3]. However, this approach suffers from two issues. Firstly, a higher cosine similarity value does not necessarily translate to a higher confidence on pseudo-labels due to existence of domain specific features. Secondly, selected data points have an intrinsic problem of being uninformative because performance of re-id task is more sensitive to different pairs with minimal distance as opposed to similar pairs with minimal distance [5]. Instead, we propose to exploit temporal continuity of images within a camera for sample selection. This proposition is based on s basic observation typical to a camera network [9]. Most people would appear in a camera view only once for a small time during a short period used for recording. This observation implies that all images of a person taken by one camera would appear in a short continuous period, assuming no reappearance, Fig. 2.

Specifically, the idea is that two images within one camera belonging to one identity should also be temporal neighbors. To put it another way, two images in one camera with similar pseudo-labels must have Frame IDs close to each other. We define a temporary cluster center for the subset of images belonging to one camera in a particular cluster. Let X_{jk} be all the images belonging to camera j and cluster k. We define $x_{min_{jk}} \in X_{jk}$ such that $sim(f_\theta(x_{min_{jk}}), c_k) \geq sim(f_\theta(x_{jk}), c_k)$ where $x_{jk} \in X_{jk}$ and $x_{jk} \neq x_{min_{jk}}$. We then define $c_{jk} = f_\theta(x_{min_{jk}})$ as a temporary cluster center for X_{jk}. Reliable samples are then selected as all those images in X_{jk} whose Frame ID is within a threshold to the Frame ID of $x_{min_{jk}}$. Mathematically, selection function for an image x_{jk} in camera j and cluster k is defined as the following indicator function:

$$S(x_{jk}) = I(\left| fid(x_{min_{jk}}) - fid(x_{jk}) \right| \leq \lambda_{fr}) \tag{3}$$

where $fid(.)$ gives the Frame ID, I is an indicator function and λ_{fr} is a positive integer for limiting the range of Frame IDs belonging to one identity.

Algorithm 1: Temporal Continuity Based Unsupervised Learning

Input   : Labelled source data S; Unlabeled target data $T = \{x_i\}_{i=1}^N$; Number of iterations N_{iter}; Number of clusters K; Selection frame range λ_{fr}; Triplet margin m; Center update coefficient λ_c

Output: Re-id model $f_\theta(.)$ for target data

Train a model f_{θ_0} on S;

k-means : initialize $\{c_k\}_{k=1}^K$;

for $iter = 1$ *to* N_{iter} **do**

    extract features: $z_i = f_{\theta_{iter-1}(x_i)}$;

    l_i : assign clusters using Eq. 2;

    D_{iter} : Select reliable data using Eq. 3;

    Fine-tune: $f_{\theta_{iter}} =$ Train model on D_{iter};

    Update Centers $\{c_k\}_{k=1}^K$ using Eq. 6;

end

3.3   Fine-Tuning

In order to fine-tune the CNN model over selected samples for T, we minimize a mini-batch based triplet loss to pull similarly labeled images closer in feature space while pushing differently labeled images away.

$$L_{trip_total} = \sum_{x_a:x_a,x_p,x_n \in B} L_{trip}(x_a, x_p, x_n) \tag{4}$$

where L_{trip} denotes triplet loss for anchor x_a, positive sample x_p and negative sample x_n in mini-batch B. Inspired by Hermans et al. [5], we perform Online Hard Mining to generate triplets that results in faster convergence and better performance. After each iteration, cluster-centers are updated using features from fine-tuned CNN model such that centers are better representative of T. Specifically, for each cluster center c_k we compute an update term as given by:

$$\Delta c_k = \frac{I(l_i = k).(c_k - f_\theta(x_i))}{\beta + \sum_{1=1}^N I(l_i = k)} \tag{5}$$

where β is a small positive number to avoid division by zero. l_i denotes the pseudo-label for image x_i, N is the size of T and $I(condition)$ is an indicator function. Cluster centers are updated by following simple equation:

$$c_{k_{updated}} = c_k - \lambda_c \Delta c_k \tag{6}$$

Table 1. Comparison with state-of-the-art methods. A→B refers to A as source dataset and B as target dataset.

Method	Duke→Market			Market→Duke			MSMT→Market			MSMT→Duke		
	rank1	rank5	mAP	rank1	rank5	mAP	rank1	rank5	mAP	rank1	rank5	mAP
Baseline	42.8	57.2	18.8	26.5	41.5	11.8	47.7	66.5	24.2	48.5	65.1	29.2
PUL [3]	44.7	59.1	20.1	30.4	44.5	16.4	–	–	–	–	–	–
TJ-AIDL [15]	58.2	74.8	26.5	44.3	59.6	23.0	–	–	–	–	–	–
ARN [12]	70.3	80.4	39.4	60.2	73.9	33.4	–	–	–	–	–	–
SPGAN + LMP [1]	58.1	76.0	26.9	49.6	62.6	26.4	–	–	–	–	–	–
TAUDL [9]	63.7	–	41.2	61.7	–	43.5	–	–	–	–	–	–
SUL (ours)	72.5	86.4	47.5	62.2	77.4	42.5	69.1	81.5	40.3	67.1	77.3	44.9
TCUL (ours)	**75.8**	**89.0**	**51.7**	**64.0**	**79.2**	**45.0**	**72.1**	**86.3**	**44.5**	**69.8**	**81.9**	**48.6**

where λ_c is regularization parameter to control the change in cluster-centers. This update method provides a smooth transition from one set of cluster-centers to the next by perturbing the original center values only by a small amount given by Eq. 5.

4 Experiments

4.1 Implementation Details

We evaluate our proposed method on three large-scale benchmark person re-id datasets, Market-1501 (Market) [18] and DukeMTMC-reID (Duke) [20] and MSMT17 (MSMT) [16]. We use ResNet-50 [4] as our backbone CNN model. During baseline model training on S, we insert two fully connected layer of size 2048 and 702 (depending upon number of identities in S). We train the baseline model using triplet loss and cross entropy loss on resized images of shape 256×128 for 150 epochs. When training on T, last fully connected layer is removed. Triplet loss along-with SGD is used for optimization. During each iteration, we perform data augmentation by randomly erasing and flipping the images. The model is fine-tuned for 20 epochs in one iteration. Other training parameters are as follows: number of identities per batch = 16, number of images per identity = 8, number of iterations = 10, learning rate = 5×10^{-4}, m = 0.5, $\lambda_c = 0.5$, $\lambda_{fr} = 100$. For evaluation purposes, we report rank-1 and 5 accuracy along with mean average precision (mAP) following Zheng et al. [18]. All experiments employ single query.

4.2 Comparison with State-of-the-art Methods

We compare our method (TCUL) with existing state-of-the-art unsupervised person re-id approaches on Market, Duke and MSMT datasets in Table 1. Firstly, TCUL improves the baseline results 32.9% and 33.2% in mAP values on

Market→Duke and Duke→Market, respectively. Similar improvements can also be observed for on MSMT→Duke and MSMT→Market[1]. These large advancements verify that TCUL is an effective framework to utilize learned representations on source dataset for learning discriminative representation on target dataset. Secondly, we observe that our approach achieves a clear improvement on Market and Duke datasets when compared with state-of-the-art methods.

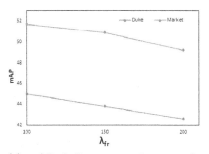

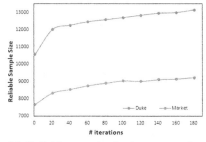

(a) mAP declines as we increase the value for λ_{fr}

(b) Reliable sample size increases after each iteration

Fig. 3. Impact of parameter changes on TCUL

4.3 Further Evaluation of TCUL

In order to highlight the effect of temporal continuity based sample selection, we follow PUL [3] to select reliable samples by measuring similarity of each feature vector from respective cluster-center. We employ the threshold value 0.85 for sample selection as used in PUL. We call this modified framework as Similarity-based Unsupervised Learning (SUL). Table 1 shows that performance of SUL has significantly reduced as compared to TCUL for all three datasets verifying the effectiveness of temporal continuity based reliable sample section. Table 1 also highlights the overall superiority of our framework, as SUL, which uses the same sample selection mechanism as PUL, still outperforms all existing methods.

As images belonging to one person in a camera should be temporally close to each other, increasing the value of λ_{fr} above an optimal value should drop reliability of selected samples. Figure 3a verifies this where increasing λ_{fr} from 150 to 200 significantly lowers the mAP values. Following the trends shown in Fig. 3a, we can deduce that performance of re-id model will further fall down if there is no sample selection step. Finally, we also compute the variance in reliable samples size after each iteration. Ideally, as the training progresses, we should have better cluster assignment producing more reliable samples. Figure 3b shows that after each iteration, there is a steady growth in the size of reliable sample set for both datasets. Therefore, we can conclude that clustering and fine-tuning steps alternatively improve each other by generating pseudo-labels that are more reliable.

---

[1] MSMT is only used as source dataset since Frame ID information is not available.

5   Conclusion

In this paper, we proposed a temporal continuity based unsupervised learning (TCUL) method for person re-id. It iterates between similarity-based clustering, reliable sample selection using temporal continuity of images within a camera and fine-tuning a CNN model on selected samples. Experimental results show that TCUL outperforms existing unsupervised person re-id methods.

Acknowledgements. This paper is supported by NSFC (No. 61772330, 61533012, 61876109), the pre-research project (no. 61403120201), Shanghai authentication Key Lab. (2017XCWZK01), and Technology Committee the interdisciplinary Program of Shanghai Jiao Tong University (YG2019QNA09).

References

1. Deng, W., Zheng, L., Ye, Q., Kang, G., Yang, Y., Jiao, J.: Image-image domain adaptation with preserved self-similarity and domain-dissimilarity for person re-identification. In: CVPR 2018 (2018)
2. Ding, G., Zhang, S., Khan, S., Tang, Z., Zhang, J., Porikli, F.M.: Feature affinity-based pseudo labeling for semi-supervised person re-identification. IEEE Trans. Multimedia **21**, 2891–2902 (2018)
3. Fan, H., Zheng, L., Yang, Y.: Unsupervised person re-identification: clustering and fine-tuning. TOMCCAP **14**, 83:1–83:18 (2018)
4. He, K., Zhang, X., Ren, S., Sun, J.: Deep residual learning for image recognition. In: CVPR (2016)
5. Hermans, A., Beyer, L., Leibe, B.: In defense of the triplet loss for person re-identification. In: CoRR (2014)
6. Khan, F.M., Brémond, F.: Unsupervised data association for metric learning in the context of multi-shot person re-identification. In: AVSS 2016 (2016)
7. Kodirov, E., Xiang, T., Fu, Z.Y., Gong, S.: Person re-identification by unsupervised l1 graph learning. In: ECCV (2016)
8. Kodirov, E., Xiang, T., Gong, S.: Dictionary learning with iterative laplacian regularisation for unsupervised person re-identification. In: BMVC (2015)
9. Li, M., Zhu, X., Gong, S.: Unsupervised person re-identification by deep learning tracklet association. In: ECCV (2018)
10. Li, W., Zhao, R., Xiao, T., Wang, X.: DeepReID: deep filter pairing neural network for person re-identification. In: CVPR (2014)
11. Li, W., Zhu, X., Gong, S.: Harmonious attention network for person re-identification. In: CVPR (2018)
12. Li, Y.J., Yang, F.E., Liu, Y.C., Yeh, Y.Y., Du, X., Wang, Y.C.F.: Adaptation and re-identification network: an unsupervised deep transfer learning approach to person re-identification. In: CVPRW (2018)
13. Ma, X., et al.: Person re-identification by unsupervised video matching. Pattern Recogn. **65**, 197–210 (2017)
14. Peng, P., et al.: Unsupervised cross-dataset transfer learning for person re-identification. In: CVPR (2016)
15. Wang, J., Zhu, X., Gong, S., Li, W.: Transferable joint attribute-identity deep learning for unsupervised person re-identification. In: CVPR (2018)

16. Wei, L., Zhang, S., Gao, W., Tian, Q.: Person trasfer GAN to bridge domain gap for person re-identification. In: CVPR (2018)

17. Yi, D., Lei, Z., Liao, S., Li, S.Z.: Deep metric learning for person re-identification. In: ICPR, August 2014

18. Zheng, L., Shen, L., Tian, L., Wang, S., Wang, J., Tian, Q.: Scalable person re-identification: a benchmark. In: ICCV (2015)

19. Zheng, L., Yang, Y., Hauptmann, A.G.: Person re-identification: past, present and future. CoRR abs/1610.02984 (2016)

20. Zheng, Z., Zheng, L., Yang, Y.: Unlabeled samples generated by GAN improve the person re-identification baseline in vitro. In: ICCV (2017)

Networking Self-Organising Maps and Similarity Weight Associations

Younjin Chung[1]([⊠]) [iD] and Joachim Gudmundsson[2]

[1] VIDEA Lab, The Australian National University, Canberra, ACT 2601, Australia
younjin.chung@anu.edu.au
[2] School of Computer Science, The University of Sydney,
Sydney, NSW 2006, Australia
joachim.gudmundsson@sydney.edu.au

Abstract. Using a Self-Organising Map (SOM), the structure of a data set can be explored when analysing patterns between data that are multivariate, nonlinear and unlabelled in nature. As a SOM alone cannot be used to explore patterns between different data sets, a similarity weighting scheme was previously introduced to associate different SOMs in a network fashion and approximate output patterns for given inputs. This approach uses a global weight association method on the combination of all SOMs specified for a network. However, there has been a difficulty in defining the association when changing the SOM network structure. Furthermore, it has always produced the same output weight distribution for different input data that have the same best matching unit. In an attempt to overcome the issues, we propose a new approach in this paper for locally associating a pair of SOMs as a basic network building block and approximating individually associated weight distribution. The experiments using ecological data demonstrate that the proposed approach effectively associates a pair of input and output SOMs for structural flexibility of the SOM network with better approximation of output weight distributions for individual input data.

Keywords: Self-Organizing Map · Network · Input and output · Pattern analysis · Weight association · Similarity weight measure

1 Introduction

Many analytical questions in the diverse study areas such as health, social and ecological sciences are associational and causal in solving their domain problems. They deal with multivariate and nonlinear data that are unlabelled in nature and different data sets from different data sources that are associated with each other through some related factors. The analysis of such complex data should allow exploring the joint factors, which are not explicitly expressed in the data

Funded by the Australian Government through the Australian Research Council (DP150101134 and DP180102870).

T. Gedeon et al. (Eds.): ICONIP 2019, CCIS 1143, pp. 779–788, 2019.
https://doi.org/10.1007/978-3-030-36802-9_82

space. Besides, the associational and causal information should not be based on deterministic analysis as our knowledge in reality is probabilistic, uncertain, graded and fuzzy [11].

There have been many methods developed for analysing associational and causal relations of complex data; the examples are Bayesian Network and Artificial Neural Networks (ANNs). Bayesian Network has been most used for structural and probabilistic analysis [4, 9]. However, its network based on probabilistic parameters has been the significant bottleneck for large data structures as the learning process is computationally complex and inflexible [6]. Among ANNs, Backpropagation Network has been most used to learn patterns between input and output data. However, its supervised and deterministic learning structure has limited providing indeterminate pattern information for unlabelled data [8]. The black-boxed classification or maximum probability adjustment of Backpropagation Network has also been discussed as insufficient and insignificant in exploring the underlying association [8]. Addressing the issues, Chung and Takatsuka [3] have introduced a network approach using Self-Organising Maps (SOMs) to analyse indeterminate associational and causal relations of complex data that are unlabelled and largely structured. However, its global weight association method between all SOMs specified for a network has showed a difficulty in defining the association when changing the SOM network structure. It has also limited approximating unique output weight distributions associated with individual input data that fall in the same best matching unit.

In an attempt to address the global weight association issues, we propose a new weight association approach in this paper. This approach is developed to locally associate a pair of SOMs as a basic building block of the SOM network. It also takes into account similarity weight measures for the approximation of individually associated weight distributions. Unlabelled ecological data are used for the experimental evaluation of the local and individual weight association approach. The experiments demonstrate that the proposed approach effectively associates a pair of input and output SOMs and makes the SOM network structure flexible. The experimental results show that the similarity weight measure of the approach also provides better approximation of unique output weight distributions associated with individual input data, regardless of having the same best matching unit.

The following section reviews the global weight association method and identifies its issues in networking SOMs and approximating association weights. Addressing the issues, our local and individual weight association approach is proposed in Sect. 3. Experimental results and discussion are given in Sect. 4, followed by the conclusion in Sect. 5.

2   Background

Self-Organising Map (SOM) [7] is an unsupervised learning algorithm and projects a high-dimensional data space in a nonlinear fashion onto a low-dimensional map space (typically 2D) in an ordered fashion. Using a SOM,

An input data,
$D = \{x_1, x_2, ..., x_n\}$

1st association by the SOMT
linking range given for the
input BMU (c) on S_I

2nd association by the SOMT
linking range given for each
neuron associated on S_{IO}

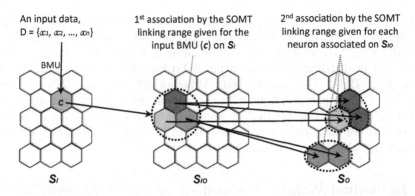

S_I S_{IO} S_O

Fig. 1. A visual process of associating similarity weights between a pair of input and output SOMs based on the global association algorithm of the SOMT [3].

the structure of either an input or an output data set can be explored when analysing patterns between data or between variables. As a SOM alone does not have the function to learn the structure between different data sets as input and output, Chung and Takatsuka [3] have utilised multiple SOMs in a network fashion to capture the indeterminate nature of associational and causal relations between multiple data sets of complex data. In the study, a weighting scheme was developed to approximate the similarity weight information, which describes how likely outputs in an output SOM are related to given inputs from an input SOM. Using its inverse Euclidean similarity measure, output neurons that are more closely associated with a given input likely have higher weights.

This approach uses the association process of the SOM Tree (SOMT) [2] when associating multiple SOMs in a network. The basic unit of the SOMT consists of one internal node SOM and two external node SOMs in a binary tree fashion. Each external SOM represents each of two data sets, and the internal SOM is trained with the combined data set of the two data sets. The role of the internal SOM is to link two external SOMs by associating their connection weight vectors. Based on the connection weight vectors between SOMs, the SOMT defines a globally optimized linking range for all links between all SOMs (e.g. three input SOMs linked to an output SOM through three internal SOMs) specified for a network. Then, the globally defined linking range is given to each link from an external SOM (an input data set) to another external SOM (an output data set) through their internal SOM. The global weight association process is illustrated in Fig. 1. An input data (D; n is the number of input variables) to the input SOM (S_I) is associated with the neurons of the internal SOM (S_{IO}) in the given global linking range to the best matching unit (BMU), c of D. Each neuron associated in S_{IO} for the input data is then associated with the neurons of the output SOM (S_O) in the given global linking range to each.

However, defining a global linking range by the SOMT for the weight association becomes a problematic process for exploratory analysis. The SOMT association process is required to define the specific linking range for a specific structure of the SOM network, whenever changing the network structure between SOMs. Another issue of using the global association method is identified in measuring similarity weight distributions for individual data. It always produces the same output weight distribution in the output SOM for different input data that fall in the same BMU of an input SOM. Although different data have the same input BMU, their joint relations with outputs can be somewhat different. This cannot be discovered by the SOMT association process. Furthermore, the output weights are distributed to only neurons in the given range, which does not consider the joint relations throughout the data space.

3   Individual Weight Association of the SOM Network

3.1   Local Weight Association Process

In an attempt to address the issues of the global weight association approach, we propose a local weight association approach to network a pair of SOMs (input and output) as a basic building block and allow own linking process for individual data. This approach defines an individual linking range at the first association process to find a group of the most similar input connection weight vectors in the internal SOM for a given input data on the input SOM. In the second association process, no linking range is defined to measure the overall similarity weight distribution, covering the whole data space, in the output SOM.

Referring to Fig. 2 compared with Fig. 1, an individual linking range is defined by the differences between BMUs of a given input data (D; n is the number of input variables) to select the most similar connection weight vectors in S_{IO} at the first step of the weight association. Denote the BMU of D in S_I as $W_I(c)$, the input connection BMU of $W_I(c)$ in S_{IO} as $W_{IO(I)}(c1)$, and the input connection BMU of D in S_{IO} as $W_{IO(I)}(c2)$. Based on those BMUs, two distances, $Dist_1$ and $Dist_2$ from $W_I(c)$ to $W_{IO(I)}(c1)$ and $W_{IO(I)}(c2)$ are measured respectively using the Euclidean distance:

$$Dist_1 = |W_I(c) - W_{IO(I)}(c1)| = \sqrt{\sum_{j=1}^{n}(wi(c)_j - wio(i)(c1)_j)^2}, \qquad (1)$$

and

$$Dist_2 = |W_I(c) - W_{IO(I)}(c2)| = \sqrt{\sum_{j=1}^{n}(wi(c)_j - wio(i)(c2)_j)^2}. \qquad (2)$$

The individual linking range (ILR) to group the most similar input connection weight vectors in S_{IO} for the given input data (D) is then defined as:

$$ILR(D) = |Dist_1 - Dist_2|. \qquad (3)$$

Once ILR is defined for the given input data (D), the distances of the input connection weight vectors in S_{IO} are measured from the input BMU, $W_I(c)$

in S_I. Based on the distance measure, a group of l input connection weight vectors in ILR are selected and given equal similarity weights (SW_1, sum to 1), as they substitute D when associating to the outputs in S_O. The l output connection weight vectors ($W_{IO(O)}(v)$, $v = 1, 2, ..., l$) selected in S_{IO} become the input vectors when associating to S_O.

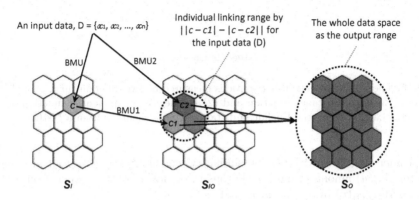

Fig. 2. A visual process of associating similarity weights between a pair of input and output SOMs based on the local association algorithm.

3.2   Negative Exponential Similarity Weight

At the second step of the weight association, the similarity weights of all neurons in S_O are measured from each of the l output connection weight vectors selected in S_{IO}. When measuring the similarity weights, we found that the inverse Euclidean distance function used for the global weight association approach brings the similarity weights down rapidly for small differences at some points, as shown in Fig. 3. The similarity weight distribution obtained by the inverse Euclidean distance function is compared with the similarity weight distribution obtained by the negative exponential function of the Shepard Law [10].

Based on Euclidean distance, Shepard proposed a universal law where similarity and distance are related via an exponential function that can describe probability. A variety of machine learning algorithms have been proposed to learn the similarity using this law [1]. According to the law, the weight of perceiving the similarity between data is a negative exponential function of the distance between data. The negative exponential function can describe the probability where data fall in a region of a SOM associated with the same response [10]. Compared with the inverse Euclidean measure, the negative exponential measure of the Shepard Law shows a gradual decline of similarity weights for the closer distances (see Fig. 3). As it is very unlikely to find the exact matched outputs to an input data, the possibly associated outputs can be more weighted in the output space. For this reason, the negative exponential function of the Shepard Law can be more robust than the inverse Euclidean distance function to be used for the similarity weight measure.

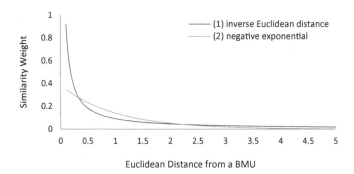

Fig. 3. The similarity weight distributions measured by two different functions: (1) the inverse Euclidean distance function and (2) the negative exponential function of the Shepard Law based on the Euclidean distance of SOM neurons from a BMU.

Under the Shepard Law, the similarity weight (SW_2) of an output weight vector $W_O(u)$ among m output neurons in S_O for an output connection weight vector $W_{IO(O)}(v)$ in S_{IO} is calculated as:

$$SW_2(W_O(u)|W_{IO(O)}(v)) = \frac{e^{-|W_{IO(O)}(v)-W_O(u)|}}{\sum_{u=1}^{m} e^{-|W_{IO(O)}(v)-W_O(u)|}}. \tag{4}$$

Each of the output weight vectors, $W_O(u)$ now holds its similarity weight of SW_2. As there can be more than one output connection weight vectors in S_{IO}, their weights (SW_1) given at the first association have to be applied to the weight, SW_2 for every output neuron in S_O at the second association process. Thus, the similarity weight, SW_3 of an output weight vector ($W_O(u)$) in S_O for the given weight (SW_1) of an output connection weight vector ($W_{IO(O)}(v)$) in S_{IO} is:

$$SW_3(W_O(u)|W_{IO(O)}(v)) = SW_1 * SW_2(W_O(u)|W_{IO(O)}(v)). \tag{5}$$

The weighted sum of all the output neurons in S_O from Eq. 5 is the same as the SW_1 of an output connection weight vector in S_{IO}. Thus, the total similarity weights for the l output connection weight vectors from S_{IO} is applied to every output neuron in S_O. The final similarity weight (SW) of each output neuron ($W_O(u)$) in S_O for the given input data (D) is:

$$SW(W_O(u)|D) = \sum_{u=1}^{m} SW_3(W_O(u)|W_{IO(O)}(v)). \tag{6}$$

The sum of the final similarity weights for all the output neurons in S_O becomes 1. The more similar output neurons in each measure have the higher weights.

4   Experiment and Discussion

The ecological domain data[1], used in the work of Chung and Takatsuka [3, 5], were used to evaluate the proposed local and individual weight association approach of networking SOMs. Among the ecological data associations in the network, the association of the biological output SOM with the physical input SOM is represented in this paper. Different physical data can have the same BMU in the physical input SOM; however, their original feature values are not the exactly same and possibly have different relations to the biological output. Based on the local association, we examined the biological output patterns for individual physical inputs presented in the same BMU.

4.1   Individual Similarity Weight Associations

Three different physical input data were separately given for their own association with the biological output. For the experiment, we compared their association results of the negative exponential measure and the inverse Euclidean measure of similarity weights. Using the individual linking range for both measures, the similarity weights were distributed over the biological SOM, covering the whole output data space. Figure 4 shows the similarity weight distributions of the biological output, associated with each of the three different physical inputs ('M7', 'M11' and 'D2') in the same BMU, obtained by the two measures.

The overall output patterns for the different input data by the negative exponential measure were similar to the result patterns by the inverse Euclidean measure. However, the negative exponential similarity weight distributions were more distinguishable than the inverse Euclidean similarity weight distributions for individual associations. Using the negative exponential measure in Fig. 4(a), the associated region including 'M7' in the biological output SOM was highlighted stronger when 'M7' was given to the physical input SOM. The weights distributed in the biological output regions around 'M11' and 'D2' became higher respectively, while the weights distributed in the biological output region around 'M7' became lower, when 'M11' and 'D2' were given to the physical input SOM. Recall that the weight difference could not be captured when measuring the similarity by the inverse Euclidean distance function. As seen in Fig. 4(b) using the inverse Euclidean measure, the similarity weight distributions for 'M11' and 'D2' showed the same pattern with 'M7'. This shows that the negative exponential measure of the local weight association approach provides unique output pattern information for individually different input data.

In order to compare the two measures regarding the similarity weight distribution, the similarity weights for 'M7', 'M11' and 'D2' were measured from their BMUs over the biological output SOM space. The graphs in Fig. 5(a), (b) and (c) show the different similarity weight distributions for the different data, 'M7', 'M11' and 'D2', respectively. In each graph, the more weights are given to the

---

[1] 130 ecological data in the physical, chemical and land-use data sets as inputs and the biological data set as output with 5 variables each data set.

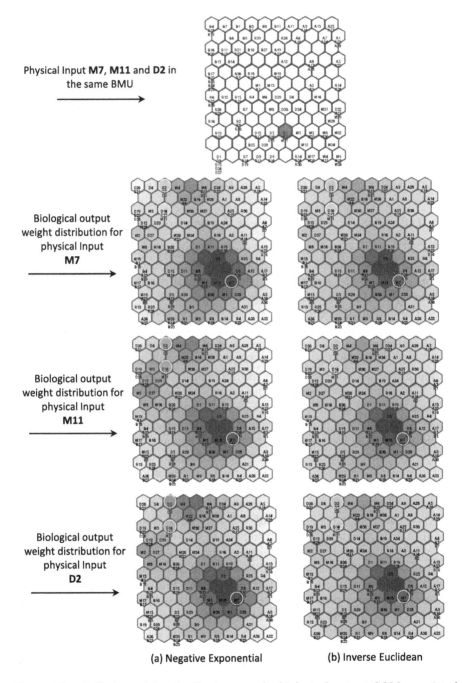

Fig. 4. The similarity weight distributions on the biological output SOM associated with the physical inputs for 'M7', 'M11' and 'D2' by (a) the negative exponential measure and (b) the inverse Euclidean measure. The biological output BMUs of the given data are circled in yellow. The darker colour indicates the higher weight. (Color figure online)

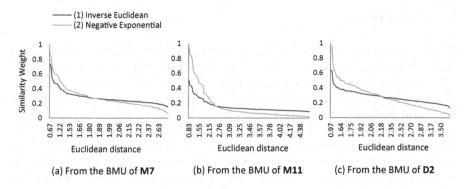

Fig. 5. The graph of the similarity weight distribution from the BMU of 'M7' in (*a*), 'M11' in (*b*) and 'D2' in (*c*) over the biological output SOM space using (1) the inverse Euclidean measure and (2) the negative exponential measure.

closer neurons to the given output BMU by the negative exponential measure than the inverse Euclidean measure. According to Euclidean distances of the biological output SOM space, the distances from the BMUs of 'M11' and 'D2' are much further than the distances from the BMU of 'M7'. This shows that the biological properties are more closely related to the data, 'M7' than 'M11' and 'D2', and explains why the stronger weights are distributed in the region around 'M7' than the regions around 'M11' and 'D2' over the biological output data space.

4.2 Discussion

The local weight association approach using the negative exponential similarity weight measure better estimated indeterminate output patterns for individual input data. An individual linking range was defined for a single association between two different input and output data sets as a basic building block of the SOM network. The individual linking process for an association could make the network modification easy for the process of exploratory analysis. The negative exponential function of the Shepard Law intensively assigned more weights to the outputs that are more closely related to a given input data. Therefore, the proposed approach can help one to explore complex and uncertain information through various associational and causal analysis processes detecting possible alternatives between multiple data sets.

However, it is difficult to compare the analytical results in a theoretical or statistical way since the weight information is based on the distance similarity of data in a given data space. This requires a wide range of empirical studies to assess the quality of the SOM network for the proposed approach by enabling application domain experts to evaluate the associational and causal information.

5   Conclusion

We presented a new weight association approach for networking SOMs when dealing with different data sets of complex data for exploring the indeterminate associational and causal information. The issues of the global association method with the inverse Euclidean similarity weight measure were identified on structural flexibility of the SOM network and unique weight approximation for individual data. Our local weight association approach in this paper used the individual linking process and the negative exponential function of the Shepard Law for the similarity weight measure. The experimental results, using the ecological domain data, showed that the possibility to be outputs for a given input data were better discovered by the local and individual weight association approach. This demonstrated that the proposed approach is highly acceptable for associational and causal information processing and provides an opportunity for the exploratory analysis of complex data in diverse application domains for solving their specific domain problems.

References

1. Ashby, F.G., Ennis, D.M.: Similarity measures. Scholarpedia **2**(12), 4116 (2007)
2. Chung, Y., Takatsuka, M.: The self-organizing map tree (SOMT) for nonlinear data causality prediction. In: Lu, B.-L., Zhang, L., Kwok, J. (eds.) ICONIP 2011. LNCS, vol. 7063, pp. 133–142. Springer, Heidelberg (2011). https://doi.org/10.1007/978-3-642-24958-7_16
3. Chung, Y., Takatsuka, M.: A causal model using self-organizing maps. In: Arik, S., Huang, T., Lai, W.K., Liu, Q. (eds.) ICONIP 2015. LNCS, vol. 9490, pp. 591–600. Springer, Cham (2015). https://doi.org/10.1007/978-3-319-26535-3_67
4. Darwiche, A.: Bayesian networks. In: Handbook of Knowledge Representation, pp. 467–508 (2008)
5. Giddings, E.M.P., et al.: Selected physical, chemical, and biological data used to study urbanizing streams in nine metropolitan areas of the united states, 1999–2004. Technical Report Data Series 423, National Water-Quality Assessment Program, U.S. Geological Survey (2009)
6. Jurgelenaite, R., Lucas, P.J.F.: Exploiting causal independence in large Bayesian networks. Knowl. Based Syst. **18**, 153–162 (2005)
7. Kohonen, T.: Self-Organizing Maps. Information Sciences, 3rd edn. Springer, Heidelberg (2001)
8. Lee, C., Rey, T., Mentele, J., Garver, M.: Structured neural network techniques for modeling loyalty and profitability. In: Data Mining and Predictive Modeling Paper 082-30, Proceedings of SAS SUGI, pp. 1–13 (2005)
9. Pearl, J.: Causal inference in statistics: an overview. Technical Report R350, Statistics Surveys (2009)
10. Shepard, R.N., et al.: Toward a universal law of generalization for psychological science. Science **237**(4820), 1317–1323 (1987)
11. Sun, R.: A neural network model of causality. IEEE Trans. Neural Networks **5**(4), 604–611 (1994)

Author Index